T0184317

Lecture Notes in Computer Science   11494

Commenced Publication in 1973
Founding and Former Series Editors:
Gerhard Goos, Juris Hartmanis, and Jan van Leeuwen

More information about this series at http://www.springer.com/series/7407

Louis-Martin Rousseau ·
Kostas Stergiou (Eds.)

Integration of Constraint Programming, Artificial Intelligence, and Operations Research

16th International Conference, CPAIOR 2019
Thessaloniki, Greece, June 4–7, 2019
Proceedings

 Springer

Editors
Louis-Martin Rousseau
Ecole Polytechnique de Montréal
Montréal, QC, Canada

Kostas Stergiou
University of Western Macedonia
Kozani, Greece

ISSN 0302-9743   ISSN 1611-3349 (electronic)
Lecture Notes in Computer Science
ISBN 978-3-030-19211-2   ISBN 978-3-030-19212-9 (eBook)
https://doi.org/10.1007/978-3-030-19212-9

LNCS Sublibrary: SL1 – Theoretical Computer Science and General Issues

This Springer imprint is published by the registered company Springer Nature Switzerland AG
The registered company address is: Gewerbestrasse 11, 6330 Cham, Switzerland

Preface

This volume contains the papers that were presented at the 16th International Conference on the Integration of Constraint Programming, Artificial Intelligence, and Operations Research (CPAIOR 2019), held in Thessaloniki, Greece, June 4–7, 2019.

The conference received a total of 111 submissions, including 99 regular paper and 12 extended abstract submissions. The regular papers reflect original unpublished work, whereas the extended abstracts contain either original unpublished work or a summary of work that was published elsewhere. Each regular paper was reviewed by at least three Program Committee members, which was followed by an author response period and a general discussion by the Program Committee. The extended abstracts were reviewed for appropriateness for the conference. At the end of the review period, 43 regular papers were accepted for presentation during the conference and publication in this volume, and all abstracts were accepted for presentation at the conference.

In addition to the regular papers and extended abstracts, four invited talks were given, by Tobias Achterberg (Gurobi Optimization), Zico Kolter (Carnegie Mellon University), Martine Labbé (Université Libre de Bruxelles), and Thomas Schiex (Université de Toulouse, INRA). The abstracts of the invited talks can also be found in this volume.

The conference program included a Master Class on the topic "CP, AI, and OR for Social Good," organized by Bistra Dilkina and Phebe Vayanos, with invited talks by Christopher Beck (University of Toronto), Bistra Dilkina (University of South California), John Hooker (Carnegie Mellon University), Marie-Éve Rancourt (HEC Montreal), Sibel Salman (Koc University), Pascal Van Hentenryck (Georgia Institute of Technology), Phebe Vayanos (University of South California), and Joann de Zegher (MIT).

The organization of this conference would not have been possible without the help of many individuals. First, we would like to thank the Program Committee members and external reviewers for their hard work. Several Program Committee members deserve additional thanks because of their help with timely reviewing of fast-track papers, shepherding regular papers, or overseeing the discussion of papers for which we had a conflict of interest. We are also particularly grateful to Bistra Dilkina and Phebe Vayanos (Master Class Chairs) and Nikolaos Ploskas (Sponsorship Chair) for their help in organizing this conference. Special thanks goes to Nikolaos Samaras (Conference Chair), whose support has been instrumental in making this event a success.

Lastly, we want to thank all sponsors for their generous contributions. At the time of writing, these include: the International Conference on Automated Planning and Scheduling (ICAPS), the Association for Constraint Programming (ACP), Springer, the Artificial Intelligence Journal, AIMMS, COSLING, Dimoulas Special Cables, the European Association for Artificial Intelligence (EurAI), FICO, General Algebraic

Modeling System (GAMS), Gurobi, Lindo, The Optimization Firm, Marathon Data Systems, the University of Macedonia, and the University of Western Macedonia.

June 2019
Louis-Martin Rousseau
Kostas Stergiou

Organization

Program Chairs

Louis-Martin Rousseau   École Polytechnique de Montréal, Canada
Kostas Stergiou   University of Western Macedonia, Greece

Conference Chairs

Nikolaos Samaras   University of Macedonia, Greece
Kostas Stergiou   University of Western Macedonia, Greece

Master Class Chairs

Bistra Dilkina   University of Southern California, USA
Phebe Vayanos   University of Southern California, USA

Sponsorship Chair

Nikolaos Ploskas   University of Western Macedonia, Greece

Program Committee

Chris Beck   University of Toronto, Canada
Nicolas Beldiceanu   IMT Atlantique (LS2N), France
David Bergman   University of Connecticut, USA
Timo Berthold   Zuse Institute Berlin/FICO, Germany
Natashia Boland   Georgia Institute of Technology, USA
Andre Augusto Cire   University of Toronto, Canada
Mathijs De Weerdt   Delft University of Technology, The Netherlands
Bistra Dilkina   University of Southern California, USA
Bernard Gendron   University of Montreal, Canada
Ambros Gleixner   Zuse Institute Berlin, Germany
Carla Gomes   Cornell University, USA
Tias Guns   Vrije Universiteit Brussel, Belgium
Emmanuel Hebrard   LAAS, CNRS, France
John Hooker   Carnegie Mellon University, USA
Matti Järvisalo   University of Helsinki, Finland
Serdar Kadioglu   Oracle Corporation, USA
Philip Kilby   Data61 and the Australian National University, Australia
Joris Kinable   Eindhoven University of Technology, The Netherlands
Philippe Laborie   IBM, France

Jeff Linderoth	University of Wisconsin-Madison, USA
Andrea Lodi	École Polytechnique de Montréal, Canada
Michele Lombardi	University of Bologna, Italy
Laurent Michel	University of Connecticut, USA
Michela Milano	University of Bologna, Italy
Ioannis Mourtos	Athens University of Economics and Business, Greece
Nina Narodytska	VMware Research, USA
Laurent Perron	Google, France
Gilles Pesant	Polytechnique Montréal, Canada
Claude-Guy Quimper	Laval University, Canada
Jean-Charles Regin	University of Nice-Sophia Antipolis/I3S/CNRS, France
Andrea Rendl	Satalia, UK
Domenico Salvagnin	University of Padova, Italy
Scott Sanner	University of Toronto, Canada
Pierre Schaus	UCLouvain, Belgium
Andreas Schutt	CSIRO and The University of Melbourne, Australia
Paul Shaw	IBM, France
Michael Trick	Carnegie Mellon University, USA
Charlotte Truchet	LINA, UMR 6241, Université de Nantes, France
Willem-Jan Van Hoeve	Carnegie Mellon University, USA
Pascal Van-Hentenryck	Georgia Institute of Technology, USA
Petr Vilím	IBM Czech, Czech Republic
Mark Wallace	Monash University, Australia
Alessandro Zanarini	ABB Corporate Research

Additional Reviewers

Babaki, Behrouz	Huang, Teng
Belov, Gleb	Kimura, Ryo
Benade, Gerdus	Kong, Shufeng
Berg, Jeremias	Lo Bianco, Giovanni
Bestuzheva, Ksenia	Miltenberger, Matthias
Bjorck, Johan	Montmirail, Valentin
Bodur, Merve	Müller, Benjamin
Borghesi, Andrea	Nannicini, Giacomo
Carbonnel, Clément	Niskanen, Andreas
Castro, Margarita	Olivier, Philippe
Daryalal, Maryam	Pelleau, Marie
Emadikhiav, Mohsen	Raghunathan, Arvind
Gmira, Maha	Refalo, Philippe
Gocht, Stephan	Repoussis, Panagiotis
Hendel, Gregor	Serra, Thiago

Serrano, Felipe
Shi, Qinru
Talbot, Pierre
Tang, Ziye
Urli, Tommaso

van den Bogaerdt, Pim
Wang, Xin
Witzig, Jakob
Živný, Stanislav

Extended Abstracts

The following extended abstracts were accepted for presentation at the conference:

- Joseph Kreimer and Eduard Ianovsky: Constrained Programming Optimization of Multiple Drones System with Shortage of Maintenance Teams
- Felix Winter and Nysret Musliu: Exact Methods for a Paint Shop Scheduling Problem from the Automotive Supply Industry
- Kevin Leo, Peter J. Stuckey, and Guido Tack: Hier-MUS: Structure-Guided MUS Enumeration using Hierarchy Maps
- Emir Demirovic, Tias Guns, Peter J. Stuckey, James Bailey, Christopher Leckie, Jeffrey Chan, and Ramamohanarao Kotagiri: A Framework for Predict+Optimise with Ranking Objectives: Learning Linear Functions for Optimisation Parameters in Exhaustive Search Fashion
- Emir Demirovic and Peter J. Stuckey: LinSBPS: A Novel Incomplete MaxSAT Approach Based on the Linear MaxSAT Algorithm and Local Search Style Techniques
- Neil Yorke-Smith: Loose Hybridisation for the Cyclic Hoist Scheduling Problem
- Sara Frimodig and Christian Schulte: Radiation Therapy Patient Scheduling
- Eleftherios Manousakis, Panagiotis Repoussis, Emmanouil Zachariadis, and Christos Tarantilis: A Hybrid Branch-and-Cut Method for the Inventory Routing Problem
- Grigoris Kasapidis, Dimitris Paraskevopoulos, Panagiotis Repoussis, and Christos Tarantilis: A Scatter Search Algorithm for Large Scale Flexible Job-Shop Scheduling Problems with Complex Precedence Constraints
- Elias Khalil, Rakshit Trivedi, and Bistra Dilkina: Neural Integer Optimization
- Elias Khalil, Amrita Gupta, and Bistra Dilkina: Combinatorial Attacks Against Binarized Neural Networks
- Michael Romer, Andre Augusto Cire, and Louis-Martin Rousseau: Dynamic Programming for Combinatorial Optimization: A Primal-Dual Approach Based on Decision Diagrams

Abstracts of Invited Talks

Products in Mixed Integer Programming

Tobias Achterberg

Gurobi Optimization
achterberg@gurobi.com

Abstract. Products of problem variables appear naturally in pseudo-Boolean programs as well as in quadratic programs. Special preprocessing, linearization and cutting plane techniques are available to deal with such products.

If at least one of the two variables in a product is binary, then the product can be modeled using a set of linear constraints. As a consequence, there are many mixed integer linear programs (MILPs) that actually contain products of variables hidden in their constraint structure. Rediscovering these product relationships between the variables enables us to exploit the solving techniques for product terms.

In this talk, we will explain how such product relationships can be detected in a given mixed integer linear program. Furthermore, we present some ideas how they can then be exploited to improve the performance of an MILP solver. In particular, we describe cuts from the Reformulation Linearization Technique (RLT) and cuts for the Boolean Quadric Polytope (BQP). These techniques have been implemented in the upcoming Gurobi version 9.0. Some preliminary computational results will be presented.

Leveraging Optimization and Convexity Within Deep Learning

Zico Kolter

Computer Science Department, Carnegie Mellon University
zkolter@cs.cmu.edu

Abstract. Deep learning is frequently seen as the "breakthrough" AI technology of recent years, revolutionizing areas spanning computer vision, natural language processing, and game playing. However, it is also often viewed as a largely heuristic-driven field, where advances occur not through rigorous analysis, but through experimentation alone. And indeed, major problems exist in modern deep learning: the systems are brittle (sensitive to adversarial manipulation and a general lack of robustness), opaque (difficult to interpret and debug their components), and expensive (often requiring vastly more data than practical in real-world settings). I this talk, I will present ways in which we can leverage techniques from optimization and convexity to overcome these problems in deep learning. First, I will discuss our approaches to designing provably robust deep networks using tools from convex relaxations and duality. I also highlight recent work on scaling these methods to much larger domains, including some initial work on provable robustness at ImageNet scale. Second, I will present our work on integrating more complex modules as interpretable layers within deep learning architectures. I show how modules such as optimization solvers, physical simulation, model-based control, and game equilibrium solvers can all be integrated as layers within a deep network, enabling more intuitive architectures that can learn from vastly less data. Last, I will highlight some additional ongoing directions and open questions in both these areas.

Bilevel Optimisation, Stackelberg Games and Pricing Problems

Martine Labbé

Université Libre de Bruxelles
mlabbe@ulb.ac.be

Abstract. Bilevel optimisation problems consist in constraint optimisation problems in which a subset of variables constitutes the optimal solution of second optimisation problem. They correspond to situations in which two groups of decisions are taken sequentially. A first part of this talk will present the main aspects, properties and algorithms for bilevel optimization problems with a particular attention to the bilevel linear ones. The second part will focus on bilevel problems with bilinear objectives and in particular on applications such as Stackelberg games and pricing problems.

Optimization in Graphical Models

Thomas Schiex

MIAT, Université de Toulouse, INRA UR 875, Castanet-Tolosan, France
thomas.schiex@inra.fr

Abstract. Graphical models (GMs) describe a function of many variables as the combination of many functions of smaller scope, or size. This idea of "decomposable" functions has been used in many areas. We restrict ourselves here to functions of discrete variables with Boolean (CSP/SAT) or numerical (Valued CSP, Cost Function Networks – CFNs) output. Interpreting cost as energy and using the Boltzmann probability law, CFNs can also describe probability distributions as Markov Random Fields or Bayesian Networks do. CFNs can also be represented as Integer Linear or Quadratic Programs. Over the last decades, the main ingredients of Constraint Programming: backtrack search, arc consistency and global constraints have been extended to "efficiently" optimize functions described as CFNs using Branch and Bound, soft arc consistencies and global cost functions, all implemented in the open source solver toulbar2. The talk will introduce soft arc consistency, show its tight relation with LP and so-called convergent Message Passing in stochastic GMs. I will also give a quick description of the many bells and whistles inside toulbar2. Ultimately, the connection between CFNs and stochastic GMs can be leveraged to learn the soft part of CFNs from data, and I will illustrate this on one hot scientific application area for toulbar2: protein design.

Contents

Constraint Programming for Dynamic Symbolic Execution of JavaScript

Roberto Amadini[1]($\boxtimes$), Mak Andrlon[1], Graeme Gange[2], Peter Schachte[1],
Harald Søndergaard[1], and Peter J. Stuckey[2]

[1] University of Melbourne, Melbourne, VIC, Australia
roberto.amadini@unimelb.edu.au
[2] Monash University, Melbourne, VIC, Australia

Abstract. Dynamic Symbolic Execution (DSE) combines concrete and symbolic execution, usually for the purpose of generating good test suites automatically. It relies on constraint solvers to solve path conditions and to generate new inputs to explore. DSE tools usually make use of SMT solvers for constraint solving. In this paper, we show that constraint programming (CP) is a powerful alternative or complementary technique for DSE. Specifically, we apply CP techniques for DSE of JavaScript, the *de facto* standard for web programming. We capture the JavaScript semantics with MiniZinc and integrate this approach into a tool we call ARATHA. We use G-STRINGS, a CP solver equipped with string variables, for solving path conditions, and we compare the performance of this approach against state-of-the-art SMT solvers. Experimental results, in terms of both speed and coverage, show the benefits of our approach, thus opening new research vistas for using CP techniques in the service of program analysis.

1 Introduction

Dynamic symbolic execution (DSE), also known as concolic execution/testing, or directed automated random testing [21,35], is a hybrid technique that integrates the *concrete* execution of a program with its *symbolic* execution [28]. The main application is the automated generation of test suites with high coverage relative to their size. In a nutshell, DSE collects the constraints (or *path conditions*) encountered at conditional statements during concrete execution; then, a constraint solver or theorem prover is used to detect alternative execution paths by systematically negating the path conditions. This process is repeated until all the feasible paths are covered or a given threshold (e.g., a timeout) is exceeded.

Key factors for the success of DSE are the efficiency and the expressiveness of the underlying constraint solver. The significant advances made by *satisfiability modulo theories* (SMT) solvers over recent years have stimulated interest in DSE and led to the development of many popular tools [11,15,36,43,45,47]. In particular, improvements in expressive power (due the ability to combine different theories) and solver performance have made SMT solvers very attractive

© Springer Nature Switzerland AG 2019
L.-M. Rousseau and K. Stergiou (Eds.): CPAIOR 2019, LNCS 11494, pp. 1–19, 2019.
https://doi.org/10.1007/978-3-030-19212-9_1

for DSE, to the point that they are considered the *de facto* standard for DSE tools. Alternatives such as *constraint programming* (CP) exist, however.

Constraint programming [40] is a declarative paradigm aimed at solving combinatorial problems consisted of variables (typically having finite domains) and constraints over those variables. CP is applied in fields like resource allocation, scheduling, and planning, but apart from some dedicated approaches [14,16,23], it has seen limited use in software analysis. Arguably, the main impediment has been lack of support for common data structures such as dynamic arrays, bit vectors, and strings.

In this paper, we show that DSE can benefit from modern CP solving. In particular, we apply CP techniques to solve the path conditions generated by the dynamic symbolic execution of *JavaScript* programs. JavaScript is nowadays the standard programming language of the web, extensively used by developers on both the client and server side, and supported by all common browsers. Its dynamic nature can easily lead to programming errors and security vulnerabilities. This makes the dynamic symbolic execution of JavaScript an important task, but also a highly challenging one. Hence, it is not surprising that only a small number of DSE tools are available for JavaScript.

To capture JavaScript semantics, we first modelled the main language constructs with the CP modelling language *MiniZinc* [38]. It is essential to note that we are using the MiniZinc extension with *string variables* defined by Amadini et al. [3]. Strings play a central role in JavaScript because each JavaScript object is a map from string keys to values, and hence coercions to strings frequently occur in JavaScript programs (notably, arrays are objects and hence array indices are converted to their corresponding string values). Moreover, JavaScript programs often use *regular expressions* to match string patterns [6].

We then developed ARATHA, a DSE tool using the JALANGI analysis framework [44]. ARATHA can generate path conditions in our MiniZinc encoding, and solve them with G-STRINGS [7], a recent extension of the CP solver GECODE [20] able to handle string variables. ARATHA is also able to generate path conditions in the form of SMT-LIB assertions, allowing us to empirically evaluate our CP approach against the state-of-the-art SMT solvers CVC4 [32] and Z3STR3 [13].

Results indicate that a CP approach can easily be competitive with SMT approaches, and in particular the techniques can be used in conjunction. We emphasize that this technique can be replicated and extended to analyze languages other than JavaScript by using different MiniZinc encodings and different solvers (MiniZinc is a solver-independent language). We are not aware of any similar existing approaches for dynamic symbolic execution.

Paper structure. Section 2 introduces the basics of CP and DSE. Section 3 explains how we use MiniZinc to model JavaScript semantics. Section 4 describes ARATHA. Section 5 presents our experimental evaluation. Section 6 discusses related work. Section 7 concludes by outlining possible future research directions.

2 Preliminaries

We begin by summarizing some basic notions related to constraint programming, string solving, DSE, and JavaScript.

For a given finite alphabet Σ, we denote by Σ^* the set of all finite strings over Σ. The length of a string $x \in \Sigma^*$ is denoted $|x|$.

2.1 Constraint Programming and String Constraint Solving

Constraint programming [40] comprises modelling and solving combinatorial problems. This often means to define and solve a *constraint satisfaction problem* (CSP), which is a triple $\langle \mathcal{X}, \mathcal{D}, \mathcal{C} \rangle$ where: $\mathcal{X} = \{x_1, \ldots, x_n\}$ is a finite set of *variables*, $\mathcal{D} = \{D(x_1), \ldots, D(x_n)\}$ is a set of *domains*, where each $D(x_i)$ is the set of the values that x_i can take, and $\mathcal{C}$ is a set of *constraints* over the variables of $\mathcal{X}$ defining the feasible assignments of values to variables. The goal is typically to find an assignment $\xi \in D(x_1) \times \cdots \times D(x_n)$ of domain values to corresponding variables that satisfies all of the constraints of $\mathcal{C}$.

Most CSPs found in the literature are defined over *finite domains*, i.e., $\mathcal{D}$ only contains finite sets. This guarantees the decidability of these problems, that are in general NP-complete. Typically, only integer variables and constraints are considered. However, some variants have been proposed. In this work, we also consider constraints over *bounded-length strings*. Fixing a finite alphabet Σ and a maximum string length $\lambda \in \mathbb{N}$, a CSP with bounded-length strings contains a number $k > 0$ of string variables $\{x_1, \ldots, x_k\} \subseteq \mathcal{X}$ such that $D(x_i) \subseteq \Sigma^*$ and $|x_i| \leq \lambda$. The set $\mathcal{C}$ contains a number of well-known string constraints, such as string length, (dis-)equality, membership in a regular language, concatenation, substring selection, and finding/replacing. In the following, we will refer to constraint solving involving string variables as *string (constraint) solving*.

Different approaches to string constraint solving have been proposed, based on: *automata* [25,31,46], *word equations* [13,32], *unfolding* (using either bit-vector solvers [27,41] or CP [42]), and *dashed strings* [7,8]. In particular, dashed strings are a recent CP approach that can be seen as "lazy" unfolding. Thanks to dedicated propagation, dashed strings enable efficient "high-level" reasoning on string constraints, by weakening the dependence on λ [5,6].

Several modelling languages have been proposed for encoding CP problems into a format understandable by constraint solvers. One of the most popular nowadays is *MiniZinc* [38], which is solver-independent (the motto is *"model once, solve anywhere"*), enabling the separation of model and data. Each MiniZinc model (together with corresponding data, if any) is translated into *FlatZinc*—the solver-specific target language for MiniZinc—in the form required by a solver. From the same MiniZinc model, different FlatZinc instances can be derived.

MiniZinc was equipped with string variables and constraints by Amadini et al. [3]. A MiniZinc model with strings can be solved "directly" by CP solvers natively supporting string variables (GECODE+S [42] and G-STRINGS [7]) or "indirectly" via the static unfolding into integer variables. Clearly, direct resolution is generally more efficient—especially as λ grows.

2.2 Dynamic Symbolic Execution

Symbolic execution is a static analysis technique that has its roots in the 1970s [28].

The idea of symbolic execution is to assume symbolic values for input and to interpret programs correspondingly, i.e., to use a concept of "value" that is in fact an expression over the variables representing possible input values. The symbolic interpreter can then explore the possible program paths by reasoning about the conditions under which execution will branch this way or that. The set of constraints leading to a particular path being taken is a *path condition*, so that a given path is feasible if and only if the corresponding constraint is satisfiable. The test for satisfiability (and the generation of a witness in the affirmative case) is delegated to a *constraint solver*.

Symbolic execution can be useful to automatically prove a given property of interest, provided that: *(i)* the whole program—including libraries—is available to the interpreter, and *(ii)* the underlying constraint solver is expressive and efficient enough to handle the generated path conditions. Unfortunately, these conditions are often not met.

Dynamic symbolic execution (DSE) is a software verification approach that performs symbolic execution along with concrete (or dynamic) execution of a given program. Concrete execution is straightforward: concrete input values are generated according to some heuristics and tested by executing the program.

DSE can mitigate the above issues by: *(i)* directly invoking unavailable functions (a complete symbolic interpreter is not required), and *(ii)* ignoring or approximating unsupported constraints. The idea is to use symbolic values alongside concrete values: during a concrete execution of the program on a given input, symbolic expressions are tracked as in the symbolic execution.

However, in general DSE cannot guarantee full coverage (e.g., in presence of loops or recursion), whereas symbolic execution tries to cover all the possible paths (although still executing a path at a time, e.g., using interpolation to collapse identical or subsumed paths).

After each run, the recorded path conditions are used to generate inputs for the next concrete execution. Indeed, negating a constraint of a path condition will generate a new set of constraints that can either be satisfiable (in which case a new input will be generated to explore the new path) or unsatisfiable (we found an unreachable path in the program). By repeating the process, we can ideally reach the maximum code coverage.

Note that the constraints of the path conditions can be unsupported or too hard to solve in a reasonable time. This can result in over-approximated solutions (when a constraint is ignored or relaxed) or timeouts. This does not compromise the soundness of DSE, however, as it only means that in the worst case, fewer paths might be explored.

2.3 JavaScript

Dynamic symbolic execution is language-independent, but the definition and the resolution of path conditions clearly depends on the semantics of the target language. Although this work only considers the JavaScript language, our approach is flexible enough to encode the semantics of other well-known languages.

Designed in mid 1990's by Brendan Eich in only ten days, JavaScript has now become a *de facto* standard for web applications. This dynamic, weakly typed language has a number of unconventional rules and pitfalls that sometimes make its behaviour surprising. For instance, there is no concept of class: JavaScript uses prototype-based inheritance between objects. Its weak typing implies a lot of—often implicit—coercions. In particular, coercions to *strings* are very common because apart from few primitive types, in JavaScript everything is an *object*, which is a dictionary with string keys. Each key-value pair is called a *property*, and the key is called a property's *name*. The syntax to access property "x" of object O is, equivalently, $O[$"x"$]$ or $O.x$. For example, JavaScript arrays are actually objects where instead of indices $0, 1, 2, \ldots$ we have properties "0","1","2", ... The same applies for strings. For instance, the value of property "1" of string "hello" is its second character (indexing is 0-based), i.e., "hello"["1"] is "e".

The weakness of the semantic rules for JavaScript is an obstacle for program analysis. Static analysis tools have been proposed [9,26,29,41], but the dynamism of the language makes static reasoning difficult and often ineffective. Dynamic techniques such as fuzzing seem more suitable for the analysis of this language. The DSE approach aims to combine the best of the static and dynamic worlds, by orchestrating the dynamic execution via symbolic reasoning.

```
1  var x = symVar();
2  var y = "length";
3  if (x[y] >= 2)
4      console.log("PC1") // path condition (1)
5  else
6      if (y[x] === "g")
7          console.log("PC2") // path condition (2)
8      else
9          console.log("PC3") // path condition (3)
```

Fig. 1. Example of JavaScript program annotated with symbolic variables.

We conclude this section by providing an example of how DSE works on a snippet of JavaScript code. In Fig. 1, variable x is symbolic, i.e., it can take any JavaScript value, while y is a concrete variable initialised to "length". When a property of a string primitive is accessed, JavaScript automatically creates a temporary String *wrapper object* to resolve the property access.[1] This wrapper

---

[1] Similarly, Booleans and numbers are wrapped into Boolean and Number objects.

inherits all the string methods (e.g., `indexOf`, `toUpperCase`, ...) and also has an immutable `length` property containing the length of the string.

Dynamic symbolic execution starts by initialising x to an arbitrary default value. Let us assume for simplicity that we start with the empty string: $\{x \leftarrow \text{""}\}$. The first concrete iteration is then executed. Line 3 checks if the `"length"` property of (the `String` object wrapping) x is at least 2. Since $|x| = 0$, this condition is false and the constraint $\neg(|x| \geq 2)$ is added to the path condition. Then, in line 6 we check if property x of `y` is equal to string `"g"`. But string `"length"` has no property named `""`, so this check also fails and constraint $\neg(\text{"length"}[x] = \text{"g"})$ is added. We thus reach line 9 by finding that path condition (3), reached with $\{x \leftarrow \text{""}\}$, is $\{\neg(|x| \geq 2), \neg(\text{"length"}[x] = \text{"g"})\}$. This path condition characterizes all inputs x that would take us along a path identical to that of `""`. It is now used to generate a new path. By negating its first constraint we get $|x| \geq 2 \wedge \neg(\text{"length"}[x] = \text{"g"})$. A suitable solver can find a feasible assignment, e.g., $\{x \leftarrow \text{"aa"}\}$. This input leads to path condition (1). Similarly, we negate the second constraint of path condition (3) to get $\neg(|x| \geq 2) \wedge \text{"length"}[x] = \text{"g"}$. The assignment $\{x \leftarrow \text{"3"}\}$ satisfies this constraint: $|x| = 1 \not\geq 2$ and the fourth character of `"length"`, i.e., the one with index 3, is the string `"g"` (of length 1). At this stage, there are no new constraints that can be generated: the set of inputs $\{\{x \leftarrow \text{""}\}, \{x \leftarrow \text{"aa"}\}, \{x \leftarrow \text{"3"}\}\}$ covers all the lines of Fig. 1.

3   Modelling JavaScript Semantics

Understanding, and then modelling, the semantics of JavaScript is not always straightforward. For example, the comparison `[] == []` between empty arrays fails because JavaScript actually compares their memory locations, which are distinct because two different temporary objects are created. Faithfully modelling the JavaScript semantics also requires the full support of data types like strings, arrays and floating-point numbers. The lack of proper support for these types is probably the main reason CP solvers are not widely used in software analysis, where SMT solvers are typically preferred. However, recent progress in CP (in particular clause learning) makes the modelling and solving tasks more feasible.

In this section we explain how we encode the path conditions generated by DSE as CP models, focusing in particular on how we handle JavaScript variables and objects. It is important to note that correctness and completeness are not strict requirements in this particular context. Indeed, the nature of JavaScript requires a compromise between a faithful mapping of the language's semantics and the complexity of the resulting CP model. Fortunately, difficult JavaScript constructs can be ignored or approximated. While this affects the correctness of the resolution, the ramifications are not dramatic for test data generation: the worst case outcome is that we fail to achieve optimal test coverage. This should be acceptable if "good enough" coverage is reached in a relatively short time.

3.1   JavaScript Variables

JavaScript is dynamic and weakly typed. Variables in JavaScript do not have statically defined types, but refer to heterogeneous values that may vary during program execution. A variable can have a *primitive* type (null, undefined, Boolean, number or string) or an *object* type. In particular, null and undefined types each have only one possible value (**null** and **undefined** respectively), a Boolean is either **true** or **false**, strings are encoded in UTF-16 format, and numbers are represented as 64-bit IEEE 754 floating-point values. Objects are collections of properties mapping a name (a string) to an arbitrary JavaScript value. For each symbolic JavaScript variable x, we define a triple $\langle type(x), sval(x), addr(x) \rangle$ of CP variables such that: (i) $type(x)$ encodes the type of x; (ii) $sval(x)$ represents the *string value* of x; (iii) $addr(x)$ models the *memory address* of x.

The domain of $type(x)$ is $\mathbb{T} = \{Null, Undef, Bool, Num, Str, Obj\}$.[2] Note that $\mathbb{T}$ can be arbitrarily extended; e.g., the current implementation also considers JavaScript global objects like **Array** and **Function** as standalone types. However, for simplicity, in this paper we do not consider extensions of $\mathbb{T}$. The string variable $sval(x)$ defines the string representation of x, i.e., the value returned by JavaScript when coercing x to a string. As aforementioned, these coercions frequently occur during JavaScript program executions. For example, we have that $type(x) \in \{Null, Bool\} \Rightarrow sval(x) \in \{\texttt{"null"}, \texttt{"true"}, \texttt{"false"}\}$, while $sval(x) = \texttt{"42"} \Rightarrow type(x) \in \{Num, Str\}$ because x can be either the number 42 or the string "42". The value of $addr(x)$ is instead a natural number that can be seen as a logical address of x. If $addr(x) = 0$ then x is a constant, primitive value; otherwise, $addr(x)$ uniquely identifies object x (see Sects. 3.2, 3.3).

$$type(x) = Undef \Rightarrow (sval(x) = \texttt{"undefined"} \wedge addr(x) = 0) \tag{1}$$

$$type(x) = Null \Rightarrow (sval(x) = \texttt{"null"} \wedge addr(x) = 0) \tag{2}$$

$$type(x) = Bool \Rightarrow sval(x) \in \{\texttt{"true"}, \texttt{"false"}\} \tag{3}$$

$$type(x) = Num \Rightarrow sval(x) \in \mathcal{NS} \tag{4}$$

$$type(x) = Obj \Rightarrow sval(x) = \texttt{"[object Object]"} \tag{5}$$

$$0 \le addr(x) \le N_{addr} \wedge (addr(x) = 0 \Rightarrow type(x) \ne Obj) \tag{6}$$

Fig. 2. Invariants for $type(x), sval(x), addr(x)$.

Figure 2 shows some of the invariants we enforce to keep $type(x)$, $sval(x)$, $addr(x)$ in a consistent state. Implications 1 and 2 handle the cases where x is undefined or null ($addr(x) = 0$ because x is a constant having no properties). If x is a Boolean variable (implication 3) then $sval(x)$ is either **"true"** or **"false"**. Note that we do not impose any condition on $addr(x)$. If $addr(x) = 0$, we refer to a Boolean constant; otherwise, to the corresponding *wrapper object*. For our

---

[2] We treat $\mathbb{T}$ as an enumeration where $Null = 1, Undef = 2, \ldots, Obj = 6$.

purposes, wrappers are only necessary when we explicitly access a property of x (e.g., $x[\texttt{"length"}]$ when $type(x) = Str$). Otherwise, we can safely treat x as a constant value. Invariant 4 says that if x is a number, then its string value must represent a number. In the current implementation, the language $\mathcal{NS}$ of numeric strings is denoted by the following regular expression:

$$(\texttt{NaN} \mid (\varepsilon \mid \texttt{-})\texttt{Infinity} \mid \texttt{0} \mid (\varepsilon \mid \texttt{-})\texttt{[1-9][0-9]}^*).$$

However, $\mathcal{NS}$ can be extended to handle exponentials, hexadecimal, and floats. Implication 5 defines the string representation of objects according to ECMAScript specifications [17]. Note that if we also consider other global objects, this invariant is no longer true (e.g., the string value of an **Array** object is the comma-separated concatenation of the array elements). Finally, invariant 6 defines the address space. The constant N_{addr} is the upper bound for each address (no greater than the number of symbolic variables involved in the path condition).

3.2   JavaScript Objects

A JavaScript object is essentially a dictionary that maps strings to JavaScript values. While SMT has a well-defined theory of arrays, parametric in the types of keys and values, CP offers no native encodings for array variables and constraints. Thus, in order to model the semantics of JavaScript objects, we devised a proper CP encoding of arrays. Assuming for the moment a fixed and finite set of attributes, a simple encoding would be to introduce a variable for the value of each attribute in each object. Inspecting the value of an attribute just returns the corresponding variable; destructive updates involve creating a new object, equal to the original in all attributes except the updated one. Unfortunately, this encoding is rather large, and tends to propagate poorly.

Francis et al. [18] invert this model: instead of storing the state of each object at each time, the model records the *history of evolution* of the attribute of interest. In this representation, encoding an attribute write $write(O, attr, val)$ simply appends a cell to the "history of writes"—basically, an array storing subsequent attribute writes. The encoding of $read(O, attr)$ must then select the most recent (if any) write to $attr$ on object O from the history array. If no matching write occurred, the read falls through to a default value.

JavaScript poses a further difficulty: since attributes are arbitrary strings, the set of indices is unbounded. Further, because of aliasing, we cannot even determine statically *which* objects are being written to. To handle non-fixed indices, Plazar et al. [39] exploit the observation that, because a finite sequence of reads and writes can only affect a bounded number of indices, it is possible to emulate an unbounded mapping with bounded arrays using an indirection.

To encode destructive update of JavaScript objects, we combine these two approaches: we record the evolution of the program as a sequence of $\langle O, attr, val \rangle$ tuples, plus a bounded number of additional entries which are read without being written (and for built-in attributes, discussed later). Then encoding a read amounts to identifying the latest $\langle O, attr, val \rangle$ tuple for a given O and $attr$.

We define five arrays of CP variables $OAddr, PName, PType, PSval$ and $PAddr$ such that: $OAddr$ stores the address of the objects to uniquely identify them, $PName$ stores the property names (i.e., the keys), while $PType, PSval$, and $PAddr$ store the property values. We model each property write $O[x] \leftarrow y$ with a predicate $write(O, x, y, i)$ such that:

$$write(O, x, y, i) \iff OAddr[i] = addr(O) \land PName[i] = sval(x) \land$$
$$PType[i] = type(y) \land PSval[i] = sval(y) \land PAddr[i] = addr(y)$$

where $i > 0$ is a *property index* necessary to handle property overwriting: if $O[x] \leftarrow y$ happens *before* a write $O[x] \leftarrow y'$ then we have two corresponding writes $write(O, x, y, i)$ and $write(O, x, y', i')$ such that $i < i'$. Hence, we have to track the *temporal order* of the writes: a property index is nothing but a sequence number identifying the time instant of a given write. Each time we have a new write, this sequence number must be incremented.

A property read $y \leftarrow O[x]$ is modelled by a function $read(O, x, T)$ returning the proper index $0 \leq i \leq T$ for $O[x]$. Formally:

$$read(O, x, T) = i \iff 0 \leq i \leq T \land type(O) \notin \{Null, Undef\} \qquad \land$$
$$O = [O, OAddr[1], \ldots, OAddr[T]][i] \qquad \land$$
$$x = [x, PName[1], \ldots, PName[T]][i] \qquad \land$$
$$\forall_{j=1,\ldots,T} : (O = OAddr[j] \land x = PName[j]) \Rightarrow j \leq i$$

where the upper bound T is needed to exclude property reads that still are to happen. Note that T is an input constant that can be pre-computed *before* the solving with a counter incremented at each property write.

Index 0 is returned if $O[x]$ is not defined: since in this case JavaScript returns undefined, we set $OAddr[0] = PAddr[0] = 0, PType[0] = Undef, PSval[0] =$ "undefined". Let us suppose, e.g., that O is a symbolic object and after i property writes the following statements are executed sequentially:

$$y \leftarrow O[x]; \quad O[x] \leftarrow z; \quad y' \leftarrow O[x]$$

This is modelled by:

$$j = read(O, x, i) \quad \land \quad write(O, x, z, i+1) \quad \land \quad j' = read(O, x, i+1) \quad \land$$
$$type(y) = PType[j] \quad \land \quad sval(y) = PSval[j] \quad \land \quad addr(y) = PAddr[j] \quad \land$$
$$type(y') = PType[j'] \quad \land \quad sval(y') = PSval[j'] \quad \land \quad addr(y') = PAddr[j']$$

It is fundamental to set a precise upper bound for reads: e.g., if the first read was $j = read(O, x, i+1)$, the above constraint would hold only if $z = y$.

JavaScript has a number of *builtin* properties (e.g., the length property for strings and arrays) that can be read and overwritten. We handle them by simulating their writing *before* the program execution, i.e., the index of a builtin property will always be lower than the index of the first property accessed in the program execution. For example, we treat the indices 0, 1, 2, ... of a symbolic string ω as builtin properties "0", "1", "2", ... of a `String` object wrapping ω.

We approximate the deletion of $O[x]$ with a write $O[x] \leftarrow$ `undefined`. This does not agree exactly with the JavaScript semantics but, as already mentioned, we need some relaxations to avoid overloading the solver. For example, although prototype chains are allowed, if object O does not have property x, the function $read(O, x, T)$ returns 0 without checking if there exists a prototype O' of O having property x.

3.3 Other JavaScript Constructs

From the encodings described above we can model most of the other JavaScript operations. For example, a common JavaScript operation is the *strict comparison* $x === y$. This relation holds if x and y have the same type, the same value (different from `NaN`) and, if one is a non-wrapper object, x and y must be exactly the same object (see the example in Fig. 1). We encode the strict comparison as:

$$
\begin{aligned}
x === y \iff{}& type(x) = type(y) \ \wedge\ sval(x) = sval(y) && \wedge \\
& (type(x) = Num \Rightarrow sval(x) \neq \text{"NaN"}) && \wedge \\
& (type(y) = Num \Rightarrow sval(y) \neq \text{"NaN"}) && \wedge \\
& ((type(x) = Obj \ \vee\ type(y) = Obj) \Rightarrow addr(x) = addr(y))
\end{aligned}
$$

We model the semantics of other JavaScript operations such as $==, !==, !=$, $<, \leq, >, \geq, +, -, /, \%$, `indexOf`, `charAt`, `concat`, `slice`, `substr`, and regular expression testing. Some of them need special attention because the semantics of the operation depends on the type of the operators. For example, $x < y$ refers to lexicographic order if $type(x) = type(y) = Str$, otherwise arithmetic comparison is performed (via coercion to numbers). Analogously, $z = x + y$ can refer to either the string concatenation or arithmetic addition. Note that in the current implementation we use channelling functions to convert strings to integers and vice versa. An alternative solution might be to use, in addition to the string value $sval(x)$, a CP integer variable $ival(x)$ to keep track of the integer value of x. For example, if $x =$ `true` then $ival(x) = 1$. The tricky part in this representation is to encode a non-integer value: if we use a special integer $v \in \mathbb{Z}$ to represent a JavaScript value x not convertible to integers, then we have to discriminate whether $ival(x) = v$ means "not an integer" or the actual number v.

The CP encoding we propose is implemented in the MiniZinc language.[3] Each solver supporting MiniZinc can therefore solve the resulting model. We remark that, due to the fundamental role played by strings in JavaScript, we are using the MiniZinc extension with string variables [3]. Clearly a dedicated string solver like G-STRINGS is currently the best candidate to solve these models, but other solvers could be used by essentially converting strings to arrays of integers.

---

[3] Publicly available at https://bitbucket.org/robama/g-strings/src/master/gecode-5. 0.0/gecode/flatzinc/javascript.

4 Implementation: ARATHA

We implemented our DSE framework into a tool we call ARATHA.[4] We followed the standard implementation strategy of DSE systems. After annotating the program with symbolic inputs, we begin by running the program with a concrete seed input. During this execution, we record which branches were taken and then construct the path condition corresponding to the input. A set of new path conditions is obtained by negating the last element of each non-empty prefix of the current path condition, which are then appended to the exploration queue. We then take the next path condition in the queue, use a constraint solver to obtain a satisfying input, and repeat the process. Figure 3 presents a graphical summary of the system.

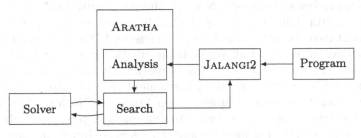

Fig. 3. The architecture of ARATHA.

4.1 Extracting Path Conditions

Branching information is extracted by running an instrumented program. This is performed via *source-to-source* rewriting using JALANGI2, the successor to JALANGI [44]. That is, instead of writing or modifying a JavaScript interpreter, we insert instrumentation into the source code itself. The instrumenter is invoked whenever new code is accessed, allowing us to analyze code executed using the eval() function. We then use the analysis interface of JALANGI2 to intercept, rewrite and record all operations involving symbolic values. All conditional branches depending on symbolic values are added to the current path condition. Apart from if-then-else statements and loops, the logical operators are also treated as conditionals, owing to short-circuit evaluation. This does cause some loss of efficiency, as it pessimistically assumes that any logical operation might involve an expression containing side effects.

In the program, symbolic variables are obtained by calling J$.readInput(), returning a pair containing the concrete value to be used in the current run, and a symbolic value representing an input variable. Essentially, any concrete value can have a symbolic expression associated to it. That expression is a record of the operations that produced the associated concrete value. If the program performs an operation we cannot trace symbolically (e.g., a call to a library function), or an operation that we cannot model properly, we return only the concrete result and we throw away the symbolic expression. This is a common approach in DSE.

---

[4] Publicly available at https://github.com/ArathaJS/aratha.

4.2   Source-to-Source Translation

Source-to-source translation brings with it two advantages. Firstly, the analysis is independent of any particular interpreter, and is therefore not sensitive to changes in system architecture. This is especially valuable, as JavaScript interpreters are a moving target, due to the never-ending search for improved performance. Secondly, this allows us to reap the efficiency benefits of those same optimized interpreters. Although instrumentation introduces some overhead, modern interpreters can nearly eliminate its impact. The cost is, however, that our analysis relies on having as much code instrumented as possible.

Though concretization allows DSE to be run on programs containing uninstrumented code, coverage quickly becomes limited as more and more symbolic expressions become concrete. Notably, though primitive values can be concretized with little impact, any object that is passed to an un-instrumented function must have its entire object graph concretized. This can result in a cascade effect that strips almost all symbolic information from that point.

One thing to note is that, regardless of method, it is difficult to fully mask the presence of instrumentation. As we instrument all JavaScript operations, we can rewrite operations such as introspection functions which could reveal the presence of instrumentation. However, a timing-sensitive program might still be disrupted by the overhead of instrumentation. As most JavaScript programs do not interrogate their execution environment, we have not attempted to handle such things in much depth.

4.3   Backend Solving and Optimizations

ARATHA can model path conditions in both the MiniZinc and the SMT-LIB [12] constraint languages. This means it is compatible with any constraint solver supporting either of those languages, as long as it also supports the string extensions. The SMT-LIB output relies on a partial axiomatization of JavaScript's semantics. This axiomatization is itself written in the SMT-LIB 2 language, and is hence independent of any particular solver.

To the best of our knowledge, the only mature SMT solvers that currently support *all* the theories it requires are Z3 and CVC4. Note that previous systems such as Kudzu [41], JALANGI [44], SymJS [30] and ExpoSE [34] were all designed for use with particular constraint solvers. As such, our implementation is the first multi-solver DSE tool. By enabling the use of both CP and SMT solvers, we can potentially benefit from the strengths of both.

Our analysis runs on Node.js, which uses the highly efficient V8 JavaScript interpreter. However, it is constraint solving rather than program execution that dominates execution time in DSE. As such, we implemented a number of optimizations in an attempt to make solving more efficient. To reduce the number of solver queries and mitigate the memory impact of storing symbolic expression trees, we perform computations concretely whenever it is possible to do so without losing precision. For instance, the unary void operator always returns `undefined`, regardless of its argument. Similarly, we eagerly simplify read operations on properties of symbolic objects. In many such cases, we can determine the property's value unambiguously and hence return just that particular value.

We also attempt to use type information to simplify expressions whenever possible. Though ARATHA fully supports JavaScript's type system, it is beneficial for both performance and understandability to simplify type-dependent operations as early as possible. As many functions only return values of a specific type, we can frequently choose which "overload" of a function to invoke. For instance, if both of the arguments of a + operation are numbers, we can use the more specific numeric addition instead of the general JavaScript addition function.

In terms of back-end optimizations, ARATHA can submit constraint queries incrementally to a supporting solver. Constraint solvers which support incremental solving can reuse previous work when answering a query, potentially yielding a result much more quickly. However, such functionality is at present generally provided only by SMT solvers because CP does not handle incremental solving.

ARATHA deals with loops and recursions by setting a parameter that limits the maximum number of iterations allowed. This is a common approach in DSE.

5   Evaluation

We now use ARATHA to assess the performance of CP and SMT solvers within our tool. We emphasize that it is not our goal to make a comparison of different DSE tools. Such a comparison would be difficult because of the limited availability and development of JavaScript DSE tools so far. Rather, we have wanted to test the hypothesis that CP is a valuable option for software analysis, when used in synergy with (not necessarily in place of) SMT or SAT solving technologies. We also underline that the performance of the individual solvers also depends on the SMT/CP encoding we chose for the JavaScript semantics: different models may lead to a different performance.

We compared four different solvers: the CP-based string solver G-STRINGS and the SMT solvers CVC4 [32], Z3 [37], together with Z3's most recent string solver extension Z3STR3 [13]. For each path condition, we set a small timeout of $T_{pc} = 10$ s. This is because DSE implies a high number of path conditions having a limited number of constraints. Moreover, setting a too high value of T_{pc} would be unnecessarily harmful given the heavy-tailed nature of solving: these problems are typically either solved in few seconds, or not solvable at all within hours of computation. We set a maximum number of $N = 1024$ DSE iterations for each problem, and also an overall timeout $T_{tot} = 300$ s, because sometimes reaching N iterations can take too long.[5] We ran all the experiments on an Ubuntu 15.10 machine with 16 GB of RAM and 2.60 GHz Intel® i7 CPU.

Unfortunately, there are no standard benchmarks for JavaScript DSE. Moreover, retrieving large JavaScript benchmarks is tedious because the source-to-source rewriting of ARATHA needs a manual instrumentation for the symbolic

---

[5] T_{tot} is also useful because CVC4 may get stuck in presolving regardless of T_{pc} limit. (see http://cvc4.cs.stanford.edu/wiki/User_Manual#Resource_limits).

input. We therefore tested ARATHA on the test suite of EXPOSE, consisting of 197 already annotated JavaScript sources. This is not a DSE benchmark in the strict sense, but it is however very useful in this context. We did not compare ARATHA against EXPOSE because ARATHA does not yet fully support complex JavaScript operations such as backreferences and greedy matching.

Table 1. Average results and cross comparisons between solvers

Solver	% statements	Time [s]	Inputs	Timeouts	G-STRINGS	CVC4	Z3	Z3STR3
G-STRINGS	**82.85**	4.74	**3.54**	0	—	**41**	**73**	**112**
CVC4	77.25	33.08	2.83	21	3	—	51	98
Z3	72.81	3.06	3.01	1	11	19	—	66
Z3STR3	62.69	**0.46**	1.80	0	0	2	11	—

The results are shown in Table 1, where we see the average coverage of statements in the entire test suite, the average time to process each problem in the suite, the average number of unique inputs generated, and the total number of times the overall timeout T_{tot} was reached. Clearly the CP approach implemented by G-STRINGS is competitive with SMT methods: it is fast and provides the best coverage. Note that being fast is not always good in this context, because a solver can terminate its execution in a few seconds without yielding significant inputs. This is the case of Z3 and Z3STR3: they are the fastest solvers, but they have the smallest coverage (in particular Z3STR3 appears unstable on these problems). However, we remark that this performance should not be taken as an absolute value because it also depends on the SMT encodings we chose.

The average coverage of CVC4 is closer to that of G-STRINGS. Its high average time is slightly misleading, as it is mainly due to the high number of timeouts. In fact, for 130 cases CVC4 is faster in reaching (at least) the same coverage as G-STRINGS. This suggests that CP and SMT solvers should not be seen as mutually exclusive, but possibly cooperating via a *portfolio* approach [4, 10] that aims to select and run the best solver(s)—possibly in parallel and by exchanging information—for a given path condition.

The second part of the table shows the number of times the solver for that row reaches a strictly better coverage than the solver for that column. On this measure, G-STRINGS has the best performance. However, there are cases where CVC4 and Z3 achieve a better coverage.

6 Related Work

The main ideas behind DSE go back to Godefroid, Klarlund and Sen's *DART* project [21]. Since then, advances in solver technology saw DSE tools improve rapidly, in some cases finding large-scale use. For example, Microsoft's *SAGE* [22]

DSE tool reportedly detected up to one third of all bugs discovered during the development of Windows 7—bugs that were missed by other testing methods.

DSE was first applied to JavaScript programs in the *Kudzu* project [41]. Existing solvers were found inadequate for the task of reasoning about JavaScript behaviour, for a number of reasons, including JavaScript's orientation towards strings as a default data structure. Hence a major part of the Kudzu project turned out to be the development of a dedicated string + bitvector solver, *Kaluza*.

SymJS [30] is a symbolic execution and fuzzing tool for JavaScript. It relies on the *PASS* [31] solver, and it uses a model of the DOM combined with an intelligent, feedback-driven event generator to automatically test web applications.

EXPOSE [34] is the first JavaScript DSE tool able to reason about string matching via (extended) regular expressions, although in a limited fashion. It uses Z3 for constraint solving and it has been applied successfully to several important `Node.js` libraries, though overall coverage is relatively low because of the limited nature of the analysis.

ARATHA is the first JavaScript DSE tool capable of reasoning about inputs without resorting to unsound heuristic type assignments or requiring the user to commit to the type of each input in advance. It allows for easy replacement of constraint solver. It is built using *Jalangi* 2 [44], a framework for implementing dynamic analyses for JavaScript.

Meaningful comparison of the JavaScript DSE tools discussed here is hampered by their limited availability. Comparison of different DSE *backends*, i.e., constraint solvers focused on the types of constraints typically generated in dynamic analysis of JavaScript, is a simpler task, provided we have a DSE tool that allows for easy backend plugging and unplugging. ARATHA does exactly that.

String solvers are still in their infancy and current solvers naturally show varying degrees of robustness. Many are based on the DPLL(T) paradigm [19], including CVC4 [33], Z3str* family [13,51,52], S3* family [48–50], and Norn [2]. More recent proposals are Sloth [24] and Trau [1]. These solvers handle constraints over strings of unbounded length; however, they are known to be incomplete. Z3STR in particular claims to be complete for the set of positive formulas in the theory of concatenation and linear integer arithmetic in length, however, its successors are of a different design and have not made such promises.

Some solvers provide finite decision procedures by stipulating an upper bound on the length of strings (e.g., HAMPI [27] and Kaluza [41]). G-STRINGS [7, 8] takes a propagation based approach to bounded string solving, where the complexity weakly depends on the length bound.

7   Conclusions

In this paper we have described how to build a dynamic symbolic execution tool for JavaScript, utilising an underlying CP solver. Critical to this approach is the ability to translate the complex object behaviour of JavaScript into a set of constraints that are handled by a CP solver. In particular for JavaScript, since strings are essential to almost any operation in the language, our tool makes use of string extensions of the MiniZinc language [3] and the efficient string constraint solving capabilities of the dashed-string solver G-STRINGS [8].

Our experiments suggest that CP solvers can be competitive with state-of-the-art SMT solvers, for the kind of constraints that arise in dynamic symbolic execution of JavaScript. In particular, a *portfolio* consisting of both SMT and CP solvers might turn out to be a good strategy for maximizing code coverage and minimising the DSE time.

Important future work to improve the CP approach is to extend CP constraints to do equality propagation, to propagate more information from object constraints. Extending the string solver to be usable in a nogood solver should also significantly improve CP solving times.

Acknowledgments. This work is supported by the Australian Research Council (ARC) through Linkage Project Grant LP140100437 and Discovery Early Career Researcher Award DE160100568.

References

1. Abdulla, P.A., et al.: Flatten and conquer: a framework for efficient analysis of string constraints. In: Proceedings of the 38th ACM SIGPLAN Conference on Programming Language Design and Implementation, PLDI 2017, Barcelona, Spain, 18–23 June 2017, pp. 602–617 (2017)
2. Abdulla, P.A., et al.: Norn: an SMT solver for string constraints. In: Kroening, D., Păsăreanu, C.S. (eds.) CAV 2015. LNCS, vol. 9206, pp. 462–469. Springer, Cham (2015). https://doi.org/10.1007/978-3-319-21690-4_29
3. Amadini, R., Flener, P., Pearson, J., Scott, J.D., Stuckey, P.J., Tack, G.: MiniZinc with strings. In: Hermenegildo, M.V., Lopez-Garcia, P. (eds.) LOPSTR 2016. LNCS, vol. 10184, pp. 59–75. Springer, Cham (2017). https://doi.org/10.1007/978-3-319-63139-4_4
4. Amadini, R., Gabbrielli, M., Mauro, J.: A multicore tool for constraint solving. In: Proceedings 24th International Joint Conference Artificial Intelligence, pp. 232–238. AAAI Press (2015)
5. Amadini, R., Gange, G., Stuckey, P.J.: Propagating LEX, FIND and REPLACE with dashed strings. In: van Hoeve, W.-J. (ed.) CPAIOR 2018. LNCS, vol. 10848, pp. 18–34. Springer, Cham (2018). https://doi.org/10.1007/978-3-319-93031-2_2
6. Amadini, R., Gange, G., Stuckey, P.J.: Propagating regular membership with dashed strings. In: Hooker, J. (ed.) CP 2018. LNCS, vol. 11008, pp. 13–29. Springer, Cham (2018). https://doi.org/10.1007/978-3-319-98334-9_2
7. Amadini, R., Gange, G., Stuckey, P.J.: Sweep-based propagation for string constraint solving. In: Proceedings 32nd AAAI Conference Artificial Intelligence, pp. 6557–6564. AAAI Press (2018)
8. Amadini, R., Gange, G., Stuckey, P.J., Tack, G.: A novel approach to string constraint solving. In: Beck, J.C. (ed.) CP 2017. LNCS, vol. 10416, pp. 3–20. Springer, Cham (2017). https://doi.org/10.1007/978-3-319-66158-2_1
9. Amadini, R., et al.: Combining string abstract domains for JavaScript analysis: an evaluation. In: Legay, A., Margaria, T. (eds.) TACAS 2017. LNCS, vol. 10205, pp. 41–57. Springer, Heidelberg (2017). https://doi.org/10.1007/978-3-662-54577-5_3

10. Amadini, R., Stuckey, P.J.: Sequential time splitting and bounds communication for a portfolio of optimization solvers. In: O'Sullivan, B. (ed.) CP 2014. LNCS, vol. 8656, pp. 108–124. Springer, Cham (2014). https://doi.org/10.1007/978-3-319-10428-7_11

11. Artzi, S., et al.: Finding bugs in web applications using dynamic test generation and explicit-state model checking. IEEE Trans. Software Eng. **36**(4), 474–494 (2010)

12. Barrett, C., Fontaine, P., Tinelli, C.: The SMT-LIB standard: Version 2.6. Technical report, Department of Computer Science, University of Iowa (2017). www.SMT-LIB.org

13. Berzish, M., Ganesh, V., Zheng, Y.: Z3str3: a string solver with theory-aware heuristics. In: Stewart, D., Weissenbacher, G. (eds.) Proceedings 17th Conference Formal Methods in Computer-Aided Design, pp. 55–59. FMCAD Inc. (2017)

14. Blanc, B., Junke, C., Marre, B., Gall, P.L., Andrieu, O.: Handling state-machines specifications with GATeL. Electr. Notes Theor. Comput. Sci. **264**(3), 3–17 (2010). https://doi.org/10.1016/j.entcs.2010.12.011

15. Cadar, C., Dunbar, D., Engler, D.: KLEE: unassisted and automatic generation of high-coverage tests for complex systems programs. In: Proceedings 8th USENIX Conference Operating Systems Design and Implementation, OSDI, vol. 8, pp. 209–224 (2008)

16. Delahaye, M., Botella, B., Gotlieb, A.: Infeasible path generalization in dynamic symbolic execution. Inf. Softw. Technol. **58**, 403–418 (2015)

17. ECMA International: Ecmascript 2018 language specification (2018). https://www.ecma-international.org/publications/files/ECMA-ST/Ecma-262.pdf

18. Francis, K., Navas, J., Stuckey, P.J.: Modelling destructive assignments. In: Schulte, C. (ed.) CP 2013. LNCS, vol. 8124, pp. 315–330. Springer, Heidelberg (2013). https://doi.org/10.1007/978-3-642-40627-0_26

19. Ganzinger, H., Hagen, G., Nieuwenhuis, R., Oliveras, A., Tinelli, C.: DPLL(T): fast decision procedures. In: Alur, R., Peled, D.A. (eds.) CAV 2004. LNCS, vol. 3114, pp. 175–188. Springer, Heidelberg (2004). https://doi.org/10.1007/978-3-540-27813-9_14

20. Gecode Team: Gecode: Generic constraint development environment (2016). http://www.gecode.org

21. Godefroid, P., Klarlund, N., Sen, K.: DART: directed automated random testing. In: Proceedings ACM SIGPLAN Conference Programming Language Design and Implementation (PLDI 2005), pp. 213–223. ACM (2005)

22. Godefroid, P., Levin, M.Y., Molnar, D.: SAGE: whitebox fuzzing for security testing. Commun. ACM **55**(3), 40–44 (2012)

23. Gotlieb, A.: TCAS software verification using constraint programming. Knowl. Eng. Rev. **27**(3), 343–360 (2012). https://doi.org/10.1017/S0269888912000252

24. Holík, L., Janku, P., Lin, A.W., Rümmer, P., Vojnar, T.: String constraints with concatenation and transducers solved efficiently. PACMPL **2**(POPL), 4:1–4:32 (2018)

25. Hooimeijer, P., Weimer, W.: StrSolve: solving string constraints lazily. Autom. Softw. Eng. **19**(4), 531–559 (2012)

26. Kashyap, V., et al.: JSAI: a static analysis platform for JavaScript. In: Proceedings 22nd ACM SIGSOFT International Symposium Foundations of Software Engineering, pp. 121–132. ACM (2014)

27. Kiežun, A., Ganesh, V., Artzi, S., Guo, P.J., Hooimeijer, P., Ernst, M.D.: HAMPI: a solver for word equations over strings, regular expressions, and context-free grammars. ACM Trans. Softw. Eng. Methodol. **21**(4) (2012). Article 25

28. King, J.C.: Symbolic execution and program testing. Commun. ACM **19**(7), 385–394 (1976)
29. Lee, H., Won, S., Jin, J., Cho, J., Ryu, S.: SAFE: formal specification and implementation of a scalable analysis framework for ECMAScript. In: Proceedings 19th International Workshop on Foundations of Object-Oriented Languages (FOOL 2012) (2012)
30. Li, G., Andreasen, E., Ghosh, I.: SymJS: automatic symbolic testing of JavaScript web applications. In: Proceedings 22nd ACM SIGSOFT International Symposium Foundations of Software Engineering, pp. 449–459. ACM (2014)
31. Li, G., Ghosh, I.: PASS: string solving with parameterized array and interval automaton. In: Bertacco, V., Legay, A. (eds.) HVC 2013. LNCS, vol. 8244, pp. 15–31. Springer, Cham (2013). https://doi.org/10.1007/978-3-319-03077-7_2
32. Liang, T., Reynolds, A., Tinelli, C., Barrett, C., Deters, M.: A DPLL(T) theory solver for a theory of strings and regular expressions. In: Biere, A., Bloem, R. (eds.) CAV 2014. LNCS, vol. 8559, pp. 646–662. Springer, Cham (2014). https://doi.org/10.1007/978-3-319-08867-9_43
33. Liang, T., Reynolds, A., Tsiskaridze, N., Tinelli, C., Barrett, C., Deters, M.: An efficient SMT solver for string constraints. Formal Methods Syst. Des. **48**(3), 206–234 (2016)
34. Loring, B., Mitchell, D., Kinder, J.: ExpoSE: practical symbolic execution of standalone JavaScript. In: Proceedings 24th ACM SIGSOFT International SPIN Symposium Model Checking of Software, pp. 196–199. ACM (2017)
35. Majumdar, R., Sen, K.: Hybrid concolic testing. In: Proceedings 29th International Conference Software Engineering (ICSE 2007), pp. 416–426. IEEE (2007)
36. Majumdar, R., Xu, R.-G.: Reducing test inputs using information partitions. In: Bouajjani, A., Maler, O. (eds.) CAV 2009. LNCS, vol. 5643, pp. 555–569. Springer, Heidelberg (2009). https://doi.org/10.1007/978-3-642-02658-4_41
37. de Moura, L., Bjørner, N.: Z3: an efficient SMT solver. In: Ramakrishnan, C.R., Rehof, J. (eds.) TACAS 2008. LNCS, vol. 4963, pp. 337–340. Springer, Heidelberg (2008). https://doi.org/10.1007/978-3-540-78800-3_24
38. Nethercote, N., Stuckey, P.J., Becket, R., Brand, S., Duck, G.J., Tack, G.: MiniZinc: towards a standard CP modelling language. In: Bessière, C. (ed.) CP 2007. LNCS, vol. 4741, pp. 529–543. Springer, Heidelberg (2007). https://doi.org/10.1007/978-3-540-74970-7_38
39. Plazar, Q., Acher, M., Bardin, S., Gotlieb, A.: Efficient and complete FD-solving for extended array constraints. In: Sierra, C. (ed.) Proceedings 26th International Joint Conference Artificial Intelligence, pp. 1231–1238 (2017). ijcai.org
40. Rossi, F., van Beek, P., Walsh, T. (eds.): Handbook of Constraint Programming. Elsevier, New York (2006)
41. Saxena, P., Akhawe, D., Hanna, S., Mao, F., McCamant, S., Song, D.: A symbolic execution framework for JavaScript. In: Proceedings 2010 IEEE Symposium Security and Privacy, pp. 513–528. IEEE Computer Socience (2010)
42. Scott, J.D., Flener, P., Pearson, J., Schulte, C.: Design and implementation of bounded-length sequence variables. In: Salvagnin, D., Lombardi, M. (eds.) CPAIOR 2017. LNCS, vol. 10335, pp. 51–67. Springer, Cham (2017). https://doi.org/10.1007/978-3-319-59776-8_5
43. Sen, K., Agha, G.: CUTE and jCUTE: concolic unit testing and explicit path model-checking tools. In: Ball, T., Jones, R.B. (eds.) CAV 2006. LNCS, vol. 4144, pp. 419–423. Springer, Heidelberg (2006). https://doi.org/10.1007/11817963_38

44. Sen, K., Kalasapur, S., Brutch, T.G., Gibbs, S.: Jalangi: a selective record-replay and dynamic analysis framework for JavaScript. In: Joint Meeting of the European Software Engineering Conference and the ACM SIGSOFT Symposium Foundations of Software Engineering, pp. 488–498 (2013)
45. Sen, K., Marinov, D., Agha, G.: CUTE: a concolic unit testing engine for C. In: Proceedings 10th European Software Engineering Conference, pp. 263–272. ACM (2005). https://doi.org/10.1145/1081706.1081750
46. Tateishi, T., Pistoia, M., Tripp, O.: Path- and index-sensitive string analysis based on monadic second-order logic. ACM Trans. Softw. Eng. Methodol. **22**(4) (2013). Article 33
47. Tillmann, N., de Halleux, J.: Pex–white box test generation for.NET. In: Beckert, B., Hähnle, R. (eds.) TAP 2008. LNCS, vol. 4966, pp. 134–153. Springer, Heidelberg (2008). https://doi.org/10.1007/978-3-540-79124-9_10
48. Trinh, M.T., Chu, D.H., Jaffar, J.: S3: a symbolic string solver for vulnerability detection in web applications. In: Proceedings 2014 ACM SIGSAC Conference Computer and Communications Security, pp. 1232–1243. ACM (2014)
49. Trinh, M.-T., Chu, D.-H., Jaffar, J.: Progressive reasoning over recursively-defined strings. In: Chaudhuri, S., Farzan, A. (eds.) CAV 2016. LNCS, vol. 9779, pp. 218–240. Springer, Cham (2016). https://doi.org/10.1007/978-3-319-41528-4_12
50. Trinh, M.-T., Chu, D.-H., Jaffar, J.: Model counting for recursively-defined strings. In: Majumdar, R., Kunčak, V. (eds.) CAV 2017. LNCS, vol. 10427, pp. 399–418. Springer, Cham (2017). https://doi.org/10.1007/978-3-319-63390-9_21
51. Zheng, Y., Ganesh, V., Subramanian, S., Tripp, O., Dolby, J., Zhang, X.: Effective search-space pruning for solvers of string equations, regular expressions and length constraints. In: Kroening, D., Păsăreanu, C.S. (eds.) CAV 2015. LNCS, vol. 9206, pp. 235–254. Springer, Cham (2015). https://doi.org/10.1007/978-3-319-21690-4_14
52. Zheng, Y., Zhang, X., Ganesh, V.: Z3-str: a Z3-based string solver for web application analysis. In: Proceedings 9th Joint Meeting on Foundations of Software Engineering, pp. 114–124. ACM (2013)

Sequential and Parallel Solution-Biased Search for Subgraph Algorithms

Blair Archibald[1] , Fraser Dunlop[2] , Ruth Hoffmann[2] ,
Ciaran McCreesh[1(✉)] , Patrick Prosser[1] , and James Trimble[1]

[1] University of Glasgow, Glasgow, Scotland
ciaran.mccreesh@glasgow.ac.uk
[2] University of St Andrews, St Andrews, Scotland

Abstract. The current state of the art in subgraph isomorphism solving involves using degree as a value-ordering heuristic to direct backtracking search. Such a search makes a heavy commitment to the first branching choice, which is often incorrect. To mitigate this, we introduce and evaluate a new approach, which we call "solution-biased search". By combining a slightly-random value-ordering heuristic, rapid restarts, and nogood recording, we design an algorithm which instead uses degree to direct the proportion of search effort spent in different subproblems. This increases performance by two orders of magnitude on satisfiable instances, whilst not affecting performance on unsatisfiable instances. This algorithm can also be parallelised in a very simple but effective way: across both satisfiable and unsatisfiable instances, we get a further speedup of over thirty from thirty-six cores, and over one hundred from ten distributed-memory hosts. Finally, we show that solution-biased search is also suitable for optimisation problems, by using it to improve two maximum common induced subgraph algorithms.

1 Introduction

The subgraph isomorphism problem is to decide whether a copy of a small "pattern" graph occurs inside a larger "target" graph. The problem is broadly applicable, arising in areas including bioinformatics [2], chemistry [46], computer vision [11,49], law enforcement [8], model checking [47], malware detection [4], compilers [1,43], pattern recognition [12], program similarity comparison [10], the design of mechanical locks [50], and graph databases [37].

Although the problem is NP-complete, by combining design techniques from artificial intelligence with careful algorithm engineering, modern subgraph isomorphism solvers can often produce exact solutions quickly even on graphs with thousands of vertices. The current single strongest subgraph isomorphism solver

This work was supported by the Engineering and Physical Sciences Research Council (grant numbers EP/P026842/1, EP/M508056/1, and EP/N007565/1). This work used the Cirrus UK National Tier-2 HPC Service at EPCC (http://www.cirrus.ac.uk) funded by the University of Edinburgh and EPSRC (EP/P020267/1).

ⓒ Springer Nature Switzerland AG 2019
L.-M. Rousseau and K. Stergiou (Eds.): CPAIOR 2019, LNCS 11494, pp. 20–38, 2019.
https://doi.org/10.1007/978-3-030-19212-9_2

uses "highest degree first" as a value-ordering heuristic to direct a constraint programming style search [25, 35, 37]. This heuristic is much better than branching randomly, but is still far from perfect. To offset mistakes made by this heuristic, this paper proposes a new perspective on value-ordering: rather than defining a search order, we use degree to direct what *proportion* of the search effort should be spent in each subproblem. By combining rapid restarts and nogood recording, and introducing a small amount of randomness into the value-ordering heuristic, we make a state-of-the-art subgraph algorithm perform two orders of magnitude better on a large number of satisfiable instances, whilst performing worse only rarely on satisfiable instances, and never on unsatisfiable instances. This strategy is also effective in an optimisation setting, producing benefits in two maximum common induced subgraph algorithms.

This new form of search can also be parallelised, with a *much* simpler implementation than conventional work-stealing. By running many threads with different random seeds but the same restart schedule, and sharing nogoods only following restarts, we can achieve aggregate speedups [20] of thirty-one from a thirty-six core machine, or over one hundred by using ten such machines.

1.1 Background

The *non-induced subgraph isomorphism problem* is to find an injective mapping from the vertices of a *pattern* graph $\mathcal{P}$ to the vertices of a *target* graph $\mathcal{T}$, such that adjacent vertices in $\mathcal{P}$ are mapped to adjacent vertices in $\mathcal{T}$ (including that vertices with loops in $\mathcal{P}$ may only be mapped to vertices with loops in $\mathcal{T}$). The *induced* problem additionally requires that non-adjacent vertices are mapped to non-adjacent vertices. The *degree* of a vertex is the number of other vertices to which it is adjacent.

This paper looks at improving the Glasgow Subgraph Solver[1], which can solve both the non-induced and the induced subgraph isomorphism problems. The solver is very closely based upon the $k\downarrow$ algorithm of Hoffmann et al. [21] with $k = 0$, and we refer the reader to that paper for full technical details; that algorithm, in turn, is a simplification and re-engineering of an older Glasgow algorithm [25, 35]. Essentially, the solver is a dedicated forward-checking constraint programming implementation specifically for subgraph problems. It works with a model having a variable per pattern graph vertex, with domains ranging over the target graph vertices, and performs a backtracking search to map each pattern vertex to a target vertex whilst propagating adjacency and injectivity constraints (together with further implied constraints based upon degrees and paths). However, it uses specialised bit-parallel data structures and algorithms, and propagates constraints in a fixed order rather than using a queue.

1.2 Experimental Setup

Our experiments are performed on the EPCC Cirrus HPC facility, on systems with dual Intel Xeon E5-2695 v4 CPUs and 256GBytes RAM, running Centos

---

[1] https://github.com/ciaranm/glasgow-subgraph-solver/.

7.3.1611. We use GCC 7.2.0 as the compiler. For parallelism, we use C++ native threads, and for distributed parallelism we also use the SGI MPT implementation of MPI. All timing measurements are steady-clock, and we use a deterministic pseudo-random number generator for reproducibility.

We use the dataset introduced by Kotthoff et al. [25] for evaluation. This dataset brings together a range of randomly-generated and application instance families from earlier papers:

BVG(r), M4D(r), and Rand are families of randomly generated graphs using different models (bounded degree, regular mesh, and uniform), where each pattern is a permuted random connected subgraph of the target (and so each instance is satisfiable) [9]. These benchmark instances are widely used, but have unusual properties and so broad conclusions should not be drawn based solely upon behaviour of these instances [37].

SF contains randomly generated scale-free graphs using a similar method [52].

LV consists of various kinds of graph gathered by Larrosa and Valiente [27] from the Stanford Graph Database. We include both the 50 small graphs, and the 50 large graphs.

Phase contains hand crafted instances that lie near the satisfiable / unsatisfiable phase transition [37].

PR contains graphs generated from segmented images, corresponding to a computer vision problem [49].

Images and Meshes contain graphs representing 2D segmented images and 3D object models, again representing a computer vision problem [11].

Other studies use a random selection of 200 of each of the instances from the "meshes" and "images" families because some earlier solvers find many of these instances extremely hard. We would like to have a larger number of satisfiable instances in our test set, and so we include all pattern/target pairs. This gives a total of 14,621 instances (rather than the original 5,725). At least 2,150 of these instances are known to be satisfiable for the non-induced problem, and at least 12,348 are unsatisfiable.

2   Improving Sequential Search

We begin with a set of baseline performance measurements. In the top two plots of Fig. 1 we show the cumulative number of instances solved over time for the non-induced and induced problems respectively. We compare the Glasgow Subgraph Solver using depth-first search (DFS) and with the modifications described in the remainder of this paper (solution-biased search, SBS), the PathLAD variation of the LAD algorithm [25, 48], VF2 [9], RI [2], and VF3 [5] (which only supports the induced problem), in each case using the original implementation provided by the algorithm's authors. The plots show that our starting point comfortably beats PathLAD, VF2, VF3 and RI, except for very low choices of timeout. For each algorithm, the y value gives the cumulative number of instances

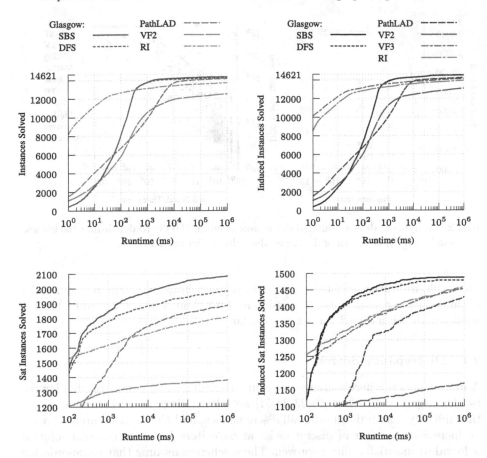

Fig. 1. On the top row, the cumulative number of instances solved over time, comparing the Glasgow Subgraph Solver (both in its basic form, and with the improvements introduced in the remainder of the paper) to other solvers, for the non-induced and induced problems. On the bottom row, the same, considering only satisfiable instances.

which (individually) can be solved in no more than x milliseconds. The vertical distance between two lines therefore shows how many more instances can be solved by one solver than another, if every instance is run separately with the chosen x timeout. The horizontal distance shows how many times longer the per-instance timeout would need to be to allow the rightmost algorithm to succeed on y out of the 14,621 instances (bearing in mind that the two sets of y instances could be different), and gives a measure called *aggregate speedup* [20].

The dataset includes many instances which are extremely easy for a good solver, and so it can be hard to see the differences between the stronger solvers at higher runtimes. This paper focusses upon improving the performance on the remaining hard satisfiable instances, and so in the bottom two plots in Fig. 1 (and in subsequent cumulative plots for sequential algorithms) we show only

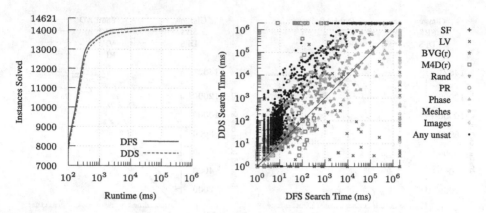

Fig. 2. Comparing depth-bounded discrepancy search (DDS) to depth-first backtracking search (DFS), both using degree as the value-ordering heuristic.

satisfiable instances, and use a reduced range on both axes. For the remainder of this paper, we show only the non-induced problem, which tends to be harder; results with the induced variant are similar.

2.1 Discrepancy Searches

A *discrepancy* is where search goes against the advice of a value-ordering heuristic. Discrepancy searches [19,23,24,51] are alternatives to backtracking search that initially search disallowing all discrepancies, and then retry search allowing an increasing number of discrepancies at each iteration until either a solution is found or unsatisfiability is proven. These schemes assume that value-ordering heuristics are usually reliable, and that most solutions can be found with only a small number of discrepancies. In such cases, the heavy commitment to early branching choices made by backtracking search can be extremely costly.

Figure 2 shows the effects of adding Walsh's [51] depth-bounded discrepancy search (DDS) to the solver (results with other discrepancy search variants are similar). On the scatter plot, each point represents the solving time for one instance—to avoid noise for easier instances, we measure only time spent during search, and exclude time spent in preprocessing and initialisation. Points below the $x - y$ diagonal are speedups, whilst points on the top and right axes represent instances which timed out after one thousand seconds with one algorithm, but not the other. For satisfiable instances, the different point styles show the different families, whilst all unsatisfiable instances are shown as dark dots. The points well below the diagonal line and along the right-hand axis on the scatter plot show that DDS can sometimes be extremely beneficial on satisfiable instances. However, on both unsatisfiable and most satisfiable instances, the overheads can be extremely large, and DDS is much worse in aggregate and is not a viable approach (even when only considering satisfiable instances). These large overheads are to be expected: discrepancy searches are aimed primarily at getting better feasible solutions in optimisation problems which are too large for a proof of optimality to be a realistic prospect, and they are not well-suited for unsatisfiable decision problems. Despite this, the extremely large gains on some satisfiable instances confirm our suspicions that we should find an alternative to heavy-commitment backtracking search.

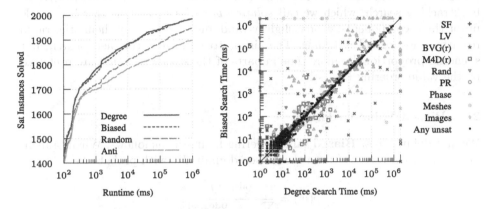

Fig. 3. On the left, the cumulative number of satisfiable instances solved over time, using four different value-ordering heuristics. On the right, an instance by instance comparison of the degree and biased heuristics on all instances. Points on the outer axes are timeouts, and point style shows instance family.

2.2   Value-Ordering Heuristics

Traditionally, value-ordering heuristics are designed to drive search towards the most promising region of the search space [14], or the most constrained [15], or the region with the highest solution density [44]. In subgraph isomorphism, this is done by selecting vertices from highest degree to lowest [37]. The left-hand plot of Fig. 3 demonstrates that this is indeed a good choice: the **Degree** heuristic's line shows much better performance on satisfiable instances than the **Random** (branch randomly) or **Anti** (branch from lowest degree to highest) heuristic lines. Meanwhile, on unsatisfiable instances, the value-ordering heuristic has no effect on performance, because a complete search must be performed.

But what happens if our value-ordering heuristic has to choose between mapping a pattern vertex to one of, for example, three target vertices of degree ten, two vertices of degree nine, or five of degree two? When driving conventional backtracking search, the degree heuristic would pick one of the vertices of degree ten, and we would commit all of our search effort to the exponentially large search tree underneath it, not considering any other choice until this tree has been fully explored and eliminated. We will show that this is not a wise choice, and that instead, we should *commit equal search effort* to each of the three subproblems found by mapping to vertices of degree ten. And similarly, should we be certain that a vertex of degree ten is so much better than a vertex of degree nine that we should commit no effort to degree nine vertices until all the degree ten subproblems have been explored? Or might it be better to commit, say, twice as much effort to each degree ten subproblems as to each degree nine subproblems, and only a very small amount of effort to the degree two subproblems? To test this hypothesis, we will now introduce a new alternative to backtracking search, which we call *solution-biased search*. This search is made up of three components: a new slightly-random value-ordering heuristic, rapid restarts, and nogood recording. The aim is to perform a complete search, but spending proportionally more time in parts of the search tree that are preferred by the value-ordering heuristic.

2.3 Biased Value-Ordering

We first define a new **Biased** value-ordering heuristic, as follows. When branching, we select a vertex v' from the chosen domain D_v with probability

$$p(v') = \frac{2^{\deg(v')}}{\sum_{w \in D_v} 2^{\deg(w)}}.$$

This heuristic is now equally likely to pick between vertices of equal degree, is twice as likely to pick a vertex of degree d as one of degree $d - 1$, and is over a thousand times more likely to pick a vertex of degree d than degree $d - 10$.

Figure 3 confirms that this heuristic, when used with backtracking search, will solve close to the same number of instances as the degree heuristic would for any given choice of timeout. In other words, we can introduce an element of randomness into the degree value-ordering heuristic without adversely affecting its performance *in aggregate*. The right-hand plot gives a detailed comparison. It shows that despite the aggregate performance being similar, on a case by case basis, the two heuristics can make a large difference to the performance for individual satisfiable instances. This justifies our belief that although degree is a good heuristic, we should perhaps not commit heavily to a single vertex of highest degree, but also consider vertices of the same or similar degree.

2.4 Restarting Search and Nogood Recording

Having introduced a new value-ordering heuristic, we must now also move away from depth-first backtracking search. We do this by using *restarts* and *nogood recording*. The general idea is to perform a certain amount of search, and then if no solution has been found (and unsatisfiability has not been proven), to abandon search and restart from the beginning. Such an approach can only be beneficial if something changes after restarting—in a constraint programming setting, this is usually the variable-ordering heuristic [13,16,28,29]. In this paper, we instead rely upon randomness in our new *value*-ordering heuristic, and continue to use smallest domain first with static tiebreaking for variable-ordering.[2] Using restarts on value-ordering heuristics is uncommon (although Razgon et al. [45] look at learning value-ordering heuristics from restarts, Chu et al. [6] use a similar scheme in the context of parallel search, and an early approach by Gomes et al. [17] does so in an optimisation context).

Preliminary experiments directed us to use the Luby scheme [33] to determine when to restart. Following convention, we multiply each item in the Luby sequence by a constant—we used the SMAC automatic parameter tuner [22] to select the value 660.

To avoid exploring portions of the search space that we have already visited, every time we restart, we add new constraints to the problem which eliminate already-explored subtrees—such a constraint is called a *nogood*. We generate simple decision nogoods. That is, upon backtracking due to a decision to restart, we post a nogood of the form $(v \mapsto v') \land (w \mapsto w') \land (x \mapsto x') \Rightarrow \bot$ for every branch to the left of the current (incomplete) branch at every level of the search tree, and when we first make a decision to restart before backtracking, we post a similar nogood eliminating the entire subtree explored. We use the two watched literals technique [42] to propagate stored nogoods. This has two benefits: the propagation complexity does not particularly depend upon the number of stored nogoods, and it does not require any work upon backtracking. Other more sophisticated nogood generation and propagation schemes exist [16,29], but these are not helpful in this setting (our solver does not maintain arc consistency or use a propagation queue).

2.5   Solution-Biased Search in Practice

In Fig. 4 we show the effects of adding restarts and nogood recording to the algorithm. With restarts and nogood recording (random search with restarts, RSR), the random value-ordering heuristic comfortably beats the degree strategy with depth-first search. In other words, although having a good value-ordering heuristic is beneficial, introducing randomness into the search is better, if it is done alongside a mechanism to avoid heavy commitment to any particular random choice. However, the biased heuristic together with restarts (solution-biased search, SBS) is better still—that is, if we are introducing restarts, then it is better to add a small amount of randomness to a tailored heuristic than it is to throw away the heuristic altogether. Indeed, the original algorithm can solve 1983 satisfiable instances by 909 s, whilst the biased and random restarting algorithms require only 12 s and 35 s respectively to solve the same number.

In the more detailed view in the right-hand plot of Fig. 4, comparing the original algorithm to solution-biased search, all of the unsatisfiable instances

---

[2] It may be possible to further improve the solver by also introducing randomness or some form of learning into its variable-ordering heuristic. However, simultaneously introducing a second change would considerably complicate the empirical analysis. Additionally, the solver's current hand-crafted variable-ordering heuristics already beat adaptive heuristics like impact or activity-based search.

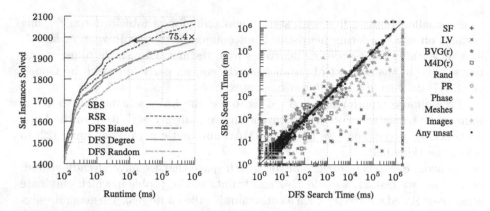

Fig. 4. On the left, the number of satisfiable instances solved over time, comparing solution-biased search (SBS), random search with restarts (RSR), and the three value-ordering heuristics with conventional depth-first search. To the right, a comparison between the original algorithm and solution-biased search.

are very close to the $x - y$ diagonal, showing that their performance is nearly unchanged. On the other hand, there are large numbers of satisfiable instances well below the diagonal line, indicating large speedups. Better yet, there are only a handful of satisfiable instances that are more than a factor of ten times worse. In other words, as well as improving performance, we have made up most of the consistency we lost by introducing randomness.

As we might expect, these properties do not hold if any of the combination of changes are disabled. In the left-hand plot of Fig. 5, we see large slowdowns on unsatisfiable instances when disabling nogood recording, and on the right-hand plot we see many more satisfiable instances above the $x - y$ diagonal when using the random value-ordering heuristic as opposed to the degree-biased heuristic.

2.6 Solution-Biased Search in Theory

Although we have shown that it provides good results, we have yet to justify where the biased formula comes from, or indeed why we call this approach "solution-biased". Our goal is to use biased randomness in a value-ordering heuristic to spend time in subproblems proportional to an estimate of their solution density [44]. Such an approach is better than committing entirely to the area of maximum solution density because estimators only give a probability— although we may estimate that one subtree has twice the solution density of another, in reality the "better" subtree may contain no solutions at all.

To estimate solution density, we need an estimate of how big different sub-problems are likely to be, and of how many solutions each subproblem is likely to contain. Of course, obtaining exact (or even approximate) values for these figures is at least as hard as solving the problem in its entirety, but we may obtain usable approximations. For pairs of Erdős-Rényi random graphs with large solution counts (i.e. chosen from within the "easy satisfiable" region [37]), we can

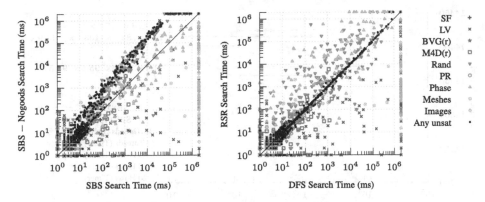

Fig. 5. On the left, not using nogood recording introduces slowdowns, particularly on unsatisfiable instances. On the right, using a random value-ordering gives much worse performance on many satisfiable instances.

observe a linear relationship between subproblem size and number of solutions. Thus, for graphs from this distribution, we need only an estimator of subproblem size. Measurements also suggest that, for pairs of Erdős-Rényi graphs, subproblems under a target vertex of degree d tend to contain a small constant times more search nodes than those under a target vertex of degree $d-1$. This empirical analysis suggests that an estimator that is exponential in d will give our method the desired behaviour, at least for Erdős-Rényi graphs. We expect it may be possible to derive better estimators for particular input classes, although over the full range of problem instances, we have verified that exponential estimators substantially outperform polynomial and factorial weightings.

3 Parallel Search

Exploiting multiple cores to speed up constraint programming solvers remains an active area of research, with no universally perfect solution being available. Four of the more common approaches are based upon decompositions [26,34], work-stealing [6,20,35,39], parallel discrepancy searches [40,41], and algorithm portfolios [32]. Decomposition approaches are unsuitable for decision problems, or problems where we have good value-ordering heuristics, because the decomposition interferes strongly with the shape of the search tree [34]. Work-stealing, traditionally, also interferes with value-ordering [36], although specially designed exceptions exist [6,20,35]. However, these have very complicated implementations. Parallel discrepancy searches are aware of value-ordering heuristics, but have other limitations: they struggles on search trees with heavy filtering, and rely upon inner search tree nodes being orders of magnitude less expensive to process than leaf nodes. Portfolios, meanwhile, typically rely upon running multiple models or heuristics simultaneously, and selecting whichever finishes first, whereas here we have a known good model and set of heuristics.

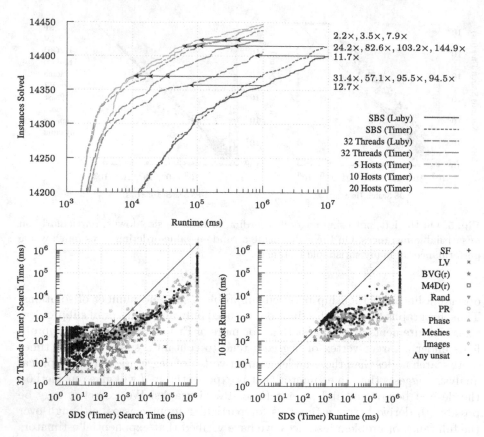

Fig. 6. Above, the cumulative number of instances solved over time, comparing the sequential algorithm to results using 32 threads on a single machine, and using five, ten or twenty distributed memory hosts. Below, instance by instance comparisons.

3.1 Shared Memory Parallelism

Solution-biased search allows for a much simpler parallel implementation. We create a number of threads, and give each thread its own random seed; otherwise each thread performs the same sequential search. Threads synchronise on restarts, a simple barrier causing each thread to wait for every other thread to also restart. Nogoods from all threads are then gathered and combined before search resumes, now with a larger set of nogoods than in a sequential run. Finally, whenever any single thread terminates, either due to having found a solution or proved unsatisfiability, then every other thread may immediately terminate.

This technique requires only limited changes to the top level search driver, and none whatsoever to the main recursive search algorithm. Notably, it does not require any locking or communication during the recursive search, aside from a single atomic boolean flag to assist early termination. A number of factors combine to make this approach feasible:

- Each thread will be run with the same restart schedule, and so will spend approximately the same amount of time between restarts. Because the only synchronisation between threads is at a restart, we expect threads to be busy doing search. (This is in contrast to an alternative method for paralellising restarting search [7], which packs together successive sequence values to produce a balanced workload.) This approach therefore avoids the irregular task size issues which usually arise in parallel combinatorial search.
- Sequentially, on non-trivial instances the algorithm will restart often (many tens of thousands, for instances that reach the thousand second timeout).
- Because the search trees we explore are exponentially large, the randomness in the value-ordering heuristic is sufficient to ensure that most of the time, threads are exploring different parts of the search tree.
- The gathering of nogoods to describe the work done so far provides an alternative to requiring either a specific mechanism to allocate work, or expensive synchronisation between threads. Notably, this completely bypasses the typical difficulties of sharing all learned nogoods in learning solvers [18].
- If sometimes threads do happen to explore part of the same subproblems, this is not a problem: if the instance is satisfiable, either thread might find a solution first, and if the instance is unsatisfiable, we merely introduce some redundancy into the proof.[3] The combination of rapid restarts and nogood recording is enough to ensure that this is only a small overhead.

Figure 6 shows how this scheme performs in practice. Sequentially, we can solve 14,357 instances within the thousand second timeout, with the last instance being solved at 939.0 s. Using thirty-six threads on machines with two eighteen core processors, we can solve 14,357 instances with a timeout of only 74.2 s, giving an aggregate speedup [20] of 12.7.

Closer inspection of the results reveals that with this many threads, a considerable proportion of the overall search time is spent with threads waiting at the barrier for synchronisation. This is because the time taken to carry out search until 660 backtracks are encountered is only *roughly* a constant (in practice it usually varies by around a factor of two). Furthermore, the Luby sequence includes occasional large multipliers, and if unsatisfiability is proved during towards the end of one of these runs, each thread will end up duplicating a large amount of work.

Because we are using nogood recording, an alternative approach is possible. Rather than using the Luby sequence for restarts, we could restart after either a constant number of backtracks, or simply after a certain time interval has passed—this bounds the maximum possible idle time that threads could spend at a barrier. Figure 6 also shows the effects of restarting every 100ms. Sequentially, this approach is slightly better than using the Luby sequence, being able to solve 14,370 instances, with the last at 996.4 s. With thirty-six threads, solving this many instances takes 31.8 s, giving an aggregate speedup of 31.4.

---

[3] For solving a counting or enumeration problem, matters become slightly more complicated, but not devastatingly so.

It is important to emphasise that there is no expectation of a linear speedup from this approach [3,30,31], particularly for satisfiable instances. Due to the biased random nature of the value-ordering heuristic, one of our extra threads may "get lucky" and find a solution much quicker than we would sequentially, leading to a superlinear speedup. Conversely, our extra threads may contribute nothing to finding a solution—or worse, due to new nogoods altering the choice made by the random branching heuristic, we could even get an absolute slowdown. We can see both of these effects in the bottom left plot of Fig. 6: although we achieve roughly a linear speedup on unsatisfiable instances, satisfiable instances show much greater variability. However, this approach does at least mirror our intuition of allocating search effort in proportion to where the value-ordering heuristic believes it will be most fruitful, and so we should not be too surprised that we see roughly a linear speedup on average on harder instances.

Additionally, for easy instances, most of the algorithm's execution time is spent in a preprocessing phase. We have not parallelised this, which is why our results are poor below the one second mark.

3.2 Distributed Memory Parallelism

To further test the scalability of this technique, we also used MPI to implement a distributed-memory parallelism layer on top of the threaded layer. In contrast to the huge difficulties of implementing work-stealing in a distributed memory setting, this required only the addition of two MPI calls: an "all gather" operation to communicate nogoods, and a "gather" to collect and combine the results of each host. Termination, meanwhile, was handled by posting an empty nogood (and so termination can only occur on a restart).

Figure 6 also shows the results of these experiments, using five, ten, and twenty hosts. Because each host has two CPU sockets, the five host results use ten MPI ranks, each with eighteen threads, and the ten and twenty host results use twenty and forty MPI ranks respectively. The supercomputing service we use is not designed for huge numbers of very short problems, and so we ran only instances whose sequential runtime was at least one second; for the sake of plotting results, we treat skipped instances as taking one second. Due to the job launcher used, it is also not possible to accurately measure "total" runtime including startup costs, and so instead we report the runtime of the rank zero process—this figure is somewhat optimistic for very easy instances. With these caveats in mind, when seeing how long a timeout is needed to solve any 14,370 instances, we get aggregate speedups of 57.1 from five hosts and 95.5 from ten hosts over a sequential baseline; using twenty hosts is slightly slower, due to increased overheads. Finally, if we look at harder instances, by allow a longer sequential timeout, we can solve 14,415 instances sequentially with the last at 8,549 s; at this difficulty level, we achieve aggregate speedups of 82.6, 103.2, and 144.9 from five, ten and twenty hosts.

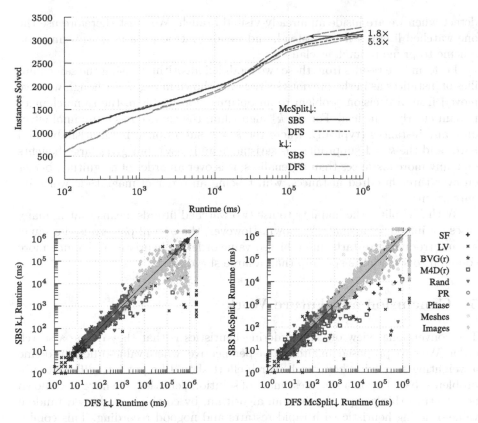

Fig. 7. Above, the cumulative number of maximum common induced subgraph instances solved by k↓ and McSplit↓ over time, with the two forms of search. Below, comparing k↓ (left) and McSplit↓ (right) on an instance by instance basis.

4   Maximum Common Subgraph Algorithms

Having looked at subgraph isomorphism in detail, we now briefly discuss the maximum common induced subgraph problem, to see whether our new approach to search has more general applicability. Two recent algorithms for this problem also make use of backtracking search with degree as a value-ordering heuristic. The k↓ algorithm [21] attempts to solve the problem by first trying to solve the induced subgraph isomorphism problem, and then if that fails, retries allowing a single unmatched vertex (and thus using weaker invariants), and so on. Due to its similarity to the Glasgow Subgraph Solver, we can introduce the same bias and restart strategy.

Meanwhile, the McSplit↓ algorithm [38] uses a constraint programming style search, but with special propagators and backtrackable data structures that exploit special properties of the problem. The unconventional domain store used by McSplit↓ precludes the use of arbitrary unit propagation, and so when introducing restarts, we cannot propagate using nogoods. Instead, we can only detect when we are inside an already-visited branch. We must therefore use the one watched literal scheme instead, and we also introduce a basic subsumption scheme to prune redundant clauses.

Performance results from these two modified algorithms, using the same families of instances as in the previous section, are shown in Fig. 7. Although we have moved from a decision problem to an optimisation problem, the same changes remain clearly beneficial. For the $k\downarrow$ algorithm, the change has a minimal effect on many instances (typically, where the $k = 0$ subproblem is unsatisfiable and hard, and the $k = 1$ subproblem is satisfiable and easy), but gives large benefits on many more instances than it penalises: it is over an order of magnitude better on over three hundred instances, whilst being an order of magnitude worse on only seven.

With McSplit$\downarrow$, the inability to use two watched literals means that in many cases we introduce a small slowdown. However, the overall pattern is the same: when introducing restarts and a biased value ordering heuristic, it is much more common to see a large speedup than a large slowdown.

5   Conclusion and Future Work

The conventional view of value-ordering heuristics is that they define a search order. We have proposed an alternative perspective, where value-orderings define a weighting specifying how much search effort should be put into different subproblems, based upon a rough estimate of solution densities. We have also shown how to turn this perspective into an algorithm, by combining a biased random value-ordering heuristic with rapid restarts and nogood recording. This combination of techniques gives us, for the first time, a practical alternative to backtracking search where we have a strong *value*-ordering heuristic, and where we care both about satisfiable *and* unsatisfiable instances.

A further benefit is the ease with which such a search can be parallelised. By having each thread carry out the same search with a different random seed, and sharing nogoods only on restarts, we remove the need for intrusive changes to the core search algorithm, eliminate the irregularity problem, and still respect the advice of the value-ordering heuristic.

We believe that these technique are broadly applicable, beyond subgraph algorithms, and we intend to implement them in a full constraint programming solver. We are also interested in making better use of statistical knowledge (either *a priori* or learned during search) to further refine the biased randomisation process. And finally, we are trying hard to work out whether our new perspective also has some relevance to *variable* ordering heuristics.

Acknowledgements. The authors would like to thank Özgür Akgün, Chris Jefferson, Sonja Kraiczy, Christophe Lecoutre, Christine Solnon, and Craig Reilly for their comments.

References

1. Hjort Blindell, G., Castañeda Lozano, R., Carlsson, M., Schulte, C.: Modeling universal instruction selection. In: Pesant, G. (ed.) CP 2015. LNCS, vol. 9255, pp. 609–626. Springer, Cham (2015). https://doi.org/10.1007/978-3-319-23219-5_42
2. Bonnici, V., Giugno, R., Pulvirenti, A., Shasha, D.E., Ferro, A.: A subgraph isomorphism algorithm and its application to biochemical data. BMC Bioinform. **14**(S-7), S13 (2013)
3. de Bruin, A., Kindervater, G.A.P., Trienekens, H.W.J.M.: Asynchronous parallel branch and bound and anomalies. In: Ferreira, A., Rolim, J. (eds.) IRREGULAR 1995. LNCS, vol. 980, pp. 363–377. Springer, Heidelberg (1995). https://doi.org/10.1007/3-540-60321-2_29
4. Bruschi, D., Martignoni, L., Monga, M.: Detecting self-mutating malware using control-flow graph matching. In: Büschkes, R., Laskov, P. (eds.) DIMVA 2006. LNCS, vol. 4046, pp. 129–143. Springer, Heidelberg (2006). https://doi.org/10.1007/11790754_8
5. Carletti, V., Foggia, P., Saggese, A., Vento, M.: Introducing VF3: a new algorithm for subgraph isomorphism. In: Foggia, P., Liu, C.-L., Vento, M. (eds.) GbRPR 2017. LNCS, vol. 10310, pp. 128–139. Springer, Cham (2017). https://doi.org/10.1007/978-3-319-58961-9_12
6. Chu, G., Schulte, C., Stuckey, P.J.: Confidence-based work stealing in parallel constraint programming. In: Gent, I.P. (ed.) CP 2009. LNCS, vol. 5732, pp. 226–241. Springer, Heidelberg (2009). https://doi.org/10.1007/978-3-642-04244-7_20
7. Ciré, A.A., Kadioglu, S., Sellmann, M.: Parallel restarted search. In: Proceedings of the Twenty-Eighth AAAI Conference on Artificial Intelligence, Québec City, Québec, Canada, 27–31 July 2014, pp. 842–848 (2014)
8. Coffman, T., Greenblatt, S., Marcus, S.: Graph-based technologies for intelligence analysis. Commun. ACM **47**(3), 45–47 (2004). https://doi.org/10.1145/971617.971643
9. Cordella, L.P., Foggia, P., Sansone, C., Vento, M.: A (sub)graph isomorphism algorithm for matching large graphs. IEEE Trans. Pattern Anal. Mach. Intell. **26**(10), 1367–1372 (2004)
10. Dalla Preda, M., Vidali, V.: Abstract similarity analysis. Electron. Notes Theoret. Comput. Sci. **331**, 87–99 (2017). https://doi.org/10.1016/j.entcs.2017.02.006, Proceedings of the Sixth Workshop on Numerical and Symbolic Abstract Domains (NSAD 2016)
11. Damiand, G., Solnon, C., de la Higuera, C., Janodet, J., Samuel, É.: Polynomial algorithms for subisomorphism of ND open combinatorial maps. Comput. Vis. Image Underst. **115**(7), 996–1010 (2011)
12. Foggia, P., Percannella, G., Vento, M.: Graph matching and learning in pattern recognition in the last 10 years. IJPRAI **28**(1) (2014). https://doi.org/10.1142/S0218001414500013
13. Gay, S., Hartert, R., Lecoutre, C., Schaus, P.: Conflict ordering search for scheduling problems. In: Pesant, G. (ed.) CP 2015. LNCS, vol. 9255, pp. 140–148. Springer, Cham (2015). https://doi.org/10.1007/978-3-319-23219-5_10
14. Geelen, P.A.: Dual viewpoint heuristics for binary constraint satisfaction problems. In: ECAI, pp. 31–35 (1992)
15. Gent, I.P., MacIntyre, E., Prosser, P., Walsh, T.: The constrainedness of search. In: Proceedings of the Thirteenth National Conference on Artificial Intelligence and Eighth Innovative Applications of Artificial Intelligence Conference, AAAI 1996, IAAI 1996, Portland, Oregon, USA, 4–8 August 1996, vol. 1, pp. 246–252 (1996)

16. Glorian, G., Boussemart, F., Lagniez, J.-M., Lecoutre, C., Mazure, B.: Combining nogoods in restart-based search. In: Beck, J.C. (ed.) CP 2017. LNCS, vol. 10416, pp. 129–138. Springer, Cham (2017). https://doi.org/10.1007/978-3-319-66158-2_9

17. Gomes, C.P., Selman, B., Kautz, H.A.: Boosting combinatorial search through randomization. In: Proceedings of the Fifteenth National Conference on Artificial Intelligence and Tenth Innovative Applications of Artificial Intelligence Conference, AAAI 1998, IAAI 1998, Madison, Wisconsin, USA, 26–30 July 1998, pp. 431–437 (1998)

18. Hamadi, Y., Jabbour, S., Sais, J.: Control-based clause sharing in parallel SAT solving. In: Hamadi, Y., Monfroy, E., Saubion, F. (eds.) Autonomous Search, pp. 245–267. Springer, Heidelberg (2011). https://doi.org/10.1007/978-3-642-21434-9_10

19. Harvey, W.D., Ginsberg, M.L.: Limited discrepancy search. In: Proceedings of the Fourteenth International Joint Conference on Artificial Intelligence, IJCAI 1995, Montréal Québec, Canada, 20–25 August 1995, 2 volumes, pp. 607–615 (1995)

20. Hoffmann, R., et al.: Observations from parallelising three maximum common (connected) subgraph algorithms. In: Integration of Constraint Programming, Artificial Intelligence, and Operations Research - 15th International Conference, CPAIOR 2018, Delft, The Netherlands, 26–29 June 2018, Proceedings, pp. 298–315 (2018). https://doi.org/10.1007/978-3-319-93031-2_22

21. Hoffmann, R., McCreesh, C., Reilly, C.: Between subgraph isomorphism and maximum common subgraph. In: Proceedings of the Thirty-First AAAI Conference on Artificial Intelligence, San Francisco, California, USA, 4–9 February 2017, pp. 3907–3914 (2017)

22. Hutter, F., Hoos, H.H., Leyton-Brown, K.: Sequential model-based optimization for general algorithm configuration. In: Coello, C.A.C. (ed.) LION 2011. LNCS, vol. 6683, pp. 507–523. Springer, Heidelberg (2011). https://doi.org/10.1007/978-3-642-25566-3_40

23. Karoui, W., Huguet, M.-J., Lopez, P., Naanaa, W.: YIELDS: a yet improved limited discrepancy search for CSPs. In: Van Hentenryck, P., Wolsey, L. (eds.) CPAIOR 2007. LNCS, vol. 4510, pp. 99–111. Springer, Heidelberg (2007). https://doi.org/10.1007/978-3-540-72397-4_8

24. Korf, R.E.: Improved limited discrepancy search. In: Proceedings of the Thirteenth National Conference on Artificial Intelligence and Eighth Innovative Applications of Artificial Intelligence Conference, AAAI 1996, IAAI 1996, Portland, Oregon, 4–8 August 1996, vol. 1, pp. 286–291 (1996)

25. Kotthoff, L., McCreesh, C., Solnon, C.: Portfolios of subgraph isomorphism algorithms. In: Festa, P., Sellmann, M., Vanschoren, J. (eds.) LION 2016. LNCS, vol. 10079, pp. 107–122. Springer, Cham (2016). https://doi.org/10.1007/978-3-319-50349-3_8

26. Kotthoff, L., Moore, N.C.A.: Distributed solving through model splitting. CoRR abs/1008.4328 (2010)

27. Larrosa, J., Valiente, G.: Constraint satisfaction algorithms for graph pattern matching. Math. Struct. Comput. Sci. **12**(4), 403–422 (2002)

28. Lecoutre, C., Sais, L., Tabary, S., Vidal, V.: Recording and minimizing nogoods from restarts. JSAT **1**(3–4), 147–167 (2007)

29. Lee, J.H.M., Schulte, C., Zhu, Z.: Increasing nogoods in restart-based search. In: Proceedings of the Thirtieth AAAI Conference on Artificial Intelligence, Phoenix, Arizona, USA, 12–17 February 2016, pp. 3426–3433 (2016)

30. Li, G., Wah, B.W.: How to cope with anomalies in parallel approximate branch-and-bound algorithms. In: Proceedings of the National Conference on Artificial Intelligence. Austin, TX, 6–10 August 1984, pp. 212–215 (1984)

31. Li, G., Wah, B.W.: Coping with anomalies in parallel branch-and-bound algorithms. IEEE Trans. Comput. **35**(6), 568–573 (1986). https://doi.org/10.1109/TC.1986.5009434
32. Lindauer, M., Hoos, H., Hutter, F.: From sequential algorithm selection to parallel portfolio selection. In: Dhaenens, C., Jourdan, L., Marmion, M.-E. (eds.) LION 2015. LNCS, vol. 8994, pp. 1–16. Springer, Cham (2015). https://doi.org/10.1007/978-3-319-19084-6_1
33. Luby, M., Sinclair, A., Zuckerman, D.: Optimal speedup of Las Vegas algorithms. Inf. Process. Lett. **47**(4), 173–180 (1993)
34. Malapert, A., Régin, J., Rezgui, M.: Embarrassingly parallel search in constraint programming. J. Artif. Intell. Res. **57**, 421–464 (2016). https://doi.org/10.1613/jair.5247
35. McCreesh, C., Prosser, P.: A parallel, backjumping subgraph isomorphism algorithm using supplemental graphs. In: Pesant, G. (ed.) CP 2015. LNCS, vol. 9255, pp. 295–312. Springer, Cham (2015). https://doi.org/10.1007/978-3-319-23219-5_21
36. McCreesh, C., Prosser, P.: The shape of the search tree for the maximum clique problem and the implications for parallel branch and bound. TOPC **2**(1), 8:1–8:27 (2015). https://doi.org/10.1145/2742359
37. McCreesh, C., Prosser, P., Solnon, C., Trimble, J.: When subgraph isomorphism is really hard, and why this matters for graph databases. J. Artif. Intell. Res. **61**, 723–759 (2018)
38. McCreesh, C., Prosser, P., Trimble, J.: A partitioning algorithm for maximum common subgraph problems. In: Proceedings of the Twenty-Sixth International Joint Conference on Artificial Intelligence, IJCAI 2017, Melbourne, Australia, 19–25 August 2017, pp. 712–719 (2017)
39. Michel, L., See, A., Van Hentenryck, P.: Parallelizing constraint programs transparently. In: Bessière, C. (ed.) CP 2007. LNCS, vol. 4741, pp. 514–528. Springer, Heidelberg (2007). https://doi.org/10.1007/978-3-540-74970-7_37
40. Moisan, T., Gaudreault, J., Quimper, C.-G.: Parallel discrepancy-based search. In: Schulte, C. (ed.) CP 2013. LNCS, vol. 8124, pp. 30–46. Springer, Heidelberg (2013). https://doi.org/10.1007/978-3-642-40627-0_6
41. Moisan, T., Quimper, C.-G., Gaudreault, J.: Parallel depth-bounded discrepancy search. In: Simonis, H. (ed.) CPAIOR 2014. LNCS, vol. 8451, pp. 377–393. Springer, Cham (2014). https://doi.org/10.1007/978-3-319-07046-9_27
42. Moskewicz, M.W., Madigan, C.F., Zhao, Y., Zhang, L., Malik, S.: Chaff: engineering an efficient SAT solver. In: Proceedings of the 38th Design Automation Conference, DAC 2001, Las Vegas, NV, USA, 18–22 June 2001, pp. 530–535 (2001)
43. Murray, A.C., Franke, B.: Compiling for automatically generated instruction set extensions. In: 10th Annual IEEE/ACM International Symposium on Code Generation and Optimization, CGO 2012, San Jose, CA, USA, 31 March–04 April 2012, pp. 13–22 (2012). https://doi.org/10.1145/2259016.2259019
44. Pesant, G., Quimper, C., Zanarini, A.: Counting-based search: branching heuristics for constraint satisfaction problems. J. Artif. Intell. Res. **43**, 173–210 (2012). https://doi.org/10.1613/jair.3463
45. Razgon, M., O'Sullivan, B., Provan, G.M.: Search ordering heuristics for restarts-based constraint solving. In: Proceedings of the Twentieth International Florida Artificial Intelligence Research Society Conference, 7–9 May, 2007, Key West, Florida, USA, pp. 182–183 (2007)
46. Régin, J.: Développement d'outils algorithmiques pour l'Intelligence Artificielle. Application à la chimie organique. Ph.D. thesis, Université Montpellier 2 (1995)

47. Sevegnani, M., Calder, M.: Bigraphs with sharing. Theor. Comput. Sci. **577**, 43–73 (2015). https://doi.org/10.1016/j.tcs.2015.02.011
48. Solnon, C.: Alldifferent-based filtering for subgraph isomorphism. Artif. Intell. **174**(12–13), 850–864 (2010)
49. Solnon, C., Damiand, G., de la Higuera, C., Janodet, J.: On the complexity of submap isomorphism and maximum common submap problems. Pattern Recogn. **48**(2), 302–316 (2015)
50. Vömel, C., de Lorenzi, F., Beer, S., Fuchs, E.: The secret life of keys: on the calculation of mechanical lock systems. SIAM Rev. **59**(2), 393–422 (2017). https://doi.org/10.1137/15M1030054
51. Walsh, T.: Depth-bounded discrepancy search. In: Proceedings of the Fifteenth International Joint Conference on Artificial Intelligence, IJCAI 97, Nagoya, Japan, 23–29 August 1997, 2 volumes, pp. 1388–1395 (1997)
52. Zampelli, S., Deville, Y., Solnon, C.: Solving subgraph isomorphism problems with constraint programming. Constraints **15**(3), 327–353 (2010)

Core-Boosted Linear Search for Incomplete MaxSAT

Jeremias Berg[1](✉), Emir Demirović[2], and Peter J. Stuckey[3,4]

[1] HIIT, Department of Computer Science, University of Helsinki, Helsinki, Finland
jeremias.berg@cs.helsinki.fi
[2] University of Melbourne, Melbourne, Australia
emir.demirovic@unimelb.edu.au
[3] Monash University, Melbourne, Australia
peter.stuckey@monash.edu
[4] Data61, CSIRO, Canberra, Australia

Abstract. Maximum Satisfiability (MaxSAT), the optimisation extension of the well-known Boolean Satisfiability (SAT) problem, is a competitive approach for solving NP-hard problems encountered in various artificial intelligence and industrial domains. Due to its computational complexity, there is an inherent tradeoff between scalability and guarantee on solution quality in MaxSAT solving. Limitations on available computational resources in many practical applications motivate the development of *complete any-time* MaxSAT solvers, i.e. algorithms that compute optimal solutions while providing intermediate results. In this work, we propose *core-boosted linear search*, a generic search-strategy that combines two central approaches in modern MaxSAT solving, namely linear and core-guided algorithms. Our experimental evaluation on a prototype combining reimplementations of two state-of-the-art MaxSAT solvers, PMRES as the core-guided approach and LinSBPS as the linear algorithm, demonstrates that our core-boosted linear algorithm often outperforms its individual components and shows competitive and, on many domains, superior results when compared to other state-of-the-art solvers for incomplete MaxSAT solving.

Keywords: Maximum Satisfiability · MaxSAT ·
SAT-based MaxSAT · Incomplete solving · Linear algorithm ·
Core-guided MaxSat

1 Introduction

Discrete optimisation problems are ubiquitous throughout society. When solving a discrete optimisation problem, the goal is to find the best solution according to

The first author is financially supported by the University of Helsinki Doctoral Program in Computer Science and the Academy of Finland (grant 312662). We thank the University of Melbourne and the Melbourne School of Engineering Visiting Fellows scheme for supporting the visit of Jeremias Berg.

© Springer Nature Switzerland AG 2019
L.-M. Rousseau and K. Stergiou (Eds.): CPAIOR 2019, LNCS 11494, pp. 39–56, 2019.
https://doi.org/10.1007/978-3-030-19212-9_3

a given objective function among a finite, but potentially large set of possibilities. Examples of such problems include scheduling, routing, timetabling, and other forms of management decision problems. The solution approaches to discrete optimisation can be divided into *complete* and *incomplete* methods. The aim of complete methods is to find the best possible solution and prove its optimality. However, in many real world applications complete solving is a difficult, and in many cases, a practically infeasible task. Hence, in practice, one might resort to incomplete solving, i.e. computing the best possible solution within a *limited time*, rather than exclusively searching for an optimal solution.

There is a wide range of technologies available for discrete optimisation. The focus of this work is on the Boolean optimisation paradigm of Maximum (Boolean) Satisfiability (MaxSAT), the optimization extension of the well-known Boolean satisfiability (SAT) problem. MaxSAT can be used to solve any NP-hard discrete optimisation problem that can be formulated as minimising a linear objective over Boolean variables subject to a set of clausal constraints. Modern MaxSAT solving technology builds on the exceptional performance improvements of SAT solvers, starting in the late 90s [39, 49]. Most MaxSAT solvers used in real-world applications are SAT-based, i.e. reduce the discrete optimisation problem into a sequence of satisfiability queries of Boolean formulas conjunctive normal form (CNF), and tackle the queries with SAT solvers. In the last decade, MaxSAT solving technology has matured significantly, leading to successful applications of MaxSAT in a wide range of AI and industrial domains, such as timetabling, planning, debugging, diagnosis, machine learning, and systems biology [3, 10, 15, 19, 20, 22, 24, 36, 51]. See [5, 6, 42] for more details.

SAT-based MaxSAT solvers can be roughly partitioned into *linear* [28, 29], *core-guided* [4, 8, 25, 41, 41, 44], and *implicit hitting-set-based* [21, 45] algorithms. The two most relevant ones for this work are the linear and core-guided algorithms. Linear algorithms are upper bounding approaches that encode the MaxSAT instance, along with its pseudo-Boolean objective function, into *conjunctive normal form* (CNF) and iteratively query a SAT solver for a solution better than the current best one. In contrast, core-guided algorithms are lower-bounding approaches that use a SAT solver to extract a series of *unsatisfiable cores*, i.e. sets of soft constraints that cannot be simultaneously satisfied, and reformulate the underlying MaxSAT instance to rule out each core as a source of unsatisfiability. Both search strategies have shown strong performance in the annual MaxSAT evaluations, linear search is particularly effective for incomplete solving while many of the best performing complete solvers are core-guided.

As our main contribution, we propose *core-boosted linear search* for incomplete MaxSAT solving, a novel search strategy that combines linear and core-guided search with the aim of achieving the best of both worlds. A core-boosted solver initially reformulates an input instance with a core-guided solver and then solves the reformulated instance with a linear search solver. The exchange of information from the core-guided phase to the linear phase tightens the gap between the lower and upper bound, allowing the use of a simpler pseudo-Boolean encoding. As a result, the approach is often more effective than either a pure linear or a pure core-guided search.

To demonstrate the potential of core-boosted linear search we report on an experimental evaluation of a prototype solver that combines reimplementations of two state-of-the-art MaxSAT solvers, PMRES [44] as the core-guided algorithm and LinSBPS [14] as the linear algorithm. We compare core-boosted linear search to its individual components on a standard set of benchmarks. Our results indicate that core-boosted linear search is indeed more effective that either core-guided or linear search for incomplete solving. An in-depth look at the search progression on three selected instances demonstrates the ability of core-boosted linear search to both avoid the worst-case executions of its components, and make use the information flow between them to more quickly find solutions of higher quality.

The rest of the paper is organised as follows. After the preliminaries in Sect. 2, we give a detailed discussion of core-guided and linear search methods for MaxSAT in Sect. 3. Core-boosted linear search is then presented in Sect. 4. We discuss related work in Sect. 5, after which we present our experimental evaluation in Sect. 6. Lastly, we give concluding remarks in Sect. 7.

2 Preliminaries

For a Boolean variable x there are two literals, the positive x and the negative $\neg x$. The negation $\neg l$ of a literal l satisfies $\neg\neg l = l$. A clause C is a disjunction ($\lor$) of literals (represented as a set of its literals), and a CNF formula F a conjunction ($\land$) of clauses (represented as a set of its clauses). The set $\mathrm{VAR}(F)$ of the variables of F contains all variables x s.t. $x \in C$ or $\neg x \in C$ for some $C \in F$. We assume familiarity with other logical connectives and denote by $\mathrm{CNF}(\phi)$ a set of clauses logically equivalent to the formula ϕ. We also assume without loss of generality, that the size of $\mathrm{CNF}(\phi)$ is linear in the size of ϕ [47].

A truth assignment τ is a function mapping Boolean variables to 1 (true) or 0 (false). A clause C is satisfied by τ (denoted by $\tau(C) = 1$) if $\tau(l) = 1$ for a positive or $\tau(l) = 0$ for a negative literal $l \in C$, otherwise C is falsified by τ (denoted $\tau(C) = 0$). A CNF formula F is satisfied by τ ($\tau(F) = 1$) if τ satisfies all clauses in the formula and falsified otherwise ($\tau(F) = 0$). If some τ satisfies a CNF formula F, then F is satisfiable, otherwise it is unsatisfiable. The NP-complete Boolean Satisfiability problem (SAT) asks to decide if a given CNF formula F is satisfiable [17].

A (weighted partial) MaxSAT instance $\mathcal{F}$ consists of two sets of clauses: the hard $\mathrm{HARD}(\mathcal{F})$, the soft $\mathrm{SOFT}(\mathcal{F})$, and a function $w^{\mathcal{F}} \colon \mathrm{SOFT}(\mathcal{F}) \to \mathbb{N}$ associating a positive integral cost to each soft clause. The set $\mathrm{VAR}(\mathcal{F})$ of the variables of $\mathcal{F}$ is $\mathrm{VAR}(\mathrm{HARD}(\mathcal{F})) \cup \mathrm{VAR}(\mathrm{SOFT}(\mathcal{F}))$. An assignment τ is a solution to $\mathcal{F}$ if $\tau(\mathrm{HARD}(\mathcal{F})) = 1$. The cost $\mathrm{COST}(\mathcal{F}, \tau)$ of a solution τ to $\mathcal{F}$ is the sum of the weights of the soft clauses it falsifies i.e. $\mathrm{COST}(\mathcal{F}, \tau) = \sum_{C \in \mathrm{SOFT}(\mathcal{F})} w^{\mathcal{F}}(C) \times (1 - \tau(C))$. A solution τ is optimal if $\mathrm{COST}(\mathcal{F}, \tau) \leq \mathrm{COST}(\mathcal{F}, \tau')$ for all solutions τ'

Algorithm 1. LIN-SEARCH

Input: A MaxSAT instance $\mathcal{F}$
Output: An optimal solution τ to $\mathcal{F}$
begin

$\quad n \leftarrow |\text{SOFT}(\mathcal{F})|, \quad \tau^* \leftarrow \text{INITIALSOLUTION}(\mathcal{F})$
$\quad \mathcal{R} \leftarrow \{r_1, \dots, r_n\}, \quad F_s^R = \{C_i \vee r_i \mid C_i \in \text{SOFT}(\mathcal{F}), r_i \notin \text{VAR}(\mathcal{F})\}$
$\quad$ **while** *true* **do**
$\quad\quad$ **if** *Resource-Out* **then** return τ^*
$\quad\quad PB \leftarrow \sum_{i=1}^{n} w^{\mathcal{F}}(C_i) \times r_i < \text{COST}(\mathcal{F}, \tau^*)$
$\quad\quad F_w \leftarrow \text{HARD}(\mathcal{F}) \cup F_s^R \cup \text{CNF}(PB)$
$\quad\quad (res, \tau) \leftarrow \text{SATSOLVE}(F_w)$
$\quad\quad$ **if** *res="satisfiable"* **then** $\tau^* \leftarrow \tau$
$\quad\quad$ **else return** τ^*

---

to $\mathcal{F}$. We denote the cost of the optimal solutions to $\mathcal{F}$ by $\text{COST}(\mathcal{F})$. The NP-hard (weighted partial) MaxSAT problem asks to compute an optimal solution to a given instance $\mathcal{F}$. In the rest of the paper we will assume that all MaxSAT instances have solutions, i.e. that $\text{HARD}(\mathcal{F})$ is satisfiable.

A central concept in many SAT-based MaxSAT algorithms is that of an *(unsatisfiable) core*. For a MaxSAT instance $\mathcal{F}$, a subset $\kappa \subseteq \text{SOFT}(\mathcal{F})$ of soft clauses is an unsatisfiable core of $\mathcal{F}$ iff $\text{HARD}(\mathcal{F}) \wedge \kappa$ is unsatisfiable.

3   Core-Guided and Linear Search for Incomplete MaxSAT

We detail two abstract MaxSAT solving algorithms, LIN-SEARCH (Algorithm 1) and CORE-GUIDED (Algorithm 2), representing linear and core-guided search, respectively. Both use SAT-solvers to reduce MaxSAT solving into a sequence of satisfiability queries. However, the manner in which the SAT solver is used differs significantly. We present both algorithms as complete *any-time* algorithms, i.e. algorithms that, given enough resources, compute the optimal solution to a MaxSAT instance while also providing intermediate solutions during search.

In the following descriptions of the MaxSAT algorithms, we abstract the use of the SAT-solver into two functions. The function SATSOLVE represents a basic SAT-solver query. Given a CNF formula F, the query $\text{SATSOLVE}(F)$ returns a tuple (res, τ), where res denotes whether the formula is satisfiable and τ is a satisfying assignment to F if one exists. The extended function $\text{EXTRACT-CORE}(\text{HARD}(\mathcal{F}), \text{SOFT}(\mathcal{F}))$ takes as input the hard and soft clauses of a MaxSAT instance $\mathcal{F}$ and returns a triplet (res, κ, τ), where res indicates if $\text{HARD}(\mathcal{F}) \wedge \text{SOFT}(\mathcal{F})$ is satisfiable, τ is a satisfying assignment for $\text{HARD}(\mathcal{F}) \wedge \text{SOFT}(\mathcal{F})$ if one exists, and $\kappa \subset \text{SOFT}(\mathcal{F})$ is a core of $\mathcal{F}$ if $\text{HARD}(\mathcal{F}) \wedge \text{SOFT}(\mathcal{F})$ is unsatisfiable. Practically all SAT-solvers used in MaxSAT solving offer a so-called *assumption interface* [43] that can be used to implement SATSOLVE and EXTRACT-CORE.

The pseudocode of LIN-SEARCH, is detailed in Algorithm 1. When solving an instance $\mathcal{F}$, LIN-SEARCH refines an upper bound on $\mathrm{COST}(\mathcal{F})$ by maintaining and iteratively improving a best known solution $\tau^\star$ to $\mathcal{F}$. Initially, $\tau^\star$ is set to any solution of $\mathcal{F}$, for example by invoking the SAT solver on $\mathrm{HARD}(\mathcal{F})$. During search, the existence of a solution τ having cost less that $\tau^\star$ is checked by querying the internal SAT solver. If no such solution is found, then $\tau^\star$ is optimal and LIN-SEARCH terminates. Otherwise $\tau^\star$ is updated and the search continues. In more detail, the existence of a solution τ for which $\mathrm{COST}(\mathcal{F}, \tau) < \mathrm{COST}(\mathcal{F}, \tau^\star)$ is checked by querying the SAT-solver for the satisfiability of a working formula $F_w = \mathrm{HARD}(\mathcal{F}) \cup F_s^R \cup \mathrm{CNF}(PB)$ consisting of the hard clauses, the soft clauses each extended with a unique *relaxation variable* r_i and a CNF-encoding of a *pseudo-Boolean* (PB) constraint $PB = \sum_{i=1}^n w^{\mathcal{F}}(C_i) \times r_i < \mathrm{COST}(\mathcal{F}, \tau^\star)$ that is satisfied by an assignment τ iff $\sum_{i=1}^n w^{\mathcal{F}}(C_i) \times \tau(r_i) < \mathrm{COST}(\mathcal{F}, \tau^\star)$. The intuition underlying F_w is that setting a relaxation variable r_i to true allows falsification of the corresponding soft clause C_i. Thus the PB constraint essentially limits the sum of the weights of the soft clauses falsified by an assignment τ to be less than the current best known upper bound $\mathrm{COST}(\mathcal{F}, \tau^\star)$ on $\mathrm{COST}(\mathcal{F})$. In other words, F_w is satisfied by an assignment τ iff τ is a solution to $\mathcal{F}$ for which $\mathrm{COST}(\mathcal{F}, \tau) < \mathrm{COST}(\mathcal{F}, \tau^\star)$.

Before proceeding with core-guided search, we make two observations regarding the effectiveness of LIN-SEARCH that are important for understanding core-boosted linear search. As the search in LIN-SEARCH is focused on decreasing the best known upper bound, we expect it to be most effective for solving an instance $\mathcal{F}$ when the difference between $\mathrm{COST}(\mathcal{F})$ and the cost $\mathrm{COST}(\mathcal{F}, \tau^\star)$ of the initial solution $\tau^\star$ is small. Thus, a high quality, i.e. low cost, initial solution can have a significant impact on the overall performance of LIN-SEARCH. The second observation concerns the PB constraint $\sum_{i=1}^n w^{\mathcal{F}}(C_i) \times r_i < \mathrm{COST}(\mathcal{F}, \tau^\star)$. Similar constraints are encountered in many different domains, as such a lot of research has been put into developing efficient CNF encodings of them [11,27,46]. Even so, the PB constraint is arguably the main bottleneck of the overall performance of LIN-SEARCH and we expect any further techniques that allow the use of simpler, and more compact (encodings) PB constraints to improve the overall performance of LIN-SEARCH.

The pseudocode of CORE-GUIDED, basic core-guided search extended with *stratification* [7,37], is detailed in Algorithm 2. Stratification is a heuristic designed to steer the core extraction of CORE-GUIDED toward cores κ for which the minimum weight of the clauses in κ is high. Stratification is a standard technique in modern core-guided solvers. Importantly for this work, stratification allows us to treat core-guided search as an any-time method for MaxSAT.

When solving an instance $\mathcal{F}$, CORE-GUIDED maintains a working instance initialised to $\mathcal{F}$ and a stratification bound b^{STRAT} initialised to the highest weight of the soft clauses in $\mathcal{F}$. During iteration i of the main search loop, the SAT solver is queried for a core κ^i of a subset of the current working instance $\mathcal{F}^i$ containing all hard clauses and STRAT, all soft clauses with weight greater than or equal to b^{STRAT}. If no such core exists, an intermediate solution τ is

---

Algorithm 2. CORE-GUIDED

Input: A MaxSAT instance $\mathcal{F}$
Output: An optimal solution τ to $\mathcal{F}$
begin

$\quad \tau^* \leftarrow$ INITIALSOLUTION$(\mathcal{F})$, $\quad b^{\text{STRAT}} \leftarrow \max\{w^{\mathcal{F}}(C) \mid C \in \text{SOFT}(\mathcal{F})\}$
$\quad \mathcal{F}^1 \leftarrow \mathcal{F}$, $\quad i \leftarrow 1$
$\quad$**while** *true* **do**
$\quad\quad$**if** *Resource-Out* **then return** τ^*
$\quad\quad$STRAT $\leftarrow \{C \mid C \in \text{SOFT}(\mathcal{F}^i), w^{\mathcal{F}^i}(C) \geq b^{\text{STRAT}}\}$
$\quad\quad(res, \kappa^i, \tau) \leftarrow$ EXTRACT-CORE(HARD$(\mathcal{F}^i)$, STRAT)
$\quad\quad$**if** *res="satisfiable"* **then**
$\quad\quad\quad$**if** COST$(\mathcal{F}, \tau) <$ COST$(\mathcal{F}, \tau^*)$ **then** $\tau^* \leftarrow \tau$
$\quad\quad\quad$**if** STRAT $=$ SOFT$(\mathcal{F}^i)$ **then return** τ
$\quad\quad\quad$**else** $b^{\text{STRAT}} \leftarrow \max\{w^{\mathcal{F}^i}(C) \mid C \in \text{SOFT}(\mathcal{F}^i), w^{\mathcal{F}^i}(C) < b^{\text{STRAT}}\}$
$\quad\quad$**else**
$\quad\quad\quad\mathcal{F}^{i+1} \leftarrow$ REFORMULATE$(\mathcal{F}^i, \kappa^i)$
$\quad\quad\quad i \leftarrow i + 1$

---

obtained and compared to the best known solution τ^*. If all soft clauses were considered in the SAT call, the obtained solution is also optimal and the algorithm terminates. If not, the bound b^{STRAT} is lowered and the search continues. When a core κ^i is extracted, the working instance is updated by the function REFORMULATE. Informally speaking, REFORMULATE reformulates $\mathcal{F}^i$ in a way that rules out κ^i as a source of unsatisfiability and allows falsifying one clause in κ^i without incurring cost. Most of the core-guided MaxSAT solvers that fit the CORE-GUIDED abstraction [4,8,25,41,44] differ mainly in the implementation of REFORMULATE. The correctness of such solvers is often established by showing that $\mathcal{F}^i$ is MaxSAT-reducible to $\mathcal{F}^{i+1}$ and that VAR$(\mathcal{F}^i) \subset$ VAR$(\mathcal{F}^{i+1})$ [6]. While a precise treatment of MaxSAT-reducibility is outside the scope of this work, the next proposition summarises the consequences of it that are important for understanding core-boosted linear search.

Proposition 1. *Let $\mathcal{F}$ be a MaxSAT instance, κ a core of $\mathcal{F}$, $w^\kappa = \min\{w^{\mathcal{F}}(C) \mid C \in \kappa\}$ and $\mathcal{F}^R =$ REFORMULATE$(\mathcal{F}, \kappa)$. Assume that $\mathcal{F}$ is MaxSAT reducible to $\mathcal{F}^R$ and that VAR$(\mathcal{F}) \subset$ VAR$(\mathcal{F}^R)$. Then the following hold: (i) any solution τ to $\mathcal{F}$ can be extended into a solution τ^R to $\mathcal{F}^R$ s.t. COST$(\mathcal{F}, \tau) =$ COST$(\mathcal{F}^R, \tau^R) + w^\kappa$ and (ii) any solution τ^R to $\mathcal{F}^R$ is a solution to $\mathcal{F}$ for which COST$(\mathcal{F}^R, \tau^R) =$ COST$(\mathcal{F}, \tau^R) - w^\kappa$.*

An alternative intuition to core-guided search offered by Proposition 1 is thus a search strategy that lowers the optimal cost of its working instance by extracting cores that witness lower bounds and reformulating the instance s.t the cost of every solution to the instance is lowered exactly by the identified lower bound. Core-guided search terminates once the optimum cost of the working instance has been lowered to 0.

Example 1. Let $\mathcal{F}$ be a MaxSAT instance having $\text{HARD}(\mathcal{F}) = \{(x_1 \vee x_2), (x_3 \vee x_4)\}$ and $\text{SOFT}(\mathcal{F}) = \{(\neg x_i) \mid i = 1 \ldots 4\}$ with $w^{\mathcal{F}}((\neg x_1)) = w^{\mathcal{F}}((\neg x_2)) = 1$ and $w^{\mathcal{F}}((\neg x_3)) = w^{\mathcal{F}}((\neg x_4)) = 2$. We sketch one possible execution of the PMRES algorithm [44], an instantiation of CORE-GUIDED, when invoked on $\mathcal{F}$. First, the initial working formula $\mathcal{F}^1$ is set to $\mathcal{F}$ and the stratification bound b^{STRAT} is set to the highest weight of soft clauses, i.e. 2. Thus $\text{STRAT} = \{(\neg x_3), (\neg x_4)\}$ in the first iteration. The formula $\text{HARD}(\mathcal{F}^1) \wedge \text{STRAT}$ is unsatisfiable, the only core obtainable at this point is $\kappa^1 = \{(\neg x_3), (\neg x_4)\}$. Using the PMRES algorithm, the next working instance $\mathcal{F}^2 = \text{REFORMULATE}(\mathcal{F}^1, \kappa^1)$ has $\text{HARD}(\mathcal{F}^2) = \text{HARD}(\mathcal{F}^1) \cup \{(\neg x_3 \vee \neg r_1), \text{CNF}(d_1 \leftrightarrow \neg x_4)\}$, $\text{SOFT}(\mathcal{F}^2) = \{(\neg x_1), (\neg x_2), (\neg r_1 \vee \neg d_1)\}$ with $w^{\mathcal{F}^2}(x_1) = w^{\mathcal{F}^2}(x_2) = 1$ and $w^{\mathcal{F}^2}((\neg r_1 \vee \neg d_1)) = 2$. The stratification bound is not altered so $\text{STRAT} = \{(\neg r_1 \vee \neg d_1)\}$ during the next iteration. Now $\text{HARD}(\mathcal{F}^2) \wedge \text{STRAT}$ is satisfiable so b^{STRAT} is lowered to 1. In the next iteration $\text{STRAT} = \text{SOFT}(\mathcal{F}^2)$ and the SAT solver obtains the core $\kappa^2 = \{(\neg x_1), (\neg x_2)\}$. The instance is again reformulated and the next working instance $\mathcal{F}^3 = \text{REFORMULATE}(\mathcal{F}^2, \kappa^2)$ has $\text{HARD}(\mathcal{F}^3) = \text{HARD}(\mathcal{F}^2) \cup \{(\neg x_1 \vee \neg r_2), \text{CNF}(d_2 \leftrightarrow \neg x_2))\}$ and $\text{SOFT}(\mathcal{F}^3) = \{(\neg r_2 \vee \neg d_2), (\neg r_1 \vee \neg d_1)\}$ with $w^{\mathcal{F}^3}((\neg r_2 \vee \neg d_2)) = 1$ and $w^{\mathcal{F}^3}((\neg r_1 \vee \neg d_1)) = 2$. In the final iteration $\text{STRAT} = \text{SOFT}(\mathcal{F}^3)$ and since $\text{HARD}(\mathcal{F}^3) \wedge \text{SOFT}(\mathcal{F}^3)$ is satisfiable, CORE-GUIDED terminates.

We conclude this section with a few observations regarding CORE-GUIDED that are important for understanding core-boosted linear search. When solving an instance $\mathcal{F}$, CORE-GUIDED focuses its search on the lower bound of $\text{COST}(\mathcal{F})$. Thus, we expect CORE-GUIDED to be effective if $\text{COST}(\mathcal{F})$ is low and, in particular, to not be significantly affected by the quality of the initial solution. The main bottleneck of CORE-GUIDED is instead the increased complexity of the core-extraction steps. Note that the core κ^i extracted during the i:th iteration of CORE-GUIDED is a core of the i:th working instance $\mathcal{F}^i$ and not necessarily of the original instance $\mathcal{F}$. While the effects of reformulation on the complexity of the EXTRACT-CORE calls are not fully understood, it has been shown that extracting a core of $\mathcal{F}^i$ can be exponentially harder than extracting a core of $\mathcal{F}$ [13].

4   Core-Boosted Linear Search for Incomplete MaxSAT

In this section, we propose and discuss *core-boosted linear search*, the main contribution of our work. The execution of a core-boosted (linear search) algorithm is split into two phases. On input $\mathcal{F}$, the algorithm begins search in a core-guided phase by invoking CORE-GUIDED on $\mathcal{F}$. If CORE-GUIDED is able to find an optimal solution within the resources allocated to it, then the core-boosted algorithm terminates. Otherwise CORE-GUIDED returns its final working instance $\mathcal{F}_w$ along with the best solution $\tau^\star$ it found. The core-boosted algorithm then moves on to its linear phase by invoking LIN-SEARCH on $\mathcal{F}_w$ using $\tau^\star$ as the initial solution. The linear phase runs until either finding the optimal solution to $\mathcal{F}_w$, or running out of computational resources. By Proposition 1, the best solution τ^* to $\mathcal{F}_w$ found by LIN-SEARCH is also a solution to $\mathcal{F}$. Specifically, an optimal solution of $\mathcal{F}_w$ is also an optimal solution to $\mathcal{F}$ implying the completeness of core-boosted linear search for MaxSAT. We emphasize that the linear component LIN-SEARCH of a core-boosted algorithm is invoked on $\mathcal{F}_w$, the final working instance of CORE-GUIDED, and not on $\mathcal{F}$, the initial input instance. As we discuss next and demonstrate in our experiments, this allows the linear phase of core-boosted linear search to benefit from the core-guided phase in a non-trivial manner.

The discussion on LIN-SEARCH and CORE-GUIDED in Sect. 3 serves as useful basis for understanding the potential benefits of core-boosted linear search. Since core-boosted linear search makes use of both core-guided and linear search, we expect it to be effective both on the same instances as linear search, and as core-guided search, or at least not significantly worse. For example, if the instance $\mathcal{F}$ being solved has low optimal cost, then we expect a core-boosted algorithm to be able to solve the instance effectively during its initial core-guided phase. Similarly, if COST($\mathcal{F}$) is close to the cost COST($\mathcal{F}, \tau^*$) of the initial solution τ^*, then COST($\mathcal{F}_w$) is also close to COST($\mathcal{F}_w, \tau^*$). Hence we expect a core-boosted algorithm to be effective during its linear phase, even factoring in the reformulations done during the core-guided phase.

The potential benefits of core-boosted linear search go beyond merely averaging out the performance of core-guided and linear search. As discussed in the previous section, one of the main drawbacks of core-guided search is the increased complexity of core extraction over time. Thus stopping the core-guided phase and solving the working instance by linear search should be beneficial. Further, the linear phase can also benefit from the reformulation steps performed by the core-guided phase. Specifically, such reformulations can decrease the size of the PB constraint $PB = \sum_{i=1}^{n} w^{\mathcal{F}}(C_i) \times r_i < \text{COST}(\mathcal{F}, \tau^*)$ that needs to be encoded during linear search. Depending on the specific encoding used, the number of clauses resulting from encoding PB into CNF depends either on the magnitudes of the weights of the soft clauses and the right-hand side [23] or on the number of unique sums that can be created from those weights [27]. The reformulation steps performed during the core-guided phase of a core-boosted algorithm can affect both of these factors. By Proposition 1 COST($\mathcal{F}_w, \tau^*$) $\leq$ COST($\mathcal{F}, \tau^*$) which implies that both the magnitude of the weights in $\mathcal{F}_w$ and the initial right hand side COST($\mathcal{F}_w, \tau^*$) of PB are smaller in the reformulated $\mathcal{F}_w$ than in the original $\mathcal{F}$. Additionally, the core-guided phase can also decrease the number of soft clauses in the instance; the second working instance of Example 1 has one less soft clause than the first one. Finally, the so-called *hardening rule* [7] commonly used in conjunction with core-guided search, can also decrease the number of soft clauses of the instance, and thus allow the linear phase of a core-boosted algorithm to use a more compact PB constraint.

5   Related Work

We begin by detailing the instantiations of LIN-SEARCH and CORE-GUIDED that we use in the prototype core-boosted linear search algorithm experimented with in the next section. As the linear search component we use the basic LIN-SEARCH extended with varying resolution and solution-based phase-saving in the style of LinSBPS, the best performing solver of the incomplete 300s track of the 2018 MaxSAT evaluation [14]. Solution-based phase-saving is a heuristic designed to steer the search towards the currently best known solution by modifying the branching heuristic of the internal SAT solver to always prefer setting the polarity of a literal it branches on to equal its polarity in the currently best known solution. Varying resolution is a heuristic designed to alleviate the issues that LIN-SEARCH has with large PB constraints. When invoked on an instance $\mathcal{F}$ a linear search algorithm using varying resolution starts its search by creating a lower resolution version of $\mathcal{F}$ by dividing all weights of soft clauses by some constant d and removing all clauses $C \in \text{SOFT}(\mathcal{F})$ for which $\lfloor w^{\mathcal{F}}(C)/d \rfloor = 0$. The low resolution version is then solved by standard linear search. When an optimal solution is found, the value of d is decreased and the search continued in higher resolution. Following LinSBPS, we used the generalized totalizer encoding (GTE) [27] to convert the PB constraints to CNF. Given a set of input literals $L = \{l_1, \ldots, l_n\}$ and their corresponding weights $\{w_1, \ldots w_n\}$ the GTE creates a set of output literals $o_1, \ldots, o_k$ s.t. each o_i corresponds to a sum s_i formable with the weights in W for which $s_i < s_j$ if $i < j$. The sum of weights of the literals in L set to true is then restricted to be less than s_i with the unit clause $(\neg o_i)$.

As the instantiation of CORE-GUIDED we use the PMRES algorithm [44] extended with weight aware core extraction (WCE) [16] and the hardening rule. Weight aware core extraction is a heuristic designed to allow CORE-GUIDED to extract multiple cores before reformulating the instance and increasing its complexity. When extracting a new core κ PMRES with WCE first computes $c^{\kappa} = \min\{w^{\mathcal{F}}(C) \mid C \in \kappa\}$, then lowers the weight of all clauses in κ by c^{κ} (removing all clauses with weight 0). When no new cores can be extracted, the REFORMULATE function is invoked on all of the found cores and the stratification bound is reset. The search continues until no new cores can be found after a reformulation step. This strategy corresponds to the S/to/WCE strategy of [16]. While an alternative strategy that prefers reformulating to lowering the stratification bound was deemed more effective for complete MaxSAT solving in [16], we found that S/to/WCE is more effective for incomplete solving. For lowering the stratification bound, we use the *diversity heuristic* [7] that balances the amount that b^{STRAT} is lowered with the number of new soft clauses introduced.

In the next section, we report on a comparison of core-boosted linear search and all of the solvers that participated in the incomplete track of the 2018 MaxSAT Evaluation: LinSBPS, maxroster, SATLike, Open-WBO and Open-WBO-Inc and their variations. Most of them implement variations of an approach where: (i) a heuristic of some kind if used to find a good initial solution to the instance being solved and (ii) that solution is used to initialise a complete any-time algorithm. In most cases, the complete algorithm is some variant of LIN-SEARCH. The solver SATLike [30] deviates from this description and instead uses local-search techniques in order to quickly traverse the search space and look for solutions of increasing quality. A more detailed description of the solvers can be found on the evaluation homepage [14].

For related work from the field of complete MaxSAT solving, the Primal-Dual MaxSAT algorithm [18] extends PMRES with a second instance reformulation used to rule out solutions that falsify the same clauses as an intermediate solution obtained during search. The main two differences between Primal-Dual and core-boosted linear search are that Primal-Dual reformulates the instance on each iteration, thus increasing the complexity of core extraction steps, and that the reformulation only rules out solutions that falsify a particular set of clauses. In contrast, lowering the bound on the PB constraint in LIN-SEARCH rules out all solutions that have higher cost than the best known solution. The WMSU3 [38] algorithm maintains a cardinality constraint over soft clauses similar to LIN-SEARCH but only relaxes a soft clause C after extracting a core κ for which $C \in \kappa$. The similar WPM3 [9] uses linear-search as a subroutine within core-guided search in order to obtain tighter bounds on the cardinality constraints.

In addition to core-guided and linear search, a third central approach to SAT-based MaxSAT solving is based on *implicit-hitting sets* [21,45]. When solving, such solvers maintain a set of cores of the input instance. During each iteration, a minimum-cost hitting set over the set of cores is computed. The clauses in the hitting set are then removed from the instance and the SAT solver invoked on the remaining clauses. If the SAT solver reports satisfiable, the obtained solution is optimal. Otherwise, a new core is obtained and the search continues. Finally, MaxSAT solvers based on branch and bound have been shown to be effective on random MaxSAT instances as well as challenging instances of smaller size. Such instances are encountered for example in combinatorics [1, 2, 31–35, 48].

6   Experimental Evaluation

Next we present the results on a experimental evaluation of a prototype core-boosted linear search algorithm that combines the instantiations of LIN-SEARCH and CORE-GUIDED discussed in Sect. 5. We refer to our implementation of LIN-SEARCH extended with varying resolution and solution-guided phase saving by Linear-Search. Similarly, we use Core-Guided to refer to our implementation of PMRES extended with WCE and hardening. Finally, Core-Boosted-XXs is the core-boosted algorithm that first runs Core-Guided until either XX seconds have passed or no more cores can be found with the stratification bound at 1, then reformulates the instance and solves the reformulated instance with Linear-Search. The state of the internal SAT solver of Core-Boosted-XXs is kept throughout the core-guided phase, but reset (that is learned clauses are eliminated and activities of all variables reset to 0) when execution is switched to the linear search phase and whenever resolution is increased during the linear phase.

All three algorithms were implemented on top of the publicly available Open-WBO system [40] using Glucose 4.1 [12] as the back-end SAT solver. The initial solution of all three algorithms is obtained by invoking the SAT solver on the hard clauses of the instance being solved. We emphasise that core-boosted linear search is a general idea applicable with all implementations and extensions of LIN-SEARCH and CORE-GUIDED that we are aware off. The goal of these experiments is to show that core-boosting can be used to improve performance of modern core-guided and linear search solvers, not to evaluate different instantiations and extensions of CORE-GUIDED and LIN-SEARCH.

Our experimental setup is similar to the 300s weighted incomplete track of the 2018 MaxSAT evaluation [14]. In most of the experiments, we use the 172 benchmarks from the weighted incomplete track of the evaluation, available from https://maxsat-evaluations.github.io/2018/benchmarks.html. We enforce a per-instance time limit of 300 s and memory limit of 32 GB. All of the experiments were run on the StarExec cluster (https://www.starexec.org) that has 2.4-GHz Intel(R) Xeon(R) E5-2609 0 quad-core machines with 128-GB RAM.

As the metric for comparing solvers we use the same incomplete score as the evaluation. For an instance $\mathcal{F}$ let BEST-COST$(\mathcal{F})$ denote the lowest cost found in 300 s by any of (the variants of) the solvers Linear-Search, Core-Guided, Core-Boosted-XXs or the solvers that participated in the evaluation. The score a solver S on $\mathcal{F}$ is defined as the ratio between BEST-COST$(\mathcal{F})$ and the cost of the best solution τ^S to $\mathcal{F}$ found by S, i.e. SCORE$(S, \mathcal{F}) = \frac{\text{BEST-COST}(\mathcal{F})+1}{\text{COST}(\mathcal{F}, \tau^S)+1}$. In other words, the score of S is the ratio between the cost of the solution of the virtual-best-strategy (VBS) among our methods and the MaxSAT Evaluation 2018 solvers, and the cost obtained by S. Hence the score difference between two solvers shows the percentage points by which the better solver is closer to the VBS.

The first experiment we report on evaluates effect of different time limits on the core-guided phase of Core-Boosted-XXs. As limits we chose 30 s (10% of the total time), 75 s (25%), 150 s (50%), 225 s (75%) and 300 s (100%), respectively. An important fact to keep in mind is that the core-guided phase can end earlier than the limit. For example, the solver Core-Boosted-150s runs its core-guided phase until no more cores can be found with the stratification bound at 1 *or* 150 s have elapsed.

Table 1 lists the average score obtained by the Core-Boosted-XXs (CB-XXs in the table) solver for different values of XX. Overall we observe a decrease in the average score when the time limit is increased, even if the effect is small in most domains. A possible explanation for this behavior is offered by Fig. 1 showing the duration of the core-guided phase of the Core-Boosted-300s solver on all benchmarks. On 107 out of the 172 benchmarks, the core-guided phase ended within 30 s and on 38 benchmarks Core-Boosted-300s did not enter its linear search phase at all. In other words, on a clear majority of the benchmarks, the duration of core-guided phase was either very short or very long, which explains the good performance of Core-Boosted-30s. For the rest of the experiments, we fix the time limit for the core-guided phase to 30 s. Table 1 also lists the average

Table 1. Average score obtained by Core-Boosted-XXs with different maximum times for the core-guided phase as well as its core-guided and linear search components. In the table CB-XXs refers to the Core-Boosted-XXs solver, Lin to the Linear-Search solver and CG to the Core-Guided solver.

Domain (#benchmarks)	CB-30s	CB-75s	CB-150s	CB-225s	CB-300s	CG	Lin
BTBNSL (16)	**0.996**	0.995	**0.996**	0.995	0.965	0.956	0.959
abstraction-refinement (2)	**1.000**	**1.000**	**1.000**	**1.000**	**1.000**	**1.000**	0.517
af-synthesis (19)	0.990	0.990	0.990	0.990	0.990	0.944	**0.991**
causal-discovery (14)	0.776	0.776	0.799	**0.803**	0.795	0.563	0.454
cluster-expansion (20)	**0.941**	**0.941**	**0.941**	**0.941**	**0.941**	**0.941**	**0.941**
correlation-clustering (12)	0.953	**0.956**	0.953	0.953	0.953	0.736	0.675
hs-timetabling (13)	0.701	0.655	0.566	0.459	0.144	0.076	**0.717**
lisbon-wedding (12)	**0.582**	**0.582**	**0.582**	**0.582**	**0.582**	0.544	**0.582**
maxcut (11)	**0.892**	**0.892**	**0.892**	**0.892**	**0.892**	0.594	0.884
min-width (16)	0.961	**0.965**	0.962	0.956	0.962	0.825	0.898
miplib (5)	**0.587**	**0.587**	0.584	0.584	0.444	0.309	0.571
power-distribution (2)	**0.704**	**0.704**	**0.704**	**0.704**	**0.704**	0.497	0.484
railway-transport (4)	0.927	0.923	0.916	0.920	**0.935**	0.708	0.906
relational-inference (2)	0.041	0.041	0.041	0.041	**0.429**	0.414	0.041
robot-nagivation (3)	**0.943**	**0.943**	**0.943**	**0.943**	0.000	0.000	**0.943**
spot5 (3)	0.990	0.990	0.990	0.990	0.990	0.914	**0.999**
staff-scheduling (10)	**0.895**	**0.895**	0.863	0.840	0.493	0.385	0.877
tcp (7)	**1.000**	0.998	0.998	**1.000**	**1.000**	0.864	0.988
timetabling (1)	0.667	0.148	0.130	0.131	0.131	0.026	**0.941**
Total (172)	**0.870**	0.864	0.857	0.847	0.785	0.680	0.807

score obtained by the two components of Core-Boosted individually. The scores clearly demonstrate the potential of core-boosted linear search. The average score of Core-Boosted-30s is higher than either Core-Guided (CG in the table) or Linear-Search (Lin in the table) on 10 out of 19 domains and equal to its better component on 3 more.

Figure 2 shows a detailed analysis on the behaviour of core-boosted linear search in the form of plots showing the evolution of the gap between the upper and lower bound (in logscale) of Core-Boosted-30s, Linear-Search and Core-Guided on three hand-picked benchmarks. The benchmark on the left shows a case where core-guided search is effective. During the first 30 s, both Core-Boosted-30s and Core-Guided rapidly decrease the gap. After 30 s, Core-Boosted-30s switches to its linear search phase, which on this benchmark slows

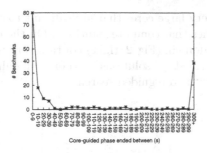

Fig. 1. Time spent in core-guided phase by Core-Boosted-300s.

its search progression. Core-Guided continues with the same search strategy, finding (and proving optimality of) a solution of cost 76250 in just under 190 s. Even if the gap of Core-Boosted-30s is larger due to a smaller lower bound, it still finds an "almost optimal" solution having cost 76251. On this benchmark Linear-Search is unable to improve on its initial solution at all and returns a solution with cost 226338. An important observation to make is that, in contrast to Linear-Search, Core-Boosted-30s did manage to improve its solution also in the linear phase. This indicates that the linear search phase of core-boosted search can indeed benefit from the reformulation steps performed and the best solution obtained during the core-guided phase.

The benchmark in the middle of Fig. 2 demonstrates the opposite behaviour to the one on the left. On this benchmark Core-Guided is unable to improve on its initial solution having cost 651, while Linear-Search continuously improves it and ends up finding one that has cost 17. Core-Boosted-30s is initially unable to make progress, but starts decreasing its gap when switching to the linear phase after 30s and ends up finding a solution of cost 23. Finally, the benchmark on the right demonstrates a best-case scenario for core-boosted search. On this benchmark Linear-Search is unable to improve at all on its initial solution that has cost 311544. Core-Guided is able to decrease the gap by increasing the lower bound to 104585, but is unable to find a single better solution and returns the initial solution of cost 311544 as well. Core-Boosted-30s is able to use the best of both worlds by first increasing the lower bound during the core-guided phase and then switching to the linear phase in order to find a solution of cost 171437, significantly better than either of its components. Notice that the initial solution given to the linear phase of Core-Boosted-30s is the same as the one found by Linear-Search, so the performance difference between the two is only due to the reformulation steps done during core-guided search.

The results shown in Fig. 2 suggest, that a more sophisticated strategy for deciding when to switch from the core-guided to the linear phase could be used to further improve the empirical performance of core-boosted linear search. Even though the instances in Fig. 2 are hand-picked, the average scores over all benchmarks in the corresponding domains listed in Table 1 support the observations. For example, the instances in the hs-timetabling domain (Fig. 2, middle) tend to contain only a few very large cores that are difficult to extract, making them well suited for approaches that compute solutions. On the other hand, instances in the causal-discovery domain (Fig. 2, right) contain very many small cores that make finding good intermediate solutions to them difficult without first ruling out some of the cores with core guided search.

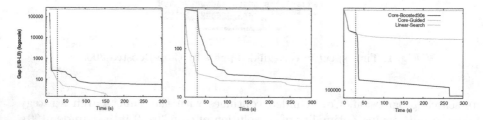

Fig. 2. Evolution of the gap between the upper and lower bound during search. The specific benchmarks shown are abstraction-refinement-downcast-antlr (left) [50], hs-timetabling-BrazilInstance5.xml (middle) [22], causal-discovery-causal_carpo_8_100 (right) [26].

Figure 3 shows a per-instance comparison of the score obtained by Core-Boosted-30s and four variants of it: (1) Core-Boosted-30s-no-reformulation that ignores the reformulated instance and invokes the linear phase on the original instance, (2) Core-Boosted-30s-no-solution that ignores the best solution obtained during the core-guided phase in the linear phase and instead initialises a new solution by invoking the SAT-solver on the hard clauses of the reformulated instance, (3) Core-Boosted-30s-keep-SAT-solver that keeps the state of the internal SAT solver throughout the entire search and (4) Core-Boosted-30s-wce-to-strat that uses of the original search strategy proposed in [16] during the core-guided phase. In all plots Core-Boosted-30s is on the y-axis, so any data points in the upper left triangle correspond to benchmarks on which the baseline performed better than the variant. We observe that the baseline solver performs better than all of its variants, justifying our design choices. The results suggest that using the reformulated instance and initialising the Linear Search with the best solution obtained during core-guided search are especially important for the overall performance.

Finally, we compare Core-Boosted-30s and its components to the other solvers that participated in the 2018 evaluation. Due to running our experiments in the same environment as the evaluation, we did not rerun the other solvers but instead compared our solvers directly to the results of the evaluation. Figure 4 demonstrates the performance of our solvers on the 300s weighted (left) and unweighted (right) tracks[1]. We observe that Core-Boosted-30s performs very well in the weighted track, improving the previous state-of-the-art (LinSBPS) by

---

[1] A consequence of the metric we use is that the scores of the other solvers we report are lower than in the evaluation. Their relative ranking is however the same.

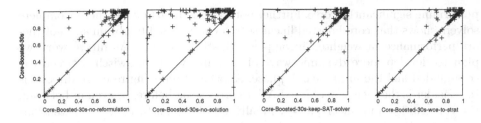

Fig. 3. The effect of different factors of Core-Boosted-30s on the overall performance.

approximately 2% while also finishing 3rd in the unweighted category. In more detail, out of the 172 weighted instances, Core-Boosted-30s and LinSBPS are equal on 63 instances (36%), Core-Boosted-30s finds a solution of strictly lower cost on 65 (37%), and LinSBPS on 44 (25%). We also evaluated our solvers in the 60s track of the evaluation, i.e. with the time out set to 60 s. In the weighted track, Core-Boosted-30s gets the average score 0.814 which is again highest of all solvers followed by Open-WBO-Inc-BMO (0.793). In the unweighted track, the average score of Core-Boosted-30s is 0.696 which is second highest after SATLike-c (0.699).

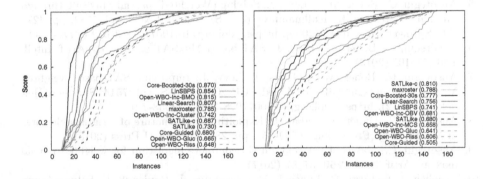

Fig. 4. Performance of Core-Boosted-30s, Linear-Search and Core-Guided compared to the results of the 300s weighted (left) and unweighted (right) track of the 2018 MaxSAT Evaluation.

7   Conclusions

We proposed core-boosted linear search, a novel search strategy for incomplete MaxSAT solving, that combines the strengths of core-guided and linear search and is, to the best of our knowledge, the first effective application of core-guided reformulation techniques in incomplete MaxSAT solving. Our experimental evaluation on a prototype implementation indicates that the information flow between the two phases of a core-boosted linear search solver often allows it to perform better than either of its individual components, while very rarely performing significantly worse. Furthermore, our comparison to other incomplete solvers shows that core-boosted linear search can be used to obtain state-of-the-art performance in weighted incomplete MaxSAT solving. As future work we plan to develop more dynamic ways of deciding when to switch between the core-guided and the linear search phase. Another interesting research directions to consider is the inclusion of MaxSAT preprocessing before, or even in-between, the core-guided and linear phases. Finally we also plan to look into extensions of core-boosted linear search to other constraint optimization paradigms.

References

1. Abramé, A., Habet, D.: AHMAXSAT: description and evaluation of a branch and bound MaxSAT solver. J. Satisf. Boolean Model. Comput. **9**, 89–128 (2015)
2. Abramé, A., Habet, D.: Learning nobetter clauses in MaxSAT branch and bound solvers. In: Proceedings of the 28th International Conference on Tools with Artificial Intelligence, pp. 452–459. IEEE Computer Society (2016)
3. Achá, R.J.A., Nieuwenhuis, R.: Curriculum-based course timetabling with SAT and MaxSAT. Ann. Oper. Res. **218**(1), 71–91 (2014)
4. Alviano, M., Dodaro, C., Ricca, F.: A MaxSAT algorithm using cardinality constraints of bounded size. In: Proceedings of IJCAI, pp. 2677–2683. AAAI Press (2015)
5. Ansótegui, C., Bonet, M.L., Levy, J.: Solving (Weighted) partial MaxSAT through satisfiability testing. In: Kullmann, O. (ed.) SAT 2009. LNCS, vol. 5584, pp. 427–440. Springer, Heidelberg (2009). https://doi.org/10.1007/978-3-642-02777-2_39
6. Ansótegui, C., Bonet, M., Levy, J.: SAT-based MaxSAT algorithms. Artif. Intell. **196**, 77–105 (2013)
7. Ansótegui, C., Bonet, M.L., Gabàs, J., Levy, J.: Improving SAT-based weighted MaxSAT solvers. In: Milano, M. (ed.) CP 2012. LNCS, vol. 7514. Springer, Heidelberg (2012). https://doi.org/10.1007/978-3-642-33558-7_9
8. Ansótegui, C., Didier, F., Gabàs, J.: Exploiting the structure of unsatisfiable cores in MaxSAT. In: Proceedings of IJCAI, pp. 283–289. AAAI Press (2015)
9. Ansótegui, C., Gabàs, J.: Wpm3: an (in)complete algorithm for weighted partial maxsat. Artif. Intell. **250**, 37–57 (2017)
10. Argelich, J., Le Berre, D., Lynce, I., Marques-Silva, J., Rapicault, P.: Solving Linux upgradeability problems using Boolean optimization. In: Proceedings of LoCoCo. Electronic Proceedings in Theoretical Computer Science, vol. 29, pp. 11–22 (2010)
11. Asín, R., Nieuwenhuis, R., Oliveras, A., Rodríguez-Carbonell, E.: Cardinality networks and their applications. In: Kullmann, O. (ed.) SAT 2009. LNCS, vol. 5584, pp. 167–180. Springer, Heidelberg (2009). https://doi.org/10.1007/978-3-642-02777-2_18
12. Audemard, G., Simon, L.: Predicting learnt clauses quality in modern SAT solvers. In: Proceedings of IJCAI, pp. 399–404. Morgan Kaufmann Publishers Inc. (2009)
13. Bacchus, F., Narodytska, N.: Cores in core based MaxSat algorithms: an analysis. In: Sinz, C., Egly, U. (eds.) SAT 2014. LNCS, vol. 8561, pp. 7–15. Springer, Cham (2014). https://doi.org/10.1007/978-3-319-09284-3_2
14. Bacchus, F., Järvisalo, M., Martins, R., et al.: MaxSat evaluation 2018 (2018). https://maxsat-evaluations.github.io/2018/. Accessed 05 Sept 2018
15. Berg, J., Järvisalo, M.: Cost-optimal constrained correlation clustering via weighted partial maximum satisfiability. Artif. Intell. **244**, 110–143 (2017)

16. Berg, J., Järvisalo, M.: Weight-aware core extraction in SAT-based MaxSAT solving. In: Beck, J.C. (ed.) CP 2017. LNCS, vol. 10416, pp. 652–670. Springer, Cham (2017). https://doi.org/10.1007/978-3-319-66158-2_42

17. Biere, A., Biere, A., Heule, M., van Maaren, H., Walsh, T.: Handbook of Satisfiability. Frontiers in Artificial Intelligence and Applications., vol. 185. IOS Press, Amsterdam (2009)

18. Bjørner, N., Narodytska, N.: Maximum satisfiability using cores and correction sets. In: Proceedings of IJCAI, pp. 246–252. AAAI Press (2015)

19. Bunte, K., Järvisalo, M., Berg, J., Myllymäki, P., Peltonen, J., Kaski, S.: Optimal neighborhood preserving visualization by maximum satisfiability. In: Proceedings of AAAI, vol. 3, pp. 1694–1700. AAAI Press (2014)

20. Chen, Y., Safarpour, S., Marques-Silva, J., Veneris, A.: Automated design debugging with maximum satisfiability. IEEE Trans. Comput. Aided Des. Integr. Circuits Syst. **29**(11), 1804–1817 (2010)

21. Davies, J., Bacchus, F.: Exploiting the power of MIP solvers in MaxSAT. In: Järvisalo, M., Van Gelder, A. (eds.) SAT 2013. LNCS, vol. 7962, pp. 166–181. Springer, Heidelberg (2013). https://doi.org/10.1007/978-3-642-39071-5_13

22. Demirović, E., Musliu, N.: MaxSAT based large neighborhood search for high school timetabling. Comput. Oper. Res. **78**, 172–180 (2017)

23. Eén, N., Sörensson, N.: Translating pseudo-boolean constraints into SAT. J. Satisf. Boolean Model. Comput. **2**(1–4), 1–26 (2006)

24. Guerra, J., Lynce, I.: Reasoning over biological networks using maximum satisfiability. In: Milano, M. (ed.) CP 2012. LNCS, pp. 941–956. Springer, Heidelberg (2012). https://doi.org/10.1007/978-3-642-33558-7_67

25. Heras, F., Morgado, A., Marques-Silva, J.: Core-guided binary search algorithms for maximum satisfiability. In: Proceedings of AAAI. AAAI Press (2011)

26. Hyttinen, A., Eberhardt, F., Järvisalo, M.: Constraint-based causal discovery: conflict resolution with answer set programming. In: Proceedings of UAI, pp. 340–349. AUAI Press (2014)

27. Joshi, S., Martins, R., Manquinho, V.: Generalized totalizer encoding for pseudo-boolean constraints. In: Pesant, G. (ed.) CP 2015. LNCS, vol. 9255, pp. 200–209. Springer, Cham (2015). https://doi.org/10.1007/978-3-319-23219-5_15

28. Koshimura, M., Zhang, T., Fujita, H., Hasegawa, R.: QMaxSat: a partial Max-Sat solver. J. Satisf. Boolean Model. Comput. **8**, 95–100 (2012)

29. Le Berre, D., Parrain, A.: The Sat4j library, release 2.2 system description. J. Satisf. Boolean Model. Comput. **7**, 59–64 (2010)

30. Lei, Z., Cai, S.: Solving (weighted) partial MaxSat by dynamic local search for SAT. In: Proceedings of IJCAI, pp. 1346–1352 (2018)

31. Li, C.M., Manyà, F., Planes, J.: Exploiting unit propagation to compute lower bounds in branch and bound Max-SAT solvers. In: van Beek, P. (ed.) CP 2005. LNCS, vol. 3709, pp. 403–414. Springer, Heidelberg (2005). https://doi.org/10.1007/11564751_31

32. Li, C.M., Manya, F., Planes, J.: New inference rules for MaxSAT. J. Artif. Intell. Res. **30**(1), 321–359 (2007)

33. Li, C.M., Quan, Z.: An efficient branch-and-bound algorithm based on MaxSAT for the maximum clique problem. In: Proceedings of AAAI, vol. 10, pp. 128–133. AAAI Press (2010)

34. Lin, H., Su, K., Li, C.M.: Within-problem learning for efficient lower bound computation in MaxSAT solving. In: Proceedings of AAAI, pp. 351–356. AAAI Press (2008)

35. Liu, Y.L., Li, C.M., He, K., Fan, Y.: Breaking cycle structure to improve lower bound for MaxSAT. In: Zhu, D., Bereg, S. (eds.) FAW 2016. LNCS, vol. 9711, pp. 111–124. Springer, Cham (2016). https://doi.org/10.1007/978-3-319-39817-4_12

36. Marques-Silva, J., Janota, M., Ignatiev, A., Morgado, A.: Efficient model based diagnosis with maximum satisfiability. In: Proceedings of IJCAI, pp. 1966–1972. AAAI Press (2015)

37. Marques-Silva, J., Argelich, J., Graça, A., Lynce, I.: Boolean lexicographic optimization: algorithms & applications. Ann. Math. Artif. Intell. **62**(3–4), 317–343 (2011)

38. Marques-Silva, J., Planes, J.: On using unsatisfiability for solving maximum satisfiability. CoRR abs/0712.1097 (2007)

39. Marques-Silva, J., Sakallah, K.A.: GRASP - a new search algorithm for satisfiability. In: Proceedings of ICCAD, pp. 220–227. IEEE Computer Society (1996)

40. Martins, R., Manquinho, V., Lynce, I.: Open-WBO: a modular MaxSAT solver'. In: Sinz, C., Egly, U. (eds.) SAT 2014. LNCS, vol. 8561, pp. 438–445. Springer, Cham (2014). https://doi.org/10.1007/978-3-319-09284-3_33

41. Morgado, A., Dodaro, C., Marques-Silva, J.: Core-duided MaxSAT with soft cardinality constraints. In: O'Sullivan, B. (ed.) CP 2014. LNCS, vol. 8656, pp. 564–573. Springer, Cham (2014). https://doi.org/10.1007/978-3-319-10428-7_41

42. Morgado, A., Heras, F., Liffiton, M.H., Planes, J., Marques-Silva, J.: Iterative and core-guided maxsat solving: a survey and assessment. Constraints **18**(4), 478–534 (2013)

43. Nadel, A., Ryvchin, V.: Efficient SAT solving under assumptions. In: Cimatti, A., Sebastiani, R. (eds.) SAT 2012. LNCS, vol. 7317, pp. 242–255. Springer, Heidelberg (2012). https://doi.org/10.1007/978-3-642-31612-8_19

44. Narodytska, N., Bacchus, F.: Maximum satisfiability using core-guided MaxSAT resolution. In: Proceedings of AAAI, pp. 2717–2723. AAAI Press (2014)

45. Saikko, P., Berg, J., Järvisalo, M.: LMHS: a SAT-IP hybrid MaxSAT solver. In: Creignou, N., Le Berre, D. (eds.) SAT 2016. LNCS, vol. 9710, pp. 539–546. Springer, Cham (2016). https://doi.org/10.1007/978-3-319-40970-2_34

46. Sinz, C.: Towards an optimal CNF encoding of boolean cardinality constraints. In: van Beek, P. (ed.) CP 2005. LNCS, vol. 3709, pp. 827–831. Springer, Heidelberg (2005). https://doi.org/10.1007/11564751_73

47. Tseitin, G.S.: On the complexity of derivation in propositional calculus. In: Siekmann, J.H., Wrightson, G. (eds.) Automation of Reasoning: 2: Classical Papers on Computational Logic 1967–1970. Symbolic Computation (Artificial Intelligence), pp. 466–483. Springer, Heidelberg (1983). https://doi.org/10.1007/978-3-642-81955-1_28

48. Xing, Z., Zhang, W.: MaxSolver: an efficient exact algorithm for (weighted) maximum satisfiability. Artif. intell. **164**(1–2), 47–80 (2005)

49. Zhang, L., Madigan, C.F., Moskewicz, M.H., Malik, S.: Efficient conflict-driven learning in a Boolean satisfiability solver. In: Proceedings of ICCAD, pp. 279–285. IEEE Computer Society (2001)

50. Zhang, X., Mangal, R., Grigore, R., Naik, M., Yang, H.: On abstraction refinement for program analyses in datalog. In: Proceedings of PLDI, PLDI 2014, pp. 239–248. ACM, New York (2014)

51. Zhu, C., Weissenbacher, G., Malik, S.: Post-silicon fault localisation using maximum satisfiability and backbones. In: Proceedings of FMCAD, pp. 63–66. FMCAD Inc. (2011)

Binary Decision Diagrams for Bin Packing with Minimum Color Fragmentation

David Bergman[1]([envelope])[iD], Carlos Cardonha[2][iD], and Saharnaz Mehrani[1]

[1] Department of Operations and Information Management,
University of Connecticut, Storrs, USA
{david.bergman,saharnaz.mehrani}@uconn.edu
[2] IBM Research, São Paulo, Brazil
carloscardonha@br.ibm.com

Abstract. Bin Packing with Minimum Color Fragmentation (**BPMCF**) is an extension of the Bin Packing Problem in which each item has a size and a color and the goal is to minimize the sum of the number of bins containing items of each color. In this work, we introduce the **BPMCF** and present a decomposition strategy to solve the problem, where the assignment of items to bins is formulated as a binary decision diagram and an optimal integrated solutions is identified through a mixed-integer linear programming model. Our computational experiments show that the proposed approach greatly outperforms a direct formulation of **BPMCF** and that its performance is suitable for large instances of the problem.

Keywords: Bin packing · Binary decision diagrams · Integer programming

1 Introduction

In this work, we investigate Bin Packing with Minimum Color Fragmentation Problem (**BPMCF**), an extension of the Bin Packing Problem in which each item is associated with a *color* and one wishes to identify assignments where items of a common color are placed in the fewest number of bins possible. The **BPMCF** provides a characterization of event seating problems (e.g., wedding) where parties of people (e.g., individuals of the same family) belong to groups (e.g., bride's relatives, groom's relatives, bride's friends, and groom's friends) and the goal is to seat parties of the same group in as few tables as possible; similarly, the goal is to maximize the number of parties of the same group sharing tables. In this application bins represent tables of fixed capacity (number of seats) and items represent parties whose sizes are the number of individuals and colors indicate the groups. Similar problems can also be observed in production planning, where the execution of jobs of a certain type enforce the availability of specific processing modules on the assigned plants, and in logistics, where transportation of certain types of goods may require specific instrumentation in the vehicles (e.g., temperature or pressure-controlling devices).

© Springer Nature Switzerland AG 2019
L.-M. Rousseau and K. Stergiou (Eds.): CPAIOR 2019, LNCS 11494, pp. 57–66, 2019.
https://doi.org/10.1007/978-3-030-19212-9_4

Colored extensions of the bin packing problem have been studied in the literature, but typically with a different objective. In the *colored bin packing problem* (**CBPP**, also investigated as the *class constrained bin packing problem*), commonly colored items are not allowed to be packed next to each other in the same bin [2,3,14,35,36,38]; Approximation results have been obtained for the variant where bins may have different sizes and the goal is to minimize the sum of the sizes of the used bins [17,38]. Jansen introduced the *bin packing problem with conflicts* (**BPPC**), a generalization of the **CBPP** where we are given a graph on the items, with edges indicating pairs of elements that cannot be placed in the same bin [23,24]. Several algorithms have been introduced in the literature to address the **BPPC** [19,20,31,34], such as a branch-and-price algorithm for general conflict graphs [34]. Another variant of the **CBPP** is the *co-printing problem*, in which bins are bounded both in terms of weight and number of colors they may contain [32]; both heuristic and exact algorithms have been proposed to solve this problem [26,27]. To the best of our knowledge, the **BPMCF** is yet to be investigated in the literature.

In this article, we introduce the **BPMCF**, show how to cast the assignment of items to bins as binary decision diagrams (BDDs) [15,16], and present a mixed-integer linear programming (MIP) formulation to solve the problem. BDDs and their multivalued extension have been successfully applied in different applications for optimization [1,12,29,30], especially discrete optimization problems [9–12,22]. Decomposition strategies relying on the combination of decision diagrams and integer programming, such as the one employed in this work, have been applied to other optimization problems [6,8,13,28]. Our experiments suggest the efficiency of the proposed algorithm, with a clear superiority over a direct MIP formulation of the **BPMCF**.

2   Problem Overview

A formal definition of the **BPMCF** is presented below. Note that the problem definition and its algorithms can be adapted to differentiate solutions based also on the total number of bins being used.

Definition 1 (Bin Packing with Minimum Color Fragmentation). *Let* $\mathcal{B}$ *denote the set of bins available for packing* ($|\mathcal{B}| = k \in \mathbb{N}$), $B \in \mathbb{N}$ *be the capacity of each bin,* $C \subseteq \mathbb{N}$ *be a set of colors, and* $\mathcal{O} = \{o_1, o_2, \ldots, o_n\}$ *be a set of indivisible items such that, for each* $o \in \mathcal{O}$, $w(o) \in \mathbb{N}$ *denotes its size and* $c(o) \in C$ *its color. A feasible solution for the problem consists of a partition of* $\mathcal{O}$ *into disjoint sets* $\mathcal{O}_1, \mathcal{O}_2, \ldots, \mathcal{O}_k$ *such that* $\forall i \in [k], \sum_{o \in \mathcal{O}_i} w(o) \leq B$. *Let* n_c *denote the number of bins containing items of color* c. *A feasible solution is said to be optimal if it minimizes* $\sum_{c \in C} n_c$.

Proposition 1. *Deciding whether an arbitrary instance of the **BPMCF** with at least 2 bins is feasible is NP-complete.*

Proof. The result follows from a reduction of the *partition problem* (**PP**), which is NP-complete [25]. Given an instance I of the **PP** with a set A of elements, with each $a \in A$ of size $s'(a)$, we create an instance I' of the **BPMCF** such that each item i in I' is associated with an element $i(a)$ of A, has size $w(i(a)) = s'(a)$, and color $c(i(a)) = 1$, i.e., all items have the same color. Moreover, I' consists of two bins, each of size $\frac{1}{2} \sum_{a \in A} s'(a)$. A feasible solution for I can be directly converted into a solution for I' and vice-versa, so the reduction follows. $\square$

Proposition 2. *If each bin can contain at most two items, an optimal solution of the **BPMCF** can be computed in polynomial time.*

Proof. For every instance admitting feasible solutions, for each item o in $\mathcal{O}$ there is at least one item o' for which $w(o) + w(o') \leq B$. A feasible solution should contain at least $q = |\mathcal{O}| - |\mathcal{B}|$ pairs of items being placed in the same bin. For each possible value of q, we create the following instance of the maximum weighted matching problem. Let $G = (V, E, w)$ be a graph where $V = V(\mathcal{O}) \cup V'$, each vertex in $V(\mathcal{O})$ is associated with an item in $\mathcal{O}$ and vertices in V' contains $|\mathcal{O}| - 2q$ artificial elements. Set E contains an edge for each pair $\{u, v\} \subseteq V(\mathcal{O})$; if $c(u) = c(v)$, $w(\{u, v\}) = 2 + |\mathcal{O}|$, and otherwise we have $w(\{u, v\}) = 1 + |\mathcal{O}|$. E also contains an edge for each pair $\{u, v\} \in V(\mathcal{O}) \times V'$, each with weight $|\mathcal{O}|^2$.

By construction, any optimal solution contains a set of edges covering all the artificial vertices, and solutions with q edges covering all the remaining elements have higher value than others with $q-1$ or less. The maximum weighted matching problem can be computed in polynomial time [18], and the number of values of q that need to be inspected is bounded by $|\mathcal{O}|$, so the result follows. $\square$

3 Direct Formulation

The following binary program is the direct formulation of the **BPMCF** used for a baseline algorithm in our computational experiments.

$$
\begin{aligned}
(\textbf{IP}) \min \quad & \sum_{(b,c) \in \mathcal{B} \times C} y_{b,c} \\
& \sum_{b \in \mathcal{B}} x_{b,o} && = 1 && \forall o \in \mathcal{O} \\
& \sum_{o \in \mathcal{O}} w(o) x_{b,o} && \leq B && \forall b \in \mathcal{B} \\
& x_{b,o} && \leq y_{b,c} && \forall (b, c, o) \in \mathcal{B} \times C \times \mathcal{O}_c \\
& x_{b,o} \in \{0, 1\} && && \forall (b, o) \in \mathcal{B} \times \mathcal{O} \\
& y_{b,c} \in \{0, 1\} && && \forall (b, c) \in \mathcal{B} \times C
\end{aligned}
$$

In **IP**, the assignment of each item o to each bin b is defined by binary decision variable $x_{b,o}$. Additionally, we use $y_{b,c}$ to indicate whether bin b contains at least one item of color c. The first family of constraints of **IP** asserts that each item is assigned to exactly one bin. The second family of constraints avoids assignments where the sum of the sizes of the selected items exceeds the capacity of the bin. The last set of constraints is used to set $y_{b,c}$; if $x_{b,o} = 1$ for some $o \in \mathcal{O}_c$, $y_{b,c} = 1$, whereas the objective function drives $y_{b,c}$ to zero otherwise.

4   Binary Decision Diagram-Based Algorithm

Our algorithm relies on a decomposition strategy in which the assignment of items to each bin b is represented as a BDD D^b, with feasible assignments being associated with paths connecting the root node to the terminal node in D^b. The structure of a BDD depends solely on the capacity B of the associated bin. The **BPMCF** can be solved through the construction of these BDDs and the identification of a mutually exclusive and collectively exhaustive collection of paths (with respect to items selected) that minimizes $\sum_{b \in B} C(p^b)$, where $C(p^b)$ equals the number of colors associated with the objects covered by p^b; this problem is known in the literature as the *consistent path problem* [8,13,28,33].

A BDD $D^b = (N^b, A^b, v^b, d^b)$ is a layered-acyclic graph composed of a set of nodes N^b, a set of arcs A^b, together with a cost function $v^b : A^b \to \mathbb{R}$ and arc-domain function $d^b : A^b \to \{0,1\}$ defined on the arcs. Nodes in N^b are partitioned into a set $L^b = \{0, 1, ..., |\mathcal{O}|, |\mathcal{O}| + 1\}$ of layers. For every node $u \in N^b$, $l^b(u)$ denotes the layer where u belongs. Layers 0 and $|\mathcal{O}| + 1$ contain only the *root node* r^b and the *terminal node* t^b of N^b, respectively. Each layer $l \in \{1, ..., |\mathcal{O}|\}$ is associated with an item $o(l)$; analogously, we define $o(u) = o(l^b(u))$ for each node $u \in N$. We assume the layers are ordered by colors first (any arbitrary ordering of C may be employed) and then arbitrarily in each color. Each arc $a \in A^b$ connects two nodes in consecutive layers, being directed from a start-node $u^s(a)$ to an end-node $u^e(a)$; the item associated with $u^e(a)$ is denoted by $o(a)$. Every node in $N^b \backslash \{t^b\}$ is the start-node of a *zero-arc* a such that $d(a) = 0$ and may be the start-node an *one-arc* a' such that $d(a') = 1$.

In our BDD formulation, each arc a represents the decision about the inclusion of the item associated with its end-node $u^s(a)$; namely, one-arcs indicate the inclusion (or coverage) of $u^s(a)$, whereas zero-arcs indicate exclusion. Every root-to-terminal arc-specified path $p = (a^1, \ldots, a^{|\mathcal{O}|})$ therefore encodes a collection of items $\mathcal{O}(p) := \{o(l) : d(a^l) = 1\}$, defined by the one-arcs on the path, and the *cost* of this assignment is given by $C(p) := \sum_{l=1}^{|\mathcal{O}|} d(a^l) v^b(a^l)$. For any *exact* BDD for a bin b of size B, there is a one-to-one mapping between each collection of items $\tilde{\mathcal{O}} \subseteq \mathcal{O}$ such that $\sum_{o \in \tilde{\mathcal{O}}} s(o) \leq B$ and root-to-terminal paths p with $\mathcal{O}(p) = \tilde{\mathcal{O}}$. The dynamic programming-based construction algorithm of a BDD representing such a set of solutions is well-known [4,5,7,37]. We adopt this algorithm, but with additional care required because of the objective function; namely, for every path p, $C(p)$ must equal the number of colors present in $\mathcal{O}(p)$ (i.e., $C(p) = |\{g : g(o) = g \text{ for } o \in \mathcal{O}(p)\}|$).

Each node u of the BDD is associated with a *state* $z(u) = (B', d) \in \mathbb{Z} \cup \mathbb{B}$. The first coordinate of $z(u)$ contains the remaining capacity in the bin for any partial solution defined by a path starting from the root node r^b and ending at u. The second coordinate indicates whether any item with the color of the object in layer $l(u)$ has been selected.

We build each BDD by assigning state $(B, 0)$ to the root node r^b and generating layer $l + 1$ iteratively by processing the nodes in layer l as follows. Given a node u with state $z(u) = (B', d)$, we create a zero-arc a_0 directed to a node

u_0 with state $s(u_0) = (B', d')$. If object $o(u_0)$ is the first of its color in the order by layers, we set $d' = 0$ (as, by construction, no other object of this color could have been covered by any path from the root to u_0); otherwise, we set $d' = d$. The cost of arc a_0 is always 0.

If $B'' = B' - s(l(o)) \geq 0$ we create a one-arc a_1 directed to a node u_1 with state $s(u_1) = (B'', 1)$; note that a one-arc represents the selection of an item, so the color of $o(u_1)$ is necessarily covered if a_1 is traversed. The cost of a_1 is 0 if u and u_1 are associated with objects of same color and $s(u) = (B', 1)$; otherwise, a_1 represent the incorporation of the first object with the color of u_1, so its cost equals 1. If $B'' < 0$, a_1 is not created, as the resulting paths would exceed the capacity of the bin.

Finally, if the state of the resulting node u_0 or u_1 equals some other state $\tilde{u}$ that was previously generated, the respective arc is directed out of u to $\tilde{u}$; we employ this technique in order to avoid the duplication of states across nodes of a layer. After constructing all layers, each node in layer $l = n+1$ is merged into a single terminal node.

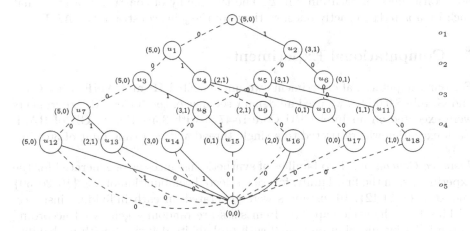

Fig. 1. Example BDD

Example: Consider an instance with 5 items of sizes 2, 3, 2, 3, 2 and colors 1, 1, 1, 2, 2, respectively, with bin capacity 5. A BDD for this bin is presented in Fig. 1. Each layer corresponds to an item. The solid/dashed arcs correspond to one-arcs/zero-arcs. The arc costs are specified next to each arc. Note that there can be one-arcs with zero cost (e.g., the arc from u_4 to u_{10}). Also, any solution corresponds to a path. For example, selecting items o_1 and o_3 is a feasible solution that corresponds to the arc-directed path $r - u_2 - u_5 - u_{11} - u_{18} - t$.

Network Flow Model: The BDDs allow us to formulate the consistent path problem through **ANF**, an *Arc-based Network Flow* MIP to solve the problem (for other examples where the same approach was employed, see [6,8]).

$$(\textbf{ANF}) \min \sum_{b \in \mathcal{B}} \sum_{a \in A^b} v^b(a) y_{b,a}$$

$$\sum_{\substack{a \in A^b; \\ u^e(a)=u}} y_{b,a} - \sum_{\substack{a \in A^b; \\ u^s(a)=u}} y_{b,a} = 0 \ \forall b \in \mathcal{B}, u \in N \backslash \{r^b, t^b\}$$

$$\sum_{\substack{a \in A^b; \\ u^s(a)=r^b}} y_{b,a} \qquad\qquad = 1 \ \forall b \in \mathcal{B}$$

$$\sum_{\substack{a \in A^b; \\ u^e(a)=t^b}} y_{b,a} \qquad\qquad = 1 \ \forall b \in \mathcal{B}$$

$$\sum_{b \in \mathcal{B}} \sum_{\substack{a \in A^b; \\ o(a)=o; \\ d(a)=1}} y_{b,a} \qquad\qquad = 1 \ \forall o \in \mathcal{O}$$

$$y_{b,a} \in \{0,1\} \qquad\qquad\qquad \forall b \in \mathcal{B}, a \in A^b$$

ANF employs binary variables $y_{b,a}$, which indicate whether arc a composes the path selected for D^b. The first three families of equalities model the network flow constraints for each bin b in $\mathcal{B}$. The last family of constraints asserts that each item is picked exactly once, so they are the joint constraints of **ANF**.

5 Computational Experiments

For our computational experiments we implemented **IP** and **ANF** using C++ and Gurobi 8.0.0 [21]; we used all default settings of the solver. All experiments were executed on an Intel CPU Core i7-4770 with 3.4 GHz, 32 GB of RAM. Each execution was restricted to a single thread and to a time limit of 30 min.

Instance Generation: Two families of synthetic instances were generated for the experiments. In the first family, for each selected combination of $k \in \{10, 20, 30\}$ and $B \in \{8, 10, 12\}$, 10 instances were generated; in each individual instance, all bins have the same capacity. Item sizes are randomly generated according to the following distribution: size 2 with probability 0.4; size 3 with probability 0.3; size 4 with probability 0.2; and size 5 with probability 0.1. This distribution was selected because of the authors' experience with group seating optimization applications. Items are generated uniformly and independently at random from the above distribution until 85% of the overall capacity is occupied. We then sequentially assign colors to the items by selecting p items to form each color class. With probability 0.6 we selected $p \in \{2, 3, 4\}$ items, and with probability 0.4 we select between $p \in \{5, 6, 7, 8\}$, in both cases sampled uniformly at random. If only one item remains, we assign it to the last color.

In order to test the scalability of **ANF**, we also generated instances with $k = 50$ and $B = 12$ as well as a second family with combinations of $k \in \{10, 20, 30, 40\}$ and $B = 20$ where items of size 2 were generated with probability 0.2 and items of size x are generated with probability 0.1 for each $x \in [3, 11]$. Finally, we restrict our experiments to scenarios where $B \geq 8$, as instances with smaller bins can be efficiently solved (see Proposition 2).

Results: The results of our experiments are shown in Table 1 in aggregation. Each row corresponds to a configuration of instances with k, B, as indicated by the first and second columns. The next eight columns report solution statistics, first for **IP** and then for **ANF**. In sequence, we report the average solution times for those instances that were solved within 1800 s, with the number of instances solved within 1800 s in superscript, the average ending lower bound, the average ending upper bound, and the average gap.

Table 1. Aggregate summary of results.

Instances		IP				ANP			
k	B	Time	LB	UB	Gap	Time	LB	UB	Gap
10	8	58.64^{10}	10.1	10.1	0	0.09^{10}	10.1	10.1	0
10	10	524.32^{8}	11.2	11.4	0.02	0.33^{10}	11.4	11.4	0
10	12	429.62^{6}	11.2	11.7	0.04	8.65^{10}	11.7	11.7	0
20	8	-	16.4	21.2	0.22	0.76^{10}	21.2	21.2	0
20	10	-	18.9	22.6	0.16	81.40^{10}	22.6	22.6	0
20	12	-	19.4	23.7	0.18	341.16^{10}	23.7	23.7	0
30	8	-	23.1	31.9	0.28	69.99^{10}	31.9	31.9	0
30	10	-	27.7	34.1	0.19	699.91^{9}	33.9	34.1	0.01
30	12	-	28.0	34.6	0.19	1357.46^{2}	30.9	34.6	0.10
50	10	-	40.9	56.2	0.27	-	50.8	56.2	0.10
10	20	232.82^{3}	11.1	11.8	0.058	1.69^{10}	11.8	11.8	0
20	20	-	18.8	23.9	0.21	682.47^{8}	23.4	23.9	0.02
30	20	-	28.1	36.1	0.22	-	32.6	36.1	0.10
40	20	-	34.1	46.5	0.27	-	41.3	46.5	0.11

We see a considerable superiority of **ANF** over **IP**, both in terms of gap and running time. **IP** solves only those instances with $k = 10$ (and only solves 24 of the 30 instances with this k) while **ANF** solves all instances with $k = 10$ and $k = 20$, and even 10 with $k = 30$. Additionally, the ending gap and quality of solutions are significantly better, even for those instances unsolved by both.

A depiction of the solution time and ending gaps is provided in the Fig. 2 through a cumulative distribution plot of performance. For both algorithms, the left half provides a plot with height equal to the cumulative number of instances solved at the time given on the horizontal axis. In the right half, the height of the plot corresponds to the number of instances with at most the optimality gap given on the horizontal axis by the time limit of 1800 s. Figure 2 more readily depicts the overall performance of **ANF**. After any amount of time, **ANF** solves more instances than **IP**, with smaller gaps at time limit.

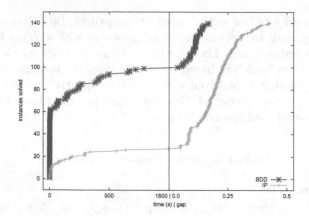

Fig. 2. Cumulative distribution plot comparing **BDD** with **IP**.

6 Conclusion and Future Work

In this work, we have introduced the *bin packing with minimum color fragmentation* and presented an algorithm consisting of the integration of decision diagrams and mixed-integer linear programming. Namely, we showed how to represent the assignment of items to individual bins as binary decision diagrams and formulated the integration of the sub-problems using a network flow model. Our computational experiments have shown that the proposed algorithm scales well and is clearly superior to a direct formulation of the **BPMCF**.

In future work, we intend to investigate the performance of the proposed algorithm in real-world scenarios. Additionally, we also would like to investigate alternative advanced solution approaches that have been successfully applied to other variants of the colored bin packing problem, such as branch and price. Finally, we believe that the present work motivates further investigation on decision diagram decomposition techniques to variants of the bin packing problem.

References

1. Andersen, H.R., Hadzic, T., Hooker, J.N., Tiedemann, P.: A constraint store based on multivalued decision diagrams. In: Bessière, C. (ed.) CP 2007. LNCS, vol. 4741, pp. 118–132. Springer, Heidelberg (2007). https://doi.org/10.1007/978-3-540-74970-7_11
2. Balogh, J., Békési, J., Dósa, G., Epstein, L., Kellerer, H., Tuza, Z.: Online results for black and white bin packing. Theor. Comput. Syst. **56**(1), 137–155 (2015)
3. Balogh, J., Békési, J., Dosa, G., Kellerer, H., Tuza, Z.: Black and white bin packing. In: Erlebach, T., Persiano, G. (eds.) WAOA 2012. LNCS, vol. 7846, pp. 131–144. Springer, Heidelberg (2013). https://doi.org/10.1007/978-3-642-38016-7_12
4. Behle, M.: On threshold BDDs and the optimal variable ordering problem. J. Comb. Optim. **16**(2), 107–118 (2008). https://doi.org/10.1007/s10878-007-9123-z
5. Bergman, D., Bodur, M., Cardonha, C., Cire, A.A.: Network models for multiobjective discrete optimization. arXiv:1802.08637 (2018)

6. Bergman, D., Cire, A.A.: Decomposition based on decision diagrams. In: Quimper, C.-G. (ed.) CPAIOR 2016. LNCS, vol. 9676, pp. 45–54. Springer, Cham (2016). https://doi.org/10.1007/978-3-319-33954-2_4

7. Bergman, D., Cire, A.A.: Multiobjective optimization by decision diagrams. In: Rueher, M. (ed.) CP 2016. LNCS, vol. 9892, pp. 86–95. Springer, Cham (2016). https://doi.org/10.1007/978-3-319-44953-1_6

8. Bergman, D., Cire, A.A.: Discrete nonlinear optimization by state-space decompositions. Manage. Sci. **64**(10), 4700–4720 (2018). https://doi.org/10.1287/mnsc.2017.2849

9. Bergman, D., Cire, A.A., van Hoeve, W.J., Hooker, J.N.: Discrete optimization with decision diagrams. INFORMS J. Comput. **28**(1), 47–66 (2016). https://doi.org/10.1287/ijoc.2015.0648

10. Bergman, D., Cire, A.A., van Hoeve, W.-J., Hooker, J.N.: Variable ordering for the application of BDDs to the maximum independent set problem. In: Beldiceanu, N., Jussien, N., Pinson, É. (eds.) CPAIOR 2012. LNCS, vol. 7298, pp. 34–49. Springer, Heidelberg (2012). https://doi.org/10.1007/978-3-642-29828-8_3

11. Bergman, D., Cire, A.A., Van Hoeve, W.J., Hooker, J.: Decision Diagrams for Optimization. Springer, Cham (2016). https://doi.org/10.1007/978-3-319-42849-9

12. Bergman, D., van Hoeve, W.-J., Hooker, J.N.: Manipulating MDD relaxations for combinatorial optimization. In: Achterberg, T., Beck, J.C. (eds.) CPAIOR 2011. LNCS, vol. 6697, pp. 20–35. Springer, Heidelberg (2011). https://doi.org/10.1007/978-3-642-21311-3_5

13. Bergman, D., Lozano, L.: Decision diagram decomposition for quadratically constrained binary optimization (2018)

14. Böhm, M., Sgall, J., Veselý, P.: Online colored bin packing. In: Bampis, E., Svensson, O. (eds.) WAOA 2014. LNCS, vol. 8952, pp. 35–46. Springer, Cham (2015). https://doi.org/10.1007/978-3-319-18263-6_4

15. Bryant, R.E.: Graph-based algorithms for Boolean function manipulation. IEEE Trans. Comput. **35**(8), 677–691 (1986). https://doi.org/10.1109/TC.1986.1676819

16. Bryant, R.E.: Symbolic Boolean manipulation with ordered binary-decision diagrams. ACM Comput. Surv. **24**(3), 293–318 (1992). https://doi.org/10.1145/136035.136043

17. Dawande, M., Kalagnanam, J., Sethuraman, J.: Variable sized bin packing with color constraints. Electron. Notes Discrete Math. **7**, 154–157 (2001)

18. Edmonds, J.: Paths, trees, and flowers. Canad. J. Math. **17**(3), 449–467 (1965)

19. Elhedhli, S., Li, L., Gzara, M., Naoum-Sawaya, J.: A branch-and-price algorithm for the bin packing problem with conflicts. INFORMS J. Comput. **23**(3), 404–415 (2011)

20. Gendreau, M., Laporte, G., Semet, F.: Heuristics and lower bounds for the bin packing problem with conflicts. Comput. Oper. Res. **31**(3), 347–358 (2004)

21. Gurobi Optimization, LLC: Gurobi optimizer reference manual (2018). http://www.gurobi.com

22. Hadzic, T., Hooker, J.N., O'Sullivan, B., Tiedemann, P.: Approximate compilation of constraints into multivalued decision diagrams. In: Stuckey, P.J. (ed.) CP 2008. LNCS, vol. 5202, pp. 448–462. Springer, Heidelberg (2008). https://doi.org/10.1007/978-3-540-85958-1_30

23. Jansen, K.: An approximation scheme for bin packing with conflicts. J. Comb. Optim. **3**(4), 363–377 (1999)

24. Jansen, K., Öhring, S.: Approximation algorithms for time constrained scheduling. Inf. Comput. **132**(2), 85–108 (1997)

25. Karp, R.M.: Reducibility among combinatorial problems. In: Miller, R.E., Thatcher, J.W., Bohlinger, J.D. (eds.) Complexity of Computer Computations. The IBM Research Symposia Series, pp. 85–103. Springer, Boston (1972). https://doi.org/10.1007/978-1-4684-2001-2_9

26. Kochetov, Y., Kondakov, A.: VNS matheuristic for a bin packing problem with a color constraint. Electron. Notes Discrete Math. **58**, 39–46 (2017)

27. Kondakov, A., Kochetov, Y.: A core heuristic and the branch-and-price method for a bin packing problem with a color constraint. In: Eremeev, A., Khachay, M., Kochetov, Y., Pardalos, P. (eds.) OPTA 2018. CCIS, vol. 871, pp. 309–320. Springer, Cham (2018). https://doi.org/10.1007/978-3-319-93800-4_25

28. Lozano, L., Bergman, D., Smith, J.C.: On the consistent path problem (2018)

29. Matsumoto, K., Hatano, K., Takimoto, E.: Decision diagrams for solving a job scheduling problem under precedence constraints. In: LIPIcs-Leibniz International Proceedings in Informatics, vol. 103. Schloss Dagstuhl-Leibniz-Zentrum fuer Informatik, Saarbrücken (2018)

30. Miller, D.M., Drechsler, R.: Implementing a multiple-valued decision diagram package. In: Proceedings of the 28th IEEE International Symposium on Multiple-Valued Logic, pp. 52–57. IEEE (1998)

31. Muritiba, A.E.F., Iori, M., Malaguti, E., Toth, P.: Algorithms for the bin packing problem with conflicts. INFORMS J. Comput. **22**(3), 401–415 (2010)

32. Peeters, M., Degraeve, Z.: The co-printing problem: a packing problem with a color constraint. Oper. Res. **52**(4), 623–638 (2004)

33. Raghunathan, A.U., Bergman, D., Hooker, J., Serra, T., Kobori, S.: Seamless multimodal transportation scheduling (2018)

34. Sadykov, R., Vanderbeck, F.: Bin packing with conflicts: a generic branch-and-price algorithm. INFORMS J. Comput. **25**(2), 244–255 (2013)

35. Shachnai, H., Tamir, T.: Polynomial time approximation schemes for class-constrained packing problems. J. Sched. **4**(6), 313–338 (2001)

36. Shachnai, H., Tamir, T.: Tight bounds for online class-constrained packing. Theoret. Comput. Sci. **321**(1), 103–123 (2004)

37. Trick, M.A.: A dynamic programming approach for consistency and propagation for knapsack constraints. Ann. Oper. Res. **118**(1), 73–84 (2003). https://doi.org/10.1023/A:1021801522545

38. Xavier, E.C., Miyazawa, F.K.: The class constrained bin packing problem with applications to video-on-demand. Theoret. Comput. Sci. **393**(1–3), 240–259 (2008)

Local Rapid Learning for Integer Programs

Timo Berthold[1(✉)], Peter J. Stuckey[2], and Jakob Witzig[3]

[1] Fair Isaac Germany GmbH, Takustr. 7, 14195 Berlin, Germany
timoberthold@fico.com
[2] Monash University and Data61, Melbourne, Australia
Peter.Stuckey@monash.edu
[3] Zuse Institute Berlin, Takustr. 7, 14195 Berlin, Germany
witzig@zib.de

Abstract. Conflict learning algorithms are an important component of modern MIP and CP solvers. But strong conflict information is typically gained by depth-first search. While this is the natural mode for CP solving, it is not for MIP solving. Rapid Learning is a hybrid CP/MIP approach where CP search is appliedat the root to learn information to support the remaining MIP solve. This has been demonstrated to be beneficial for binary programs. In this paper, we extend the idea of Rapid Learning to integer programs, where not all variables are restricted to the domain $\{0, 1\}$, and rather than just running a rapid CP search at the root, we will apply it repeatedly at local search nodes within the MIP search tree. To do so efficiently, we present six heuristic criteria to predict the chance for local Rapid Learning to be successful. Our computational experiments indicate that our extended Rapid Learning algorithm significantly speeds up MIP search and is particularly beneficial on highly dual degenerate problems.

1 Introduction

Constraint programming (CP) and integer programming (IP) are two complementary ways of tackling discrete optimization problems. Hybrid combinations of the two approaches have been used for many years, see, e.g., [2,9,10,17,22,37,42]. Both technologies have incorporated *conflict learning* capabilities [1,21,27,35,38] that derive additional valid constraints from the analysis of infeasible subproblems extending methods developed by the SAT community [33].

Conflict learning is a technique that analyzes infeasible subproblems encountered during a tree search algorithm. In a tree search, each subproblem can be identified by its local variable bounds, i.e., by local bound changes that come from branching decisions and propagation at the current node and its ancestors. If propagation detects infeasibility, conflict learning will traverse this chain of decisions and deductions reversely, reconstructing which bound changes led to which other bound changes. In this way, conflict learning identifies explanations for the infeasibility. If it can be shown that a small subset of the bound changes

© Springer Nature Switzerland AG 2019
L.-M. Rousseau and K. Stergiou (Eds.): CPAIOR 2019, LNCS 11494, pp. 67–83, 2019.
https://doi.org/10.1007/978-3-030-19212-9_5

suffices to prove infeasibility, a so-called conflict constraint is generated that can be exploited in the remainder of the search to prune parts of the tree.

In the context of constraint programming, conflict constraints are also referred to as *no-goods*. For binary programs (BPs), i.e., mixed integer (linear) programs for which all variables have domain $\{0, 1\}$, conflict constraints will have the form of *set covering* constraints. These are linear constraints of the form "sum of variables (or their negated form) is greater than or equal to one".

Rapid Learning [13] is a heuristic algorithm for BPs that searches for valid conflict constraints, global bound reductions, and primal solutions. It is based on the observation that a CP solver can typically perform an incomplete search on a few thousand nodes in a fraction of the time that a MIP solver needs for processing the root node. In addition, CP solvers make use of depth-first search, as opposed to the hybrid best-first/depth-first search of MIP solvers, which more rapidly generates strong no-goods. Typically CP solvers do not differentiate the root node from other nodes. They apply fast (at least typically) propagation algorithms to infer new information about the possible values variables can take, and then take branching decisions. In contrast, a MIP solver invests a substantial amount of time at the root node to gather global information about the problem and to initialize statistics that can help for the search. A significant portion of root node processing time comes from the computational effort needed to solve the initial LP relaxation from scratch. Further aspects are the LP resolves during cutting plane generation, strong branching [7] for branching statistic evaluation, and primal heuristics, see, e.g., [11].

The idea of Rapid Learning is to apply a fast CP depth-first branch-and-bound search for a few hundred or thousand nodes, generating and collecting valid conflict constraints at the root node of a MIP search. Using this, the MIP solver is already equipped with the valuable information of which bound changes will lead to an infeasibility, and can avoid them by propagating the derived constraints. Just as important, the partial CP search might find primal solutions, thereby acting as a primal heuristic. Furthermore, the knowledge of conflict constraints can be used to initialize branching statistics, just like strong branching. In this paper, we will extend Rapid Learning to integer programs and to nodes beyond the root.

The remainder of the paper is organized as follows. In Sect. 2, we provide more background on conflict learning for MIPs, in particular the extension to general integer variables, which is important for our extended Rapid Learning algorithm. In Sect. 3, we describe details of the Rapid Learning algorithm for general integer programs, extending the work of Berthold et al. [13]. In Sect. 4, we discuss what special considerations have to be taken when applying Rapid Learning repeatedly at local subproblems during the MIP tree search instead of using it as a onetime global procedure. We introduce six criteria to predict the benefit of local Rapid Learning. Section 5 presents our computational study, in which we apply our extended Rapid Learning algorithm to a set of integer programs from the well-known benchmark sets of MIPLIB 3, MIPLIB 2003, and MIPLIB 2010 [28]. The experiments have been conducted with the constraint integer programming solver SCIP [24] and indicate that a significant speed-up can be achieved for (pure) integer programs, when using Rapid Learning locally. In Sect. 6, we conclude.

2 Conflict Learning in Integer Programming

A mixed integer program is a mathematical optimization problem defined as follows.

Definition 1 (mixed integer program). *Let $m, n \in \mathbb{Z}_{\geq 0}$. Given a matrix $A \in \mathbb{R}^{m \times n}$, a right-hand-side vector $b \in \mathbb{R}^m$, an objective function vector $c \in \mathbb{R}^n$, a lower and an upper bound vector $l \in (\mathbb{R} \cup \{-\infty\})^n$, $u \in (\mathbb{R} \cup \{+\infty\})^n$ and a subset $\mathcal{I} \subseteq \mathcal{N} = \{1, \ldots, n\}$, the corresponding* mixed integer program (MIP) *is given by*

$$
\begin{aligned}
\min \quad & c^\mathsf{T} x \\
s.t. \quad & Ax \leq b \\
& l_j \leq x_j \leq u_j \quad \text{for all } j \in \mathcal{N} \\
& x_j \in \mathbb{R} \quad \text{for all } j \in \mathcal{N} \setminus \mathcal{I} \\
& x_j \in \mathbb{Z} \quad \text{for all } j \in \mathcal{I}.
\end{aligned}
\tag{1}
$$

Mixed integer programs can be categorized by the classes of variables that are part of their formulation:

- If $\mathcal{N} = \mathcal{I}$, problem (1) is called a *(pure) integer program (IP)*.
- If $\mathcal{N} = \mathcal{I}$, $l_j = 0, j \in \mathcal{N}$ and $u_j = 1, j \in \mathcal{N}$, problem (1) is called a *(pure) binary program (BP)*.
- If $\mathcal{I} = \emptyset$, problem (1) is called a *linear program (LP)*.

Conflict analysis techniques were originally developed by the artificial intelligence research community [40] and, later extended by the SAT community [33]; they led to a huge increase in the size of problems modern SAT solvers can handle [31,33,43]. The most successful SAT learning approaches use so-called *one-level first unique implication point (1-UIP)* [43] learning which in some sense captures the conflict constraint "closest" to the infeasibility. Conflict analysis also is successfully used in the CP community [25,26,35] (who typically refer to it as no-good learning) and the MIP world [1,21,38,41]. Nowadays, commercial MIP solvers like FICO Xpress [23] employ conflict learning by default.

Constraint programming and mixed integer programming are two complementary ways of tackling discrete optimization problems. Because they have different strengths and weaknesses hybrid combinations are attractive. One notable example, the software SCIP [3], is based on the idea of *constraint integer programming (CIP)* [2,6]. CIP is a generalization of MIP that supports the notion of general constraints as in CP. SCIP itself follows the idea of a very low-level integration of CP, SAT, and MIP techniques. All involved algorithms operate on a single search tree and share information and statistics through global storage of, e.g., solutions, variable domains, cuts, conflicts, the LP relaxation and so on. This allows for a very close interaction amongst CP and MIP (and other) techniques.

There is one major difference between BPs and IPs in the context of Rapid Learning: in IP, the problem variables are not necessarily binary. To deal with this, the concept of a *conflict graph* needs to be extended. A conflict graph gets constructed whenever infeasibility is detected in a local search node; it represents the logic of how the set of branching decisions led to the detection of infeasibility.

More precisely, the conflict graph is a directed acyclic graph in which the vertices[1] represent bound changes of variables, e.g., $x_i \leq \lambda_i$ or $x_i \geq \mu_i$. The conflict graph is built such that when the solver infers a bound change v as a consequence of a set of existing bound changes U, i.e., $U \rightarrow v$, then we have an arc (u, v) from each $u \in U$ to v. Bound changes caused by branching decisions are vertices without incoming edges. Finally the conflict graph includes a dummy vertex *false* representing failure which is added when the solver infers unsatisfiability.

Given a conflict graph, each cut that separates the branching decisions from the artificial infeasibility vertex *false* gives rise to a valid conflict constraint. A *unique implication point (UIP)* is an (inner) vertex of the conflict graph which is traversed by all paths from the branching vertices to the conflict vertex. Or, how Zhang et al. [43] describe it: "Intuitively, a UIP is the *single* reason that implies the conflict at [the] current decision level." UIPs are natural candidates for finding small cuts in the conflict graph. The *1-UIP* is the first cut separating the conflict vertex from the branching decisions when traversing in reverse assignment order.

For integer programs, conflict constraints can be expressed as so-called *bound disjunction* constraints:

Definition 2. *For an IP, let $\mathcal{L} \subseteq \mathcal{I}, \mathcal{U} \subseteq \mathcal{I}$ be disjoint index sets of variables, let $\lambda \in \mathbb{Z}^{\mathcal{L}}$ with $l_i \leq \lambda_i \leq u_i$ for all $i \in \mathcal{L}$, and $\mu \in \mathbb{Z}^{\mathcal{U}}$ with $l_i \leq \mu_i \leq u_i$ for all $i \in \mathcal{U}$. Then, a constraint of the form*

$$\bigvee_{i \in \mathcal{L}} (x_i \geq \lambda_i) \vee \bigvee_{i \in \mathcal{U}} (x_i \leq \mu_i)$$

is called a bound disjunction *constraint.*

For details on bound disjunction constraints, see Achterberg [1]. If all involved conflict values λ, μ correspond to global bounds of the variables, the bound disjunction constraint can be equivalently expressed as a knapsack constraint of form

$$\sum_{i \in \mathcal{U}} x_i - \sum_{i \in \mathcal{L}} x_i \leq \sum_{i \in \mathcal{U}} u_i - \sum_{i \in \mathcal{L}} l_i - 1. \tag{2}$$

Note that for BPs all conflicts only involve global bounds.

---

[1] For disambiguation, we will use the term *vertex* for elements of the conflict graph, as opposed to *nodes* of the search tree.

The power of conflict learning arises because often branch-and-bound based algorithms implicitly repeat the same search in a slightly different context in another part of the tree. Conflict constraints help to avoid redundant work in such situations. As a consequence, the more search is performed by a solver and the earlier conflicts are detected, the greater the chance for conflict learning to be beneficial. Note that conflict generation has a positive interaction with depth-first search. Depth-first search leads to the creation of no-goods that explain why a whole subtree contains no solutions, and hence the no-goods generated by depth-first search are likely to prune more of the subsequent search.

3 Rapid Learning for Integer Programs

The principle motivation for Rapid Learning [13] is the fact that a CP solver can typically search hundreds or thousand of nodes in a fraction of the time that a MIP solver needs for processing the root node of the search tree. Rapid Learning applies a fast CP search[2] for a few hundred or thousand nodes, before starting the MIP search. Using this approach, conflict constraints can be learnt before, and not only during, MIP search. Very loosely speaking: while the aim of conflict learning is to avoid making mistakes a second time, Rapid Learning tries to avoid making them the first time (during MIP search).

Rapid Learning is related to large neighborhood search heuristics, such as RINS and RENS [12,20]. But, rather than doing an incomplete search on a sub-problem using the same (MIP search) algorithm, Rapid Learning performs an incomplete search on the same problem using a much faster algorithm (CP search). Rapid Learning differs from primal heuristics in that it aims at improving the dual bound by collecting information on infeasibility rather than searching for feasible solutions.

Each piece of information collected in a rapid CP search can be used to guide the MIP search or even deduce further reductions during root node processing. Since the CP solver is solving the same problem as the MIP solver

- each generated conflict constraint is valid for the MIP search,
- each global bound change can be applied at the MIP root node,
- each feasible solution can be added to the MIP solver's solution pool,
- the branching statistics can initialize a hybrid MIP branching rule, see [4], and
- if the CP solver completely solves the problem, the MIP solver can abort.

All five types of information may be beneficial for a MIP solver, and are potentially generated by our algorithm which we now describe more formally.

The Rapid Learning algorithm is outlined in Fig. 1. Here, $l(P)$ and $u(P)$ are lower and upper bound vectors, respectively, of the problem at hand, P. For the moment we assume P is the root problem, in the next section we will examine the use of Rapid Learning at subproblem nodes. The symbol C refers to a single globally valid conflict constraint explaining the infeasibility of the current

---

[2] By CP search we mean applying a depth-first search using only propagation for reasoning, no LP relaxation is solved during the search.

subproblem. Rapid Learning is an incomplete CP search: a branch-and-bound algorithm which traverses the search space in a depth-first manner (Line 3), using propagation (Line 4) and conflict analysis (Line 7), but no LP relaxation. Instead, the *pseudo-solution* [2], i.e., an optimal solution of a relaxation consisting only of the variable bounds (Line 5), is used for the bounding step.

Propagation of linear constraints is conducted by the bound strengthening technique of Brearley et al. [18] which uses the residual activity of linear constraints within the local bounds. For special cases of linear constraints, SCIP implements special, more efficient propagators. Knapsack constraints use efficient integer arithmetic instead of floating point arithmetic, and sort by coefficient values to propagate each variable only once. SCIP also features methods to extract clique information about the binary variables of a problem. A clique is a set of binary variables of which at most one variable can take the value 1 in a feasible solution. Clique information can be used to strengthen the propagation of knapsack constraints. Set cover constraints are propagated by the highly efficient two-watched literal scheme [33], which is based on the fact that the only domain reduction to be inferred from a set cover constraint is to fix a variable to 1 if all other variables have already been fixed to 0.

Variable and value selection takes place in Line 14; inference branching [2] is used as branching rule. Inference branching maintains statistics about how often the fixing of a variable led to fixings of other variables, i.e., it is a history rule, its essentially a MIP equivalent of impact-based search [29,36]. Since history rules are often weak in the beginning of the search, we seed the CP solver with statistics that the MIP solver has collected in probing [39] during MIP presolving.

We assume that the propagation routines in Line 4 may also deduce global bound changes and modify the global bound vectors $l(P)$ and $u(P)$. Single-clause conflicts are automatically upgraded to global bound changes in Line 9. Note that it suffices to check constraint feasibility in Line 11, since the pseudo-solution $\bar{x}$ (see Line 5) will always take the value of one of the (integral) bounds for each variable.

Our implementation of the Rapid Learning heuristic uses a secondary SCIP instance to perform the CP search. Only a few parameters need to be altered from their default values to turn SCIP into a CP solver, an overview is given in Table 1. Most importantly, we disabled the LP relaxation and use a pure depth-first search with inference branching (but without any additional tie breakers). Further, we switch from All-UIP to 1-UIP in order to generate only one conflict per infeasibility. This is a typical behavior of CP solvers, but not for MIP solvers. Expensive feasibility checks and propagation of the objective function as a constraint are also avoided.

In order to avoid spending too much time in Rapid Learning, the number of nodes explored during the CP search is limited to at most 5000. The actual number of allowed nodes is determined by the number of simplex iterations iter_{LP} performed so far in the main SCIP but at least 500, i.e.,

$$\lim_{\text{node}} = \min\{5000, \max\{500, \text{iter}_{\text{LP}}\}\}.$$

Table 1. Settings for Rapid Learning sub-SCIP.

Parameter name	Value	Effect
lp/solvefreq	-1	Disable LP
conflict/fuiplevels	1	Use 1-UIP
nodeselection/dfs/stdpriority	INT_MAX/4	Use DFS
branching/inference/useweightedsum	FALSE	Pure inference, no VSIDS
constraints/disableenfops	TRUE	No extra checks
propagating/pseudoobj/freq	-1	No objective propagation
conflict/maxvarsfac	0.05	Only short conflicts
history/valuebased	TRUE	Extensive branch. Statistics

Input : IP P as in (1) (with $\mathcal{R} = \emptyset$),
  node limit $\lim_{\text{node}}$,
  primal bound $\bar{c}$ for P (might be ∞)

Output: set of valid conflict constraints $\mathcal{L}_C$ for P,
  valid global domain box $[l, u]$ for P,
  feasible solution $\tilde{x}$ for P or $\emptyset$

1   $\mathcal{L} \leftarrow \{P\}$, $n_{\text{node}} \leftarrow 0$, $\mathcal{L}_C \leftarrow \emptyset$, $\tilde{x} \leftarrow \emptyset$;
2   **while** $\mathcal{L} \neq \emptyset \wedge n_{\text{node}} < \lim_{\text{node}}$ **do**
3     $\tilde{P} \leftarrow \texttt{select\_dfs}(\mathcal{L})$, $\mathcal{L} \leftarrow \mathcal{L} \setminus \tilde{P}$, $n_{\text{node}} \leftarrow n_{\text{node}} + 1$;
4     $[l(\tilde{P}), u(\tilde{P})] \leftarrow \texttt{propagate}([l(\tilde{P}), u(\tilde{P})])$;
5     $\tilde{x} \leftarrow \arg\min\{c^T x \mid x \in [l(\tilde{P}), u(\tilde{P})]\}$;

    /* analyze infeasible subproblem, potentially store globally
      valid conflict constraint   */
6   **if** $[l(\tilde{P}), u(\tilde{P})] = \emptyset$ **or** $c(\tilde{x}) \geq \bar{c}$ **then**
7       $C \leftarrow \texttt{analyze}(\tilde{P})$;
8   **if** $C \neq \emptyset$ **then** $\mathcal{L}_C \leftarrow \mathcal{L}_C \cup \{C\}$;
9   **if** $|C| = 1$ **then** $\texttt{tighten}([l(P), u(P)])$;
10   **continue**;

    /* check for new incumbent solution   */
11   **if** $A\tilde{x} \leq b$ **and** $c^T \tilde{x} < \bar{c}$ **then**
12       $\tilde{x} \leftarrow \tilde{x}$, $\bar{c} \leftarrow c^T \tilde{x}$;
13   **continue**;

14     $(x_i, v) \leftarrow \texttt{select\_infer}(\tilde{P}, \tilde{x})$;
15     $\tilde{P}_l \leftarrow \tilde{P} \cup \{x_i \leq v\}$, $\tilde{P}_r \leftarrow \tilde{P} \cup \{x_i \geq v\}$;
16     $\mathcal{L} \leftarrow \mathcal{L} \cup \{\tilde{P}_l, \tilde{P}_r\}$;
17 **return** $(\mathcal{L}_C, [l(P), u(P)], \tilde{x})$;

Fig. 1. Rapid Learning algorithm

The idea is to restrict Rapid Learning more rigorously for problems where processing of a single MIP node is cheap already. The number of simplex iterations is a deterministic estimate for node processing cost.

We aim to generate short conflict constraints, since these are most likely to frequently trigger propagations in the upcoming MIP search. Thus, we only collect conflicts that contain at most 5% of the problem variables. Finally, we adapt the collection of branching statistics such that history information on general integer variables are collected per value in the domain rather than having one counter for down- and one for up-branches regardless of the value on which was branched. This can be essential for performing an efficient CP search on general integer variables, and was a building block that enabled us to use Rapid Learning on IPs rather than solely on BPs, as in [13].

In addition to the particular parameters listed in Table 1, we set the emphasis[3] for presolving to "fast". Emphasis settings for cutting are not necessary, since no LP relaxation is solved, from the armada of primal heuristics only a few are applied that do not require an LP relaxation, see [5]. Note that since Rapid Learning will be called at the end of the MIP root node, or even locally, see next Section, the problem that the CP solver considers has already been presolved, might contain cutting planes as additional linear constraints and have an objective cutoff constraint if a primal solution has been found by a primal heuristic during root node processing.

4   Local Rapid Learning

The original Rapid Learning algorithm [13] was used as part of a root preprocessing, i.e., for every instance it was run exactly once at the end of the root node. But only running Rapid Learning at the root limits its effectiveness. We now discuss the factors that arise when we allow Rapid Learning to be run at local nodes inside the search tree.

When running in the root only all information returned by the CP solver is globally valid, and the overhead to maintain the information gathered by Rapid Learning is negligible [13]. In contrast, when applying Rapid Learning at a local node within the tree conflicts and bound changes will only be locally valid in general. Since Rapid Learning uses a secondary SCIP instance to perform the CP search, all local information of the current node becomes part of the initial problem formulation for the CP search. Thus, conflicts gathered by Rapid Learning do not include bound changes made along the path from the root to the current node, they are simply considered as valid for this local node. As a consequence, these conflicts will only be locally valid and hence only applied to the current node of the MIP search. Using an assumption interface [34], local conflicts could be lifted to be globally valid. However, this is subject to future investigation and not considered in the current implementation of Rapid Learning.

---

[3] In SCIP, emphasis settings correspond to a group of individual parameters being changed.

In practice, all local information needs to be maintained when switching from one node of the tree to another. In CP solvers, switching nodes is typically very cheap, because depth-first search is used. However, a MIP solver frequently "jumps" within the tree. Therefore, two consecutively processed nodes can be quite different. In what follows, we will refer to the time spent for moving from one node to another node as *switching time*. The switching time can be used as an indicator to quantify the overhead introduced by all locally added information found by Rapid Learning.

To ensure that the amount of locally added information does not increase the switching time too much, we apply Rapid Learning very rarely by using a exponentially decreasing frequency of execution. Rapid Learning is executed at every node of depth d with

$$\log_\beta(d/f) \in \mathbb{Z}, \tag{3}$$

where β and f are two parameters to control the speed of decrease. For example, if $\beta = 1$ Rapid Learning is executed at every depth $d = i \cdot f$ with $i \in \mathbb{Z}_+$.

Unfortunately, the amount of locally valid information produced by Rapid Learning still leads to an increase of switching time by 21%. Consequently, the overall performance decreased by 20% in our first experiments. At the same time the number of explored branch-and-bound nodes decreased by 16%. This indicates the potential gains possible using local Rapid Learning.

To control at which subproblem Rapid Learning is applied we propose six criteria to forecast the potential of Rapid Learning. These criteria aim at identifying one of two situations. The first is to estimate whether the (sub)problem is infeasible or a pure feasibility problem. In these cases propagating conflicts is expected to be particularly beneficial. The second is to estimate the dual degeneracy of a problem. In this case, VSIDS branching statistics are expected to be particularly beneficial. The VSIDS [31] (variable state independent decaying sum) statistics takes the contribution of every variable (and its negated complement) in conflict constraints found so far into account. For every variable, the number of clauses (in MIP speaking: conflict constraints) the variable is part of is counted. In the remainder of the search the VSIDS are periodically scaled by a predefined constant. By this, the weight of older clauses is reduced over time and more recent observations have a bigger impact.

A basic solution of an LP is called *dual degenerate*, when it has nonbasic variables with zero reduced costs. One can define the dual degeneracy of a MIP as the average number of nonbasic variables with zero reduced costs appearing in a basic solution of its LP relaxation. The higher the dual degeneracy, the higher the chance that the LP objective will not change by branching and hence many of the costs involved in the pseudo-cost computation are zero. Therefore, for highly dual degenerate problems, using other branching criteria, such as VSIDS or inference scores, is crucial for solving the problem.

We now describe the six criteria we use to identify infeasible or dual degenerate problems, already using the criteria abbreviations from the tables in Sect. 5.

Criterion I: Dual Bound Improvement. During the tree search a valid lower bound for each individual subproblem is given by the respective LP solution. A globally valid lower bound is given by the minimum over all individual lower bounds. This global bound is called the *dual bound*. If the dual bound has not changed after processing a certain number of nodes, i.e., the dual bound is equal to the lower bound of the root node, it might be the case that the MIP lies inside a level plane of the objective, i.e., all feasible LP (and MIP) solutions will have the same objective. In other words, the instance might be a feasibility instance for which Rapid Learning was already shown to be very successful [13]. Feasibility instances are typically highly dual degenerate. The `dualbound` criterion means to call local Rapid Learning if the dual bound never changed during the MIP search.

Criterion II: Leaves Pruned by Infeasibility or Exceeding the Cutoff Bound. During the tree search every leaf node either provides a new incumbent solution (the rare case), is proven to be infeasible or to exceed the current cutoff bound which is given by the incumbent solution. The ratio of the latter two cases is used in SCIP's default branching rule. *Hybrid branching* [4] combines pseudo-costs, inference scores, and conflict information into one single branching score. The current implementation in SCIP puts a higher weight on conflict information, e.g., VSIDS [31], and a lower weight on pseudo-costs when the ratio of infeasible and cutoff nodes is larger than a predefined threshold. The `leaves` criterion means to call local Rapid Learning if ratio of infeasible leaves over those exceeding the cutoff bound is larger than 10. The rationale is that we expect (local) conflicts to be most beneficial, when infeasibility detection appears to be the main driver for pruning the tree.

Criterion III: LP Degeneracy. As mentioned above, the more nonbasic variables are dual degenerate, the less information can be gained during strong branching or pseudo-cost computation. As a consequence, Berthold et al. [14] introduced a modification to strong branching that considers the dual degeneracy of the LP solution. In rough terms, if either the share of dual degenerate nonbasic variables or the variable-constraint ratio of the optimal face exceed certain thresholds, strong branching will be deactivated. We adapt this idea of using the dual degeneracy of the current LP solution. The `degeneracy` criterion means to call local Rapid Learning if more than 80% of the nonbasic variables are degenerate or the variable-constraint ratio of the optimal face is larger than 2, as proposed in [14]. In both cases we expect that "strong conflict generation" will be useful.

Criterion IV: (Local) Objective Function. If all variables with non-zero objective coefficients are fixed at the local subproblem, i.e., the objective is constant, Criteria I and II will apply: every LP solution is fully dual degenerate and the only possibility to prune a leaf node is by infeasibility. If there are only very few unfixed variables with nonzero objective are left, the criteria might not apply. However, it is likely that the targeted situations occur frequently in the tree rooted at the current subproblem, at the latest, when all the variables occurring in the objective are fixed. The `obj` criterion means to call local Rapid Learning once the objective support is small enough, in anticipation of the current subproblem turning into a feasibility problem. In our implementation we apply this criterion very conservatively, and call Rapid Learning only if the local objective is zero.

Criterion V: Number of Solutions. The most obvious evidence, and indeed a necessary one, that a MIP instance is infeasible, is that no feasible solution has been found during the course of the MIP search. Note that for most (feasible) MIP instances, primal heuristics find a feasible solution at the root node [11] or at the latest during the first dive in the branch-and-bound. The `nsols` criterion means to call local Rapid Learning if no feasible solution has been found so far.

Criterion VI: Strong Branching Improvements. In the beginning of the tree search it is very unlikely that enough leaf nodes are explored to reliably guess whether the actual MIP is a feasibility instance. Therefore, we consider the subproblems evaluated during strong branching, which are concentrated at the top of the search tree. Similarly to Criterion II, we compute the ratio between the number of strong branching problems that gave no improvement in the objective or went infeasible to the number of strong branching problems where we observed an objective change. The `sblps` criterion means to call local Rapid Learning if this ratio exceeds a threshold of 10, hence strong branching does not appear to be efficient for generating pseudo-cost information.

In addition to the exponentially decreasing frequency and the six criteria above, we applied the following three changes to the original implementation of Rapid Learning used in [13].

- We limited the number of conflict constraints transferred from Rapid Learning back to the original search tree to ten. This corresponds to the SCIP parameter `conflict/maxconss` for the maximal allowed number of added conflicts per call of conflict analysis. We greedily use the shortest conflicts.
- We prefer conflict constraints that have a linear representation over bound disjunction constraints (see Definition 2).
- To exploit performance variability [19,30] every CP search is initialized with a different pseudo-random seed.

5 Computational Results

To evaluate how local Rapid Learning impacts IP solving performance we used the academic constraint integer programming solver SCIP 6.0 [24] (with SoPlex 4.0 as LP solver) and extended the existing code of Rapid Learning. The original implementation of Rapid Learning was already shown to significantly improve the performance of SCIP 1.2.0.5 on pure binary instances [13]. In this setting, Rapid Learning was applied exactly once at the root node. However, during the last eight years SCIP has changed in many places. In SCIP 6.0, Rapid Learning is deactivated by default, since it led to a big performance variability.

Therefore, we use SCIP without Rapid Learning (as it is the current default) as a baseline. We will refer to this setting as `default`. In our computational experiments we evaluate the impact of local Rapid Learning if one or more of the criteria described in Sect. 4 are fulfilled. In the following, we will refer to the criteria I–VI as `dualbound`, `leaves`, `degeneracy`, `obj`, `nsols`, and `sblps`, respectively. Within the tree, Rapid Learning is applied with an exponentially decreasing frequency (see Sect. 4). In our experiments, we used $f = 5$ and $\beta = 4$, i.e., Rapid Learning is called at depths d with $log_4(d/5) \in \mathbb{Z}$, i.e., $d = 0, 5, 20, 80, 320 \ldots$, if one of the six criteria is fulfilled.

As a test set we used all pure integer problems of MIPLIB 3 [16], MIPLIB 2003 [8] and the MIPLIB 2010 [28] benchmark set. This test set consists of 71 publicly available instances, which we will refer to as `MMM-IP`. The experiments were run on a cluster of identical machines, each with an Intel Xeon E5-2690 with 2.6 GHz and 128 GB of RAM; a time limit of 3600 s was set.

In a first experiment we evaluated the efficacy of each individual criterion and global Rapid Learning as published in [13]. Aggregated results are shown in Table 2, section Exp.1. For detailed results we refer to the appendix of [15]. For every setting, the table shows the number of solved instances out of 71 (**solved**), shifted geometric means [3] of the absolute solving time in seconds (**time**, shift = 1) and number of explored nodes (**nodes**, shift = 100), as well as the relative solving time (**time$_Q$**) and number of nodes (**nodes$_Q$**) w.r.t. to `default` as a baseline. Local Rapid Learning without any of the presented criteria (`nochecks`) leads to a performance decrease of 21% on the complete test set `MMM-IP` compared to `default`. Always applying Rapid Learning only at the root (`onlyroot`), which corresponds to Rapid Learning as published in [13], leads to slowdown of 10% but solves one instance more. For this settings, we could observe a performance decrease of 29% on the group of instances that are not affected[4] by Rapid Learning. To avoid a computational overhead and performance variability on instances where Rapid Learning is not expected to be beneficial, we apply the criteria `degeneracy`, `obj`, and `nsols` at the root node, too. Afterwards, the performance decrease of global Rapid Learning reduced to 3%. The computational results indicate that almost all individual criteria are useful on their own. The solving time and generated nodes can be reduced by up to 7% and 14%, respectively, on the complete test set of 71. The exception is the `obj` criterion, which leads to a slowdown of 2%, but solves one more instance than `default`. These results can be confirmed when repeating the experiments with five different random seeds [15]. On the group of affected instances the solving time can be reduced by up to 21%, using the `leaves` criterion. The number of generated nodes can be reduced by up to 39% (for `degeneracy`) on the same group of instances.

The impact of the individual criteria on the solving time is illustrated in Fig. 2. For each criterion, the box plot [32] shows the median (dashed line), and the 1st and 3rd quartile (shaded box) of all observations. For all criteria the median time ratio is at most one; only for `degeneracy` and `leaves` the median is

---

[4] An instance is called affected when the solving path changes.

strictly smaller than one. Hence, these two settings improve the performance on more than 50% of the affected instances. Furthermore, degeneracy and leaves have by far the smallest 1st and 3rd quartile, indicating that the corresponding settings often improve performance and rarely deteriorate it.

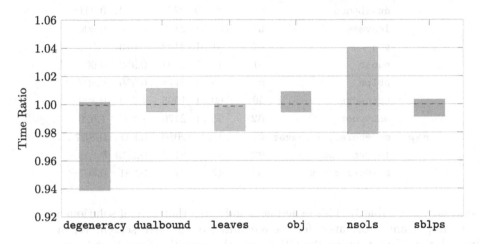

Fig. 2. Box-plot of the performance ratios of the individual criteria compared to default on the set of affected instances.

Grouping all instances of MMM-IP based on the degeneracy at the end of the root node shows the importance of this criterion. On the group of instances where at least 1% of the variables is dual degenerate at the end of the root node Rapid Learning leads to a performance improvement of 9.1%. On all instances where at least 80% of the variable are dual degenerate at the root node, we could observe a reduction of solving time by 28.8%. Note that this was one of the two thresholds for the degeneracy criterion.

In a second experiment (Table 2, section Exp.2) we combined all individual criteria. Combining two or more criteria leads to more aggressive version of Rapid Learning since it runs if at least one of the chosen criteria is satisfied. The two (out of fifteen) best pairwise combinations as well as the (most aggressive) combination of all six criteria are shown in Table 2. Interestingly, no combined setting is superior to degeneracy. The combination of degeneracy and leaves, which were the two outstanding criteria in the individual test, performs almost the same as the degeneracy criterion alone. These results can be confirmed when repeating the experiments with five different random seeds [15].

For a final experiment we choose degeneracy as the best criterion, since it was one of two criteria that solved an additional instance, clearly showed the best search reduction, and was a close second to leaves with respect to running time. Our final experiment evaluates the impact of the individual information gained from local Rapid Learning. To this end, we individually deactivated transferring variable bounds, conflict constraints, inference information, and primal feasible

Table 2. Computational results for every individual heuristic criterion on `MMM-IP`.

		solved	time	nodes	time$_Q$	nodes$_Q$
Exp.1	`default`	61	50.34	2428	–	–
	`degeneracy`	**62**	47.23	2078	0.938	**0.856**
	`dualbound`	61	49.00	2284	0.973	0.941
	`leaves`	61	46.88	2199	**0.931**	0.906
	`obj`	**62**	51.51	2432	1.023	1.002
	`nsols`	61	47.95	2194	0.952	0.904
	`sblps`	61	48.14	2178	0.956	0.897
	`nochecks`	59	60.81	1995	1.208	0.822
	`onlyroot`	**62**	55.71	2476	1.107	1.020
Exp.2	`degeneracy + leaves`	**62**	47.37	2080	0.941	**0.857**
	`leaves + obj`	**62**	47.26	2167	**0.939**	0.893
	`all6criterion`	**62**	47.88	2104	0.951	0.867

solutions (see Table 3). This experiment indicates that primal solutions are the most important information for the remainder of the MIP search. When ignoring solutions found during the CP search, the overall solving time increased by 10.4% (`primsols`). When ignoring conflict constraints, the original motivation of Rapid Learning, solving time increased by 2.4% (`conflicts`). Both transferring variable bounds and inference information proved beneficial, with a 2.1% (`variablebounds`) and 2.8% (`infervals`) impact on performance, respectively. To take performance variability into account, we repeated the experiment with five different random seeds, see [15] for detailed results. This experiment indicated that conflict constraints are the second most important criterion. Over five seeds the solving time increased by 9.9% (`primsols`), 4.4% (`conflicts`), 1.4% (`variablebounds`), and 0.6% (`infervals`). It is not surprising that finding primal solutions has the largest effect. Firstly, they are applied globally, in contrast to bound changes and conflicts. Secondly, highly dual degenerate problems are known to be cumbersome not only for MIP branching but also for primal heuristics [11], which means that solution-generating procedures that do not rely on solving LPs are particularly promising for such problems.

Table 3. Performance impact of individual gained information on `MMM-IP`.

		solved	time	nodes	time$_Q$	nodes$_Q$
Exp.3	`degeneracy`	62	47.23	2078	–	–
	`variablebounds`	62	48.23	2180	1.021	1.049
	`conflicts`	63	48.38	2213	1.024	1.065
	`infervals`	62	48.53	2230	1.028	1.073
	`primsols`	62	52.15	2400	1.104	1.155

6 Conclusion

In this paper, we extended the idea of Rapid Learning [13]. Firstly, we generalized Rapid Learning to integer programs and described the details that were necessary for doing so: value-based inference branching, additional propagators and generalized conflict constraints, most of which were already available in SCIP. Secondly, we applied Rapid Learning repeatedly during the search. This generates a true hybrid CP/MIP approach, with two markedly different search strategies communicating information forth and back. To this end, we introduced six heuristic criteria to decide when to start local Rapid Learning. Those criteria are based on degeneracy information, branch-and-bound statistics, and the local structure of the problem. Our computational experiments showed a speed-up of up to 7% when applying local Rapid Learning in SCIP. Calling local Rapid Learning depending on the local degree of dual degeneracy is the best strategy found in our experiments.

Interesting future work in this direction includes: extending the CP search to generate global conflicts at local nodes using an assumption interface, running the CP search in a parallel thread where whenever the MIP solver moves to a new node the CP search restarts from that node, and extending the method to handle problems that include continuous variables.

Acknowledgments. The work for this article has been partly conducted within the Research Campus MODAL funded by the German Federal Ministry of Education and Research (BMBF grant number 05M14ZAM). We thank the anonymous reviewers for their valuable suggestions and helpful comments.

References

1. Achterberg, T.: Conflict analysis in mixed integer programming. Discrete Optim. 4(1), 4–20 (2007)
2. Achterberg, T.: Constraint integer programming. Ph.D. thesis, Technische Universität Berlin (2007)
3. Achterberg, T.: SCIP: solving constraint integer programs. Math. Program. Comput. 1(1), 1–41 (2009)
4. Achterberg, T., Berthold, T.: Hybrid branching. In: van Hoeve, W.-J., Hooker, J.N. (eds.) CPAIOR 2009. LNCS, vol. 5547, pp. 309–311. Springer, Heidelberg (2009). https://doi.org/10.1007/978-3-642-01929-6_23
5. Achterberg, T., Berthold, T., Hendel, G.: Rounding and propagation heuristics for mixed integer programming. In: Klatte, D., Lüthi, H.-J., Schmedders, K. (eds.) Operations Research Proceedings 2011. ORP, pp. 71–76. Springer, Heidelberg (2012). https://doi.org/10.1007/978-3-642-29210-1_12
6. Achterberg, T., Berthold, T., Koch, T., Wolter, K.: Constraint integer programming: a new approach to integrate CP and MIP. In: Perron, L., Trick, M.A. (eds.) CPAIOR 2008. LNCS, vol. 5015, pp. 6–20. Springer, Heidelberg (2008). https://doi.org/10.1007/978-3-540-68155-7_4
7. Achterberg, T., Koch, T., Martin, A.: Branching rules revisited. Oper. Res. Lett. 33(1), 42–54 (2005)

8. Achterberg, T., Koch, T., Martin, A.: Miplib 2003. Oper. Res. Lett. **34**(4), 361–372 (2006)
9. Althaus, E., Bockmayr, A., Elf, M., Jünger, M., Kasper, T., Mehlhorn, K.: SCIL— symbolic constraints in integer linear programming. In: Möhring, R., Raman, R. (eds.) ESA 2002. LNCS, vol. 2461, pp. 75–87. Springer, Heidelberg (2002). https:// doi.org/10.1007/3-540-45749-6_11
10. Aron, I., Hooker, J.N., Yunes, T.H.: SIMPL: a system for integrating optimization techniques. In: Régin, J.-C., Rueher, M. (eds.) CPAIOR 2004. LNCS, vol. 3011, pp. 21–36. Springer, Heidelberg (2004). https://doi.org/10.1007/978-3-540-24664-0_2
11. Berthold, T.: Heuristic algorithms in global MINLP solvers. Ph.D. thesis, Technische Universität Berlin (2014)
12. Berthold, T.: RENS - the optimal rounding. Math. Program. Comput. **6**(1), 33–54 (2014)
13. Berthold, T., Feydy, T., Stuckey, P.J.: Rapid learning for binary programs. In: Lodi, A., Milano, M., Toth, P. (eds.) CPAIOR 2010. LNCS, vol. 6140, pp. 51–55. Springer, Heidelberg (2010). https://doi.org/10.1007/978-3-642-13520-0_8
14. Berthold, T., Gamrath, G., Salvagnin,D.: Cloud Branching (in Preparation)
15. Berthold, T., Stuckey, P.J., Witzig, J.: Local rapid learning for integer programs. Technical report 18–56, ZIB, Berlin (2018)
16. Bixby, R.E., Ceria, S., McZeal, C.M., Savelsbergh, M.W.: An updated mixed integer programming library: MIPLIB 3.0. Technical report (1998)
17. Bockmayr, A., Kasper, T.: Branch-and-infer: a unifying framework for integer and finite domain constraint programming. INFORMS J. Comput. **10**(3), 287–300 (1998)
18. Brearley, A., Mitra, G., Williams, H.: Analysis of mathematical programming problems prior to applying the simplex algorithm. Math. Program. **8**, 54–83 (1975)
19. Danna, E.: Performance variability in mixed integer programming. In: Presentation Slides from MIP 2008 Workshop in New York City (2008). http://coral.ie.lehigh. edu/~jeff/mip-2008/program.pdf
20. Danna, E., Rothberg, E., Pape, C.L.: Exploring relaxation induced neighborhoods to improve MIP solutions. Math. Program. **102**(1), 71–90 (2004)
21. Davey, B., Boland, N., Stuckey, P.J.: Efficient intelligent backtracking using linear programming. INFORMS J. Comput. **14**(4), 373–386 (2002)
22. Davies, T., Gange, G., Stuckey, P.J.: Automatic logic-based benders decomposition with MiniZinc. In: Proceedings of the 31st AAAI Conference on Artificial Intelligence (AAAI-2017), pp. 787–793. AAAI Press (2017). https://aaai.org/ocs/index. php/AAAI/AAAI17/paper/view/14489
23. FICO Xpress Optimizer. http://www.fico.com/en/Products/DMTools/xpress-overview/Pages/Xpress-Optimizer.aspx
24. Gleixner, A., et al.: The SCIP Optimization Suite 6.0. Technical report 18–26, ZIB, Berlin (2018)
25. Jussien, N., Barichard, V.: The PaLM system: explanation-based constraint programming. In: Proceedings of TRICS: Techniques for Implementing Constraint Programming Systems, A Post-Conference Workshop of CP 2000, pp. 118–133 (2000)
26. Katsirelos, G., Bacchus, F.: Generalised nogoods in CSPs. In: Proceedings of AAAI-2005, pp. 390–396 (2005)

27. Katsirelos, G., Bacchus, F.: Generalized nogoods in CSPs. In: Proceedings of the Twentieth National Conference on Artificial Intelligence and the Seventeenth Innovative Applications of Artificial Intelligence Conference, 9–13 July 2005, Pittsburgh, Pennsylvania, USA, pp. 390–396. AAAI Press/The MIT Press (2005)
28. Koch, T., et al.: MIPLIB 2010. Math. Program. Comput. **3**(2), 103–163 (2011)
29. Li, C.M., Anbulagan: Look-ahead versus look-back for satisfiability problems. In: Smolka, G. (ed.) Principles and Practice of Constraint Programming-CP97. CP 1997. LNCS, vol. 1330, pp. 341–355 (1997). https://doi.org/10.1007/BFb0017450
30. Lodi, A., Tramontani, A.: Performance variability in mixed-integer programming. In: Theory Driven by Influential Applications, pp. 1–12. INFORMS (2013)
31. Marques-Silva, J.P., Sakallah, K.A.: GRASP: a search algorithm for propositional satisfiability. IEEE Trans. Comput. **48**, 506–521 (1999)
32. McGill, R., Tukey, J.W., Larsen, W.A.: Variations of box plots. Am. Stat. **32**(1), 12–16 (1978)
33. Moskewicz, M.H., Madigan, C.F., Zhao, Y., Zhang, L., Malik, S.: Chaff: engineering an efficient SAT solver. In: Proceedings of DAC 2001, pp. 530–535 (2001)
34. Nadel, A., Ryvchin, V.: Efficient SAT solving under assumptions. In: Cimatti, A., Sebastiani, R. (eds.) SAT 2012. LNCS, vol. 7317, pp. 242–255. Springer, Heidelberg (2012). https://doi.org/10.1007/978-3-642-31612-8_19
35. Ohrimenko, O., Stuckey, P.J., Codish, M.: Propagation via lazy clause generation. Constraints **14**(3), 357–391 (2009)
36. Refalo, P.: Impact-based search strategies for constraint programming. In: Wallace, M. (ed.) CP 2004. LNCS, vol. 3258, pp. 557–571. Springer, Heidelberg (2004). https://doi.org/10.1007/978-3-540-30201-8_41
37. Rodosek, R., Wallace, M.G., Hajian, M.T.: A new approach to integrating mixed integer programming and constraint logic programming. Ann. Oper. Res. **86**(1), 63–87 (1999)
38. Sandholm, T., Shields, R.: Nogood learning for mixed integer programming. In: Workshop on Hybrid Methods and Branching Rules in Combinatorial Optimization, Montréal (2006)
39. Savelsbergh, M.W.P.: Preprocessing and probing techniques for mixed integer programming problems. ORSA J. Comput. **6**, 445–454 (1994)
40. Stallman, R.M., Sussman, G.J.: Forward reasoning and dependency-directed backtracking in a system for computer-aided circuit analysis. Artif. Intell. **9**(2), 135–196 (1977)
41. Witzig, J., Berthold, T., Heinz, S.: Experiments with conflict analysis in mixed integer programming. In: Salvagnin, D., Lombardi, M. (eds.) CPAIOR 2017. LNCS, vol. 10335, pp. 211–220. Springer, Cham (2017). https://doi.org/10.1007/978-3-319-59776-8_17
42. Yunes, T.H., Aron, I.D., Hooker, J.N.: An integrated solver for optimization problems. Oper. Res. **58**(2), 342–356 (2010)
43. Zhang, L., Madigan, C.F., Moskewicz, M.H., Malik, S.: Efficient conflict driven learning in a Boolean satisfiability solver. In: Proceedings of the 2001 IEEE/ACM International Conference on Computer-Aided Design, pp. 279–285. IEEE Press (2001)

A Status Report on Conflict Analysis in Mixed Integer Nonlinear Programming

Jakob Witzig[1], Timo Berthold[2(✉)], and Stefan Heinz[2]

[1] Zuse Institute Berlin, Takustr. 7, 14195 Berlin, Germany
witzig@zib.de
[2] Fair Isaac Germany GmbH, Takustr. 7, 14195 Berlin, Germany
{timoberthold,stefanheinz}@fico.com

Abstract. Mixed integer nonlinear programs (MINLPs) are arguably among the hardest optimization problems, with a wide range of applications. MINLP solvers that are based on linear relaxations and spatial branching work similar as mixed integer programming (MIP) solvers in the sense that they are based on a branch-and-cut algorithm, enhanced by various heuristics, domain propagation, and presolving techniques. However, the analysis of infeasible subproblems, which is an important component of most major MIP solvers, has been hardly studied in the context of MINLPs. There are two main approaches for infeasibility analysis in MIP solvers: *conflict graph analysis*, which originates from artificial intelligence and constraint programming, and *dual ray analysis*.

The main contribution of this short paper is twofold. Firstly, we present the first computational study regarding the impact of dual ray analysis on convex and nonconvex MINLPs. In that context, we introduce a modified generation of infeasibility proofs that incorporates linearization cuts that are only locally valid. Secondly, we describe an extension of conflict analysis that works directly with the nonlinear relaxation of convex MINLPs instead of considering a linear relaxation. This is work-in-progress, and this short paper is meant to present first theoretical considerations without a computational study for that part.

1 Introduction

In this paper, we consider *mixed integer nonlinear programs* (MINLPs) of the form

$$\min\{c^{\mathsf{T}}x \mid Ax \geq b, \; g_k(x) \leq 0 \; \forall k \in \mathcal{K}, \; \ell \leq x \leq u, \; x_j \in \mathbb{Z} \; \forall j \in \mathcal{I}\} \quad (1)$$

with objective coefficient vector $c \in \mathbb{R}^n$, linear constraint matrix $A \in \mathbb{R}^{m \times n}$, nonlinear constraint functions $g_k \colon \mathbb{R}^n \mapsto \mathbb{R}$, $k \in \mathcal{K} := \{1, \ldots, p\}$, continuously differentiable, and possibly nonconvex, and variable bounds $\ell, u \in \overline{\mathbb{R}}^n$, where $\overline{\mathbb{R}} := \mathbb{R} \cup \{\pm\infty\}$. Furthermore, let $\mathcal{N} = \{1, \ldots, n\}$ be the index set of all variables and $\mathcal{I} \subseteq \mathcal{N}$ the set of variables that need to be integral in every feasible solution. Without loss of generality, we assume the objective function to be linear. A

© Springer Nature Switzerland AG 2019
L.-M. Rousseau and K. Stergiou (Eds.): CPAIOR 2019, LNCS 11494, pp. 84–94, 2019.
https://doi.org/10.1007/978-3-030-19212-9_6

nonlinear objective function can be transformed into a constraint bounded by an artificial variable z that needs to be minimized. We call an MINLP *convex* when all of its constraint functions g_k are convex. Otherwise, we call the MINLP *nonconvex*. When omitting the integrality requirements, we obtain the *nonlinear programming* (NLP) relaxation of (1)

$$\min\{c^\mathsf{T}x \mid Ax \geq b, \ g_k(x) \leq 0 \, \forall k \in \mathcal{K}, \ \ell \leq x \leq u, \ x \in \mathbb{R}^n\}. \tag{2}$$

The *mixed integer programming* (MIP) relaxation of (1) is given by omitting all nonlinear constraints g_k for all $k \in \mathcal{K}$

$$\min\{c^\mathsf{T}x \mid Ax \geq b, \ \ell \leq x \leq u, \ x_i \in \mathbb{Z} \, \forall i \in \mathcal{I}\}. \tag{3}$$

Omitting both, integrality requirements and nonlinear constraints, yields the *linear programming* (LP) relaxation of (1)

$$\min\{c^\mathsf{T}x \mid Ax \geq b, \ \ell \leq x \leq u, \ x \in \mathbb{R}^n\}. \tag{4}$$

All three relaxations provide a lower bound on the optimal solution value of the MINLP (1). MINLP combines discrete decisions and nonlinear functions that are potentially nonconvex. In theory, linear and convex smooth nonlinear programs are solvable in polynomial time [27, 48]. In practice, both classes can be solved very efficiently [10, 42]. In contrast to that, nonconvexities as imposed by discrete variables or nonconvex nonlinear functions easily lead to problems that are both $\mathcal{NP}$-hard in theory and computationally demanding in practice [49].

Commonly used methods to solve convex MINLPs (1) include the extended cutting plane algorithm (ECP) [52], the extended supporting hyperplane algorithm [31], outer approximation (OA) [17,19], NLP-based branch-and-bound [23], and LP/NLP-based branch-and-bound [45]. The most commonly used method to solve nonconvex MINLPs is a combination of OA [29,50] and spatial branch-and-bound [24,34,35]. Different MINLP solvers either use LP or MIP relaxations or both during the tree search. For example, Couenne [14] and SCIP [49] derive valid lower bounds by solving LP relaxations only, whereas BARON [5,28] and BONMIN [11,12] solve both LP and MIP relaxations. In contrast to that, only a handful of MINLP solvers provide the possibility to exclusively use NLP relaxations, e.g., BONMIN and FICO Xpress Optimizer [18]. For a detailed overview of MINLP solvers that can handle convex and/or nonconvex MINLPs and the implemented algorithm, we refer to [30].

In the following, we will focus on MINLP solvers that use a combination of OA and spatial branch-and-bound. Spatial branch-and-bound is – analogous to LP-based branch-and-bound [15,33] – a divide-and-conquer method which splits the search space sequentially into smaller subproblems that are intended to be easier to solve. Additionally, convex relaxations are used to compute lower bounds on the individual subproblems. Based on the computed lower bound, a subproblem can be pruned earlier if the lower bound already exceeds the currently best-known solution. To divide the search space into smaller pieces, spatial branch-and-bound branches on discrete variables with a fractional solution value in the relaxation solution. In addition to that, spatial branch-and-bound uses continuous variables for branching if they appear in nonconvex terms of nonlinear constraints that are violated by the current relaxation solution. During this procedure, infeasible subproblems may be encountered. Infeasibility can either be detected by contradicting variable bounds, derived by domain propagation, or by an infeasible convex relaxation. In contrast to modern MIP solvers that can refer to a variety of well-studied techniques, e.g., [2,16,46], to 'learn' from infeasible subproblems, similar techniques for MINLPs exist for certain special cases only.

2 Conflict Analysis in MINLP

In this section, we will briefly describe conflict analysis for MIPs of type (3) and the drawbacks when applying these techniques to general MINLP.

2.1 Technical Background: Conflict Analysis in MIP

Conflict analysis for MIP has a long history and has its origin in artificial intelligence [47] and solving satisfiability problems (SAT) [36]. Similar ideas are used in constraint programming (CP), see, e.g., [21,25]. Integrations of these techniques into MIP were independently suggested by [2,16,46].

If infeasibility is encountered by domain propagation, modern SAT and MIP solvers construct a directed acyclic graph which represents the logic of how the set of branching decisions led to the detection of infeasibility. This graph is called the *conflict graph*. Valid *conflict constraints* can be derived from cuts in the graph that separate the branching decisions from an artificial vertex representing the infeasibility. Based on such a cut, a conflict constraint consists of a set of variables with associated bounds, requiring that in each feasible solution at least one of the variables has to take a value outside these bounds.

If the LP relaxation of a subproblem with local bounds ℓ' and u' turns out to be infeasible, it is necessary to identify a set of variables and bound changes that are sufficient to render the infeasibility. Such a set, the so-called *Farkas proof* [44,53], can be constructed by using LP duality theory that states that exactly one of the systems

$$Ax \geq b, \ \ell' \leq x \leq u' \tag{5}$$

$$y^\mathsf{T} A + r^\mathsf{T}\{\ell', u'\} = 0, \ y^\mathsf{T}b + r^\mathsf{T}\{\ell', u'\} > 0, \ y \geq 0 \tag{6}$$

where $r^\mathsf{T}\{\ell', u'\} := \sum_{j \in \mathcal{N}:\, r_j > 0} r_j \ell'_j + \sum_{j \in \mathcal{N}:\, r_j < 0} r_j u'_j$, can be satisfied. System (6) implies a proof of infeasibility w.r.t. to the local bounds

$$0 < y^\mathsf{T}b + r^\mathsf{T}\{\ell', u'\} = y^\mathsf{T}b - (y^\mathsf{T}A)\{\ell', u'\} \iff (y^\mathsf{T}A)\{\ell', u'\} < y^\mathsf{T}b. \tag{7}$$

Consequently, every feasible solution has to satisfy

$$(y^\mathsf{T}A)x \geq y^\mathsf{T}b, \tag{8}$$

which is called Farkas proof; it is a globally valid constraint because it is a nonnegative combination of all globally valid constraints. Thereby, Farkas proofs are a special case of Benders cuts [6]. The Farkas proof is used as a starting point for conflict graph analysis or dual ray analysis. Note, in MIP conflict graph analysis yields at least one conflict that does not need to be linear, whereas dual ray analysis yields exactly one linear constraint.

2.2 Conflict Analysis in MINLP

Only a few publications are dealing with infeasibility in MINLP. Most of the literature is restricted to a certain class of MINLPs, e.g., conic certificates for convex MINLPs [13] which has been proven to be very successful on *mixed-integer second-order cone* (MISOCP) problems. Purely theoretical results for *mixed integer semidefinite programs* (MISDP) were recently published in [26]. Both publications deal with MINLPs that are infeasible as a whole, and not with the analysis of infeasible subproblems to learn information.

For MINLP algorithms that are based on solving LP relaxations, in particular, for OA- and ECP-based solvers, conflict analysis methods for MIP can be applied under certain conditions. To this end, let us first recap the idea of constructing an LP relaxation for an MINLP.

During the tree search, nonlinear functions are approximated by linear functions if they are violated by a relaxation solution. Let $\tilde{x}$ be a relaxation solution with $g_k(\tilde{x}) > 0$. If g_k is convex, a so-called *gradient cut*

$$g_k(\tilde{x}) + \nabla g_k(\tilde{x})(x - \tilde{x}) \leq 0$$

is added. If g_k is nonconvex, convex underestimators are added, see, e.g., [49]. For quadratic functions, e.g., these are the so-called McCormick underestimators [37]. More general nonlinear functions are typically decomposed into functions of a single variable, for which explicit underestimators are known. Note that gradient cuts are globally valid, while underestimators for non-convex functions typically involve the local bounds and are hence not globally valid.

For a subproblem s during the tree search, let $\mathcal{G}^s := \{l_1^s, \ldots, l_q^s\}$ be the index set of all linear approximations of all g_k with $k \in \mathcal{K}$ that have been added at the node corresponding to s or any of its ancestors. Hence, it is the current set of (local) *linear relaxation cuts*; all are valid at s. Let G^s be the matrix containing all of these linearizations and d^s be the corresponding right-hand sides. Thus, the LP relaxation solved for subproblem s reads as

$$\min\{c^\mathsf{T}x \mid Ax \geq b, \ G^s x \geq d^s, \ \ell \leq x \leq u\}. \tag{9}$$

We denote the set of linearizations added at the root node by $\mathcal{G}^0$. During the (spatial) branch-and-bound the set of linearizations expands along each path of the tree: It holds that $\mathcal{G}^0 \subseteq \mathcal{G}^{s_1} \subseteq \ldots \subseteq \mathcal{G}^{s_p} \subseteq \mathcal{G}^s$ for each path $(0, s_1, \ldots, s_p, s)$. In analogy to solving MIPs, if (9) is infeasible each ray (y, w, r) in its dual can be used to construct a proof of local infeasibility. Here, y_i are the dual variables corresponding to $A_i.$, w_l are the dual variables corresponding to $G_l^s.$ for all $l \in \mathcal{G}^s$, and r_j denotes the reduced costs (the duals of the bound constraints) of every variable x_j. Note that $r_j = c_j - y^\mathsf{T} A_{\cdot j} - w^\mathsf{T} G^s_{\cdot j}$.

Hence, a local infeasibility proof w.r.t. the local bounds ℓ' and u' is given by

$$y^\mathsf{T} b + w^\mathsf{T} d^s + r^\mathsf{T}\{\ell', u'\} > 0, \tag{10}$$

In contrast to (8) the constraint $y^\mathsf{T} Ax + w^\mathsf{T} G^s x \geq y^\mathsf{T} b + w^\mathsf{T} d^s$ is not globally valid in general because linearizations of nonlinear constraints might rely on intermediate local bounds. Conflict analysis as introduced in [1,53] only considers globally valid reasons of infeasibility. Therefore, every local certificate of infeasibility (10) needs to be relaxed to consider $\mathcal{G}^0$ only

$$y^\mathsf{T} b + \bar{w}^\mathsf{T} d^s + \bar{r}^\mathsf{T}\{\ell', u'\} > 0, \tag{11}$$

where $\bar{w}_l := w_l$, if $l \in \mathcal{G}^0$, and $\bar{w}_l := 0$, otherwise, and $\bar{r}_j := c_j - y^\mathsf{T} A_{\cdot j} - \bar{w}^\mathsf{T} G^s_{\cdot j}$. As a consequence, the relaxed certificate (11) might not provide an infeasibility proof anymore and cannot be used to generate a conflict constraint. If, however, (11) is a valid proof of local infeasibility, all conflict analysis techniques known from MIP can be applied.

2.3   Locally Valid Certificates of Infeasibility

In MIP both conflict graph analysis and dual ray analysis rely on globally valid proofs. In most MIP solvers, local cuts are applied rarely, if at all. This is very different for non-convex MINLP solvers which rely on local linearization cuts. A computational study within the constraint integer programming and MINLP solver SCIP showed that the impact of conflict graph analysis for general MINLPs is almost negligible [49]. A computational study regarding the impact of dual ray analysis on an MINLP solver has – to the best of our knowledge – never been conducted before. We present such a computational study in Sect. 3.

The observation that conflict graph analysis on MINLP instances has a much smaller impact than on MIP instances [8,49] led to the assumption that a substantial amount of infeasibility proofs of form (11) were not globally valid. Hence, they are not suitable for conflict graph analysis as known from the literature and implemented in SCIP. These results indicate that locally added linearization cuts are, non-surprisingly, important to render infeasibility w.r.t. local bounds.

To incorporate local linearizations of nonlinear constraints we propose to generalize dual infeasibility proofs of subproblem s with local bounds ℓ' and u' as described in Sect. 2.1 to locally valid certificates of form

$$y^\mathsf{T} b + \hat{w}^\mathsf{T} d^s + \hat{r}^\mathsf{T}\{\ell', u'\} > 0, \tag{12}$$

incorporating linearizations $\hat{\mathcal{G}}$ with $\mathcal{G}^0 \subseteq \hat{\mathcal{G}} \subseteq \mathcal{G}^{s_p}$, $\hat{w}_l := w_l$, if $l \in \hat{\mathcal{G}}$, and $\hat{w}_l := 0$, otherwise, and $\hat{r}_j := c_j - y^\mathsf{T} A_{\cdot j} - \hat{w}^\mathsf{T} G^s_{\cdot j}$. The certificate (12) is valid for the search tree induced by subproblem q, where q is chosen to satisfy

$$q = \min_{q \in \{1,\ldots,s_p\}} \{\mathcal{G}^{q-1} \subseteq \hat{\mathcal{G}},\ \hat{\mathcal{G}} \cap (\mathcal{G}^{q+1} \setminus \mathcal{G}^q) = \emptyset\}. \tag{13}$$

Table 1. Aggregated results on MINLPLIB

	#	solved	time	nodes	time$_Q$	nodes$_Q$	confs$_{glb}$	confs$_{loc}$
all								
noconflict	1170	689	79.11	3014.25	1.000	1.000	–	–
confgraph	1170	694	77.94	2952.07	0.985	0.979	9679.01	–
dualray	1170	695	76.78	2871.86	0.970	0.953	1359.92	–
dualray-loc	1170	**698**	76.35	2841.90	**0.965**	**0.943**	1338.65	3192.50
[100,tilim]								
noconflict	99	83	638.34	86860.54	1.000	1.000	–	–
confgraph	99	88	563.06	74251.69	0.882	0.855	23653.88	–
dualray	99	89	458.28	62890.08	0.718	0.724	2019.46	–
dualray-loc	99	**92**	429.31	59629.05	**0.673**	**0.686**	2086.62	3177.67

Hence, the infeasibility proof might be lifted to an ancestor q of the subproblem s it was created for, if all local information used for the proof were already available at q. Note that it would be possible to apply conflict graph analysis to (12), too. However, this would introduce a computational overhead because the order of locally apply bound changes and separated local linearizations needs to be tracked and maintained. Since conflict graph analysis already comes with an overhead due to maintaining the so-called delta-tree, i.e., complete information about bound deductions and its reasons within the tree, we omit applying conflict graph analysis on locally valid infeasibility certificates.

3   Computational Study

For our computational study, we implemented the generation of locally valid infeasibility certificates in the academic constraint integer programming solver SCIP [22]. In the following, we refer to SCIP with (global) conflict graph analysis as confgraph and SCIP with (global) dual ray analysis as dualray. Moreover, we refer to dualray extended by locally valid infeasibility proofs as dualray-loc. As a baseline we use SCIP with deactivated conflict analysis (noconflict). As a test set we use the MINLPLIB [40] without instances for which at least one setting finished with numerical violations. This yields a test set of 1170 instances. The experiments were run on a cluster of Intel Xeon E5-2690 2.6 GHz machines with 128 GB of RAM; a time limit of 3600 s was set.

Aggregated results of all four settings are shown in Table 1. Here, [100,tilim] denotes the set of instances for which all settings need at least 100 s and are solved by at least one setting [4].

All settings with activated conflict analysis improve both the running time of SCIP, the number of branch-and-bound nodes, and the number of solved instances. Moreover, there seems to be a clear ordering: dualray-loc is superior to dualray which in turn is superior to confgraph. Further, the harder the instances are, the more performance is gained by `dualray` and `dualray-loc` compared to `confgraph`. The number of locally added conflict constraints (**confs**$_{\text{loc}}$) by `dualray-loc` is on average larger than the amount of globally added conflict constraints (**confs**$_{\text{glb}}$) but in the same order of magnitude. On the set of nonconvex MINLPs, however, `dualray-loc` constructs 11.08 times more locally than globally valid conflict constraints. These results indicate that locally added linearizations of nonlinear constraints are important to render local infeasibility.

When looking into the generation of local proofs into detail, we could observe that in 5% of all analyzed infeasible LPs no local cut was needed to construct a valid infeasibility certificate, i.e., we could lift the local conflict to a global one. For 14% of all local proofs we found a set of local cuts such that $q = \lfloor s/2 \rfloor$, the conflict could be lifted up at least half of the depth. 78% of the local proofs could not be lifted. Since a lot of infeasibility information is lost, we propose to use a nonlinear relaxation instead. The theoretical base for nonlinear conflict analysis will be discussed in the following section, whereas the implementation and a computational study is future work.

4   Outlook and Theoretical Thoughts

In this final section, we will discuss theoretical considerations how conflict analysis can be directly applied to a nonlinear relaxation of convex MINLPs. The content described in the following is work-in-progress. At the beginning of this paper, we argued that after LP/MIP-based branch-and-bound, another common method to solve MINLPs is NLP-based branch-and-bound. We will briefly sketch how a generalization of LP infeasibility analysis can be derived from the KKT-conditions of convex NLPs. Given a convex MINLP of form

$$\min_{x \in X}\{f(x) \mid g_k(x) \le 0 \; \forall k \in \mathcal{K}, \; h_e(x) = 0 \; \forall e \in \mathcal{E}\}, \tag{14}$$

where f, g_k are convex, continuously differentiable functions over $\mathbb{R}^n$ and h_e are affine functions. For every optimal solution $x^\star$ of (14) of the (convex) NLP relaxation of (14) there exist $\lambda \ge 0$ such that it holds that

$$\nabla f(x^\star) + \sum_{k \in \mathcal{K}} \lambda_k \nabla g_k(x^\star) + \sum_{e \in \mathcal{E}} \mu_e \nabla h_e(x^\star) = 0, \quad \lambda_k g_k(x^\star) = 0. \tag{15}$$

These conditions raise from the so-called *Karush-Kuhn-Tucker-Conditions* [32]. Equality (15) is the gradient of the *Lagrangian dual* that reads as

$$\mathcal{L}(x, \lambda, \mu) := f(x) + \sum_{k \in \mathcal{K}} \lambda_k g_k(x) + \sum_{e \in \mathcal{E}} \mu_e h_e(x), \tag{16}$$

with $\lambda \ge 0$ and $\mu \in \mathbb{R}^{|\mathcal{E}|}$. By duality theory, the *Lagrangian dual function* which reads as $q(\lambda, \mu) := \sup_{\lambda, \mu} \mathcal{L}(x, \lambda, \mu)$ yields a lower bound on the optimal value of (14). Maximizing $q(\lambda, \mu)$ would give the tightest lower bound of (14), and strict duality of convex optimization tells us that this is equivalent to the optimal value of (14). Consequently, if there exists $(\lambda^\star, \mu^\star)$ such that $\sum_{k \in \mathcal{K}} \lambda_k^\star g_k(x) + \sum_{e \in \mathcal{E}} \mu_e^\star h_e(x) > 0$, then the dual is unbounded and thus $(\lambda^\star, \mu^\star)$ proofs infeasibility of (14). Even though Slater regularity does not hold for infeasible points[1],

$$\sum_{k \in \mathcal{K}} \lambda_k^\star g_k(x) + \sum_{e \in \mathcal{E}} \mu_e^\star h_e(x) \leq 0 \qquad (17)$$

is a valid inequality for (14); it is a convex combination (defined by the dual multipliers) of the constraints of (14). Inequality (17) is the convex optimization equivalent of the Farkas proof (8).

Assume that constraint (17) is given as proof of infeasibility for a subproblem within an NLP-based branch-and-bound. If no local cuts are involved in the infeasibility proof, inequality (17) is a globally valid convex nonlinear constraint. Note in this context that gradient cuts are globally valid.

Clearly, inequality (17) holds for all non-negative $\lambda^\star$. The following observation makes the concrete $(\lambda^\star, \mu^\star)$ from the infeasibility proof interesting to use as global information inside a branch-and-bound tree search for convex MINLP. Consider the linearization at an infeasible point $x^\star$

$$\nabla g_k(x^\star)^\mathsf{T}(x - x^\star) \leq 0 \quad \Leftrightarrow \quad \nabla g_k(x^\star)^\mathsf{T} x \leq \nabla g_k(x^\star)^\mathsf{T} x^\star \ \forall k \in \mathcal{K}. \qquad (18)$$

Then, the corresponding dual multipliers $\lambda^\star$ give the (linear) Farkas proof

$$\sum_{k \in \mathcal{K}} \lambda_k^\star \nabla g_k(x^\star) + \sum_{e \in \mathcal{E}} \mu_e^\star \nabla h_e(x^\star) = 0 \qquad (19)$$

$$\sum_{k \in \mathcal{K}} \lambda_k^\star \nabla g_k(x^\star)^\mathsf{T} x^\star + \sum_{e \in \mathcal{E}} \mu_e^\star \nabla h_e(x^\star)^\mathsf{T} x^\star < 0. \qquad (20)$$

Hence, as in the case of dual ray analysis for MIP, inequality (17) is a single inequality that would have provided the infeasibility proof from its derivative. The hope (which is true for MIP) is that it is a good candidate to detect infeasibility by propagation (under the use of integrality information) in other parts of the search tree, and might be a meaningful aggregation of problem constraints to create cuts from.

For many NLP solvers, in particular *dual active set methods* [20,41,43] and barrier algorithms [38,39,51], dual multipliers will be readily available. The added advantage of active set methods is that they typically yield a sparse dual weight-vector (λ, μ). This might come in handy when the local bounds involved in the infeasibility proof should be used to seed a conflict graph analysis. Like in the linear case, the problem is that the initial reason will typically be too large to be meaningful.

All of this is subject to further investigation. We plan to implement NLP-based conflict analysis into the academic constraint integer programming solver

---

[1] If one wanted to assume regularity on the constraint functions of (14), linear independence constraint classification would be applicable.

SCIP and to study its impact on solver behavior. As in the MIP case, infeasibility information might be used in several other contexts, consider hybrid branching [3], conflict-driven diving heuristics [54], and also rapid learning [7,9].

Acknowledgments. We thank Zsolt Csizmadia for his valuable comments on Sect. 4. The work for this article has been conducted within the Research Campus Modal funded by the German Federal Ministry of Education and Research (fund number 05M14ZAM). We thank three anonymous reviewers for their valuable suggestions and helpful comments.

References

1. Achterberg, T.: Conflict analysis in mixed integer programming. Discrete Optim. 4(1), 4–20 (2007)
2. Achterberg, T.: Constraint integer programming (2007)
3. Achterberg, T., Berthold, T.: Hybrid branching. In: van Hoeve, W.-J., Hooker, J.N. (eds.) CPAIOR 2009. LNCS, vol. 5547, pp. 309–311. Springer, Heidelberg (2009). https://doi.org/10.1007/978-3-642-01929-6_23
4. Achterberg, T., Wunderling, R.: Mixed integer programming: analyzing 12 years of progress. In: Jünger, M., Reinelt, G. (eds.) Facets of Combinatorial Optimization: Festschrift for Martin Grötschel, pp. 449–481. Springer, Heidelberg (2013). https://doi.org/10.1007/978-3-642-38189-8_18
5. BARON. https://minlp.com/baron
6. Benders, J.F.: Partitioning procedures for solving mixed-variables programming problems. Numerische Mathematik 4(1), 238–252 (1962)
7. Berthold, T., Feydy, T., Stuckey, P.J.: Rapid learning for binary programs. In: Lodi, A., Milano, M., Toth, P. (eds.) CPAIOR 2010. LNCS, vol. 6140, pp. 51–55. Springer, Heidelberg (2010). https://doi.org/10.1007/978-3-642-13520-0_8
8. Berthold, T., Gleixner, A.M., Heinz, S., Vigerske, S.: Analyzing the computational impact of MIQCP solver components. Numer. Algebra Control Optim. 2(4), 739–748 (2012)
9. Berthold, T., Stuckey, P.J., Witzig, J.: Local rapid learning for integer programs. Technical report 18–56, ZIB, Takustr. 7, 14195 Berlin (2018)
10. Bixby, R.E.: Solving real-world linear programs: a decade and more of progress. Oper. Res. 50(1), 3–15 (2002)
11. Bonami, P., et al.: An algorithmic framework for convex mixed integer nonlinear programs. Discrete Optim. 5(2), 186–204 (2008)
12. Bonmin. https://projects.coin-or.org/Bonmin
13. Coey, C., Lubin, M., Vielma, J.P.: Outer approximation with conic certificates for mixed-integer convex problems. arXiv preprint arXiv:1808.05290 (2018)
14. Couenne. https://www.coin-or.org/Couenne/
15. Dakin, R.J.: A tree-search algorithm for mixed integer programming problems. Comput. J. 8(3), 250–255 (1965)
16. Davey, B., Boland, N., Stuckey, P.J.: Efficient intelligent backtracking using linear programming. INFORMS J. Comput. 14(4), 373–386 (2002)
17. Duran, M.A., Grossmann, I.E.: An outer-approximation algorithm for a class of mixed-integer nonlinear programs. Math. Program. 36(3), 307–339 (1986)
18. FICO Xpress Optimizer. https://www.fico.com/de/products/fico-xpress-optimization

19. Fletcher, R., Leyffer, S.: Solving mixed integer nonlinear programs by outer approximation. Math. Program. **66**(1–3), 327–349 (1994)
20. Forsgren, A., Gill, P.E., Wong, E.: Primal and dual active-set methods for convex quadratic programming. Math. Program. **159**(1–2), 469–508 (2016)
21. Ginsberg, M.L.: Dynamic backtracking. J. Artif. Intell. Res. **1**, 25–46 (1993)
22. Gleixner, A., et al.: The SCIP Optimization Suite 6.0. Technical report 18–26, ZIB, 18–56, ZIB, Takustr. 7, 14195 Berlin (2018)
23. Gupta, O.K., Ravindran, A.: Branch and bound experiments in convex nonlinear integer programming. Manage. Sci. **31**(12), 1533–1546 (1985)
24. Horst, R., Tuy, H.: Global Optimization: Deterministic Approaches. Springer, Heidelberg (2013). https://doi.org/10.1007/978-3-662-03199-5
25. Jiang, Y., Richards, T., Richards, B.: Nogood backmarking with min-conflict repair in constraint satisfaction and optimization. In: Borning, A. (ed.) PPCP 1994. LNCS, vol. 874, pp. 21–39. Springer, Heidelberg (1994). https://doi.org/10.1007/3-540-58601-6_87
26. Kellner, K., Pfetsch, M.E., Theobald, T.: Irreducible infeasible subsystems of semidefinite systems. arXiv preprint arXiv:1804.01327 (2018)
27. Khachiyan, L.G.: A polynomial algorithm in linear programming. Doklady Academii Nauk SSSR **244**, 1093–1096 (1979)
28. Kılınç, M.R., Sahinidis, N.V.: Exploiting integrality in the global optimization of mixed-integer nonlinear programming problems with BARON. Optim. Methods Softw. **33**(3), 540–562 (2018)
29. Kocis, G.R., Grossmann, I.E.: Global optimization of nonconvex mixed-integer nonlinear programming (MINLP) problems in process synthesis. Ind. Eng. Chem. Res. **27**(8), 1407–1421 (1988)
30. Kronqvist, J., Bernal, D., Lundell, A., Grossmann, I.: A review and comparison of solvers for convex MINLP. Optim. Eng. (2018)
31. Kronqvist, J., Lundell, A., Westerlund, T.: The extended supporting hyperplane algorithm for convex mixed-integer nonlinear programming. J. Glob. Optim. **64**(2), 249–272 (2016)
32. Kuhn, H.W., Tucker, A.W.: Nonlinear programming. In: Giorgi, G., Kjeldsen, T.H. (eds.) Traces and Emergence of Nonlinear Programming, pp. 247–258. Springer, Basel (2014). https://doi.org/10.1007/978-3-0348-0439-4_11
33. Land, A.H., Doig, A.G.: An automatic method of solving discrete programming problems. Econometrica **28**(3), 497–520 (1960)
34. Land, A.H., Doig, A.G.: An automatic method for solving discrete programming problems. In: Jünger, M., et al. (eds.) 50 Years of Integer Programming 1958–2008, pp. 105–132. Springer, Heidelberg (2010). https://doi.org/10.1007/978-3-540-68279-0_5
35. Liberti, L., Pantelides, C.C.: Convex envelopes of monomials of odd degree. J. Glob. Optim. **25**(2), 157–168 (2003)
36. Marques-Silva, J.P., Sakallah, K.: GRASP: a search algorithm for propositional satisfiability. IEEE Trans. Comput. **48**(5), 506–521 (1999)
37. McCormick, G.P.: Computability of global solutions to factorable nonconvex programs: part I—convex underestimating problems. Math. Program. **10**(1), 147–175 (1976)
38. Mehrotra, S.: On the implementation of a primal-dual interior point method. SIAM J. Optim. **2**(4), 575–601 (1992)
39. Mészáros, C.: The BPMPD interior point solver for convex quadratic problems. Optim. Meth. Softw. **11**(1–4), 431–449 (1999)

40. MINLPLib: Githash 033934c0. http://www.minlplib.org/
41. Murty, K.G., Yu, F.-T.: Linear Complementarity, Linear and Nonlinear Programming, vol 3. Citeseer (1988)
42. Nocedal, J., Wright, S.: Numerical Optimization. Springer Series in Operations Research and Financial Engineering, 2nd edn. Springer, Heidelberg (2006). https://doi.org/10.1007/978-0-387-40065-5
43. Nocedal, J., Wright, S.J.: Nonlinear equations. In: Numerical Optimization. Springer Series in Operations Research and Financial Engineering, pp. 270–302. Springer, New York (2006). https://doi.org/10.1007/978-0-387-40065-5_11
44. Pólik, I.: Some more ways to use dual information in MILP. In: International Symposium on Mathematical Programming, Pittsburgh, PA (2015)
45. Quesada, I., Grossmann, I.E.: An LP/NLP based branch and bound algorithm for convex MINLP optimization problems. Comput. Chem. Eng. **16**(10–11), 937–947 (1992)
46. Sandholm, T., Shields, R.: Nogood learning for mixed integer programming. In: Workshop on Hybrid Methods and Branching Rules in Combinatorial Optimization, Montréal (2006)
47. Stallman, R.M., Sussman, G.J.: Forward reasoning and dependency-directed backtracking in a system for computer-aided circuit analysis. Artif. Intell. **9**(2), 135–196 (1977)
48. Vavasis, S.A.: Complexity issues in global optimization: a survey. In: Horst, R., Pardalos, P.M. (eds.) Handbook of Global Optimization, pp. 27–41. Springer, New York (1995). https://doi.org/10.1007/978-1-4615-2025-2_2
49. Vigerske, S., Gleixner, A.: SCIP: global optimization of mixed-integer nonlinear programs in a branch-and-cut framework. Optim. Meth. Softw. **33**(3), 563–593 (2018)
50. Viswanathan, J., Grossmann, I.E.: A combined penalty function and outer-approximation method for MINLP optimization. Comput. Chem. Eng. **14**(7), 769–782 (1990)
51. Wächter, A.: Short tutorial: getting started with Ipopt in 90 minutes. In: Dagstuhl Seminar Proceedings. Schloss Dagstuhl-Leibniz-Zentrum für Informatik (2009)
52. Westerlund, T., Pettersson, F.: An extended cutting plane method for solving convex MINLP problems. Comput. Chem. Eng. **19**, 131–136 (1995)
53. Witzig, J., Berthold, T., Heinz, S.: Experiments with conflict analysis in mixed integer programming. In: Salvagnin, D., Lombardi, M. (eds.) CPAIOR 2017. LNCS, vol. 10335, pp. 211–220. Springer, Cham (2017). https://doi.org/10.1007/978-3-319-59776-8_17
54. Witzig, J., Gleixner, A.: Conflict-driven heuristics for mixed integer programming. Technical report 19–08, ZIB, Takustr. 7, 14195 Berlin (2019)

Generating Compound Moves
in Local Search by Hybridisation
with Complete Search

Gustav Björdal$^{(\boxtimes)}$ ⓘ, Pierre Flener ⓘ, and Justin Pearson ⓘ

Department of Information Technology, Uppsala University, 751 05 Uppsala, Sweden
{Gustav.Bjordal,Pierre.Flener,Justin.Pearson}@it.uu.se

Abstract. A black-box local-search *backend* to a solving-technology-independent modelling language, such as MiniZinc, automatically infers from the structure of a declarative model for a satisfaction or optimisation problem a combination of a neighbourhood, heuristic, and meta-heuristic. These ingredients are then provided to a local-search *solver*, but are manually designed in a handcrafted local-search *algorithm*. However, such a backend can perform poorly due to model structure that is inappropriate for local search, for example when it considers moves modifying only variables that represent auxiliary information. Towards overcoming such inefficiency, we propose *compound-move generation*, an extension to local-search solvers that uses a complete-search solver in order to augment moves modifying non-auxiliary variables so that they also modify auxiliary ones. Since compound-move generation is intended to be applied to such models, we discuss how to identify them.

We present several refinements of compound-move generation and show its very positive impact on several third-party models. This helps reduce the unavoidable gap between black-box local search and local-search algorithms crafted by experts.

1 Introduction

The aim of technology-independent modelling languages for satisfaction and optimisation problems is to allow many solvers to run for a single problem model and hopefully avoid too early commitment to a solving technology. MiniZinc [13] is such a language where a user designs a model and can then solve the problem using a wide range of backends that call solvers from technologies such as constraint programming (CP), lazy clause generation (LCG), (constraint-based) local search (LS and CBLS), mixed-integer programming (MIP), Boolean satisfiability (SAT), and satisfiability modulo theories (SMT). Given a MiniZinc model, a backend should infer a representation and search strategy that are suitable for its solver and technology.

This work is supported by the Swedish Research Council (VR) through Project Grant 2015-04910.

L.-M. Rousseau and K. Stergiou (Eds.): CPAIOR 2019, LNCS 11494, pp. 95–111, 2019.
https://doi.org/10.1007/978-3-030-19212-9_7

A *black-box local-search backend* automatically infers from a MiniZinc model, which is purely declarative, a representation required to compute efficiently a cost function as well as a combination of a neighbourhood, heuristic, and meta-heuristic, which form the search strategy and are then provided to a CBLS solver: these ingredients are manually designed in a handcrafted local-search algorithm, which processes no model. A drawback of technology-independent modelling is that backends of some solving technologies can be sensitive to model structure (which backends of other technologies may be unaffected by). For all these reasons, there will always be a gap between black-box local-search backends and local-search algorithms handcrafted by experts for specific problems.

To help reduce this gap, we here explore a model structure where black-box LS performs poorly, namely when its inferred neighbourhood has moves modifying only variables representing auxiliary information. This model structure can appear when such auxiliary variables are (or seem) not functionally determined by the other variables. The intuition for why this can degrade performance is that the search strategy should consider new values for the non-auxiliary variables and infer (possibly when generating the considered moves) new values for the auxiliary variables from those considered values, and not vice versa.

Towards improving the performance of black-box local search, our contributions are as follows, after defining all required background in Sect. 2. In Sect. 3, we propose *compound-move generation* (CMG), an extension to local-search solvers that uses a complete-search solver (based on CP in our implementation) in order to try to augment moves modifying non-auxiliary variables so that they also modify auxiliary ones. In Sect. 4, we present two approaches for detecting the model structure that CMG is intended for. In Sect. 5, we experimentally demonstrate the very positive impact of CMG. We discuss related work in Sect. 6 and conclude in Sect. 7.

2 Background

We define the relevant concepts of MiniZinc, (CP-style) complete search, and (constraint-based) local search.

2.1 MiniZinc, FlatZinc, Models, and Instances

The constraint-based modelling language MiniZinc [13] for satisfaction and optimisation problems is independent of solving technologies, such as CP, LCG, LS, CBLS, MIP, SAT, and SMT. Its open-source toolchain contains a *flattener*, which translates a model and instance data into a sub-language called FlatZinc, which is amenable to interpretation and analysis by a *backend* that calls a targeted *solver*. We now present a MiniZinc model for our running example.

Example 1. Consider the travelling salesperson problem with time windows (TSPTW). Given are n locations, a travel-time matrix T (where T[i,j] is the travel time from location i to location j plus the service time at i), and a matrix

W of time windows for each location (where W[i,1] is the earliest arrival time and W[i,2] the latest arrival time for location i). The goal is to find a shortest route that visits each location exactly once and within its time window.

Figure 1 shows a MiniZinc model that is good for most backends. Since the arrival time at a location depends on the departure time at the previous location, the route is modelled using a predecessor array pred, where variable pred[i] denotes the location visited before location i. The circuit constraint in line 5 requires pred to represent a Hamiltonian circuit. Location 1 is assumed to be the start of the route. The arrival times are modelled using the array A, where variable A[i] denotes the arrival time at location i. Each arrival time is constrained, in lines 8 to 11, to be either the arrival time at the preceding location plus the travel time or the start of its time window, whichever is greater, and at most the end of its time window. The objective is to minimise the travel time of the entire route, which is stated in line 12.

```
1  int: n; set of int: Loc = 1..n; % (number of) locations
2  array[Loc,Loc] of int: T; % travel times
3  array[Loc,1..2] of int: W; % time windows
4  array[Loc] of var Loc: pred; % predecessor locations
5  constraint circuit(pred);
6  int: depot = 1; % location 1 is the depot
7  array[Loc] of var int: A; % arrival times
8  constraint A[depot] = W[depot,1];
9  constraint forall(i in Loc where i != depot)(
10    A[i] = max(A[pred[i]] + T[pred[i],i], W[i,1]));
11 constraint forall(i in Loc)(A[i] <= W[i,2]);
12 solve minimize sum(i in Loc)(T[pred[i],i]);
```

Fig. 1. A MiniZinc model for TSP with time windows (TSPTW)

Each variable A[i] represents auxiliary information and seems at first sight to be functionally determined by lines 8 and 10. However, since A[pred[i]] on the right-hand side of the equality in line 10 defines A[i] possibly in terms of itself, a backend should infer that A[i] cannot be functionally determined by lines 8 and 10 alone. This is an example of a model that, somewhat unexpectedly, a backend can see as having non-functionally determined variables representing auxiliary information.[1] Furthermore, the equality constraints in lines 8 and 10 are here only correct because in a minimal solution the salesperson always arrives as early as possible. If this assumption cannot be made, possibly due to additional side constraints, then lines 8 and 10 have to be expressed using inequalities, thus making the auxiliary A[i] variables *necessarily* non-functionally determined. □

---

[1] Note that upon *also* considering the semantics of the circuit constraint in line 5, a backend that *only* explores assignments satisfying that constraint can infer that the A[i] are in fact functionally determined by line 10. However, to the best of our knowledge, no backend to MiniZinc performs such a semantic analysis. Also, doing so would not address all cases where a model can be seen as having non-functionally determined variables representing auxiliary information.

Without loss of generality, we explain everything for minimisation problems: a maximisation or satisfaction problem can be transformed into a minimisation problem by minimising the negated objective function or a constant, respectively. In order to emphasise the independence from MiniZinc of our method, a Flat-Zinc model for a minimisation problem *instance* is here abstracted as a tuple $\langle \mathcal{V}, \mathcal{D}, \mathcal{C}, o \rangle$, where $\mathcal{V}$ is the set of variables; $\mathcal{D}$ is the function mapping each variable to its set of possible values, called its *domain*; $\mathcal{C}$ is the set of constraints over variables in $\mathcal{V}$; and $o \in \mathcal{V}$ is the variable that is constrained in $\mathcal{C}$ to take the value of the objective function, which is to be minimised.

2.2 Complete Search and Constraint Programming

Given a minimisation instance $\langle \mathcal{V}, \mathcal{D}, \mathcal{C}, o \rangle$, a *solution* is an assignment of all variables $\mathcal{V}$ to values allowed by the domain mapping $\mathcal{D}$ such that all the constraints $\mathcal{C}$ are satisfied. We denote the value assigned to a variable v in a solution by $\mathrm{sol}(v)$ and a solution by $\mathrm{sol}(\mathcal{V}) := \{v \mapsto \mathrm{sol}(v) \mid v \in \mathcal{V}\}$.

Given enough time, a *complete* solver is guaranteed either to return a proven *minimal solution*, which is a solution where $\mathrm{sol}(o)$ is minimal, or to prove unsatisfiability otherwise. If a complete solver is stopped early , then it returns either the *best-found solution*, without proof of minimality, or nothing, meaning it is not known if the instance is satisfiable or not.

Many solving technologies offer complete solvers, such as CP, LCG, MIP, SAT, and SMT. Our main ideas are independent of which complete technology is used, but some refinements exploit features of CP solvers, defined next.

A *CP solver* builds a search tree by interleaving propagation and search. It modifies the *current domain*, which maps each variable $v \in \mathcal{V}$ to a set $\mathrm{dom}(v)$, initialised to $\mathcal{D}(v)$. *Propagation* computes the fixpoint of the propagators, one for each constraint in $\mathcal{C}$: a *propagator* for a constraint c deletes (not necessarily all) values from the current domain of each variable in c that are impossible under c. The current domain for the *root node* of the search tree is computed by a first run of propagation. If the current domain of some variable becomes empty at some node, then there is a *failure* and backtracking occurs. If the current domain of each variable becomes a singleton at some node, then the instance is proven satisfiable, under the assignment $\{v \mapsto d \mid v \in \mathcal{V} \wedge \mathrm{dom}(v) = \{d\}\}$, which is $\mathrm{sol}(\mathcal{V})$, and the constraint $o < \mathrm{sol}(o)$ is added to $\mathcal{C}$ before backtracking in order to search for a better solution. If at least one current domain has at least two values, then a *child node* is created for each part of a partition of $\mathrm{dom}(v)$ into at least two non-empty disjoint subsets for some variable v, guided by a *branching strategy*. Solving (by propagation and search) recursively continues for each child node, under usually a depth-first exploration order. Solving either returns a minimal solution or reports unsatisfiability.

2.3 (Constraint-Based) Local Search

Other solving technologies offer non-complete solvers. For example, *local search* (LS), say [11], initialises and iteratively modifies the *current assignment*, which maps each variable $v \in \mathcal{V}$ to a value val(v), called its *current value*, in its domain $\mathcal{D}(v)$. The current assignment need not satisfy all the constraints $\mathcal{C}$. The *initial assignment* is built under some amount of randomisation. At every *iteration*, a two-step search *heuristic* is followed. First, a set of *candidate moves* is considered, each being a set of reassignments $v \mapsto d$ for at least one variable $v \in \mathcal{V}$ and value $d \in \mathcal{D}(v)$. We assume that each candidate move is *probed* by (i) tentatively performing its reassignments, (ii) estimating the proximity $\hat{p}$ of the resulting *tentative assignment* to some assignment satisfying $\mathcal{C}$ and computing the resulting value $\hat{o}$ of the objective variable o, and (iii) undoing the tentatively performed reassignments and returning the pair $\langle \hat{p}, \hat{o} \rangle$. The set of probed candidate moves is called the *neighbourhood*, which is said to be *explored*, and its elements are called *neighbours*. Second, among the candidate moves, the heuristic *selects* one based on a *cost function* applied to each pair $\langle \hat{p}, \hat{o} \rangle$ and actually *commits* it, yielding the new current assignment. A *meta-heuristic*, such as tabu search, say [10], can be used to escape local optima of the cost function. Together, the neighbourhood, heuristic, and meta-heuristic form a *local-search strategy*.

In *constraint-based local search* (CBLS) [19], a declarative model is coupled with either a user-defined LS strategy, yielding a *white-box LS solver* (such as Comet [19] and OscaR.cbls [6]), or a solver-inferred LS strategy, yielding a *black-box LS solver* (such as LocalSolver [1] and fzn-oscar-cbls [2]). For each built-in constraint c, a predefined *violation function* viol(c), which returns the value 0 when c is satisfied and otherwise a positive value, can be used for estimating the proximity of a tentative assignment to an assignment satisfying c. One can then estimate the proximity $\hat{p}$ as the *violation* viol($\mathcal{C}$) := $\sum_{c \in \mathcal{C}}$ viol(c). *Note that objective function, cost function, and violation function are here not synonyms.*

A CBLS model has two categories of *explicit constraints*. *Soft constraints* have a violation function and may be violated during search but must be satisfied in a solution. *One-way constraints*, such as z <== x * y in OscaR.cbls syntax and called *invariants* in Comet, are impossible to violate by candidate moves: in z <== x * y, the functionally determined variable z cannot undergo a move, since its value is maintained by the solver to be the product of the variables x and y, which can undergo moves. An *implicit constraint* in a CBLS model is satisfied by the initial assignment and preserved by all committed moves: this can be done by using a *constraint-specific neighbourhood* [2].

For each constraint of a problem, a CBLS modeller must choose whether to make it soft, one-way, or implicit. *Note that implicit and one-way constraints do not exist as such in MiniZinc and FlatZinc.*

2.4   A Local-Search Backend to MiniZinc

Our fzn-oscar-cbls [2] LS backend to MiniZinc conceptually performs three steps. First, the constraints of a given flattened MiniZinc model are categorised into the three CBLS constraint categories (soft, one-way, and implicit) by using a structure identification scheme (see [2] for full details). Second, an LS strategy for the CBLS solver OscaR.cbls [6] is inferred: the neighbourhood is the union of the constraint-specific neighbourhoods for all identified implicit constraints and a default neighbourhood for all variables that are not part of any constraint-specific neighbourhood (note that variables identified as functionally determined, and thus maintained by one-way constraints, are not in any neighbourhood); the heuristic selects a random best candidate move from the neighbourhood; and the meta-heuristic is a variation of tabu search [10]. Third, OscaR.cbls is invoked. *Note that backend and solver are here not synonyms.*

Example 2. For the model in Fig. 1, fzn-oscar-cbls categorises the `circuit` constraint in line 5 as implicit, since a constraint-specific neighbourhood (namely 3-opt) is available in fzn-oscar-cbls. The `A[i]` variables are mistakenly conjectured not to be functionally determined, as the structure identification scheme does not take the semantics of `circuit` into account, hence the `A[i]` seem defined possibly in terms of themselves and are not maintained by one-way constraints: a default neighbourhood is inferred for them. The objective variable (introduced by line 12) is maintained by a one-way constraint. The soft constraints are the equality constraints in lines 8 and 10.   □

Two major burdens for an LS backend to a technology-independent modelling language such as MiniZinc are the identification of an LS-appropriate structure of a model, which is non-trivial as models need not be written with LS in mind, and the ensuing neighbourhood inference, which depends on the identified structure. We now address these two burdens by trying to make LS backends more robust to models without an identifiable LS-appropriate structure and by making the moves of the inferred neighbourhoods more suitable to such models.

3   Compound-Move Generation

We present compound-move generation (CMG), an extension to local search (LS) that hybridises LS with complete solving and is geared for models where an LS solver is forced to make moves over what we will call *auxiliary variables*, which we will demonstrate in Sect. 5.2 to greatly degrade performance. The main idea is to use a complete solver, in our implementation a CP solver, to try to augment each move probed by the LS solver in order to generate what we will call a *compound move* that also reassigns auxiliary variables. We first explain the basic CMG algorithm and then discuss implementation-specific refinements.

3.1   Basic Algorithm

Consider a flattened MiniZinc model $\langle \mathcal{V}, \mathcal{D}, \mathcal{C}, o \rangle$, partitioned a priori such that $\mathcal{V} = \mathcal{V}_c \cup \mathcal{V}_a$, where the variables of $\mathcal{V}_c$ are called *core variables* and those of $\mathcal{V}_a$ are called *auxiliary variables*; and $\mathcal{C} = \mathcal{C}_c \cup \mathcal{C}_a \cup \mathcal{C}_\ell$ where the constraints of $\mathcal{C}_c$ are called *core constraints* and are all the constraints over only variables in $\mathcal{V}_c$, those of $\mathcal{C}_a$ are called *auxiliary constraints* (also known as *side constraints*) and are all the constraints over only variables in $\mathcal{V}_a$, and those of $\mathcal{C}_\ell$ are called *linking constraints*. Note that o need not be in $\mathcal{V}_c$.

In Example 1, the `pred[i]` variables are ideally in $\mathcal{V}_c$ and the `A[i]` in $\mathcal{V}_a$: the constraint sets $\mathcal{C}_c$, $\mathcal{C}_a$, and $\mathcal{C}_\ell$ follow by their definitions. We discuss in Sect. 4 how to guess this partition automatically.

A model for $\langle \mathcal{V}, \mathcal{D}, \mathcal{C}, o \rangle$ is created for the CBLS solver and a neighbourhood is inferred for the non-functionally determined variables in $\mathcal{V}_c$ but *not* for any variables in $\mathcal{V}_a$; the values of all the variables in $\mathcal{V}_c$ identified as functionally determined are maintained by one-way constraints (see Sect. 2.3) in the CBLS model.

Further, a model for $\langle \mathcal{V}', \mathcal{D}', \mathcal{C}'_a \cup \mathcal{C}'_\ell, o' \rangle$ is created for the complete solver: we add the prime symbol to the corresponding objects for the CBLS solver.

The probing (recall that we assume it consists of (i) tentatively performing a candidate move m; (ii) computing the resulting value of the cost function; and (iii) undoing the candidate move) in the CBLS solver is modified such that between (i) and (ii) an extra step of calling the complete solver is added, divided into three sub-steps:

1. Each variable v' in $\mathcal{V}'_c$ of the model in the complete solver is fixed to the tentative value of its corresponding variable v in the CBLS solver by adding the constraint $v' = \text{val}(v)$ to the model in the complete solver. The search of the complete solver is then launched in order to find an assignment of the variables $\mathcal{V}'_a$ that satisfies all constraints in $\mathcal{C}'_a \cup \mathcal{C}'_\ell \cup \{v' = \text{val}(v) \mid v' \in \mathcal{V}'_c\}$.
2. There are two possible outcomes: either (a) the complete solver reports unsatisfiability, whether at the root node or through search, and the normal probing of m continues; or (b) the complete solver returns a minimal solution $\text{sol}(\mathcal{V}')$ and the normal probing continues for the candidate move $m \cup \{v \mapsto \text{sol}(v') \mid v \in \mathcal{V}_a\}$, which we call a *compound move*, instead of m.
3. Sub-step 1 is undone in the complete solver.

The other aspects of the search in the CBLS solver, such as the heuristic and the meta-heuristic, remain unchanged.

3.2 Refinements and Implementation

We have implemented CMG for our black-box local-search backend fzn-oscar-cbls [2] to MiniZinc, calling the OscaR.cbls solver [6], by using as the complete solver the OscaR.cp solver of the same OscaR framework [14], thereby exploiting the felicitous co-existence of CP and CBLS solvers within the OscaR toolkit.

In its basic form, CMG can be very slow or memory-intensive. We here describe several refinements that improve the performance of CMG, sometimes modifying parts of fzn-oscar-cbls. Some refinements have parameters (denoted by Greek letters), for which we propose values in Sect. 5. We refer to "the complete solver" when refinements or concepts are technology-agnostic, and to "the CP solver" when refinements or concepts are dependent on CP technology.

A. Incomplete Solving. Since finding a minimising assignment for the variables in $\mathcal{V}_a$ can be NP-hard, we can limit the complete solver in the total runtime τ, the number ϕ of failures (if it is a CP solver), and the number σ of intermediate solutions.

B. Always Modifying Auxiliary Variables. If, for a large number of consecutive committed moves, the complete solver has failed to augment them into compound ones, then the auxiliary variables V_a remain unchanged in the CBLS solver during those iterations. Since the current assignment of V_a contributes to the violation in the CBLS solver, as $C_a \cup C_\ell$ is part of its model, and thus can affect which candidate move is committed, this can result in the local search getting stuck in some region of the search space. So fzn-oscar-cbls can be modified to infer also a neighbourhood for V_a: whenever the CBLS solver commits a move for which the complete solver has failed, a move from the neighbourhood of V_a is also committed, in the same iteration.

C. Calling the Complete Solver Again Before Committing a Move. If Refinement A is used, then the assignment of the auxiliary variables V_a returned by the complete solver may not be minimal with respect to the objective function. So, between selecting and committing a candidate move, the complete solver can be run again for that move in order to get a possibly better assignment of V_a. This second solving can be either complete or, as done in our experiments in Sect. 5, a deeper incomplete solving, depending on the new parameters $\tau^\uparrow$, $\phi^\uparrow$, and $\sigma^\uparrow$, which have the semantics of their counterparts in Refinement A.

D. Only Calling the Complete Solver Before Committing a Move. Refinement C can be taken to the extreme where the complete solver is not used at all while probing, but only after selecting a candidate move: instead of modifying the probing step, we can call the complete solver as a post-processing step to selecting a move. For large neighbourhoods, this will significantly speed up the probing, but at the cost of possibly missing good candidate moves.

E. Only Returning the Objective Value. The cost function minimised by fzn-oscar-cbls is $\alpha \cdot \mathrm{val}(o) + \beta \cdot \mathrm{viol}(C)$, where o is the objective variable, $\mathrm{viol}(C)$ is the violation of the constraints in $C = C_c \cup C_a \cup C_\ell$, and α, β are non-negative weights that are tuned during search. An assignment $\mathrm{sol}(V_a')$ returned by the complete solver satisfies all constraints in $C_a' \cup C_\ell'$ and $\mathrm{sol}(o')$ is the same value as o will have in the CBLS solver if the corresponding reassignment is made. Therefore, if using Refinement C, then it is enough, *while probing*, for the complete solver to return $\mathrm{sol}(o')$ since $\alpha \cdot \mathrm{val}(o) + \beta \cdot \mathrm{viol}(C) = \alpha \cdot \mathrm{sol}(o') + \beta \cdot \mathrm{viol}(C_c)$ in this case. By maintaining $\mathrm{viol}(C_c)$ in a separate constraint system [19] in the underlying CBLS solver, we can compute the cost function faster while probing.

F. Exploiting Conflicting Assignments. We say that the current assignment of some variables in the CBLS solver is *conflicting* if they cause the current domain of at least one variable to become empty in the CP solver due to root-node propagation. For efficiency reasons, we limit this definition to root-node failure, but one can generalise it to any failure.

For example, consider the constraints $x < a$ and $a < y$, where $x, y \in V_c$ and $a \in V_a$, and the conflicting assignment $\{x \mapsto 2, y \mapsto 1\}$. Until x or y is reassigned by a CBLS move, the CP solver empties the domain of its variable a' and fails at the root node upon adding the constraints $\{x' = 2, \ y' = 1\}$ in sub-step 1 of the basic CMG. However, for this conflicting assignment and most values of a, one of the two inequality constraints is always satisfied in the CBLS solver and the other one might only make a small contribution to $\text{viol}(\mathcal{C})$. So there might be no strong indication for the LS strategy that x or y needs to be reassigned.

So one should try to identify which variables in $\mathcal{V}'_c$ have caused a root-node failure in the CP solver, and then force the CBLS solver to commit moves on its corresponding core variables. This can be done with a solver that provides explanations for failures, such as any LCG solver, say Chuffed [3]. Most CP solvers (e.g., OscaR.cp, which we use in our implementation) do not provide explanations for failures, so we try to identify which variables have caused such a failure by extending the basic CMG algorithm:

a. The constraints $v' = \text{val}(v)$ for each $v' \in \mathcal{V}'_c$ are, in a random order, iteratively added to the model of the CP solver in sub-step 1, triggering root-node propagation each time. If a failure occurs, then the last variable that was fixed (i.e., that triggered the failure) is returned, say u'. Otherwise, there is no conflicting assignment.

b. Each such variable $u' \in \mathcal{V}'_c$ is recorded in a map that maintains its number of triggered failures, which we call its *conflict count*. The counter is reset to zero for a variable in $\mathcal{V}'_c$ whenever a move reassigning its corresponding variable is committed by the CBLS solver.

If at least one variable has a conflict count that is at least a parameter ω, we force the search heuristic in the CBLS solver to make a move on one or more variables in $\mathcal{V}_c$ with a conflict count of at least another parameter $\underline{\omega}$, by exploiting the tabu search of fzn-oscar-cbls: we make all variables in $\mathcal{V}_c$ with a conflict count under $\underline{\omega}$ tabu. We recommend $\omega > \underline{\omega}$ to avoid making too many variables tabu.

4 Partitioning a Model for CMG

To use compound-move generation (CMG), one must first partition a model instance in order to get the sets $\mathcal{V}_c$, $\mathcal{V}_a$, $\mathcal{C}_c$, $\mathcal{C}_a$, and $\mathcal{C}_\ell$. By $\mathcal{V} = \mathcal{V}_c \cup \mathcal{V}_a$ and the definitions of $\mathcal{C}_c$, $\mathcal{C}_a$, $\mathcal{C}_\ell$, all these sets can be inferred given either $\mathcal{V}_c$ or $\mathcal{V}_a$. We do not impose any semantics to $\mathcal{V}_c$ and $\mathcal{V}_a$: CMG can be applied to any such partition, as done at the end of Sect. 5.1. However, we conjecture that CMG is most efficient when $\mathcal{V}_c$ is the set of variables that model the combinatorial sub-structure of the problem, and $\mathcal{V}_a$ has the variables whose values can easily be determined (ideally by CP-style propagation) given an assignment of $\mathcal{V}_c$.

We present two ways of making this partition, namely by user-provided hints in a MiniZinc model and automatically through a heuristic: in a black-box setting, a user should not have to provide a hint to use CMG and most third-party MiniZinc models are not written with a method such as CMG in mind.

4.1 Hint-Based Partitioning

MiniZinc allows modellers to provide hints to a backend through *annotations* to parts of a model. We introduce the `search_variables(array of var int: V)` annotation, which is attached to the solve statement of a model and indicates that the modeller wants search to be performed on the variables in V: an LS backend with CMG can then use the V[i] as the core variables.

In Example 2 we saw that both the pred[i] and A[i] variables of the model in Fig. 1 are searched on by fzn-oscar-cbls. Upon annotating the solve statement in line 12 with search_variables(pred), fzn-oscar-cbls with CMG can compute that $\mathcal{V}_a = \{A[i] \mid i \text{ in Loc}\}$ because $\mathcal{V}_c = \{pred[i] \mid i \text{ in Loc}\}$.

MiniZinc officially supports search annotations for CP and LCG solvers in order to specify branching strategies. Unfortunately, in general, those search annotations cannot be used in place of our here introduced search_variables annotation, as their semantics does not hint at distinguishing core and auxiliary variables. One could make the (often incorrect) assumption that all variables appearing in a branching strategy are core variables: however, in practice, many MiniZinc modellers specify a branching strategy for *all* variables of a model, and our aim includes good performance on *third-party* models.

4.2   Heuristic-Based Partitioning

Based on our conjecture that $\mathcal{V}_c$ should have the variables that model the combinatorial sub-structure of the problem, we can try to detect such model structure automatically by using a heuristic to guess a partition. Since the global constraints in MiniZinc (such as circuit in Fig. 1) capture combinatorial sub-structures of a problem and fzn-oscar-cbls has constraint-specific neighbourhoods for some global constraints, we can use the following heuristic to decide which variables belong to $\mathcal{V}_c$: if fzn-oscar-cbls infers that constraint-specific neighbourhoods can be used, then all variables that belong to those neighbourhoods are guessed to be in $\mathcal{V}_c$, and all other variables (which would have been put into a default neighbourhood) are therefore in $\mathcal{V}_a$. Otherwise, the heuristic will decide that CMG cannot be used.

This heuristic leads to the same partition for the model in Fig. 1 as when annotating its solve statement by search_variables(pred), but without having to modify the MiniZinc model. Furthermore, this heuristic is able to guess a good partition for some third-party models used in the MiniZinc Challenges, as we will see in Sect. 5.1. However, the heuristic can guess bad partitions, so the modeller currently has to say at the command line if CMG should be used.

5   Experimental Evaluation

We believe the strength of CMG lies in dealing with solver-independent models, say in MiniZinc, where non-functionally-determined auxiliary variables can appear naturally and where the modeller need not be familiar with local search (LS). Therefore, we evaluate CMG on third-party MiniZinc models to see its impact on the robustness of an LS backend to MiniZinc across a variety of models in Sect. 5.1. In order to see how other LS solvers are affected by the presence of non-functionally-determined auxiliary variables, we modify Example 1 to force their presence and evaluate the impact in Sect. 5.2.

5.1 Benchmark Problems

We compare two configurations of CMG in fzn-oscar-cbls with our original fzn-oscar-cbls and Yuck,[2] which is also a CBLS backend to MiniZinc. As a point of reference, we also run the LCG backend Chuffed [3].

The first configuration, called config1, uses Refinements A, B, C, E, F of Sect. 3.2, while the second one, called config2, uses A, B, D, F, but not E, which is meaningless with D. Indeed, initial experiments showed that using D instead of C can both improve and degrade performance, depending on the model, while each other refinement individually seems to improve performance. For both configurations, we set the parameters $\tau = \tau^\uparrow = 30$ s, $\phi = 10000$ failures, $\sigma = 2$ solutions, $\omega = 3$ conflicts, $\underline{\omega} = 1$ conflict, $\phi^\uparrow = 100000$ failures, and $\sigma^\uparrow = \infty$ solutions: initial experiments showed that all those are good values.

We do not compare with the basic CMG algorithm: initial experiments showed that it is too slow. We do not compare with the black-box local-search solver LocalSolver [1] as it offers no backend to MiniZinc. Reformulating models in LocalSolver's modelling language LSP would not yield a meaningful performance comparison as it does *not* have all the global constraints of MiniZinc and as LSP has features that MiniZinc does *not* have.

We evaluate CMG on models and instances for a capacitated vehicle routing problem (CVRP) and a time-dependent travelling salesperson problem (TDTSP), which are taken from the MiniZinc Challenges [18] of 2015 and 2017, as well as our model in Fig. 1 for our running example, the travelling salesperson problem with time windows (TSPTW). Furthermore, we run CMG on all instances of all models of the MiniZinc Challenge 2018 where the heuristic-based partitioning of Sect. 4.2 detects that CMG can be used (in a competition setting, the original fzn-oscar-cbls would run on the other instances). Finally, to showcase the hint-based partitioning of Sect. 4.1, we also perform experiments on a group scheduling problem (GFD), used in the MiniZinc Challenge 2018 with a model where the partitioning heuristic of Sect. 4.2 does not detect that CMG can be used. All models except the one of Fig. 1 and all instances are third-party.

For the local-search backends, we made 10 independent runs with a 600-second-timeout each. For the complete-search backend Chuffed, which is deterministic, we report the objective value of one run with the same timeout. The results are reported in Table 1: note that all problems are minimisation or satisfaction problems and that all chosen instances happen to be satisfiable.

TSPTW. The heuristic-based partitioning of Sect. 4.2 detects auxiliary variables in our model in Fig. 1. We selected five ".001" instances of the GendreauDumasExtended benchmark[3] around the instance size where Chuffed and the original fzn-oscar-cbls stopped establishing satisfiability. We see in Table 1 that both CMG configurations improved the best-found and median values of the original fzn-oscar-cbls. Both configurations established satisfiability for all instances in at least 50% of the runs, whereas the original fzn-oscar-cbls only did so in at most

---

[2] https://github.com/informarte/yuck.

[3] http://lopez-ibanez.eu/tsptw-instances.

20% of the runs. The best-found objective value by config1 is for each instance equal to the best-known objective value reported at the benchmark site. Yuck was not able to establish satisfiability for any instance.

CVRP. The heuristic-based partitioning detects auxiliary variables in the model cvrp. We used all except the toy instance of the MiniZinc Challenge 2015. We see in Table 1 that config1 performed worse than the original fzn-oscar-cbls, except for winning on the P-n16-k8 instance, whereas config2 otherwise outperformed all other backends. On all but P-n16-k8, the best-found solutions by config2 are better than those at the MiniZinc Challenge 2015 for any challenge category.

TDTSP. The heuristic-based partitioning detects auxiliary variables in the model tdtsp. We used the four largest instances among MiniZinc Challenges 2015 and 2017. In Table 1 we see that both Yuck and Chuffed outperformed the original fzn-oscar-cbls, but that config1 outperformed all other backends.

MiniZinc Challenge 2018. The heuristic-based partitioning of Sect. 4.2 finds CMG to be applicable to 3 models and 14 instances of the 20 models and 100 instances in the MiniZinc Challenge 2018, namely elitserien, soccer-computational, and vrplc. For elitserien and vrplc, neither the original fzn-oscar-cbls, nor config1, nor config2, nor Yuck found any solution within the given timelimit to any instance, whereas Chuffed solved all five elitserien instances and two vrplc instances to optimality within the given timelimit.

For soccer-computational, which is the only satisfaction problem in our evaluation, the heuristic-based partitioning determines that CMG is not applicable to the xIGData_22_12_22_5 instance, as a global constraint for which fzn-oscar-cbls has a neighbourhood is removed during flattening. For the other four instances, Yuck found a solution in all runs, while the original fzn-oscar-cbls found a solution in at most half the runs, and Chuffed found a solution to only one instance. Both configurations of CMG had a negative impact on fzn-oscar-cbls, as they did not find any solution in any run.

GFD Schedule. The model gfd-schedule2 is for a scheduling problem with multiple levels of decisions: items are allocated to groups, groups are allocated to factories, and groups are scheduled to be processed on a day. The heuristic-based partitioning does not detect auxiliary variables since the model does not use any global constraint for which fzn-oscar-cbls has a neighbourhood. We therefore use this model to showcase the hint-based partitioning of Sect. 4.1 by annotating the variables representing the allocation of groups to factories as core variables: it is then inferred that all other variables are auxiliary, and their values will be sought by the complete solver instead. Note that not all of the here inferred auxiliary variables are actually auxiliary, and that CMG will here behave similarly to a decomposition where a master problem of allocating groups to factories is solved by LS and a sub-problem of allocating items to groups and scheduling groups is solved by CP. We use the five instances of the MiniZinc Challenge 2018. We see in Table 1 that both config1 and config2 greatly improved over the original fzn-oscar-cbls. Chuffed found and proved minimal solutions to three instances,

Table 1. Comparison on MiniZinc models and third-party benchmark instances between our original LS backend fzn-oscar-cbls, two configurations enriching it with CMG, the LS backend Yuck, and the complete-search backend Chuffed: best-found objective value over 10 runs (column 'best'), boldface indicating overall best performance for the instance of that row, flagged by '+' if equal to the best-known value, and flagged by '*' if proven optimal by Chuffed; median of the best-found objective values over these runs (column 'median'), a superscript indicating the number of runs establishing satisfiability before timing out, a '−' indicating no such run.

	fzn-oscar-cbls original		fzn-oscar-cbls CMG config1		fzn-oscar-cbls CMG config2		Yuck		Chuffed
TSPTW	Best	Median	Best	Median	Best	Median	Best	Median	Best
n20w180	377	377^1	**+253**	253^{10}	261	263^{10}	−	−	*253
n20w200	347	373^2	**+233**	233^{10}	**+233**	234^{10}	−	−	*233
n40w120	−	−	**+434**	439^5	437	464^9	−	−	536
n40w140	−	−	**+328**	334^7	367	388^{10}	−	−	−
n40w160	−	−	**+348**	349^8	362	393^{10}	−	−	−
CVRP									
A-n37-k5	2614	2870^9	2925	2934^{10}	**875**	983^9	−	−	1570
A-n64-k9	5431	5659^4	5518	5661^9	**2868**	3472^8	−	−	3667
B-n45-k5	3638	4121^6	4201	4207^{10}	**972**	1182^{10}	−	−	2466
P-n16-k8	489	503^2	**450**	523^6	481	481^1	−	−	502
TDTSP									
20_14_10	22546	22883^9	**12556**	13506^{10}	13390	15706^{10}	17446	17446^{10}	17024
20_25_00	22924	22961^2	**14888**	16024^{10}	15014	16114^8	18646	18646^{10}	22328
20_26_00	22901	22930^2	**12926**	13917^{10}	13723	14711^8	20790	20790^{10}	19076
20_36_10	22611	22946^6	**12809**	14559^{10}	13859	16752^{10}	17247	17247^{10}	17054
GFD schedule									
n65f2d50...	12741	18547^8	446	552^{10}	343	645^{10}	6861	6861^{10}	*19
n80f7d30...	9952	13338^{10}	665	1062^{10}	**660**	857^{10}	10578	10578^{10}	2023
n90f5d40...	8655	15865^9	473	967^{10}	655	880^{10}	15488	15488^{10}	*11
n100f7d5...	29967	40071^4	1187	2384^9	885	1474^{10}	26397	26397^{10}	*14
n200f5d5...	−	−	−	−	−	−	−	−	−

though config2 found the best objective value for the second instance. On the largest instance, none of the backends found any solution.

5.2 Impact of Auxiliary Variables on Local-Search Solvers

We now show the negative impact on CMG-free black-box local search of committing moves on non-functionally-determined variables that represent auxiliary information. Towards this, we reformulate the TSPTW model in Fig. 1 such that it *can* be written in LocalSolver's modelling language LSP and, unlike in Example 2, the A[i] variables *can* be detected to be functionally determined without a semantic analysis of the entire model. We replace

the predecessor array `pred` in line 4 by the array `order`, where variable
`order[i]` denotes the ith location visited. The `circuit(pred)` constraint in
line 5 then becomes `alldifferent(order)`, and the earliest-arrival-time constraint in line 10 becomes `A[i] = max(A[i-1] + T[order[i-1],order[i]],`
`W[order[i],1])`: this means that the `A[i]` *can* now be determined by one-way
constraints and there is no need for CMG. The main drawbacks of the new
model, called the `alldifferent` model, are that an LS backend can infer a more
suitable neighbourhood for the old model, called the `circuit` model, namely 3-
opt in the case of fzn-oscar-cbls, and that the `circuit` model is better suited for
complete solvers as it captures the combinatorial sub-structure of the problem
better.

However, in the `alldifferent` model, we can now artificially make the `A[i]`
variables non-functionally determined by replacing the equality constraint above
by an `>=` inequality constraint, which will *not* change the minimal objective
value. This allows us to measure the negative impact on CMG-free black-box
local search of committing moves on non-functionally-determined auxiliary variables. Recall from Example 1 that using an inequality is not only correct but
may also be necessary in TSPTW variants.

We examined the impact on LocalSolver, Yuck, and the original fzn-oscar-cbls
of the `alldifferent` model with either equality (model variant `eq`) or inequality
(model variant `ineq`) constraints in the modified line 10. Since LocalSolver does
not have a backend to MiniZinc, we wrote equivalent models in LSP.

In Table 2 we see the negative impact of having auxiliary variables that are
not functionally determined and thus must undergo moves: both fzn-oscar-cbls
and LocalSolver did not find any solutions to `ineq`, where the auxiliary variables
are not functionally determined, though Yuck found solutions but with worse
objective values than for `eq`. On model `eq`, where the auxiliary variables need
not undergo moves, both fzn-oscar-cbls and Yuck found solutions, as opposed to
when running the `circuit` model for TSPTW used in Table 1.

This shows that other black-box local-search solvers are adversely affected
when moves must be made on non-functionally-determined auxiliary variables.

6   Related Work

In the hybridisation context, [8] discusses two categories of hybrids between local
search (LS) and constraint programming (CP):

– **Augmenting LS with CP:** Examples include using only the root-node
   propagation of a CP solver to try to check the feasibility of side constraints
   of capacitated vehicle routing problems and thereby try to find values for the
   auxiliary variables when probing an LS candidate move [5]; modelling an LS
   neighbourhood as an optimisation problem and using a CP solver to find a
   best neighbour to the current assignment [15]; and *large-neighbourhood search*
   (LNS) [17], where some variables in a feasible current LS assignment are fixed
   for a CP solver to look for a best assignment to the other variables, thereby
   building an LS move to another feasible current assignment.

Table 2. Best-found objective values over 5 independent runs for variants of the `alldifferent` model of TSPTW, showing the negative impact on the original fzn-oscar-cbls, on Yuck, and on LocalSolver (neither of them using CMG) when reformulating so that functionally-determined auxiliary variables (column 'eq') become non-functionally-determined ones (column 'ineq'). Boldface indicates overall best performance for the instance of that row, flagged by '+' if equal to the best-known value.

instance\model	fzn-oscar-cbls		Yuck		LocalSolver	
	eq	ineq	eq	ineq	eq	ineq
n40w120	+**434**	–	436	468	+**434**	–
n40w140	+**328**	–	+**328**	391	+**328**	–
n40w160	352	–	+**348**	411	+**348**	–

- **Augmenting CP with LS:** Examples include performing LS starting from the leaf nodes of the CP search tree in order to improve solutions; performing LS at the internal nodes of the CP search tree in order to repair or improve a node [16]; and using LS in order to guide the CP branching strategy [12].

We have here augmented LS with CP in a manner most similar to [5] and LNS. Unlike [5], we allow the CP solver to *search* for a *best* assignment to the auxiliary variables when *feasible* values cannot be inferred through only *root-node propagation*; furthermore, we refine CMG and make it available in a problem-independent context. Like LNS, a partial assignment is here fixed for a CP solver, and complete search is made over the remaining variables. However, unlike LNS, the complete search is here on a subset of the constraints, always the same variables are here fixed, and we allow moves to *infeasible* current assignments.

In the MiniZinc context, LS-CP hybrids exist for LNS, namely the GELATO framework [4], combining their LS solver EasyLocal++ with the CP solver Gecode [9], and Mini-LNS [7], which is solver-independent, but neither of these are black-box and therefore both require a search strategy to be specified.

7   Conclusion and Future Work

We presented compound-move generation (CMG), an extension to black-box local search, geared for model structure that may cause local-search solvers to make moves reassigning variables representing auxiliary information. We have outlined two methods for detecting such model structure, which can appear naturally, for example for routing problems with side constraints. This means that such solvers without CMG might perform unexpectedly worse than complete solvers and considerably worse than handcrafted local-search algorithms on such problems. Our experiments show that several black-box local-search solvers are adversely affected in the presence of that model structure, and that CMG greatly improves the performance of our fzn-oscar-cbls backend to MiniZinc on such models, without requiring any model reformulation.

Future work includes extracting more information from the complete solver in order to help guide local search. For example, if an LCG solver is used as the complete solver for CMG, then learned clauses could be used to construct the next local-search move. Furthermore, making some constraints soft for the complete solver could help when infeasible solutions are explored and would improve on refinement B in Sect. 3.2.

References

1. Benoist, T., Estellon, B., Gardi, F., Megel, R., Nouioua, K.: LocalSolver 1.x: a black-box local-search solver for 0-1 programming. 4OR Q. J. Oper. Res. **9**(3), 299–316 (2011). LocalSolver is available at https://www.localsolver.com
2. Björdal, G., Monette, J.N., Flener, P., Pearson, J.: A constraint-based local search backend for MiniZinc. Constraints **20**(3), 325–345 (2015). https://doi.org/10.1007/s10601-015-9184-z, the fzn-oscar-cbls backend is available at http://optimisation.research.it.uu.se/software
3. Chu, G.: Improving Combinatorial Optimization. Ph.D. thesis, Department of Computing and Information Systems, University of Melbourne, Australia (2011). The Chuffed solver and MiniZinc backend are available at https://github.com/chuffed/chuffed
4. Cipriano, R., Di Gaspero, L., Dovier, A.: A multi-paradigm tool for large neighborhood search. In: Talbi, E.G. (ed.) Hybrid Metaheuristics. SCI, vol. 434, pp. 389–414. Springer, Heidelberg (2013). https://doi.org/10.1007/978-3-642-30671-6_15
5. De Backer, B., Furnon, V., Shaw, P., Kilby, P., Prosser, P.: Solving vehicle routing problems using constraint programming and metaheuristics. J. Heuristics **6**(4), 501–523 (2000)
6. De Landtsheer, R., Ponsard, C.: OscaR.cbls: an open source framework for constraint-based local search. In: ORBEL-27, the 27th Annual Conference of the Belgian Operational Research Society (2013). http://www.orbel.be/orbel27/pdf/abstract293.pdf, the OscaR.cbls solver is available at https://bitbucket.org/oscarlib/oscar/branch/CBLS
7. Dekker, J.J., de la Banda, M.G., Schutt, A., Stuckey, P.J., Tack, G.: Solver-independent large neighbourhood search. In: Hooker, J. (ed.) CP 2018. LNCS, vol. 11008, pp. 81–98. Springer, Cham (2018). https://doi.org/10.1007/978-3-319-98334-9_6
8. Focacci, F., Laburthe, F., Lodi, A.: Local search and constraint programming. In: Glover, F.W., Kochenberger, G.A. (eds.) Handbook of Metaheuristics, ORMS, vol. 57, chap. 13, pp. 369–403. Springer, Boston (2003). https://doi.org/10.1007/978-1-4419-8917-8_9
9. Gecode Team: Gecode: A generic constraint development environment (2018). The Gecode solver and MiniZinc backend are available at http://www.gecode.org
10. Glover, F., Laguna, M.: Tabu search. In: Reeves, C.R. (ed.) Modern Heuristic Techniques for Combinatorial Problems, pp. 70–150. Wiley, New York (1993)
11. Hoos, H.H., Stützle, T.: Stochastic Local Search: Foundations & Applications. Elsevier/Morgan Kaufmann, San Francisco (2004)
12. Jussien, N., Lhomme, O.: Local search with constraint propagation and conflict-based heuristics. Artif. Intell. **139**(1), 21–45 (2002)

13. Nethercote, N., Stuckey, P.J., Becket, R., Brand, S., Duck, G.J., Tack, G.: MiniZinc: towards a standard CP modelling language. In: Bessière, C. (ed.) CP 2007. LNCS, vol. 4741, pp. 529–543. Springer, Heidelberg (2007). https://doi.org/10.1007/978-3-540-74970-7_38
14. OscaR Team: OscaR: Scala in OR (2012). https://oscarlib.bitbucket.io
15. Pesant, G., Gendreau, M.: A constraint programming framework for local search methods. J. Heuristics 5(3), 255–279 (1999)
16. Prestwich, S.: A hybrid search architecture applied to hard random 3-SAT and low-autocorrelation binary sequences. In: Dechter, R. (ed.) CP 2000. LNCS, vol. 1894, pp. 337–352. Springer, Heidelberg (2000). https://doi.org/10.1007/3-540-45349-0_25
17. Shaw, P.: Using constraint programming and local search methods to solve vehicle routing problems. In: Maher, M., Puget, J.-F. (eds.) CP 1998. LNCS, vol. 1520, pp. 417–431. Springer, Heidelberg (1998). https://doi.org/10.1007/3-540-49481-2_30
18. Stuckey, P.J., Feydy, T., Schutt, A., Tack, G., Fischer, J.: The MiniZinc challenge 2008–2013. AI Mag. 35(2), 55–60 (2014). (Summer), https://www.minizinc.org/challenge.html
19. Van Hentenryck, P., Michel, L.: Constraint-Based Local Search. The MIT Press, London (2005)

SAT Encodings of Pseudo-Boolean Constraints with At-Most-One Relations

Miquel Bofill$^{(\boxtimes)}$, Jordi Coll, Josep Suy, and Mateu Villaret

Universitat de Girona, Girona, Spain
{miquel.bofill,jordi.coll,josep.suy,mateu.villaret}@imae.udg.edu

Abstract. Pseudo-Boolean (PB) constraints appear often in a large variety of constraint satisfaction problems. Encoding such constraints to SAT has proved to be an efficient approach in many applications. However, most of the existing encodings in the literature do not take profit from side constraints that often occur together with the PB constraints. In this work we introduce specialized encodings for PB constraints occurring together with at-most-one (AMO) constraints over subsets of their variables. We show that many state-of-the-art SAT encodings of PB constraints from the literature can be dramatically reduced in size thanks to the presence of AMO constraints. Moreover, the new encodings preserve the propagation properties of the original ones. Our experiments show a significant reduction in solving time thanks to the new encodings.

Keywords: SAT · Pseudo-Boolean · Encoding

1 Introduction

Linear equations are ubiquitous in Constraint Satisfaction Problems (CSP). A particular case are Pseudo-Boolean (PB) constraints, which are linear expressions of the form $\sum_{i=1}^{n} q_i x_i \# K$, where $\# \in \{<, \leq, =, \geq, >\}$, $q_1, \ldots, q_n$ and K are integer constants, and $x_1, \ldots, x_n$ are 0/1 variables. A successful approach to solve CSPs is to encode them to propositional Boolean formulas, which are then solved using off-the-shelf SAT solvers. Therefore, there exist many works on encoding PB constraints to SAT [20]. State-of-the-art encodings are based on Binary Decision Diagrams [1,12], Sequential Weight Counters [15], Generalized Totalizers [16], and Polynomial Watchdog schemes [7,18].

In [10] it is proposed a Multi-Decision Diagram (MDD) based SAT encoding of PB constraints, under the assumption that there exist some at-most-one (AMO) relations on disjoint subsets of variables. The AMO relations let to erase certain interpretations from decision diagrams, and to represent the PB as an MDD instead of as a Binary Decision Diagram (BDD). This way, the resulting encoding is notably smaller than the one of an equivalent BDD, and solving time is significantly reduced. This technique has been used in [9,10] to formulate particular kinds of scheduling problems. Similar techniques are also applied in [2] in

© Springer Nature Switzerland AG 2019
L.-M. Rousseau and K. Stergiou (Eds.): CPAIOR 2019, LNCS 11494, pp. 112–128, 2019.
https://doi.org/10.1007/978-3-030-19212-9_8

the field of Mixed Integer Linear Programming, where AMO relations between a set of 0/1 variables are used to substitute these variables by an integer variable.

Efficient encodings of conjunctions of PB and AMO constraints can have a high impact in a wider range of CSPs. Such combination of constraints appear in settings where it has to be chosen one option among a set of incompatible options, and this decision has an associated cost. A non exhaustive list of applications where this happens are logistics [8], resource allocation [17], capital budgeting [21], telecommunications [22], combinatorial auctions [11] and routing [19]. In short, any problem which is essentially a multi-choice knapsack problem is likely to contain both PB and AMO constraints. Therefore, it is of big interest to provide new and better encodings for this combination constraints.

In this paper we address the question of whether other SAT encodings of PB constraints different than decision diagram-based ones can be improved in the presence of AMOs. We revisit many state-of-the-art SAT encodings of PB constraints and propose improved versions of those encodings for conjunctions of PB and AMO constraints. More precisely, we provide modifications of the Sequential Weight Counter, Generalized Totalizer, and Global Polynomial Watchdog encodings. We also show that the new encodings preserve the propagation properties of the original ones. Our experimental results show that the size of the SAT encodings of PB constraints can be dramatically reduced thanks to taking the AMO constraints into account, and that there can be a huge time performance improvement when using the new encodings. We provide datasets which contain AMO constraints and PB constraints with different configurations, and we show that some encodings are better than others for particular kinds of PB.

2 Preliminaries

A *Boolean variable* is a variable than can take truth values 0 (false) and 1 (true). A *literal* is a Boolean variable x or its negation $\overline{x}$. A *clause* is a disjunction of literals. A *propositional formula in conjunctive normal form* (CNF) is a conjunction of clauses. Clauses are usually seen as sets of literals, and formulas as sets of clauses. A *Boolean function* is a function of the form $f : \{0,1\}^n \to \{0,1\}$. In this paper we will only consider constraints which are defined on a finite set of Boolean variables, i.e., Boolean functions. An *assignment* is a mapping of Boolean variables to truth values; it can also be seen as a set of literals (e.g., $\{x = 1, y = 0, z = 0\}$ is usually denoted $\{x, \overline{y}, \overline{z}\}$). A *satisfying assignment* of a Boolean function f is an assignment that makes it evaluate to 1. In particular, an assignment A satisfies a formula F in CNF if at least one literal l of each clause in F belongs to A. Such an assignment is called a *model* of the formula. In this paper we will assume that all propositional formulas are in CNF. Given two Boolean functions F and G, we say that G is a logical consequence of F, written $F \models G$, iff every model of F is also a model of G.

Definition 1. *An* at-most-one *(AMO) constraint is a Boolean function of the form $\sum_{i=1}^{n} x_i \leq 1$, where all x_i are 0/1 variables.*

Definition 2. *A pseudo-Boolean (PB) constraint is a Boolean function of the form $\sum_{i=1}^{n} q_i x_i \# K$ where K and all q_i are integer constants, all x_i are $0/1$ variables, and $\# \in \{<, \leq, =, \geq, >\}$.*

It is usually assumed that $\#$ is $\leq$, and that K and all q_i are non-negative, since the other cases can be easily reduced to this one [12].

Definition 3. *By PB(AMO) constraint we refer to a constraint of the form $P \wedge M_1 \wedge \cdots \wedge M_m$, where P is a PB constraint, and $M_1, \ldots, M_m$ are AMO constraints.*

Unit propagation (UP) is a propagation mechanism used in modern SAT solvers. It is based on the principle that if a clause contains a single literal (i.e., under a given assignment, all literals but one are false), then every model must make that literal true. Hence, the assignment can be extended with this literal.

We say that a formula G is an *encoding* of a Boolean function F if the following holds: given an assignment A over the variables of F, A satisfies F iff A can be extended to a satisfying assignment of G.

An encoding E of a constraint C is said to *UP-maintain GAC* if it satisfies the following property: given a partial assignment A, if a variable x of C is true (respectively false) in every extension of A satisfying C, then unit propagating A on E will extend A to $A \cup \{x\}$ (respectively $A \cup \{\overline{x}\}$) [7].

3   New Encodings of Monotonic Decreasing PB(AMO) Constraints

In this section we present three different SAT encodings for PB(AMO) constraints. Given a PB(AMO) of the form $P \wedge M_1 \wedge \cdots \wedge M_m$, a straightforward approach to encode it is to generate a formula F of the form $G \wedge H_1 \wedge \cdots \wedge H_m$, where G is an encoding of P, and each H_i is an encoding of M_i. Instead, similarly to the MDD-based approach of [10], we propose to encode PB(AMO)s in a combined way. On the one hand we encode the conjunction of AMO constraints in the usual way, i.e., we encode each AMO separately and use the conjunction of all the resulting clauses. On the other hand we encode the PB constraint assuming that the accompanying AMO constraints are already enforced in some way. This is precisely what will let us reduce the size of the encoding of the PB constraint. We do not restrict to a particular encoding for the AMO constraints. Even more, in the context of a bigger formula, if the AMO constraints are logically implied by the formula at hand, then the encoding of the PB constraint will suffice to obtain a correct encoding of the PB(AMO) constraint.

We start each subsection giving an intuitive explanation of an already existing encoding of PB constraints. Then, we propose a generalized version of it in order to encode PB(AMO) constraints. Since PB(AMO) constraints generalize PB constraints,[1] we follow the convention of naming the new encodings after

---

[1] A PB constraint of the form $\sum_{i=1}^{n} q_i x_i \leq K$ corresponds to a PB(AMO) constraint of the form $\sum_{i=1}^{n} q_i x_i \leq K \wedge x_1 \leq 1 \wedge \cdots \wedge x_n \leq 1$.

the original encoding, prefixing them with the word *Generalized*, e.g., from the Sequential Weight Counter (SWC) encoding we provide the Generalized Sequential Weight Counter (GSWC) encoding.

We make the following assumptions on the PB(AMO) constraint $P \wedge M_1 \wedge \cdots \wedge M_m$ at hand: (i) every variable in P occurs at least in one M_i (a variable x can always be included in a single-variable AMO constraint of the form $x \leq 1$); (ii) unless otherwise stated, we assume that P is of the form $\sum_{i=1}^{n} q_i x_i \leq K$, with $q_i \geq 0$, i.e. it is monotonic decreasing; (iii) P is neither trivially true nor false (i.e., we assume that $0 \leq K < \sum_{i=1}^{n} q_i$); (iv) no variable is trivially removable (i.e., $0 < q_i \leq K$).

Input. Given a PB(AMO) constraint $\mathcal{P}$ of the form $P \wedge M_1 \wedge \cdots \wedge M_m$, the encodings of the following subsections receive as input the PB constraint P and a partition $\mathcal{X} = \{X_1, \ldots, X_N\}$ of the variables of P, such that any assignment satisfying $M_1 \wedge \cdots \wedge M_m$ also satisfies the AMO constraints $\sum_{x_{i_j} \in X_i} x_{i_j} \leq 1$, for all $X_i \in \mathcal{X}$. Note that there may be more than one possible partition, and that for every partition we will have $n = \sum_{i=1}^{N} |X_i|$, where n is the number of variables in the scope of P. A simple way to obtain such a partition is to start with a list of sets $X_1, \ldots, X_m$, where each set X_i contains the variables in the scope of M_i. Then, remove every variable from all X_i but one, and finally remove all the empty sets.

3.1 Sequential Weight Counter Encoding

The *Sequential Weight Counter* (SWC) encoding for PB constraints was introduced in [15]. The idea is to encode the PB constraint by a circuit that sequentially sums from left to right the coefficients (a.k.a. weights) q_i whose variable x_i is set to true. Specifically, given a PB constraint $\sum_{i=1}^{n} q_i x_i \leq K$, there is a sequence of n counters of K inputs and K outputs, where the i-th counter is associated to the variable x_i. Each counter receives as input a vector of Boolean variables, which is the unary representation of an integer value, and adds the weight q_i to the output if the associated variable x_i is set to true. Therefore, the i-th counter receives as input $\sum_{j=1}^{i-1} q_j x_j$ and outputs $\sum_{j=1}^{i} q_j x_j$. Note that the output of the counter number $i - 1$ is the input of the i-th counter.

An example of a sequence of counters is shown in Fig. 1. The encoding introduces $n \cdot K$ variables, denoted $s_{i,j}$, with $1 \leq i \leq n$, $1 \leq j \leq K$, where $s_{i,j}$ is the j-th output of the i-th counter and also the j-th input of the $(i+1)$-th counter. The encoding introduces the clauses

$$\overline{s_{i-1,j}} \vee s_{i,j} \qquad 2 \leq i < n, 1 \leq j \leq K \qquad (1)$$

$$\overline{x_i} \vee s_{i,j} \qquad 1 \leq i < n, 1 \leq j \leq q_i \qquad (2)$$

$$\overline{s_{i-1,j}} \vee \overline{x_i} \vee s_{i,j+q_i} \qquad 2 \leq i < n, 1 \leq j \leq K - q_i \qquad (3)$$

$$\overline{s_{i-1,K+1-q_i}} \vee \overline{x_i} \qquad 2 \leq i \leq n \qquad (4)$$

where $s_{0,j}$ is the constant 0 for all j, to represent the input of the first counter which is the empty sum. Clauses (1) state that $\sum_{j=1}^{i} q_j x_j \geq \sum_{j=1}^{i-1} q_j x_j$. Clauses

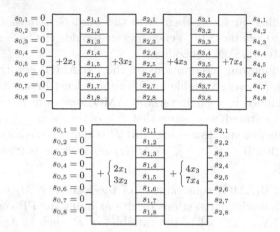

Fig. 1. At the top: high level circuit representation of $SWC(2x_1 + 3x_2 + 4x_3 + 7x_4 \leq 8)$. At the bottom: high level circuit representation of $GSWC(2x_1 + 3x_2 + 4x_3 + 7x_4 \leq 8, \{\{x_1, x_2\}, \{x_3, x_4\}\})$.

(2) and (3) enforce that if a variable x_i is true then its coefficient is added to the input of the next counter. Finally, Clauses (4) enforce that the sum never exceeds K.

Generalized Sequential Weight Counter (GSWC). We define the GSWC encoding by, instead of associating a single product $q_i x_i$ from the PB constraint to each counter, associating a set of products to each of them. In our generalization, given a partition $\mathcal{X} = \{X_1, \ldots, X_N\}$ of the variables of the PB constraint, the resulting formulation will have just N counters, where the i-th counter will handle all the products $q_l x_l$ for the variables x_l in X_i. If the variables in each set X_i are subject to an AMO constraint then, given an assignment satisfying those constraints, at most one coefficient q_l will be added by each counter, and the output of the whole circuit will correspond to the value of the left hand side sum of the PB constraint, i.e., $\sum_{i=1}^{n} q_i x_i$. Analogously as in the original encoding, we will enforce that it is not reached a sum that exceeds K. The GSWC encoding introduces the following clauses:

$$\overline{s_{i-1,j}} \vee s_{i,j} \qquad 2 \leq i < N, 1 \leq j \leq K \tag{5}$$

$$\overline{x_l} \vee s_{i,j} \qquad 1 \leq i < N, x_l \in X_i, 1 \leq j \leq q_l \tag{6}$$

$$\overline{s_{i-1,j}} \vee \overline{x_l} \vee s_{i,j+q_l} \qquad 2 \leq i < N, x_l \in X_i, 1 \leq j \leq K - q_l \tag{7}$$

$$\overline{s_{i-1,K+1-q_l}} \vee \overline{x_l} \qquad 2 \leq i \leq N, x_l \in X_i \tag{8}$$

Clauses (5) propagate the accumulated sum in the same way as Clauses (1). Clauses (6) and (7) enforce $S_i \geq S_{i-1} + q_l x_l$, for all $x_l \in X_i$, where S_{i-1} and S_i are respectively the input and output value of the i-th counter. Clauses (8) enforce that the sum never exceeds K. A high level circuit representation of a GSWC encoding is shown in Fig. 1.

The main difference between the SWC and GSWC encodings is that the latter has only N counters, instead of n, and therefore introduces less fresh variables (assuming $N < n$). Also, the number of Clauses (5) in the GSWC encoding is smaller than the number of Clauses (1) in the SWC encoding. The SWC encoding requires $O(nK)$ auxiliary variables and $O(nK)$ clauses, while the GSWC encoding requires $O(NK)$ auxiliary variables and $O(nK)$ clauses.

By $GSWC(P, \mathcal{X})$ we denote the set of clauses derived from a PB constraint P and a partition $\mathcal{X}$, as described above.

Lemma 1. *Let $\mathcal{P}$ be a PB(AMO) of the form $P \wedge M_1 \wedge \cdots \wedge M_m$ and $\mathcal{X}$ be a partition of the variables of P such that $M_1 \wedge \cdots \wedge M_m \models \sum_{x_{i_j} \in X_i} x_{i_j} \leq 1$, for all $X_i \in \mathcal{X}$. The conjunction of $GSWC(P, \mathcal{X})$ with an encoding of $M_1 \wedge \cdots \wedge M_m$ is an encoding of $\mathcal{P}$.*

In [15] it is proved that the SWC encoding UP-maintains GAC. The GSWC encoding preserves this property.

Theorem 1. *Let $\mathcal{P}$ be a PB(AMO) of the form $P \wedge M_1 \wedge \cdots \wedge M_m$ and $\mathcal{X}$ be a partition of the variables of P such that $M_1 \wedge \cdots \wedge M_m \models \sum_{x_{i_j} \in X_i} x_{i_j} \leq 1$, for all $X_i \in \mathcal{X}$. The conjunction of $GSWC(P, \mathcal{X})$ with an UP-maintaining GAC encoding of $M_1 \wedge \cdots \wedge M_m$ is UP-maintaining GAC.*

Proof. Let S denote the conjunction of $GSWC(P, \mathcal{X})$ with an UP-maintaining GAC encoding of $M_1 \wedge \cdots \wedge M_m$. Let A be a partial assignment to the variables of S, which is extendible to a satisfying assignment of $\mathcal{P}$. Therefore, no AMO constraint M_i is violated under A. We need to show that for every variable x of $\mathcal{P}$ such that x is not assigned in A, if $A \cup \{x\}$ cannot be extended to a satisfying assignment of $\mathcal{P}$, then x is set to false by unit propagating A on S (note that $A \cup \{\bar{x}\}$ can always be extended to a satisfying assignment, so we don't need to consider this case). W.l.o.g., assume that $x_1 \in X_1$ is such variable. If $A \cup \{x_1\}$ cannot be extended to a satisfying assignment of $M_1 \wedge \cdots \wedge M_m$ then, by the assumption that S contains an UP-maintaining GAC encoding of $M_1 \wedge \cdots \wedge M_m$, we have that x_1 is set to false by unit propagation. Assume now the contrary, i.e., that $A \cup \{x_1\}$ can be extended to an assignment satisfying the AMOs. In this case, the reason why UP should set x_1 to false is that $A \cup \{x_1\}$ cannot be extended to satisfy P. Since $A \cup \{x_1\}$ does not violate $M_1 \wedge \cdots \wedge M_m$, at most one variable in X_i is true in A, for $2 \leq i \leq N$, and no variable in X_1 is true in A. Let us construct a PB constraint P' from P by picking one variable x_{j_i} from each set X_i, $2 \leq i \leq N$, as follows: if X_i contains a variable which is true in A, then this is the variable to be picked up from X_i, otherwise pick up any variable. We define $P' : q_1 x_1 + \sum_{i=2}^{N} q_{j_i} x_{j_i} \leq K$. Since P' contains all the variables which are true in A, and due to the monotonicity of P, we have that $q_1 x_1 + \sum_{i=2}^{N} q_{j_i} x_{j_i}$ is equisatisfiable to $\sum_{i=1}^{n} q_i x_i$ under the assignment $A \cup \{x_1\}$, and therefore $A \cup \{x_1\}$ can neither be extended to a model of P'. It is not hard to see that $GSWC(P, \mathcal{X})$ contains all the clauses of $SWC(P')$, and it is already proved that the SWC encoding UP-maintains GAC. Therefore, all the clauses required to set x_1 to false by UP are contained in S.

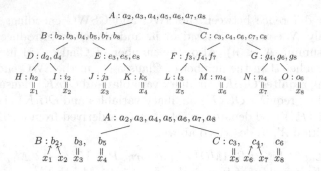

Fig. 2. At the top: binary tree of $GT(2x_1 + 2x_2 + 3x_3 + 5x_4 + 3x_5 + 4x_6 + 4x_7 + 6x_8 \leq 7)$. At the bottom: binary tree of $GGT(2x_1 + 2x_2 + 3x_3 + 5x_4 + 3x_5 + 4x_6 + 4x_7 + 6x_8 \leq 7, \{\{x_1, x_2, x_3, x_4\}, \{x_5, x_6, x_7, x_8\}\})$.

3.2 Generalized Totalizer Encoding

The *Generalized Totalizer* (GT) encoding was presented in [16] as a generalization of the Totalizer encoding for cardinality constraints [6]. The overall idea of GT is to represent a PB constraint $\sum_{i=1}^{n} q_i x_i \leq K$ as a binary tree. Every node of the tree has associated a distinct label and an attribute *vars* which consists of a set of Boolean variables. Each variable x_i of the PB constraint is placed into the attribute *vars* of a different leaf node, and is renamed after the label of the node and its associated coefficient q_i (e.g., given the product $3x_1$, if the variable x_1 is inserted into a leaf node labelled by letter O, then the variable is named o_3). The attribute *vars* of any non-leaf node labelled O contains a variable o_w for every subset of the underlying leaves which sums exactly w, for values of w in the range $[1, K]$, taking i for the value of each leaf node L with variable l_i. Also, *vars* contains a variable o_{K+1} iff any of the sums is larger than K. Figure 2 illustrates an example binary tree.

The clauses of the encoding enforce that each non-leaf variable o_w is set to true if the underlying variables which sum w (or more than K for $w = K + 1$) are set to true. Moreover it is enforced, at the root node, that the variable representing a sum larger than K is false. The GT encoding introduces the following clauses for each non-leaf node O with children L and R:

$$\overline{l_{w_1}} \vee \overline{r_{w_2}} \vee o_{w_3} \quad l_{w_1} \in L.vars, \; r_{w_2} \in R.vars, \; w_3 = \min(w_1 + w_2, K + 1) \quad (9)$$

$$\overline{t_w} \vee o_w \quad t_w \in L.vars \cup R.vars \quad (10)$$

It also introduces the unary clause

$$\overline{a_{k+1}} \quad (11)$$

where A is the root node of the tree and $a_{k+1} \in A.vars$. Note that variable a_{k+1} will exist, since otherwise the constraint would be trivially satisfied. Clauses (9) enforce that the variable o_{w_3} will be set to true by UP if there exists a pair of variables l_{w_1}, r_{w_2} from the children nodes that are set to true and such that

$w_3 = \min(w_1 + w_2, K + 1)$. Clauses (10) enforce that the variable o_w will be set to true by UP if some child has a variable t_w set to true. Finally, Clause (11) states that the sum of the tree (i.e., the value of the left hand side expression of the PB constraint) cannot be greater than K.

Generalized Generalized Totalizer (GGT). In our generalization of the GT encoding, we will use the same definition of the binary tree, but the leafs will be instantiated differently. Instead of introducing a leaf node for each variable of the PB constraint, what we do is to introduce a leaf node for each of the sets in the partition $\mathcal{X}$. The leaf node O associated to set X_i will contain a variable o_{q_l} in its *vars* attribute for each different coefficient q_l such that $x_l \in X_i$. If there is a single coefficient q_l, then x_l is renamed as o_{q_l} and placed in $O.vars$, as in the GT encoding. If there are multiple occurrences of a coefficient q_l, we introduce a fresh variable o_{q_l}. The following clauses relate the fresh leaf variables with the variables of the PB constraint:

$$\overline{x_l} \vee o_{q_l} \qquad X_i \in \mathcal{X},\, x_l \in X_i,\, |\{x_{l'} \in X_i \mid q_{l'} = q_l\}| \geq 2 \qquad (12)$$

The GGT encoding introduces Clauses (9), (10) and (11) as in the GT encoding, and Clauses (12). Figure 2 depicts the binary tree of a GGT encoding. Note that assuming that an AMO constraint over each set X_i is satisfied, at most one of the variables in each leaf node will be true, and therefore the encoding correctly evaluates $\sum_{i=1}^{n} q_i x_i \leq K$.

The GT encoding requires $O(nK)$ auxiliary variables and $O(nK^2)$ clauses, while GGT encoding requires $O(NK)$ auxiliary variables and $O(NK^2)$ clauses.

By $GGT(P, \mathcal{X})$ we denote the set of clauses derived from a PB constraint P and a partition $\mathcal{X}$, as described above.

Lemma 2. *Let $\mathcal{P}$ be a PB(AMO) of the form $P \wedge M_1 \wedge \cdots \wedge M_m$ and $\mathcal{X}$ be a partition of the variables of P such that $M_1 \wedge \cdots \wedge M_m \models \sum_{x_{i_j} \in X_i} x_{i_j} \leq 1$, for all $X_i \in \mathcal{X}$. The conjunction of $GGT(P, \mathcal{X})$ with an encoding of $M_1 \wedge \cdots \wedge M_m$ is an encoding of $\mathcal{P}$.*

In [16] it is proved that the GT encoding UP-maintains GAC. The GGT encoding preserves this property.

Theorem 2. *Let $\mathcal{P}$ be a PB(AMO) of the form $P \wedge M_1 \wedge \cdots \wedge M_m$ and $\mathcal{X}$ be a partition of the variables of P such that $M_1 \wedge \cdots \wedge M_m \models \sum_{x_{i_j} \in X_i} x_{i_j} \leq 1$, for all $X_i \in \mathcal{X}$. The conjunction of $GGT(P, \mathcal{X})$ with an UP-maintaining GAC encoding of $M_1 \wedge \cdots \wedge M_m$ is UP-maintaining GAC.*

The proof is analogous to the proof of Theorem 1.

3.3 Global Polynomial Watchdog Encoding

The *Global Polynomial Watchdog* (GPW) encoding was presented in [7]. It uses as basis a *polynomial watchdog* formula, denoted by $PW(P)$, which is associated

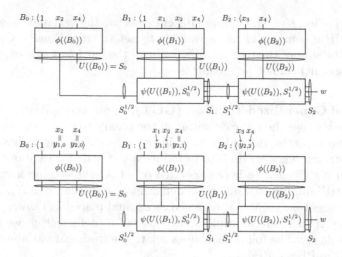

Fig. 3. At the top: circuit representation of $PW(2x_1 + 3x_2 + 4x_3 + 7x_4 < 9)$. At the bottom: circuit representation of $PW(2x_1 + 3x_2 + 4x_3 + 7x_4 < 9, \{\{x_1, x_2\}, \{x_3, x_4\}\})$.

with a PB constraint P. The formula $PW(P)$ has a variable named the *output* variable, denoted w, which is set to 1 by UP as soon as P is falsified.

The GPW encoding is defined for PB constraints of the form $\sum_{i=1}^{n} q_i x_i < K$, i.e., with a strict inequality instead of a non-strict one. The first step is to normalize the constraint to the form $T + \sum_{i=1}^{n} q_i x_i < m2^p$, where p, T and m are defined as follows: $p = \lfloor \log_2(\max_{i=1..n}(q_i)) \rfloor$ is the index of the most significant bit in the binary representation of the largest coefficient q_i, being 0 the index of the least significant bit. In other words, $p + 1$ is the number of bits needed to represent q_i in binary notation; T is the smallest non-negative integer such that $K + T$ is a multiple of 2^p; $m = (K + T)/2^p$.

Once the constraint is expressed in this form, it is computed a set B_r of variables of P (called *bucket*) for each bit $0 \le r \le p$. We denote by $b_r(q_i)$ the r-th bit of the binary representation of the integer q_i. Bucket B_r contains all the variables x_i such that $b_r(q_i) = 1$. Bucket B_r also contains a 1 constant if $b_r(T) = 1$.

Example 1. The following is the transformation to apply to the PB constraint $2x_1 + 3x_2 + 4x_3 + 7x_4 < 9$. We have that $p = 2$, and $T = 3$ is the smallest integer such that $T + K = 12$ is a multiple of 2^p, with $m = 3$. Therefore, the constraint is expressed as $3 + 2x_1 + 3x_2 + 4x_3 + 7x_4 < 12$. The content of buckets B_0, B_1 and B_2 is illustrated in Fig. 3.

The idea is to decompose each coefficient in its binary representation an sum each bit having the same weight.

The formula $PW(P)$ can be represented as a circuit, as can be seen in Fig. 3 corresponding to Example 1. We denote by $\langle B_r \rangle$ a vector with an arbitrary order containing the elements of bucket B_r. The formula $PW(P)$ uses two main components: the formulas $\phi(V)$ and $\psi(V_1, V_2)$. The formula $\phi(V)$ has as input a vector of Boolean variables V, and has as output a vector of $|V|$ variables named $U(V)$. The formula $\phi(V)$ enforces that $U(V)$ is the unary representation of the sum of the input variables. The formula $\psi(V_1, V_2)$, has as input two vectors of variables V_1 and V_2, which are the unary representation of two integers, and has as output a vector of $|V_1|+|V_2|$ variables named S. The formula $\psi(V_1, V_2)$ enforces that S is the unary representation of $V_1 + V_2$. In the definition of $PW(P)$, we denote by S_r the output of the ψ formula related with bucket B_r, for $1 \leq r \leq p$, and we define $S_0 = U(\langle B_0 \rangle)$. Half of the value of S_k, for a weight 2^k, is computed with operator $\frac{1}{2}$ and integrated in the sum for weight 2^{k+1}. Then, the formula $PW(P)$ is defined as the conjunction of these two formulas:

$$\phi(\langle B_r \rangle) \qquad\qquad\qquad 0 \leq r \leq p \qquad\qquad (13)$$

$$\psi(U(\langle B_r \rangle), S_{r-1}^{1/2}) \qquad\qquad 1 \leq r \leq p \qquad\qquad (14)$$

The GPW encoding is then defined as

$$PW(P) \wedge \overline{w} \qquad\qquad\qquad\qquad (15)$$

where $PW(P)$ encodes ϕ with a totalizer, and ψ with an adder of unary numbers. Due space restrictions we cannot elaborate more on the correctness of this encoding, so we refer the reader to [7]. The basic idea is that the m-th bit of S_p, represented with variable w, is set to 1 by UP if the sum of the constraint is greater or equal than $m2^p = T + K$. If w is set to 1 the encoding is not satisfied.

Generalized Global Polynomial Watchdog (GGPW). We define the GGPW encoding by using a *generalized polynomial watchdog formula* $PW(P, \mathcal{X})$ instead of the original polynomial watchdog formula. Again, P has to be normalized to the form $T + \sum_{i=1}^{n} q_i x_i < m2^p$ in the same way as in $PW(P)$. For each set X_i, $PW(P, \mathcal{X})$ will contain a vector of variables $Y_i = \langle y_{i,p}, y_{i,p-1}, \ldots, y_{i,0} \rangle$.

Y_i is interpreted as binary number, where (at least) the bits corresponding to the binary representation of q_l, for all $x_l \in X_i$ such that x_l is true, are set to one. Therefore, when exactly one x_l is true, Y_i will be greater than or equal to q_l. The following clauses define the variables Y_i:

$$\overline{x_l} \vee y_{i,r} \qquad 0 \leq r \leq p,\ 1 \leq i \leq N,\ x_l \in X_i,\ b_r(q_l) = 1 \qquad (16)$$

In this case the bucket B_r, for each bit $0 \leq r \leq p$, will contain the variables $y_{1,r}, y_{2,r}, \ldots, y_{N,r}$. Bucket B_r will also contain a 1 constant if $b_r(T) = 1$.

The formula $PW(P, \mathcal{X})$ is defined as the conjunction of (13), (14) and (16). Some considerations can be taken into account on Clauses (16) in order to optimize the encoding:

– If there is no $x_l \in X_i$ such that $b_r(q_l) = 1$, and therefore the variable $y_{i,r}$ does not appear in any clause of (16), then this variable is not created nor included in any bucket.

– If there is only one variable $x_l \in X_i$ such that $b_r(q_l) = 1$, then the variable $y_{i,r}$ is the variable x_l itself, and Clause (16) is not added for $y_{i,r}$.
– Otherwise, $y_{i,r}$ is indeed a fresh variable and Clause (16) is added.

Figure 3 contains a circuit representation of $PW(P, \mathcal{X})$.

The GGPW encoding is defined as

$$PW(P, \mathcal{X}) \wedge \overline{w} \tag{17}$$

where $PW(P, \mathcal{X})$ encodes ϕ with a totalizer, and ψ with an adder of unary numbers. Similarly as in the other newly introduced encodings, given an assignment that satisfies an AMO constraint over each $X_i \in \mathcal{X}$, this encoding represents the PB constraint $\sum_{i=1}^{n} q_i x_i < K$ in a more compact way.

The GPW encoding introduces $O(n \log(n) \log(q_{max}))$ fresh variables and $O(n^2 \log(n) \log(q_{max}))$ clauses, while the GGPW encoding introduces $O(N \log(N) \log(q_{max}))$ fresh variables and $O(N^2 \log(N) \log(q_{max}))$ clauses, where $q_{max} = \max_{i=1}^{n} q_i$. This follows from the fact that a totalizer ϕ with n input variables requires $O(n \log(n))$ auxiliary variables and $O(n^2 \log(n))$ clauses, and an adder ψ of unary numbers with n input variables requires $O(n)$ auxiliary variables and $O(n^2)$ clauses; see [7].

Lemma 3. *Let $\mathcal{P}$ be a PB(AMO) of the form $P \wedge M_1 \wedge \cdots \wedge M_m$, with P of the form $\sum_{i=1}^{n} q_i x_i < K$, and $\mathcal{X}$ be a partition of the variables of P such that $M_1 \wedge \cdots \wedge M_m \models \sum_{x_{i_j} \in X_i} x_{i_j} \leq 1$, for all $X_i \in \mathcal{X}$. The conjunction of $GGPW(P, \mathcal{X})$ with an encoding of $M_1 \wedge \cdots \wedge M_m$ is an encoding of $\mathcal{P}$.*

In [7] it is shown that the GPW encoding does not UP-maintain GAC. As stated earlier, a PB(AMO) constraint with only AMO constraints of one variable is indeed a PB constraint. In this case the GGPW and GPW encodings would be identical. Therefore, the GGPW encoding does neither UP-maintain GAC.

Binary Merger (BM). The *Binary Merger* (BM) encoding was introduced in [18]. This encoding is essentially another implementation of the GPW encoding, in which the formulas ϕ and ψ are respectively implemented using sorters and odd-even mergers [4]. This way, the BM encoding is asymptotically smaller in the number of clauses and slightly bigger in the number of variables than the GPW encoding. The BM encoding can be generalized to deal with PB(AMO) constraints in the same way as GPW encoding. However, in our experiments we have not observed significant differences between the BM and GPW based encodings, and therefore we do not provide detailed results for BM encoding.

4   Normalization of PB(AMO) Constraints

Most existing encodings of PB constraints to SAT are designed for constraints of the form $\sum_{i=i}^{n} q_i x_i \leq K$, with non-negative q_i, since other cases can be easily transformed to this one. Also the encoding that we have presented in Sect. 3

requires this normalization. The usual way of getting rid of negative coefficients [12] is by using the equality $x = 1 - \overline{x}$, e.g. $-2x_1 + 6x_2 \leq 5 \equiv 2\overline{x_1} + 6x_2 \leq 7$. Then, if we want to encode a constraint of the form $\sum_{i=i}^{n} q_i x_i \geq K$, we can simply replace it by $-\sum_{i=i}^{n} q_i x_i \leq -K$ and get rid of the negative coefficients. However, this rewriting might not be applicable to PB(AMO) constraints. Consider the PB(AMO) constraint $P \wedge x_1 + x_2 + x_3 \leq 1 \wedge x_4 + x_5 + x_6 \leq 1$, with $P : -x_1 - 3x_2 - 4x_3 - 2x_4 - 3x_5 - 5x_6 \leq -6$. If we remove the negative coefficients we obtain $P' : \overline{x_1} + 3\overline{x_2} + 4\overline{x_3} + 2\overline{x_4} + 3\overline{x_5} + 5\overline{x_6} \leq 12$. Notice that $x_1 + x_2 + x_3 \leq 1$ (similarly $x_4 + x_5 + x_6 \leq 1$) no longer impose AMO constraints over the literals of P', and therefore $P' \wedge x_1 + x_2 + x_3 \leq 1 \wedge x_4 + x_5 + x_6 \leq 1$ is not a PB(AMO) constraint. We could still use any existing PB encoding to encode P' without taking the AMO constraints into account, but we would not be using the simplification potential of PB(AMO) constraints.

We present another rewriting procedure to get rid of negative coefficients, which does not require to negate the literals of the original PB constraint, and hence still allows us to take into account the AMO constraints. Moreover, this procedure let choose the polarity in the PB of any variable by using $x = 1 - \overline{x}$, even if the substitution introduces a negative coefficient, because it will be dealt with in the rewriting. Hence it is possible to make use not only of AMOs between the variables of the PB, but also of AMOs between literals with any polarity.

The first step is, for each AMO constraint of the form $x_{i_1} + \cdots + x_{i_l} \leq 1$, define a fresh variable y_i as $y_i \leftrightarrow \overline{x_{i_1}} \wedge \cdots \wedge \overline{x_{i_l}}$. Then, all the auxiliary y_i variables can be included in the PB constraint with coefficient 0. In previous example, from:

$$x_1 - 3x_2 - 4x_3 - 2x_4 - 3x_5 - 5x_6 \leq -6 \wedge x_1 + x_2 + x_3 \leq 1 \wedge x_4 + x_5 + x_6 \leq 1$$

we get:
$$0y_1 - x_1 - 3x_2 - 4x_3 + 0y_2 - 2x_4 - 3x_5 - 5x_6 \leq -6$$
$$\wedge \, x_1 + x_2 + x_3 \leq 1 \, \wedge \, x_4 + x_5 + x_6 \leq 1$$
$$\wedge \, (y_1 \leftrightarrow \overline{x_1} \wedge \overline{x_2} \wedge \overline{x_3}) \wedge (y_2 \leftrightarrow \overline{x_4} \wedge \overline{x_5} \wedge \overline{x_6})$$

After that, for each AMO constraint $x_{i_1} + \cdots + x_{i_l} \leq 1$, choose any integer I such that $I_i \geq -q_{i_j}$, for all $1 \leq j \leq l$. We have that $x_{i_1} + \cdots + x_{i_l} \leq 1 \wedge y_i \leftrightarrow \overline{x_{i_1}} \wedge \cdots \wedge \overline{x_{i_l}} \models y_i + x_{i_1} + \cdots + x_{i_l} = 1$, and therefore $I_i y_i + I_i x_{i_1} + \cdots + I_i x_{i_l} = I_i$. By adding these equalities to the PB constraint, all the negative coefficients become non-negative, due to the values of I_i that we have chosen. The size of the constraint will not increase if we choose $I = -\min_{i=1}^{n}(q_i)$, since we will be cancelling at least one coefficient of each AMO. In our example, we get:

$$(4+0)y_1 + (4-1)x_1 + (4-3)x_2 + (4-4)x_3$$
$$+ (5+0)y_2 + (5-2)x_4 + (5-3)x_5 + (5-5)x_6 \leq -6 + 4 + 5$$
$$\wedge \, x_1 + x_2 + x_3 \leq 1 \, \wedge \, x_4 + x_5 + x_6 \leq 1$$
$$\wedge \, (y_1 \leftrightarrow \overline{x_1} \wedge \overline{x_2} \wedge \overline{x_3}) \wedge (y_2 \leftrightarrow \overline{x_4} \wedge \overline{x_5} \wedge \overline{x_6})$$

that is:
$$4y_1 + 3x_1 + x_2 + 5y_2 + 3x_4 + 2x_5 \leq 3$$
$$\wedge\, x_1 + x_2 + x_3 \leq 1 \,\wedge\, x_4 + x_5 + x_6 \leq 1$$
$$\wedge\, (y_1 \leftrightarrow \overline{x_1} \wedge \overline{x_2} \wedge \overline{x_3}) \,\wedge\, (y_2 \leftrightarrow \overline{x_4} \wedge \overline{x_5} \wedge \overline{x_6})$$

Finally, since $y_i + x_{i_1} + \cdots + x_{i_l} = 1 \models y_i + x_{i_1} + \cdots + x_{i_l} \leq 1$, we can use any of the presented PB(AMO) encodings, for instance GSWC:

$$GSWC(4y_1 + 3x_1 + x_2 + 5y_2 + 3x_4 + 2x_5 \leq 3, \{\{y_1, x_1, x_2\}, \{y_2, x_4, x_5\}\})$$
$$\wedge\, x_1 + x_2 + x_3 \leq 1 \,\wedge\, x_4 + x_5 + x_6 \leq 1$$
$$\wedge\, (y_1 \leftrightarrow \overline{x_1} \wedge \overline{x_2} \wedge \overline{x_3}) \,\wedge\, (y_2 \leftrightarrow \overline{x_4} \wedge \overline{x_5} \wedge \overline{x_6})$$

5   Experiments

In this section we report a clean comparison between the different encodings for PB(AMO) constraints, and also between those and the classical encodings for PB constraints. For this purpose, we use benchmark sets of problems consisting on conjunctions of AMO constraints and PB constraints. Each instance is defined by four parameters: L is the number of PB constraints, N is the number of AMO constraints, M is the number of Boolean variables in each AMO constraint, and Q is the maximum coefficient of a variable in a PB constraint. The variables of the AMO constraints will be disjoint, so there is a total of $n = N \cdot M$ Boolean variables in each instance. The PB constraints contain all n variables. The j-th variable in the i-it AMO constraint is named $x_{i,j}$. The coefficients in the PB constraints are generated uniformly and independently at random in the range $[1, Q]$. The resulting instance has the following constraints:

$$\sum_{i=1}^{N} \sum_{j=1}^{M} q_{i,j,k} \cdot x_{i,j} \leq K_k \qquad\qquad 1 \leq k \leq L \qquad\qquad (18)$$

$$\sum_{j=1}^{M} x_{i,j} \leq 1 \qquad\qquad 1 \leq i \leq N \qquad\qquad (19)$$

$$\sum_{j=1}^{M} x_{i,j} \geq 1 \qquad\qquad 1 \leq i \leq N \qquad\qquad (20)$$

The conjunction of PB and AMO constraints (18) and (19) is not a hard problem, since a trivial solution is to set all the variables $x_{i,j}$ to 0. For this reason we add *at-least-one* Constraints (20), which require that at least one variable in each AMO group is set to true. Essentially, this set of constraints is a decision version of the Multi-Choice Multidimensional Knapsack Problem (MMKP), which is NP-complete. We have generated the benchmarks using the MMKP instance generator from [14].

We provide three different datasets with different parameters, with the aim of showing which kind of PB constraints are better suited for the different encodings. The instances in a dataset are distributed in families, and every family has values of K_k randomly distributed around a different mean in the range $[1, M \cdot Q]$. The values of K_k are proportional to the values of the coefficients, because otherwise the PB constraints would be trivially satisfied or unsatisfied. We choose different values of K_k to ensure that in the datasets there are instances of different hardness, and that approximately half the instances are satisfiable:

Set1 100 families of 5 instances, with $L = 10$, $N = 15$, $M = 10$, $Q = 1000$. The families have increasing K_k values from family 1 (capacities of about 1000) to family 100 (capacities of about 14000).

Set2 100 families of 5 instances, with $L = 10$, $N = 15$, $M = 10$, $Q = 60$. The families have increasing K_k values from family 1 (capacities of about 100) to family 100 (capacities of about 800).

Set3 20 families of 20 instances, with $L = 50$, $N = 15$, $M = 5$, $Q = 10$. The values of K_k increase in each family, ranging between 65 and 100.

All the instances have been encoded to SAT using the presented encodings for PB(AMO) constraints. The AMO Constraints (19) have been encoded with the well-known UP-maintaining GAC encoding referred as *regular* in [3] and *ladder* in [13], which only introduces a number of clauses and variables linear in the number of variables of the AMO constraint. Constraints (20) have been encoded with clauses $x_{i,1} \lor \cdots \lor x_{i,M}$, for all $1 \leq i \leq N$. We have used the Glucose 4.1 SAT solver [5] to solve the instances, on a 8 GB, 3.10 GHz Intel® Xeon® E3-1220v2.

The results are contained in Table 1. The evaluated encodings are the ones introduced in this paper, and their counterpart original ones. For completeness we also report results on the MDD-based encoding from [10] (MDD), and its version without taking AMOs into account (BDD), from [1]. Both MDD and BDD UP-maintain GAC. For each encoding we report solving times on each dataset and the average size required to encode a PB(AMO) constraint. The size results do not include the number of variables and clauses introduced by AMO constraints (19), which is the same for all encodings and negligible.

In summary, it can be observed a dramatic decrease in size, and hence in generation time, as well a significant decrease in solving time, in all the generalized encodings for PB(AMO)s w.r.t. the original encodings for PBs. In most of cases the size and solving time reduction is of one order of magnitude. Even in Set3, which is the one with smallest AMO constraints (only 5 variables per AMO) the reduction is notable. The GPW encoding is the smallest, and the one which is less reduced when using the AMO constraints.

In Set1, which contains instances with large coefficients, the best approach is GGPW. Although this encoding do not UP-maintain GAC, the number of clauses and variables is remarkably small compared to the other encodings, whose size is proportional to K. In particular this dataset is prohibitive for GT and GGT encodings, which require a number of clauses quadratic in the value of K.

Table 1. From left to right: first quartile (Q1), median (med) and third quartile (Q3) of solving times (in seconds); average solving time in seconds (avg) counting time outs as 600 s; number of instances that timed out before being solved (t.o.); in thousands, average number of auxiliary variables (v.) and clauses (cl.) needed to encode one of the Constraints (18); in seconds, average time required to generate the CNF formula of an instance (g.t.). Solving time out (t.o.) is set to 600 s. Long dash (—) means that the resulting formulas are too large and their generation either run out of memory or did not finish in less than 600 s (in these instances we have been able to identify constraints requiring 33,000,000 clauses).

	enc.	Q1	med	Q3	avg	t.o.	v.	cl.	g.t.	enc.	Q1	med	Q3	avg	t.o.	v.	cl.	g.t.
Set1	BDD	14.00	17.59	t.o.	219	158	857	1714	35.6	MDD	3.89	14.78	73	131	87	25	266	3.71
	SWC	10.51	14.12	t.o.	199	144	1100	2177	17.2	GSWC	4.50	5.92	277	158	112	105	1076	10.01
	GT	—	—	—	—	—	—	—	—	GGT	—	—	—	—	—	—	—	—
	GPW	0.93	0.97	23	114	85	5.9	77	0.8	GGPW	0.04	0.04	5.54	93	67	1.0	4.4	0.05
Set2	BDD	4.29	5.65	133	141	96	57	115	2.0	MDD	0.21	0.41	1.42	74	53	2.1	21	0.28
	SWC	4.10	5.41	138	140	95	68	135	1.3	GSWC	0.58	0.62	1.09	71	52	6.4	66	0.62
	GT	5.33	6.94	182	154	110	10	1640	18.0	GGT	2.42	8.83	53	132	95	1.9	120	1.53
	GPW	0.46	0.48	11	108	77	3.5	42	0.4	GGPW	0.02	0.03	3.36	89	65	0.6	2.5	0.03
Set3	BDD	215	t.o.	t.o.	423	218	4.8	9.6	0.7	MDD	16.2	78.6	525	221	97	0.4	2.6	0.17
	SWC	247	t.o.	t.o.	429	227	6.0	12	0.6	GSWC	17.5	87.3	597	225	100	1.1	6.5	0.32
	GT	240	t.o.	t.o.	427	223	1.3	31	1.6	GGT	70.8	281	t.o.	322	152	0.4	4.6	0.25
	GPW	172	t.o	t.o	415	229	0.8	5.1	0.3	GGPW	133	t.o	t.o	407	226	0.3	1.2	0.07

Set2 is similar to Set1 but it contains instances with small coefficients. In this case, the best approaches are MDD and GSWC, whose sizes are reasonably smaller than the ones in Set1. The GGT encoding introduces the largest number of clauses in this dataset, and has the worst time performance.

Instances in Set3 contain more PB constraints than the other datasets, and the values of K are distributed around the transition value from unsatisfiable instances to satisfiable instances. We have observed empirically that it is in this transition where the instances become harder. In this case, GGPW has the worst time performance although it still has a smaller size than the other encodings. This may be because GGPW is the only one which does not UP-maintain GAC.

6   Conclusions and Further Work

In this work we have provided different SAT encodings of PB(AMO)s, i.e., conjunctions of PB and AMO constraints. These new encodings have been defined by generalizing existing state-of-the-art SAT encodings of PB constraints, in a way that the size is highly reduced thanks to assuming that the AMO constraints are already enforced. Moreover, the propagation properties of the original encodings are preserved in the new ones. Our results show that all the new encodings are dramatically smaller and more efficient than their counterpart PB encodings. We have observed size reductions of an order of magnitude and solving time improvements of 1 or 2 orders of magnitude in many cases.

We have also shown that there is no best encoding for PB(AMO)s but it depends on the characteristics of the instances at hand. The datasets that we provide expose some strengths and weaknesses of the different encodings.

The PB(AMO) constraints have application in a large number of problems, and our new encodings should be taken into account as an alternative to tackle them. It is matter of a future work to see how this SAT-based approach compares with other solving approaches in different specific domains. Finally, other new encodings for PB(AMO)s might be studied, either based on existing ones or brand new ones.

Acknowledgements. Work supported by grants TIN2015-66293-R (MINECO/ FEDER, UE), and *Ayudas para Contratos Predoctorales 2016* (grant number BES-2016-076867, funded by MINECO and co-funded by FSE).

References

1. Abío, I., Nieuwenhuis, R., Oliveras, A., Rodríguez-Carbonell, E., Mayer-Eichberger, V.: A new look at BDDs for Pseudo-Boolean constraints. J. Artif. Intell. Res. **45**, 443–480 (2012)
2. Achterberg, T., Bixby, R.E., Gu, Z., Rothberg, E., Weninger, D.: Presolve reductions in mixed integer programming. Zuse Institute Berlin (2016)
3. Ansótegui, C., Manyà, F.: Mapping problems with finite-domain variables to problems with Boolean variables. In: Hoos, H.H., Mitchell, D.G. (eds.) SAT 2004. LNCS, vol. 3542, pp. 1–15. Springer, Heidelberg (2005). https://doi.org/10.1007/11527695_1
4. Asín, R., Nieuwenhuis, R., Oliveras, A., Rodríguez-Carbonell, E.: Cardinality networks: a theoretical and empirical study. Constraints **16**(2), 195–221 (2011). https://doi.org/10.1007/s10601-010-9105-0
5. Audemard, G., Simon, L.: On the glucose SAT solver. Int. J. Artif. Intell. Tools **27**(1), 1–25 (2018)
6. Bailleux, O., Boufkhad, Y.: Efficient CNF encoding of Boolean cardinality constraints. In: Rossi, F. (ed.) CP 2003. LNCS, vol. 2833, pp. 108–122. Springer, Heidelberg (2003). https://doi.org/10.1007/978-3-540-45193-8_8
7. Bailleux, O., Boufkhad, Y., Roussel, O.: New encodings of Pseudo-Boolean constraints into CNF. In: Kullmann, O. (ed.) SAT 2009. LNCS, vol. 5584, pp. 181–194. Springer, Heidelberg (2009). https://doi.org/10.1007/978-3-642-02777-2_19
8. Basnet, C., Wilson, J.: Heuristics for determining the number of warehouses for storing non-compatible products. Int. Trans. Oper. Res. **12**(5), 527–538 (2005)
9. Bofill, M., Coll, J., Suy, J., Villaret, M.: An efficient SMT approach to solve MRCPSP/max instances with tight constraints on resources. In: Beck, J.C. (ed.) CP 2017. LNCS, vol. 10416, pp. 71–79. Springer, Cham (2017). https://doi.org/10.1007/978-3-319-66158-2_5
10. Bofill, M., Coll, J., Suy, J., Villaret, M.: Compact MDDs for Pseudo-Boolean constraints with at-most-one relations in resource-constrained scheduling problems. In: Proceedings of the Twenty-Sixth International Joint Conference on Artificial Intelligence - IJCAI 2017, pp. 555–562. ijcai.org (2017)
11. De Vries, S., Vohra, R.V.: Combinatorial auctions: a survey. INFORMS J. Comput. **15**(3), 284–309 (2003)

12. Eén, N., Sorensson, N.: Translating Pseudo-Boolean constraints into SAT. J. Satisf. Boolean Model. Comput. **2**, 1–26 (2006)
13. Gent, I.P., Nightingale, P.: A new encoding of all different into SAT. In: 3rd International Workshop on Modelling and Reformulating Constraint Satisfaction - CP 2004, pp. 95–110 (2004)
14. Han, B., Leblet, J., Simon, G.: Hard multidimensional multiple choice knapsack problems, an empirical study. Comput. Oper. Res. **37**(1), 172–181 (2010)
15. Hölldobler, S., Manthey, N., Steinke, P.: A compact encoding of Pseudo-Boolean constraints into SAT. In: Glimm, B., Krüger, A. (eds.) KI 2012. LNCS (LNAI), vol. 7526, pp. 107–118. Springer, Heidelberg (2012). https://doi.org/10.1007/978-3-642-33347-7_10
16. Joshi, S., Martins, R., Manquinho, V.: Generalized totalizer encoding for Pseudo-Boolean constraints. In: Pesant, G. (ed.) CP 2015. LNCS, vol. 9255, pp. 200–209. Springer, Cham (2015). https://doi.org/10.1007/978-3-319-23219-5_15
17. Ma, P.R., Lee, E.Y.S., Tsuchiya, M.: A task allocation model for distributed computing systems. IEEE Trans. Comput. **31**(1), 41–47 (1982)
18. Manthey, N., Philipp, T., Steinke, P.: A more compact translation of Pseudo-Boolean constraints into CNF such that generalized arc consistency is maintained. In: Lutz, C., Thielscher, M. (eds.) KI 2014. LNCS (LNAI), vol. 8736, pp. 123–134. Springer, Cham (2014). https://doi.org/10.1007/978-3-319-11206-0_13
19. Miller, C.E., Tucker, A.W., Zemlin, R.A.: Integer programming formulation of traveling salesman problems. J. ACM **7**(4), 326–329 (1960)
20. Philipp, T., Steinke, P.: PBLib – a library for encoding Pseudo-Boolean constraints into CNF. In: Heule, M., Weaver, S. (eds.) SAT 2015. LNCS, vol. 9340, pp. 9–16. Springer, Cham (2015). https://doi.org/10.1007/978-3-319-24318-4_2
21. Pisinger, D.: Budgeting with bounded multiple-choice constraints. Eur. J. Oper. Res. **129**(3), 471–480 (2001)
22. Watson, R.: Packet networks and optimal admission and upgrade of service level agreements: applying the utility model. MA Sc. Ph.D. thesis, Department of ECE, University of Victoria (2001)

A Constraint Programming Approach to Electric Vehicle Routing with Time Windows

Kyle E. C. Booth[✉] and J. Christopher Beck

Department of Mechanical and Industrial Engineering, University of Toronto, Toronto, Ontario M5S 3G8, Canada
{kbooth,jcb}@mie.utoronto.ca

Abstract. The Electric Vehicle Routing Problem with Time Windows (EVRPTW) extends traditional vehicle routing to address the recent development of electric vehicles (EVs). In addition to traditional VRP problem components, the problem includes consideration of vehicle battery levels, limited vehicle range due to battery capacity, and the presence of vehicle recharging stations. The problem is related to others in emissions-conscious routing such as the Green Vehicle Routing Problem (GVRP). We propose the first constraint programming (CP) approaches for modeling and solving the EVRPTW and compare them to an existing mixed-integer linear program (MILP). Our initial CP model follows the alternative resource approach previously applied to routing problems, while our second CP model utilizes a single resource transformation. Experimental results on various objectives demonstrate the superiority of the single resource transformation over the alternative resource model, for all problem classes, and over MILP, for the majority of medium-to-large problem classes. We also present a hybrid MILP-CP approach that outperforms the other techniques for distance minimization problems over long scheduling horizons, a class that CP struggles with on its own.

Keywords: Electric Vehicle Routing · Green Vehicle Routing · Constraint programming · Mixed-integer linear programming · Optimization

1 Introduction

Fueled by emission regulations, government subsidies, and the benefits of a more eco-friendly image, electric vehicle (EV) utilization in logistics has seen significant growth in recent years [9]. Outside of logistics, EVs have experienced a growing adoption within the consumer automotive industry [28] and have shown promise in car sharing pilot projects [26]. While not currently cost-competitive with internal combustion engines due to high acquisition costs and limited operational range [14], the benefits of EVs coupled with an increasing number of socially and environmentally-aware consumers are driving the adoption of the

© Springer Nature Switzerland AG 2019
L.-M. Rousseau and K. Stergiou (Eds.): CPAIOR 2019, LNCS 11494, pp. 129–145, 2019.
https://doi.org/10.1007/978-3-030-19212-9_9

technology. The industry has also seen significant investment in the development of required recharging infrastructure. As with traditional fleets, the case for EVs can be significantly bolstered via effective route planning.

The vehicle routing literature has recently addressed this emerging technology through the introduction of the Electric Vehicle Routing Problem with Time Windows (EVRPTW) [29], building on previous work conducted on green logistics, including the Green Vehicle Routing Problem (GVRP) [12]. The problem involves routing a fleet of vehicles to satisfy customer demands while adhering to the battery capacity and range of the fleet EVs. The EVRPTW literature has seen considerable research activity, including the development of sophisticated exact approaches [7,10], metaheuristics [13,29], and the introduction of increasingly rich problem definitions driven by real-world logistics use cases [17,27]. There have been, however, no efforts thus far to explore the use of constraint programming (CP) to model and solve the problem.

Recognizing EV routing as a strategic area for methodological development, we investigate the use of monolithic (i.e., non-decomposed) MILP and CP models to solve the problem. The contributions of this paper are as follows:

  i. We propose the first CP approaches for the EVRPTW.
 ii. We introduce a single resource transformation for CP formulations that use optional interval, sequence, and cumulative function expression variables. The transformation significantly extends the size of problems that can be solved with CP, and can be applied to other homogeneous VRP and multi-machine scheduling problems.
iii. We demonstrate, through empirical evaluation, that our single resource CP approach significantly outperforms the alternative resource CP model, and outperforms MILP for nearly all medium-to-large problem classes.
 iv. Following the observation that MILP excels at quickly finding high quality solutions to distance minimization problems with large scheduling horizons, we propose a hybrid MILP-CP technique that outperforms the individual approaches on this problem class.

This paper is organized as follows. Section 2 defines the EVRPTW problem and presents an existing MILP model. Section 3 details related work for the problem studied. Section 4 presents two CP models, alternate modeling strategies, and an initial empirical evaluation with accompanying analysis. Section 5 illustrates a hybrid CP-MILP approach, motivated by the strength of MILP for larger, long horizon problems, and presents hybrid experimental results with detailed analysis. Finally, Sect. 6 provides concluding remarks.

2 Problem Definition

The Electric Vehicle Routing Problem with Time Windows (EVRPTW) is a static optimization problem that aims to route a fleet of electric vehicles to satisfy customer requests [29]. Following existing notation, we let $V' = V \cup F'$ be the set of N vertices where $V = \{v_1, \ldots, v_n\}$ is the set of customer requests, F

is the set of recharging stations, and $F' = \{v_{n+1}, \ldots, v_N\}$ is the set of augmented recharging stations that includes dummy vertices to allow multiple visits to each of the stations in F. We let vertices v_0 and v_{N+1} correspond to start and end instances of the vehicle depot, where each vehicle starts and ends. Sets with depot subscripts include the indicated instances of the depot (i.e., $V'_{N+1} = V' \cup \{v_{N+1}\}$ and $V'_{0,N+1} = V' \cup \{v_0, v_{N+1}\}$). The problem is then defined on a graph with vertices $V'_{0,N+1}$ and undirected arcs $A = \{(i,j)|i,j \in V'_{0,N+1}, i \neq j\}$. Each arc is assigned a distance, travel time, and energy consumption, d_{ij}, t_{ij}, and $h \cdot d_{ij}$, respectively, where h is a constant energy consumption rate. Vehicles are initially positioned at the depot and start with maximum capacity C, while customer vertices, $i \in V$, are assigned a positive demand, $q_i \leq C$, and a time window, $[e_i, l_i]$.[1] The time window of the start depot is $[0, 0]$, and the end depot is $[H, H]$, where H is the problem horizon. Each recharging station has a time window of the entire horizon, namely $[e_i = 0, l_i = H], \forall i \in F'$. Customer vertices, $i \in V$, have a service time s_i. The depot instances each have a null service time and the service time at recharging stations is a variable. Vehicles have maximum battery capacity Q and recharge linearly at rate g. The problem then minimizes an objective function, often a combination of fleet size and travel distance.

An existing two-index MILP model from the literature [29] is detailed by Eq. (1) through (12). Binary variable x_{ij} is 1 if arc $(i,j) \in A$ is traveled and 0 otherwise. Continuous variables τ_i, u_i, and y_i represent the arrival time, remaining cargo, and remaining energy, respectively, at vertex $i \in V'_{0,N+1}$. This formulation assumes an unlimited number of homogeneous vehicles are available and only permits full vehicle recharges (i.e., if a vehicle visits a recharge station vertex, the service time is the difference between its maximum energy capacity and current energy level, divided by the recharge rate). The augmented recharge station set, F', is constructed such that the number of dummy vertices associated with each recharge station, n_f, represents the number of times the associated recharge station can be visited across all vehicles (with $|F'| = n_f \cdot |F|$). Following the literature, n_f is set to be relatively small, to reduce the network size, but large enough to not restrict multiple beneficial visits [12]. We note that heuristically choosing a value for n_f, as in the literature, can potentially remove optimal solutions.

Objective (1) details the weighted objective function, where $\alpha \in [0, 1]$ identifies the emphasis on fleet size minimization and $\beta \in [0, 1]$ on travel distance minimization. Constraint (2) ensures each customer request is satisfied, while Constraint (3) restricts each recharge station in the augmented set to be visited at most once (due to the augmented vertices, each recharge station can be visited at most n_f times). Constraint (4) enforces the flow for non-depot nodes. Constraints (5)–(6) prevent the formation of subtours, with disjunctive constant $M = (l_0 + g \cdot Q)$. Constraint (7) ensures demand fulfillment at customer vertices and Constraints (8)–(9) constrain energy levels to be feasible. Constraint (10) requires customer visits to satisfy the time windows and Constraints (11)–(12) identify binary and continuous variable domains.

---

[1] Service must start within the time window.

$$\min \ \alpha \sum_{j \in V'} x_{0j} + \beta \sum_{i \in V'_0} \sum_{j \in V'_{N+1}, i \neq j} d_{ij} x_{ij} \tag{1}$$

$$\text{s.t.} \ \sum_{j \in V'_{N+1}, i \neq j} x_{ij} = 1 \qquad\qquad \forall i \in V, \tag{2}$$

$$\sum_{j \in V'_{N+1}, i \neq j} x_{ij} \leq 1 \qquad\qquad \forall i \in F', \tag{3}$$

$$\sum_{i \in V'_{N+1}, i \neq j} x_{ji} - \sum_{i \in V'_0, i \neq j} x_{ij} = 0 \qquad\qquad \forall j \in V', \tag{4}$$

$$\tau_i + (t_{ij} + s_i) x_{ij} - l_0 (1 - x_{ij}) \leq \tau_j \qquad \forall i \in V_0, j \in V'_{N+1}, i \neq j, \tag{5}$$

$$\tau_i + t_{ij} x_{ij} + g(Q - y_i) - M(1 - x_{ij}) \leq \tau_j \qquad \forall i \in F', j \in V'_{N+1}, i \neq j, \tag{6}$$

$$0 \leq u_j \leq u_i - q_i x_{ij} + C(1 - x_{ij}) \qquad \forall i \in V'_0, j \in V'_{N+1}, i \neq j, \tag{7}$$

$$0 \leq y_j \leq y_i - (h \cdot d_{ij}) x_{ij} + Q(1 - x_{ij}) \qquad \forall j \in V'_{N+1}, i \in V, i \neq j, \tag{8}$$

$$0 \leq y_j \leq Q - (h \cdot d_{ij}) x_{ij} \qquad \forall j \in V'_{N+1}, i \in F'_0, i \neq j, \tag{9}$$

$$e_j \leq \tau_j \leq l_j \qquad\qquad \forall j \in V'_{0,N+1}, \tag{10}$$

$$x_{ij} \in \{0, 1\} \qquad\qquad \forall i \in V'_0, j \in V'_{N+1}, i \neq j, \tag{11}$$

$$\tau_0 = 0, u_0 = C, y_0 = Q. \tag{12}$$

Two Index Formulation. With the exception of x_{ij}, the variables are continuous, modeling visit time, load, and energy level via sequencing constraints. The model represents multiple vehicles by relaxing the unitary out- and in-flow on the start and end depot vertices, respectively. This modeling technique is effective as it does not multiply the number of variables by the number of (symmetric) vehicles.

Problem Variants. The fixed fleet variant can be modeled with the inclusion of a constraint of the form: $\sum_{j \in V'} x_{0j} \leq m$, where m is the fleet size. A variant with heterogeneous vehicles can be modeled for k different vehicle types by adding an index for the vehicle type to the arc, cargo, and energy consumption variables (i.e., x_{ij}^k, u_i^k, and y_i^k), with similar adjustments to the parameters [17]. Additional problem variants, such as partial recharges, can also be considered through the inclusion of additional constraints/variables [7,10].

3 Related Work

Research on energy-aware, environmentally conscious vehicle routing is a relatively new area with a flurry of research in recent years [24]. In response to a growing commitment within the United States to investigate alternative fuel sources, the vehicle routing literature introduced the GVRP, detailing a MILP formulation, construction heuristic, and clustering algorithm [12]. Since the introduction of the GVRP, EV routing has grown dramatically, with initial EVRPTW work for homogeneous fleets and full re-charges including a MILP model and a hybrid variable neighborhood search/tabu search solution technique [29]. Subsequent research was developed with approaches for heterogeneous vehicles [17], and partial recharging problem variants [10,13,19]. Recent work has also been conducted on modeling non-linear energy consumption [27] as well as incorporating richer, industry-driven problem constraints [1].

Although the literature for mathematical programming-based approaches (e.g., MILP, branch-and-price, branch-and-cut) for the GVRP and the EVRPTW is abundant [10, 12, 17, 29], to the authors' knowledge, the use of CP for solving these problems has not yet been investigated. While the performance achieved by sophisticated branch-and-price-and-cut algorithms for EVRPTW [10] is unlikely to be surpassed by monolithic modeling methods utilizing off-the-shelf solvers, the practicality and flexibility of such approaches, including MILP and CP, often translate to more widespread adoption.

In general vehicle routing, CP has been offered as an alternative to mathematical programming approaches for quite some time [2, 30]. Recent applications include work on the multiple traveling salesman problem [31], team orienteering [15], dynamic dial-a-ride routing [3], bike share balancing [11], joint vehicle and crew routing [23], and patient transportation [8, 25], though these efforts do not consider fuel constraints. While CP has not been explicitly proposed for GVRP nor EVRPTW as of yet, previous work on snow plow routing [20] and robot task allocation and scheduling [5, 6] propose CP models with consideration for energy consumption and replenishment.

4   Constraint Programming Approaches

In this section we present two CP formulations for the EVRPTW. Our models are posed as scheduling formulations with optional activities [21, 22]. As is becoming increasingly common in CP-based approaches [5, 8, 25], the proposed models make use of three primary decision variable types, namely: optional interval variables, sequence variables, and cumulative function expressions.

Optional Interval Variables. Formally, optional interval variables are decision variables whose possible values are a convex interval: $\{\perp\} \cup \{[s, \epsilon) | s, \epsilon \in \mathbb{Z}, s \le \epsilon\}$, where s and ϵ are the start and end values of the interval and $\perp$ is a special value indicating the variable is not present in the solution. The presence (binary), start time, and length of an optional interval variable, var, can be expressed within a CP model using $\text{Pres}(var)$, $\text{Start}(var)$, and $\text{Length}(var)$, respectively. We use the notation $\text{optIntervalVar}(p, [s, \epsilon])$ to define these variables in our models, where p is the processing time of the task (and can be variable). Model constraints are only enforced over present interval variables.

Sequence Variables. This variable type is useful for expressing model constraints over a permutation of present (i.e., $\text{Pres}(var) = 1$) interval variables. Given the definition of a sequence variable, π, various constraints can be expressed, including those on the interval variable previous to var in the sequence, $\text{Prev}_\pi(var)$, and temporal constraints such as the $\text{NoOverlap}(\pi)$ constraint, which ensures the interval variables in the sequence do not interfere temporally.

Cumulative Function Expressions. It is often useful to represent the usage of a renewable resource as the sum of individual interval variable contributions over time. Given a cumulative function expression variable, f, we can express impact on the expression using the $f \pm \texttt{StepAtStart}(var, impact)$ expression, specifying that at the start of interval variable var, function f has an increment (or decrement) of $impact$. The constraint $\texttt{AlwaysIn}(f, [s, \epsilon], [min, max])$ ensures that $min \leq f \leq max$ holds for all time points in s to ϵ and a similar constraint $\texttt{AlwaysIn}(f, var, [min, max])$ ensures that $min \leq f \leq max$ holds during the processing of interval variable var. Cumulative expression variables are useful in representing both the vehicle load and energy constraints, and have been used for similar problems [5, 20].

4.1   Alternative Resource Model

Our first CP model follows the traditional alternative resource model for formulating VRPs in CP [8, 20], and, in contrast to the two-index MILP presented in Sect. 2, explicitly represents the vehicles. We define an upper bound on the number of vehicles to be equal to the number of customer requests, $|K| = |V|$, representing the worst-case where each customer is serviced by a separate vehicle. For each customer request, $i \in V$, we introduce a mandatory interval variable, $\bar{x}_i$. We create an optional interval variable, x_i^k, for each vertex, $i \in V'$, for each vehicle, $k \in K$. We also introduce start and end interval variables, x_0^k and x_{N+1}^k, with null duration for each vehicle to represent the depot.

The model considers a set of $|V'_{0,N+1}|$ interval variables and a sequence variable, π^k, for each vehicle $k \in K$. Each interval variable x_i^k, for all $i \in V'_{0,N+1}$, represents the time period in which the vehicle visits i. Thus, expressions $\texttt{StartOf}(x_i^k)$ and $\texttt{EndOf}(x_i^k)$ correspond to the arrival and departure time of vehicle k at location i, respectively. The expression $\texttt{Pres}(x_i^k) = 1$ if vehicle $k \in K$ visits location i (i.e., the interval variable is present in the solution), and 0 otherwise. Sequence variable π^k is defined over the set of interval variables involving vehicle k, and represents the sequence of visits. Vehicle load consumption and energy level consumption/replenishment are modeled with cumulative function expressions. We let C^k and Q^k be cumulative function expressions representing the load and energy level of vehicle $k \in K$ throughout its route.

Our alternative resource CP model is detailed by Eqs. (13)–(27). Objective (13) represents the minimization of fleet size and distance traveled. Constraint (14) ensures that each customer is serviced by one vehicle and Constraint (15) enforces that tasks assigned to a vehicle, represented by the sequence variable π^k, do not interfere temporally, including travel times. Constraints (16)–(17) ensure that the vehicle load does not fall below zero over the planning horizon, represented as a cumulative function expression with negative impact for served customers. Constraints (18)–(19) ensure vehicle energy stays within permissible limits, also represented as a cumulative function expression with negative impact for travel between locations and a positive impact for vehicle recharging. We note that the impact for energy replenishment tasks includes the negative contribution of the travel to the recharge station. Constraint (20) ensures that during a recharge task, the energy of the vehicle is set to its capacity; whenever a vehicle recharges, it does so fully. Constraint (22) ensures each recharge station is used at most n_f times across the fleet, where $F'(i)$ represents all dummy recharge stations associated with real recharge station $i \in F$. Constraint (21) enforces that the start and end depot instances for a vehicle be first and last in the sequence variable for that vehicle, while Constraints (23)–(27) provide the definitions of the interval and sequence variables.

As the alternative resource formulation explicitly represents each vehicle, the number of variables can become unwieldy for larger problems. Specifically, the formulation has $|V| + |K| \cdot |V'_{0,N+1}|$ interval variables, $|K|$ sequence variables, and $2|K|$ cumulative function expression variables.

$$\min \sum_{k \in K} \left(\alpha \mathrm{Pres}(x_0^k) + \beta \sum_{i \in V'_{N+1}} \mathrm{Pres}(x_i^k) \cdot d_{\mathrm{Prev}_{\pi^k}(i),i} \right) \tag{13}$$

$$\text{s.t.} \quad \mathtt{Alternative}(\bar{x}_i, \{x_i^1, \ldots, x_i^{|V|}\}) \qquad\qquad \forall i \in V, \tag{14}$$

$$\mathtt{NoOverlap}(\pi^k, \{t_{ij} : (i,j) \in A\}) \qquad\qquad \forall k \in K, \tag{15}$$

$$C^k = \mathtt{StepAtStart}(x_0^k, C)$$
$$\quad - \sum\nolimits_{i \in V} \mathtt{StepAtStart}(x_i^k, q_i) \qquad\qquad \forall k \in K, \tag{16}$$

$$\mathtt{AlwaysIn}(C^k, [0, H], [0, C]) \qquad\qquad \forall k \in K, \tag{17}$$

$$Q^k = \mathtt{StepAtStart}(x_0^k, Q)$$
$$\quad - \sum\nolimits_{i \in V'_{N+1}} \mathtt{StepAtStart}(x_i^k, h \cdot d_{\mathrm{Prev}_{\pi^k}(i),i})$$
$$\quad + \sum\nolimits_{i \in F'} \mathtt{StepAtStart}(x_i^k, g \cdot \mathtt{Length}(x_i^k)) \qquad\qquad \forall k \in K, \tag{18}$$

$$\mathtt{AlwaysIn}(Q^k, [0, H], [0, Q]) \qquad\qquad \forall k \in K, \tag{19}$$

$$\mathtt{AlwaysIn}(Q^k, x_i^k, [Q, Q]) \qquad\qquad \forall i \in F', k \in K, \tag{20}$$

$$\mathtt{First}(\pi^k, x_0^k), \ \mathtt{Last}(\pi^k, x_{N+1}^k) \qquad\qquad \forall k \in K, \tag{21}$$

$$\sum_{k \in K} \sum_{j \in F'(i)} \mathrm{Pres}(x_j^k) \leq n_f \qquad\qquad \forall i \in F, \tag{22}$$

$$x_i^k : \mathtt{optIntervalVar}([0, Q \cdot g^{-1}], [0, H]) \qquad\qquad \forall i \in F', k \in K, \tag{23}$$

$$x_i^k : \mathtt{optIntervalVar}(s_i, [e_i, l_i]) \qquad\qquad \forall i \in V, k \in K, \tag{24}$$

$$\bar{x}_i : \mathtt{intervalVar}(s_i, [e_i, l_i]) \qquad\qquad \forall i \in V, \tag{25}$$

$$x_0^k : \mathtt{intervalVar}(0, [0, 0]), x_{N+1}^k : \mathtt{intervalVar}(0, [H, H]) \qquad \forall k \in K, \tag{26}$$

$$\pi^k : \mathtt{sequenceVar}(\{x_0^k, \ldots, x_{N+1}^k\}) \qquad\qquad \forall k \in K. \tag{27}$$

Model Strengthening

Cumulative Resource Constraint. Similar to a previous CP formulation for patient transportation [8], we strengthen the baseline formulation with a cumulative resource constraint. We define an auxiliary integer variable representing the number of vehicles in the fleet, $z = \sum_{k \in K} \mathrm{Pres}(x_0^k)$. The cumulative constraint is then $\mathtt{Cumulative}(\bar{x} \cup \{x_i : i \in F'\}, z)$, which expresses that at any time point in the horizon, the total number of customer interval variables, $\bar{x}$, and present recharge interval variables is bounded by the number of vehicles in the fleet.

Symmetry Breaking Constraints. Due to the large number of homogeneous vehicles, the use of symmetry breaking can be effective. We introduce a constraint of the form $\texttt{Pres}(x_0^k) \geq \texttt{Pres}(x_0^{k+1})$, ensuring vehicles are used in a lexicographic order. We then specify that if a vehicle depot task is not present, it cannot be assigned any other activities via $\texttt{Pres}(x_0^k) \geq \texttt{Pres}(x_i^k), \forall i \in V_{N+1}'$.

Energy Expression Tightening. While the energy impact of a customer visit on vehicle energy level is a variable, we can tighten the domain of its impact by reasoning about minimum and maximum travel consumptions to the considered customer location. More specifically, we add the constraint $\min_{i \in V_0'}(h \cdot d_{ij}) \leq \texttt{HeightAtStart}(x_j^k) \leq \max_{i \in V_0'}(h \cdot d_{ij}), \forall j \in V, k \in K$, where the $\texttt{HeightAtStart}(var, f)$ expression evaluates the individual contribution of an interval variable, var, to a cumulative function expression, f.

4.2   Single Resource Model

Our second CP model, inspired by the modeling efficiency of the two-index MILP for homogeneous vehicles, utilizes a single resource transformation to significantly reduce the number of variables. The transformation represents the problem as an interval variable sequence over an augmented horizon and, like the MILP, does not explicitly represent the vehicles. This modeling strategy, while common in MILP models for VRPs, has been rarely used in CP. In previous work on joint vehicle and crew routing, a similar strategy was used to artificially join the end of one route to the beginning of another when using the Circuit global constraint [23], which prevents the formation of subtours among a set of integer variables. However, to our knowledge, the single resource transformation has never been proposed for scheduling-based CP models involving interval, sequence, and cumulative function expression variables. The transformation using these formalisms is challenging as the modeling paradigm does not permit the "resetting" of time as in [23]; we detail how this is accomplished in the remainder of this section. The described transformation can also be applied to homogeneous machine scheduling problems, which we leave to future work.

We visualize the single resource model in Fig. 1. The transformation augments the problem horizon from H to $|V| \cdot H$, generating a horizon for each potential vehicle used. In addition to the start and end depot instances, v_0 and v_{N+1}, we define a set of auxiliary depot instances, $\mathcal{H} = \{v_{N+2}, \ldots, v_{N+|V|}\}$, representing the end depots of the additional horizon segments. We define the notation $V_{0,N+1,\mathcal{H}}' = V_{0,N+1}' \cup \mathcal{H}$ and undirected arcs $A' = \{(i,j) | i, j \in V_{0,N+1,\mathcal{H}}', i \neq j\}$. Similarly, we define $\mathcal{H}_{N+1} = v_{N+1} \cup \mathcal{H}$ and $\mathcal{H}_{0,N+1} = \{v_0, v_{N+1}\} \cup \mathcal{H}$. A depot instance, represented as an interval variable, x_i, is assigned with null duration for $i \in \mathcal{H}_{0,N+1}$. These interval variables have start time σ_i, such that $\sigma_0 = 0, \sigma_{N+1} = H, \sigma_{N+2} = 2H$, and so forth. We then create a mandatory interval variable, x_i, for each customer request, $i \in V$, and an optional interval variable for each recharge station instance in the augmented set, $i \in F'$. Our model uses a single sequence variable, π, defined over the set of all interval variables, and a single cumulative function expression to model vehicle load, $\mathcal{C}$, with

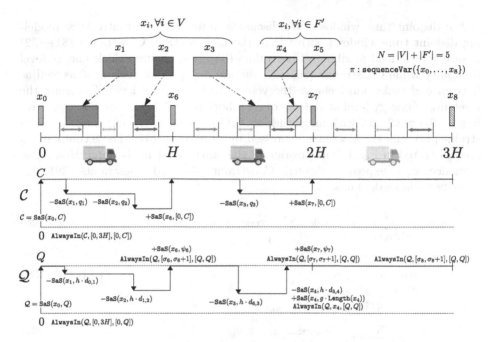

Fig. 1. Single resource transformation for problem with $|V| = 3$ and a single recharge station, $|F| = 1$, with $n_f = 2$ (such that $|F'| = 2$). A horizon segment is created for each potential vehicle and time windows are duplicated. All customer tasks are mandatory with disjoint start time domains and energy tasks (optional) have start time domain of $[0, 3H]$. A cumulative function expression represents vehicle load and energy level (where notation SaS corresponds to StepAtStart in models). Vehicle assignments can then be inferred by the start times of the tasks themselves. The last horizon segment is not used (set as absent).

another for energy level, Q. Additionally, at the start of each end depot instance, $i \in \mathcal{H}_{N+1}$, the state of the vehicle must be reset to initial conditions. Thus, the cumulative function expressions for vehicle load and energy have auxiliary positive impacts bringing them to their maximum capacity states. The start time domain for customer requests, $i \in V$, becomes a set of disjoint time windows, where each request time window is replicated over each of the horizon segments. The start domain for customer requests is the entire augmented horizon and the disjoint time windows are enforced with constraints. The start time domain for recharge tasks, $i \in F'$, becomes the entire augmented horizon.

Our single resource CP model is detailed by Eqs. (28)–(42). Objective (28) is our fleet and distance minimization objective function. Constraint (29) enforces temporal feasibility of the interval variable sequence, π, including travel times. To make sure customers are serviced during a valid time window, we use Constraint (30), where $\phi_i = \{0, \ldots, |V| \cdot H\} \setminus \bigcup_{\delta \in \{0, \ldots, |V|\}} \{\delta H + e_i, \ldots, \delta H + l_i + s_i\}$. The ForbidExtent constraint prevents an interval variable, x_i, from being scheduled during any time point within the augmented horizon that is not also within one of the disjoint time windows. We discuss a number of alternatives for modeling disjoint time windows with CP in the next section. Constraints (31)–(32) ensure vehicle load feasibility. Constraints (33)–(35) ensure vehicle energy level feasibility while Constraints (36)–(37) dictate any present recharges, as well as horizon end tasks, must charge the vehicle to full energy level. To ensure the resetting of energy level at the end of each horizon, $i \in \mathcal{H}_{N+1}$, we use a positive impact `StepAtStart` with magnitude in $[0, Q - h \cdot d_{\text{Prev}_\pi(i),i}]$ expressed by Constraint (34), and the `AlwaysIn` expressed by Constraint (37). These components are illustrated in Fig. 1. The position of the start depot in the interval variable sequence, π, is expressed through Constraint (38) and Constraints (39)–(42) identify variable domains.

$$\min \quad \alpha \sum_{i \in \mathcal{H}_{N+1}} \text{Pres}(x_i) + \beta \sum_{i \in V'_{\mathcal{H}, N+1}} \text{Pres}(x_i) \cdot d_{\text{Prev}_\pi(i),i} \tag{28}$$

$$\text{s.t.} \quad \text{NoOverlap}(\pi, \{t_{ij} : (i,j) \in A'\}) \tag{29}$$

$$\text{ForbidExtent}(x_i, \phi_i) \qquad \forall i \in V, \tag{30}$$

$$\mathcal{C} = \text{StepAtStart}(x_0, C)$$
$$\quad - \sum_{i \in V} \text{StepAtStart}(x_i, q_i)$$
$$\quad + \sum_{i \in \mathcal{H}_{N+1}} \text{StepAtStart}(x_i, [0, C]) \tag{31}$$

$$\text{AlwaysIn}(\mathcal{C}, [0, |V| \cdot H], [0, C]) \tag{32}$$

$$\mathcal{Q} = \text{StepAtStart}(x_0, Q)$$
$$\quad - \sum_{i \in V'} \text{StepAtStart}(x_i, h \cdot d_{\text{Prev}_\pi(i),i})$$
$$\quad + \sum_{i \in F'} \text{StepAtStart}(x_i, g \cdot \text{Length}(x_i))$$
$$\quad + \sum_{i \in \mathcal{H}_{N+1}} \text{StepAtStart}(x_i, \psi_i) \tag{33}$$

$$0 \leq \psi_i \leq Q - h \cdot d_{\text{Prev}_\pi(i),i} \qquad \forall i \in \mathcal{H}_{N+1}, \tag{34}$$

$$\text{AlwaysIn}(\mathcal{Q}, [0, |V| \cdot H], [0, Q]) \tag{35}$$

$$\text{AlwaysIn}(\mathcal{Q}, x_i, [Q, Q]) \qquad \forall i \in F', \tag{36}$$

$$\text{AlwaysIn}(\mathcal{Q}, [\sigma_i, \sigma_i + 1], [Q, Q]) \qquad \forall i \in \mathcal{H}_{N+1}, \tag{37}$$

$$\text{First}(\pi, x_0) \tag{38}$$

$$x_i : \text{optIntervalVar}([0, Q \cdot g^{-1}], [0, H \cdot |V|]) \qquad \forall i \in F', \tag{39}$$

$$x_i : \text{intervalVar}(s_i, [0, H \cdot |V|]) \qquad \forall i \in V, \tag{40}$$

$$x_i : \text{intervalVar}(0, \sigma_i) \qquad \forall i \in \mathcal{H}_{0,N+1}, \tag{41}$$

$$\pi : \text{sequenceVar}(\{x_0, \ldots, x_{N+|V|}\}). \tag{42}$$

The single resource transformation requires only $|V'_{0,N+1}| + |\mathcal{H}|$ interval variables, one sequence variable, and two cumulative function expression variables, a significant reduction from the alternative resource model.

Model Strengthening

Optional Horizon Segments. Initially, a horizon segment must be created for each potential vehicle, recalling that the upper bound used is the number of customer requests, $|V|$. This augmented horizon significantly increases the start time domain of the recharge tasks, even though most high quality solutions only use a small fraction of the vehicles allotted. To improve upon this, we develop a technique, similar to the symmetry breaking in the alternate resource model, where horizon segments can be set absent. First, we set all auxiliary end depot instances, $x_i, \forall i \in \mathcal{H}$, as optional interval variables. Next, we introduce an integer variable for each of the end depot instances, w_i, and constrain its value to be the start time of the interval variable (0 if the variable is set as absent), via $\text{StartOf}(x_i) = w_i, \forall i \in \mathcal{H}_{N+1}$. We then constrain the end time of the set of customer and recharge visit tasks to be bounded by the maximum w_i value, $\text{EndOf}(x_j) \leq \max_{i \in \mathcal{H}_{N+1}} w_i, \forall j \in V'$. Finally, we impose an ordering on the present depot instances using: $\text{Pres}(x_i) \geq \text{Pres}(x_{i+1}), \forall i \in \mathcal{H} \setminus v_{N+|V|}$.

Energy Expression Tightening. Similar to the technique presented for the alternative resource model, we introduce energy impact tightening constraints for the single resource model as well, namely: $\min_{i \in V'_0}(h \cdot d_{ij}) \leq \text{HeightAtStart}(x_j) \leq \max_{i \in V'_0}(h \cdot d_{ij}), \forall j \in V$.

4.3   Alternate Modeling Strategies

We investigated a number of alternate modeling strategies that were found, through initial experiments, to under-perform the proposed models.

Vehicle Energy and Load. The modeling of energy and vehicle load can also be accomplished via auxiliary tracking variables [20] similar to those used in the MILP. The idea is to introduce a numeric variable for each interval variable representing the load or energy level in the sequence after that particular task. This technique is advantageous in that the exact vehicle load or energy level can be accessed at any point along the route, whereas current implementations of cumulative function expressions, as within CP Optimizer, do not support this.

Disjoint Time Windows. The single resource model results in a set of disjoint time windows for customer tasks.[2] The model uses the $\text{ForbidExtent}(var, T)$ constraint, restricting an interval variable var from executing at any time point within the restricted set of time points T. This relationship can also be expressed using interval variables by generating a set of fixed interval variables that occupy all the time points in T. Then, a $\text{NoOverlap}(\pi'_i)$ is added to the model for each customer request task, $i \in V$, where the sequence variable π'_i contains the set of all customer interval variables and the auxiliary fixed interval variables. Finally, similar to the alternative resource CP model, one can generate an alternative

---

[2] There also exist VRP variants posed with multiple disjoint time windows [18].

task for each of the time windows. Although these alternate techniques perform moderately well on smaller problems, the increased model size was detrimental for larger problems.

4.4 Experimental Analysis

We present an empirical assessment of our models on the benchmark data in Table 2. We explore three different objective functions: fleet distance minimization ($\alpha = 0$, $\beta = 1$) as reported in [10,29], fleet size minimization ($\alpha = 1$, $\beta = 0$), and fleet size minimization with distance minimization as a secondary objective ($\alpha = 1, \beta = \xi$), where ξ is a sufficiently small number to lexicographically order the objective components. We reiterate that the intent of this work is to investigate the performance of off-the-shelf optimization models for EVPRTW; state-of-the-art results for distance minimization are found in [10] using sophisticated branch-price-and-cut techniques bolstered by customized labeling algorithms.

Set-Up. All experiments are implemented in C++ on an Intel Xeon CPU E5-2690 v4 2.60 GHz processor and 16 GB of RAM running Ubuntu 14.04. We use CP Optimizer for the CP models and CPLEX for the MILP model from the IBM ILOG CPLEX Optimization Studio version 12.8. All experiments are single-threaded with default search and inference settings. A five minute time limit is used for all experiments.

Table 1. Problem instances. Each value represents the number of instances for a given size/characteristic combination. $|V| \leq 15$ are small instances containing 5, 10, and 15 customers. Clustered, random, and mix refer to the geographical distribution of customer vertices. $|F|$ values are averages across the instances.

$|V|$	$|F|$	Total	Short horizon			Long horizon		
			Clustered	Random	Mix	Clustered	Random	Mix
≤ 15	4.2	36	6	6	6	6	6	6
25	21	56	9	12	8	8	11	8
50	21	56	9	12	8	8	11	8

Instances and Implementation. We conduct our analysis on problem instances taken from the literature [10,29]. Instances vary w.r.t. the number of customer and recharge station vertices, the length of the scheduling horizon (short and long) and the geographical distribution of the customer vertices (random, clustered, and a mixture of both). The benchmark utilized contains a total of 148 instances summarized in Table 1.

Following the procedure outlined in previous work on the same instances [10], we transform floating point parameter values to integer values such that the problems are amenable to CP modeling. As with most integer transformations, the scaling involved in this process results in much larger variable domain ranges which can have a negative impact on CP approaches. Additionally, we heuristically set the number of visits allowable to each recharge station as $n_f = \lceil |V| \cdot 0.2 \rceil$ (i.e., problems with five customers allow a single visit to each recharge station).

During testing, we found that the CP solver, for both formulations, had a difficult time producing initial feasible solutions for the larger problems, $|V| \in \{25, 50\}$. To mitigate this, we seed the CP search with the initial solution found by the MILP presolve routine. We note that we could have used any initial heuristic here to yield the same result; the presolve results used were often trivial (i.e., each customer serviced by a separate vehicle).

Table 2. Experimental results. The best result of each column for each objective function in bold. 'M': method ran out of memory before entering the search.

	Short Horizon									Long Horizon								
	# Feasible			# Best			MRE (%)			# Feasible			# Best			MRE (%)		
Method	≤15	25	50	≤15	25	50	≤15	25	50	≤15	25	50	≤15	25	50	≤15	25	50
$\alpha = 0; \beta = 1$																		
MILP	18	29	29	17	4	9	**7.3**	63.4	68.1	18	27	27	18	16	25	**3.7**	**50.3**	**51.7**
CP_{AR}	18	29	M	8	0	M	12.6	76.4	M	18	27	M	3	0	M	17.9	71.3	M
CP_{SR}	18	29	29	9	**25**	**20**	9.9	**54.9**	**64.5**	18	27	27	5	11	2	11.0	**50.3**	67.0
$\alpha = 1; \beta = 0$																		
MILP	18	29	18	14	0	7	24.8	83.5	88.3	18	27	27	15	0	2	17.6	82.0	93.6
CP_{AR}	18	29	M	7	0	M	35.4	89.3	M	18	27	M	8	3	M	41.7	84.4	M
CP_{SR}	18	29	23	**18**	**29**	**17**	22.8	**71.7**	**88.2**	18	27	27	**18**	**27**	**25**	**11.1**	**52.2**	**78.9**
$\alpha = 1; \beta = \xi$																		
MILP	18	29	29	14	3	11	**22.8**	68.6	66.2	18	27	27	17	0	8	**14.0**	65.5	59.5
CP_{AR}	18	29	M	5	0	M	48.6	88.8	M	18	27	M	2	0	M	56.5	94.4	M
CP_{SR}	18	29	29	**14**	**26**	**18**	23.7	**58.8**	**64.0**	18	27	27	5	**27**	**19**	25.0	**40.0**	**57.0**

Results. The results are illustrated in Table 2. The mean relative error (MRE) compares the best solution found by a given technique to the best bound found across all techniques; the results in the table take the average of this across all instances solved by the technique.

The MILP displayed fairly strong performance on the distance minimization objective function, particularly for small problems, where the strong bound is able to effectively direct the search, and on long horizon problems, where CP inference is less effective. While the MILP approach is often able prove optimality for small instances when minimizing travel distance, it struggles to produce high quality solutions and meaningful bounds for the fleet minimization objective functions for medium-to-large problems, with optimality gaps close to 100%.

The alternative resource CP model, CP_{AR}, encountered memory issues for $|V| = 50$, while the single resource transformation model, CP_{SR}, was able to initiate the search for all problems.[3] Overall, it was found that the single resource model outperformed the alternative model for all problem classes. Additionally, CP_{SR} outperforms the MILP formulation on almost all classes of larger problems ($|V| \in \{25, 50\}$), with the exception of distance minimization over long

---

[3] In fact, the single resource model required significantly less memory for $|V| = 50$ problems than the alternative resource model did for $|V| = 25$ problems.

horizons. The horizon augmentation of the single resource model results in large domain sizes for problems with larger initial scheduling horizons, resulting in weaker inference throughout the search. Both CP approaches tend to produce more meaningful bounds for the fleet minimization problems, where the MILP approach has difficulty producing non-trivial lower bounds.

5   A Hybrid Approach

From the experiments, it is evident that MILP outperforms CP_{SR} for distance minimization for large problems over long scheduling horizons. This finding is similar to that from previous research that demonstrated scheduling-based CP models containing optional activities can suffer from poor inference as the problem scales without good upper bounds on horizon length [4]; the authors of this previous work found that seeding CP with high quality solutions found by a different solver can be significantly beneficial.

Given these findings, we construct a hybrid approach that passes the best solution found by the MILP solver to the CP solver as a starting point. Following previous work [4], we allocate half of the runtime to MILP and half to CP noting that the MILP solution improvement diminishes with time. We apply this hybrid to the larger problem instances, $|V| \in \{25, 50\}$. The remainder of the experimental set-up remains as described in the previous section.

Table 3. Hybrid results, large problems. Best result of column for each objective function in bold. 'M': method ran out of memory before entering the search.

	Short Horizon						Long Horizon					
	# Feas.		# Best		MRE (%)		# Feas.		# Best		MRE (%)	
Method	25	50	25	50	25	50	25	50	25	50	25	50
$\alpha = 0; \beta = 1$												
MILP	29	29	2	5	63.4	68.1	**27**	**27**	5	8	50.3	51.7
CP_{SR}	29	29	14	14	54.9	**64.5**	**27**	**27**	8	2	50.3	67.0
MILP→CP_{SR}	29	29	13	10	**54.6**	**64.5**	**27**	**27**	14	17	**47.3**	**47.3**
$\alpha = 1; \beta = 0$												
MILP	29	18	0	7	83.7	87.6	**27**	**27**	0	2	81.8	92.8
CP_{SR}	29	23	23	15	**72.0**	87.5	**27**	**27**	27	25	**51.2**	**77.5**
MILP→CP_{SR}	29	16	11	3	74.0	86.7	**27**	**27**	24	13	53.3	82.3
$\alpha = 1; \beta = \xi$												
MILP	29	29	0	4	68.9	66.2	**27**	**27**	0	5	65.5	59.5
CP_{SR}	29	29	19	14	**59.1**	64.0	**27**	**27**	16	17	**40.0**	57.0
MILP→CP_{SR}	29	29	10	11	59.8	**62.9**	**27**	**27**	11	5	47.9	**55.3**

Results. We present the results for our hybrid approach, denoted MILP→CP_{SR}, in Table 3, alongside the original MILP and CP_{SR} results. It is apparent that the hybrid approach is beneficial for the distance minimization objective over long horizons, outperforming the other approaches by a wide margin and improving over MILP MRE values by up to 4.4% ($|V| = 50$, long horizon). However, outside of the large, long horizon distance minimization problems, the hybrid provides improvement in few other areas, and is commonly outperformed by the standalone CP method. Based on this observation, we can conclude that it makes sense to hybridize MILP and CP in particular circumstances, but often standalone CP will produce the best result. In these experiments, a CP-based approach (either hybrid or standalone) provides the best results for every problem class, and the hybrid approach outperforms standalone MILP across nearly all problem classes, with the exception of fleet minimization on short horizons.

6   Conclusion and Perspective

In this paper we presented the first approaches for solving the Electric Vehicle Routing Problem with Time Windows (EVRPTW) using constraint programming (CP). We present two scheduling-based CP formulations: the initial model uses an alternative resource technique previously applied to other routing problems, while the second uses a single resource transformation for CP models using optional activities, sequence variables, and cumulative function expressions. We detail techniques used to strengthen the formulations and discuss alternate modeling strategies.

Numerical results indicate the superiority of the single resource CP model over the alternative resource model, for all problems, and the MILP formulation, for the majority of medium-to-large problem classes. Recognizing the ability of MILP to quickly produce good quality solutions for large distance minimization problems with long scheduling horizons, we also investigate a hybrid MILP-CP approach where the best solution from the mathematical programming solver is used to seed the CP search. Results indicate the hybrid approach outperforms both of the standalone techniques for the problems that motivated the effort, but is not beneficial overall.

Given the growth of electric vehicle (EV) adoption in the logistics and consumer automotive industries, we believe the study of these problems in the context of CP modeling and solving is a strategic direction. Outside of EVs and transportation, there is considerable opportunity in the highly related field of multi-robot task allocation (MRTA) [5,16]. Future work will investigate the applicability of the techniques developed in this paper to problems found in MRTA.

Acknowledgment. We would like to thank the anonymous reviewers whose detailed feedback helped improve the paper.

References

1. Afroditi, A., Boile, M., Theofanis, S., Sdoukopoulos, E., Margaritis, D.: Electric vehicle routing problem with industry constraints: trends and insights for future research. Transp. Res. Procedia **3**, 452–459 (2014)
2. Beck, J.C., Prosser, P., Selensky, E.: Vehicle routing and job shop scheduling: What's the difference? In: International Conference on Automated Planning and Scheduling, pp. 267–276 (2003)

3. Berbeglia, G., Cordeau, J.F., Laporte, G.: A hybrid tabu search and constraint programming algorithm for the dynamic dial-a-ride problem. INFORMS J. Comput. **24**(3), 343–355 (2012)
4. Booth, K.E.C., Do, M., Beck, J.C., Rieffel, E., Venturelli, D., Frank, J.: Comparing and integrating constraint programming and temporal planning for quantum circuit compilation. In: International Conference on Automated Planning and Scheduling, pp. 366–374. AAAI Press (2018)
5. Booth, K.E.C., Nejat, G., Beck, J.C.: A constraint programming approach to multi-robot task allocation and scheduling in retirement homes. In: Rueher, M. (ed.) CP 2016. LNCS, vol. 9892, pp. 539–555. Springer, Cham (2016). https://doi.org/10. 1007/978-3-319-44953-1_34
6. Booth, K.E.C., Tran, T.T., Nejat, G., Beck, J.C.: Mixed-integer and constraint programming techniques for mobile robot task planning. IEEE Robot. Autom. Lett. **1**(1), 500–507 (2016)
7. Bruglieri, M., Pezzella, F., Pisacane, O., Suraci, S.: A variable neighborhood search branching for the electric vehicle routing problem with time windows. Electron. Notes Discret. Math. **47**, 221–228 (2015)
8. Cappart, Q., Thomas, C., Schaus, P., Rousseau, L.-M.: A constraint programming approach for solving patient transportation problems. In: Hooker, J. (ed.) CP 2018. LNCS, vol. 11008, pp. 490–506. Springer, Cham (2018). https://doi.org/10.1007/ 978-3-319-98334-9_32
9. Dekker, R., Bloemhof, J., Mallidis, I.: Operations research for green logistics-an overview of aspects, issues, contributions and challenges. Eur. J. Oper. Res. **219**(3), 671–679 (2012)
10. Desaulniers, G., Errico, F., Irnich, S., Schneider, M.: Exact algorithms for electric vehicle-routing problems with time windows. Oper. Res. **64**(6), 1388–1405 (2016)
11. Di Gaspero, L., Rendl, A., Urli, T.: Balancing bike sharing systems with constraint programming. Constraints **21**(2), 318–348 (2016)
12. Erdoğan, S., Miller-Hooks, E.: A green vehicle routing problem. Transp. Res. Part E: Logist. Transp. Rev. **48**(1), 100–114 (2012)
13. Felipe, Á., Ortuño, M.T., Righini, G., Tirado, G.: A heuristic approach for the green vehicle routing problem with multiple technologies and partial recharges. Transp. Res. Part E: Logist. Transp. Rev. **71**, 111–128 (2014)
14. Feng, W., Figliozzi, M.: An economic and technological analysis of the key factors affecting the competitiveness of electric commercial vehicles: a case study from the usa market. Transp. Res. Part C: Emerg. Technol. **26**, 135–145 (2013)
15. Gedik, R., Kirac, E., Milburn, A.B., Rainwater, C.: A constraint programming approach for the team orienteering problem with time windows. Comput. Ind. Eng. **107**, 178–195 (2017)
16. Gerkey, B.P., Matarić, M.J.: A formal analysis and taxonomy of task allocation in multi-robot systems. Int. J. Robot. Res. **23**(9), 939–954 (2004)
17. Hiermann, G., Puchinger, J., Ropke, S., Hartl, R.F.: The electric fleet size and mix vehicle routing problem with time windows and recharging stations. Eur. J. Oper. Res. **252**(3), 995–1018 (2016)
18. de Jong, C., Kant, G., Van Vlient, A.: On finding minimal route duration in the vehicle routing problem with multiple time windows. Department of Computer Science, Utrecht University, Holland, Manuscript (1996)
19. Keskin, M., Çatay, B.: Partial recharge strategies for the electric vehicle routing problem with time windows. Transp. Res. Part C: Emerg. Technol. **65**, 111–127 (2016)

20. Kinable, J., van Hoeve, W.-J., Smith, S.F.: Optimization models for a real-world snow plow routing problem. In: Quimper, C.-G. (ed.) CPAIOR 2016. LNCS, vol. 9676, pp. 229–245. Springer, Cham (2016). https://doi.org/10.1007/978-3-319-33954-2_17

21. Laborie, P.: IBM ILOG CP optimizer for detailed scheduling illustrated on three problems. In: van Hoeve, W.-J., Hooker, J.N. (eds.) CPAIOR 2009. LNCS, vol. 5547, pp. 148–162. Springer, Heidelberg (2009). https://doi.org/10.1007/978-3-642-01929-6_12

22. Laborie, P., Rogerie, J., Shaw, P., Vilím, P.: IBM ILOG CP Optimizer for scheduling. Constraints 23(2), 210–250 (2018)

23. Lam, E., Van Hentenryck, P., Kilby, P.: Joint vehicle and crew routing and scheduling. In: Pesant, G. (ed.) CP 2015. LNCS, vol. 9255, pp. 654–670. Springer, Cham (2015). https://doi.org/10.1007/978-3-319-23219-5_45

24. Lin, C., Choy, K.L., Ho, G.T., Chung, S.H., Lam, H.: Survey of green vehicle routing problem: past and future trends. Expert Syst. Appl. 41(4), 1118–1138 (2014)

25. Liu, C., Aleman, D.M., Beck, J.C.: Modelling and solving the senior transportation problem. In: van Hoeve, W.-J. (ed.) CPAIOR 2018. LNCS, vol. 10848, pp. 412–428. Springer, Cham (2018). https://doi.org/10.1007/978-3-319-93031-2_30

26. Luè, A., Colorni, A., Nocerino, R., Paruscio, V.: Green move: an innovative electric vehicle-sharing system. Procedia-Soc. Behav. Sci. 48, 2978–2987 (2012)

27. Montoya, A., Guéret, C., Mendoza, J.E., Villegas, J.G.: The electric vehicle routing problem with nonlinear charging function. Transp. Res. Part B: Methodol. 103, 87–110 (2017)

28. Rezvani, Z., Jansson, J., Bodin, J.: Advances in consumer electric vehicle adoption research: a review and research agenda. Transp. Res. Part D: Transport Environ. 34, 122–136 (2015)

29. Schneider, M., Stenger, A., Goeke, D.: The electric vehicle-routing problem with time windows and recharging stations. Transp. Sci. 48(4), 500–520 (2014)

30. Shaw, P.: Using constraint programming and local search methods to solve vehicle routing problems. In: Maher, M., Puget, J.-F. (eds.) CP 1998. LNCS, vol. 1520, pp. 417–431. Springer, Heidelberg (1998). https://doi.org/10.1007/3-540-49481-2_30

31. Vali, M., Salimifard, K.: A constraint programming approach for solving multiple traveling salesman problem. In: The Sixteenth International Workshop on Constraint Modelling and Reformulation (2017)

A Sampling-Free Anticipatory Algorithm for the Kidney Exchange Problem

Danuta Sorina Chisca[1](✉), Michele Lombardi[2](✉), Michela Milano[2](✉), and Barry O'Sullivan[1](✉)

[1] Insight Centre for Data Analytics, University College Cork, Cork, Ireland
{sorina.chisca,barry.osullivan}@insight-centre.org
[2] DISI, University of Bologna, Bologna, Italy
{michele.lombardi2,michela.milano}@unibo.it

Abstract. Kidney exchange programs try to improve accessibility to kidney transplants by allowing incompatible patient-donor pairs to swap donors. Running such a program requires to solve an optimization problem (the Kidney Exchange Problem, or KEP) as new pairs arrive or, unfortunately, drop-off. The KEP is a stochastic online problem, and can greatly benefit from the use of anticipatory algorithms. Unfortunately, most such algorithms suffer from scalability issues due to the reliance on scenario sampling, limiting their practical applicability. Here, we recognize that the KEP allows for a sampling-free probabilistic model of future arrivals and drop-offs, which we capture via a so-called Abstract Exchange Graph (AEG). We show how an AEG-based approach can outperform sampling-based algorithms in terms of quality, while being comparable to a myopic algorithm in terms of scalability. While our current experimentation is preliminary and limited in scale, these qualities make our technique one of the few that can hope to address nation-wide programs with thousands of enrolled pairs.

Keywords: Kidney Exchange Problem ·
Stochastic online optimization · Probabilistic model

1 Introduction

For many patients suffering of organ failure, transplants are the most effective solution [11]. However, transplants are also difficult to access, due to the lack of donors and biological compatibility issues: as a result, patients often remain in waiting lists for a few years. In the case of kidneys, transplant accessibility may be improved by resorting to exchanges: these arise when a patient-donor pair is incompatible, but a match can be found by looking at other incompatible pairs.

The Insight Centre for Data Analytics is supported by Science Foundation Ireland under Grant Number SFI/12/RC/2289, which is co-funded under the European Regional Development Fund.

L.-M. Rousseau and K. Stergiou (Eds.): CPAIOR 2019, LNCS 11494, pp. 146–162, 2019.
https://doi.org/10.1007/978-3-030-19212-9_10

Centralized kidney exchange programs exist in many countries, including the US, the Netherlands and the UK [18]: in the US at the moment 95,345 patients are waiting for a kidney, while 17,575 transplants have been performed in the course of 2018[1]. Similar numbers can be found worldwide, and grow every year. Running a large-scale exchange program requires to regularly solve an optimization problem (the Kidney Exchange Problem – KEP) to choose which transplants should be performed. The KEP is typically defined over a compatibility graph, whose nodes represent patient-donor pairs (or individual "altruistic" donors), and arcs correspond to viable transplants.

The KEP is inherently a stochastic online problem, since arrivals (and drop-offs) occur over time and cannot be predicted with certainty. Problems in this class can greatly benefit from the use of anticipatory algorithms (see e.g. [14]). However, most such algorithms rely on scenario sampling to handle uncertainty, leading to scalability issues. As a result, in practice the KEP is often solved myopically, i.e. by taking into account only the pairs that are currently in the program.

In this paper we recognize that, with a few reasonable assumptions, the KEP allows for the construction of a sampling-free probabilistic model of future arrivals and drop-offs. We capture this information via a so-called Abstract Exchange Graph (AEG), whose nodes represent "types" of pairs, rather than individual pairs, and are associated to probability values. An AEG can be easily (and efficiently) obtained from medical or historical data.

By relying on the AEG, we describe how to enrich a given optimization model for the myopic KEP with an anticipatory component, often with very limited impact on its scalability. By doing this, the model becomes capable of taking into account both future arrivals and drop-offs. For the sake of simplicity, we show how to apply our technique to a specific KEP approach, i.e. the cycle formulation.

In an experimentation on instances obtained via a realistic simulator, we show how sampling-based anticipatory algorithms quickly run into scalability issues, and even fail to provide high quality solutions in case of insufficient sample numbers. Conversely, an AEG-augmented method is consistently able to outperform its myopic counterpart, while having a comparable run-time. While the current experimentation is preliminary and limited in scale, these qualities make our technique one of the few that can hope to address nation-wide programs with thousands of enrolled pairs

The paper is organized as follows: in Sect. 2 we present the background and related literature, in Sect. 3 we give a detailed description of the AEG model. In Sect. 4 we show some experimental results, while concluding remarks are in Sect. 5.

2   Background and Related Work

Formally, the offline (i.e. myopic) KEP is defined on a directed graph $D = \langle N, A(N) \rangle$. Nodes in N correspond to patient-donor pairs or "altruistic" donors (i.e. people who are willing to donate a kidney to the program). A is a function

---

[1] https://unos.org/data/transplant-trends/.

that maps a set of nodes to a corresponding set of directed arcs, represented as pairs of indices, i.e. $A : 2^N \rightarrow 2^{N \times N}$. The mapping is based on the biological properties of the nodes: in particular, an arc (i, j) from node i to j appears in $A(N)$ iff the donor at node i is compatible with patient at node j. We refer to such data structure as *compatibility graph*.

A cycle corresponds to a viable set of exchanges, and cycles starting from an altruistic donor are called chains. A cycle/chain involving k nodes is also known as a k-exchange. A limit on k is often enforced, since transplants in the same exchange need to be performed simultaneously to avoid donor withdrawal. The goal is to maximize a utility function, usually the total number of transplants. The KEP is known (see [1]) to be NP-complete for $k \geq 3$.

The problem can be stated in an abstract fashion by referring as $\mathbf{x}$ to the set of exchanges $\mathbf{x_j}$ that should be performed. Then the KEP corresponds to:

$$\max z = \sum_{\mathbf{x_j} \in \mathbf{x}} value(\mathbf{x_j}) \qquad \textbf{(KEP)} \qquad (1)$$

$$\text{s.t. } usage_i(\mathbf{x}) \leq 1 \qquad \forall i \in N \qquad (2)$$

$$valid(\mathbf{x_j}) \qquad \forall \mathbf{x_j} \in \mathbf{x} \qquad (3)$$

where $value(x_j)$ is the utility of performing exchange $\mathbf{x_j}$ (e.g. the number of involved transplants), and $usage_i(\mathbf{x})$ denotes the number of times that vertex i is used by all exchanges in $\mathbf{x}$. No patient-donor pair can be selected twice, leading to Constraints (2). Finally, $valid(\mathbf{x_j})$ is a predicate that is true iff exchange $\mathbf{j}$ is feasible (e.g. corresponds to a cycle/chain and has the correct length).

Approaches for the Offline KEP: In the past years, the Kidney Exchange Problem received considerable attention from the medical, economics and computer science communities. Most works have focused on finding the best matching for a given graph, i.e., on the *offline* version of the problem. In the Operations Research community the problem has been considered in [3,7,17,18] and more recently in [2,8], via a variety of models. The KEP has also been linked to the cycle roommates problem in [15], and to barter-exchange markets in [1,4].

A popular Mathematical Programming model for the KEP is the so-called *cycle formulation*, which is an almost direct translation of the abstract model. Let $\mathcal{C} = \{C_1 \dots C_{n_c}\}$ be the set of all cycles/chains of valid length. Each cycle is associated to a 0-1 variable x_j such that $x_j = 1$ iff the corresponding exchange is selected. Similarly, each cycle is associated to a weight w_j, corresponding to its value. The cycle formulation is then given by the following Integer Program:

$$\max z = \sum_{C_j \in \mathcal{C}} w_j x_j \qquad \textbf{(CYF)} \qquad (4)$$

$$\text{s.t. } \sum_{C_j \in \mathcal{C}, i \in C_j} x_j \leq 1 \qquad \forall i \in N \qquad (5)$$

$$x_j \in \{0, 1\} \qquad \forall C_j \in \mathcal{C} \qquad (6)$$

The model can be seen as a set packing formulation: it is simple and capable to easily capture complex constraints on the formation of cycles, including their size.

The main drawback is its limited scalability, since the formulation requires enumerating all (valid) cycles in the graph. This issue has been addressed effectively via column generation in [1,10,13,16,22]. Alternatively, compact models have been proposed in [2,8] (and references therein) to improve scalability without resorting to column generation. Constraint generation has been employed to speed-up the edge formulation of the KEP in [1], while a KEP model based on the Traveling Salesman Problem has been proposed in [3].

Approaches for the Online KEP: As mentioned in our introductory section, the KEP is actually an online problem: choosing to perform a transplant may have unintended side effects in the future. Anticipatory algorithms typically deal with this source of uncertainty by sampling *scenarios* (possible outcomes) and optimizing the (estimated) expected value of current decisions.

A few of the main anticipatory algorithms from [14] have been adapted to the KEP in [5], but the task was not straightforward and forced the authors to introduce heuristic approximations. We proposed one more sampling-based anticipatory algorithm in [6]. These algorithms will be described in our results section, since they are considered in our experimentation. Here, we simply observe that they all rely on scenario sampling, a scenario being a set of nodes that may enter in the next few steps. The main two algorithms from [5] require the solution of a multiple (smaller) off-line KEPs, while the method from [6] solves a single modified KEP obtained via the Sample Average Approximation [20].

The only sampling-free anticipatory methods for the online KEP to date are those from [9,12]. They are both based on the idea of discounting the value of each cycle with the lost "potential" of the involved nodes. Formally, the new weight of each cycle is given by $w_j - \sum_{i \in C_j} v_i$, where v_i is the estimated potential of $i \in N$. A set of exchanges can then be obtained by solving an off-line KEP with the modified weights. In [9] the potentials are estimated via probability considerations, while [12] employs a parameter tuning algorithm. Even if the technique proposed here does not rely on sampling, from a mathematical point of view it is more akin to the sampling-based algorithms than to these sampling-free methods. For this reason, a comparison with either [9] or [12] is missing in this paper, but we still consider that a priority for future research.

To the best of our knowledge, these are all the methods for the online KEP in the literature. Works [2,10,21] have considered the effect of potential failures, but not of entering pairs. A simulator for the online KEP is presented in [24], but the authors still rely on periodic execution of an offline approach.

3   The Abstract Exchange Graph

Here we present our main data structure, i.e. the Abstract Exchange Graph, and its potential applications. Our contributions stem from two simple assumptions, which we state initially in their basic form.

First, all sampling-based anticipatory methods from the literature attempt to account for the arrival of individual pairs/altruistic donors. However, pairs having the same connectivity lead to equivalent sets of cycles: in the absence of an external criterion for favoring certain pairs over others, we have that:

Assumption 1. *Nodes having the same connectivity are equivalent.*

Assumption 2. *Arrivals are independent and identically distributed (i.i.d.).*

Second, banning mass-sensitization campaigns or similar actions, the arrival of a pair should provide no information on which pair is arriving next. Moreover, population changes occur slowly over time. Hence, during regular operation we apply Assumption 2.

Both assumptions are imposed for technical feasibility and can actually be relaxed, and we will show how this can be done at the end of this section.

Formally, the AEG is an annotated directed graph $\langle N, A(N), p \rangle$. Rather than to individual pairs, the nodes correspond to classes ("types") of equivalent pairs. The arcs correspond to "types" of transplants, and can be obtained via the same function A used for the classical KEP. Each node can be associated to a probability value $p_i \in [0, 1]$. Overall, the AEG specifies in a compact fashion both the arrival probabilities of future nodes and the cycles they can form.

Obtaining an AEG: From a mathematical standpoint, the AEG is a simple extension of the compatibility graph used by most KEP approaches. In particular, a "concrete" compatibility graph can be seen as an AEG where $p_i = 1$ for all nodes. This observation allows to easily obtain an AEG from historical data.

Kidney exchange programs usually keep a record of the participating pairs, which can be used to populate a set N. This set can form the bases for a "concrete" graph. Starting from this graph, an AEG can be obtained by iteratively merging nodes with the same connectivity, as described in Algorithm 1. Once the graph can no longer be contracted in this fashion, all p values are normalized so that they represent frequencies of occurrence, i.e. estimated probabilities.

Algorithm 1. AEG extraction

Require: A graph $\langle N, A(N), p \rangle$, with $p_i = 1 \; \forall i \in N$
  loop
    Search for two nodes $i, j \in N$ with the same outgoing/ingoing arcs, i.e.:
    A) $(i, h) \in A(N) \Leftrightarrow (j, h) \in A(N)$ and
    B) $(h, i) \in A(N) \Leftrightarrow (h, i) \in A(N)$
    if such nodes exist **then**
      $N = N \setminus j$ (i.e. remove node j)
      $p_i = p_i + p_j$ (i.e. compute aggregated count)
    else
      set $p_i = p_i / \sum_i p_i$ (normalize counts)
      break loop
  return $\langle N, A(N), p \rangle$

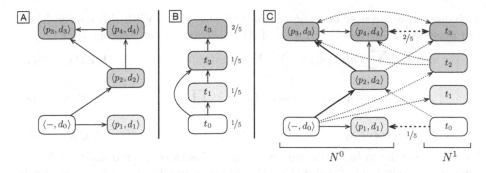

Fig. 1. (A) A "concrete" compatibility graph; (B) The corresponding AEG; (C) The AEG-base graph used in optimization

This process always yields a valid AEG. As an example, assume that we start from the compatibility graph in Fig. 1A: nodes are labeled with the corresponding patient donor pair, and node 0 corresponds to an altruistic donor. Nodes having the same connectivity (i.e. (p_3, d_3) and (p_4, d_4)) are colored the same shade of gray. Figure 1B shows the AEG produced by Algorithm 1, where t_0 is associated to the altruistic donor, t_1 to (p_1, d_1) and t_2 to (p_2, d_2), and t_3 is an aggregated node merging (p_3, d_3) and (p_4, d_4), and with the estimated probabilities reported next to each node.

Using the AEG in Optimization: The AEG can be employed to enrich a model for the off-line KEP with an anticipatory component. We will describe the process using the abstract KEP from Sect. 2.

We start by augmenting the problem graph with h instances of all the AEG nodes, so as to represent h future arrival events. Formally, let N^0 be the original set of nodes, N^k be the k-th instance of the AEG nodes, and N be the set of all nodes (i.e. $N = N^0 \cup N^1 \cup \ldots N^h$). We then construct the graph:

$$\langle N, A(N) \setminus A(N^1) \setminus \ldots A(N^h), p^0 \cup p^1 \cup \ldots p^h \rangle$$

Arcs from $A(N^1)$ to $A(N^k)$ are removed since multiple nodes from the same N^k (with $k > 0$) correspond to different possible types *for a single arrival event*. The p_i values associated to nodes in N^0 are all equal to 1, while other p_i values are those from the AEG. An example of such a graph for $h = 1$ is shown in Fig. 1C, where for the sake of simplicity we assume that N^0 is given by the graph from Fig. 1A. Dashed arcs connect nodes in N^0 with nodes in N^1.

Exchanges defined exclusively over nodes in N^0 can be interpreted as usual. Exchanges involving even a single node from N^k with $k > 0$ represent *potential future exchanges*, and are associated to a stochastic strategy, similar to those employed in Markov Decision Processes.

Formally, *all* future exchanges (not just the selected ones) are collected in a set $\mathbf{x^F}$. Each of such exchanges is associated to a 0-1 random variable with a Bernoulli distribution. The probability that the variable is equal to 1 captures the odds that (1) the involved nodes arrive; and (2) we choose to perform the exchange: *what we need to decide are these probabilities.* With this approach, we can provide an abstract description of AEG-based the anticipatory KEP:

$$\max z = \sum_{x_i \in x} value(x_i) + \sum_{x_i \in x^F} E\left[value(x_i)\right] \qquad \text{(KEP}_A\text{)} \qquad (7)$$

$$\text{s.t. } usage_S(x) + E\left[usage_S(x^F)\right] \leq \prod_{i \in S} p_i \qquad \forall S \subseteq N \qquad (8)$$

$$valid(x_i) \qquad \forall x_i \in x \cup x^F \qquad (9)$$

We maximize the utility of current exchanges, plus the expected utility of future exchanges. Constraints (8) state that the expected number uses of each subset of nodes S cannot exceed the expected number of its occurrences. For set of nodes in N^0, all p_i values are 1 and the formula boils down to the one in the KEP abstract model. For exchanges that involve future nodes, we take into account the probability that all involved nodes are present. All exchanges should be valid. In principle there is an exponential number of Constraints (8): in practice, however, the number is usually polynomial due to restrictions on the exchange size. Moreover, many subset S lead to redundant constraints that can be eliminated.

The thick arcs in Fig. 1C show the exchanges that would be selected in a (sub-optimal) solution of KEP_A formulation: the solid ones correspond to (selected) deterministic exchanges, while the dashed ones to stochastic exchanges with a non-zero probability (reported as a label on the arc). The main appeal of the technique is that the abstract formulation KEP_A is anticipatory, *and yet it does not rely on sampling.* Provided that our assumptions hold, the probabilistic model we employ gives a characterization of future uncertainty that is *limited in accuracy only by our estimates of the p_i values.*

Grounding Based on the Cycle Formulation: Our technique is general, but particularly easy to ground on the cycle model. Let $\mathcal{C}$ be the set of all cycles of valid length in the (augmented) graph. Let $\mathcal{C}^0$ be the subset of cycles that involve only nodes in N^0, and let $\mathcal{C}^F = \mathcal{C} \setminus \mathcal{C}^0$ be the set of "future" cycles, i.e. those involving at least one node in N^k with $k > 0$. We can then introduce: (1) a variable $x_j \in \{0, 1\}$ for each $C_j \in \mathcal{C}^0$, such that $x_j = 1$ iff the cycle is chosen in the solution; plus (2) a variable $x_j \in [0, 1]$ for each cycle $C_j \in \mathcal{C}^F$, representing the probability that the involved nodes arrive and the cycle is chosen. We can than state the new formulation as a Mixed Integer Linear Program:

$$\max z = \sum_{C_j \in \mathcal{C}} w_j x_j \qquad \text{(CYF}_A\text{)} \qquad (10)$$

$$\text{s.t. } \sum_{C_j \in \mathcal{C}, S \subseteq C_j} x_j \leq \prod_{i \in S} p_i \qquad \forall S \in \mathcal{S} \qquad (11)$$

$$x_j \in \{0, 1\} \qquad \forall C_j \in \mathcal{C}^0 \qquad (12)$$

$$x_j \in [0, 1] \qquad \forall C_j \in \mathcal{C}^F \qquad (13)$$

Since for cycles in $\mathcal{C}^F$ the value of x_j naturally represents a probability, summing the variables in the objective and in Constraints (8) is enough to obtain the expected values from of Objective (7) and Constraints (8).

The family $\mathcal{S}$ contains all subsets of nodes such that lead to non-redundant constraints. Formally, $\mathcal{S}$ contains all $S \subseteq N$ such that: (1) the cardinality is not greater than the allowed cycle length, i.e. $|S| \leq L$; (2) all nodes S appear together in at least one cycle, i.e. $\exists C_j \in \mathcal{C}$ s.t. $S \subseteq C_j$, since subsets that do not appear in any cycle are irrelevant. Finally, (3) if all nodes in S are in N^0, then the cardinality of S is exactly 1, since for subsets entirely in N^0 posting the constraints for each individual node is sufficient.

There are three key facts to observe: first, the CYF_A model has the same structure of the original cycle formulation, meaning that most techniques employed to increase its scalability (e.g. column generation) should still be applicable. Second, the lack of arcs between nodes in the same N^k acts as a mitigation factor for the number of cycles. Third, all the variables related to the anticipatory components are real-valued, and therefore much easier to handle for the solver. Conversely, all sampling-based approaches require 0-1 variables to handle the cycles appearing in scenarios. The main drawback is the increased number of constraints, whose adverse effects will need to be empirically evaluated.

Handling Drop-Offs: Additionally, the AEG-based formulation provides a natural framework for taking into account drop-offs, which may arise as a consequence of pairs leaving the program. For the sake of simplicity, we will initially make the assumption that each pair may drop-off the program between consecutive arrivals with a fixed probability. In other words:

Assumption 3. *Drop-offs for each pair i between arrivals follow a Bernoulli process, with probability r_i.*

A discussion on how to relax the assumption appears at the end of this section.

We can now show how to extend the CYF_A model to take into account the effect of drop-offs (the same reasoning applies to the abstract KEP_A formulation). In this case, it becomes important to understand *when* a given exchange can be performed. Let τ_i be the index of the time step when a node enters the program, i.e. $\tau_i = k$ iff $i \in N^k$. An exchange involving a set of nodes S can be performed only once the last of the involved nodes arrive. We extend our notation so that, for a set $S \subseteq N$, we have $\tau_S = \max\{\tau_i \mid i \in S\}$.

For an exchange to be performed, all involved nodes should arrive *and remain in the program* long enough. Formally, Constraints (11) should be rewritten as:

$$\sum_{C_j \in \mathcal{C}, S \subseteq C_j} x_j \leq \prod_{i \in S} r_i^{\tau_S - \tau_i} p_i \qquad \forall S \in \mathcal{S} \qquad (14)$$

where $r_i^{\tau_S - \tau_i}$ is the probability that node i remains in the program until τ_S.

Transplants vs Survivors: Taking into account drop-off events enables one further extension. In an online setting, the number of transplants is important, but does not take into account that *having a large pool of participants is also a value*: it means more people have survived and may still receive a transplant.

We can take this into account by tracking the *expected* number of instances of each node i at time step k via an additional set of variables $q_i^k \geq 0$. Formally we have that:

$$q_i^k = \begin{cases} p_i & \text{if } k = \tau_i \\ r_i(q_i^{k-1} - y_i^k) & \text{if } k > \tau_i \end{cases} \qquad \forall i \in N, \forall k \in \{\tau_i \ldots h+1\} \qquad (15)$$

where y_i^k is the expected number of instances of node i that received a transplant at step k, and there is no need to introduce q_i^k variables for $k < \tau_i$. Equation (15) follows directly from Assumption 3. The value of y_i^k is given by:

$$y_i^k = \sum_{\substack{C_j \in \mathcal{C}, i \in C_j, \\ \tau_{C_j} = k}} x_j \qquad \forall i \in N, \forall k \in \{\tau_i \ldots h\} \qquad (16)$$

Incorporating everything in the CYF_A formulation leads to the following model:

$$\max z = \sum_{i \in N} r_i(q_i^h - y_i^h) + \sum_{C_j \in \mathcal{C}} w_j x_j \qquad (\textbf{CYF}_{\textbf{A},\textbf{D}}) \qquad (17)$$

$$\text{s.t.} \quad \sum_{C_j \in \mathcal{C}, S \subseteq C_j} x_j \leq \prod_{i \in S} r_i^{\tau_S - \tau_i} p_i \qquad \forall S \in \mathcal{S} \qquad (18)$$

$$q_i^{\tau_i} = p_i \qquad \forall i \in N \qquad (19)$$

$$q_i^k = r_i(q_i^{k-1} - y_i^k) \qquad \forall i \in N, k \in \{\tau_i + 1 \ldots h\} \qquad (20)$$

$$y_i^k = \sum_{\substack{C_j \in \mathcal{C}, i \in C_j, \\ \tau_{C_j} = k}} x_j \qquad \forall i \in N, \forall k \in \{\tau_i \ldots h\} \qquad (21)$$

$$x_j \in \{0, 1\} \qquad \forall C_j \in \mathcal{C}^0 \qquad (22)$$

$$x_j \in [0, 1] \qquad \forall C_j \in \mathcal{C}^F \qquad (23)$$

$$y_i^k, q_i^k \geq 0 \qquad \forall i \in N, \forall k \in \{\tau_i \ldots h\} \qquad (24)$$

The goal is to maximize the number of transplants, *plus the expected number of survivors* at the end of the look-ahead horizon h. Performing a transplant is still be more beneficial than waiting, since the q_i^{h+1} values are discounted by at least one factor r_i. Taking into account drop-offs results only in marginal modifications of the original CYF_A, while keeping track of survivors requires more extensive changes. None of the additional variables is subject to integrality constraints, however, which is good for the scalability of the method.

Limitations and Workarounds: Here we discuss some of the limitations of our approach, together with some means for addressing them. At the moment, however, none of these solutions has been experimentally evaluated.

Assumption 1 may be violated if there are additional criteria that differentiate the nodes, e.g. different patient survival probabilities. Violations of this kind are easily accounted for by including such factors in Algorithm 1, when searching for equivalent nodes. The price to pay is an increased number of nodes, which in general may be an issue with AEG-based approaches. A possible mitigation measure would be to use ideas from clustering algorithms to merge in Algorithm 1 nodes that are sufficiently similar, even if they are not exactly equivalent.

A violation of Assumption 2 prevents the definition of a vector p of arrival probabilities. However, our methods still work by replacing the product of probabilities in Constraints (8) and (11) with a joint arrival probability $P(S)$, i.e:

$$\sum_{C_j \in \mathcal{C}, i \in C_j} x_j \leq \prod_{i \in S} p_i \longleftrightarrow \sum_{C_j \in \mathcal{C}, i \in C_j} x_j \leq P(S) \tag{25}$$

Non-stationary drop-off probabilities, i.e. time dependent r_i values, require to rewrite the $r_i^{\tau_S - \tau_i}$ expressions in Constraints (18):

$$r_i^{\tau_S - \tau_i} \longleftrightarrow \prod_{k=\tau_i}^{\tau_S} r_i^k \tag{26}$$

where r_i^k is the drop-off probability of node i at step k. We focused on maximizing the number of transplants/survivors, but different objective functions may be employed as long as they can be handled in the problem models. In particular, the objectives considered in [12] based on fairness and expected organ failures should be manageable without much trouble.

4 Experiments

In this section we present our experimentation. We start with a brief survey of the main approaches considered in our comparison in Sect. 4.1. We then present our instances, experimental methods, and results in Sect. 4.2.

4.1 Other Anticipatory Algorithms in the Experimentation

From a mathematical standpoint our approach is closest to the algorithms from [5] and [6], which we therefore chose for our comparison, together with a myopic approach and an oracle (used respectively as baseline and optimistic bound).

Conversely, our method is more distantly related to the ones from [9,12], despite being also sampling-free and with the same practical use cases. Such methods rely on adjusting (e.g. via parameter tuning) the weights of an off-line KEP to take into account the impact of current decisions on the future. The AEG is arguably more accurate from a formal point of view, and does not require a computationally expensive fitting step. However, the approach from [12] can use an off-line KEP formulation with virtually no modification, and provides a bit more flexibility in terms of the supported problem objectives. An empirical comparison of the two approaches is planned as part of future research.

All the considered algorithms treat a scenario as a set of nodes that may enter in the next h steps (no drop-offs are considered the scenarios). The two main algorithms from [5] estimate the expected impact of choosing a cycle, and then solve a KEP using such impacts as weights. The first is referred to as APST1, it can be considered an adaptation of the REGRETS method from [14], and computes cycle scores by solving an offline KEP *for each scenario*. The second is referred to as APST2, it shares ideas with the EXPECTATION method from [14], and computes scores by solving an offline KEP *for each cycle and scenario*. For further details, the reader may refer to [5] or [6].

The CSBA algorithm from [6] solves a single modified KEP on a graph constructed using all scenarios, obtained by using the Sample Average Approximation [20]. The final matching is given by all exchanges in the solution that are defined solely over nodes from the current time step (i.e. that do not include nodes entering in any scenario). In the objective, exchanges related to the current time step are summed exactly, whereas future exchanges are considered in expectation. For further details, the reader may refer to [6].

4.2 Methods and Instances

Some anticipatory algorithms for the KEP may occasionally suggest to delay an exchange, if they estimate that such actions is going to be beneficial on the long term. This may happen even if the delayed exchange could be performed in the current time interval without conflicting with other selected exchanges. For a human decision maker, this kind of behavior is hard to justify, especially when personal health is at stake.

Luckily, in the cycle formulation such a behavior can be prevented or at least discouraged via a simple pre-processing technique. Namely, we can remove from the graph all cycles that include no node currently in the program: as a result, delaying an exchange that can be performed immediately becomes far less likely. In our experimentation, we have applied this technique to all methods that may benefit from that, in particular the AEG-based approach and the CSBA algorithm. Neither APST1 nor APST2 have such need since, they eventually solve a KEP including only the current nodes.

Table 1. 31-months setup sample algorithms (batch 5)

Lookahead/samples	Pairs	Altr.	CSBA		APST2		APST1	
			Lives	WaitL	Lives	WaitL	Lives	WaitL
h2_s5	149	11	61.0 ± 4.08	64.5	60.6 ± 4.32	64.7	62.0 ± 3.52	63.4
h2_s10	149	11	60.6 ± 4.59	65.1	$\mathbf{60.7 \pm 4.63}$	64.5	61.6 ± 4.29	63.8
h2_s15	149	11	60.6 ± 4.50	65.0	60.5 ± 4.70	64.6	62.2 ± 4.39	63.2
h3_s5	149	11	61.2 ± 4.46	64.3	60.5 ± 4.45	64.8	62.1 ± 3.84	63.2
h3_s10	149	11	60.8 ± 4.36	65.0	60.7 ± 4.73	64.5	62.8 ± 3.82	62.4
h3_s15	149	11	60.3 ± 4.52	65.4	60.3 ± 4.70	64.7	61.7 ± 3.91	63.5
h4_s5	149	11	60.7 ± 4.44	64.7	60.5 ± 4.45	64.8	62.2 ± 3.70	62.8
h4_s10	149	11	$\mathbf{61.1 \pm 4.35}$	64.5	60.7 ± 4.73	64.6	62.0 ± 3.52	63.1
h4_s15	149	11	60.8 ± 4.75	64.8	60.6 ± 4.74	64.6	$\mathbf{63.2 \pm 4.08}$	62.2

Instances: We've generated two main instances using real world probabilities: each instance is a (concrete) exchange graph which represents a population. At each time step (corresponding to one month), arrival events are modeled by sampling from the graph a number of nodes (referred to as batch size). Unlike in all works in the literature about the online KEP, pairs are sampled *with reinsertion*: this means that the graphs represent stable or slowly varying populations, and captures much better what happens in the real world compared to sampling without reinsertion. Scenarios are sampled from the same graphs.

Table 2. 31-months setup AEG (batch 5)

Lookahead	Pairs	Altr.	AEG		Myopic		Oracle
			Lives	WaitL	Lives	WaitL	Lives
h2	149	11	**76.1 ± 3.87**	83.9	53.5 ± 2.75	66.5	88.3 ± 7.46
h3	149	11	75.0 ± 3.43	85.0	53.5 ± 2.75	66.5	88.3 ± 7.46
h4	149	11	75.6 ± 3.43	84.4	53.5 ± 2.75	66.5	88.3 ± 7.46

In detail, we generated a small instance with 160 pairs (11 altruistic), used for experiments running over 31 months and batch size 5, and a larger one with 1029 pairs (29 altruistic), for experiments over 12 months and batch size 15 and 20. For each setup, the generated pairs have blood types, PRA values, and ages following the distributions from [23]. Tissue compatibility has been approximated by suppressing a fraction of the arcs, chosen uniformly at random according to statistics reported in the same paper: this process disregards patterns that may arise in the real world and may therefore lead to larger number of types. The AEG graph was generated as described in Sect. 3: in our settings, we detected 22 node types for the small instance and 26 for the big instance, i.e. very small numbers compared to the size of the original graphs. The death rate is adjusted to match the reality that 12% of kidney patients survive 10 years[2].

We implemented all the algorithms in Python, using Numberjack [19] as a modeling front-end and CPLEX as a back-end. As we mentioned, we also include in our comparison an oracle that solves a single offline KEP including all pairs entering the program (and disregarding drop-off dates), which provides an optimistic bound on the performance of any online algorithm. We consider all transplants equally worthy.

4.3 Results

We report the results for the sampling based algorithms (i.e. CSBA, APST1, and APST2) in one set of tables, while the results for the (sampling-free) AEG, myopic, and oracle methods are in a second set of tables.

---

[2] United States Renal Data System (USRDS), 2007: http://www.usrds.org/.

Table 3. 12-months setup sample algorithms (batch 15)

Lookahead/samples	Pairs	Altr	CSBA		APST2		APST1	
			Lives	WaitL	Lives	WaitL	Lives	WaitL
h2_s5	171	9	82.8 ± 6.00	67.8	82.8 ± 7.11	67.8	82.4 ± 7.29	67.1
h2_s10	171	9	$\mathbf{83.2 \pm 6.20}$	68.0	$\mathbf{83.5 \pm 7.06}$	66.9	$\mathbf{84.6 \pm 7.92}$	66.0
h2_s15	171	9	83.0 ± 6.68	67.8	–	–	86.0 ± 7.51	65.0
h3_s5	171	9	82.9 ± 6.59	68.1	82.5 ± 7.06	65.9	82.7 ± 8.21	67.4
h3_s10	171	9	82.3 ± 6.20	68.3	–	–	83.3 ± 7.07	67.5
h3_s15	171	9	82.6 ± 7.65	67.6	–	–	84.0 ± 6.31	67.5
h4_s5	171	9	82.8 ± 6.66	67.9	–	–	79.7 ± 8.64	70.3
h4_s10	171	9	82.4 ± 6.17	68.3	–	–	83.7 ± 4.65	68.5
h4_s15	171	9	82.6 ± 7.75	68.0	–	–	81.6 ± 8.24	69.1

Table 4. 12-months setup AEG (batch 15)

Lookahead	Pairs	Altr.	AEG		Myopic		Oracle
			Lives	WaitL	Lives	WaitL	Lives
h2	171	9	$\mathbf{94.8 \pm 8.59}$	85.2	67.9 ± 5.49	71.0	113.4 ± 2.67
h3	171	9	94.5 ± 8.34	85.5	67.9 ± 5.49	71.0	113.4 ± 2.67
h4	171	9	94.8 ± 8.78	85.2	67.9 ± 5.49	71.0	113.4 ± 2.67

In particular, Tables 1, 3, and 5 show the results of the scenario sampling algorithms: each row is labeled as $\{hx_1\_sx_2\}$, where $x_1 = \{2, 3, 4\}$ is the look-ahead horizon and $x_2 = \{5, 10, 15\}$ is the number of scenarios. Tables 2, 4, and 6 show instead the results of the sampling-free algorithms. Each row is labeled in this case as $\{hx_1\}$, where $x_1 = \{2, 3, 4\}$ is the look-ahead horizon. Each algorithm, except for the Oracle, has two columns: one called "Lives" for the number of transplants, and one called "WaitL" and for the number of pairs still in the waiting list. Each cell reports an average over 10 runs, and for the "Lives" column we also report the standard deviation. The solver timelimit is set to 600 s for each run, therefore for each time step each algorithm has 600/Months setup overall. If an algorithm requires to solve multiple KEPs (say k) in a single time step, then the timelimit for each attempt is set to 600/(Months * k) seconds. A "dash" in a cell means that the algorithm was not able to find a solution for all time steps.

Effect of the Batch Size: In Tables 1, 3, and 5, we show the results of a long run setup with small batch (5) size vs a short run setup with larger batch size (15, 20). As it can be seen in Table 1, $APST1$ outperforms the other scenario-sampling algorithms, but increasing the batch size to 20 is enough for the method to hit the time limit, due to the large number of KEPs that the algorithms needs to solve. The $APST2$ method runs out of time even with batch size 10, since it needs to loop over all scenarios and also all cycles; the algorithm didn't scale for batch size 20. Both the CSBA and AEG algorithms manage to complete all experiments within the time limit.

Table 5. 12-months setup sample algorithms (batch 20)

Lookahead/samples	Pairs	Altr.	CSBA		APST1	
			Lives	WaitL	Lives	WaitL
h2_s5	228	12	106.7 ± 9.68	92.6	107.2 ± 8.83	91.9
h2_s10	228	12	106.8 ± 9.43	92.9	108.2 ± 7.98	90.6
h2_s15	228	12	107.6 ± 10.16	92.1	$\mathbf{108.4 \pm 8.80}$	90.2
h3_s5	228	12	106.8 ± 9.90	92.4	103.9 ± 8.46	94.0
h3_s10	228	12	$\mathbf{107.7 \pm 10.77}$	91.7	106.7 ± 9.21	91.4
h3_s15	228	12	106.7 ± 10.24	93.0	–	–
h4_s5	228	12	106.3 ± 10.04	92.6	103.8 ± 9.27	93.3
h4_s10	228	12	107.4 ± 8.77	92.6	103.83 ± 7.41	95
h4_s15	228	12	106.0 ± 8.91	93.1	–	–

Table 6. 12-months setup AEG (batch 20)

Lookahead	Pairs	Altruistic	AEG		Myopic		Oracle
			Lives	WaitL	Lives	WaitL	Lives
h2	228	12	121.6 ± 9.29	118.4	92.5 ± 8.05	92.3	137.4 ± 13.12
h3	228	12	$\mathbf{122.2 \pm 9.85}$	117.8	92.5 ± 8.05	92.3	137.4 ± 13.12
h4	228	12	121.8 ± 9.41	118.2	92.5 ± 8.05	92.3	137.4 ± 13.12

Effect of Look-Ahead and Number of Scenarios. Increasing the look-ahead and the number of scenarios does not have a consistent effect on the algorithms, which reach the best results for different configurations (in bold in the tables). Unlike [6] we notice that the $CSBA$ algorithm doesn't outperform the other scenario-based algorithms, probably due to the fact that we are sampling pairs with reinsertion.

Trends: The AEG model improves the number of transplants over the myopic algorithm by a factor 26.9% in Table 3 and by a factor 29.7% in Table 5. These results are in line with the improvements obtained by the sampling-free method from [12], and interestingly they are reached *even if the AEG method in fact optimise the number of survivors.*

In Fig. 2 we compare the results for the best configuration: our method consistently (and significantly) outperforms all sampling-based algorithms, in terms of both the number of transplants and that of survivors (transplants + waiting list). Figure 3 shows the evolution of the cumulative (average) number of transplants for each algorithm, which is initially very similar for all methods, until month 3–4 where the AEG models start to outpace the competitors.

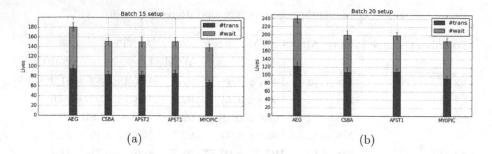

Fig. 2. Comparison of #transplants and #waiting list of the algorithms (a) 15 batch setup, (b) 20 batch setup.

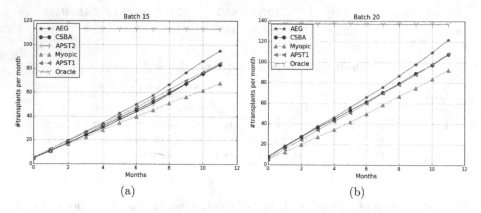

Fig. 3. Trend of the algorithms (a) batch 15 setup, (b) batch 20 setup.

Solution Times: In Fig. 4 we show the solution times of the anticipatory algorithms for the best configuration of 5, 10 and 15 batch size. Each value is measured in seconds and represents the average of 10 runs. The $APST2$ algorithm has only 2 points since it doesn't scale further than 10 batch size, and the solution time for algorithm $APST1$ grows very fast with the batch size. Meanwhile the AEG model remains very scalable even increasing the batch size.

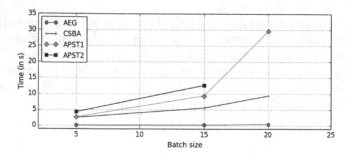

Fig. 4. Times of the algorithms

5 Conclusions and Future Research

Informally speaking, our main contribution allows to build an anticipatory model for the KEP based on probabilistic, rather than statistical, considerations, and therefore to forgo sampling. Additionally, our AEG-based technique allows to take into account the effect of drop-offs. When grounded on the cycle formulation, our method does not require the introduction of additional *integer* variables. Overall, the resulting approach manages to outperform significantly both sampling-based anticipatory algorithms and a myopic approach, while having a scalability similar to the latter. Plans for future research include an experimental comparison with other sampling-free methods, and tackling large-scale, real world, instances. This will likely require the use of column generation, which intuitively could prove particularly effective, given that most of our problem variables are not subject to integrality constraints.

References

1. Abraham, D.J., Blum, A., Sandholm, T.: Clearing algorithms for barter exchange markets: enabling nationwide kidney exchanges. In: Proceedings of EC, pp. 295–304 (2007)
2. Alvelos, F., Klimentova, X., Rais, A., Viana, A.: A compact formulation for maximizing the expected number of transplants in kidney exchange programs. J. Phys. Conf. Ser. **616**, 012011 (2015)
3. Anderson, R., Ashlagi, I., Gamarnik, D., Roth, A.E.: Finding long chains in kidney exchange using the traveling salesman problem. Proc. Natl. Acad. Sci. **112**(3), 663–668 (2015)
4. Ashlagi, I., Roth, A.E.: New challenges in multihospital kidney exchange. Am. Econ. Rev. **102**(3), 354–59 (2012)
5. Awasthi, P., Sandholm, T.: Online stochastic optimization in the large: application to kidney exchange. In: Proceedings of IJCAI, vol. 9, pp. 405–411 (2009)
6. Chisca, D.S., O'Sullivan, B., Lombardi, M., Milano, M.: From off-line to on-line kidney exchange optimization. In: 2018 IEEE 29th International Conference on Tools with Artificial Intelligence (ICTAI). IEEE (2018)
7. Constantino, M., Klimentova, X., Viana, A., Rais, A.: New insights on integer-programming models for the kidney exchange problem. Eur. J. Oper. Res. **231**(1), 57–68 (2013)
8. Dickerson, J.P., Manlove, D.F., Plaut, B., Sandholm, T., Trimble, J.: Position-indexed formulations for kidney exchange. In: Proceedings of EC, pp. 25–42. ACM (2016)
9. Dickerson, J.P., Procaccia, A.D., Sandholm, T.: Dynamic matching via weighted myopia with application to kidney exchange. In: Proceedings of AAAI 2012, pp. 98–100 (2012)
10. Dickerson, J.P., Procaccia, A.D., Sandholm, T.: Failure-aware kidney exchange. In: Proceedings of EC, pp. 323–340. ACM (2013)
11. Dickerson, J.P., Sandholm, T.: Liver and multi-organ exchange. Am. J. Transplant. **13**, 272–273 (2013)
12. Dickerson, J.P., Sandholm, T.: Futurematch: combining human value judgments and machine learning to match in dynamic environments. In: Proceedings of AAAI, pp. 622–628 (2015)

13. Glorie, K.M., van de Klundert, J.J., Wagelmans, A.P.M.: Kidney exchange with long chains: an efficient pricing algorithm for clearing barter exchanges with branch-and-price. Manuf. Serv. Oper. Manage. **16**(4), 498–512 (2014)
14. Hentenryck, P.V., Bent, R.: Online Stochastic Combinatorial Optimization. The MIT Press, Cambridge (2009)
15. Irving, R.W.: The cycle roommates problem: a hard case of kidney exchange. Inf. Process. Lett. **103**(1), 1–4 (2007)
16. Klimentova, X., Alvelos, F., Viana, A.: A new branch-and-price approach for the kidney exchange problem. In: Murgante, B., et al. (eds.) ICCSA 2014. LNCS, vol. 8580, pp. 237–252. Springer, Cham (2014). https://doi.org/10.1007/978-3-319-09129-7_18
17. Mak-Hau, V.: On the kidney exchange problem: cardinality constrained cycle and chain problems on directed graphs: a survey of integer programming approaches. J. Comb. Optim. **33**(1), 35–59 (2017)
18. Manlove, D.F., O'Malley, G.: Paired and altruistic kidney donation in the UK: algorithms and experimentation. ACM J. Exp. Algorithmics **19**(1), 271–282 (2014)
19. Version 1.1.0 Numberjack. https://github.com/eomahony/numberjack
20. Pagnoncelli, B.K., Ahmed, S., Shapiro, A.: Sample average approximation method for chance constrained programming: theory and applications. J. Optim. Theory Appl. **142**(2), 399–416 (2009)
21. Pedroso, J.P.: Maximizing expectation on vertex-disjoint cycle packing. In: Murgante, B., et al. (eds.) ICCSA 2014. LNCS, vol. 8580, pp. 32–46. Springer, Cham (2014). https://doi.org/10.1007/978-3-319-09129-7_3
22. Plaut, B., Dickerson, J.P., Sandholm, T.: Fast optimal clearing of capped-chain barter exchanges. In: Proceedings of AAAI, pp. 601–607 (2016)
23. Saidman, S.L., Roth, A.E., Sönmez, T., Ünver, M.U., Delmonico, F.L.: Increasing the opportunity of live kidney donation by matching for two-and three-way exchanges. Transplantation **81**(5), 773–782 (2006)
24. Santos, N., Tubertini, P., Viana, A., Pedroso, J.P.: Kidney exchange simulation and optimization. J. Oper. Res. Soc. **68**(12), 1521–1532 (2017)

Evaluating Ising Processing Units
with Integer Programming

Carleton Coffrin$^{(\boxtimes)}$, Harsha Nagarajan, and Russell Bent

Los Alamos National Laboratory, Los Alamos, NM, USA
cjc@lanl.gov

Abstract. The recent emergence of novel computational devices, such as adiabatic quantum computers, CMOS annealers, and optical parametric oscillators, present new opportunities for hybrid-optimization algorithms that are hardware accelerated by these devices. In this work, we propose the idea of an Ising processing unit as a computational abstraction for reasoning about these emerging devices. The challenges involved in using and benchmarking these devices are presented and commercial mixed integer programming solvers are proposed as a valuable tool for the validation of these disparate hardware platforms. The proposed validation methodology is demonstrated on a D-Wave 2X adiabatic quantum computer, one example of an Ising processing unit. The computational results demonstrate that the D-Wave hardware consistently produces high-quality solutions and suggests that as IPU technology matures it could become a valuable co-processor in hybrid-optimization algorithms.

Keywords: Discrete optimization · Ising model ·
Quadratic unconstrained binary optimization · Integer programming ·
Large Neighborhood Search · Adiabatic quantum computation

1   Introduction

As the challenge of scaling traditional transistor-based Central Processing Unit (CPU) technology continues to increase, experimental physicists and high-tech companies have begun to explore radically different computational technologies, such as adiabatic quantum computers (AQCs) [1], gate-based quantum computers [2–4], CMOS annealers [5–7], neuromorphic computers [8–10], memristive circuits [11,12], and optical parametric oscillators [13–15]. The goal of all of these technologies is to leverage the dynamical evolution of a physical system to perform a computation that is challenging to emulate using traditional CPU technology (e.g., the simulation of quantum physics) [16]. Despite their entirely disparate physical implementations, AQCs, CMOS annealers, memristive circuits, and optical parametric oscillators are unified by a common mathematical abstraction known as the Ising model, which has been widely adopted by the physics community for the study of naturally occurring discrete optimization

© Springer Nature Switzerland AG 2019
L.-M. Rousseau and K. Stergiou (Eds.): CPAIOR 2019, LNCS 11494, pp. 163–181, 2019.
https://doi.org/10.1007/978-3-030-19212-9_11

processes [17]. Furthermore, this kind of "Ising machine" [13,14] is already commercially available with more than 2000 decision variables in the form of AQCs developed by D-Wave Systems [18].

The emergence of physical devices that can quickly solve Ising models is particularly relevant to the constraint programming, artificial intelligence and operations research communities, because the impetus for building these devices is to perform discrete optimization. As this technology matures, it may be possible for this specialized hardware to rapidly solve challenging combinatorial problems, such as Max-Cut [19] or Max-Clique [20]. Preliminary studies have suggested that some classes of Constraint Satisfaction Problems may be effectively encoded in such devices because of their combinatorial structure [21–24]. Furthermore, an Ising model coprocessor could have significant impacts on solution methods for a variety of fundamental combinatorial problem classes, such as MAX-SAT [25–27] and integer programming [28]. At this time, however, it remains unclear how established optimization algorithms should leverage this emerging technology. This paper helps to address this gap by highlighting the key concepts and hardware limitations that an algorithm designer needs to understand to engage in this emerging and exciting computational paradigm.

Similar to an arithmetic logic unit (ALU) or a graphics processing unit (GPU), this work proposes the idea of an Ising processing unit (IPU) as the computational abstraction for wide variety of physical devices that perform optimization of Ising models. This work begins with a brief introduction to the IPU abstraction and its mathematical foundations in Sect. 2. Then the additional challenges of working with real-world hardware are discussed in Sect. 3 and an overview of previous benchmarking studies and solution methods are presented in Sect. 4. Finally, a detailed benchmarking study of a D-Wave 2X IPU is conducted in Sect. 5, which highlights the current capabilities of such a device. The contributions of this work are as follows,

1. The first clear and concise introduction to the key concepts of Ising models and the limitations of real-world IPU hardware, both of which are necessary for optimization algorithm designers to effectively leverage these emerging hardware platforms (Sects. 2 and 3).
2. Highlighting that integer programming has been overlooked by recent IPU benchmarking studies (Sect. 4), and demonstrating the value of integer programming for filtering easy test cases (Sect. 5.1) and verifying the quality of an IPU on challenging test cases (Sect. 5.2).

Note that, due to the maturity and commercial availability of the D-Wave IPU, this work often refers to that architecture as an illustrative example. However, the methods and tools proposed herein are applicable to all emerging IPU hardware realizations, to the best of our knowledge.

2   A Brief Introduction to Ising Models

This section introduces the notations of the paper and provides a brief introduction to Ising models, the core mathematical abstraction of IPUs. The Ising model refers to the class of graphical models where the nodes, $\mathcal{N}$, represent *spin* variables (i.e., $\sigma_i \in \{-1, 1\}\ \forall i \in \mathcal{N}$) and the edges, $\mathcal{E}$, represent *interactions* of spin variables (i.e., $\sigma_i \sigma_j\ \forall i, j \in \mathcal{E}$). A local *field* $h_i\ \forall i \in \mathcal{N}$ is specified for each node, and an interaction strength $J_{ij}\ \forall i, j \in \mathcal{E}$ is specified for each edge. Given these data, the *energy* of the Ising model is defined as,

$$E(\sigma) = \sum_{i,j \in \mathcal{E}} J_{ij} \sigma_i \sigma_j + \sum_{i \in \mathcal{N}} h_i \sigma_i \tag{1}$$

Applications of the Ising model typically consider one of two tasks. First, some applications focus on finding the lowest possible energy of the Ising model, known as a *ground state*. That is, finding the globally optimal solution of the following binary quadratic optimization problem:

$$\min : E(\sigma)$$
$$\text{s.t.}: \sigma_i \in \{-1, 1\}\ \forall i \in \mathcal{N} \tag{2}$$

Second, other applications are interested in sampling from the Boltzmann distribution of the Ising model's states:

$$Pr(\sigma) \propto e^{\frac{-E(\sigma)}{\tau}} \tag{3}$$

where τ is a parameter representing the *effective temperature* of the Boltzmann distribution [29]. It is valuable to observe that in the Boltzmann distribution, the lowest energy states have the highest probability. Therefore, the task of sampling from a Boltzmann distribution is similar to the task of finding the lowest energy of the Ising model. Indeed, as τ approaches 0, the sampling task smoothly transforms into the aforementioned optimization task. This paper focuses exclusively on the mathematical program presented in (2), the optimization task.

Frustration: The notion of frustration is common in the study of Ising models and refers to any instance of (2) where the optimal solution, σ^*, satisfies the property,

$$E(\sigma^*) > \sum_{i,j \in \mathcal{E}} -|J_{ij}| - \sum_{i \in \mathcal{N}} |h_i| \tag{4}$$

A canonical example is the following three node problem:

$$h_1 = 0,\ h_2 = 0,\ h_3 = 0,\ J_{12} = -1,\ J_{23} = -1,\ J_{13} = 1 \tag{5}$$

Observe that, in this case, there are a number of optimal solutions such that $E(\sigma^*) = -2$ but none such that $E(\sigma) = \sum_{i,j \in \mathcal{E}} -|J_{ij}| = -3$. Note that frustration has important algorithmic implications as greedy algorithms are sufficient for optimizing Ising models without frustration.

Gauge Transformations: A valuable property of the Ising model is the gauge transformation, which characterizes an equivalence class of Ising models. For illustration, consider the optimal solution of Ising model S, $\boldsymbol{\sigma}^{s*}$. One can construct a new Ising model T where the optimal solution is the same, except that $\boldsymbol{\sigma}_i^{t*} = -\boldsymbol{\sigma}_i^{s*}$ for a particular node $i \in \mathcal{N}$ is as follows:

$$J_{ij}^t = -J_{ij}^s \ \forall i, j \in \mathcal{E}(i) \tag{6a}$$
$$h_i^t = -h_i^s \tag{6b}$$

where $\mathcal{E}(i)$ indicates the neighboring edges of node i. This S-to-T manipulation is referred to as a gauge transformation. Given a complete source state $\boldsymbol{\sigma}^s$ and a complete target state $\boldsymbol{\sigma}^t$, this transformation is generalized to all of σ by,

$$J_{ij}^t = J_{ij}^s \sigma_i^s \sigma_j^s \sigma_i^t \sigma_j^t \ \forall i, j \in \mathcal{E} \tag{7a}$$
$$h_i^t = h_i^s \sigma_i^s \sigma_i^t \ \forall i \in \mathcal{N} \tag{7b}$$

It is valuable to observe that by using this gauge transformation property, one can consider the class of Ising models where the optimal solution is $\sigma_i^* = -1 \ \forall i \in \mathcal{N}$ or any arbitrary vector of $-1, 1$ values without loss of generality.

Bijection of Ising and Boolean Optimization: It is also useful to observe that there is a bijection between Ising optimization (i.e., $\sigma \in \{-1, 1\}$) and Boolean optimization (i.e., $x \in \{0, 1\}$). The transformation of σ-to-x is given by,

$$\sigma_i = 2x_i - 1 \ \forall i \in \mathcal{N} \tag{8a}$$
$$\sigma_i \sigma_j = 4x_i x_j - 2x_i - 2x_j + 1 \ \forall i, j \in \mathcal{E} \tag{8b}$$

and the inverse x-to-σ is given by,

$$x_i = \frac{\sigma_i + 1}{2} \ \forall i \in \mathcal{N} \tag{9a}$$
$$x_i x_j = \frac{\sigma_i \sigma_j + \sigma_i + \sigma_j + 1}{4} \ \forall i, j \in \mathcal{E} \tag{9b}$$

Consequently, any results from solving Ising models are also immediately applicable to the following class of Boolean optimization problems:

$$\min : \sum_{i,j \in \mathcal{E}} c_{ij} x_i x_j + \sum_{i \in \mathcal{N}} c_i x_i$$
$$\text{s.t.: } x_i \in \{0, 1\} \ \forall i \in \mathcal{N} \tag{10}$$

The Ising model provides a clean mathematical abstraction for understanding the computation that IPUs perform. However, in practice, a number of hardware implementation factors present additional challenges for computing with IPUs.

3 Features of Analog Ising Processing Units

The core inspiration for developing IPUs is to take advantage of the natural evolution of a discrete physical system to find high-quality solutions to an Ising model [1,6,11,13]. Consequently, to the best of our knowledge, all IPUs developed to date are analog machines, which present a number of challenges that the optimization community is not accustomed to considering.

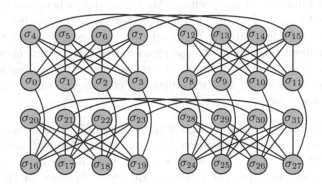

Fig. 1. A 2-by-2 chimera graph illustrating the variable product limitations of a D-Wave 2X IPU.

Effective Temperature: The ultimate goal of IPUs is to solve the optimization problem (2) and determine the globally optimal solution to the input Ising model. In practice, however, a variety of analog factors preclude IPUs from reliably finding globally optimal solutions. As a first-order approximation, current IPUs behave like a Boltzmann sampler (3) with some hardware-specific effective temperature, τ [30]. It has also been observed that the effective temperature of an IPU can vary around a nominal value based on the Ising model that is being executed [31]. This suggests that the IPU's performance can change based on the structure of the problem input.

Environmental Noise: One of the primary contributors to the sampling nature of IPUs are the environmental factors. All analog machines are subject to faults due to environmental noise; for example, even classical computers can be affected by cosmic rays. However, given the relative novelty of IPUs, the effects of environmental noise are noticeable in current hardware. The effects of environmental noise contribute to the perceived effective temperature τ of the IPU.

Coefficient Biases: Once an Ising model is input into an IPU, its coefficients are subject to at least two sources of bias. The first source of bias is a model programming error that occurs independently each time the IPU is configured for a computation. This bias is often mitigated by programming the IPU multiple times with an identical input and combining the results from all executions. The second source of bias is a persistent coefficient error, which is an artifact of the IPU manufacturing and calibration process. Because this bias is consistent across multiple IPU executions, this source of bias is often mitigated by performing multiple gauge transformations on the input and combining the results from all executions.

Problem Coefficients: In traditional optimization applications, the problem coefficients are often rescaled to best suit floating-point arithmetic. Similarly, IPUs have digital-to-analog converters that can encode a limited number of values; typically these values are represented as numbers in the range of -1 to 1. Some IPUs allow for hundreds of steps within this range, [1,6] whereas others support only the discrete set of $\{-1, 0, 1\}$ [13]. In either case, the mathematical Ising model must be rescaled into the IPU's operating range. However, this mathematically equivalent transformation can result in unexpected side effects because the coefficients used in the IPU hardware are perturbed by a constant amount of environmental noise and hardware bias, which can outweigh small rescaled coefficient values.

Topological Limitations: Another significant feature of IPUs is a restricted set of variable products. In classical optimization (e.g., (2)), it is assumed that every variable can interact with every other variable, that is, an Ising model where an edge connects every pair of variables. However, because of the hardware implementation of an IPU, it may not be possible for some variables to interact. For example, the current D-Wave IPUs are restricted to the *chimera* topology, which is a two-dimensional lattice of *unit cells*, each of which consist of a 4-by-4 bipartite graph (e.g., see Fig. 1). In addition to these restrictions, fabrication errors can also lead to random failures of nodes and edges in the IPU hardware. Indeed, as a result of these minor imperfections, every D-Wave IPU developed to date has a unique topology [32–34]. Research and development of algorithms for embedding various kinds of Ising models into a specific IPU topology is still an active area of research [21,35–37].

3.1 Challenges of Benchmarking Ising Processing Units

These analog hardware features present unique challenges for benchmarking IPUs that fall roughly into three categories: (1) comparing to established benchmark libraries; (2) developing Ising model instance generators for testing and; (3) comparing with classical optimization methods.

Benchmark Libraries: Research and development in optimization algorithms has benefited greatly from standardized benchmark libraries [38–40]. However, direct application of these libraries to IPUs is out of scope in the near term for the following reasons: (1) the Ising model is a binary quadratic program, which is sufficiently restrictive to preclude the use of many standard problem libraries; (2) even in cases where the problems of interest can be mapped directly to the Ising model (e.g., Max-Cut, Max-Clique), the task of embedding given problems onto the IPU's hardware graph can be prohibitive [41]; and (3) even if an embedding can be found, it is not obvious that the problem's coefficients will be amenable to the IPU's operating range.

Instance Generation Algorithms: Due to these challenges, the standard practice in the literature is to generate a collection of instances for a given IPU and use these cases for the evaluation of that IPU [33,34,42,43]. The hope being that these instances provide a reasonable proxy for how real-world applications might perform on such a device.

Comparison with Classical Algorithms: Because of the radically different hardware of CPUs vs IPUs and the stochastic nature of the IPUs, conducting a fair comparison of these two technologies is not immediately clear [43–45]. Indeed, comparisons of D-Wave's IPU with classical algorithms have resulted in vigorous discussions about what algorithms and metrics should be used to make such comparisons [34,46,47]. It is widely accepted that IPUs do not provide optimality guarantees and are best compared to heuristic methods (e.g. local search) in terms of runtime performance. This debate will most likely continue for several years. In this work, our goal is not to answer these challenging questions but rather to highlight that commercial mixed integer programming solvers are valuable and important tools for exploring these questions.

4   A Review of Ising Processing Unit Benchmarking Studies

Due to the challenges associated with mapping established optimization test cases to specific IPU hardware [41], the IPU benchmarking community has adopted the practice of generating Ising model instances on a case-by-case basis for specific IPUs [33,34,42,43] and evaluating these instances on a variety of solution methods. The following subsections provide a brief overview of the instance generation algorithms and solution methods that have been used in various IPU benchmarking studies. The goals of this review are to: (1) reveal the lack of consistency across current benchmarking studies; (2) highlight the omission of integer programming methods in all of the recent publications and; (3) motivate the numerical study conducted in this work.

4.1   Instance Generation Algorithms

The task of IPU instance generation amounts to finding interesting values for h and J in (1). In some cases the procedures for generating these values are elaborate [33,48] and are designed to leverage theoretical results about Ising models [42]. A brief survey reveals five primary problem classes in the literature, each of which is briefly introduced. For a detailed description, please refer to the source publication of the problem class.

Random (RAN-k and RANF-k): To the best of our knowledge, this general class of problem was first proposed in [27] and was later refined into the RAN-k problem in [34]. The RAN-k problem consists simply of assigning each value of **h** to 0 and each value of **J** uniformly at random from the set

$$\{-k, -k+1, \ldots, -2, -1, 1, 2, \ldots, k-1, k\} \tag{11}$$

The RANF-k problem is a simple variant of RAN-k where the values of **h** are also selected uniformly at random from (11). As we will later see, RAN-1 and RANF-1, where $h, J \in \{-1, 1\}$, are an interesting subclass of this problem.

Frustrated Loops (FL-k and FCL-k): The frustrated loop problem was originally proposed in [42] and then later refined to the FL-k problem in [48]. It consists of generating a collection of random cycles in the IPU graph. In each cycle, all of the edges are set to -1 except one random edge, which is set to 1 to produce *frustration*. A scaling factor, α, is used to control how many random cycles should be generated, and the parameter k determines how many cycles each edge can participate in. A key property of the FL-k generation procedure is that two globally optimal solutions are maintained at $\sigma_i = -1 \; \forall i \in \mathcal{N}$ and $\sigma_i = 1 \; \forall i \in \mathcal{N}$ [48]. However, to obfuscate this solution, a gauge transformation is often applied to make the optimal solution a random assignment of σ.

A variant of the frustrated loop problem is the frustrated *cluster* loop problem, FCL-k [43]. The FCL-k problem is inspired by the chimera network topology (i.e., Fig. 1). The core idea is that tightly coupled variables (e.g., $\sigma_0 \ldots \sigma_7$ in Fig. 1) should form a *cluster* where all of the variables take the same value. This is achieved by setting all of the values of **J** within the cluster to -1. For the remaining edges between clusters, the previously described frustrated cycles generation scheme is used. Note that a polynomial time algorithm is known for solving the FCL-k problem class on chimera graphs [45].

It is worthwhile to mention that the FL-k and FCL-k instance generators are solving a cycle packing problem on the IPU graph. Hence, the randomized algorithms proposed in [42,43] are not guaranteed to find a solution if one exists. In practice, this algorithm fails for the highly constrained settings of α and k.

Weak-Strong Cluster Networks (WSCNs): The WSCN problem was proposed in [33] and is highly specialized to the chimera network topology. The basic building block of a WSCN is a pair of spin clusters in the chimera graph (e.g., $\sigma_0 \ldots \sigma_7$ and $\sigma_8 \ldots \sigma_{15}$ in Fig. 1). In the *strong* cluster the values of **h** are set to the strong force parameter *sf* and in the *weak* cluster the values of **h** are set to the weak force parameter *wf*. All of the values of **J** within and between this cluster pair are set to -1. Once a number of weak-strong cluster pairs have been placed, the strong clusters are connected to each other using random values of $J \in \{-1, 1\}$. The values of $sf = -1.0$ and $wf = 0.44$ are recommended by [33]. The motivation for the WSCN design is that the clusters create deep local minima that are difficult for local search methods to escape.

4.2   Solution Methods

Once a collection of Ising model instances have been generated, the next step in a typical benchmarking study is to evaluate those instances on a variety of solution methods, including the IPU, and compare the results. A brief survey reveals five primary solution methods in the literature, each of which is briefly introduced. For a detailed description, please refer to the source publications of the solution method.

Simulated Annealing: The most popular staw-man solution method for comparison is Simulated Annealing [49]. Typically the implementation only considers a neighborhood of single variable flips and the focus of these implementations is on computational performance (e.g. using GPUs for acceleration). The search is run until a specified time limit is reached.

Large Neighborhood Search: The state-of-the-art meta-heuristic for solving Ising models on the chimera graphs is a Large Neighborhood Search (LNS) method called the Hamze-Freitas-Selby (HFS) algorithm [50,51]. The core idea of this algorithm is to extract low treewidth subgraphs of the given Ising model and then use dynamic programming to compute the optimal configuration of these subgraphs. This extract and optimize process is repeated until a specified time limit is reached. A key to this method's success is the availability of a highly optimized open-source C implementation [52].

Integer Programming: Previous works first considered integer quadratic programming [27] and quickly moved to integer linear programming [53,54] as a solution method. The mathematical programming survey [55] provides a useful overview of the advantages and dis-advantages of various integer programming (IP) formulations.

Based on some preliminary experiments with different formulations, this work focuses on the following integer linear programming formulation of the Ising model, transformed into the Boolean variable space:

$$\min : \sum_{i,j \in \mathcal{E}} c_{ij} x_{ij} + \sum_{i \in \mathcal{N}} c_i x_i + c \tag{12a}$$

s.t.:

$$x_{ij} \geq x_i + x_j - 1, \ \ x_{ij} \leq x_i, \ \ x_{ij} \leq x_j \ \forall i,j \in \mathcal{E} \tag{12b}$$
$$x_i \in \{0,1\} \ \forall i \in \mathcal{N}, \ \ x_{ij} \in \{0,1\} \ \forall i,j \in \mathcal{E}$$

where the application of (8) leads to,

$$c_{ij} = \sum_{i,j \in \mathcal{E}} 4 J_{ij} \ \forall i,j \in \mathcal{E} \tag{13a}$$

$$c_i = \sum_{i,j \in \mathcal{E}(i)} 2 J_{ij} + \sum_{i \in \mathcal{N}} 2 h_i \ \forall i \in \mathcal{N} \tag{13b}$$

$$c = \sum_{i,j \in \mathcal{E}} J_{ij} - \sum_{i \in \mathcal{N}} h_i \tag{13c}$$

In this formulation, the binary quadratic program defined in (10) is converted to a binary linear program by lifting the variable products $x_i x_j$ into a new variable x_{ij} and adding linear constraints to capture the $x_{ij} = x_i \wedge x_j \; \forall i, j \in \mathcal{E}$ conjunction constraints. Preliminary experiments of this work confirmed the findings of [55], that this binary linear program formulation is best on sparse graphs, such as the hardware graphs of current IPUs.

Table 1. A chronological summary of IPU benchmarking studies

Publication	Problem classes					Solution methods				
	RAN	RANF	FL	FCL	WSCN	IP	SA	LNS	QMC	AQC
[27]	✓					✓				
[53]	✓					✓				
[54]		✓				✓				
[42]			✓				✓	✓		✓
[48]			✓				✓	✓		✓
[60]	✓		✓					✓	✓	✓
[33]					✓		✓		✓	✓
[43]				✓			✓	✓	✓	✓
This work	✓	✓	✓	✓	✓	✓		✓		✓

Adiabatic Quantum Computation: An adiabatic quantum computation (AQC) [56] is a method for solving an Ising model via a quantum annealing process [57]. This solution method has two notable traits: (1) the AQC dynamical process features quantum tunneling [58], which can help it to escape from local minima; (2) it can be implemented in hardware (e.g. the D-Wave IPU).

Quantum Monte Carlo: Quantum Monte Carlo (QMC) is a probabilistic algorithm that can be used for simulating large quantum systems. QMC is a very computationally intensive method [33,59] and thus the primary use of QMC is not to compare runtime performance but rather to quantify the possible value of an adiabatic quantum computation that could be implemented in hardware at some point in the future.

4.3   Overview

To briefly summarize a variety of benchmarking studies, Table 1 provides an overview of the problems and solution methods previous works have considered. Although there was some initial interest in integer programming models [27, 53,54], more recent IPU benchmark studies have not considered these solution methods and have focused exclusively on heuristic methods. Furthermore, there are notable inconsistencies in the type of problems being considered. As indicated by the last row in Table 1, the goal of this work is revisit the use of IP methods for benchmarking IPUs and to conduct a thorough and side-by-side study of all problem classes and solution methods proposed in the literature. Note that, because this paper focuses exclusively on the quality and runtime of the Ising model optimization task (2), the study of SA and QMC are omitted as they provide no additional insights over the LNS [48] and AQC [33] methods.

5   A Study of Established Methods

This section conducts an in-depth computational study of the established instance generation algorithms and solution methods for IPUs. The first goal of this study is to understand what classes of problems and parameters are the most challenging, as such cases are preferable for benchmarking. The second goal is to conduct a validation study of a D-Wave 2X IPU, to clearly quantify its solution quality and runtime performance. This computational study is divided into two phases. First, a broad parameter sweep of all possible instance generation algorithms is conducted and a commercial mixed-integer programming solver is used to filter out the easy problem classes and parameter settings. Second, after the most challenging problems have been identified, a detailed study is conducted to compare and contrast the three disparate solution methods IP, LNS, and AQC.

Throughout this section, the following notations are used to describe the algorithm results: UB denotes the objective value of the best feasible solution produced by the algorithm within the time limit, LB denotes the value of the best lower bound produced by the algorithm within the time limit, T denotes the algorithm runtime in seconds[1], TO denotes that the algorithm hit a time limit of 600 s, $\mu(\cdot)$ denotes the mean of a collection of values, $sd(\cdot)$ denotes the standard deviation of a collection of values, and $max(\cdot)$ denotes the maximum of a collection of values.

Computation Environment: The classical computing algorithms are run on HPE ProLiant XL170r servers with dual Intel 2.10 GHz CPUs and 128 GB memory. After a preliminary comparison of CPLEX 12.7 [61] and Gurobi 7.0 [62], no significant difference was observed. Thus, Gurobi was selected as the commercial Mixed-Integer Programming (MIP) solver and was configured to use one thread. The highly specialized and optimized HFS algorithm [52] is used as an LNS-based heuristic and also uses one thread.

The IPU computation is conducted on a D-Wave 2X [63] adiabatic quantum computer (AQC). This computer has a 12-by-12 chimera cell topology with random omissions; in total, it has 1095 spins and 3061 couplers and an effective temperature of $\tau \in (0.091, 0.053)$ depending on the problem being solved [64, 65]. Unless otherwise noted, the AQC is configured to produce 10,000 samples using a 5-µs annealing time per sample and a random gauge transformation every 100 samples. The best sample is used in the computation of the upper bound value. The reported runtime of the AQC reflects the amount of time used on the IPU hardware; it does not include the overhead of communication or scheduling of the computation, which adds an overhead of about three seconds.

---

[1] For MIP solvers, the runtime includes the computation of the optimally certificate.

Table 2. Parameter settings of various problems.

Problem	First param.	Second param.
RAN-k	$k \in (1..5 : 1)$	NA
RANF-k	$k \in (1..5 : 1)$	NA
FL-k	$k \in (1..5 : 1)$	$\alpha \in (0..1 : 0.1)$
FCL-k	$k \in (1..5 : 1)$	$\alpha \in (0..1 : 0.1)$
WSCN	$wf \in (-1..1 : 0.2)$	$sf \in (-1..1 : 0.2)$

Table 3. MIP runtime on various IPU benchmark problems (seconds)

Problem	Cases	$\mu(|\mathcal{N}|)$	$\mu(|\mathcal{E}|)$	$\mu(T)$	$sd(T)$	$max(T)$
RAN	1250	1095	3061	TO	—	TO
RANF	1250	1095	3061	TO	—	TO
FL	6944	1008	2126	1.82	1.06	16.80
FCL	8347	888	2282	4.19	2.81	41.40
WSCN	30250	949	2313	0.25	0.87	17.90

All of the software used in this benchmarking study is available as open-source via: BQPJSON, a language-independent JSON-based Ising model exchange format designed for benchmarking IPU hardware; DWIG, algorithms for IPU instance generation; BQPSOLVERS, tools for encoding BQPJSON data into various optimization formulations and solvers.[2]

5.1   Identifying Challenging Cases

Broad Parameter Sweep: In this first experiment, we conduct a parameter sweep of all the inputs to the problem generation algorithms described in Sect. 4.1. Table 2 provides a summary of the input parameters for each problem class. The values of each parameter are encoded with the following triple: (start..stop : step size). When two parameters are required for a given problem class, the cross product of all parameters is used. For each problem class and each combination of parameter settings, 250 random problems are generated in order to produce a reasonable estimate of the average difficulty of that configuration. Each problem is generated using all of the decision variables available on the IPU. The computational results of this parameter sweep are summarized in Table 3.

The results presented in Table 3 indicate that, at this problem size, all variants of the FL, FCL, and WSCN problems are easy for modern MIP solvers. This is a stark contrast to [33], which reported runtimes around 10,000 s when applying Simulated Annealing to the WSCN problem. Furthermore, this result

---

[2] The source code is available at https://github.com/lanl-ansi/ under the repository names BQPJSON, DWIG and BQPSOLVERS.

Table 4. MIP runtime on RAN-k and RANF-k IPU benchmark problems (seconds)

k	Cases	$\mu(|\mathcal{N}|)$	$\mu(|\mathcal{E}|)$	$\mu(T)$	$sd(T)$	$max(T)$	$\mu(T)$	$sd(T)$	$max(T)$
Problems of increasing k				RAN-k			RANF-k		
1	250	194	528	340.0	195.0	TO	14.10	15.20	82.70
2	250	194	528	89.3	64.3	481	2.97	3.41	22.70
3	250	194	528	64.8	28.3	207	1.67	1.48	10.70
4	250	194	528	58.0	29.5	250	1.25	0.83	6.10
5	250	194	528	49.0	23.0	131	1.12	0.77	6.98
6	250	194	528	49.0	22.4	119	1.05	0.59	4.47
7	250	194	528	45.0	22.8	128	1.04	0.75	7.60
8	250	194	528	44.8	23.7	121	1.01	0.62	5.43
9	250	194	528	42.3	22.3	110	0.98	0.60	5.08
10	250	194	528	39.8	22.1	107	0.91	0.43	3.09

suggests that these problems classes are not ideal candidates for benchmarking
IPUs. In contrast, the RAN and RANF cases consistently hit the runtime limit of
the MIP solver, suggesting that these problems are more useful for benchmarking. This result is consistent with a similar observation in the SAT community,
where random SAT problems are known to be especially challenging [66,67]. To
get a better understanding of these RAN problem classes, we next perform a
detailed study of these problems for various values of the parameter k.

The RAN and RANF Problems: In this second experiment, we focus on the
RAN-k and RANF-k problems and conduct a detailed parameter sweep of $k \in$
$(1..10 : 1)$. To accurately measure the runtime difficulty of the problem, we also
reduce the size of the problem from 1095 variables to 194 variables so that the
MIP solver can reliably terminate within a 600 s time limit. The results of this
parameter sweep are summarized in Table 4.

The results presented in Table 4 indicate that (1) as the value of k increases,
both the RAN and RANF problems become easier; and (2) the RANF problem is
easier than the RAN problem. The latter is not surprising because the additional
linear coefficients in the RANF problem break many of the symmetries that exist
in the RAN problem. These results suggest that it is sufficient to focus on the
RAN-1 and RANF-1 cases for a more detailed study of IPU performance. This
is a serendipitous outcome for IPU benchmarking because restricting the problem coefficients to $\{-1, 0, 1\}$ reduces artifacts caused by noise and the numeral
precision of the analog hardware.

5.2   An IPU Evaluation Using RAN-1 and RANF-1

Now that the RAN-1 and RANF-1 problem classes have been identified as the
most interesting for IPU benchmarking, we perform two detailed studies on these

problems using all three algorithmic approaches (i.e., AQC, LNS, and MIP). The first study focuses on the scalability trends of these solution methods as the problem size increases, whereas the second study focuses on a runtime analysis of the largest cases that can be evaluated on a D-Wave 2X IPU hardware.

Scalability Analysis: In this experiment, we increase the problem size gradually to understand the scalability profile of each of the solution methods (AQC, LNS, and MIP). The results are summarized in Table 5. Focusing on the smaller problems, where the MIP solver provides an optimality proof, we observe that both the AQC and the LNS methods find the optimal solution in all of the sampled test cases, suggesting that both heuristic solution methods are of high quality.

Table 5. A comparison of solution quality and runtime as problem size increases on RAN-1 and RANF-1.

			AQC		LNS		MIP		
Cases	$\mu(|\mathcal{N}|)$	$\mu(|\mathcal{E}|)$	$\mu(UB)$	$\mu(T)$	$\mu(UB)$	$\mu(T)$	$\mu(UB)$	$\mu(LB)$	$\mu(T)$
RAN-1 problems of increasing size									
250	30	70	−44	3.53	−44	10	−44	−44	0.05
250	69	176	−110	3.57	−110	10	−110	−110	0.48
250	122	321	−199	3.60	−199	10	−199	−199	15.90
250	194	528	−325	3.64	−325	10	−325	−327	340.00
250	275	751	−462	3.68	−462	10	−461	−483	TO
250	375	1030	−633	3.73	−633	10	−629	−673	TO
250	486	1337	−821	3.77	−822	10	−814	−881	TO
250	613	1689	−1038	3.77	−1039	10	−1021	−1116	TO
250	761	2114	−1296	3.76	−1297	10	−1262	−1401	TO
250	923	2578	−1574	3.77	−1576	10	−1525	−1713	TO
250	1095	3061	−1870	3.80	−1873	10	−1806	−2045	TO
RANF-1 problems of increasing size									
250	30	70	−53	3.53	−53	10	−53	−53	0.02
250	69	176	−127	3.56	−127	10	−127	−127	0.13
250	122	321	−229	3.61	−229	10	−229	−229	0.67
250	194	528	−370	3.66	−370	10	−370	−370	14.10
250	275	751	−526	3.71	−526	10	−526	−527	128.00
250	375	1030	−719	3.76	−719	10	−719	−727	471.00
250	486	1337	−934	3.81	−934	10	−933	−954	588.00
250	613	1689	−1179	3.82	−1179	10	−1178	−1211	TO
250	761	2114	−1472	3.82	−1472	10	−1470	−1520	TO
250	923	2578	−1786	3.82	−1787	10	−1778	−1856	TO
250	1095	3061	−2121	3.86	−2122	10	−2110	−2212	TO

Focusing on the larger problems, we observe that, in just a few seconds, both AQC and LNS find feasible solutions that are of higher quality than what the MIP solver can find in 600 s. This suggests that both methods are producing high-quality solutions at this scale. As the problem size grows, a slight quality discrepancy emerges favoring LNS over AQC; however, this discrepancy in average solution quality is less than 1% of the best known value.

Detailed Runtime Analysis: Given that both the AQC and the LNS solution methods have very similar solution qualities, it is prudent to perform a detailed runtime study to understand the quality vs. runtime tradeoff. To develop a runtime profile of the LNS algorithm, the solver's runtime limit is set to values ranging from 0.01 to 10.00 s. In the case of the AQC algorithm, the number of requested samples is set to values ranging from 10 to 10,000, which has the effect of scaling the runtime of the IPU process. The results of this study are summarized in Fig. 2. Note that the stochastic sampling nature of the IPU results in some noise for small numbers of samples. However, the overall trend is clear.

The results presented in Fig. 2 further illustrate that (1) the RAN problem class is more challenging than the RANF problem class, and (2) regardless of the runtime configuration used, the LNS heuristic slightly outperforms the AQC; however, the average solution quality is always within 1% of each other. Combining all of the results from this section provides a strong validation that even if the D-Wave 2X IPU cannot guarantee a globally optimal solution, it produces high quality solutions reliably across a wide range of inputs.

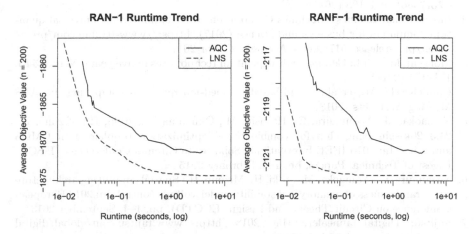

Fig. 2. Detailed runtime analysis of the AQC (D-Wave 2X) and LNS heuristic (HFS) on the RAN-1 (left) and RANF-1 (right) problem classes.

6   Conclusion

This work introduces the idea of Ising processing units (IPUs) as a computational abstraction for emerging physical devices that optimize Ising models.

It highlights a number of unexpected challenges in using such devices and proposes commercial mixed-integer programming solvers as a tool to help improve validation and benchmarking.

A baseline study of the D-Wave 2X IPU suggests that the hardware specific instance generation is a reasonable strategy for benchmarking IPUs. However, finding a class of challenging randomly generated test cases is non-trivial and an open problem for future work. The study verified that at least one commercially available IPU is already comparable to current state-of-the-art classical methods on some classes of problems (e.g. RAN and RANF). Consequently, as this IPU's hardware increases in size, one would expect that it could outperform state-of-the-art classical methods because of its parallel computational nature and become a valuable co-processor in hybrid-optimization algorithms.

Overall, we find that the emergence of IPUs is an interesting development for the optimization community and warrants continued study. Considerable work remains to determine new challenging classes of test cases for validating and benchmarking IPUs. We hope that the technology overview and the validation study conducted in this work will assist the optimization research community in exploring IPU hardware platforms and will accelerate the development of hybrid-algorithms that can effectively leverage these emerging technologies.

References

1. Johnson, M.W., et al.: Quantum annealing with manufactured spins. Nature **473**(7346), 194–198 (2011)
2. International Business Machines Corporation: IBM building first universal quantum computers for business and science (2017). https://www-03.ibm.com/press/us/en/pressrelease/51740.wss. Accessed 28 Apr 2017
3. Mohseni, M., et al.: Commercialize quantum technologies in five years. Nature **543**, 171–174 (2017)
4. Chmielewski, M., et al.: Cloud-based trapped-ion quantum computing. In: APS Meeting Abstracts (2018)
5. Yamaoka, M., Yoshimura, C., Hayashi, M., Okuyama, T., Aoki, H., Mizuno, H.: 24.3 20k-spin Ising chip for combinational optimization problem with CMOS annealing. In: 2015 IEEE International Solid-State Circuits Conference - (ISSCC) Digest of Technical Papers, pp. 1–3, February 2015
6. Yoshimura, C., Yamaoka, M., Aoki, H., Mizuno, H.: Spatial computing architecture using randomness of memory cell stability under voltage control. In: 2013 European Conference on Circuit Theory and Design (ECCTD), pp. 1–4, September 2013
7. Fujitsu: Digital annealer, May 2018. http://www.fujitsu.com/global/digital annealer/. Accessed 26 Feb 2019
8. Modha, D.S.: Introducing a brain-inspired computer (2017). http://www.research.ibm.com/articles/brain-chip.shtml. Accessed 28 Apr 2017
9. Davies, M., et al.: Loihi: a neuromorphic manycore processor with on-chip learning. IEEE Micro **38**(1), 82–99 (2018)
10. Schuman, C.D., et al.: A survey of neuromorphic computing and neural networks in hardware (2017). arXiv preprint: arXiv:1705.06963
11. Caravelli, F.: Asymptotic behavior of memristive circuits and combinatorial optimization (2017)

12. Traversa, F.L., Di Ventra, M.: MemComputing integer linear programming (2018)
13. McMahon, P.L., et al.: A fully-programmable 100-spin coherent Ising machine with all-to-all connections. Science **354**, 614–617 (2016). https://doi.org/10.1126/science.aah5178
14. Inagaki, T., et al.: A coherent Ising machine for 2000-node optimization problems. Science **354**(6312), 603–606 (2016)
15. Kielpinski, D., et al.: Information processing with large-scale optical integrated circuits. In: 2016 IEEE International Conference on Rebooting Computing (ICRC), pp. 1–4, October 2016
16. Feynman, R.P.: Simulating physics with computers. Int. J. Theor. Phys. **21**(6), 467–488 (1982)
17. Brush, S.G.: History of the lenz-ising model. Rev. Mod. Phys. **39**, 883–893 (1967)
18. D-Wave Systems Inc.: Customers (2017). https://www.dwavesys.com/our-company/customers. Accessed 28 Apr 2017
19. Haribara, Y., Utsunomiya, S., Yamamoto, Y.: A coherent Ising machine for MAX-CUT problems: performance evaluation against semidefinite programming and simulated annealing. In: Yamamoto, Y., Semba, K. (eds.) Principles and Methods of Quantum Information Technologies. LNP, vol. 911, pp. 251–262. Springer, Tokyo (2016). https://doi.org/10.1007/978-4-431-55756-2_12
20. Lucas, A.: Ising formulations of many NP problems. Frontiers Phys. **2**, 5 (2014)
21. Bian, Z., Chudak, F., Israel, R.B., Lackey, B., Macready, W.G., Roy, A.: Mapping constrained optimization problems to quantum annealing with application to fault diagnosis. Frontiers ICT **3**, 14 (2016)
22. Bian, Z., Chudak, F., Israel, R., Lackey, B., Macready, W.G., Roy, A.: Discrete optimization using quantum annealing on sparse Ising models. Frontiers Phys. **2**, 56 (2014)
23. Rieffel, E.G., Venturelli, D., O'Gorman, B., Do, M.B., Prystay, E.M., Smelyanskiy, V.N.: A case study in programming a quantum annealer for hard operational planning problems. Quantum Inf. Process. **14**(1), 1–36 (2015)
24. Venturelli, D., Marchand, D.J.J., Rojo, G.: Quantum annealing implementation of job-shop scheduling (2015)
25. de Givry, S., Larrosa, J., Meseguer, P., Schiex, T.: Solving Max-SAT as weighted CSP. In: Rossi, F. (ed.) CP 2003. LNCS, vol. 2833, pp. 363–376. Springer, Heidelberg (2003). https://doi.org/10.1007/978-3-540-45193-8_25
26. Morgado, A., Heras, F., Liffiton, M., Planes, J., Marques-Silva, J.: Iterative and core-guided maxsat solving: a survey and assessment. Constraints **18**(4), 478–534 (2013)
27. McGeoch, C.C., Wang, C.: Experimental evaluation of an adiabatic quantum system for combinatorial optimization. In: Proceedings of the ACM International Conference on Computing Frontiers, CF 2013, pp. 23:1–23:11. ACM (2013)
28. Nieuwenhuis, R.: The IntSat method for integer linear programming. In: O'Sullivan, B. (ed.) CP 2014. LNCS, vol. 8656, pp. 574–589. Springer, Cham (2014). https://doi.org/10.1007/978-3-319-10428-7_42
29. Zdeborova, L., Krzakala, F.: Statistical physics of inference: thresholds and algorithms. Adv. Phys. **65**(5), 453–552 (2016)
30. Bian, Z., Chudak, F., Macready, W.G., Rose, G.: The Ising model: teaching an old problem new tricks (2010). https://www.dwavesys.com/sites/default/files/weightedmaxsat_v2.pdf. Accessed 28 Apr 2017
31. Benedetti, M., Realpe-Gómez, J., Biswas, R., Perdomo-Ortiz, A.: Estimation of effective temperatures in quantum annealers for sampling applications: a case study with possible applications in deep learning. Phys. Rev. A **94**, 022308 (2016)

32. Boixo, S., et al.: Evidence for quantum annealing with more than one hundred qubits. Nat. Phys. **10**(3), 218–224 (2014)
33. Denchev, V.S., et al.: What is the computational value of finite-range tunneling? Phys. Rev. X **6**, 031015 (2016)
34. King, J., Yarkoni, S., Nevisi, M.M., Hilton, J.P., McGeoch, C.C.: Benchmarking a quantum annealing processor with the time-to-target metric (2015). arXiv preprint: arXiv:1508.05087
35. Boothby, T., King, A.D., Roy, A.: Fast clique minor generation in chimera qubit connectivity graphs. Quantum Inf. Process. **15**(1), 495–508 (2016)
36. Cai, J., Macready, W.G., Roy, A.: A practical heuristic for finding graph minors (2014)
37. Klymko, C., Sullivan, B.D., Humble, T.S.: Adiabatic quantum programming: minor embedding with hard faults. Quantum Inf. Process. **13**(3), 709–729 (2014)
38. Koch, T., et al.: MIPLIB 2010: mixed integer programming library version 5. Math. Program. Comput. **3**(2), 103–163 (2011)
39. Gent, I.P., Walsh, T.: CSPLib: a benchmark library for constraints. In: Jaffar, J. (ed.) CP 1999. LNCS, vol. 1713, pp. 480–481. Springer, Heidelberg (1999). https://doi.org/10.1007/978-3-540-48085-3_36
40. Hoos, H.H., Stutzle, T.: SATLIB: An online resource for research on SAT (2000)
41. Coffrin, C., Nagarajan, H., Bent, R.: Challenges and successes of solving binary quadratic programming benchmarks on the DW2X QPU. Technical report, Los Alamos National Laboratory (LANL) (2016)
42. Hen, I., Job, J., Albash, T., Rønnow, T.F., Troyer, M., Lidar, D.A.: Probing for quantum speedup in spin-glass problems with planted solutions. Phys. Rev. A **92**, 042325 (2015)
43. King, J., et al.: Quantum annealing amid local ruggedness and global frustration (2017)
44. Mandrà, S., Zhu, Z., Wang, W., Perdomo-Ortiz, A., Katzgraber, H.G.: Strengths and weaknesses of weak-strong cluster problems: a detailed overview of state-of-the-art classical heuristics versus quantum approaches. Phys. Rev. A **94**, 022337 (2016)
45. Mandrà, S., Katzgraber, H.G., Thomas, C.: The pitfalls of planar spin-glass benchmarks: raising the bar for quantum annealers (again) (2017)
46. Aaronson, S.: D-wave: Truth finally starts to emerge, May 2013. http://www.scottaaronson.com/blog/?p=1400. Accessed 28 Apr 2017
47. Aaronson, S.: Insert d-wave post here, March 2017. http://www.scottaaronson.com/blog/?p=3192. Accessed 28 Apr 2017
48. King, A.D., Lanting, T., Harris, R.: Performance of a quantum annealer on range-limited constraint satisfaction problems (2015). arXiv preprint: arXiv:1502.02098
49. Kirkpatrick, S., Gelatt, C.D., Vecchi, M.P.: Optimization by simulated annealing. Science **220**(4598), 671–680 (1983)
50. Hamze, F., de Freitas, N.: From fields to trees. In: Proceedings of the 20th Conference on Uncertainty in Artificial Intelligence, UAI 2004, Arlington, Virginia, United States, pp. 243–250. AUAI Press (2004)
51. Selby, A.: Efficient subgraph-based sampling of Ising-type models with frustration (2014)
52. Selby, A.: Qubo-chimera (2013). https://github.com/alex1770/QUBO-Chimera
53. Puget, J.F.: D-wave vs cplex comparison. Part 2: Qubo (2013). https://www.ibm.com/developerworks/community/blogs/jfp/entry/d_wave_vs_cplex_comparison_part_2_qubo. Accessed 28 Nov 2018

54. Dash, S.: A note on qubo instances defined on chimera graphs (2013). arXiv preprint: arXiv:1306.1202
55. Billionnet, A., Elloumi, S.: Using a mixed integer quadratic programming solver for the unconstrained quadratic 0-1 problem. Math. Program. **109**(1), 55–68 (2007)
56. Farhi, E., Goldstone, J., Gutmann, S., Sipser, M.: Quantum computation by adiabatic evolution (2018)
57. Kadowaki, T., Nishimori, H.: Quantum annealing in the transverse Ising model. Phys. Rev. E **58**, 5355–5363 (1998)
58. Farhi, E., Goldstone, J., Gutmann, S., Lapan, J., Lundgren, A., Preda, D.: A quantum adiabatic evolution algorithm applied to random instances of an NP-complete problem. Science **292**(5516), 472–475 (2001)
59. Nightingale, M.P., Umrigar, C.J. (eds.): Quantum Monte Carlo Methods in Physics and Chemistry. Nato Science Series C, vol. 525. Springer, Netherlands (1998)
60. Parekh, O., Wendt, J., Shulenburger, L., Landahl, A., Moussa, J., Aidun, J.: Benchmarking adiabatic quantum optimization for complex network analysis (2015)
61. IBM ILOG CPLEX Optimizer. https://www.ibm.com/analytics/cplex-optimizer. Accessed 2010
62. Gurobi Optimization, Inc.: Gurobi optimizer reference manual (2014). http://www.gurobi.com
63. D-Wave Systems Inc.: The D-wave 2X quantum computer technology overview (2015). https://www.dwavesys.com/sites/default/files/D-Wave%202X%20Tech%20Collateral_0915F.pdf. Accessed 28 Apr 2017
64. Vuffray, M., Misra, S., Lokhov, A., Chertkov, M.: Interaction screening: efficient and sample-optimal learning of Ising models. In: Lee, D.D., Sugiyama, M., Luxburg, U.V., Guyon, I., Garnett, R. (eds.) Advances in Neural Information Processing Systems, vol. 29, pp. 2595–2603. Curran Associates, Inc. (2016)
65. Lokhov, A.Y., Vuffray, M., Misra, S., Chertkov, M.: Optimal structure and parameter learning of Ising models (2016)
66. Mitchell, D., Selman, B., Levesque, H.: Hard and easy distributions of sat problems. In: Proceedings of the Tenth National Conference on Artificial Intelligence, AAAI 1992, pp. 459–465. AAAI Press (1992)
67. Balyo, T., Heule, M.J.H., Jarvisalo, M.: Sat competition 2016: recent developments. In: Proceedings of the Thirty-First National Conference on Artificial Intelligence, AAAI 2017, pp. 5061–5063. AAAI Press (2017)

Using Cost-Based Solution Densities from TSP Relaxations to Solve Routing Problems

Pierre Coste[✉], Andrea Lodi, and Gilles Pesant

Polytechnique Montréal, Montreal, Canada
{pierre.coste,andrea.lodi,gilles.pesant}@polymtl.ca

Abstract. The Traveling Salesman Problem, at the heart of many routing applications, has a few well-known relaxations that have been very effective to compute lower bounds on the objective function or even to perform cost-based domain filtering in constraint programming models. We investigate other ways of using such relaxations based on computing the frequency of edges in near-optimal solutions to a relaxation. We report early empirical results on symmetric instances from TSPLIB.

1 Introduction

The Traveling Salesman Problem (TSP) is certainly one of the most well-studied combinatorial optimization problems. It is of theoretical interest as a prominent representative of the class of $\mathcal{NP}$-hard problems but also of great practical importance in routing and other application areas. Several relaxations of this problem have long been investigated as part of the efforts to solve it by computing lower bounds on the objective for search-tree pruning and more recently for cost-based domain filtering given an upper bound. This short paper examines whether recent work related to counting-based branching heuristics in Constraint Programming (CP) may offer new and effective ways of exploiting these relaxations.

Counting-based search [19] represents a family of branching heuristics in CP that guide the search for solutions by identifying likely variable-value assignments in each constraint. Originally introduced for satisfaction problems it was later extended to optimization problems [18]. Given a constraint $c(x_1, \ldots, x_k)$ on finite-domain variables $x_i \in \mathcal{D}_i$ $1 \leq i \leq k$, let $f : \mathcal{D}_1 \times \cdots \times \mathcal{D}_k \to \mathbb{N}$ associate a cost to each k-tuple t of values for the variables appearing in that constraint and z be a finite-domain cost variable. An *optimization constraint* $c^\star(x_1, x_2, \ldots, x_k, z, f)$ holds if $c(x_1, x_2, \ldots, x_k)$ is satisfied and $z = f(x_1, x_2, \ldots, x_k)$. Let $\epsilon \geq 0$ be a small real number and $z^\star = \min_{t\,:\,c(t)} f(t)$ (without loss of generality consider that we are minimizing). We call

L.-M. Rousseau and K. Stergiou (Eds.): CPAIOR 2019, LNCS 11494, pp. 182–191, 2019.
https://doi.org/10.1007/978-3-030-19212-9_12

$$\sigma^*(x_i, d, c^*, \epsilon) = \frac{\sum_{t=(x_1,...,x_{i-1},d,x_{i+1},...,x_k)\; :\; c^*(t,z,f) \wedge z \leq (1+\epsilon)z^*} \omega(z, z^*, \epsilon)}{\sum_{t=(x_1,...,x_k)\; :\; c^*(t,z,f) \wedge z \leq (1+\epsilon)z^*} \omega(z, z^*, \epsilon)}$$

the *cost-based solution density* of variable-value pair (x_i, d) in c^* given ϵ. Its value, between 0 and 1, measures how often assignment $x_i = d$ appears in "good" satisfying assignments to c^*. If $\epsilon = 0$ this corresponds to the solution density restricted to the optimal solutions to the constraint with respect to f. A positive ϵ gives a margin to include close-to-optimal solutions, but at a discount proportional to their distance from z^* (the minimum value of f over solutions to c), as given by generic weight function $\omega(z, z^*, \epsilon) \in [0, 1]$ (for example, $\omega(z, z^*, \epsilon) = 1 - \frac{z-z^*}{\epsilon z^*}$) whose definition may vary depending on the constraint. We use cost-based solution densities in CP to favour branching decisions, in the form of a variable assignment, that retain many good-quality solutions from the individual perspective of constraints, keeping in mind that each constraint in CP tends to represent a large combinatorial substructure of the problem.

In the rest of the paper, Sect. 2 reviews the TSP and in particular its common relaxations, Sect. 3 sketches the way cost-based solution densities are computed for the few constraints that are relevant here, Sect. 4 presents how cost-based solution densities can filter out unpromising edges as a preprocessing step or offer insightful branching heuristics, and Sect. 5 provides an empirical evaluation of these ideas on standard benchmark instances from the TSPLIB.

2   TSP

We are given a complete and undirected graph $G = (V, E)$, where V is called the vertex set and E is called the edge set. We are also given some costs associated with the edges of the graph, namely $t_{ij}, \forall (i, j) \in E$, and we assume that the triangle inequalities hold, i.e., $t_{ij} + t_{jk} \geq t_{ik}$ for every triplet of vertices $i, j, k \in V$. Then, the TSP calls for finding a unique tour visiting each vertex $i \in V$ exactly once. By associating a binary variable x_{ij} that takes value 1 if edge $(i, j) \in E$ belongs to the tour, the Integer Programming model with the most effective linear programming relaxation reads as follows

$$\min \sum_{(i,j)\in E} t_{ij} x_{ij} \tag{1}$$

$$\sum_{(i,j)\in\delta(i)} x_{ij} = 2, \quad i \in V \tag{2}$$

$$\sum_{(i,j)\in\delta(S)} x_{ij} \geq 2, \quad S \subseteq V, 2 \leq |S| \leq |V| - 2 \tag{3}$$

$$x_{ij} \in \{0, 1\}, \quad (i, j) \in E \tag{4}$$

where $\delta(i)$ (resp. $\delta(S)$) denotes the set of edges incident to vertex i (resp. with one endpoint in set S). Degree constraints (2) establish that each vertex needs exactly two incident edges in a tour, while (3) forbid any subtour, i.e., a cycle of length smaller than V. As stated, the TSP is $\mathcal{NP}$-hard and a number of relaxations have been investigated. The strongest is obtained by optimizing (1) over the so-called *subtour elimination polytope*, that resulting from relaxing the integrality requirement (4) to nonnegativity. However, this relaxation is exponentially large and requires some care computationally (a sequence of min-cut/max-flow problems has to be solved on carefully constructed graphs). Nowadays, linear-programming based algorithms have taken over the solution of extremely large-scale TSP instances and CONCORDE [1] is the state-of-the-art solver. However, exploiting the combinatorial structure of the TSP on more heterogeneous problems and incorporating its solution within modular programming paradigms like CP is still extremely relevant in practice and more combinatorial TSP relaxations that have been known for decades can be extremely useful. More precisely, in this paper we consider three of them.

1-tree relaxation. For a given vertex, say vertex 1, a 1-tree is a tree spanning the vertices in $V \setminus \{1\}$ plus two edges incident with vertex 1. It is easy to see that any 1-tree has at most one cycle and if the tree is computed by solving a minimum-cost spanning tree and the two edges are those of minimum cost, then the cost of the resulting 1-tree provides a lower bound on the optimal TSP cost. The 1-tree relaxation is computable in polynomial time by solving a minimum-cost spanning tree ($O(|V|^2)$ complexity) and has been used in one of the first breakthrough algorithm for the TSP by Held and Karp [15, 16].

2-matching relaxation. The 2-matching relaxation is obtained from model (1)–(4) by dropping constraints (3) (but keeping the integrality requirements (4)). The resulting integer programming problem is a perfect 2-matching, i.e., a collection of minimum-cost disjoint cycles covering all vertices and can be solved in polynomial time by the famous algorithm of Edmonds [13].

n-path relaxation. This relaxation was introduced by Christofides et al. [7] and generally used with a path representation of the problem where a vertex, say 1, is considered the starting one of the tour and duplicated (vertex $n + 1$), so as to transform a tour in a path of $n + 1$ edges from vertex 1 to itself, i.e., $n + 1$. The idea is to relax the degree constraint of each vertex and, at the same time, imposing that $|V| := n$ edges need to be selected in the resulting path. Through dynamic programming, such a shortest but not-necessarily-elementary path can be computed in polynomial time and this relaxation has been especially used in the column generation approaches for more complex routing problems like the Capacitated Vehicle Routing and its variant.

We end the section by noting that there have been several previous attempts in CP to filter edges by using relaxations. Two of the most relevant ones in this context are Benchimol et al. [3] and Ducomman et al. [12]. In [3], the authors consider the 1-tree relaxation and, in an additive way, the 2-matching one to empower the so-called `weightedCircuit` constraint so as to remove edges from the variables' domain through cost-based domain filtering [14]. The computational experiments on the TSP show strong size reductions and improved computing times with respect to less sophisticated CP models. In [12], the authors

Table 1. A basic CP model for the TSP

$$\min z = \sum_{i=1}^{n} \gamma_{is_i} \quad \text{s.t.}$$
$$\texttt{minWeightAlldifferent}(\{s_1, \ldots, s_n\}, z, \Gamma)$$
$$\texttt{noCycle}(\{s_1, \ldots, s_n\})$$
$$s_i \in \{2, 3, \ldots, n+1\} \qquad\qquad 1 \le i \le n$$
$$z \in \mathbb{N}$$

extend the work in [3] for the `weightedCircuit` constraint by showing that among the three bounds above, 1-tree, 2-matching and n-path, there is no dominance in terms of filtering and successfully apply the framework to one of the time-constrained TSP variant, namely TSP with time windows.

3   Solution Densities of CP Optimization Constraints

The combinatorial structure of each relaxation presented in the previous section happens to be captured by some existing optimization constraint in CP. In this section we review them and outline the ways solution densities are computed for them. But first, consider the following basic CP model for the TSP as given at Table 1. Without loss of generality we start a tour at vertex 1 and end it at vertex $n + 1$, which is a duplicate of 1. We define successor variables s_i so that $s_i = j$ corresponds to using edge (i, j). Γ represents the distance matrix. The `minWeightAlldifferent` optimization constraint [6] is an `alldifferent` constraint to which we add costs for each variable-value assignment and an extra variable representing the sum of the assignments. Here it enforces the assignment part for the successor variables s_i and links them to the objective variable z but does not apply any cost-based domain filtering. Constraint `noCycle` is the subtour elimination constraint.

2-matching relaxation. The collection of minimum-cost disjoint cycles for the former IP model of Sect. 2 corresponds to a minimum-cost assignment for the latter CP model, captured by the `minWeightAlldifferent` constraint. As described in [18], to derive cost-based solution densities from that constraint we first compute a minimum-weight bipartite matching, say of cost $z^\star$, using the Hungarian algorithm. As a by-product of this computation we get a reduced-cost matrix $R = (r_{ij})$ whose non-negative entries r_{ij} tell how much of an increase in cost we can expect if we assign value j to variable i instead of the value from the computed matching. Next we define related matrix $R' = (r'_{ij})$ as

$$r'_{ij} = \max\left(0, \frac{(\epsilon z^\star + 1) - r_{ij}}{\epsilon z^\star + 1}\right).$$

Each entry r'_{ij} lies in the real interval $[0, 1]$, with value 1 corresponding to a reduced cost r_{ij} of 0 and value 0 corresponding to any reduced cost signalling a variable-value assignment whose cost would exceed the ϵ margin.

The permanent of a $n \times n$ matrix $A = (a_{ij})$ is defined as

$$\text{per}(A) = \Sigma_{p \in P} \Pi_{i=1}^{n} a_{i,p(i)}$$

where P denotes the set of all permutations of $\{1, 2, \ldots, n\}$. In the case of a binary matrix representing the domain of each variable (variables on the rows, values on the columns, and a "1" entry if and only if the value appears in the domain of the variable) this represents the number of solutions to the constraint: we sum over all possible assignments; the inner product is equal to 1 if each variable has the corresponding value in its domain and 0 otherwise. In the case of R' an optimal assignment counts as 1, any assignment whose cost exceeds $(1 + \epsilon)z^*$ or that is simply infeasible counts as 0, and any other assignment is counted at a discount proportional to how far it is from the optimum. We thus achieve a weighted counting of feasible assignments that are within our ϵ margin.

The cost-based solution density of a variable-value pair will be computed as the ratio of two permanents. Because computing the permanent is #P-complete, several computationally-tractable upper bounds have been proposed: we use bound U^1 from [23].

1-*tree relaxation.* There has been considerable work on (weighted) tree structures in CP about domain filtering [2,10,11,20,21], failure explanation [8], and (cost-based) solution densities [4,9]. We recall below the latter work that is relevant here. Note that since our goal is to compute edge frequencies in low-weight trees and not a lower bound on a tour, we consider spanning trees instead of 1-trees.

We define matrix $M = (m_{ij})$ whose elements are univariate polynomials built from edge set E and weight function w defined over E:

$$m_{ij} = \begin{cases} -x^{w(e)}, & i \neq j & e = (v_i, v_j) \in E \\ 0, & i \neq j & (v_i, v_j) \notin E \\ \displaystyle\sum_{e=(v_i,v_k) \in E} x^{w(e)}, & i = j \end{cases}$$

A remarkable result [5] states that any minor of M (i.e. the determinant of the submatrix obtained by removing from M row and column i for any i) yields a polynomial $\sum_k a_k x^k$ in which monomial $a_k x^k$ indicates the number a_k of spanning trees of weight k. Instead of computing the determinant of a matrix of polynomials, a potentially time-consuming process, we instantiate x to a real value between 0 and 1 (e.g. 0.7 in our experiments), yielding a matrix of scalars. Its effect is to apply an exponential decay to the number a_k of spanning trees of weight k according to the difference between that weight and that of a minimum spanning tree, thus giving more importance to close-to-optimal trees. Setting $x = 1$ does not apply any decay and so all trees are counted equally—the closer x gets to 0, the more aggressive the decay.

To compute the cost-based solution density of an edge we exploit the fact that the matrix M of a graph without that edge is almost identical to the original one, leading to an efficient computation by matrix inversion, as described in [4]. To consider this relaxation we add a `minWeightSpanningTree` constraint to the CP model at Table 1.

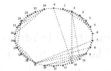 

Fig. 1. From left to right: top 15% edges for each vertex according to the 2-matching and 1-tree relaxations, their union, and the 35% cheapest edges per vertex.

n-path relaxation. We can represent an n-path with a `regular` constraint on variables $\langle x_1, x_2, \ldots, x_n \rangle$ where x_i corresponds to the vertex in ith position on the path and using the input graph as the automaton. This allows us to use `cost-regular`, its optimization variant, for which an algorithm computing cost-based solution densities has already been proposed [18]. The domain-filtering algorithm for `regular` builds and maintains a layered digraph built by unfolding the automaton over the sequence of variables. Each path from the first layer to the last in the layered digraph corresponds to a solution (here, an n-path). The cost-based solution density algorithm needs to restrict its attention to paths of cost at most $(1 + \epsilon)z^\star$ where $z^\star$ is the cost of the shortest n-path. It therefore computes at each node of the digraph the number of incoming and outgoing partial paths of each cost up to $(1+\epsilon)z^\star$. The number of relevant paths featuring a given variable-value pair is computed as the sum over all corresponding arcs of the products of number of partial incoming/outgoing paths at their endpoints, provided their composition makes an n-path of cost at most $(1 + \epsilon)z^\star$ (see [18] for details—note that contrary to that reference, here we weigh paths according to their cost, similarly to the other two relaxations). To consider this relaxation we add a `cost-regular` constraint to the CP model at Table 1.

4   Exploiting Cost-Based Solution Densities

4.1   Preprocessing

We investigate using cost-based solution densities computed from the relaxations as a preprocessing step that discards unpromising edges in an optimal tour. From each relaxation, we consider the $k\%$ highest solution densities for each vertex, with the choice of k depending on how aggressive we wish to be, and then combine that information by keeping the union of the corresponding edges. In this way we only discard edges that are considered unpromising *by all relaxations*.

As an illustration consider instance *fri26* from the TSPLIB. Each graph in Fig. 1 shows the vertices in clockwise order of its optimal tour and the edges that are kept in each case. We observed that the n-path relaxation (not shown here) does not provide discriminating information unless some variables are fixed and so will be left out for preprocessing. Note that the graph from the 2-matching (and of course its union with the 1-tree) includes all edges of the optimal tour. To preserve all the optimal edges while simply keeping the cheapest edges from each vertex we would need to increase k to 35%, yielding a much denser graph (extreme right).

4.2 Branching Heuristic

Beyond preprocessing, cost-based solution densities provide insightful information for optimistic branching in a search tree. The question of how best to combine such information from each relaxation arises again—one simple combination that has worked well in general, called $maxSD^\star$ [18], identifies the highest cost-based solution density over all constraints, variables, and values, makes the corresponding assignment in the left branch and forbids that same assignment in the right branch. Another combination that we will evaluate considers the arithmetic mean of the cost-based solution densities from each relaxation for a given variable-value pair and selects the highest one.

Table 2. Average performance (10 runs) of CONCORDE to solve to optimality symmetric instances of size 150 to 400 from the TSPLIB, with and without preprocessing.

instance	original graph				graph with discarded edges			
	# bbnodes	total time (s)	Branching (s)	# cuts	# bbnodes	total time (s)	branching (s)	# cuts
ch150	1	0.49	0	111.5	2.4	0.35	0.21	88.2
kroA150	1	0.88	0	164.3	1	0.19	0	32.9
kroB150	1.2	0.84	0.01	171.3	1.2	0.42	0.02	58.8
si175	2.8	3.42	0.24	294.2	1	0.25	0	53.8
brg180	1	0.71	0	5.0	1	0.09	0	1.8
rat195	5.6	5.49	3.14	314.8	4.6	4.07	2.67	325.6
d198	3.2	2.29	0.31	193.0	1.6	1.12	0.28	95.1
kroA200	1	0.77	0	250.1	1.4	0.39	0.10	103.7
kroB200	1	0.44	0	136.6	1.2	0.21	0.02	33.1
ts225	1.5	8.01	0.23	875.0	1	0.44	0	197.9
pr226	1	0.51	0	101.9	1	0.24	0	42.1
pr264	1	0.44	0	49.7	1	0.28	0	17.9
a280	1	1.18	0	107.0	3	1.04	0.23	111.4
pr299	1.8	3.24	0.09	446.9	2.2	2.11	0.84	387.4
lin318	1	1.77	0	237.0	2.4	3.53	0.62	306.9
rd400	11.2	18.20	12.30	754.1	11.0	17.97	14.62	814.2

5 Empirical Evaluation

5.1 State-of-the-Art Exact Solver

As discussed in Sect. 4.1, one algorithmic idea for using the density of an edge in the solutions of the TSP relaxations is to sparsify the instance accordingly, i.e., discard edges of low density. To test the computational effect of this idea, we preprocess 16 classical TSP instances with number of nodes between 150 and 400 and we run CONCORDE both on the original instance and the sparsified one (keeping the 1% highest densities, which notably still happens to include the optimal solution for all 16 instances). Some significant characteristics of the two types of run are reported in Table 2, namely the number of branch-and-bound nodes, the overall computing time, the time for branching and the number of generated cuts. All numbers are averages over 10 runs.

Of course, we know that instances of that size are not challenging for CON-CORDE. Indeed the largest tree size for the classical version of the instances is only 11.2 nodes and 9 out of the 16 instances do not require branching at all. Therefore the potential impact is limited. Nevertheless there is an overall improvement in the solving process whose most significant characteristic seems to be the reduction in the number of cutting planes CONCORDE needs to separate to reach optimality. A more detailed computational analysis on larger and more challenging instances is needed to confirm this trend and, hopefully, to observe a large impact, which in order to be useful needs to compensate the time spent in computing the edge densities, which currently takes several seconds.

5.2   CP

Next we consider both opportunities mentioned in Sect. 4—namely, discarding some edges from the input graph and branching on the s_i variables using cost-based solution densities—and evaluate their impact on solving the CP model at Table 1 for small instances from TSPLIB. Such a model is clearly not competitive with the state of the art but serves our evaluation need.

Table 3 compares several branching heuristics: maximum regret [17] (a); $maxSD^*$ on 2-matching (b), 2-matching & 1-tree (c), 2-matching & 1-tree & n-path (d); arithmetic mean of 2-matching & 1-tree (e) and of 2-matching & 1-tree & n-path (f). Instances were preprocessed using $k = 15\%$. We do not show explicitly the results on the original graphs but every heuristic performed much worse on them, thus indicating that discarding edges in this way is beneficial. We generally observe a marked improvement when using our proposed branching heuristics with respect to maximum regret. However those which involve the n-path relaxation (d,f) are too computationally expensive and are only reported on the three smallest instances.

Table 3. Performance of IBM ILOG CP 1.6 on 9 small sparsified instances (30 min timeout).

	Opt		Best	Time(s)	Fails		Opt		Best	Time(s)	Fails		Opt		Best	Time(s)	Fails
gr21	2707	a	2707	0.11	173	gr24	1272	a	1272	0.26	359	fri26	937	a	937	0.43	475
		b	2707	0.04	48			b	1272	0.03	25			b	937	1.71	2163
		c	2707	0.14	27			c	1272	0.07	2			c	937	16.54	3156
		d	2707	34.90	60			d	1272	76.67	182			d	937	486.00	1367
		e	2707	0.15	31			e	1272	0.31	47			e	937	12.37	2393
		f	2707	35.23	70			f	1272	30.39	51			f	937	1684.26	3254
bays29	2020	a	2020	125.36	118869	dantzig42	699	a	699	0.3	0	swiss42	1273	a	1464	1537.4	1349413
		b	2020	7.57	9341			b	722	114.7	40888			b	1298	127.3	42503
		c	2020	8.37	1179			c	722	1147.1	73431			c	1397	24.9	1316
		e	2020	9.03	1287			e	718	72.8	3505			e	1410	25.3	1266
gr48	5046	a	5898	1097.1	423271	hk48	11461	a	14734	1486.8	970394	berlin52	7542	a	10434	1162.1	702850
		b	5055	548.6	195503			b	12466	591.6	277333			b	8224	104.6	123179
		c	5174	860.1	31897			c	12039	765.1	29543			c	8193	621.6	18781
		e	-	-	-			e	12032	1379.1	54948			e	8016	558.9	15548

6   Conclusion

We introduced new ways of exploiting known relaxations of the TSP and presented some preliminary empirical results to evaluate their usefulness. We believe this line of research requires further investigation and is particularly interesting for a CP approach to solve routing problems with additional constraints that make it difficult to apply other exact approaches directly.

References

1. Concorde TSP Solver. https://en.wikipedia.org/wiki/Concorde_TSP_Solver
2. Beldiceanu, N., Flener, P., Lorca, X.: The *tree* constraint. In: Barták, R., Milano, M. (eds.) CPAIOR 2005. LNCS, vol. 3524, pp. 64–78. Springer, Heidelberg (2005). https://doi.org/10.1007/11493853_7
3. Benchimol, P., van Hoeve, W.J., Régin, J.-C., Rousseau, L.-M., Rueher, M.: Improved filtering for weighted circuit constraints. Constraints **17**(3), 205–233 (2012)
4. Brockbank, S., Pesant, G., Rousseau, L.-M.: Counting spanning trees to guide search in constrained spanning tree problems. In: Schulte, C. (ed.) CP 2013. LNCS, vol. 8124, pp. 175–183. Springer, Heidelberg (2013). https://doi.org/10.1007/978-3-642-40627-0_16
5. Broder, A.Z., Mayr, E.W.: Counting minimum weight spanning trees. J. Algorithms **24**(1), 171–176 (1997)
6. Caseau, Y., Laburthe, F.: Solving various weighted matching problems with constraints. In: Smolka, G. (ed.) CP 1997. LNCS, vol. 1330, pp. 17–31. Springer, Heidelberg (1997). https://doi.org/10.1007/BFb0017427
7. Christofides, N., Mingozzi, A., Toth, P.: State space relaxation procedures for the computation of bounds to routing problems. Networks **11**, 145–164 (1981)
8. de Uña, D., Gange, G., Schachte, P., Stuckey, P.J.: Weighted spanning tree constraint with explanations. In: Quimper, C.-G. (ed.) CPAIOR 2016. LNCS, vol. 9676, pp. 98–107. Springer, Cham (2016). https://doi.org/10.1007/978-3-319-33954-2_8
9. Delaite, A., Pesant, G.: Counting weighted spanning trees to solve constrained minimum spanning tree problems. In: Salvagnin, D., Lombardi, M. (eds.) CPAIOR 2017. LNCS, vol. 10335, pp. 176–184. Springer, Cham (2017). https://doi.org/10.1007/978-3-319-59776-8_14
10. Dooms, G., Katriel, I.: The *Minimum Spanning Tree* constraint. In: Benhamou, F. (ed.) CP 2006. LNCS, vol. 4204, pp. 152–166. Springer, Heidelberg (2006). https://doi.org/10.1007/11889205_13
11. Dooms, G., Katriel, I.: The "Not-Too-Heavy Spanning Tree" constraint. In: Van Hentenryck, P., Wolsey, L. (eds.) CPAIOR 2007. LNCS, vol. 4510, pp. 59–70. Springer, Heidelberg (2007). https://doi.org/10.1007/978-3-540-72397-4_5
12. Ducomman, S., Cambazard, H., Penz, B.: Alternative filtering for the weighted circuit constraint: comparing lower bounds for the TSP and solving TSPTW. In: Schuurmans and Wellman [22], pp. 3390–3396
13. Edmonds, J.: Maximum matching and a polyhedron with 0,1-vertices. J. Res. Natl. Bur. Stand. **69B**, 125–130 (1965)

14. Focacci, F., Lodi, A., Milano, M.: Cost-based domain filtering. In: Jaffar, J. (ed.) CP 1999. LNCS, vol. 1713, pp. 189–203. Springer, Heidelberg (1999). https://doi. org/10.1007/978-3-540-48085-3_14
15. Held, M., Karp, R.M.: The traveling-salesman problem and minimum spanning trees. Oper. Res. **18**, 1138–1162 (1970)
16. Held, M., Karp, R.M.: The traveling-salesman problem and minimum spanning trees: Part II. Math. Program. **1**, 6–25 (1970)
17. Kilby, P., Shaw, P.: Vehicle routing. In: Rossi, F., van Beek, P., Walsh, T. (eds.) Handbook of Constraint Programming. Foundations of Artificial Intelligence, vol. 2, pp. 801–836. Elsevier, New York (2006)
18. Pesant, G.: Counting-Based Search for Constraint Optimization Problems. In: Schuurmans and Wellman [22], pp. 3441–3448
19. Pesant, G., Quimper, C.-G., Zanarini, A.: Counting-based search: branching heuristics for constraint satisfaction problems. J. Artif. Int. Res. **43**(1), 173–210 (2012)
20. Régin, J.-C.: Simpler and incremental consistency checking and arc consistency filtering algorithms for the weighted spanning tree constraint. In: Perron, L., Trick, M.A. (eds.) CPAIOR 2008. LNCS, vol. 5015, pp. 233–247. Springer, Heidelberg (2008). https://doi.org/10.1007/978-3-540-68155-7_19
21. Régin, J.-C., Rousseau, L.-M., Rueher, M., van Hoeve, W.-J.: The weighted spanning tree constraint revisited. In: Lodi, A., Milano, M., Toth, P. (eds.) CPAIOR 2010. LNCS, vol. 6140, pp. 287–291. Springer, Heidelberg (2010). https://doi.org/ 10.1007/978-3-642-13520-0_31
22. Schuurmans, D., Wellman, M.P. (eds.): Proceedings of the Thirtieth AAAI Conference on Artificial Intelligence, 12–17 February 2016. AAAI Press, Phoenix (2016)
23. Soules, G.W.: New permanental upper bounds for nonnegative matrices. Linear Multilinear A. **51**(4), 319–337 (2003)

A Counting-Based Approach to Scalable Micro-service Deployment

Waldemar Cruz, Fanghui Liu, and Laurent Michel[✉]

Computer Science and Engineering Department, School of Engineering,
University of Connecticut, Storrs, CT 06269-4155, USA
{waldemar.cruz,fanghui.liu,laurent.michel}@uconn.edu

Abstract. Deploying a cloud-based distributed application created from the composition of micro-services is a challenging problem. It mandates the resolution of a resource allocation problem accounting for resource utilization and network load. But it also imposes security requirements such as the selection of suitable technology stacks to protect the communication channels. Both sets of decisions are intimately related as hosting decisions affect the cost or feasibility of security measures under consideration. This paper revisits the problem and focuses on a *scalable approach* suitable to deploy *large* distributed applications. Specifically, it introduces a counting-based model to deliver solutions for hundreds of services within short computation times. The essence is to side-step some of the difficulties by focusing first and foremost on deciding how many services of each type need to be deployed at each location and postponing the instance connectivity problem to a post-optimization phase. Empirical results demonstrate the scope of the improvements and illustrate the performance to expect as a function of instance sizes.

1 Introduction

A modern cloud-based application forms a distributed system, i.e., a set of communicating and reusable micro-services delivering a software stack that meets application requirements. Efficient deployment of large scale distributed software stacks is difficult as service placement has an influence on resource usage.

Application developers often over-provision with respect to actual needs to ensure that the application requirements are met. This results in waste that hinders fault-tolerance and load-balancing of the system [7]. Finding an efficient deployment scheme is important to keep costs of implementation down and reduce waste of resources.

There are many prior works related to service deployment in data center, however, they do not consider communication and security policies between services. Micro-services deployed across multiple hosts may require communication over a public channel. For instance, an application formed out of two services may deploy one of them on company servers in its own data center and the other on an AWS cloud. Communications over such public channels mandate that appropriate security technology be enabled to meet security requirements. This problem

© Springer Nature Switzerland AG 2019
L.-M. Rousseau and K. Stergiou (Eds.): CPAIOR 2019, LNCS 11494, pp. 192–207, 2019.
https://doi.org/10.1007/978-3-030-19212-9_13

was first addressed in [5]. It considered the deployment of micro-services across different hosts and security requirements from services while communicating over public channels. The strategies included IP, CP with Large Neighborhood Search (LNS) and a hybrid method. The authors in [5] showed that it is possible to deliver high-quality solutions within a short runtime but without an optimality proof. Yet, application sizes exceeding 50 services proved to be difficult and challenged scalability.

This paper introduces a counting-based approach with a two-phase method that delivers optimal deployment schemes for larger scale instances. The two-phase method reduces the model to the components that contribute to the objective function which greatly reduces the size of the model. Empirical results indicate that this approach scales up to 1000 services and can deliver an optimal deployment strategy within a reasonable amount of time.

The remainder of the paper is organized as follows. Section 2 introduces related work. Section 3 presents the counting based model and problem formulation. Section 4 explains how to match service instances to setup the pairings. Section 5 shows the empirical results and Sect. 6 concludes.

2   Related Work

Many approaches have been developed in recent years that attempt to manage service deployment while minimizing the total cost of implementation across data centers. These include Bin Packing (BP) [14] as well as several variations discussed below. Cambazard et al. [2] introduce Bin Packing with Usage Cost (BPUC) to minimize the energy consumption which is represented as linear cost. The Temporal Bin Packing (TBP) [3] considers the lifespan of each task and aims at minimizing the allocated resources (CPU cycles) in a data center. Armant et al. [1] formulate the workload consolidation problem as a semi-online Bin Packing problem (also known as Batched Bin Packing [8]) where tasks are allocated to servers in real-time. In [13], the authors use CP with a resizable decomposition method to solve the Dynamic Cache Distribution Problem (DCDP). The DCDP is modeled as a variant of BP and the allocation found by the solver delivers load balancing and fault tolerance. Hermenier et al. [9] introduced the Bin Repacking Scheduling Problem which considers allocation and reconfiguration of VMs placement.

Other works that relate to service deployment and energy/cost management include [4,10,11]. In [10], the authors explore a service deployment from an energy/cost management standpoint. That paper offers a scheme that can minimize the migration costs by dynamically reallocating VMs to hosts. In [4], Chisca et al. address the balance between workload and cooling needs for a data center while minimizing the energy utilization. The problem is represented as a non-linear energy use function and solved with local search. Kadioglu et al. introduce in [11] a Core Group Placement Problem (CGPP) to allocate heterogeneous resources in cloud centers. The goal is to minimize the maximum load for the deployment of heterogeneous services (Fig. 2).

3 Supply-Demand Model

The premise of this model is that each service instance generates a number of ports to connect to other service instances. Each pairing represents the bi-directional communication channel between two services. In each pairing, the service with greater demand requires more resources from its connection partner. The service with the highest demand is said to be demand-side whereas the lowest one is said to be supply-side. Given that the supply-side has the lowest cardinality, supply-side instances must open up ports equal to the number of demand-side instances. In other words, supply-side instances may support multiple demand-side instances. Load-balancing is applied to the instances to distribute the connections across supply-side services. Since load-balancing of the demands may not be divided equally among instances, a number of excess ports must be opened to meet the remaining demand.

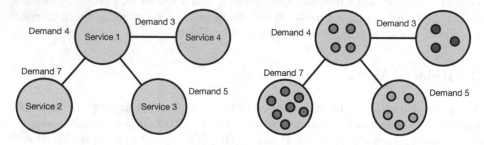

Fig. 1. Left: connectivity graph representing connections between service types Right: decomposition of service types into individual instances to be deployed on available hosts. The demand requirements dictate the number of instances needed.

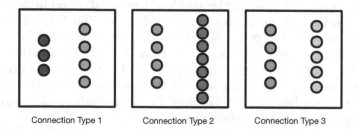

Fig. 2. Service Connections: square boxes encapsulate collections of *pairings* between service instances. Service instances appear in two columns showing supply (left) and demand (right). Node colors match those from Fig. 1.

Example: Consider a pair of connecting services s and t with demands 4 and 9 respectively where s is the supply-side service and t is the demand-side service. Load-balancing requires that each supply-side service open up $\lfloor \frac{9}{4} \rfloor$ ports each with an excess port of 1. Namely, [2,3] ports on each instance for a total of 9 ports.

The cost of deploying a micro-service application is influenced by the number of pairings across host boundaries. To support inter-host communications, service instances must include the appropriate security adapters to secure the channels. Pairings involving services on the same host rely on *private* channels and do not mandate additional adapters. Therefore, the internal wiring of co-located instances does not affect the cost. Adapters (e.g., software modules responsible for encryption and decryption of the messages pushed into the channel) increase the total cost of the implementation and impact the required resources for deployment.

Cruz et al. [5] introduced an approach that focuses on finding a global solution in one-step by providing the minimal cost deployment with the *wiring scheme* across instances. While workable, this method relies on a large model that negatively impacts the ability to scale. By separating the wiring component, a solution to the problem can be obtained with less efforts.

This method follows a two-part approach for solving for an optimal deployment. First, the optimal cost of deployment is determined by finding *how many* service instances of each type should be deployed on each host as well as *how many* inter-host connections are required to satisfy the application connectivity requirements. Once these cardinalities have been established, the second phase delivers an actual deployment of service instances by producing the wiring in the form of all required service pairings.

The remainder of this section describes the model in details starting with a lexicon of its parameters, decision variables and finally its constraints. Section 4 focuses on the actual deployment.

3.1 Constants

- T: Set of services types.
- M: Set of available physical hosts.
- Z: Set of security options required to ensure security requirements.
- D_t: Demand for service type t. $t \in T$.
- C: Set of unordered pairs of connecting service types. Namely, $\forall c \in C$: $D_{td(c)} \geq D_{ts(c)}$.
- $td(c) \in T$: The demanding service with the highest demand for connection $c \in C$.
- $ts(c) \in T$: The supplying service with the lowest demand for connection $c \in C$.
- $z(c)$: The required security protocol for securing connection $c \in C$.
- LB_c: The lower bound support on the supply-side for connection. c. LB_c is $\lfloor \frac{D_{td(c)}}{D_{ts(c)}} \rfloor$.
- FB_t: Fixed bandwidth cost for service t.
- FM_t: Fixed memory cost for service t.
- SB_z: Scaling factor to bandwidth cost for security option z.
- SM_z: Scaling factor to memory cost for security option z.
- AB_z: Fixed bandwidth cost for enabling security option z.
- AM_z: Fixed memory cost for enabling security option z.

- VM: Virtual Machine overhead introduced for each instance.
- $pin(m,t)$: Number of instances of service type t pinned on host m.
- $zone(m,t)$: Boolean indicating whether service t is permitted on host m.

3.2 Variables

- MM_m: Total memory usage for host m. $MM_m \in [0, Available\ Memory]$
- MB_m: Total bandwidth usage on host m. $MB_m \in [0, Available\ Bandwidth]$
- $Q_{m,t}$: Set of instances of service type t deployed on host m.
- $q_{m,t}$: Number of instances of service type t on host m, i.e., $|Q_{m,t}|$.
- $eq_{m,t,z}$: Number of instances of service type t on host m with an existing inter-host connection adopting security protocol z.
- $es_{m,c}$: Number of supply-side instances with inter-host connections of type c.
- $ed_{m,c}$: Number of demand-side instances with inter-host connections of type c.
- $SP_{m,c}$: Set of supply-side ports generated on host m for a connection type c.
- $X_{m,c}$: Set of excess supply-side ports generated on host m for a connection type c.
- $RX_{m,c}$: Subset of $X_{m,c}$ involved in an inter-host connection deployed on host m for connection type c.
- $RSP_{m,c}$: Subset of $SP_{m,c}$ involved in an inter-host connection deployed on host m for connection type c.
- $RDP_{m,c}$: Subset of demand-side ports involved in an inter-host connection deployed on host m for connection type c.

3.3 Constraints

Each service instance offers a set of ports. Specifically, instances on the supply-side of a connection (the smaller cardinality side) offer supply-side ports while instances on the demand-side offer demand-side ports. To correctly setup the connectivity between supply and demand, pairings must be established between supply-side ports and demand-side ports. Figure 3 illustrates this idea. Supply-side services are shown in green while demand-side services are brown. In total, 7 pairings must be established to meet this demand and one supplier ($\#2$) will need three ports while the other two ($\{0,1\}$) offer only two supply-side ports. The right-hand side of the Figure uses squares to represent ports and the service instance number is repeated in the square for clarity. For instance, the brown edge pairing services 0 and 3 on the left-hand side appears inside host n where service 3 and 0 execute. Two square rectangles represent the consumed ports for those services and the brown edge connects the ports. It is essential for the deployment model to create a hosting with enough ports to guarantee the existence of such a matching between all suppliers and consumers.

As shown in Fig. 3, each service instance generates a number of ports representing connections to supply and demand ports (boxes). The blue shaded area represents $SP_{m,c}$ the *set of supply-side ports* generated by the host for that connection c. The green shaded area represents $RDP_{m,c}$ the *set of demand-side ports* that are involved in inter-host pairings for connection c. The red shaded area

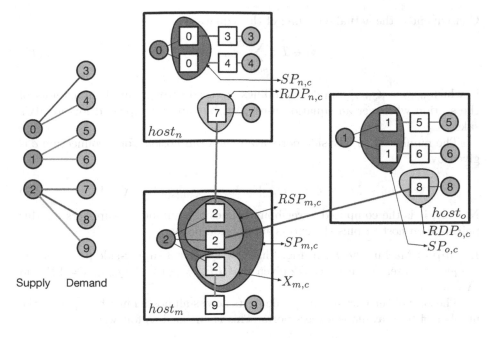

Fig. 3. Multi-host connectivity (Color figure online)

represents $RSP_{m,c}$ the *set of supply-side ports* involved in inter-host pairings for connection c. The light-blue shaded area represents $X_{m,c}$ the set of excess supply-side ports generated on host m for connection c.

$RSP_{m,c}$, $X_{m,c}$ and $RX_{m,c}$ are subsets of $SP_{m,c}$ which is defined as the set of all supply ports on host m and connection c. In Fig. 3, $host_m$ contains three supply-side service ports generated by service instance 2 and one demand-side port generated by service instance 9. The local demand (from service 9) can be satisfied by one supply port (from $X_{m,c}$) leaving two ports available to meet *external demands* emanating from other hosts (n and o). In the Figure, inter-host pairings are established between ports from service instances 2 and 7 and service instances 2 and 8 to satisfy connection requirements.

It is, perhaps, essential to note that the identity of the ports used to create those pairings is immaterial. In essence, all ports of a service are equivalent and therefore interchangeable. This insight is essential to produce the mathematical formulation. Indeed, it is not necessary to talk about individual ports in those sets, but instead, only about the cardinality of those sets to have enough ports of each "kind". In the following, equations that refer to a set will each time refer to the *cardinality* of the set which is the actual variable created in the model.

Quantity Constraints. The set of all the deployed instances for type t must be equal to the total demand of service t.

$$\forall t \in T : |\bigcup_{m \in M} Q_{m,t}| = D_t \tag{1}$$

Consequently, the actual encoding of this equation is

$$\forall t \in T : \sum_{m \in M} q_{m,t} = D_t \tag{2}$$

in which $q_{m,t} = |Q_{m,t}|$. Similar encodings are used throughout the remainder of the section whenever an equation uses a set but are not repeated for brevity's sake.

The number of supply-side ports available from host m for a connection c is given by:

$$\forall m \in M, c \in C : |SP_{m,c}| = LB_c \cdot |Q_{m,ts(c)}| + |X_{m,c}| \tag{3}$$

Namely, it is the computed lower bound times the number of suppliers for this pairing c on host m plus the excess supply-side ports.

Example: Machine $host_m$ in Fig. 3 has $2 \times 1 + 1 = 3$ supply-side ports for the unique green service instance #2 given that $LB_c = \lfloor \frac{7}{3} \rfloor = 2$, $Q_{m,ts(c)} = \{2\}$ and $|X_{m,c}| = 1$.

The sum of the excess supply-side ports across all hosts must be equal to the number of total available excess ports. This is encoded as follows:

$$\forall c \in C : \sum_{m \in M} |X_{m,c}| = D_{td(c)} \bmod D_{ts(c)} \tag{4}$$

If there exists any excess supply-side ports offered by $host_m$, then there must be at least one supply-side instance to support them.

$$\forall m \in M, c \in C : (|X_{m,c}| \geq 1) \rightarrow (|Q_{m,ts(c)}| \geq 1) \tag{5}$$

Zoning Constraints. Service instances working with sensitive information must be restricted within a particular zone. If a service type t is not permitted on host m, then the following constraint is encoded as follows:

$$\forall m \in M, t \in T : (zone(m,t) = 0) \rightarrow |Q_{m,t}| = 0 \tag{6}$$

Pinning Constraints. An existing infrastructure may have already deployed service instances that may be difficult to migrate due to strict policies. For re-deployment strategies, previously deployed service instances remain on their respective hosts and a new deployment strategy is configured around the current one. Pinned instances occupy resources on their respective hosts. This is encoded with:

$$\forall m \in M, t \in T : |Q_{m,t}| \geq pin(m,t) \tag{7}$$

where $pin(m,t)$ is the number of service instance of type t on host m.

Connectivity Constraints. It is necessary to ensure that supply meets demand for each host. If a host does not have the necessary ports to meet supply-demand requirements, then it is necessary to connect to another host with those available ports. This is achieved with:

$$\forall m \in M, c \in C : |SP_{m,c}| + |RDP_{m,c}| = |Q_{m,td(c)}| + |RSP_{m,c}| \tag{8}$$

The total number of available supply-side ports is equal to the total number of available demand-side ports:

$$\forall c \in C : |\bigcup_{m \in M} RSP_{m,c}| = |\bigcup_{m \in M} RDP_{m,c}| \tag{9}$$

Example: In Fig. 3, the union of the two green sets ($RDP_{n,c}$ and $RDP_{o,c}$) from hosts n and o has cardinality 2 while the red set of suppliers on host m ($RSP_{m,c}$) also has cardinality 2 and therefore satisfy Eq. 9. Likewise, note how, on host m, $SP_{m,c}$ contains three ports for service instance #2 and its $RDP_{m,c}$ set is empty. Meanwhile, $Q_{m,td(c)} = Q_{m,brown} = \{\#9\}$, i.e., a set of cardinality 1. Finally, $RSP_{m,c}$ is the red set of ports connected to external instances and its cardinality is 2. Overall, Eq. 8 reduces to $3 + 0 = 1 + 2$ which is satisfied.

Supply-Side Connectivity Constraints. For each supply-side service, the following constraints are encoded to determine the composition and the number of service instances required to satisfy the pairings:

The set of excess supply-side ports $RX_{m,c}$ is a subset of the set $X_{m,c}$.

$$\forall m \in M, c \in C : RX_{m,c} \subseteq X_{m,c} \tag{10}$$

If there is an excess supply-side port with an inter-host connection, then there must be at least one supply-side instance to support them.

$$\forall m \in M, c \in C : (|RX_{m,c}| \geq 1) \rightarrow (es_{m,c} \geq 1) \tag{11}$$

The total number of inter-host pairings initiated by host m cannot exceed the number of external ports generated by host m.

$$\forall m \in M, c \in C : |RSP_{m,c}| \leq LB_c \cdot es_{m,c} + |RX_{m,c}| \tag{12}$$

The number of supply-side service instances connected to inter-host pairings are bounded by the total number of service instances on host m:

$$\forall m \in M, c \in C : es_{m,c} \leq |Q_{m,ts(c)}| \tag{13}$$

Example: Suppose there exists three inter-host connections, $|RSP_{h,c}| = 3$, to host h connecting to supply-side services on host h. A supply-side service can only open $LB_c = 2$ ports for each instance. Therefore, a combination of supply-side ports or excess ports is required to establish the connections.

Demand-Side Connectivity Constraints. The following constraints determine the number of service instances required to satisfy the inter-host pairings for demand-side services:

$$\forall m \in M, c \in C : |RDP_{m,c}| = ed_{m,c} \tag{14}$$

The number of demand-side service instances connected to inter-host pairings is bounded by the total number of service instances on host m:

$$\forall m \in M, c \in C : ed_{m,c} \leq |Q_{m,td(c)}| \tag{15}$$

Example: Suppose there exists two inter-host connections, $|RDP_{h,c}| = 2$, to host h connecting to demand-side instances on host h. A demand-side instance can be connected to at most one supply-side instance and therefore to satisfy the condition $|RDP_{h,c}| = 2$, there must be at least two demand-side instances deployed on host h.

Memory Consumption Constraints. The following constraints determine the total required service instances for a particular service type that pair services deployed on different hosts and requiring a security protocol z:

$$\begin{aligned} \forall m \in M, \forall z \in Z, \forall c \in C \text{ such that } z(c) = z : eq_{m,ts(c),z} \geq es_{m,c} \\ \forall m \in M, \forall z \in Z, \forall c \in C \text{ such that } z(c) = z : eq_{m,td(c),z} \geq ed_{m,c} \end{aligned} \tag{16}$$

Finally, the constraint below determines the memory usage for each host:

$$\begin{aligned} \forall m \in M : MM_m = {} & \sum_{t \in T} \left[(VM + FM_t) \cdot |Q_{m,t}| \right. \\ & \left. + \sum_{z \in Z} \left((FM_t \cdot SM_z) \cdot eq_{m,t,z} + AM_z \cdot eq_{m,t,z} \right) \right] \end{aligned} \tag{17}$$

which is driven by the fixed overhead for a virtual machine for each service (VM), the memory usage for each service type FM_t as well as the memory cost of the adapter needed to implement the security protocol z (under the assumption that the protocol is indeed used $eq_{m,t,z}$).

Bandwidth Consumption Constraints. Each inter-host pairing induces a bandwidth consumption. Given that the link is across different physical hosts, the connection must be secured via the most stringent security policy.

$$\begin{aligned} \forall m \in M : MB_m = {} & \sum_{c \in C} (1 + SB_{z(c)}) \cdot [FB_{ts(c)} \cdot |RDP_{m,c}| + FB_{td(c)} \cdot |RSP_{m,c}|] \\ & + \sum_{z \in Z} AB_{z(c)} \cdot (\sum_{c \in C \, st. \, z(c) = z} [|RSP_{m,c}| + |RDP_{m,c}|] > 0) \end{aligned} \tag{18}$$

The bandwidth cost is driven by the number of inter-host pairings $|RDP_{m,c}|$ and $|RSP_{m,c}|$ established on each host. The existence of any inter-host pairings established on behalf of the host induces a usage cost AB_z associated with the appropriate security technology. A fixed bandwidth cost FB_t and scaling factor SB_z is applied to each inter-host pairings.

3.4 Objective

The objective is to minimize the total memory and bandwidth usage across all of the available hosts.

$$\min : \sum_{m \in M} MM_m + MB_m \qquad (19)$$

3.5 Redundant Constraints

The number of additional supply-side ports involved in an inter-host connection is bounded from above by the total number of additional ports for connection c.

$$\forall c \in C : \sum_{m \in M} |RX_{m,c}| \le D_{td(c)} \ mod \ D_{ts(c)} \qquad (20)$$

The total number of supply-side ports generated by all hosts must be equal to the number of demand ports:

$$\forall c \in C : \sum_{m \in M} |SP_{m,c}| = D_{td(c)} \qquad (21)$$

Given that there are two ports involved in each pairing, all inter-host connections must be established between two external ports. Therefore, the total number of external ports are bounded from above by the total number of ports generated for connection c.

$$\forall c \in C : \sum_{m \in M} (|RSP_{m,c}| + |RDP_{m,c}|) \le 2 \cdot D_{td(c)} \qquad (22)$$

3.6 Symmetry Breaking Constraints

First, assume that every zone contain a set of identical hosts with the same specifications (memory and bandwidth resources). This is, of course, typical in data centers where machine racks contain identical hosts. An ordering is applied to the hosts to reduce the number of symmetric solutions. For hosts located in the same zone, an ordering based on the number of external connections is applied to the hosts. The following constraint encodes this property:

$$\forall i \in \{1, ..., |M| - 1\} \ such \ that \ zone(i) = zone(i + 1) :$$
$$\sum_{t \in T} (|RDP_{i,t}| + |RSP_{i,t}|) \ge \sum_{t \in T} (|RDP_{i+1,t}| + |RSP_{i+1,t}|) \qquad (23)$$

4 Deployment Scheme

Once the optimal solution is found, the actual deployment (instances and links placement) can be computed in polynomial time. The "wiring" of service instances can be treated as a *perfect matching* problem where each supply-side port must connect to a demand-side port. Generating a bipartite graph for each pairing where edge capacities of 1 allows for the use of max-flow algorithms to achieve a perfect matching in polynomial time [12].

For each connection type, construct a bipartite graph connecting the service instances as follows:

1. Create a node for each port generated by a service instance. (square)
2. Connect an edge between each external supply-side and demand-side port node. (solid black edges)
3. Connect an edge between each internal supply-side and demand-side port node. (light-blue edges)

Additionally, if a host is given excess ports decided by the first phase, then the excess ports are given to an instance that has external connections.

By finding a perfect matching in each bipartite graph, a feasible matching between service instances is achieved. The conditions set on the first phase (Eq. 9) ensures that such a perfect matching exists between each supply-side and demand-side ports.

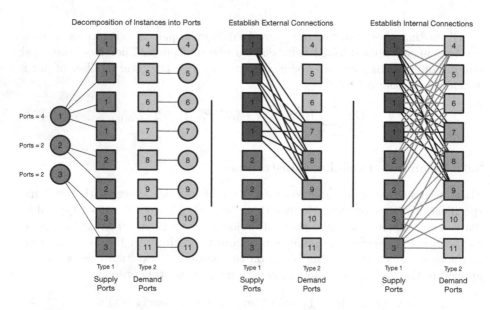

Fig. 4. Construction of bipartite graph for a pair of connecting services.

Consider the example in Fig. 4. Suppose a pair of services of type 1 and type 2 are connected with demands of 3 and 8. The set of instances $\{1, 2, 4, 5, 6\}$ are deployed on host 1 and $\{3, 7, 8, 9, 10, 11\}$ are deployed on host 2. Service instances, represented by the circles, are decomposed into individual ports represented as the colored boxes in the Figure. Suppose service instances 1, 7, 8 and 9 are decided to be the instances with external connections and two excess ports are assigned to host 1, then the two excess ports are given to service instance 1. The sum of the ports generated by service type 1 is equal to the demand of service type 2.

The light-blue and orange ports in the second column represent ports that may have external connections. Since instance 1 is the instance with external connection on host 1, all the ports on instance 1 are colored as light-blue. Similarly, instance 7, 8, 9 are instances with external connection on host 2, all the ports on instance 7, 8, 9 are orange. The black lines represent inter-host connections between ports. In the final column, connections between ports located on the same host are connected to each other with light-blue lines. Once the graph is generated, a max-flow algorithm is applied to the graph to find a perfect matching and establish all the pairings.

5   Results

To investigate the performance of the proposed model, a series of synthetic instances were created to emulate small to large sized applications. Benchmarks were solved on a Xeon(R) CPU E5-2640 v4 @ 2.40 GHz on a single core running Linux kernel 4.4.0-119- generic. Gurobi 7.5.2 was used as the MIP Solver. The MIP implementation relies on Objective-CP.

Table 1. MIP Performance for full and counting models.

Benchmark	Full MIP			Full LNS		Counting MIP		
	T	UB	LB	μ_Q	σ_Q	T	UB	LB
1	294.97	1,804.00	1,804.00	1,804.00	0.00	0.01	1,804.00	1,804.0
2	146.58	1,810.00	1,810.00	1,810.00	0.00	0.01	1,810.00	1,810.0
3	370.49	1,732.00	1,732.00	1,735.00	0.00	0.01	1,735.00	1,735.0
4	121.88	1,804.00	1,804.00	1,804.00	0.00	0.01	1,804.00	1,804.0
5	590.64	3,280.00	3,280.00	3,280.00	0.00	0.02	3,280.00	3,280.0
6	5,400.00		3,280.00	3,435.13	163.06	0.02	3,280.00	3,280.0
7	5,400.00	3,376.00	3,280.00			0.02	3,280.00	3,280.0
8	2,391.28	3,280.00	3,280.00			0.01	3,280.00	3,280.0

Model Comparison. The full model presented in [5] is capable to handle small to midsized applications with 50 service instances. Table 1 shows the performance comparison between the counting approach and the full approach on the set of benchmarks presented in [5]. Each benchmark was executed with a timeout of 5400 s. While the full model can find the optimal solution for smaller application sizes, application sizes with 50 service instances are difficult to close with many timing out. The counting approach is able to find the same optimal solution within a fraction of a second.

Symmetry Breaking Performance. The impact of symmetry breaking constraints on performance can be quite significant. On small instances (≤200 micro-services) the overhead can negate the benefits of the symmetry breaking. On larger instances (>200 micro-services), the reduction in the size of the search tree can reach a factor of 3. Without symmetry breaking, the model retains the ability to produce the optimal solution (i.e., for instances with 1000 micro-services), yet it is no longer capable to prove optimality within the time limit.

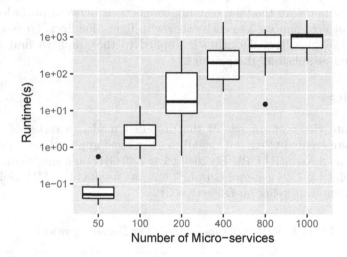

Fig. 5. Comparison of runtime for increasing application demands.

Scalability. Figure 5 shows a set of benchmarks based on a distributed application with a few services, security policies and zones meant to capture the impact of application demands (replication level) on the solving runtime. A set of 20 benchmarks are generated for application sizes of $\{50, 100, 200, 400, 800, 1000\}$ service instances. Each benchmark was executed with a timeout of 5400 s. Figure 5 represents the total run time required to reach the optimal solution *and close the problem*. The box plot chart highlights that the counting-based model delivers an order of magnitude improvement in scalability over [5]. Indeed, the largest benchmarks with 1000 service instances often terminate in 1000 s. That being said, out of the 20 benchmarks at each size, the solver timed out without an optimality proof 7 times on size 1000, 6 times on size 800 and 4 times on 400 (it never timed out on the smaller ones).

Early Termination. Solving to optimality for large benchmarks may prove excessively demanding. This experiment considers an early termination condition that stops the branch and bound if the optimality gap drops below 1%. Figure 6 shows the solving time. Comparing the results from Fig. 5, it is clear that high-quality solutions are delivered *early on* during the search process and can lead to a significant reduction in the runtime.

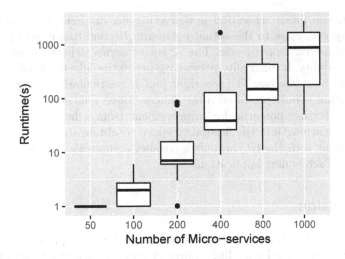

Fig. 6. Runtimes for finding a solution within a 1% optimality gap.

Trajectories. Figure 7 shows the trajectories of the lower bound and upper bound on a few benchmarks. It highlights how quickly the gap evolves as a function of time. The red line captures the lower bound and the black line the incumbent solution. Comparing the four graphs in Fig. 7, a few observation seem in

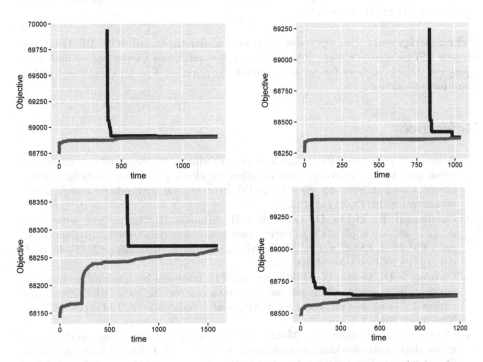

Fig. 7. Comparison of the performance gap between four benchmarks. (Color figure online)

order. First, the linear relaxation is delivering an *excellent* lower bound that often rapidly gets close to the actual optimum. Second, the curves for feasible upper bound are far more diverse. The IP solver seems to have mixed success in attaining feasibility. Incidentally, alternative heuristics, like the feasibility pump, were tried without success. The top right plot is particularly telling as the first feasible solution is only produced after almost 800 s. This observation indicates that there is further potential for improvement from a hybrid technique that leverages CP's strength in the feasibility space to obtain solutions much earlier and share them with the IP through a parallel composition as discussed in [6] for the case of scheduling applications.

6   Conclusion

This paper revisited the secure micro-service deployment problem introduced in [5] and offered a novel modeling approach based on counting that eliminates symmetries and improves the quality of the linear relaxation used by an IP solver. The impact is significant and translates into order of magnitude improvements in the runtime for the original instances and delivers an order of magnitude improvement in scalability. The approach appears capable to handle industrial scale instances with order of 1000 micro-services. The empirical results emphasize the strength of the lower-bound but also reveal additional potential for hybridization with a strong primal technique capable of delivering high-quality upper-bounds in short runtime.

Acknowledgment. This work was supported under the award SOW BL 11568 and project CSI Selected Projects 2018: *Securing Virtualization Configuration and Managing the Attack Surfaces* funded by Comcast Corporation. Special thanks to Vaibhav Garg from Comcast.

References

1. Armant, V., Cauwer, M.D., Brown, K.N., O'Sullivan, B.: Semi-online task assignment policies for workload consolidation in cloud computing systems. Future Gener. Comput. Syst. **82**, 89–103 (2018). http://www.sciencedirect.com/science/article/pii/S0167739X17319143
2. Cambazard, H., Mehta, D., O'Sullivan, B., Simonis, H.: Bin packing with linear usage costs – an application to energy management in data centres. In: Schulte, C. (ed.) CP 2013. LNCS, vol. 8124, pp. 47–62. Springer, Heidelberg (2013). https://doi.org/10.1007/978-3-642-40627-0_7
3. Cauwer, M.D., Mehta, D., O'Sullivan, B.: The temporal bin packing problem: an application to workload management in data centres. In: 2016 IEEE 28th International Conference on Tools with Artificial Intelligence (ICTAI), pp. 157–164, November 2016
4. Chisca, D.S., Castineiras, I., Mehta, D., OSullivan, B.: On energy- and cooling-aware data centre workload management. In: 2015 15th IEEE/ACM International Symposium on Cluster, Cloud and Grid Computing, pp. 1111–1114, May 2015

5. Cruz, W., Liu, F., Michel, L.: Securely and automatically deploying micro-services in an hybrid cloud infrastructure. In: Hooker, J. (ed.) CP 2018. LNCS, vol. 11008, pp. 613–628. Springer, Cham (2018). https://doi.org/10.1007/978-3-319-98334-9_40

6. Fontaine, D., Michel, L., Van Hentenryck, P.: Parallel composition of scheduling solvers. In: Quimper, C.-G. (ed.) CPAIOR 2016. LNCS, vol. 9676, pp. 159–169. Springer, Cham (2016). https://doi.org/10.1007/978-3-319-33954-2_12

7. Greenberg, A., Hamilton, J., Maltz, D.A., Patel, P.: The cost of a cloud: research problems in data center networks. SIGCOMM Comput. Commun. Rev. **39**(1), 68–73 (2008). https://doi.org/10.1145/1496091.1496103

8. Gutin, G., Jensen, T., Yeo, A.: Batched bin packing. Discret. Optim. **2**(1), 71–82 (2005). http://www.sciencedirect.com/science/article/pii/S1572528605000058

9. Hermenier, F., Demassey, S., Lorca, X.: Bin repacking scheduling in virtualized datacenters. In: Lee, J. (ed.) CP 2011. LNCS, vol. 6876, pp. 27–41. Springer, Heidelberg (2011). https://doi.org/10.1007/978-3-642-23786-7_5

10. Hermenier, F., Lawall, J., Muller, G.: BtrPlace: a flexible consolidation manager for highly available applications. IEEE Trans. Dependable Secure Comput. **10**(5), 273–286 (2013)

11. Kadioglu, S., Colena, M., Sebbah, S.: Heterogeneous resource allocation in cloud management. In: 2016 IEEE 15th International Symposium on Network Computing and Applications (NCA), pp. 35–38. IEEE (2016)

12. King, V., Rao, S., Tarjan, R.: A faster deterministic maximum flow algorithm. In: Proceedings of the Third Annual ACM-SIAM Symposium on Discrete Algorithms, SODA 1992, pp. 157–164. Society for Industrial and Applied Mathematics, Philadelphia, PA, USA (1992). http://dl.acm.org/citation.cfm?id=139404.139438

13. Sebbah, S., Bagley, C., Colena, M., Kadioglu, S.: Availability optimization in cloud-based in-memory data grids. In: Rueher, M. (ed.) CP 2016. LNCS, vol. 9892, pp. 666–679. Springer, Cham (2016). https://doi.org/10.1007/978-3-319-44953-1_42

14. Srikantaiah, S., Kansal, A., Zhao, F.: Energy aware consolidation for cloud computing. In: Proceedings of the 2008 Conference on Power Aware Computing and Systems, HotPower 2008, pp. 10–10. USENIX Association, Berkeley, CA, USA (2008). http://dl.acm.org/citation.cfm?id=1855610.1855620

An Optimization Approach to the Ordering Phase of an Attended Home Delivery Service

Günther Cwioro[1], Philipp Hungerländer[2], Kerstin Maier[1], Jörg Pöcher[1], and Christian Truden[1(✉)]

[1] Department of Mathematics, Alpen-Adria-Universität Klagenfurt, Klagenfurt, Austria
{guenther.cwioro,kerstin.maier,joerg.poecher,christian.truden}@aau.at
[2] Laboratory for Information and Decision Systems, Massachusetts Institute of Technology, Cambridge MA, 02139, USA
philipp.hungerlaender@aau.at

Abstract. Attended Home Delivery (AHD) systems are used whenever a supplying company offers online shopping services which require that customers must be present when their deliveries arrive. Therefore, the supplying company and the customer must both agree on a time window, which ideally is rather short, during which delivery is guaranteed. Typically, a capacitated Vehicle Routing Problem with Time Windows forms the underlying optimization problem of the AHD system. In this work we consider an AHD system that runs the online grocery shopping service of an international grocery retailer.

The ordering phase, during which customers place their orders through the web service, is the computationally most challenging part of the AHD system. The delivery schedule must be build dynamically as new orders are placed. We propose a solution approach that allows to determine which delivery time windows can be offered to potential customers. We split the computations of the ordering phase into four key steps. For performing these basic steps we suggest both a heuristic approach and a hybrid approach employing Mixed-Integer Linear Programs. In an experimental evaluation we demonstrate the efficiency of our approaches.

Keywords: Attended home delivery ·
Capacitated vehicle routing with time windows · Heuristics

1 Introduction

In recent years, online grocery shopping has gained increased popularity in several countries, such as the United Kingdom where about 6.3% [16] of all grocery shopping is bought online. Nowadays, all major supermarket chains provide online shopping services where customers select groceries as well as a delivery

© Springer Nature Switzerland AG 2019
L.-M. Rousseau and K. Stergiou (Eds.): CPAIOR 2019, LNCS 11494, pp. 208–224, 2019.
https://doi.org/10.1007/978-3-030-19212-9_14

time window on the supermarket's website. This provides several benefits to the customers, such as 24-h opening hours of the online store, quicker shopping times, the avoidance of traveling times, no carrying of heavy or bulky items, and facilitated access for citizens with reduced mobility. Despite of the benefits for the customers, e-grocery shopping services pose several interrelated logistic and optimization challenges to the supplying companies. Especially the *Ordering Phase*, during which customers place their orders, imposes a computationally challenging problem.

In this paper we tackle this challenge in the context of a large international supermarket chain that offers online grocery shopping. E-grocery services are a paradigm for *Attended Home Delivery* (AHD) problems [1–4] where the customers must be present for their deliveries. In order to ensure customer satisfaction and to minimize undeliverable orders, it is crucial that the supplying company provides a wide selection of rather narrow delivery time windows. Hence, in this work, we aim to provide a framework that determines the available time windows and dynamically builds the delivery schedule.

This paper is organized as follows. First we provide an overview of the logistic process behind the considered AHD system and discuss the *Ordering Phase* in detail. In Sect. 2 we introduce the related optimization problem and suggest algorithmic strategies for solving it. In Sect. 3 we demonstrate the efficiency of our solution approaches on benchmark instances related to an online grocery shopping system. Finally, Sect. 4 concludes the paper.

1.1 The Attended Home Delivery Process

Let us start with giving a short overview of the overall planning and fulfillment process behind the *Attended Home Delivery* service, by describing the actions taken by the supplying e-grocery retailer in order to fulfill the deliveries of a single day.

Tactical Planning Phase - (several months/weeks before delivery):

- A fleet of vehicles is set up and operation times of those vehicles are defined.
- Drivers are assigned to the vehicles in accordance to the legal regulations concerning drive and rest times.
- The supplying company defines the set of possible delivery time windows that will later be offered to the customers through the web service.

Ordering Phase - (several weeks up to days/hours before delivery): This phase begins once the web service starts to allow booking of delivery time windows for the specific day of delivery. Hence, the system must handle the following tasks:

- Customers use the web service through a web site or a mobile app to place their orders.
- The system must decide which delivery time windows can be offered to a specific customer, such that the delivery can be fulfilled within the time window. Only the resources that have been assigned during the tactical planning phase are available.

– Once a customer has booked a delivery time window, the system must adapt the existing delivery schedule to accommodate the respective order. Furthermore, the system periodically tries to improve the current schedule.

During this phase, the objective is to accept as many customers as possible, while offering as many time windows as possible to each potential customer in order to achieve a high degree of customer satisfaction and to also ensure good resource utilization, which translates to the overall logistics operations being cost efficient.

Preparation Phase - (Days/Hours Before Delivery): This phase is triggered once the system does no longer accept new orders through the web service. The objective function is now changed to minimization of the transportation costs (overall fulfillment costs). Another relevant aspect to be considered is the traffic flow at the depot and the vehicle loading bays. Hence, the following tasks must be performed:

– The delivery schedule is improved regarding the new objective function.
– Meanwhile, at the depot, the ordered goods are fetched from storage and consolidated accordingly to the customer orders.

Delivery Phase: In this phase the vehicles are first packed with the consolidated orders and prepared to leave the depot. Then the vehicles visit the customers according to the delivery schedule, which was generated by the system, such that the customers receive their orders within the selected time windows.

1.2 Related Work

First, let us give a short overview of related work.

– Campbell and Savelsbergh [2] describe a *Home Delivery* system that decides if a customer order is accepted. Furthermore, the system assigns accepted orders to a time window under consideration of the opportunity costs of the orders. In contrast to that, in our setup the customer takes the decision to which delivery time window their order is assigned to.
– Parts of the *Ordering Phase* are tackled by Hungerländer et al. [7] using two Mixed-Integer Linear Programs (MILPs).
– The *Slot Optimization Problem* is introduced by Hungerländer et al. [8]. It describes the problem of determining the maximal number of available delivery time windows for a new customer.
– Hungerländer and Truden [9] focus on providing competitive MILP formulations for the *Traveling Salesperson Problem with Time Windows*.

The contribution of this work is given as followed:

– A general description of an AHD system based on the use-case of an online grocery shopping service is given and a detailed description of the *Ordering Phase* is provided.

- A heuristic solution approach for the introduced problem is presented. Further, the authors propose a hybrid approach that applies a Mixed-Integer Linear Program.
- Finally, novel benchmarks instances are introduced and computational experiments that evaluate the performance of the proposed approaches are presented.

1.3 Challenges and Key Steps of the Ordering Phase

In this paper we focus on the *Ordering Phase* and suggest solution approaches to deal with the computational challenges arising during this phase. In particular, the runtime requirements for the optimization approaches applied during this phase are much more severe than in any other phase.

All decisions taken in the foregoing *Tactical Planning Phase* are considered as input variables. During the *Ordering Phase* customers can book their grocery deliveries through a web service. Figure 1 illustrates a generic example website and its main features. Clearly, the web service should respond to the customer requests with as little delay as possible. Fetching and providing the input data for the booking process requires communication across several services and many data base queries. As this already takes a significant amount of time, there is even less time to solve the actual optimization problem.

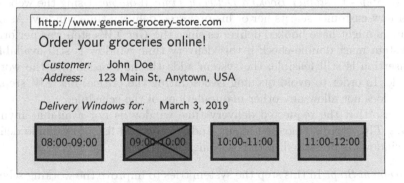

Fig. 1. Illustration of a generic example website of an AHD service for grocery online shopping. Based on the customer's address, the system determines the availability of the predefined delivery time windows. Non-available delivery time windows, e.g., 09:00–10:00, are crossed out.

During the *Ordering Phase*, the web service must repeatedly solve an online variant of a *capacitated Vehicle Routing Problem with Time Windows* (cVRPTW). The cVRPTW is concerned with finding optimal tours for a fleet of vehicles with given capacity constraints to deliver goods to customers within assigned time windows. As the cVRPTW is known to be NP-hard [13], the naive approach of solving a new cVRPTW instance from scratch for each new customer order is far from being applicable in an online environment, even when using fast Meta-heuristics [5], due to the tight restrictions regarding runtime.

For clarity of exposition, we further split the *Ordering Phase* into the following *four key steps:*

Initialization Step: The system sets up an empty delivery schedule, i.e., a cVRPTW instance with a fixed number of vehicles where each vehicle has defined operation times that were determined during the *Tactical Planning Phase*, but not having any customers assigned to yet.

Get TWs Step - The System Determines Available Delivery Time Windows: Based on the current delivery schedule the system determines which delivery time windows are available to a new customer. During times with high customer request rates this step has to be performed within milliseconds. The available time windows are then presented to the customer through the web service. Note that the customer has to provide a delivery address such that a routing system can estimate the travel times between all pairs of customers.

Optionally, for reasons of profit maximizing, some available time windows can be hidden from the customer or be offered at different rates. However, we do not consider any kind of slot pricing in this work. For related and recent publications on pricing in the context of AHD systems we refer the reader to [10–12,18].

Set TW Step - Customer Books a Delivery Time Window: Using the website or app, a new customer selects her or his preferred delivery time window. As other customers might have booked deliveries since the *Get TWs* step was performed, the system must double-check if this delivery time window is still available. If the insertion is still feasible, the system adds the new order into the working schedule. In order to avoid queuing issues during the critical *Set TW* step, the system does not allow any other manipulations of the schedule.

In case that the requested delivery time window is not available anymore, the *Get TWs* step is triggered again, and an updated list of available delivery time windows is presented to the customer.

Improvement Step: In this step the system tries to improve the working schedule such that as many delivery time windows as possible can be offered to potential future customers and therefore more customers can place their orders. Choosing the total travel time as objective function has proven to be a reasonable choice to achieve this. While the fleet and the assignment of customers to time windows is fixed, the assignment of customers to delivery vehicles as well as the sequences in which the vehicles visit the customers can be altered. Typically, the *Improvement* step may take several seconds, but during times with high customer request rates the step can be omitted or only be triggered after a certain number of *Set TW* steps. Note that at any time there is exactly one working schedule in the system.

In the following section we formally introduce the cVRPTW as the underlying optimization problem of the *Ordering Phase* and propose algorithmic strategies for dealing with the cVRPTW during each of the four steps described above.

2   Algorithms

2.1   Formal Definition of the cVRPTW

In this section we now formally introduce the cVRPTW and some further required notations.

Basic Definitions: A cVRPTW instance is typically defined by the following input data:

- A set of *time windows* $\mathcal{W} = \{w_1, \dots, w_q\}$, where each time window $w \in \mathcal{W}$ is defined through its *start time* s_w and its *end time* e_w. We assume that the time windows are unique. Hence, there do not exist time windows $w_a, w_b, \in \mathcal{W}, w_a \neq w_b$ with $s_{w_a} = s_{w_b}$ and $e_{w_a} = e_{w_b}$.
- A set of *customers* $\mathcal{C}$, $|\mathcal{C}| = p$, with corresponding *order weight* function $c : \mathcal{C} \rightarrow \mathbb{R}^{>0}$, a *service time* function $s : \mathcal{C} \rightarrow \mathbb{R}^{>0}$, and a *travel time* function $t : \mathcal{C} \times \mathcal{C} \rightarrow \mathbb{R}^{\geq 0}$ where we set the travel time from a customer a to itself to 0, i.e., $t(a, a) = 0$, $a \in \mathcal{C}$.
- A function $w : \mathcal{C} \rightarrow \mathcal{W}$ that assigns to each customer a time window, during which the delivery vehicle has to arrive at the customer.
- A *schedule* $\mathcal{S} = \{\mathcal{A}, \mathcal{B}, \dots\}$, consisting of $|\mathcal{S}| = m$ tours with assigned capacities C_k, $k \in \mathcal{S}$, where C_k corresponds to the *capacity* of the vehicle that operates tour k.

A *tour* $\mathcal{A} = \{a_1, a_2, \dots, a_n\}$ contains n customers, where the indices of the customers display the sequence in which the customers are visited. To improve clarity of exposition, we sometimes additionally use upper indices, i.e., $\mathcal{A} = \{a_1^{w(a_1)}, a_2^{w(a_2)}, \dots, a_n^{w(a_n)}\}$, which indicate the time windows assigned to the customers. Furthermore, each tour $\mathcal{A}$ has assigned *start* and *end times* that we denote as $start_{\mathcal{A}}$ and $end_{\mathcal{A}}$ respectively. Hence, the vehicle executing tour $\mathcal{A}$ can leave from the start depot no earlier than $start_{\mathcal{A}}$ and must return to the end depot no later than $end_{\mathcal{A}}$.

Structured Time Windows: Two time windows w_a and w_b are *non-overlapping* if and only if $e_{w_a} \leq s_{w_b}$ or $e_{w_b} \leq s_{w_a}$. Therefore, w_a and w_b do overlap if and only if $s_{w_b} < e_{w_a}$ and $s_{w_a} < e_{w_b}$. We speak of *structured* time windows, if all time windows in $\mathcal{W}$ are pair-wise non-overlapping and if the number of customers $|\mathcal{C}| = p$ is much larger than the number of time windows $|\mathcal{W}| = q$, i.e., $n \gg q$, and therefore typically several customers are assigned to the same time window. We denote the corresponding variant of the cVRPTW as the *capacitated Vehicle Routing Problem with structured Time Windows* (cVRPsTW). Structured time windows are a specialty that arises in the Attended Home Delivery use case, as well as some other modern routing applications. Note that the corresponding assumptions do not impose severe restrictions to the supplier nor the customers, but allow for a more efficient optimization of the corresponding logistics operations.

2.2   Arrival Times and Feasibility

Next, let us give formal, recursive definitions of the earliest and latest arrival times that are needed to define feasibility of a schedule and of the insertion of a new customer.

Earliest and Latest Arrival Times: We consider a fixed tour $\mathcal{A} = \{a_0, a_1, \ldots, a_n, a_{n+1}\}$, where a_0 is the start depot, a_{n+1} is the end depot, and $\{a_1, \ldots, a_n\}$ is the set of customers assigned to tour $\mathcal{A}$. Note that all our approaches do not move the depots. Hence, customers can only be inserted after the start depot and before the end depot. The earliest (latest) arrival time α_{a_i} (β_{a_i}) gives the earliest (latest) time at which the vehicle may arrive at a_i, who is the ith customer on the tour, while not violating time window and travel time constraints on the preceding (subsequent) tour:

$$\alpha_{a_0} := start_{\mathcal{A}}, \quad \alpha_{a_{j+1}} := \max\left\{s_{w(a_{j+1})}, \alpha_{a_j} + s(a_j) + t(a_j, a_{j+1})\right\}, \; j \in [n-1],$$

$$\alpha_{a_{n+1}} := \alpha_{a_n} + s(a_n) + t(a_n, a_{n+1}),$$

$$\beta_{a_{n+1}} = end_{\mathcal{A}}, \quad \beta_{a_{j-1}} := \min\left\{e_{w(a_{j-1})}, \beta_{a_j} - t(a_{j-1}, a_j) - s(a_{j-1}), a_j)\right\},$$

$$j \in [n] \setminus \{1\}, \quad \beta_{a_0} := \beta_{a_1} - t(a_0, a_1).$$

Feasibility of a Schedule: Now we can concisely define the feasibility of a tour and a schedule with the help of the earliest arrival times. A schedule S is *feasible*, if all its tours are feasible. A tour $\mathcal{A}$ is *feasible*, if it satisfies both of the following conditions:

$$s_{w(a_i)} \le \alpha_{a_i} \le e_{w(a_i)}, \; i \in [n], \; \wedge \; \alpha_{a_{n+1}} \le end_{\mathcal{A}}, \qquad \text{(TFEAS)},$$

$$\sum_{i \in [n]} c(a_i) \le C_{\mathcal{A}}, \qquad \text{(CFEAS)}.$$

While **TFEAS** ensures that the arrival times at each customer are within their assigned time windows, **CFEAS** ensures that the capacity of $\mathcal{A}$ is not exceeded.

Feasibility of an Insertion: Now we further use the concepts of earliest and latest arrival time to facilitate and algorithmically speed up feasibility checks of tours after inserting an additional customer. A new customer $\tilde{a}^w$ can be feasibly inserted *with respect to time* between customers a_i and a_{i+1}, $i \in [n_0]$, into a feasible tour $\mathcal{A}$ if the following condition holds:

$$\alpha_{\tilde{a}^w} \le \beta_{\tilde{a}^w}, \qquad \text{TFEAS}(\tilde{a}^w, i+1, \mathcal{A}), \qquad (1)$$

where,

$$\alpha_{\tilde{a}^w} := \max\{s_w, \alpha_{a_i} + s(a_i) + t(a_i, \tilde{a}^w)\},$$

$$\beta_{\tilde{a}^w} := \min\{e_w, \beta_{a_{i+1}} - s(\tilde{a}^w) - t(\tilde{a}^w, a_{i+1})\}.$$

Condition (1) ensures that we arrive at customer $\tilde{a}^w$ early enough, such that we can leave from $\tilde{a}^w$ early enough, to handle all subsequent customers of $\mathcal{A}$ within their assigned time windows. We refer to Fig. 2 for an illustration of the above condition.

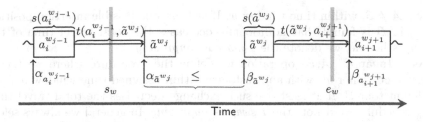

Fig. 2. Depiction of a feasible insertion with respect to time of $\tilde{a}^{w_j}$ between $a_i^{w_j-1}$ and $a_{i+1}^{w_j+1}$, i.e., $\texttt{TFEAS}(\tilde{a}^{w_j}, i+1, \mathcal{A})$ holds.

Additionally, we have to check that the sum of the weights of the customer orders assigned to tour $\mathcal{A}$ does not exceed the capacity $C_\mathcal{A}$. The insertion of $\tilde{a}^w$ into tour $\mathcal{A}$ is feasible *with respect to capacity*, if the following condition holds:

$$\sum_{i\in[n]} c(a_i) + c(\tilde{a}^w) \leq C_\mathcal{A}, \qquad \texttt{CFEAS}(\tilde{a}, \mathcal{A}). \tag{2}$$

Assuming that all earliest and latest arrival times and the sum of capacities have already been calculated, Conditions (1) and (2) allow to check the feasibility of an insertion of a new customer into a given time window in $\mathcal{O}(1)$.

If we conduct an insertion that is feasible with respect to time and capacity and decide to insert $\tilde{a}^w$, we receive a new tour $\tilde{\mathcal{A}} = \{a_0, a_1, \ldots, a_i, \tilde{a}^w, a_{i+1}, \ldots, a_n, a_{n+1}\}$. Customer $\tilde{a}^w$ is then assigned index $i + 1$ and the indices of all succeeding customers are increased by one. Clearly, earliest and latest arrival times and the sum of capacities of the modified tour must be updated, which can be done in $\mathcal{O}(n)$.

Note that, in the context of an offline *Traveling Salesperson Problem with Time Windows*, the Generalized Insertion heuristic proposed by Gendreau et al. [6] uses concepts analog to our earliest and latest arrival times. The authors also check the feasibility of possible insertions with two conditions that resemble Condition (1). Due to their efficient computation, Conditions (1) and (2) form the basic building blocks of our Local Search heuristic that we describe in the following subsections. Moreover, the concept is flexible enough such that valuable extensions, such as time-dependent travel times or the integration of driving breaks into the schedule, can be employed without major changes.

2.3 Local Search Heuristic

We consider a Local Search heuristic that uses two neighborhoods for exchanging customer orders between two tours:

1. The *1-move* neighborhood moves a customer from one tour to another tour.
2. The *1-swap* neighborhood swaps two customers between two different tours.

Accordingly we define the $\texttt{1move}(\tilde{a}^w, \mathcal{A}, \mathcal{B})$ operation as the procedure where we *remove* customer $\tilde{a}^w$ from tour $\mathcal{A} \in \mathcal{S}$ and try to feasible *insert* it into tour

$\mathcal{B} \in \mathcal{S}$, $\mathcal{A} \neq \mathcal{B}$, within time window w. If at least one feasible insertion position for $\tilde{a}^w$ in $\mathcal{B}$ is found that additionally decreases the total travel time of the delivery schedule, we denote the *1-move* as *improving*.

As a $\mathtt{1swap}(\tilde{a}^w, \mathcal{A}, \mathcal{B})$ operation we define the procedure where we try to *exchange* customer $\tilde{a}^w$ with any customer within assigned time window w from a different tour $\mathcal{B}$. If at least one such exchange decreases the total travel time of the schedule, we denote the *1-swap* as *improving*. In general we always select the exchange of an improving *1-swap* that results in the largest decrease of the total travel time of the delivery schedule. In Fig. 3 we provide an illustration of an improving $\mathtt{1swap}(a_3^{w_j}, \mathcal{A}, \mathcal{B})$.

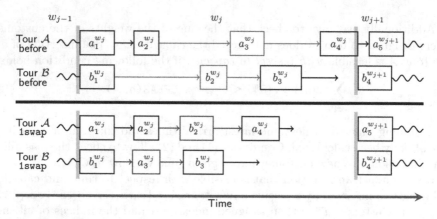

Fig. 3. Reduction of the total travel time of a schedule induced by an improving $\mathtt{1swap}(a_3^{w_j}, \mathcal{A}, \mathcal{B})$ operation.

2.4 Algorithmic Strategies

In this subsection we describe how to combine the Local Search heuristics presented in the previous subsection in order to conduct sufficiently fast *Get TWs*, *Set TW* and *Improvement* steps.

The Get TWs Step: In this step we aim to quickly identify all time windows during which a new customer $\tilde{a}$ can be inserted into (at least one of) the current tours. We suggest to use the following procedure:

- **Simple-insertion.** For each time window $w \in \mathcal{W}$ iterate over all tours $\mathcal{A} \in \mathcal{S}$ and all possible insertion points within w and check Conditions (1) and (2). A time window w is considered as being *available*, if both conditions hold for at least one insertion point. In this case we add w to the set of available time windows $\mathcal{T}_{\tilde{a}} \subseteq \mathcal{W}$. This procedure is computationally very cheap and runs within 1 ms for all benchmark instances considered in our computational study.

Then the time windows $\mathcal{T}_{\tilde{a}}$ are offered to the customers through the web service. Note that [2] proposes a similar procedure for the VRP, i.e., without considering time windows.

Additionally to this *Simple-insertion* heuristic we introduced an Adaptive Neighborhood Search (ANS) in [8] that is especially tailored to the *Get TWs* step. The ANS applies *1-move* and *1-swap* operations to free up time during a specific time window on a selected tour in order to enable the insertion of the new customer. ANS has proven to find more available time windows than *Simple-insertion*, while still being fast enough for most applications, as long as the customer request rate is moderate. However, as this strategy has already been discussed and computationally compared [8], we do not benchmark it in this paper.

The Set TW Step: Once the customer has selected a time window $\tilde{w}$ from the set $\mathcal{T}_{\tilde{a}}$, we double check its availability in the same manner as in the *Get TWs* step, and then we immediately insert $\tilde{a}$ into $\tilde{w}$ at the best found insertion point.

The Improvement Step: In this step we aim to reduce the total travel time of the delivery schedule by using one of the following two procedures:

- **Local-improvement.** Our computationally cheap, yet quite effective Local Search heuristic builds the foundation of the *Improvement* step. We combine *1-move* and *1-swap* operations, where we focus on the *1-move* operations when possible, as they are computationally cheaper and in general more effective than the *1-swap* operations. We stop our Local Search heuristic once we reach a local minimum of our objective function with respect to our neighborhoods.
- **Local+TSPTW-improvement.** After the Local Search heuristic we additionally use MILPs proposed in our previous paper [9] for optimizing all single tours that have changed since the last improvement step. In [9] we motivated and analyzed the *Traveling Salesperson Problem with Time Windows* (TSPTW) that is a subproblem of the cVRPTW as each tour of the delivery schedule corresponds to a TSPTW instance. Optimizing the single tours of a schedule to optimality has been proven to be critical to ensure driver satisfaction. Hence, it ensures that drivers do not encounter any obvious loops on their routes. Also note that we use the current tours of our delivery schedule for warm starting the TSPTW MILPs.

During the *Ordering Phase* our Local Search heuristic only performs improving operations. However, the algorithms can be simply altered into a *Simulated Annealing* approach by allowing also non-improving operations, which is more appropriate for the *Preparation Phase* when there is more time available for optimization.

3 Computational Experiments

In this section we present computational results on a set of benchmark instances that are motivated by an online shopping service of an international grocery retailer. We restrict our experiments to a setup with structured time windows as it has proven to be more computationally efficient than having overlapping time windows and therefore is better suited for the use in an AHD system. Accordingly we consider the *Traveling Salesperson Problem with structured Time Windows* (TSPsTW) [9], which is a special case of the cVRPsTW as each tour of the delivery schedule corresponds to a TSPsTW instance.

3.1 Benchmark Instances

In order to provide meaningful computational experiments we created a benchmark set that resembles real-world data focusing on urban settlement structures. Moreover, the benchmark instances are designed to reflect instances as they arise in an online grocery shopping service of a major international supermarket chain, regarding travel times, length of time windows, duration of service times, customer order weights, and their proportions to vehicle capacities. All instances can be downloaded from http://tinyurl.com/vrpstw. Note that the well-known VRPTW benchmark instances proposed by Solomon [15] do not comply with our considered use-case.

In more detail, our benchmark instances have the following characteristics:

- **Grid Size.** We consider a 20 km × 20 km square grid, which is roughly of the size of Vienna as well as a smaller grid of size 10 km × 10 km that corresponds to smaller cities.
- **Placing of Customers.** In order to achieve varying customer densities, only 20 % of the customer locations have been sampled from a two-dimensional uniform distribution. The remaining 80 % of the customer locations have been randomly assigned to 10 clusters. The centers resp. shapes of those clusters have been randomly sampled as well. Finally, the customer locations have been sampled from the assigned cluster.
- **Depot Location.** We consider two different placements of the depot: At the center of the grid, and at the center of the top left quadrant. In each test setup there are equally many instances for both variants.
- **Travel Speeds.** As proposed by Pan et al. [14] we assume a travel speed of 20 km/h. This number can be further supported by a recent report by Vienna Public Transport [17], where an average travel speed for their fleet of buses of 17.7 km/h during the day, 17.2 km/h at peak times, and 20.0 km/h during evening hours has been reported.

  For the sake of simplicity, the distance between two locations is calculated as the Euclidean distance between them. Travel times are calculated proportional to the Euclidean distances, using the assumed travel speed.
- **Order Weights.** The order weights of customers have been sampled from a truncated normal distribution with mean of 7 and standard deviation of 2, where the lower bound is 1 and the upper bound is 15.
- **Customer Choice Model.** A customer choice model simulates the decisions that are usually taken by the customers. We choose a simple model, where every customer has just one desired delivery time window that has been set

beforehand in the benchmark instance. If the preferred time window is not offered to the customer, we assume that the customer does not place an order and leaves the website. We simulate this by a random assignment, following a uniform distribution, of each customer to one time window out of a set of 5 resp. 10 consecutive time windows, where each is one hour long. The later reflects a situation where customers choose from one-hour delivery time windows between 08:00 and 18:00. Note that this is in contrast to real-world applications, where usually certain time windows are more prominent among customers than others. However, we chose a uniform distribution to obtain unbiased results that allow for an easier identification and clearer interpretation of the key findings.

– **Service Times.** We assume the service time at each customer to be 5 min.

In summary, our assumptions were chosen in order to find a good compromise between realistic real-world instances and enabling a concise description and interpretation of the experimental set-up.

Typically, in the online grocery delivery use-case no more than 500 customer are served from the same depot on a given day. Hence, for our computational experiments we consider two benchmark sets that contain 500 customers each:

1. A benchmark set with *many short* (30–40) tours having a capacity of 100 and 5 time windows each.
2. A benchmark set with *fewer long* (10–20) tours having a capacity of 200 and 10 time windows each.

Moreover, we simulate sparsely such as densely populated delivery regions by using grids of size 20 km × 20 km resp. 10 km × 10 km.

3.2 Experimental Setup

All experiments were performed on an Ubuntu 14.04 machine equipped with an Intel Xeon E5-2630V3 @ 2.4 GHz 8 core processor and 132 GB RAM. We implemented all algorithms in Java version 8 and use Gurobi 8.0.1 as IP-solver in single thread mode. We compare the algorithmic strategies presented in the previous section for both the *Get TWs* and the *Improvement* step.

In all our experiments we iteratively insert new customers into the schedule, simulating customers placing orders online following the customer choice model. Hence, we assume that if the preferred time window is not offered to the customer she or he refuses to place an order and hence, the customer is not inserted into the schedule. Due to the iterative setup we can omit the *Set TW* step and insert the new customer without double-checking the availability of the selected delivery time slot. We determine the following metrics averaged over 100 instances each:

– *Get TWs* step:
  • Average number of feasible time windows determined for each customer: Corresponds to the number of time windows in which the order can be inserted.
  • Average runtime of the *Get TWs* step.

– *Improvement* step:
  • Average improvement over *Insertion* step: Given as the average reduction of the sum of travel times over all tours relative to the total travel time of the schedule after inserting the new customer (given as percentage).
  • Average improvement of the cost of *Insertion*: Given as the average reduction of the sum of travel times over all tours relative to the increase of travel time caused by the insertion of the new customer (given as percentage).
  • Average number of TSPsTW MILPs solved.
  • Average runtime of each *Improvement* step.

3.3   Results

Now let us present the results of our computational evaluation. We examine the performance of our approaches on instances with 500 customers. The results for the sparse resp. dense benchmark sets are summarized in Table 1 resp. Table 2 for the *Get TWs* step, and in Table 3 resp. Table 4 for the *Improvement* step.

Table 1. Results for the *Get TWs step* for our sparse benchmark scenarios.

Get TWs step Simple-insertion	500 customers 100 capacity units 5 time windows			500 customers 200 capacity units 10 time windows		
Tours	30	35	40	10	15	20
Average runtime (sec:ms)	0:001	0:001	0:001	0:001	0:001	0:001
Number of time windows offered (avg.)	4.29	4.93	5.00	5.76	8.59	10.00
Total customers inserted (avg.)	428.90	493.40	500.00	288.30	429.20	500.00

Table 2. Results for the *Get TWs step* for our dense benchmark scenarios.

Get TWs step Simple-insertion	500 customers 100 capacity units 5 time windows			500 customers 200 capacity units 10 time windows		
Tours	30	35	40	10	15	20
Average runtime (sec:ms)	0:001	0:001	0:001	0:001	0:001	0:001
Number of time windows offered (avg.)	4.28	4.94	5.00	5.76	8.59	10.00
Total customers inserted (avg.)	428.30	493.70	500.00	287.80	429.50	500.00

The first benchmark set with many short tours corresponds to the left column, and the results for the second benchmark set with few long tours are displayed in the right column. First, we observe that the runtimes for both the *Get TWs* and the *Improvement* step are very low despite of the large instances,

Table 3. Results for the *Improvement step* for our sparse benchmark scenarios.

Improvement step	500 customers 100 capacity units 5 time windows			500 customers 200 capacity units 10 time windows		
Tours	30	35	40	10	15	20
Avg. runtime (sec:ms)						
Local	1:024	1:453	1:504	0:653	1:512	1:966
Local+TSPsTW	1:115	1:547	1:600	0:729	1:602	2:058
Avg. improvement over *Insertion* step (%)						
Local	0.96	0.90	0.90	0.71	0.65	0.60
Local+TSPsTW	1.00	0.94	0.93	0.78	0.72	0.65
Avg. improvement of cost of *Insertion* (%)						
Local	66.12	67.97	68.49	50.13	56.73	58.96
Local+TSPsTW	68.81	78.55	71.05	55.16	61.27	63.42
Avg. number of TSPsTW MILPs solved						
Local+TSPsTW	3.48	3.76	3.81	1.82	2.25	2.41

Table 4. Results for the *Improvement step* for our dense benchmark scenarios.

Improvement step	500 customers 100 capacity units 5 time windows			500 customers 200 capacity units 10 time windows		
Tours	30	35	40	10	15	20
Avg. runtime (sec:ms)						
Local	1:468	1:882	1:912	0:807	2:035	2:707
Local+TSPsTW	1:564	1:992	2:014	0:886	2:140	2:807
Avg. improvement over *Insertion* step (%)						
Local	1.00	0.95	0.95	0.68	0.65	0.62
Local+TSPsTW	1.05	0.99	0.98	0.76	0.70	0.66
Avg. improvement of cost of *Insertion* (%)						
Local	67.37	69.59	70.20	49.01	57.18	60.65
Local+TSPsTW	70.43	72.48	73.09	54.56	62.23	65.66
Avg. number of TSPsTW MILPs solved						
Local+TSPsTW	3.59	3.89	3.94	1.93	2.37	2.56

which demonstrates that our solving approaches scale very well. It is worth pointing out that the *Get TWs* step stays below 1 ms even for a large number of customers. This is crucial in order to deal with high customer request rates at peak times. Considering that between two *Improvement* steps the schedule is altered only by insertion of one customer, a reduction of our objective function by 0.60 % to 1.05 % per step is remarkable. This can be further underlined by the reported average reduction of the cost of inserting the new customer which ranges from 50.13 % to 78.55 %. Furthermore, we notice a moderate improvement of the hybrid heuristics over the Local-improvement heuristics.

When comparing dense to sparse instances, we notice that the average runtimes for the *Improvement* step are significantly higher on dense instances. Further, we notice higher average runtimes on the benchmark set with fewer long tours. The reason for both observations lies in the larger number of customers per tour which makes these instances more difficult to solve. We also observe that the reduction of the objective function achieved by the hybrid heuristics, compared to the Local-improvement heuristics, is similar on both benchmark sets.

In summary, our suggested algorithms perform very good an both benchmark sets as they are able to produce delivery schedules on large-scale instances within the tight runtime restrictions imposed by the considered application.

4   Conclusion

In this work, we considered an Attended Home Delivery (AHD) system in the context of an online grocery shopping service offered by an international grocery retailer. AHD systems are used whenever a supplying company offers online shopping services that require that customers must be present when their deliveries arrive. Therefore, the supplying company and the customer must both agree on a delivery time window, which ideally is rather short, during which delivery is guaranteed.

Especially, we considered the overall fulfillment process of the AHD system that can be described by four consecutive phases: (1) Tactical Planning, (2) Ordering, (3) Preparation, and (4) Delivery. We focused on the Ordering phase, during which customers place their orders through the web service. Generally, this phase is the most challenging phase of an AHD system from a computational point of view. As for most AHD approaches in the literature, we considered a capacitated Vehicle Routing Problem with Time Windows as the underlying optimization problem of the ordering phase. The online characteristic of this phase requires that the delivery schedule is built dynamically as new orders are placed. We split the computations of the ordering phase into four key steps and proposed a solution approach that allows to (non-stochastically) determine which delivery time windows can be offered to potential customers. Furthermore, we employed a Local Search heuristic to improve the delivery schedule and we also suggested a hybrid approach that additionally to the Local Search heuristic employs MILPs, which optimize single tours.

Finally, in an experimental evaluation, we demonstrated the efficiency of our approaches on benchmark sets that are motivated by an online grocery shopping service. We considered the capacitated Vehicle Routing Problem with structured Time Windows (cVRPsTW) for our benchmarking experiments. The special feature of the cVRPsTW is the additional structure of the time windows which does not impose severe restrictions neither to the supplying company nor to the customers. Our computational study showed that the suggested algorithms can solve the considered cVRPsTW instances fast enough to comply with the very strict runtime restrictions as they arise in AHD systems with high customer request rates.

For future research it would be interesting to integrate time-dependent travel times as well as driving breaks into the approach.

References

1. Agatz, N., Campbell, A.M., Fleischmann, M., Savelsbergh, M.W.P.: Challenges and opportunities in attended home delivery. In: Golden, B., Raghavan, S., Wasil, E. (eds.) The Vehicle Routing Problem: Latest Advances and New Challenges, pp. 379–396. Springer, US (2008). https://doi.org/10.1007/978-0-387-77778-8_17
2. Campbell, A.M., Savelsbergh, M.W.P.: Decision support for consumer direct grocery initiatives. Transp. Sci. **39**(3), 313–327 (2005)
3. Ehmke, J.F.: Attended home delivery. In: Integration of Information and Optimization Models for Routing in City Logistics, pp. 23–33. Springer, Boston (2012). https://doi.org/10.1007/978-1-4614-3628-7_3
4. Ehmke, J.F., Campbell, A.M.: Customer acceptance mechanisms for home deliveries in metropolitan areas. Eur. J. Oper. Res. **233**(1), 193–207 (2014). https://doi.org/10.1016/j.ejor.2013.08.028
5. El-Sherbeny, N.A.: Vehicle routing with time windows: an overview of exact heuristic and metaheuristic methods. J. King Saud Univ. **22**, 123–131 (2010)
6. Gendreau, M., Hertz, A., Laporte, G., Stan, M.: A generalized insertion heuristic for the traveling salesman problem with time windows. Oper. Res. **46**(3), 330–335 (1998)
7. Hungerländer, P., Maier, K., Pöcher, J., Rendl, A., Truden, C.: Solving an online capacitated vehicle routing problem with structured time windows. In: Fink, A., Fügenschuh, A., Geiger, M.J. (eds.) Operations Research Proceedings 2016. ORP, pp. 127–132. Springer, Cham (2018). https://doi.org/10.1007/978-3-319-55702-1_18
8. Hungerländer, P., Rendl, A., Truden, C.: On the slot optimization problem in online vehicle routing. Transp. Res. Procedia **27**, 492–499 (2017). https://doi.org/10.1016/j.trpro.2017.12.046
9. Hungerländer, P., Truden, C.: Efficient and easy-to-implement mixed-integer linear programs for the traveling salesperson problem with time windows. Transp. Res. Procedia **30**, 157–166 (2018). https://doi.org/10.1016/j.trpro.2018.09.018. EURO Mini Conference on "Advances in Freight Transportation and Logistics"
10. Klein, R., Mackert, J., Neugebauer, M., Steinhardt, C.: A model-based approximation of opportunity cost for dynamic pricing in attended home delivery. OR Spectrum (2017). https://doi.org/10.1007/s00291-017-0501-3
11. Klein, R., Neugebauer, M., Ratkovitch, D., Steinhardt, C.: Differentiated time slot pricing under routing considerations in attended home delivery. Transp. Sci. (2017). https://doi.org/10.1287/trsc.2017.0738

12. Köhler, C., Ehmke, J.F., Campbell, A.M.: Flexible time window management for attended home deliveries. Omega (2019). https://doi.org/10.1016/j.omega.2019.01.001
13. Lenstra, J.K., Kan, A.H.G.R.: Complexity of vehicle routing and scheduling problems. Networks **11**(2), 221–227 (1981). https://doi.org/10.1002/net.3230110211
14. Pan, S., Giannikas, V., Han, Y., Grover-Silva, E., Qiao, B.: Using customer-related data to enhance e-grocery home delivery. Ind. Manag. Data Syst. **117**(9), 1917–1933 (2017). https://doi.org/10.1108/IMDS-10-2016-0432
15. Solomon, M.M.: Algorithms for the vehicle routing and scheduling problems with time window constraints. Oper. Res. **35**(2), 254–265 (1987). https://doi.org/10.1287/opre.35.2.254
16. Syndy: The state of online grocery retail in Europe 2015 (2015). http://www.syndy.com/report-the-state-of-online-grocery-retail-2015/
17. Vienna public transport (Wiener Linien): 2017 - Facts and Figures (2018). https://www.wienerlinien.at/media/files/2018/facts_and_figures_2017_243486.pdf
18. Yang, X., Strauss, A.K., Currie, C.S.M., Eglese, R.: Choice-based demand management and vehicle routing in E-fulfillment. Transp. Sci. **50**(2), 473–488 (2016)

Consistency for 0–1 Programming

Danial Davarnia[1] and J. N. Hooker[2(✉)]

[1] Iowa State University, Ames, USA
davarnia@iastate.edu
[2] Carnegie Mellon University, Pittsburgh, USA
jh38@andrew.cmu.edu

Abstract. Concepts of consistency have long played a key role in constraint programming but never developed in integer programming (IP). Consistency nonetheless plays a role in IP as well. For example, cutting planes can reduce backtracking by achieving various forms of consistency as well as by tightening the linear programming (LP) relaxation. We introduce a type of consistency that is particularly suited for 0–1 programming and develop the associated theory. We define a 0–1 constraint set as LP-consistent when any partial assignment that is consistent with its linear programming relaxation is consistent with the original 0–1 constraint set. We prove basic properties of LP-consistency, including its relationship with Chvátal-Gomory cuts and the integer hull. We show that a weak form of LP-consistency can reduce or eliminate backtracking in a way analogous to k-consistency. This work suggests a new approach to the reduction of backtracking in IP that focuses on cutting off infeasible partial assignments rather than fractional solutions.

Keywords: Consistency · Resolution · Constraint satisfaction ·
Integer programming · Backtracking · Cutting planes

1 Introduction

Consistency is a fundamental concept of constraint programming (CP) and an essential tool for the reduction of backtracking during search [1]. Curiously, the concept never explicitly developed in mathematical programming, even though solvers rely on a similar type of branching search. In fact, the cutting planes of integer programming can reduce backtracking by achieving various forms of consistency as well as by tightening the linear programming (LP) relaxation.

This suggests that it may be useful to investigate the potential role of consistency concepts in mathematical programming. We do so for 0–1 integer programming in particular. We study how consistency relates to such integer programming ideas as the LP relaxation, Chvátal-Gomory cutting planes [3], and the integer hull, as well as how consistency can be achieved for 0–1 inequalities. Our main contribution is to introduce a type of consistency, *LP-consistency*, that seems particularly relevant to 0–1 programming, and to develop the underlying

© Springer Nature Switzerland AG 2019
L.-M. Rousseau and K. Stergiou (Eds.): CPAIOR 2019, LNCS 11494, pp. 225–240, 2019.
https://doi.org/10.1007/978-3-030-19212-9_15

theory. We show that achieving a form of partial LP-consistency can reduce backtracking in ways that traditional cutting planes do not.

One way to reduce backtracking is to identify partial assignments to the variables that are *inconsistent* with the constraint set, meaning that they cannot occur in a feasible solution of the constraints. Branching decisions that result in such partial assignments can then be avoided, thus removing infeasible subtrees from the search. Unfortunately, it is generally hard to identify inconsistent partial assignments in advance.

The essence of consistency is that it makes it easier to identify inconsistent partial assignments. Full consistency allows one to recognize an inconsistent partial assignment by the fact that it violates a constraint that contains only the variables in the partial assignment. Because full consistency is very hard to achieve, CP solvers rely on *domain consistency* (generalized arc consistency) [1,4,10,11], which reduces variable domains to the point that every value in them occurs in some feasible solution. If domain consistency is obtained at the current node of the search tree, branching on any value in a variable's domain can lead to a feasible solution. Domain consistency is itself hard to achieve for the entire constraint set, but can often be achieved, or partially achieved, for individual global constraints in the CP model, and this reduces backtracking significantly [14].

Our approach is based on the idea that consistency can be defined with respect to a *relaxation* of the constraint set. Specifically, we interpret consistency as making it possible to identify inconsistent partial assignments by checking whether they are consistent with a certain type of relaxation. This perspective allows us to propose alternative types of consistency by using various types of relaxation. For traditional consistency, the relaxation is obtained simply by dropping constraints that contain variables that are not in the partial assignment. We define LP-consistency by replacing this relaxation with the LP relaxation. Thus LP-consistency ensures that any partial assignment that is consistent with the LP relaxation is consistent with the original constraint set. Fortunately, one can easily check consistency with an LP relaxation simply by solving the LP problem that results from adding the partial assignment to the LP relaxation.

This poses the question of whether it is practical to achieve LP-consistency for a 0–1 problem. There is no known practical method for achieving full LP-consistency, but we take a cue from the concept of k-consistency in CP [5,15,16], which is weaker than full consistency but sufficient to avoid backtracking if the constraints are not too tightly coupled by common variables. We define a similar property, *sequential LP k-consistency*, that can avoid some backtracking that traditional cutting planes may permit, because it focuses on identifying inconsistent partial assignments rather than cutting off fractional solutions of the LP relaxation.

A method for obtaining sequential LP k-consistency is suggested by our practice of defining consistency concepts in terms of projection, as proposed in [9]. One can define sequential LP k-consistency, in particular, in terms of the results of lifting a problem from $k-1$ dimensions to k dimensions, and then projecting it back into $k-1$ dimensions. A modified form of the well-known lift-and-project technique of IP [2] achieves sequential LP k-consistency.

We begin below by defining and illustrating basic consistency concepts and showing how they can be cast in terms of projection. We also indicate how consistency can eliminate or reduce backtracking. We review some prior work showing that an inference method of propositional logic, resolution, can achieve consistency for 0–1 problems, and that a weak form of resolution, input resolution, can generate all Chvátal-Gomory cuts for a set of logical clauses.

At this point we introduce LP-consistency and show some elementary properties, namely that consistency implies LP-consistency, and a constraint set that describes the integer hull is necessarily LP-consistent. Yet LP-consistency is a concept that does not occur in polyhedral theory, and an LP-consistent constraint set need not describe the integer hull. While the facet-defining inequalities that describe the integer hull are generally regarded as the strongest valid inequalities, we show that they can be weaker than a non-facet-defining inequality that achieves LP-consistency, in the sense that they exclude fewer inconsistent 0–1 (partial) assignments. We further elaborate on connections with cutting plane theory by showing that a 0–1 partial assignment is consistent with the LP relaxation if and only if it violates no logical clause that is a Chvátal-Gomory (C-G) cut, and a 0–1 problem is LP-consistent if and only if all of its implied logical clauses are C-G cuts. We also note that while input resolution derives C-G cuts, it does not achieve LP-consistency.

The remainder of the paper defines and develops the concept of sequential LP k-consistency. It shows that achieving sequential LP k-consistency for $k = 1, \ldots, n$ (where n is the number of variables) avoids backtracking altogether for branching order $x_1, x_2, \ldots, x_n$. In practice, one would achieve sequential LP k-consistency for a few small values of k. We then prove that one step of the lift-and-project procedure [2] achieves sequential k-consistency for a given k. Finally, we illustrate how achieving sequential LP k-consistency even for $k = 2$ can avoid backtracking that is permitted by traditional separating cuts.

2   Consistency and Projection

To define consistency, it is convenient to adopt basic terminology as follows. The *domain* D_j of a variable x_j is the set of values that can be assigned to x_j. A *constraint* C is an object that *contains* some set $\{x_1, \ldots, x_k\}$ of variables, such that any given assignment of values to $(x_1, \ldots, x_k)$ either *satisfies* or *violates* C. Thus a constraint is satisfied or violated only when all of its variables have been assigned values. An assignment to x satisfies a constraint set S when it satisfies all the constraints in S. A list of symbols defined hereafter appears in Table 1.

Let x_J be the tuple containing the variables in $\{x_j \mid j \in J\}$ for $J \subseteq N = \{1, \ldots, n\}$. A *partial assignment* to x is an assignment of values to x_J for some $J \subseteq N$. We can now define a consistent partial assignment and a consistent constraint set.

Table 1. List of symbols.

x_J	tuple of variables x_j for $j \in J$
N	the set $\{1, \ldots, n\}$
J_k	the set $\{1, \ldots, k\}$
D_j	domain of x_j
D_J	cartesian product of D_j for $j \in J$
$D(\mathcal{S})$	set of assignments* to x that satisfy $\mathcal{S}$
$D_J(\mathcal{S})$	set of assignments* to x_J that are consistent with $\mathcal{S}$
$D(\mathcal{S})|_J$	projection of $D(\mathcal{S})$ onto x_J
$\mathcal{S}_J$	set of constraints in $\mathcal{S}$ that contain only variables in x_J
$\mathcal{S}_{\mathrm{LP}}$	LP relaxation of 0–1 constraint set $\mathcal{S}$
$D_J(\mathcal{S}_{\mathrm{LP}})$	set of 0–1 assignments to x_J that are consistent with $\mathcal{S}_{\mathrm{LP}}$
$\mathcal{S}_{\mathrm{C}}$	set of clausal inequalities implied by individual constraints of $\mathcal{S}$

*an assignment $x = v$ or $x_J = v_J$ assumes that $v \in D$, $v_J \in D_J$.

Definition 1. *Given a constraint set $\mathcal{S}$, a partial assignment $x_J = v_J$ is consistent with $\mathcal{S}$ if $\mathcal{S} \cup \{x_J = v_J\}$ is feasible.*

Since it is hard in general to determine whether $\mathcal{S} \cup \{x_J = v_J\}$ is feasible, it is hard to identify which partial assignments are consistent with $\mathcal{S}$. Consistent constraint sets are defined so that it is easy to identify which partial assignments are consistent with them.

Definition 2. *A constraint set $\mathcal{S}$ is* consistent *if every partial assignment to x that violates no constraint in $\mathcal{S}$ is consistent with $\mathcal{S}$.*

The contrapositive is perhaps more intuitive: $\mathcal{S}$ is consistent when every partial assignment that is inconsistent with $\mathcal{S}$ violates some individual constraint in $\mathcal{S}$. Thus a consistent constraint set can be viewed as one in which implied constraints are made explicit, in the sense that every inconsistent partial assignment is explicitly ruled out by some constraint in the set.

Since full consistency is generally hard to achieve, the constraint programming community has found various weaker forms of consistency to be more useful. By far the most popular is domain consistency, also known as generalized arc consistency [1,4,10,11].

Definition 3. *A constraint set $\mathcal{S}$ is* domain consistent *if $x_j = v_j$ is consistent with $\mathcal{S}$ for all $v_j \in D_j$ and all variables x_j.*

That is, every value in the domain of a variable x_j is assigned to x_j in some feasible solution of $\mathcal{S}$. A consistent constraint set is necessarily domain consistent.

Example 1. Suppose that $\mathcal{S}$ is the constraint set

$$
\begin{aligned}
x_1 + x_2 \quad\ + x_4 &\geq 1 \\
x_1 - x_2 + x_3 \quad &\geq 0 \\
x_1 \quad\quad\ - x_4 &\geq 0 \\
x_j \in \{0, 1\}, \text{ all } j
\end{aligned}
$$

The domains are $D_j = \{0,1\}$ for $j = 1, \ldots, 4$. The feasible solutions $(x_1, \ldots, x_4)$ of S are listed below:

$$
\begin{array}{lll}
(0,1,1,0) & (1,0,1,0) & (1,1,0,1) \\
(1,0,0,0) & (1,0,1,1) & (1,1,1,0) \\
(1,0,0,1) & (1,1,0,0) & (1,1,1,1)
\end{array}
$$

Set S is not consistent because, for instance, the partial assignment $(x_1, x_2) = (0,0)$ violates no constraint in S but is inconsistent with S due to the fact that $(x_1, x_2) = (0,0)$ in none of the feasible solutions. On the other hand, S is domain consistent because $x_j = 0$ and $x_j = 1$ occur in some feasible solution for each j.

The various consistency concepts are more easily defined in terms of projection, as proposed in [9]. Let D_J be the cartesian product of D_j for $j \in J$, and let $D = D_N$. When we speak of an assignment $x = v$ or a partial assignment $x_J = v_J$, we assume $v \in D$ and $v_J \in D_J$. Let $D(S)$ be the set of assignments to x that satisfy S, and let $D_J(S)$ be the set of assignments to x_J that are consistent with S. Thus

$$
D_J(S) = \{v_J \in D_J \mid S \cup \{x_J = v_J\} \text{ is feasible}\}
$$

The *projection* of $D(S)$ onto x_J, which we may write $D(S)|_J$, is $\{x_J \mid x \in D(S)\}$.

We can now define consistency in terms of projection. Let S_J be the set of constraints in S whose variables belong to x_J. Then $D_J(S_J)$ is the set of assignments to x_J that violate no constraints in S.

Proposition 1. *A constraint set S is consistent if and only if $D_J(S_J) = D(S)|_J$ for all $J \subseteq N$. In addition, S is domain consistent if and only if $D_j = D(S)|_{\{j\}}$ for all $j \in N$.*

3 Consistency and Backtracking

It is well known that consistency is closely related to backtracking. We note first that branching can find a feasible solution for a fully consistent constraint set without backtracking, assuming of course that the constraints have a solution. Suppose we branch on variables $x_1, \ldots, x_n$ in that order. Each node in level j of the branching tree corresponds to a partial assignment $(x_1, \ldots, x_{j-1}) = (v_1, \ldots, v_{j-1})$. We branch on x_j at the node by assigning to x_j each value $v_j \in D_j$ for which the partial assignment $(x_1, \ldots, x_j) = (v_1, \ldots, v_j)$ violates no constraint in S. Due to the consistency of S, this partial assignment is consistent with S for at least one value $v_j \in D_j$. Thus branching can continue to the bottom of the tree with no need to backtrack.

A weaker form of consistency, k-consistency, avoids backtracking if there is limited coupling of variables [6]. More relevant to our purposes is a still weaker form of consistency that assumes the branching order is given, namely $x_1, \ldots, x_n$. Let $J_k = \{1, \ldots, k\}$.

Definition 4. *A constraint set S is* sequentially k-consistent *if $D_{J_{k-1}}(S_{J_{k-1}}) = D_{J_k}(S_{J_k})|_{J_{k-1}}$.*

Thus S is sequentially k-consistent if for every partial assignment $(x_1, \ldots, x_{k-1}) = (v_1, \ldots, v_{k-1})$ that violates no constraint in S, there is a value v_k in D_k such that $(x_1, \ldots, x_k) = (v_1, \ldots, v_k)$ violates no constraint in S. The following is easy to show.

Proposition 2. *If the branching order is $x_1, \ldots, x_n$, constraint set S can be solved without backtracking if S is sequentially k-consistent for $k = 1, \ldots, n$.*

Example 2. Let $S = \{3x_1 + 2x_2 \geq 1, \ -x_1 + 2x_2 \geq 0, \ x \in \{0,1\}^2\}$. Proposition 2 implies that we can avoid backtracking by branching in the order x_1, x_2, because S is sequentially 1-consistent and sequentially 2-consistent.

4 Consistency and Resolution

Previous research has shown that the resolution procedure of propositional logic achieves consistency for a 0–1 constraint set. First, some definitions. A *literal* ℓ_j is a proposition of the form x_j or $\neg x_j$. A logical *clause* is a disjunction $\bigvee_{j \in J} \ell_j$ of literals. A clause C_1 implies C_2 when C_1 *absorbs* C_2, meaning that all the literals of C_1 are in C_2. There is a resolution proof of any clause that is logically implied by a clause set C [12,13].

Now let S be 0–1 constraint set $\{Ax \geq b, \ x \in \{0,1\}^n\}$, where the domains are $D_j = \{0,1\}$ for all j. S *logically implies* 0–1 constraint set S' when all 0–1 points that satisfy S also satisfy S'. S and S' are *logically equivalent* when they logically imply each other. A logical clause $\bigvee_{j \in J^+} x_j \vee \bigvee_{j \in J^-} \neg x_j$ is *represented* by the 0–1 inequality

$$\sum_{j \in J^+} x_j + \sum_{j \in J^-} (1 - x_j) \geq 1$$

A 0–1 inequality is *clausal* when it represents a clause. It is clear that a 0–1 inequality is logically equivalent to the set of clausal inequalities it implies. Thus if we let S_C be the set of clausal inequalities that are implied by some inequality in S, then S is logically equivalent to S_C. It is shown in [8] that resolution on clausal inequalities achieves consistency.

The following example illustrates how a traditional cutting plane can serve the dual purpose of tightening the linear programming (LP) relaxation and achieving consistency. Let the LP relaxation of $S = \{Ax \geq b, \ x \in \{0,1\}^n\}$ be $S_{\text{LP}} = \{Ax \geq b, \ x \in [0,1]^n\}$.

Example 3. Suppose that S is the constraint set of Example 1. In this case, S and S_C are identical. Resolution yields two additional clausal inequalities, $x_1 + x_2 \geq 1$ and $x_1 + x_3 \geq 1$. Adding these inequalities to S achieves consistency. These inequalities are also traditional cutting planes for S, in particular Chvátal-Gomory (C-G) cuts. The first cuts off two fractional vertices $(x_1, \ldots, x_4) = (\frac{1}{3}, \frac{1}{3}, 0, \frac{1}{3}), (\frac{1}{2}, 0, 0, \frac{1}{2})$ of the polytope described by S_{LP}, and the second cuts off the vertex $(\frac{1}{2}, \frac{1}{2}, 0, 0)$ as well. The inequalities therefore serve the dual purpose of achieving consistency and tightening the LP relaxation. As it happens, adding both resolvents yields an integral polytope, but we will see that a consistent constraint does not in general describe an integral polytope.

A special class of resolution proofs, namely input proofs, derive all clausal C–G cuts for S_C [7].

5   LP-consistency

While resolution can always achieve consistency, it is not a practical method for the reduction of backtracking. Resolution proofs tend to explode rapidly in length and complexity. However, the LP relaxation of S provides an additional tool for this purpose. Specifically, it provides a more useful test for consistency than whether a partial assignment violates a constraint.

Consistency of S implies that any partial assignment $x_J = v_J$ that is consistent with S_J (i.e., violates no constraint in S) is consistent with S. We want a type of consistency that ensures that any partial assignment consistent with S_{LP} is consistent with S. We can achieve this by defining consistency with respect to the LP relaxation S_{LP} rather than the relaxation S_J. Recall that classical consistency is defined so that $D_J(S_J) = D(S)|_J$. We therefore define *LP-consistency* as follows.

Definition 5. *A 0–1 constraint set S is* LP-consistent *if $D_J(S_{LP}) = D(S)|_J$ for all $J \subseteq N$.*

Note that $D_J(S_{LP})$ refers to the set of *0–1 assignments* to x_J that are consistent with S_{LP}, since the domains are $D_j = \{0, 1\}$ for all j. Thus S is *LP-consistent* if $S_{LP} \cup \{x_J = v_J\}$ is infeasible for any 0–1 partial assignment $x_J = v_J$ that is inconsistent with S.

Example 4. Consider the 0–1 constraint set $S = \{4x_1 + 4x_2 \geq 1,\ 2x_1 - 4x_2 \geq -3,\ x \in \{0, 1\}^2\}$ (Fig. 1). The partial assignment $x_1 = 0$ is consistent with S_{LP} but not with S, because both $(x_1, x_2) = (0, 0)$ and $(x_1, x_2) = (0, 1)$ violate S. So S is not LP-consistent.

Two elementary properties of LP-consistency follow.

Proposition 3. *A consistent 0–1 constraint set is LP-consistent.*

Proof. Consider any 0–1 partial assignment $x_J = v_J$ that is consistent with S_{LP}. We claim that $x_J = v_J$ is consistent with S, which suffices to show that S is LP-consistent. Since $S_{LP} \cup \{x_J = v_J\}$ is feasible, $x_J = v_J$ violates no constraints in S. Now since S is consistent, this means that $x_J = v_J$ is consistent with S, as claimed.   □

In addition, a 0–1 constraint set that describes the integer hull (the convex hull of feasible 0–1 points) is LP-consistent.

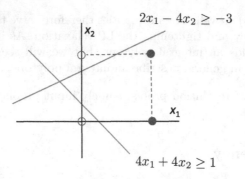

Fig. 1. Illustration of Example 4.

Proposition 4. *Given 0–1 constraint set S, if S_{LP} describes the integer hull of $D(S)$, then S is LP-consistent.*

Proof. Suppose that $S \cup \{x_J = v_J\}$ is infeasible for a given 0–1 partial assignment $x_J = v_J$. Then $x_J = v_J$ describes a face of the unit hypercube that is disjoint from $D(S)$. This implies that the face is disjoint from the convex hull of $D(S)$, which is described by S_{LP}. Thus $S_{LP} \cup \{x_J = v_J\}$ is infeasible, and it follows that S is LP-consistent. □

It is essential to observe that a convex hull model is not necessary to achieve LP-consistency, a fact that will be exploited in later sections. This can be seen in an example.

Example 5. Consider the following two constraint sets (Fig. 2), which have the same feasible set:

$$S^1 = \{x_1 + x_2 \le 1,\ x_2 + x_3 \le 1,\ x \in \{0,1\}^3\}$$
$$S^2 = \{x_1 + 2x_2 + x_3 \le 2,\ x \in \{0,1\}^3\}$$

The LP relaxation S^1_{LP} describes the integer hull of $D(S^1) = D(S^2)$, and so S^1 is LP-consistent by Proposition 4. Yet the constraint set S^2 is also LP-consistent, even though S^2_{LP} does not describe the integer hull, but describes a polytope with fractional extreme points $(x_1, x_2, x_3) = (0, \frac{1}{2}, 1), (1, \frac{1}{2}, 0)$. Interestingly, the inequality $x_1 + 2x_2 + x_3 \ge 2$ in S^2 is the sum of the two nontrivial facet-defining inequalities in S^1 and is therefore weaker than either of them from a polyhedral point of view. Yet it cuts off more infeasible 0–1 points than either of the facet-defining inequalities and is therefore stronger in this sense. Indeed, the purpose of achieving LP-consistency is to cut off infeasible 0–1 (partial) assignments, not to cut off fractional vertices of the LP relaxation.

6   Characterizing LP-Consistency

The following result gives a necessary condition for consistency based on clausal inequalities.

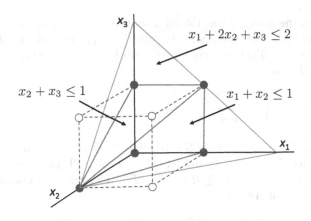

$$x_1 + 2x_2 + x_3 \leq 2$$

$$x_2 + x_3 \leq 1$$

$$x_1 + x_2 \leq 1$$

Fig. 2. Illustration of Example 5

Proposition 5. *If a constraint set S is consistent, then all of its implied clausal inequalities are in S_C.*

Proof. Suppose that S is consistent, and let C be any clausal inequality implied by S. Then the assignment $x_J = v_J$ violates C, where x_J are the variables in C and v_j is 1 when x_j is negated in C and 0 otherwise. This means $x_J = v_J$ is inconsistent with S, which implies by the consistency of S that $x_J = v_J$ violates an inequality $\alpha x \geq \beta$ in S. As a result, C must be implied by $\alpha x \geq \beta$, showing that $C \in S_C$.   □

LP-consistency allows us to derive a stronger argument on the relation between an LP-consistent set and its implied clausal inequalities, as it provides both necessary and sufficient conditions. In particular, a 0–1 constraint set S is LP-consistent if and only if all of its implied clauses are C-G cuts for S_{LP}. This is due to the following fact.

Proposition 6. *Given a 0–1 constraint set S, a 0–1 partial assignment is consistent with S_{LP} if and only if the assignment violates no clausal C-G cut for S_{LP}.*

Proof. It suffices to show that a given 0–1 partial assignment $x_J = v_J$ violates a clausal C-G for S_{LP} if and only if $S_{LP} \cup \{x_J = v_J\}$ is infeasible. Suppose first that $x_J = v_J$ violates a clausal inequality $ax \geq \beta$ that is a C-G cut for S_{LP}, where S_{LP} is the system $Ax \geq b$. Since $x_J = v_J$ violates $ax \geq \beta$, we can write the inequality as $a_J x_J \geq \beta$, where $a_J v_J \leq \beta - 1$. Now since $ax \geq \beta$ is a C-G cut, there is a tuple $u \geq 0$ of multipliers such that $uA = a$ and $\beta - 1 < ub \leq \beta$. We therefore have $(uA)_J v_J = a_J v_J \leq \beta - 1 < ub$. This implies that $x_J = v_J$ violates $uAx \geq ub$, and so $S_{LP} \cup \{x_J = v_J\}$ must be infeasible.

For the converse, suppose that $S_{LP} \cup \{x_J = v_J\}$ is infeasible, which means that the face of the unit hypercube defined by $x_J = v_J$ lies outside the polytope defined by S_{LP}. Let $J^+ = \{j \in J \mid v_j = 0\}$ and $J^- = \{j \in J \mid v_j = 1\}$. Then some inequality of the form $\sum_{j \in J^+} x_j + \sum_{j \in J^-} (1 - x_j) \geq \bar{\pi}$ for some

$\bar{\pi} > 0$ separates the face just mentioned from the polytope; i.e., $x_J = v_J$ violates this inequality. Since this inequality is valid for $\mathcal{S}_{\mathrm{LP}}$, it is dominated by some surrogate of $Ax \geq b$. That is there exists a tuple $u \geq 0$ of multipliers such that $uA \geq ub$ is of the form

$$\sum_{j \in J^+} x_j + \sum_{j \in J^-} (1 - x_j) \geq \pi \tag{1}$$

where $\pi \geq \bar{\pi}$, and $\pi \leq |J|$ because $\mathcal{S}$ is feasible. Now pick any subset $\hat{J} \subseteq J$ with $|\hat{J}| = \lceil \pi \rceil - 1$, let $\hat{J}^+ = J^+ \cap \hat{J}$, and let $\hat{J}^- = J^- \cap \hat{J}$. Take the sum of (1) with $-x_j \geq -1$ for $j \in \hat{J}^+$ and $x_j \geq 0$ for $j \in \hat{J}^-$. This yields a clausal inequality that is a surrogate of $Ax \geq b$:

$$\sum_{j \in J^+ \setminus \hat{J}^+} x_j + \sum_{j \in J^- \setminus \hat{J}^-} (1 - x_j) \geq 1 + \pi - \lceil \pi \rceil$$

Rounding up the right-hand side (if necessary) yields a clausal C–G cut violated by $x_J = v_J$. Thus $x_J = v_J$ violates a clausal C–G cut for $\mathcal{S}_{\mathrm{LP}}$, as claimed. □

Example 6. Consider again the constraint set $\mathcal{S}$ of Example 3. The partial assignment $(x_1, x_3) = (0,0)$ is inconsistent with $\mathcal{S}_{\mathrm{LP}}$ and violates a clausal C–G cut, namely $x_1 + x_3 \geq 1$. The cut is obtained by assigning multipliers $\frac{1}{4}, \frac{1}{2}, \frac{1}{4}, \frac{1}{4}, \frac{1}{2}$ to the three constraints of $\mathcal{S}$, $x_2 \geq 0$, and $x_3 \geq 0$, respectively. The partial assignment $(x_1, x_3) = (0,1)$ is consistent with $\mathcal{S}_{\mathrm{LP}}$ and therefore violates no clausal C–G cut.

Corollary 1. *A constraint set $\mathcal{S}$ is LP-consistent if and only if all of its implied clausal inequalities are C–G cuts for $\mathcal{S}_{\mathrm{LP}}$.*

Proof. Suppose first that $\mathcal{S}$ is LP-consistent, and let C be any clausal inequality implied by $\mathcal{S}$. Then the assignment $x_J = v_J$ violates C, where x_J are the variables in C and v_j is 1 when x_j is negated in C and 0 otherwise. This means $x_J = v_J$ is inconsistent with $\mathcal{S}$, which implies by the LP-consistency of $\mathcal{S}$ that $x_J = v_J$ is inconsistent with $\mathcal{S}_{\mathrm{LP}}$. By Proposition 6, $x_J = v_J$ violates some clausal C–G cut C' of $\mathcal{S}_{\mathrm{LP}}$. Then C' must absorb C, which means C is likewise a C–G cut of $\mathcal{S}_{\mathrm{LP}}$.

Conversely, suppose all clausal inequalities implied by $\mathcal{S}$ are C–G cuts for $\mathcal{S}_{\mathrm{LP}}$, and consider any partial assignment $x_J = v_J$ that is consistent with $\mathcal{S}_{\mathrm{LP}}$. By Proposition 6, $x_J = v_J$ violates no clausal C–G cut of $\mathcal{S}_{\mathrm{LP}}$. This means that it violates no clause implied by $\mathcal{S}$, which means that $x_J = v_J$ is consistent with $\mathcal{S}$, as desired. □

Example 7. The constraint set $\mathcal{S}$ of Example 1 is LP-consistent because its implied clausal inequalities are all implied by the inequalities in the set $\mathcal{S} \cup \{x_1 + x_2 \geq 1, \ x_1 + x_3 \geq 1\}$, and these are all C–G cuts for $\mathcal{S}_{\mathrm{LP}}$.

7   LP-Consistency and Backtracking

Like full consistency in CP, full LP-consistency is difficult to achieve. We there-fore follow the lead of the CP community and consider a weaker form of con-sistency, namely an analog of k-consistency. While even k-consistency is hard to achieve in practice, and the CP community focuses on domain consistency instead, a form of LP-consistency analogous to sequential k-consistency may offer possibilities to 0–1 programming.

Recall that S is sequentially k-consistent if $D_{J_{k-1}}(S_{J_{k-1}}) = D_{J_k}(S_{J_k})|_{J_{k-1}}$, and that sequential k-consistency for $k = 1, \ldots, n$ suffices to avoid backtracking when the branching order is $x_1, \ldots, x_n$. A parallel definition that relates to linear programming is as follows.

Definition 6.   *A 0–1 constraint set S is sequentially LP k-consistent if* $D_{J_{k-1}}(S_{\mathrm{LP}}) = D_{J_k}(S_{\mathrm{LP}})|_{J_{k-1}}$.

Equivalently, we can say that S is sequentially LP k-consistent if for every 0–1 partial assignment $x_{J_{k-1}} = v_{J_{k-1}}$ that is consistent with S_{LP}, there is a 0–1 assignment $x_k = v_k$ for which $x_{J_k} = v_{J_k}$ is consistent with S_{LP}. Thus sequential LP k-consistency is analogous to sequential k-consistency but based on the S_{LP} relaxation rather than the $S_{J_{k-1}}$ relaxation.

This form of consistency can also allow us to avoid backtracking, if we are will-ing to solve appropriate LP problems. Specifically, suppose that at a given node in the branching tree, prior branching has fixed $(x_1, \ldots, x_{k-1}) = (v_1, \ldots, v_{k-1})$. For the next branch, we select a value $v_k \in \{0, 1\}$ for which the partial assign-ment $(x_1, \ldots, x_k) = (v_1, \ldots, v_k)$ is consistent with S_{LP}; that is, for which the LP problem $S_{\mathrm{LP}} \cup \{(x_1, \ldots, x_k) = (v_1, \ldots, v_k)\}$ is feasible. We then set $x_k = v_k$ and continue to the next level of the tree. The following theorem guarantees that the LP problem will be feasible for at least one value of v_k, and that this process avoids backtracking.

Proposition 7. *If S is a feasible 0–1 constraint set over x and the branch-ing order is $x_1, \ldots, x_n$, achieving sequential LP k-consistency for $k = 1, \ldots, n$ suffices to solve S without backtracking.*

Proof. Since S is feasible, S_{LP} is feasible at the root node of the branching tree, and so the empty assignment is consistent with S_{LP}. Arguing inductively, suppose the partial assignment $(x_1, \ldots, x_{k-1}) = (v_1, \ldots, v_{k-1})$ that reflects the branching decisions down to the node at level k is consistent with S_{LP}. Since S is sequentially LP k-consistent, there exists a 0–1 value v_k of x_k for which the partial assignment $(x_1, \ldots, x_k) = (v_1, \ldots, v_k)$ is consistent with S_{LP}. By induction, $S_{\mathrm{LP}} \cup \{(x_1, \ldots, x_n) = (v_1, \ldots, v_n)\}$ is feasible at the terminal node of the tree for some tuple $(v_1, \ldots, v_n)$ of 0–1 values. But in this case, $(x_1, \ldots, x_n) = (v_1, \ldots, v_n)$ satisfies S, and we have solved the problem without backtracking. $\square$

Example 8. Consider the constraint set S of Example 4. S is not sequentially LP 2-consistent because $x_1 = 0$ is consistent with S_{LP}, but neither $(x_1, x_2) = (0, 0)$ nor $(x_1, x_2) = (0, 1)$ is consistent with S_{LP}. Also, backtracking is possible, because if we set $x_1 = 0$ at the root node because $x_1 = 0$ is consistent with $\mathcal{S}_{\text{LP}}$, we cannot find a consistent value for x_2 at the child node and must backtrack. Now suppose we add the clause $x_1 + x_2 \geq 1$ to $\mathcal{S}$ to obtain a constraint set $\mathcal{S}'$ that is sequentially LP 2-consistent. At the root node we must branch on $x_1 = 1$, because $x_1 = 0$ is not consistent with $\mathcal{S}'_{\text{LP}}$. At the child node, branching on $x_2 = 1$ yields an assignment $(x_1, x_2) = (1, 1)$ that is consistent with $\mathcal{S}_{\text{LP}}$ and, in fact, solves $\mathcal{S}$ without backtracking.

8 Achieving LP Consistency

We can achieve sequential LP k-consistency by using one step of a modified lift-and-project method [2]. Given $\mathcal{S} = \{Ax \geq b, \ x \in \{0, 1\}^n\}$ where $0 \leq x_i \leq 1$ is included in $Ax \geq b$, we generate the nonlinear system

$$(Ax - b)x_k \geq 0$$
$$(Ax - b)(1 - x_k) \geq 0$$

We next linearize the system by replacing each x_k^2 with x_k, and each product $x_i x_k$ with y_{ik}. Let the resulting system be $R_k(\mathcal{S}_{\text{LP}})$. Adding the constraints in this system to $\mathcal{S}_{\text{LP}}$ yields a sequentially LP k-consistent constraint set.

Proposition 8. *Given a 0–1 constraint set $\mathcal{S}$, augmenting $\mathcal{S}$ with the constraints in $R_k(\mathcal{S}_{\text{LP}})$ yields a constraint set that is sequentially LP k-consistent.*

Proof. For a given 0–1 partial assignment $x_J = v_J$, suppose that $\mathcal{S}_{\text{LP}} \cup \{(x_{J_k}) = (v_{J_k})\}$ is infeasible for $v_k = 0, 1$. It suffices to show that $R_k(\mathcal{S}_{\text{LP}})|_N \cup \{x_{J_{k-1}} = v_{J_{k-1}}\}$ is infeasible. It follows from Theorem 2.1 in [2] that $R_k(\mathcal{S}_{\text{LP}})|_N$ describes the convex hull of the union of $D(\mathcal{S}_{\text{LP}} \cup \{x_k = v_k\})$ over $v_k = 0, 1$. We claim that $x_{J_{k-1}} = v_{J_{k-1}}$ does not satisfy $R_k(\mathcal{S}_{\text{LP}})|_N$. Assume to the contrary. Then there exists a point $w = (v_{J_{k-1}}, \tilde{v}_k, \tilde{v}_K, \tilde{y})$ that satisfies $R_k(\mathcal{S}_{\text{LP}})$, where $K = N \setminus J_k$. This point must be representable as a convex combination of two points of the form $(v_{J_{k-1}}, 0, \dot{v}_K, \dot{y})$ and $(v_{J_{k-1}}, 1, \ddot{v}_K, \ddot{y})$, since the components of $v_{J_{k-1}}$ are integral and cannot be represented as the convex combination of other points. However, by assumption such points do not exist because $\mathcal{S}_{\text{LP}} \cup \{x_{J_k} = v_{J_k}\}$ is infeasible for $v_k = 0, 1$. This yields the desired contradiction. □

If desired, $R_k(\mathcal{S}_{\text{LP}})$ can be projected onto x, before adding its constraints to $\mathcal{S}_{\text{LP}}$, to obtain a sequentially LP k-consistent system in the space of original variables. Alternatively, $R_k(\mathcal{S}_{\text{LP}})$ can be projected onto $x_{J_{k-1}}$ to obtain sparse cuts that are nonetheless sufficient to achieve sequential LP k-consistency. Every partial assignment $x_{J_{k-1}} = v_{J_{k-1}}$ that is inconsistent with $\mathcal{S}_{\text{LP}}$ violates some individual cut in $R_k(\mathcal{S}_{\text{LP}})|_{J_{k-1}}$.

Example 9. Consider again Example 4, in which

$$S = \{2x_1 - 4x_2 \geq -3,\ 4x_1 + 4x_2 \geq 1,\ x_1, x_2 \in \{0,1\}\}$$

Recall that S is not LP 2-consistent because $x_1 = 0$ is consistent with S_{LP} and $(x_1, x_2) = (0, v_2)$ is inconsistent with S_{LP} for $v_2 = 0, 1$. We wish to achieve sequential LP 2-consistency by applying the modified lift-and-project procedure. First generate the constraints

$$\begin{array}{ll}
(2x_1 - 4x_2 + 3)x_2 \geq 0 & x_1 x_2 \geq 0 \\
(2x_1 - 4x_2 + 3)(1 - x_2) \geq 0 & x_1(1 - x_2) \geq 0 \\
(4x_1 + 4x_2 - 1)x_2 \geq 0 & (1 - x_1)x_2 \geq 0 \\
(4x_1 + 4x_2 - 1)(1 - x_2) \geq 0 & (1 - x_1)(1 - x_2) \geq 0
\end{array}$$

After linearizing and writing y_{12} simply as y, we obtain the system $R_2(S_{\text{LP}})$:

$$\begin{array}{ll}
-x_2 + 2y \geq 0 & y \geq 0 \\
2x_1 - 3x_2 - 2y + 3 \geq 0 & x_1 - y \geq 0 \\
3x_2 + 4y \geq 0 & x_2 - y \geq 0 \\
4x_1 + x_2 - 4y - 1 \geq 0 & -x_1 - x_2 + y + 1 \geq 0
\end{array} \qquad (2)$$

The third constraint on the left can be omitted because it is implied by $x_2, y \geq 0$. Adding the constraints in (2) to S_{LP} yields a sequentially LP 2-consistent set, and it is clear on inspection that $x_1 = 0$ is inconsistent with (2). If we wish to obtain a consistent constraint set in the original variables, we can project (2) onto (x_1, x_2). This yields a constraint $4x_1 - x_2 \geq 1$ that can be added to S_{LP} to obtain a sequentially LP 2-consistent set, as illustrated in Fig. 3(a). It is evident in the figure that $x_1 = 0$ is inconsistent with this set. Finally, we can obtain a sparse cut that achieves sequential LP 2-consistency by projecting (2) onto x_1. This yields the cut $x_1 \geq \frac{1}{4}$, which likewise excludes $x_1 = 0$.

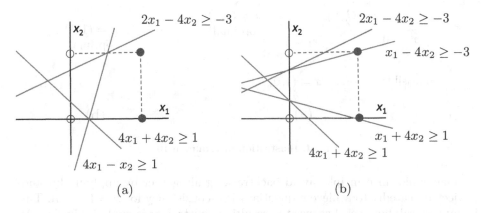

Fig. 3. Illustration of Examples 9 and 10.

An advantage of sequential LP-consistency is that it can avoid branching that traditional cutting planes do not avoid, because it focuses on excluding inconsistent partial assignments, rather than on tightening the LP relaxation by cutting off fractional points. This can be illustrated in a very simple context as follows.

Example 10. Suppose we wish to maximize $3x_2 - x_1$ subject to the constraint set S in the previous example. We first apply a traditional branch-and-cut procedure that generates separating lift-and-project cuts at the root node (Fig. 4(a)). The solution of the LP relaxation at the root node is $(x_1, x_2) = (\frac{1}{2}, 1)$. Lift and project yields the cuts $x_1 - 4x_2 \geq -3$ and $x_1 + 4x_2 \geq 1$ (corresponding to the disjunction $x_1 = 0 \lor x_1 = 1$) as illustrated in Fig. 3(b), and the cut $4x_1 - x_2 \geq 1$ (corresponding to $x_2 = 0 \lor x_2 = 1$) as illustrated in Fig. 3(a). Only the first cut is generated, because only it cuts off the fractional solution $(\frac{1}{2}, 1)$. This results in a new LP solution $(x_1, x_2) = (0, \frac{3}{4})$. The procedure then branches on the fractional variable x_2. The $x_2 = 0$ branch yields the fractional LP solution $(x_1, x_2) = (\frac{1}{2}, 0)$, and it is necessary to branch on x_1. The $x_2 = 1$ branch yields the integer LP solution $(x_1, x_2) = (1, 1)$, which solves the problem. The resulting search tree has 5 nodes.

Suppose now that we achieve sequential LP 2-consistency as described in Example 9 by generating the inequality $4x_1 - x_2 \geq 1$, even though it does not cut off the fractional LP solution (Fig. 4(b)). Since the partial assignment $x_1 = 0$ is inconsistent with the LP relaxation, we immediately branch on $x_1 = 1$, which yields the integer LP solution $(x_1, x_2) = (1, 1)$. The problem is solved with only 2 nodes in the search tree, even though we used no traditional separating cuts at all.

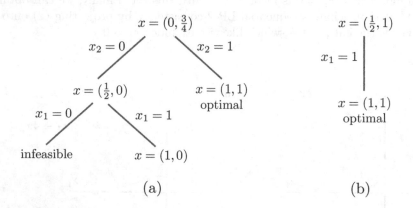

Fig. 4. Illustration of Example 10

One could, in principle, avoid backtracking altogether by applying lift-and-project repeatedly to achieve sequential LP k-consistency for $k = 1, \ldots, n$. This is impractical, however, because the resulting constraint set explodes in size. An alternative is to achieve k-consistency for a few small values of k. This can be accomplished in three ways. (a) Obtain k-consistency by applying the lift-and-project step to the set S_{LP} obtained from computing $(k-1)$-consistency in the previous step, where the procedure is modified to account for the y-variables in S_{LP}. This causes the number of variables to double in each step, but no projection is required. (b) Project $R_k(S_{LP})$ onto x before moving to the next step. The number of variables remains constant, but a time-consuming projection operation must be carried out. (c) Project $R_k(S_{LP})$ onto $x_{J_{k-1}}$ before moving to the next step. This adds only sparse cuts to the constraint set but requires more computation to carry out the projection.

These methods become computationally prohibitive in an IP solver as k increases, unless a heuristic is used to identify generated inequalities that are likely to play a role in achieving sequential LP k-consistency—much as separation algorithms are used to identify useful cutting planes. This remains an issue for future research.

9 Conclusion

We provided a theoretical foundation for a new type of consistency, LP-consistency, that is particularly suited to 0–1 programming. It is based on the idea that consistency can, in general, be defined with respect to a type of relaxation. LP-consistency is obtained by replacing the relaxation used for traditional consistency concepts with the LP relaxation. It brings a novel approach to 0–1 programming by identifying cuts that exclude infeasible partial assignments rather than fractional solutions. To our knowledge, no such concept has been proposed in the IP literature, even though it is directly relevant to the amount of backtracking that occurs. We also showed how a non-facet-defining inequality can be stronger than a facet-defining inequality in an interesting sense, and how traditional cutting planes can reduce branching even if no LP relaxation is used, because they can help achieve consistency.

We also defined sequential LP k-consistency, a weaker form of LP-consistency that nonetheless reduces backtracking. Sequential LP k-consistency for a given k can be obtained by one step of the lift-and-project process of integer programming. We showed that achieving even sequential LP 2-consistency can avoid backtracking that traditional separating cuts allow.

This work points to at least three further research programs. One is to extend the concepts introduced here to general mixed integer/linear programming (MILP), which appears to be straightforward. A second is to investigate the computational usefulness of sequential LP k-consistency for MILP solvers, in particular by achieving sequential LP k-consistency for small k near the top of the search tree. A third is to conduct a systematic study of the ability of traditional cutting planes to achieve consistency, both traditional forms and LP-consistency, in an MILP problem. This could allow one to make better use of known cutting planes by generating cuts that do not separate fractional solutions but enhance the consistency properties of the constraint set.

References

1. Apt, K.R.: Principles of Constraint Programming. Cambridge University Press, Cambridge (2003)
2. Balas, E., Ceria, S., Cornuéjols, G.: A lift-and-project cutting plane algorithm for mixed 0–1 programs. Math. Program. **58**, 295–324 (1993)
3. Chvátal, V.: Edmonds polytopes and a hierarchy of combinatorial problems. Discrete Math. **4**, 305–337 (1973)
4. Davis, E.: Constraint propagation with intervals labels. Artif. Intell. **32**, 281–331 (1987)
5. Freuder, E.C.: Synthesizing constraint expressions. Commun. ACM **21**, 958–966 (1978)
6. Freuder, E.C.: A sufficient condition for backtrack-free search. Commun. ACM **29**, 24–32 (1982)
7. Hooker, J.N.: Input proofs and rank one cutting planes. ORSA J. Comput. **1**, 137–145 (1989)
8. Hooker, J.N.: Integrated Methods for Optimization, 2nd edn. Springer, Heidelberg (2012). https://doi.org/10.1007/978-1-4614-1900-6
9. Hooker, J.N.: Projection, consistency, and George Boole. Constraints **21**, 59–76 (2016)
10. Mackworth, A.: Consistency in networks of relations. Artif. Intell. **8**, 99–118 (1977)
11. Montanari, U.: Networks of constraints: fundamental properties and applications to picture processing. Inf. Sci. **7**, 95–132 (1974)
12. Quine, W.V.: The problem of simplifying truth functions. Am. Math. Mon. **59**, 521–531 (1952)
13. Quine, W.V.: A way to simplify truth functions. Am. Math. Mon. **62**, 627–631 (1955)
14. Régin, J.C.: Global constraints: a survey. In: Milano, M., Van Hentenryck, P. (eds.) Hybrid Optimization: The Ten Years of CPAIOR, pp. 63–134. Springer, New York (2010). https://doi.org/10.1007/978-1-4419-1644-0_3
15. Tsang, E.: Foundations of Constraint Satisfaction. Academic Press, London (1983)
16. Van Hentenryck, P.: Constraint Satisfaction in Logic Programming. MIT Press, Cambridge (1989)

An Investigation
into Prediction + Optimisation
for the Knapsack Problem

Emir Demirović[1]([✉]), Peter J. Stuckey[2], James Bailey[1], Jeffrey Chan[3],
Chris Leckie[1], Kotagiri Ramamohanarao[1], and Tias Guns[4]([✉])

[1] University of Melbourne, Melbourne, Australia
{emir.demirovic,baileyj,kotagiri,caleckie}@unimelb.edu.au
[2] Monash University and Data61, Melbourne, Australia
peter.stuckey@monash.edu
[3] RMIT University, Melbourne, Australia
jeffrey.chan@rmit.edu.au
[4] Vrije Universiteit Brussel, Brussels, Belgium
tias.guns@vub.be

Abstract. We study a *prediction + optimisation* formulation of the
knapsack problem. The goal is to predict the profits of knapsack items
based on historical data, and afterwards use these predictions to solve
the knapsack. The key is that the item profits are not known beforehand
and thus must be estimated, but the quality of the solution is evaluated
with respect to the true profits. We formalise the problem, the goal of
minimising expected regret and the learning problem, and investigate
different machine learning approaches that are suitable for the optimisation problem. Recent methods for linear programs have incorporated the
linear relaxation directly into the loss function. In contrast, we consider
less intrusive techniques of changing the loss function, such as standard
and multi-output regression, and learning-to-rank methods. We empirically compare the approaches on real-life energy price data and synthetic
benchmarks, and investigate the merits of the different approaches.

Combinatorial optimisation is crucial in today's society and used throughout
many industries. In this paper, we work with the fundamental knapsack problem,
which has been studied for over a century and is well understood [9,17]. It is
studied in fields such as combinatorics, computer science, complexity theory,
cryptography, and applied mathematics. It has numerous applications, including
resource allocation problems where the aim is to select as many resources as
possible under given financial constraints. The knapsack problem is NP-hard,
though highly efficient solution methods exist for reasonably sized instances [9].

In traditional optimisation, it is assumed that all parameters, e.g. the profits
and weights in a knapsack, are precisely known beforehand. In practice, these are
often crude estimates based on domain expertise or historic data. As we enter the
age of big data, large amounts of data is available and thus parameters can be

L.-M. Rousseau and K. Stergiou (Eds.): CPAIOR 2019, LNCS 11494, pp. 241–257, 2019.
https://doi.org/10.1007/978-3-030-19212-9_16

estimated with greater precision. For example, ongoing promotion and current weather might influence the demand. The question that arises is whether such contextual data, together with historical data, can be used to improve decision making, i.e. solve the underlying optimisation problem more effectively.

Such problems are encountered in load shifting [10], where the aim is to create an energy-aware day-head schedule based on predicted hourly energy prices. Traditional approaches to this problem consist of two phases: (1) use machine learning to estimate the problem parameters, and (2) optimise over the estimated parameters. However, Grimes et al. [10] and Mathaba et al. [16] have separately shown that in energy-aware scheduling, learning accurate values of the parameters by minimising the mean-square error of the predictions, a commonly used metric in machine learning, does not necessarily lead to solutions of better quality for the optimisation problem. The reason is that not all errors on estimated energy prices have an equal effect on the optimisation problem, but the machine learning algorithm does not take this into account.

The challenge is to incorporate information from the optimisation problem into the learning. The main difficulty is that learning techniques typically assume that the loss function is convex. However, given the combinatorial optimisation component, this is no longer holds. It is also not differentiable as this would require computing the gradient over the *argmax* of the optimisation problem. Intuitively, such a gradient would capture the direction in which the predicted values should change to lead to a solution that is closer to the true solution obtainable under perfect knowledge.

The discussed methods have been evaluated on combinatorial problems which are solvable in polynomial time. However, we direct our attention to *NP-hard* combinatorial problems, i.e. difficult problems for which the existence of a polynomial algorithm is not known. We investigate their use on the *knapsack*, a fundamental combinatorial problem, both unit-weighted (polynomially solvable) and weighted (NP-hard). The knapsack was chosen as it has a simple constraint, yet captures a difficult combinatorial optimisation problem, suitable for exploring the use of prediction + optimisation techniques in constraint optimisation.

This work falls into the wider research theme of combining machine learning and constraint optimisation [18]. Most research has focussed on using machine learning to improve the solving process, e.g. algorithm selection and hyperparameter optimisation [14] and using machine learning to improve MIP solvers [4]. This is different from our setting, where the aim is to develop machine learning algorithms specifically designed for use with combinatorial optimisation problems where the parameters, e.g. profit values for items in the knapsack problem, are estimated with machine learning rather than given precisely. In terms of modeling, constraint acquisition [1] uses machine learning techniques to learn structural constraints from data, while other works are concerned with finding the most likely parameters of given hard constraints [20]. Closely related, as predictions are used in the objective, is the emerging topic of constructive machine learning [7,22], where the goal is to learn to synthesize structured objects from data, e.g. by interactively learning the preferences of a user and searching for the most preferred object. In contrast, our work is concerned with learning the weights of the objective on a per-instance basis. This has been done with a two-phase approach in practice, e.g. in energy-aware load shifting [10,16].

We formalise the problem of minimising the *expected regret* and draw the relation to stochastic optimisation. We then investigate multiple approaches to formulate the machine learning problem for the knapsack: indirectly - a two-stage method (predict then optimise); directly - by incorporating the optimisation objective as the loss function; and semi-directly - using domain-specific knowledge of the optimisation problem in the loss function, but without requiring to solve the constraint optimisation problem at each step.

To summarise, our contributions are as follows:

- We formalise the problem formulation in terms of *regret*;
- We investigate the relation between regret and surrogate loss functions;
- We propose two semi-direct methods based on appropriate semi-direct loss functions specifically designed for the knapsack problem;
- We empirically evaluate different strategies for prediction + optimisation on the knapsack problem with artificial and real-life energy-price data. Contrary to previous work, we show that direct methods do not outperform simple two-stage approaches on these benchmarks, demonstrating the difficulty that arises when NP-hard problems are introduced.

1   Formalisation

A combinatorial optimisation problem is a tuple $CSP(X, D, C, o)$, where X is the set of variables, D the domain of each variable, C a set of constraints over subsets of the variables, and o an objective function over X that needs to be maximized. A CSP is often a parameterized representation of a class of problems.

For example, the knapsack problem consists of selecting a number of items from a set, such that the total value is maximized. Each item has a weight and the sum of the weights of the selected items may not exceed a given capacity threshold. Let there be n items, then the knapsack problem can be formalised as a CSP with $X = \{x_1, \ldots, x_n\}$ variables, domain $D(x_i) = \{0, 1\}, \forall i$ representing whether an item is in or out and constraint set $C = \{\sum_i w_i x_i \leq c\}$ and objective $o = \sum_i v_i x_i$. Let $V = (v_1, \ldots, v_n)$ be the set of profits, $W = (w_1, \ldots, w_n)$ the set of weights and c the capacity. Any given assignment of parameters (V, W, c) is an instance of the parameterized knapsack CSP.

In the above example, we may not know the profits V of the items in advance, but we may have attributes a such as what temperature it is, how popular the items are, whether they are in promotion etc. At the same time, we have a set of historical $\{(a, V)\}$ data of tuples, with values for the same attributes as well as the (post-hoc) profits of the items that day. This historical data can be used to predict the most likely item profits V given today's attributes.

More formally, let us define θ as the set of parameters of a CSP. Then, in a prediction + optimisation setting, this consists of two disjoint sets $\theta = \theta^p \cup \theta^y$ where the θ^p parameters are assumed given and the θ^y parameters will need to be predicted. A problem instance hence does not consist of a tuple (θ^p, θ^y), but rather a tuple (θ^p, a) with a the attributes of this problem instance. From a set of historical data $\{(a, \theta^y)\}$ one can then *learn* a function f that outputs an estimated set of parameters $f(a) \approx \theta^y$ such that one obtains the problem parameters $(\theta^p, f(a))$ of the parameterized CSP.

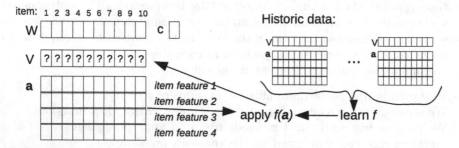

Fig. 1. Schematic figure of the components for a knapsack with 10 items, weights W, capacity c, unknown profits V and 4 attributes per item.

In the knapsack case we consider here, $\theta^p = (W, c)$ consists of the weights and the capacity, while $\theta^y = V$ are the item profits. Figure 1 shows a graphical example where it is assumed that the attributes a consist of 4 features per item (the columns), with a the union of them (one long sequence).

Solution Quality. In a standard machine learning setting, one assumes a training set $\{(a, \theta^y)\}_{train}$ and an independent test set $\{(a, \theta^y)\}_{test}$. One can then evaluate the quality of a model f trained on $\{(a, \theta^y)\}_{train}$ by measuring how it performs on $\{(a, \theta^y)\}_{test}$ through a loss function $loss(f(a), \theta^y)$ over instances (a, θ^y). Following the risk minimisation framework [23], the goal of machine learning is then to minimise the expected loss:

$$\mathbb{E}[loss(f(a), \theta^y)]$$

In our prediction + optimisation setting, the predictions are merely *intermediary results* and the true goal is to minimise the error of the optimisation procedure when using the predictions:

$$\mathbb{E}[loss(f(a), \theta^y, \theta^p)]$$

Such a loss is called the *task* loss in [5]. A natural task loss in our setting is to consider the *regret* of using the predictions rather than using the (apriori unknown) actual profits, that is, the difference in *true* solution quality obtained when optimising with the predictions as opposed to the real profits. Let $s(\theta^y, \theta^p)$ be an optimal solution to a (θ^y, θ^p) parameterized CSP, where we write $s(\theta^y)$ when θ^p is clear from the context. We can formalise the regret of f for a single instance (a, θ^y, θ^p) as:

$$regret(f(a), \theta^y, \theta^p) = o(s(\theta^y, \theta^p), \theta^y, \theta^p) - o(s(f(a), \theta^p), \theta^y, \theta^p) \quad (1)$$

where $o(X, \theta^y, \theta^p)$ denotes the value of the CSP's objective function $o()$ with solution X when using θ^y and θ^p.

In case of knapsack:

$$regret(f(\boldsymbol{a}), V, (W, c)) = V * s(V, (W, c)) - V * s(f(\boldsymbol{a}), (W, c)) \qquad (2)$$

Observe how, when computing the quality $o(X^y, \theta^y, \theta^p)$ and $o(X^f, \theta^y, \theta^p)$, in both cases the real parameters θ^y are used, e.g. profits V in case of knapsack. The regret hence quantifies the difference in the objective value when using the estimated profits compared to the ideal case, i.e. using perfect information.[1]

The goal of prediction + optimisation is hence to devise prediction and optimisation methods such that the expected regret is minimised, i.e.

$$\mathbb{E}[regret(f(\boldsymbol{a}), \theta^y, \theta^p)]$$

We assume that the parameters θ^p are independent of $\boldsymbol{a}$ and fixed, e.g. the weights in case of knapsack. The **learning problem** for a given θ^p is hence that of learning an f:

$$\min_f \mathbb{E}[regret^{\theta^p}(f(\boldsymbol{a}), \theta^y)]$$

where we write the θ^p in superscript to make clear that it is constant.

The main difficulty here is that one function evaluation of $regret()$ requires solving a discrete optimisation problem. This is different from the usual setting in machine learning, where the loss is a (differentiable) function rather than the result of a set of non-decomposable discrete optimisation problems.

2   Relation to Stochastic Optimisation

Prediction + optimisation is a setting in which one has a large sample of data from an unknown distribution, one feature vector from that distribution and an optimisation problem where we want to minimise expected regret. We discuss the setting in a stochastic optimisation problem (see e.g. [21]).

Assume, as we have done so far, that the stochasticity is only in the parameters of the objective function of the optimisation problem. Let this function be denoted by $o(X^y, \theta^y)$. It can be written as a stochastic problem as follows:

$$\max_X \mathbb{E}[o(X, \theta^y)] \; s.t. \; C(X)$$

This is a simple stochastic problem, and as the uncertainty is only on the objective, not the constraints, there are no second stage decisions or recourse. The goal is to find one X that is good in expectation, over some distribution of θ^y.

One key difference in prediction + optimisation, is that we assume the presence of a single feature vector $\boldsymbol{a}$ of observed variables which are correlated with

---

[1] Note the different problem where profits are known and the weights are learned is more complicated, since we may need some form of recourse mechanism to repair inconsistent decisions X^f.

the parameters θ^y. Hence, the knowledge of a changes the probability distribution over θ^y that we should optimise over. Indeed, we should optimise over the conditional distribution:

$$\max_X \mathbb{E}[o(X, \theta^y)|a] \; s.t. \; C(X)$$

When given a finite sample of observations $\{(a', \theta^y)\}$ we can empirically approximate it by the following:

$$\max_X \sum_{(a', \theta^y)} o(X, \theta^y) * P[(a', \theta^y)|a] \; s.t. \; C(X)$$

that is, a probability weighted sum of the objective. The remaining question is now: what is the value of $P[(a', \theta^y)|a]$, namely the probability of an historical θ^y with features a' given the feature vector a? Stochastic optimisation does not provide an answer, as it assumes that probability of each scenario is given.

From a machine learning point of view, we can take inspiration from case-based reasoning and more specifically *nearest neighbor* methods that use distance information. More specifically, we can replace the probability $P[(a', \theta^y)|a]$ by the (inverse of the) Euclidean distance between a and a'.

In fact, the k-nearest neighbor (k-NN) classification method [3], takes the k nearest neighbors and assign them a probability of $1/k$ while assigning all other instances a probability of 0. Let $kn(a)$ denote k-$nearest(\{(a', \theta^y)\}, a)$, that is, the k instances most near to a. The predicted value is then the weighted average over the samples: $\theta^f = \sum_{(a', \theta^y) \in kn(a)} \frac{1}{k} * \theta^y$. A weighted (k) nearest neighbors weights each instance by the inverse distance $\theta^f = \sum_{(a', \theta^y) \in kn(a)} \frac{1}{d(a', a)} * \theta^y$. When θ^y is a list, this is done for each component individually.

In our stochastic problem, we can also use the inverse distance as probability estimate, leading to the following:

$$\max_X \sum_{(a', \theta^y) \in kn(a)} o(X, \theta^y) * \frac{1}{d(a', a)} \; s.t. \; C(X)$$

When θ^y is a vector of values (as we assume), and the objective is *linear* wrt θ, then we can write the objective in its decomposed form $o(X, \theta^y) = \sum_i o_i(X) * \theta_i^y$. E.g. in case of knapsack where $o_i(X) = X_i$ and $\theta_i^y = v_i$ we have $o(X, \theta^y) = X * V = \sum_i X_i * v_i$. We can then do the following rewriting:

$$\sum_{(a', \theta^y) \in kn(a)} o(X, \theta^y) * \frac{1}{d(a', a)} \tag{3}$$

$$= \sum_{(a', \theta^y) \in kn(a)} \sum_i o_i(X) * \theta_i^y * \frac{1}{d(a', a)} \tag{4}$$

$$= \sum_i o_i(X) * \left(\sum_{(a', \theta^y) \in kn(a)} \frac{1}{d(a', a)} * \theta_i^y \right) \tag{5}$$

$$= \sum_i o_i(X) * \theta_i^f \tag{6}$$

where $\theta_i^f = \sum_{(a',\theta^y) \in kn(a)} \frac{1}{d(a',a)} * \theta_i^y$, the prediction of the distance-weighted k-nearest neighbor method for component i.

Hence, in the simple stochastic setting corresponding to prediction + optimisation where the objective is linear, doing a *stochastic* optimisation over multiple distance-weighted scenarios, coincides with one standard *deterministic* optimisation over the distance-weighted kNN predictions θ^f.

3 Machine Learning Formulations

As shown in the previous section, stochastic optimisation is not sufficient for solving a prediction + optimisation problem. Thus, we now turn to machine learning methods. We consider three different learning approaches:

Indirect methods use a standard learning method and loss function that is independent of the optimisation problem;

Direct methods do the learning using a convex surrogate of the regret function as loss function, which requires solving the optimisation problem repeatedly;

Semi-direct methods, which we introduce in this paper for knapsack, that use a convex surrogate of the regret which takes key properties of the optimisation problem into account but which does not require repeatedly solving it.

3.1 Indirect Learning Formulations

Following the formalisation of the problem, the most natural setting is to consider the learning as a problem of mapping the feature vector a to the list of values θ^y. This is known as **multi-output regression** [2] and many standard regression methods can be extended to multi-output regression, where the loss function is the sum of the losses of each individual prediction:

$$\mathbb{E}[loss_{mo}(f(a), \theta^y)] = \mathbb{E}[\frac{1}{n} \sum_i loss(f(a)_i, \theta_i^y)]$$

The assumption of multi-output regression, compared to pointwise regression, is that the values should be learned with respect to other items in the same group. For example, because there are correlations between the values that are always present but can get lost when predicting them independently.

Standard regression can also be used, when a set of attributes a_i is given for *each* item i, e.g. $a = \cup_i a_i$. Ideally, any correlation present between items can be indirectly accounted for through the values of the attributes, for example that sales of icecream products rise with hot weather.

The data is in a form where existing regression methods can be used to estimate the $f(a_i)$'s. The expected loss is the average loss over all items:

$$\mathbb{E}[loss(f(a_i), \theta_i^y)]$$

Regression will predict the values, and good predictions should lead to good optimisation solutions. However, predictions are estimates that have errors, and the errors can interact in the optimisation. This is not captured in the loss functions of regression.

Learning-to-Rank. Following the observation of [10] that for load-shifting problems the *correlation* between the rankings of the predictions and the true values is indicative for the optimisation quality, we may opt to learn to *rank* the items consistently. This is studied in machine learning as learning-to-rank.

The learning-to-rank problem has its roots in information retrieval [15]. In this setting, one assumes a set of *queries*, such as keywords entered in a search engine, and a set of *documents* that need to be ranked according to their relevance with the query. A relevance grade is given to each query-document pair. The goal is to rank the documents in decreasing order of relevance.

Learning-to-rank methods do this by learning a function f over a feature vector representing a query+document instance $f(a_i)$. Typically some features are related to the query and some to the document. The value the function returns has no connection to the actual value v_i other than if $v_i > v_j$ then it should be that $f(a_i) > f(a_j)$. The loss function is a 0-1-like loss defined over each pair of instances where one has a higher relevance than the other:

$$\mathbb{E}[\sum_{(i,j), v_i > v_j} loss(I_{\{f(a_i) > f(a_j)\}}, 1)] \tag{7}$$

where $I_{\{.\}}$ is 1 if the condition inside is true, and 0 otherwise. Furthermore, ranking can be decomposed into pairs and hence it is assumed the elements only interact in pairwise ways. An optimisation problem typically trades off different decision variables to each other. This can go beyond pairwise interactions, so a pairwise loss is just another problem-independent surrogate.

One can use learning to rank methods in a prediction + optimisation setting by treating each optimisation instance (a, θ^y) as one query with $|\theta^y|$ documents, each with relevance θ_i^y. Thus, in our setting, for a given knapsack instance (query), we wish to order the items (documents) by profit (relevance).

SVMRank. In [12], the author considers learning a linear function over feature vectors, aiming to learn a ranking function as defined in Eq. 7. However computing the optimal coefficients for the resulting linear function is NP-hard. As an alternative, the problem can be approximated [12]:

$$min \; \frac{1}{2}\overrightarrow{w} \cdot \overrightarrow{w} + K\sum \xi_{i,j,k} \tag{8}$$

$$\text{where } \overrightarrow{w} \cdot \overrightarrow{a_{i,j}} \geq \overrightarrow{w} \cdot \overrightarrow{a_{i,k}} + 1 - \xi_{i,j,k} \qquad q_i \in Q, \forall(j,k) \in q_i \tag{9}$$

$$\xi_{i,j,k} \geq 0, \tag{10}$$

where the first term in Eq. 8 is the regularisation term, Q is the set of queries with each query being a partial order, $\xi_{i,j,k}$ is the error variable for the j-th and k-th item in the i-th query, K is a trade-off coefficient between regularisation and training-error, and $a_{i,j}$ is the feature vector for the j-th item in the i-th query. The resulting problem is a convex quadratic program and thus the solution can be obtained using generic continuous optimisation methods. Equations 8–10 are solved to obtain the weight vector w for a linear ranking function based on item attributes, i.e. $f(a) = \sum(w_i * a_i)$, such that $f(a_1) > f(a_2)$ indicates that the first item is more profitable. Intuitively, the resulting function aims at capturing the rankings as faithfully as possible for the historical data: each benchmark defines an ordering of items based on their profitability. We note that specialised techniques have been devised for solving quadratic programs of this particular form [11,13].

3.2 Direct Learning Formulations

We consider two direct learning formulations that were recently proposed [8,24]. These methods use historical data to compute the gradient for combinatorial problems, to minimise the regret by gradient descent.

The aim of *gradient descent* is to compute the minimising point x_{min} for a function f, i.e. $\forall x : f(x_{min}) \leq f(x)$. Starting from an initial point, the algorithm iteratively moves the point towards a local minima according to the direction of its gradient.

Smart Predict then Optimise. In this approach [8] the authors aim to directly optimise the regret of an optimisation problem that has a linear objective function. For the case of knapsack, this is $\sum_{(a,V)} V * s(V) - V * s(f(a))$ where we recall that $s(L)$ returns the optimal solution when solving the CSP with values L. They derive a clever surrogate loss function using upper bounds motivated by duality, a scaling approximation and a first-order approximation of the optimal cost with respect to the predictions [8]. The resulting surrogate loss for linear objectives is the following:

$$loss_{SPO}(f(a), \theta^y) = (\theta^y - 2f(a)) * s(\theta^y - 2f(a)) + 2f(a) * s(\theta^y) - \theta^y * s(\theta^y)$$

The authors show that this indeed an upper-bounding surrogate loss to regret, and that it is convex in $f(a)$. This means that one can optimise over this loss function, for example using stochastic gradient descent methods as used in neural networks. Detailed information on what the (sub)gradients are in this case is given in [8].

For completeness we note that to avoid degenerate cases, the result of the solving method $s(\cdot)$ used should be a valid upper bound, so in case multiple optimal solutions to $s(\cdot)$ exist, the one with the best regret should be chosen. We avoid this issue for the knapsack case by solving the following greedy relaxation that has a unique optimal solution, namely it orders all items by profitability (v_i/w_i) and selects all items with the highest profitability not yet selected and distributes capacity *equally* among them, this procedure is repeated until no more capacity is left.

Quadratic Programming Task Loss (QPTL). Another approach recently proposed is not limited to linear objectives, but encompasses convex optimisation [24]. To derive gradients they apply the chain rule to a task loss, such as regret, as follows:

$$\frac{dregret(f(a), \theta^y)}{df(a)} = \frac{dregret(f(a), \theta^y)}{ds(f(a))} \frac{ds(f(a))}{df(a)}$$

The first component is the gradient of the regret with respect to the solution $s(f(a))$. The difficult part is the second part, which requires differentiation over the optimisation. They do this by differentiating through the Karush-Kuhn-Tucker conditions around the optimal point. With X, λ being the primal and dual solutions of solving the convex optimisation problem $s(f(a))$ with linear equalities represented by $BX = c$, and $\hat{f}$ the shorthand for $f(a)$ of a specific a, one needs to solve the following set of differential equations to obtain the gradients $\frac{dX}{d\hat{f}} = \frac{ds(f(a))}{df(a)}$:

$$\begin{bmatrix} \nabla_X^2 o(X, \hat{f}) & B^T \\ diag(\lambda)B & diag(BX - c) \end{bmatrix} \begin{bmatrix} \frac{dX}{d\hat{f}} \\ \frac{d\lambda}{d\hat{f}} \end{bmatrix} = \begin{bmatrix} \frac{d\nabla_X o(X, \hat{f})}{d\hat{f}} \\ 0 \end{bmatrix}$$

As described in [24] this can also be applied to linear relaxations of combinatorial optimisation problems. One difficulty is that $\nabla_X^2 o(X, \hat{f})$ is always zero for linear programs. The authors suggest to add a weighted quadratic term to the linear program to overcome this, e.g. solve:

$$max\ \hat{f}^T X - \gamma \|X\|_2^2\ s.t.\ BX = c, GX \leq h$$

for some small-valued parameter γ. As explained in [24], in this case the differential equations become, with I the identity matrix:

$$\begin{bmatrix} \gamma I & B^T \\ diag(\lambda)B & diag(BX - c) \end{bmatrix} \begin{bmatrix} \frac{dX}{d\hat{f}} \\ \frac{d\lambda}{d\hat{f}} \end{bmatrix} = \begin{bmatrix} I \\ 0 \end{bmatrix}$$

The resulting gradient can be used in the chain rule and backpropagated with stochastic gradient descent [24].

3.3   Semi-direct Learning Formulations

We now introduce a new class of techniques that are in between indirect and direct methods: *semi-direct methods*. Semi-direct methods do not require solving the optimisation problem repeatedly like direct methods, but do use information from the optimisation problem in the loss function.

Profitability-Learning. A first example is in case of weighted knapsacks, one can use regression to **predict the profitability** V_i/W_i of items rather than the profit V_i. The greedy approach to solving weighted knapsacks is to sort them by profitability and iteratively select as many as capacity allows. While we can assume that the weights W are independent of the features a, they rescale the values V_i and hence also its errors: errors on items with larger weights (and hence smaller profitability) will be relatively smaller than equal errors on items with smaller weights. In particular for weighted knapsacks we can use learning-to-rank methods on profitability as a surrogate for regret.

Simplifying QPTL. When implementing the QPTL approach [24] we observed that for knapsack, with its simple constraint of $WX \leq c$ with $B = W$ the weight vector and c the capacity, the result of solving the linear equations is often non-informative: if the solution maximizes the capacity constraint then $diag(BX - c) = 0$ and $diag(\lambda)B\frac{dX}{df} = 0$ forces the $\frac{dX}{df}$ gradients to zero, or to a below-precision small value. If $diag(BX - c) > 0$ then $\frac{d\lambda}{df}$ may be forced to zero leading to the $\frac{dX}{df}$ having the huge value of $1/\gamma$ Both cases are not meaningful and have to be guarded against, for example by replacing the gradients with a small negative constant.

In fact, we can go as far as cutting out the QP and the solving of the differential equations. The gradient is then simply $\frac{dregret(f(a),\theta^y)}{ds(f(a))}$ which is θ^y for knapsack. The gradient will hence push the predictions in the direction of the true values, *independent of the actual prediction it gives*. While unusual, the motivation is two-fold: (1) the instability of solving the differential equations has this effect in many cases and (2) the magnitude of the gradient for each item is proportional to the true value and hence the gradient updates are also proportional to it and so will the predictions be over time. For linear objective functions, which are scale invariant, that is, the same optimal solution is found when rescaling all weights, these 'proportional to true value' updates are desirable.

Specialising SVMRank for Knapsack. A key observation is that not all pairs (i, j) in Eq. 7 contribute equally towards minimising the expected regret in the case of the knapsack problem. This is illustrated in the following example.

Example 1. Consider a unit-weighted knapsack problem with four items of true profits $[10, 20, 30, 40]$ and capacity 2, and three ranking functions giving the following ranking values: $f = [-10, 0, 5, 10]$, $g = [10, 0, 30, 20]$, and $h = [1, 3, 2, 4]$. According to Eq. 7, the functions f, g, and h have 0, 2, 1 violations. The function f, as it has zero violations, captures the ranking perfectly and thus achieves the optimal solution $[0, 0, 1, 1]$. However, function g allows us to obtain the optimal solution as well, despite having two violations, because they do not affect the two highest items being ranked first. In contrast, function h only has one violation, but it misranks two critical items leading to a worse solution $[0, 1, 0, 1]$ with regret $(30 + 40) - (20 + 40) = 10$.

Based on the observation above, we modified the SVMRank objective function (Eq. 8) by adding constant weights $e_{i,j,k}$ to the error variables to account for the differences among error variables:

$$\frac{1}{2}\vec{w} \cdot \vec{w} + K \sum e_{i,j,k} \cdot \xi_{i,j,k}.$$

We propose a weight-scheme where items within a query are partitioned based on whether their profitability exceeds a given threshold. Weights for error variables within the same query are set one if the two items are not in the same partition, and zero otherwise. The threshold is taken as the value of the profitability of the least profitable item in the solution given by the linear relaxation of the query. Formally, let $p_{q_i,j}$ be the profitability for the j-th item in the i-th query, and T_i be the threshold for the i-th query, the weight-scheme is given as:

$$e_{i,j,k} = \begin{cases} 1, (p_{q_i,j} \geq T_i \wedge p_{q_i,k} < T_i) \vee (p_{q_i,j} < T_i \wedge p_{q_i,k} \geq T_i) \\ 0, (p_{q_i,j} \geq T_i \wedge p_{q_i,k} \geq T_i) \vee (p_{q_i,j} < T_i \wedge p_{q_i,k} < T_i) \end{cases}$$

Example 1. (continued) The four items with profits $(p_1, p_2, p_3, p_4) = (10, 20, 30, 40)$ compose a query q. Given capacity 2, the threshold value is set to $T = 30$ and the items are partitioned into $\{1, 2\}$ and $\{3, 4\}$. Therefore, $e_{q,1,2} = 0$ and $e_{q,3,4} = 0$, and the remaining weights are set to one. Note that if all weights were set to one, we would obtain standard SVMRank.

4   Experiments

The aim of the experimental section is to investigate the different machine learning approaches, namely indirect, direct, and semi-direct methods, for the knapsack problem. Even for a *"simple"* problem, such as the knapsack, there are multiple approaches to the solution process that warrant being investigated. We note that the benchmarks and code are available online: https://github.com/vub-dl/predopt_knapsack.

Benchmarks and Data. We perform experiments with artificially generated datasets and real-life energy price data.

The *artificial datasets* are constructed such that the profits can not be easily learned. Each item is represented by a 2-dimensional attribute vector (i, j), where $i, j \in [0, 360]$. The profit is set to $profit((i, j)) = 10^3 * sin(i) * sin(j)$. This constitutes the initial set of items, which is filtered as follows: to obtain a bijection, if multiple pairs (i, j) map to the same value, we only keep one such pair, e.g. (pi, j) and $(0, j)$ both lead to zero profit and thus only one of these pairs is kept. To ensure each profit value is positive, we add a positive constant to each profit. A weight $w \in \{3, 5, 7\}$ is assigned to each pair (i, j) and its profit is multiplied accordingly, hence preserving the profit-weight ratio. Knapsack instances are generated using these items, such that in each instance there are 16 items of each weight (total of 48 items), and special measures are taken to ensure that the distribution of item ratios is similar for each benchmark. To increase the difficulty of learning, we draw a random integer from $[1, 5]$ for each benchmark and multiply its profits.

The *real-life datasets* contain two years of historical energy price data from the day-ahead market of SEM-O, the Irish Single Electricity Market operator. The data was used in the ICON energy-aware scheduling competition and a number of publications (e.g. see [6,10]). For every half hour, the data consists of seven calendar features: whether it was a holiday, the day of the week, the week of the year, day, month, year and the half-hour-of-the-day. In addition, it includes three weather features, namely, the estimates of the temperature, windspeed, and $CO2$ intensity in the Irish city of Cork, and three key energy-related day-ahead forecasts by SEM-O itself: the forecasted wind production, the forecasted system load, and the forecasted energy price. The goal is to predict the real energy price, as determined post-hoc two days later. The task is hence not to predict energy prices from scratch, but rather learn to use the SEM-O predictions together with weather and calendar information to improve on their predictions. As is common in energy price predictions, it is difficult to derive accurate estimates and due to price swings there is a large variance in the prices and prediction errors. In our benchmarks, we consider that one physical day, consisting of 48 half-hour slots, is the range of one problem instance, i.e. it is used in a day-ahead planning setting and each benchmark contains 48 items.

The data is used in a weighted and unit-weighted setting. The weights are generated as in the artificial data and all benchmark contain 48 items.

Methodology and Implementation. The data is divided into training and test sets at a 70%–30% ratio. On the training data, we perform for each learning method a 5-fold cross-validation grid search over a small range of hyperparameters with regret as a measure to do the selection. We discuss the results of both datasets at 10%, 30% and 50% capacity of the sum of weights. The results are presented in tables, where an entry (x, y) represents the average regret for the training (x) and testing set (y), respectively.

Regarding the implementation, we use *scikit-learn* [19] and *torch* Python libraries, Gurobi as the quadratic linear program solver, and the dedicated knapsack solver of or-tools (http://developers.google.com/optimization/).

We investigate solution-quality rather than runtime. Direct methods are always more computationally expensive given their use of optimisation, and implementations have not been optimised in terms of execution time in any case.

Learning Methods. We experiment with the methods detailed in Sect. 3: **(indirect)** *kNN*, k-nearest neighbours regression; *kNN-mo*, k-nearest neighbours multi-output regression; *Ridge*, ridge regression which has shown good accuracy on the energy dataset in the past; *Ridge-mo*, multi-output ridge regression; *SVMRank*, SVMrank; **(direct)** *QPTL*, quadratic programming task loss [24]; *SPO*, smart predict then optimise [8]; **(semi-direct)** *Ridge-p*, ridge regression on profitability v_i/w_i rather than profit v_i *QPTL-s*, our quadratic programming task loss simplification as described in Sect. 3.3; and *SVMRank-s*, our weighted SVMrank modification as described in Sect. 3.3.

Table 1. Regret values for the artificial dataset: unit-weighted (top) and (bottom) weighted. (training, test) values in thousands.

Capacity	Indirect					Direct		Semi-direct	
	kNN	kNN-mo	ridge	ridge-mo	SVMR	SPO	QPTL	QPTL-s	SVMR-s
5 (10%)	(⊥; 1.6)	(⊥; 27.0)	(12.0; 12.0)	(22.0; 28.0)	(1.12; 0.99)	(20.49; 20.36)	(23.01; 19.17)	(24.22; 23.68)	(0.08; 0.41)
15 (30%)	(⊥; 1.5)	(⊥; 66.0)	(11.0; 12.0)	(61.0; 75.0)	(2.29; 2.54)	(30.31; 29.75)	(33.66; 27.62)	(35.34; 34.76)	(0.5; 0.27)
25 (50%)	(⊥; 0.7)	(⊥; 76.0)	(1.1; 0.9)	(52.0; 66.0)	(4.71; 6.07)	(0.38; 0.13)	(0.09; 0.05)	(0.1; 0.0)	(0.04; 0.03)

Capacity	Indirect					Direct		Semi-direct		
	kNN	kNN-mo	ridge	ridge-mo	SVMR	SPO	QPTL	QPTL-s	ridge-p	SVMR-s
25 (10%)	(⊥; 26)	(⊥; 119)	(62; 63)	(93; 127)	(4; 1)	(72; 67)	(62; 73)	(72; 61)	(26; 29)	(2; 0)
75 (30%)	(⊥; 64)	(⊥; 287)	(104; 102)	(218; 282)	(9; 8)	(238; 214)	(206; 235)	(240; 203)	(24; 27)	(1; 1)
125 (50%)	(⊥; 25)	(⊥; 324)	(42; 43)	(259; 347)	(23; 23)	(166; 151)	(274; 318)	(322; 270)	(8; 8)	(6; 8)

Table 2. Regret values for the energy-pricing dataset: unit-weighted (top) and (bottom) weighted.

Capacity	Indirect					Direct		Semi-direct	
	kNN	kNN-mo	ridge	ridge-mo	SVMR	SPO	QPTL	QPTL-s	SVMR-s
5 (10%)	(⊥; 88)	(⊥; 99)	(43; 51)	(67; 96)	(39; 44)	(40; 55)	(50; 64)	(50; 64)	(41; 42)
15 (30%)	(⊥; 108)	(⊥; 112)	(59; 67)	(80; 117)	(55; 53)	(59; 72)	(76; 105)	(76; 105)	(55; 51)
25 (50%)	(⊥; 68)	(⊥; 83)	(42; 48)	(51; 89)	(38; 41)	(36; 45)	(49; 70)	(49; 70)	(39; 45)

Capacity	Indirect					Direct		Semi-direct		
	kNN	kNN-mo	ridge	ridge-mo	SVMR	SPO	QPTL	QPTL-s	ridge-p	SVMR-s
25 (10%)	(⊥; 78)	(⊥; 91)	(85; 72)	(86; 88)	(97; 97)	(126; 106)	(177; 142)	(177; 142)	(84; 97)	(88; 89)
75 (30%)	(⊥; 89)	(⊥; 95)	(60; 73)	(76; 97)	(145; 155)	(259; 223)	(260; 237)	(261; 236)	(144; 144)	(143; 155)
125 (50%)	(⊥; 80)	(⊥; 82)	(47; 60)	(56; 84)	(117; 121)	(136; 113)	(236; 191)	(236; 191)	(112; 105)	(124; 126)

Results. We discuss the results of both datasets in the unit-weighted and weighted case at 10%, 30% and 50% capacity of the total weight to vary the combinatorial component in the problem instances. The results are presented in tables, where an entry (x, y) represents the average regret for the training (x) and testing set (y), respectively. We omit the training results of kNN as it simply stores all training samples in its knowledge base.

Artificial data, unit-weighted, Table 1 (top). Among the indirect methods, the multi-output regression variants are outperformed by the individual learning settings, which is an indication that learning from a joint feature representation (for all items at once) is more difficult then when learning for each item specifically and that the ability to account for correlation in the output does not compensate that. The data is generated to be difficult for pointwise methods, and indeed for small and medium capacities, Ridge regression performs poorly while SVMRank does well. The direct and semi-direct methods capture the case of 50% capacity all really well. For tighter capacities, the direct methods do not seem to capture the essential parts of the predictions better. The specialised SVMRank-s on the other hand clearly improves on vanilla SVMRank and ridge.

Artificial data, weighted, Table 1 (bottom). In the weighted case, the weights add more combinatorial effects in the optimisation as well as varying the importance of the prediction errors. Among the indirect methods the rank-learning SVMRank is the clear winner. The semi-direct SVMRank-s modification again improves it and has the best results. The direct methods perform well at 10% capacity but much less at higher capacities where more items are involved. The semi-direct method of learning profitability rather than profit shows very strong results, demonstrating the benefit of including limited information of the optimisation problem.

Energy data, unit-weighted, Table 2 (top). The data is much more noisy in this case which can be seen at the worse results of kNN versus ridge regression. The SVMRank method that learns to rank pairs of items rather than individual values performs better than ridge regression. The direct methods are more effective on this data, especially SPO. The semi-direct SVMRank-s modification is again improving over SVMRank leading to the best results overall. Ranking methods are clearly the best for our unit-weighted sets.

Energy data, weighted, Table 2 (bottom). In contrast to the unit-weighted case, learning the profits with Ridge regression leads to the lowest expected regret. Ranking is less advantageous it seems, as an optimal solution cannot be computed merely by ordering items based on their profit ratios as in the unit-weighted case. In contrast to the artificial data, learning the profitability rather than profit does not guarantee better results either and somewhat disappointingly the direct methods do not gain much by solving the optimisation problem and learning from that at each gradient step.

Note that in some cases there is a difference in regret for test and training. This can be explained by the difficulty of our setting: the machine learning function that minimises regret is highly nonlinear and nondifferentiable, as it depends on a combinatorial optimisation problem.

Discussion. Our investigation demonstrate that we have a long way to go to know how best to predict+optimise. Direct methods may be very promising, but non-gradient-descent methods like ridge regression may provide more robust predictions in general. Surprisingly, QPTL-s, which omits the QP part, is even on par or better than QPTL again demonstrating the difficulty of deriving gradients over a combinatorial optimisation problem. The only case when the reverse holds were in simple versions of the knapsack which were close to convex problems. The alternative approach, semi-direct, of including some knowledge of the optimisation problem without having to solve it repeatedly has shown its potential, but this needs to be done for each optimisation problem specifically. The ranking based methods, which have not been considered in prediction + optimisation works so far, perform well here since they do not learn on items individually, but over the relation between pairs of items. But when we examine the most difficult class of instances: weighted energy; it appears that existing indirect methods have the capability to more accurately learn the profits and override the methods which attempt to take into account the actual regret as loss function.

5 Conclusion

We study the prediction + optimisation knapsack problem. We provided insight into the relation to stochastic optimisation and the suitability of different learning methods. We compare indirect methods, standard learning approaches, versus direct methods which combine learning with the optimisation problem, and introduce semi-direct methods which combine learning with the optimisation problem while avoiding solving the optimisation problem using a form of surrogate for regret. We show that direct methods can outperform alternative indirect methods, however their utility seems limited to cases when the optimisation problem is near convex. Hence, the best approach for prediction + optimisation problems is still an open question. Some of the challenges include further exploiting learning to rank methods, improving direct methods and to explore more automatic ways to create semi-direct approaches to a new optimisation problem.

References

1. Bessiere, C., Koriche, F., Lazaar, N., O'Sullivan, B.: Constraint acquisition. Artif. Intell. **244**, 315–342 (2017). https://doi.org/10.1016/j.artint.2015.08.001, http://www.sciencedirect.com/science/article/pii/S0004370215001162, combining Constraint Solving with Mining and Learning
2. Borchani, H., Varando, G., Bielza, C., Larrañaga, P.: A survey on multi-output regression. Wiley Interdisc. Rev. Data Min. Knowl. Dis. **5**(5), 216–233 (2015)
3. Cover, T., Hart, P.: Nearest neighbor pattern classification. IEEE Trans. Inform. Theory **13**(1), 21–27 (1967)
4. Di Liberto, G., Kadioglu, S., Leo, K., Malitsky, Y.: Dash: dynamic approach for switching heuristics. Eur. J. Oper. Res. **248**(3), 943–953 (2016)
5. Donti, P.L., Amos, B., Kolter, J.Z.: Task-based end-to-end model learning in stochastic optimization. In: Proceedings of the 31st Conference on Neural Information Processing Systems (NIPS 2017), pp. 5484–5494 (2017)
6. Dooren, D.V.D., Sys, T., Toffolo, T.A.M., Wauters, T., Berghe, V.: Multi-machine energy-aware scheduling. EURO J. Comput. Optim. **5**(1–2), 285–307 (2017). https://doi.org/10.1007/s13675-016-0072-0
7. Dragone, P., Teso, S., Passerini, A.: Pyconstruct: constraint programming meets structured prediction. In: Proceedings of the Twenty-Seventh International Joint Conference on Artificial Intelligence, IJCAI 2018, pp. 5823–5825. International Joint Conferences on Artificial Intelligence Organization, July 2018
8. Elmachtoub, A.N., Grigas, P.: Smart "predict, then optimize". Technical report (2017). https://arxiv.org/pdf/1710.08005.pdf
9. Gilmore, P., Gomory, R.E.: The theory and computation of knapsack functions. Oper. Res. **14**(6), 1045–1074 (1966)
10. Grimes, D., Ifrim, G., O'Sullivan, B., Simonis, H.: Analyzing the impact of electricity price forecasting on energy cost-aware scheduling. Sustain. Comput. Inform. Syst. **4**(4), 276–291 (2014). https://doi.org/10.1016/j.suscom.2014.08.009,http://www.sciencedirect.com/science/article/pii/S221053791400050X, special Issue on Energy Aware Resource Management and Scheduling (EARMS)

11. Joachims, T.: Making large-scale svm learning practical. Technical report, SFB 475: Komplexitätsreduktion in Multivariaten Datenstrukturen, Universität Dortmund (1998)
12. Joachims, T.: Optimizing search engines using click through data. In: Proceedings of the Eighth ACM SIGKDD International Conference on Knowledge Discovery and Data Mining, KDD 2002, pp. 133–142. ACM, New York (2002). https://doi.org/10.1145/775047.775067, http://doi.acm.org/10.1145/775047.775067
13. Joachims, T.: Training linear SVMs in linear time. In: Proceedings of the 12th ACM SIGKDD International Conference on Knowledge Discovery and Data Mining, pp. 217–226. ACM (2006)
14. Kotthoff, L.: Algorithm selection for combinatorial search problems: a survey. AI Mag. **35**(3), 48–60 (2014). http://www.aaai.org/ojs/index.php/aimagazine/article/view/2460
15. Liu, T.Y., et al.: Learning to rank for information retrieval. Found. Trends Inf. Retrieval **3**(3), 225–331 (2009)
16. Mathaba, T., Xia, X., Zhang, J.: Analysing the economic benefit of electricity price forecast in industrial load scheduling. Electric Power Syst. Res. **116**, 158–165 (2014).https://doi.org/10.1016/j.epsr.2014.05.008, http://www.sciencedirect.com/science/article/pii/S0378779614001886
17. Matthews, G.: On the partition of numbers. Proc. Lond. Math. Soc. **28**, 486–490 (1897)
18. Passerini, A., Tack, G., Guns, T.: Introduction to the special issue on combining constraint solving with mining and learning. Artif. Intell. **244**, 1–5 (2017). https://doi.org/10.1016/j.artint.2017.01.002
19. Pedregosa, F., et al.: Scikit-learn: machine learning in Python. J. Mach. Learn. Res. **12**, 2825–2830 (2011)
20. Picard-Cantin, É., Bouchard, M., Quimper, C.-G., Sweeney, J.: Learning the parameters of global constraints using branch-and-bound. In: Beck, J.C. (ed.) CP 2017. LNCS, vol. 10416, pp. 512–528. Springer, Cham (2017). https://doi.org/10.1007/978-3-319-66158-2_33
21. Spall, J.: Introduction to Stochastic Search and Optimization. Wiley, New York (2003)
22. Teso, S., Passerini, A., Viappiani, P.: Constructive preference elicitation by setwise max-margin learning. In: Proceedings of the Twenty-Fifth International Joint Conference on Artificial Intelligence, IJCAI 2016, New York, NY, USA, 9–15 July 2016, pp. 2067–2073 (2016)
23. Vapnik, V.: Principles of risk minimization for learning theory. In: Advances in Neural Information Processing Systems, pp. 831–838 (1992)
24. Wilder, B., Dilkina, B., Tambe, M.: Melding the data-decisions pipeline: Decision-focused learning for combinatorial optimization. In: Proceedings of the Thirty-Third AAAI Conference on Artificial Intelligence (AAAI 2019) (2019, to appear). https://arxiv.org/pdf/1809.05504.pdf

The Maximum Weighted Submatrix Coverage Problem: A CP Approach

Guillaume Derval[(✉)][iD], Vincent Branders[iD], Pierre Dupont[iD],
and Pierre Schaus[iD]

UCLouvain - ICTEAM/INGI, Louvain-la-Neuve, Belgium
{guillaume.derval,vincent.branders,
pierre.dupont,pierre.schaus}@uclouvain.be

Abstract. The objective of the maximum weighted submatrix coverage problem (MWSCP) is to discover K submatrices that together cover the largest sum of entries of the input matrix. The special case of $K = 1$ called the maximal-sum submatrix problem was successfully solved with CP. Unfortunately, the case of $K > 1$ is more difficult to solve as the selection of the rows of the submatrices cannot be decided in polynomial time solely from the selection of K sets of columns. The search space is thus substantially augmented compared to the case $K = 1$. We introduce a complete CP approach for solving this problem efficiently composed of the major CP ingredients: (1) filtering rules, (2) a lower bound, (3) dominance rules, (4) variable-value heuristic, and (5) a large neighborhood search. As the related biclustering problem, MWSCP has many practical data-mining applications such as gene module discovery in bioinformatics. Through multiple experiments on synthetic and real datasets, we provide evidence of the practicality of the approach both in terms of computational time and quality of the solutions discovered.

Keywords: Constraint programming ·
Maximum weighted submatrix coverage problem · Data mining

1 Introduction

Constraint Programming (CP) has received an increasing interest for solving unsupervised (clustering) data-mining problems [1,3,5,7,12,14,18]. This article is interested into the mining of a numerical matrix to discover submatrices (also called biclusters) that capture a high total value. More exactly we consider an input matrix $\mathcal{M}$ with m rows and n columns where element $\mathcal{M}_{i,j}$ is a given real value. The matrix is associated with a set of rows $R = \{r_1, \ldots, r_m\}$ and a set of columns $C = \{c_1, \ldots, c_n\}$. We use $(R; C)$ to denote matrix $\mathcal{M}$. If $I \subseteq R$ and $J \subseteq C$ are subsets of the rows and of the columns, respectively, $\mathcal{M}_{I,J} = (I; J)$ denotes the submatrix $\mathcal{M}_{I,J}$ of $\mathcal{M}$ that contains only the elements $\mathcal{M}_{i,j}$ belonging to the submatrix with set of rows I and set of columns J.

© Springer Nature Switzerland AG 2019
L.-M. Rousseau and K. Stergiou (Eds.): CPAIOR 2019, LNCS 11494, pp. 258–274, 2019.
https://doi.org/10.1007/978-3-030-19212-9_17

The maximal sum submatrix problem introduced in [4] is to discover a subset of rows and columns of an input matrix that maximizes the sum of the covered entries. An example is provided in Fig. 1.

Definition 1. *The Maximal-Sum Submatrix Problem. Given a matrix $\mathcal{M} \in \mathbb{R}^{m \times n}$. Let $R = \{1, \ldots, m\}$ and $C = \{1, \ldots, n\}$ be index sets for rows and for columns, respectively. The maximal-sum submatrix is the submatrix $(I^*; J^*)$, with $I^* \subseteq R$ and $J^* \subseteq C$, such that:*

$$(I^*; J^*) = \operatorname*{argmax}_{I,J} f(I, J) = \operatorname*{argmax}_{I,J} \sum_{i \in I, j \in J} \mathcal{M}_{i,j} \tag{1}$$

The objective function rewards the selection of positive values and penalizes selection of negative values. In case of positive input matrices, the domain expert can subtract a constant threshold θ from all entries. The choice of this threshold is not discussed here. Therefore, the problem matrix is assumed to contain both positive and negative values in order to be interesting and challenging to solve.

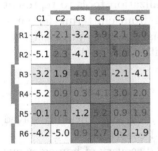

The submatrix $(\{R_1, R_2, R_4, R_5\}; \{C_2, C_4, C_5, C_6\})$, in red, is of maximal sum as the value of the objective function is 27.3.

For $K = 2$, the two submatrices depicted in red, $(\{R_1, R_2, R_4, R_5\}; \{C_2, C_4, C_5, C_6\})$, and blue, $(\{R_3, R_4, R_6\}; \{C_3, C_4\})$, are of maximal sum. The objective value equals 38.6.

Fig. 1. Example of matrix and associated submatrices of maximal sum. (Color figure online)

The maximum weighted submatrix coverage problem, that we study in this work, generalizes the maximal-sum submatrix problem to K submatrices. An example is provided in Fig. 1.

Definition 2. *The Maximum Weighted Submatrix Coverage Problem. Given a matrix $\mathcal{M} \in \mathbb{R}^{m \times n}$ and a parameter K, the maximum weighted submatrix coverage problem is to select a set of submatrices $(\mathcal{R}_k, \mathcal{C}_k)$ with $k = 1, \ldots, K$ such that the sum of the cells covered by at least one submatrix is maximal:*

$$(\mathcal{R}_1^*; \mathcal{C}_1^*), \ldots, (\mathcal{R}_K^*; \mathcal{C}_K^*) = \operatorname*{argmax}_{(\mathcal{R}_1;\mathcal{C}_1),\ldots,(\mathcal{R}_K;\mathcal{C}_K)} \sum_{i \in R, j \in C} \mathcal{M}_{i,j} \times \mathbb{1}_{cover}((i,j)) \tag{2}$$

where $\mathbb{1}_{cover}$ is the indicator function over the set $cover = \bigcup_{k \in 1..K} \mathcal{R}_k \times \mathcal{C}_k$.

1.1 Applications

The maximum weighted submatrix coverage problem has many practical data mining applications where one is interested to discover K strong relations between two groups of variables (rows and columns) represented as a matrix:

- In gene expression analysis, rows correspond to genes and columns to samples and the value in $\mathcal{M}_{i,j}$ is the measurement of the expression of gene i in sample j. One is typically interested in finding subsets of genes that present high expression in a subset of the samples as it would indicate that a particular biological pathway made of these genes is active in these samples.
- In migration data, value $\mathcal{M}_{i,j}$ represents the number of persons that moved from location i to j. The goal is the to identify groups of locations that together migrate to other groups of locations.
- A sports journalist could also be interested in Olympic games to discover group of countries that together obtained similar strong performances on the same subset of sports. The matrix value $\mathcal{M}_{i,j}$ then represents the number of medals obtained by the country i in sport j.
- Dendrograms and Sankey plots are standard visualization tools to represent relations. Unfortunately those plots quickly suffer from cluttering for large matrices. The MWSCP can be used as a preliminary step to preselect submatrices that can then be analyzed more easily with those plots.

1.2 Related Work

The maximal-sum submatrix problem was introduced in [4] and efficiently solved using constraint programming with a dedicated global constraint.

The biclustering problems are concerned with the discovery of homogeneous submatrices (called biclusters in this context) rather than maximizing the sum of the covered entries. A comprehensive review can be found in [15]. Common approaches are heuristic based and greedily selects the next bicluster after randomization of entries covered by the previously discovered biclusters.

The maximum subarray problem introduced by [2] is looking for a maximal-sum submatrix with contiguous subsets of rows and contiguous subset of columns.

The maximum ranked tile mining problem has been introduced in [14]. This is a special case of the maximal-sum submatrix problem for which the matrix entries are discrete ranks, corresponding to a permutation of column indices on each row. Another relevant difference is the constraint that sets of entries covered by the submatrices are disjoint. This restriction is more convenient for solving the problem efficiently but unnatural for the applications motivating this work.

1.3 Contributions

Our contributions are:

- The introduction of the maximum weighted submatrix coverage problem (MWSCP) as a generalization of the maximal-sum submatrix problem.
- A CP approach for solving MWSCP including filtering, lower-bound, dominance rules, a variable heuristic, and a large neighborhood search.
- An evaluation of the performances of the CP approach as compared to a greedy baseline approach (using the maximal-sum submatrix problem as subroutine) and two mathematical programming models on synthetic and real datasets.

2 CP Approach

Constraint programming (CP) is a flexible programming paradigm for solving (discrete) optimization problems. A CP model is a triplet (V, D, C) where V is the set of variables, D their domains and C is a set of constraints. In constraint programming the set domain bounds representation [8] is used to approximate the domain of a set variable S by a closed interval denoted $[S^\in, S^\in \cup S^\perp]$ where $S^\in$ are the mandatory elements and $S^\perp$ are the possible additional ones ($S^\in \cap S^\perp = \emptyset$). Such an interval represents all the sets in between those two bound sets according to the inclusion relation $\{S \mid S^\in \subseteq S \subseteq (S^\in \cup S^\perp)\}$. A set variable is bound (or assigned) whenever it contains a single set in its domain. This situation (called an assignment) happens when set interval bounds are equal, that is the possible set is empty: $S^\perp = \emptyset$.

For a set variable, the domain's update operations are:

- The inclusion of an item j in the mandatory set, denoted require(j, S), which implies that $S^\in \leftarrow S^\in \cup \{j\}$ and $S^\perp \leftarrow S^\perp \setminus \{j\}$.
- The exclusion of an item j from the possible set, denoted exclude(j, S), which implies that $S^\perp \leftarrow S^\perp \setminus \{j\}$ (and $j \notin S^\in$).

For each submatrix k, a set variable $\mathcal{R}_k$ (resp. $\mathcal{C}_k$) is introduced to represent the possible rows (resp. columns) selections in submatrix k.

Preliminary Notations. We define $\mathcal{R}_k^{\in, +j}$ (resp. $\mathcal{R}_k^{\in, -j}$) as the subset of $\mathcal{R}_k^\in$ whose matrix value in column j is positive (resp. strictly negative):

$$\mathcal{R}_k^{\in, +j} = \{i \in \mathcal{R}_k^\in \mid \mathcal{M}_{i,j} \geq 0\} \qquad \mathcal{R}_k^{\in, -j} = \{i \in \mathcal{R}_k^\in \mid \mathcal{M}_{i,j} < 0\} \qquad (3)$$

Similar notations hold for $\mathcal{C}_k$ and $\perp$. The sum of the elements in a given row i (resp. column j) and in a column (resp. row) set S is noted as:

$$\operatorname*{sum}_{\text{row } i}(S) = \sum_{j \in S} \mathcal{M}_{i,j} \qquad \operatorname*{sum}_{\text{col } j}(S) = \sum_{i \in S} \mathcal{M}_{i,j} \qquad (4)$$

The set of cells selected by at least one submatrix is denoted $\text{Cover}^{\in}$. The set of cells excluded by all submatrices is denoted $\text{Cover}^{\notin}$:

$$\text{Cover}^{\in} = \{(i,j) \mid \exists k : i \in \mathcal{R}_k^{\in} \wedge j \in \mathcal{C}_k^{\in}\} \tag{5}$$

$$\text{Cover}^{\notin} = \{(i,j) \mid \forall k : i \notin (\mathcal{R}_k^{\in} \cup \mathcal{R}_k^{\perp}) \vee j \notin (\mathcal{C}_k^{\in} \cup \mathcal{C}_k^{\perp})\} \tag{6}$$

The CP resolution is made via a Depth-First-Search (DFS) exploration. The following subsections discuss the search space, sketch the algorithm and its key components.

2.1 Search Space

As explained in [4], the search space of MWSCP with $K = 1$ can be limited to searching on a single dimension, for instance $\mathcal{C}_1$. Indeed, the variable $\mathcal{R}_1$ can be fixed optimally in polynomial time by a simple inspection argument: $\forall i \in \mathcal{R}_1^{\perp} : \underset{\text{row } i}{\text{sum}}(\mathcal{C}_1) > 0 \implies i \in \mathcal{R}_1^{\in}$.

For $K > 1$, once all the columns set variables are fixed ($\mathcal{C}_k \; \forall k \in [1..K]$) it remains to decide for each row i and each submatrix k whether i should be part of $\mathcal{R}_k$ or not. Those K decisions per row does not enjoy the monotonicity or the anti-monotonicity properties as illustrated on the next example.

Example 1. Let us consider $K = 2$ with column selection $\mathcal{C}_1 = \{1,3\}$, $\mathcal{C}_2 = \{2,3\}$. For the 1×3 input matrix $\mathcal{M} = [[2,2,-3]]$. Individually for each submatrix, the sum of entries that would be covered by selecting this row in both $\mathcal{R}_1$ and $\mathcal{R}_2$ would be negative (-1). But since weights of covered elements count only once, the value -3 is added only once and the objective value obtained is 1. Now consider the matrix $\mathcal{M} = [[-2,-2,3]]$. Individually for each submatrix, the sum of entries that would be covered by selecting this row in both $\mathcal{R}_1$ and $\mathcal{R}_2$ would be positive (1). But since weights of covered elements count only once, the value 3 is added only once and the final objective value is -1.

Actually, those K decisions per row cannot be optimally taken in polynomial time anymore as stated in Theorem 1. As a consequence, the CP search will have to branch both on the rows and columns variables rather than branching on the columns only.

Theorem 1. *For fixed variables* $\mathcal{C}_k \; \forall k \in [1..K]$, *fixing optimally* $\mathcal{R}_k \; \forall k \in [1..K]$ *is NP-Hard.*

Proof. We reduce the NP-Hard Set Cover Problem [11] to our problem: Given a universe $U = \{1, \ldots, n\}$ and a set $\{S_1, \ldots, S_K\}$ of K subsets of U, the Set Cover Problem is to find the minimum number of sets such that their union covers the universe. We construct a matrix with a single row and $n + K$ columns. The unique row values of this matrix are given by the regular expression $[K + 1]\{n\}[-1]\{K\}$ (value $K + 1$ repeated n times followed by -1 repeated K times).

The column variables are fixed to $\mathcal{C}_k = S_k \cup \{n + k\}$. In this reduction, S_k is selected if and only if $\mathcal{R}_k = \{1\}$ for every set k. A first observation is that any optimal solution covers the universe otherwise it could be improved by K by selecting any additional set that contains an uncovered element. The optimal objective function can thus be written as $n \cdot (K + 1) - |\{k \mid \mathcal{R}_k = \{1\}\}|$. As $n \cdot (K + 1)$ is fixed, maximizing this expression amounts at minimizing $|\{k \mid \mathcal{R}_k = \{1\}\}|$ which is exactly the set cover objective. □

2.2 Resolution via Depth-First-Search

The CP resolution through Depth-First-Search (DFS) exploration is sketched in Algorithm 1. All the procedures are assumed to take the decision variables $\{\mathcal{R}_1, \ldots, \mathcal{R}_K, \mathcal{C}_1, \ldots, \mathcal{C}_K\}$ and the input matrix $\mathcal{M}$ as parameters.

Algorithm 1. Sketch of the DFS resolution algorithm

```
function SOLVEDFS( )
    if !ALLVARIABLESBOUND( ) then
        S ← SELECTUNBOUNDSETVAR( )
        i ← SELECTVALUE(S⊥)
        for action ∈ [require(i, S), exclude(i, S)]  do
            SAVESTATE( )
            POST(action)
            PROPAGATEDOMINANCERULE( )
            (lb, cb, ub) ← UPDATEBOUNDS( )
            best ← max(best, cb)
            if ub > best then
                SOLVEDFS( )
            end if
            RESTORESTATE( )
        end for
    end if
end function
```

The procedure SELECTUNBOUNDSETVAR chooses a not yet bound set variable among $\{\mathcal{R}_1^\perp, \ldots, \mathcal{R}_K^\perp, \mathcal{C}_1^\perp, \ldots, \mathcal{C}_K^\perp\}$. The subsequent line chooses for the selected row/column set of some submatrix k, the specific row/column i (among the possible ones) to be included on the left branch and to be excluded on the right branch. The explored search tree is thus binary. Once the constraint is posted, and the previous state saved for later backtracking, the procedure PROPAGATEDOMINANCERULE can include (exclude) rows or columns in every submatrix that can be proven to (not) participate in any optimal solution. The UPDATEBOUNDS function updates and returns the lower, current and upper bounds for the state. The current bound is obtained by transforming the partial assignment into a complete feasible solution that excludes all rows/columns in $\perp$. If the current bound cb is better than the best value found so far (stored in variable best), the current state $(\mathcal{R}_1^\in, \ldots, \mathcal{R}_K^\in, \mathcal{C}_1^\in, \ldots, \mathcal{C}_K^\in)$ is a better solution and the value of the variable best (storing the best objective found so far) is updated (and the solution is logged). Once this is done, a check is made to ensure that there may still be a better solution below this tree node, by verifying that the upper bound is greater than the best objective value found so far; if that is the case, the DFS continues recursively. Once these steps are done, the state is backtracked and the next state visited.

Efficient backtracking is achieved through trailing, which is a state management strategy that facilitates the restoration of the computation state to an earlier version. Trailing enables the design of reversible objects. We refer to MiniCP [13] for a detailed description of trail-based solvers and to [17] for a trailed based implementation of set domains with sparse-sets.

The following subsections are dedicated to the four main functions of our algorithm: SELECTUNBOUNDSETVAR, SELECTVALUE, PROPAGATEDOMINANCE-RULE and UPDATEBOUNDS.

2.3   Functions SELECTUNBOUNDSETVAR and SELECTVALUE

SELECTUNBOUNDSETVAR chooses, at each step of the DFS, the next (unbounded) row/column interval set $\mathcal{S}$ to branch on, while SELECTVALUE selects the value $l \in \mathcal{S}^\perp$ to include/exclude from this set when branching. That is, when a pair $(\mathcal{S}, l)$ has been chosen, the DFS branches on the left, by setting require $(l, \mathcal{S})$, and on the right, by setting exclude $(l, \mathcal{S})$. The decision of the interval set and of the value are not done independently. To choose the next (set, value) pair to branch on, our algorithm maintains two (reversible) counters per row or column and per submatrix:

- $t_{k,i}^{\text{row}}$ contains the sum of cell values that will be immediately added to the objective value if row i is included in $\mathcal{R}_k$:

$$t_{k,i}^{\text{row}} = \operatorname*{sum}_{\text{row } i} \left(\{ j \mid j \in \mathcal{C}_k^{\in} \ \wedge (i,j) \notin \text{Cover}^{\in} \} \right) \tag{7}$$

- $p_{k,i}^{\text{row}}$ contains the sum of positive values in the line i that *could* be taken by submatrix k, i.e. whose columns have not been excluded:

$$p_{k,i}^{\text{row}} = \operatorname*{sum}_{\text{row } i} \left(\{ j \mid j \in (\mathcal{C}_k^{\in} \cup \mathcal{C}_k^{\perp}) \ \wedge (i,j) \notin \text{Cover}^{\in} \} \right) \tag{8}$$

$t_{k,j}^{\text{col}}$ and $p_{k,j}^{\text{col}}$ are defined similarly. The algorithm then selects the (submatrix, row) (or (submatrix, column)) pair (k, i) (or (k, j)) that maximizes $t_{k,i}^{\text{row}}$ (or $t_{k,j}^{\text{col}}$). Ties are broken by maximizing $p_{k,i}^{\text{row}}$ (or $p_{k,j}^{\text{col}}$). The selected interval set and value are then $\mathcal{R}_k$ and i (or $\mathcal{C}_k$ and j).

Recomputing these counters at each iteration is costly, as this operation is in $\mathcal{O}(Knm + K(n+m))$ for the MWSCP with an $m \times n$ matrix and K submatrices. We propose here to maintain these counters using the finite state machine (FSM) shown in Fig. 2. The algorithm we propose virtually maintains a FSM for each (row, column, submatrix) triplet. The FSMs are updated each time a row/column is added to/excluded from a submatrix:

- When a row i is included in/removed from the submatrix k, at most n FSMs must be updated (one for each cell in the row).

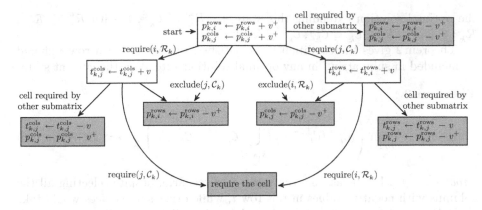

Fig. 2. FSM maintained for each (row, column, submatrix) i, j, k in the variable/value selection algorithm. For simplicity, $v = M_{i,j}$, $v^+ = \max(v, 0)$ and $v^- = \min(v, 0)$. FSMs states in blue are terminal states.

- When a column j is included in/removed from the submatrix k, at most m FSMs must be updated (one for each cell in the column).
- Updating a cell is $\mathcal{O}(1)$, if it does not become *selected* by a submatrix (i.e. the row and column of the cell are both in the mandatory sets of the submatrix).
- If a cell becomes *selected*, $K - 1$ other cells must be updated.

Given that Δ_{rows}, Δ_{cols} and Δ_{selected} are respectively the number of added or excluded (submatrix, row) tables, added/excluded (submatrix, column) tables and selected cells between two calls of the algorithm, this update runs in $\mathcal{O}(\Delta_{\text{rows}}n + \Delta_{\text{cols}}m + \Delta_{\text{selected}}K)$. To this update process must be added the verification of the counters to select the best set/value pair, which is in $\mathcal{O}(K(m + n))$.

Over a complete branch of the DFS tree (which has a maximum depth of $K(m + n)$), we have that:

$$\sum_{\text{branch}} \Delta_{\text{rows}} \leq K \cdot m \qquad \sum_{\text{branch}} \Delta_{\text{cols}} \leq K \cdot n \qquad \sum_{\text{branch}} \Delta_{\text{selected}} \leq n \cdot m \qquad (9)$$

Over a complete branch, the FSM-based algorithm maintains the states and returns the best set/value pair in $\mathcal{O}(K^2(m+n)^2)$, which is a significant improvement over the recomputation-based algorithm which runs in $\mathcal{O}(K^2(n^2m+nm^2))$ over a complete branch.

2.4 Dominance Rules

In some cases, given a partial assignment with some rows and columns already included in the set variables $\mathcal{C}_k$ and $\mathcal{R}_k$, dominance rules permit to detect additional rows or columns that must be included in any optimal solution extending this partial assignment, or rows or columns that never participate in an optimal solution. The current state is defined by $(\mathcal{R}_k^\in, \mathcal{R}_k^\perp, \mathcal{C}_k^\in, \mathcal{C}_k^\perp)$, and we denote the optimal solution extending this state as $(\mathcal{R}_k^{*\in}, \emptyset, \mathcal{C}_k^{*\in}, \emptyset)$ with $\mathcal{R}_k^{\in} \subseteq \mathcal{R}_k^{*\in}$, $\mathcal{R}_k^{*\in} \subseteq (\mathcal{R}_k^{\in} \cup \mathcal{R}_k^{\perp})$, $\mathcal{C}_k^{\in} \subseteq \mathcal{C}_k^{*\in}$, $\mathcal{C}_k^{*\in} \subseteq (\mathcal{C}_k^{\in} \cup \mathcal{C}_k^{\perp})$.

Theorem 2 gives the condition to be satisfied to detect that a row i should be included in submatrix l in any optimal solution extending the current state.

Theorem 2

$$\forall i \in \mathcal{R}_l^{\perp} \ : \ \underset{row\ i}{sum} \left((\mathcal{C}_l^{\in} \cup \mathcal{C}_l^{\perp,-i}) \setminus (\bigcup_{k|k \neq l} \mathcal{C}_k^{\in,+i} \cup \mathcal{C}_k^{\perp,+i}) \right) > 0 \Rightarrow i \in \mathcal{R}_l^{*\in} \quad (10)$$

Proof (sketch). Let us assume the worst-case scenario: despite selecting all the columns with negative values in this row i, while other submatrices would take the columns with positive values, the submatrix still has a positive sum contribution for this row i. Therefore this row must be included in submatrix l in any optimal solution extending the current state.   □

Theorem 3 gives the condition to be satisfied to detect that a row i will never be included submatrix l in any optimal solution extending the current state, using the best-case scenario.

Theorem 3

$$\forall i \in \mathcal{R}_l^{\perp} \ : \ \underset{row\ i}{sum} \left((\mathcal{C}_l^{\in} \cup \mathcal{C}_l^{\perp,+i}) \setminus (\bigcup_{k|k \neq j} \mathcal{C}_k^{\in,-i} \cup \mathcal{C}_k^{\perp,-i}) \right) < 0 \Rightarrow j \notin \mathcal{R}_l^{*\in} \quad (11)$$

These two properties (and their symmetric counterparts for columns) can be used in any node of the search tree to reduce the search space.

2.5   PROPAGATEDOMINANCERULE: Dominance Rules Check

Dominance rules from Eqs. (10) and (11) (and their symmetric counterparts for the columns) can be used to reduce the search space. As in the previous subsections, recomputing the rules at each call to PROPAGATEDOMINANCERULE is expensive ($\mathcal{O}(Kmn)$ at each call, $\mathcal{O}(K^2(m^2 n + mn^2))$ over a complete branch of the DFS). We describe below how to maintain the rules on rows. Of course, the method is symmetric for columns.

As in SELECTUNBOUNDSETVAR and SELECTVALUE, we maintain virtual FSMs for each triplet (row, column, submatrix), as shown in shown Fig. 3. The FSMs collectively maintain two reversible values, shared between FSMs, for each (submatrix k, row i) table:

- $lb_{k,i}$ is the value of the worst-case scenario for submatrix k and row i (the left part of Eq. (10))
- $ub_{k,i}$ is the value of the best-case scenario for submatrix k and row i (the left part of Eq. (11)).

The FSMs also maintain the number of *supports* of each cell (i, j), i.e. the number of submatrices that could still select the cell:

$$\text{support}_{i,j} = \left|\{k \mid i \in (\mathcal{R}_k^{\in} \cup \mathcal{R}_k^{\perp}) \wedge j \in (\mathcal{C}_k^{\in} \cup \mathcal{C}_k^{\perp})\}\right| \tag{12}$$

Each $\text{support}_{i,j}$, shared across all FSMs, is maintained as reversible integer by the solver: its state can then be backtracked.

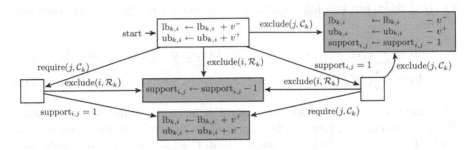

Fig. 3. FSM maintained for each (row, column, submatrix) i, j, k in PROPAGATEDOM-INANCERULE. For simplicity, $v = M_{i,j}$, $v^+ = \max(v, 0)$ and $v^- = \min(v, 0)$. FSMs states in blue are terminal states. (Color figure online)

The transition and update operations of our FSMs are the following:

- When a row i (resp. column j) is excluded from a submatrix k, at most n (resp. m) cells' FSMs must be updated. The contribution of the cell (i, j) to $\text{ub}_{k,i}$ and $\text{lb}_{k,i}$ are removed and the support of the cell is decremented. Each of these operations are in constant time, and overall takes $\mathcal{O}(n)$ (resp. $\mathcal{O}(m)$).
- When a cell (i, j) becomes supported by only one remaining submatrix k ($\text{support}_{i,j} = 1$), and the column j is included in this submatrix k ($j \in \mathcal{C}_k^{\in}$, and since $\text{support}_{i,j} = 1$, it implies that $i \in (\mathcal{R}_k^{\in} \cup \mathcal{R}_k^{\perp})$), the value of lb and ub for this submatrix k is updated by the cell's value. This operation is also in constant time, and thus $\mathcal{O}(K)$ for all submatrices.
- When a row i (resp. column j) is included in a submatrix k, a check on all columns j (resp. rows i) must be performed to see if a cell (i, j) with $\text{support}_{i,j} = 1$ and $i \in \mathcal{R}_k^{\in}$ and $j \in \mathcal{C}_k^{\in}$ exists. If that is the case, $\text{lb}_{k,i}$ and $\text{ub}_{k,i}$ are updated to include the value of the cell. Overall, this operation is $\mathcal{O}(n)$ (resp. $\mathcal{O}(m)$).

Once the update of the FSMs is done, each (row, submatrix) pair is verified w.r.t. the rules, in $\mathcal{O}(Km)$. A call to PROPAGATEDOMINANCERULE is in $\mathcal{O}(Km + \Delta_{\text{rows}}n + \Delta_{\text{cols}}m + \Delta_{\text{required}}K + \Delta_{\text{support}=1}K)$. Over a complete branch, the number of operations required is in $\mathcal{O}(Km^2 + Kmn)$. If the rules are applied symmetrically on columns, the overall running time is in $\mathcal{O}(K \max(m, n)^2)$.

2.6   UPDATEBOUNDS: Efficient Lower and Upper Bounds Computations

In order to run the Branch & Bound, upper bounds on the objective for the current tree node must be computed efficiently. The chosen method also provides a lower bound, with no additional (asymptotic) computational cost.

The upper bound ub is the sum of every cell that is either selected in a submatrix or that is positive and could still be selected. The lower bound lb is similarly defined, but keeping negative-valued cells. Formally, they are computed as follows:

$$\text{ub} = \sum \{ \mathcal{M}_{i,j} \mid (i,j) \in \text{Cover}^\in \vee (\mathcal{M}_{i,j} > 0 \wedge (i,j) \notin \text{Cover}^{\notin}) \} \qquad (13)$$

$$\text{lb} = \sum \{ \mathcal{M}_{i,j} \mid (i,j) \in \text{Cover}^\in \vee (\mathcal{M}_{i,j} < 0 \wedge (i,j) \notin \text{Cover}^{\notin}) \} \qquad (14)$$

Recomputing these bounds from scratch in each node is again costly: $\mathcal{O}(Knm)$. The running time can be improved by maintaining incrementally the number of submatrices supporting each cell, in the same way as previously done in PROPAGATEDOMINANCERULE.

These bounds, stored as reversible floating point numbers, can then be maintained easily:

- When a row i is included in a submatrix k, check if any column j is already in $\mathcal{C}_k^\in$, and that $(i,j) \notin \text{Cover}^\in$ yet. If that is the case and that $\mathcal{M}_{i,j} > 0$ (resp. < 0), increase ub (resp. lb) by $\mathcal{M}_{i,j}$. This operation runs in $\mathcal{O}(n)$.
- The similar operation must be performed when a column is included in a submatrix. Each of these operations runs in $\mathcal{O}(m)$.
- When a row i is excluded from a submatrix k, check if any column j is not already excluded ($j \notin (\mathcal{C}_k^\in \cup \mathcal{C}_k^\perp)$). If that is the case, decrease $\text{support}_{i,j}$ by one. This operation runs in $\mathcal{O}(n)$.
- The same operation goes for excluded columns in $\mathcal{O}(m)$.
- When the $\text{support}_{i,j}$ is reduced to zero, if $\mathcal{M}_{i,j} > 0$ (resp. < 0), then decrease ub (resp. lb) by $\mathcal{M}_{i,j}$. This operation runs in $\mathcal{O}(1)$.

The whole maintenance process for the bounds behaves in $\mathcal{O}(\Delta_{rows}n + \Delta_{cols}m)$. Over a complete branch, the incremental method is in $\mathcal{O}(Knm)$, while the one based on recomputations is in $\mathcal{O}(K^2(n^2m + nm^2))$.

2.7   The Large Neighborhood Search

The exhaustive approach presented above eventually finds and proves the optimum value provided enough time is given. Unfortunately, the search space is so large that even for small matrices and a limited number of submatrices, it tends to quickly find a good solution but is not able to improve it. To overcome this limitation, we propose to embed the exhaustive CP search into a Large Neighborhood Search (LNS) [19]. LNS is a local search approach using CP to discover improvements around the current best solution:

- First the CP exhaustive search is used during a limited time, to discover an initial solution.
- For a given number of iterations, the CP exhaustive search is used again but this time with some variables partially fixed (fragment) as in the current best solution.

In addition, to limit the risk of having an iteration stuck for too long, we limit the DFS to 1000 *failures*.

The current best solution at iteration t has the form $((\mathcal{R}^{*\in}_{1,t}, \ldots, \mathcal{R}^{*\in}_{K,t}); (\mathcal{C}^{*\in}_{1,t}, \ldots, \mathcal{C}^{*\in}_{K,t}))$. We propose three different fragment selection heuristics (part of the solution to constrain when restarting the LNS for next iteration):

1. Select uniformly at random a subset of rows and columns in the set of lines and columns used by some submatrix: $R^p \subseteq (\bigcup_{k \in M^p} \mathcal{R}^{*\in}_{k,t})$, $C^p \subseteq (\bigcup_{k \in M^p} \mathcal{C}^{*\in}_{k,t})$, then for each submatrix, include the set of rows and columns intersecting with those sets: $\mathcal{R}^{\in}_{k,t+1} = \mathcal{R}^{\in}_{k,t} \cap R^p$, $\mathcal{R}^{\perp}_{k,t+1} = R \setminus \mathcal{R}^{\in}_{k,t+1}$ and similarly for columns.
2. A similar operator is defined with rows and columns selected inside the whole matrix: $R^p \subseteq R$, $C^p \subseteq C$. This allows for greater diversification, notably by allowing discovery of previously unselected rows/columns.
3. Selecting uniformly at random a subset of submatrices $M^p \subseteq \{1, \ldots, K\}$. For each of these submatrices, select at random different subsets of rows and columns $R^p_k \subseteq \mathcal{R}^{*\in}_{k,t}$, $C^p_k \subseteq \mathcal{C}^{*\in}_{k,t}$ that is constrained: $\mathcal{R}^{\in}_{k,t+1} = \mathcal{R}^{\in}_{k,t} \cap R^p_k$, $\mathcal{R}^{\perp}_{k,t+1} = R \setminus \mathcal{R}^{\in}_{k,t+1}$ and similarly for columns.

Empirical observations show that these three operators are complementary.

3   Experiments

This section describes experiments conducted to assess the performances of the proposed algorithms and to provide guidance on the selection of the appropriate solution. We first evaluate the methods on synthetic datasets, where the optimum is known, then on real datasets.

We compare our exhaustive CP and LNS methods against a greedy baseline approach, CP-Greedy, that solves at each step the maximal-sum submatrix ($K = 1$) problem using the CP approach from [4]. This approach iteratively selects the next best submatrix, on a modified matrix in which the previously selected entries are set to 0 such that there is no incentive to select several times the same (positive) entries. Each iteration is performed within $\frac{t^{\max}}{K}$ with $t^{\max}$ the allocated budget of time.

The implementation has been carried out on OscaR [16], using Java 1.8.0 (Hotspot VM) on an AMD Bulldozer clocked at 2.1 GHz; one core and 3 Go of RAM per instance.

The source code is available here: https://github.com/GuillaumeDerval/MWSCP.

3.1 Synthetic Datasets

A synthetic dataset composed of 1,617 instances have been generated using a Python script (available on Zenodo [9]). For those, the optimal solution is known as they were all generated by implanting randomly K submatrices before adding some noise[1]. Table 1 describes parameter values considered in the generation. The parameters used to generate the instances are described in Table 1.

Approaches are compared using any-time profiles as described in Definition 3.

Definition 3 *Any-Time Profile.* *Let $f(a, i, t)$ be the objective value of the best solution found so far by an algorithm a for an instance i at time t. Let $t^{\max}$ be the provided budget of time before interrupting a run. Let f_i^* be the optimal solution for i if known (as is the case for synthetic data). The any-time profile of a is the solution quality $Q_a(t)$ of a on all instances as a function of time:*

$$Q_a(t) = \frac{1}{|i|} \sum_i \frac{f(a, i, t)}{\max(f(a_i^*, i, t^{\max}), f^*)} \ with \ a_i^* = \operatorname*{argmax}_a f(a, i, t^{\max}). \quad (15)$$

Table 1. Parameters for the synthetic dataset generation

Parameter	Description	Values used
m, n	Size of the matrix $\mathcal{M} \in \mathbb{R}^{m \times n}$	$(800, 200), (640, 250), (400, 400)$
K	Number of submatrices	2, 4, 8
o	Minimum overlap between submatrices (in % of cells)	0, 0.3, 0.6
σ	Background noise variance (mean is 0)	0, 0.5, 1.0
r, s	Size of submatrices (noisy, Gaussian with $\sigma = \frac{r \text{ or } s}{20}$)	$(35, 70), (50, 50)$
seed	Seed for matrix generation	$[0, 9]$

Figure 4 gives the any-time profiles of the CP-Greedy baseline method, along with CP-Exhaustive (the exhaustive process presented above) and CP-LNS. The results clearly illustrates the overall better performances of the CP-LNS whenever the computation time exceeds roughly 20 s.

Table 2a presents, for each parameter value considered in the synthetic data generation, the performances of the algorithms. Reported performances are computed as the average performance of each algorithm obtained before a certain limit of computation time.

Through analysis of the performances with respect to parameters' values, we observed that the major parameters are, in decreasing order of influence, the following: (1) the submatrices overlap, (2) K = the number of submatrices. The difficulty of reaching good solution increases quickly as the minimum overlap parameter increases until 50%, after which it decreases. Similarly, as the number of implanted submatrices increases, good solution quality becomes harder to grasp.

---

[1] Notice that the optimal solution may be slightly different than the implanted submatrices because of the noise addition.

3.2 Real Datasets

We also experiment with non-synthetic datasets of several types (*olympic, migration, genes*) described in Sect. 1.1. The results, presented in Table 2b, are similar to those obtained for synthetic datasets. CP-LNS is the best method on most datasets given 10 s of computation time, with two notable exceptions (alizadeh and garber datasets), in which case LNS did not find the optimum in the 20 min allowed for each dataset.

Table 2. Comparison between CP-Greedy (GRE), CP-Exhaustive (EX) and CP-LNS (LNS). The table shows the $Q_a(t)$ for each algorithm a given a certain amount of time t (see Eq. (3)).

(a) Synthetic dataset

Parameters	10s			20s			100s			1080s		
	GRE	EX	LNS	GRE	EX	LNS	GRE	EX	LNS	GRE	EX	LNS
$\{m = 400, n = 400\}$	**0.70**	0.33	0.37	0.74	0.57	**0.76**	0.76	0.75	**0.95**	0.77	0.75	**0.97**
$\{m = 640, n = 250\}$	**0.71**	0.34	0.32	0.75	0.48	**0.79**	0.77	0.74	**0.95**	0.77	0.75	**0.97**
$\{m = 800, n = 200\}$	**0.73**	0.34	0.29	**0.77**	0.48	0.61	0.79	0.77	**0.94**	0.79	0.78	**0.96**
$K = 2$	0.85	0.78	0.32	0.85	**0.88**	0.83	0.85	0.90	**0.96**	0.85	0.91	**0.97**
$K = 4$	**0.72**	0.20	0.30	**0.77**	0.51	0.72	0.78	0.74	**0.94**	0.78	0.75	**0.96**
$K = 8$	**0.57**	0.03	0.36	**0.64**	0.13	0.61	0.68	0.62	**0.94**	0.68	0.62	**0.97**
$o = 0\%$	**0.58**	0.27	0.34	0.67	0.45	**0.71**	0.71	0.66	**0.97**	0.71	0.66	**0.98**
$o = 30\%$	**0.71**	0.34	0.31	**0.73**	0.50	0.69	0.75	0.75	**0.93**	0.75	0.76	**0.95**
$o = 60\%$	**0.85**	0.40	0.34	**0.86**	0.57	0.77	0.86	0.86	**0.94**	0.86	0.86	**0.97**
$\sigma = 0.0$	0.73	0.34	**0.78**	0.78	0.63	**0.80**	0.81	0.77	**0.98**	0.81	0.78	**1.00**
$\sigma = 0.5$	**0.72**	0.33	0.04	**0.75**	0.44	0.67	0.78	0.74	**0.94**	0.78	0.74	**0.97**
$\sigma = 1.0$	**0.69**	0.33	0.16	**0.73**	0.44	0.68	0.73	0.75	**0.93**	0.73	0.75	**0.94**
$\{r = 50, s = 50\}$	**0.71**	0.34	0.34	**0.75**	0.52	0.73	0.77	0.76	**0.94**	0.77	0.77	**0.96**
$\{r = 35, s = 70\}$	**0.71**	0.32	0.32	**0.76**	0.50	0.71	0.78	0.75	**0.95**	0.78	0.75	**0.97**

(b) Real datasets

$K = 4$		1s			5s			20s		
Type	Dataset	GRE	EX	LNS	GRE	EX	LNS	GRE	EX	LNS
migration	migration_0.001 [6]	**0.96**	0.92	**0.96**	0.96	0.92	**0.99**	0.96	0.92	**1.00**
migration	migration_0.003 [6]	0.87	0.89	**0.93**	0.87	0.89	**0.99**	0.87	0.89	**1.00**
migration	migration_0.005 [6]	0.83	0.79	**0.96**	0.83	0.79	**1.00**	0.83	0.79	**1.00**
olympic	olympic_0.01 [10]	0.88	0.69	**0.92**	0.88	0.91	**0.97**	0.91	0.91	**1.00**
olympic	olympic_0.02 [10]	0.79	0.69	**0.87**	0.84	0.84	**0.97**	0.84	0.84	**1.00**
olympic	olympic_0.04 [10]	0.62	0.81	**0.91**	0.76	0.82	**0.96**	**0.93**	0.82	**1.00**
olympic	olympic_0.06 [10]	0.80	0.92	**0.93**	0.97	0.92	**0.98**	0.97	0.92	**0.99**
$K = 4$		10s			20s			100s		
Type	Dataset	GRE	EX	LNS	GRE	EX	LNS	GRE	EX	LNS
gene	alizadeh-2000-v1_095 [20]	**1.00**	0.48	0.82	**1.00**	0.48	0.82	**1.00**	0.48	0.92
gene	armstrong-2002-v1_095 [20]	0.73	0.60	**0.92**	0.73	0.60	**0.99**	0.73	0.60	**1.00**
gene	bhattacharjee-2001_095 [20]	0.82	0.31	**0.98**	0.91	0.86	**0.99**	0.91	0.96	**1.00**
gene	bittner-2000_095 [20]	**0.96**	0.53	0.86	**0.96**	0.53	**0.98**	**0.96**	0.53	**0.98**
gene	bredel-2005_095 [20]	0.98	0.86	**1.00**	0.98	0.86	**1.00**	0.98	0.86	**1.00**
gene	chen-2002_095 [20]	0.74	0.80	**1.00**	0.89	0.80	**1.00**	0.89	0.80	**1.00**
gene	chowdary-2006_095 [20]	0.82	0.83	**1.00**	0.82	0.83	**1.00**	0.87	0.83	**1.00**
gene	dyrskjot-2003_095 [20]	0.97	0.94	**0.99**	0.97	0.94	**1.00**	0.97	0.94	**1.00**
gene	garber-2001_095 [20]	**0.59**	0.24	0.58	**0.82**	0.32	0.58	**1.00**	0.50	0.86
gene	golub-1999-v1_095 [20]	0.86	0.88	**0.92**	0.86	0.88	**0.95**	0.86	0.88	**0.96**

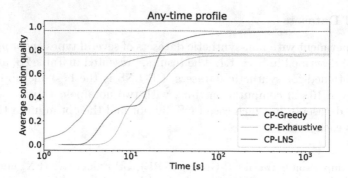

Fig. 4. Comparison between CP-Greedy, CP-Exhaustive and CP-LNS on 1,617 matrices generated as described in Sect. 3.1. The graph presents the any-time profile described in Eq. (3). For each instance, 18 min were allocated for computations.

3.3 Comparison Against Mixed Integer Linearly and Quadratically Constrained Programming

We tested our methods against MIP (linear) and MIQCP (quadratic terms in the constraints) methods. As these two methods do not perform well on bigger instances, we do not integrate them in our experiments on large matrices, presented above.

$$
\begin{array}{ll}
\textbf{MIP model} & \\
\max \sum_{i,j} \mathcal{M}_{i,j} \cdot s_{i,j} & \\
s_{i,j} \geq e_{i,j,k} & \forall i, j, k \\
s_{i,j} \leq \sum_k e_{i,j,k} & \forall i, j \\
e_{i,j,k} + 1 \geq r_{k,i} + c_{k,j} & \forall i, j, k \\
2 \cdot e_{i,j,k} \leq r_{k,i} + c_{k,j} & \forall i, j, k
\end{array}
\qquad
\begin{array}{ll}
\textbf{MIQCP model} & \\
\max \sum_{i,j} \mathcal{M}_{i,j} \cdot s_{i,j} & \\
K \cdot s_{i,j} \geq \sum_k r_{k,i} \cdot c_{k,j} & \forall i, j \\
s_{i,j} \leq \sum_k r_{k,i} \cdot c_{k,j} & \forall i, j
\end{array}
$$

$$\text{All variables} \in \{0, 1\}$$

MIP and MIQCP methods are plagued by the number of variables, that is in $\mathcal{O}(Knm)$ for MIP and $\mathcal{O}(K(n+m))$ for MIQCP, and by the number of constraints, which is $\mathcal{O}(Knm)$ for MIP and $\mathcal{O}(nm)$ for MIQCP. Tables 3a and b show that both models are slow compared to our LNS method, and are heavily affected by matrix size, number of submatrices to find and noise. For bigger submatrices, such as the synthetic and real ones presented in the previous section, both methods timeout either without returning solutions or with comparatively poor solutions.

Table 3. Comparison between CP-LNS, MIP and MIQCP, on a synthetic dataset (generated as described in Sect. 3.1). All methods were given a fixed time limit of 300 s. The metric used is the any-time profile at the time limit (see Definition 3). CP-LNS finds the optimum on each dataset. The time when the best found solution was found is indicated inside parentheses. Experiments made on Gurobi 8.1.0.

(a) Varying number of submatrices and noise, with matrices of size 50×50 and submatrices of size 16×16.

K	σ	CP-LNS	MIP	MIQCP
2	0.0	**1.00** (**1s**)	1.00 (0s)	1.00 (1s)
2	0.5	**1.00** (**1s**)	1.00 (7s)	1.00 (7s)
2	1.0	**1.00** (**1s**)	0.89 (233s)	0.79 (57s)
3	0.0	**1.00** (**2s**)	1.00 (1s)	1.00 (2s)
3	0.5	**1.00** (**3s**)	1.00 (140s)	1.00 (138s)
3	1.0	**1.00** (**3s**)	0.74 (254s)	0.48 (256s)
4	0.0	**1.00** (**2s**)	1.00 (1s)	1.00 (62s)
4	0.5	**1.00** (**3s**)	1.00 (252s)	0.88 (290s)
4	1.0	**1.00** (**6s**)	0.64 (260s)	0.69 (225s)
5	0.0	**1.00** (**4s**)	1.00 (79s)	1.00 (275s)
5	0.5	**1.00** (**5s**)	0.82 (257s)	0.69 (237s)
5	1.0	**1.00** (**6s**)	0.77 (24s)	0.36 (38s)

(b) Varying size of the matrix and noise, with matrices of size $m \times m$ and $K = 2$ submatrices of size $\lfloor \frac{m}{3} \rfloor \times \lfloor \frac{m}{3} \rfloor$.

m	σ	CP-LNS	MIP	MIQCP
50	0.0	**1.00** (**0s**)	1.00 (1s)	1.00 (3s)
50	0.5	**1.00** (**1s**)	1.00 (5s)	1.00 (7s)
50	1.0	**1.00** (**1s**)	0.95 (207s)	0.82 (204s)
100	0.0	**1.00** (**4s**)	1.00 (1s)	1.00 (33s)
100	0.5	**1.00** (**1s**)	0.86 (293s)	1.00 (45s)
100	1.0	**1.00** (**3s**)	0.65 (269s)	0.82 (191s)
200	0.0	**1.00** (**17s**)	1.00 (8s)	1.00 (135s)
200	0.5	**1.00** (**21s**)	0.37 (191s)	3% (81s)
200	1.0	**1.00** (**6s**)	0% (0s)	5% (134s)
400	0.0	**1.00** (**1s**)	1.00 (31s)	1.00 (54s)
400	0.5	**1.00** (**1s**)	0% (1s)	0% (0s)
400	1.0	**1.00** (**1s**)	0% (1s)	4% (301s)

4   Conclusions

We presented a generalization of the Maximal-Sum Submatrix Problem [4] to multiple submatrices, called the Maximum Weighted Submatrix Coverage Problem (MWSCP), along with a method to solve this problem based on constraint programming and large neighborhood search. Experiments on both synthetic and real datasets show that our CP-LNS method finds consistently better solutions (when more than 10 s are allocated) than both MIP/MIQCP, an exhaustive CP method and a greedy approach using the method from [4].

Acknowledgments. Computational resources have been provided by the Consortium des Équipements de Calcul Intensif (CÉCI), funded by the Fonds de la Recherche Scientifique de Belgique (F.R.S.-FNRS) under Grant No. 2.5020.11.

References

1. Aoga, J.O.R., Guns, T., Schaus, P.: An efficient algorithm for mining frequent sequence with constraint programming. In: Frasconi, P., Landwehr, N., Manco, G., Vreeken, J. (eds.) ECML PKDD 2016. LNCS (LNAI), vol. 9852, pp. 315–330. Springer, Cham (2016). https://doi.org/10.1007/978-3-319-46227-1_20
2. Bentley, J.: Programming pearls: algorithm design techniques. Commun. ACM **27**(9), 865–873 (1984)
3. Bessiere, C., De Raedt, L., Kotthoff, L., Nijssen, S., O'Sullivan, B., Pedreschi, D. (eds.): Data Mining and Constraint Programming. LNCS (LNAI), vol. 10101. Springer, Cham (2016). https://doi.org/10.1007/978-3-319-50137-6
4. Branders, V., Schaus, P., Dupont, P.: Mining a sub-matrix of maximal sum. In: Proceedings of the 6th International Workshop on New Frontiers in Mining Complex Patterns in Conjunction with ECML-PKDD 2017 (2017)

5. Chabert, M., Solnon, C.: Constraint programming for multi-criteria conceptual clustering. In: Beck, J.C. (ed.) CP 2017. LNCS, vol. 10416, pp. 460–476. Springer, Cham (2017). https://doi.org/10.1007/978-3-319-66158-2_30
6. Dao, T., Docquier, F., Maurel, M., Schaus, P.: Global migration in the 20th and 21st centuries: the unstoppable force of demography (2018)
7. Duong, K.C., Vrain, C., et al.: Constrained clustering by constraint programming. Artif. Intell. **244**, 70–94 (2017)
8. Gervet, C.: Interval propagation to reason about sets: definition and implementation of a practical language. Constraints **1**(3), 191–244 (1997)
9. Guillaume, D., Vincent, B., Pierre, D., Pierre, S.: Synthetic dataset used in the maximum weighted submatrix coverage problem: A CP approach, November 2018. https://doi.org/10.5281/zenodo.1688740
10. IOC Research and Reference Service, The Guardian: Olympic sports and medals 1896–2014. https://www.kaggle.com/the-guardian/olympic-games
11. Karp, R.M.: Reducibility among combinatorial problems. In: Miller, R.E., Thatcher, J.W., Bohlinger, J.D. (eds.) Complexity of Computer Computations, pp. 85–103. Springer, Boston (1972). https://doi.org/10.1007/978-1-4684-2001-2_9
12. Kuo, C.T., Ravi, S., Vrain, C., Davidson, I., et al.: Descriptive clustering: ILP and CP formulations with applications. In: IJCAI-ECAI 2018, the 27th International Joint Conference on Artificial Intelligence and the 23rd European Conference on Artificial Intelligence (2018)
13. Michel, L., Schaus, P., Van Hentenryck, P.: MiniCP: a lightweight solver for constraint programming (2018). https://minicp.bitbucket.io
14. Le Van, T., van Leeuwen, M., Nijssen, S., Fierro, A.C., Marchal, K., De Raedt, L.: Ranked tiling. In: Calders, T., Esposito, F., Hüllermeier, E., Meo, R. (eds.) ECML PKDD 2014. LNCS (LNAI), vol. 8725, pp. 98–113. Springer, Heidelberg (2014). https://doi.org/10.1007/978-3-662-44851-9_7
15. Madeira, S.C., Oliveira, A.L.: Biclustering algorithms for biological data analysis: a survey. IEEE/ACM Trans. Comput. Biol. Bioinform. (TCBB) **1**(1), 24–45 (2004)
16. OscaR Team: OscaR: Scala in OR (2012). https://bitbucket.org/oscarlib/oscar
17. de Saint-Marcq, V.L.C., Schaus, P., Solnon, C., Lecoutre, C.: Sparse-sets for domain implementation. In: CP workshop on Techniques for Implementing Constraint Programming Systems (TRICS), pp. 1–10 (2013)
18. Schaus, P., Aoga, J.O.R., Guns, T.: CoverSize: a global constraint for frequency-based itemset mining. In: Beck, J.C. (ed.) CP 2017. LNCS, vol. 10416, pp. 529–546. Springer, Cham (2017). https://doi.org/10.1007/978-3-319-66158-2_34
19. Shaw, P.: Using constraint programming and local search methods to solve vehicle routing problems. In: Maher, M., Puget, J.-F. (eds.) CP 1998. LNCS, vol. 1520, pp. 417–431. Springer, Heidelberg (1998). https://doi.org/10.1007/3-540-49481-2_30
20. de Souto, M.C., Costa, I.G., de Araujo, D.S., Ludermir, T.B., Schliep, A.: Clustering cancer gene expression data: a comparative study. BMC Bioinform. **9**(1), 497 (2008)

Learning MILP Resolution Outcomes
Before Reaching Time-Limit

Martina Fischetti[1], Andrea Lodi[2], and Giulia Zarpellon[2(✉)]

[1] Vattenfall, Kolding, Denmark
martina.fischetti@vattenfall.com
[2] Polytechnique Montréal, Montréal, Canada
{andrea.lodi,giulia.zarpellon}@polymtl.ca

Abstract. The resolution of some Mixed-Integer Linear Programming (MILP) problems still presents challenges for state-of-the-art optimization solvers and may require hours of computations, so that a time-limit to the resolution process is typically provided by a user. Nevertheless, it could be useful to get a sense of the optimization trends after only a fraction of the specified total time has passed, and ideally be able to tailor the use of the remaining resolution time accordingly, in a more strategic and flexible way. Looking at the evolution of a partial branch-and-bound tree for a MILP instance, developed up to a certain fraction of the time-limit, we aim to predict whether the problem will be solved to proven optimality before timing out. We exploit machine learning tools, and summarize the development and progress of a MILP resolution process to cast a prediction within a classification framework. Experiments on benchmark instances show that a valuable statistical pattern can indeed be learned during MILP resolution, with key predictive features reflecting the know-how and experience of field's practitioners.

Keywords: MILP resolution · Branch and Bound · Machine learning

1 Introduction

Within the realm of discrete optimization, we consider Mixed-Integer Linear Programming (MILP) problems, of the form

$$\min\{c^T x : Ax \geq b, x \geq 0, \ x_i \in \mathbb{Z} \ \forall i \in \mathcal{I}\}, \tag{1}$$

where $A \in \mathbb{R}^{m \times n}$, $b \in \mathbb{R}^m$, $c, x \in \mathbb{R}^n$ and $\mathcal{I} \subseteq \{1, \ldots, n\}$ is the set of indices of variables that are required to be integral. We do not assume A, b having any special structure (as it is, e.g., for Traveling Salesman Problem instances). Models like (1) can be used to mathematically describe a number of different real-world problems, and are daily deployed across a wide spectrum of applications – network, scheduling, planning and finance, just to mention a few.

Despite being $\mathcal{NP}$-hard problems, MILPs are nowadays solved in very reliable and effective ways, ultimately based on the divide-and-conquer paradigm

© Springer Nature Switzerland AG 2019
L.-M. Rousseau and K. Stergiou (Eds.): CPAIOR 2019, LNCS 11494, pp. 275–291, 2019.
https://doi.org/10.1007/978-3-030-19212-9_18

of Branch and Bound (B&B) [23]. State-of-the-art optimization solvers, such as IBM-CPLEX [10], experienced a dramatic performance improvement over the past decades, due to both hardware and software advances (see, e.g., [2,26]). Nonetheless, the resolution of some MILPs can prove to be challenging for solvers, and may require hours of computations, so that the experimental practice of imposing a time-limit (TL) to the MILP resolution process is not only very reasonable, but well established too. However, it would be useful to get a sense of the optimization trends after only a fraction of the specified TL has passed, and ideally be able to tailor the usage of the remaining resolution time in a more strategic and flexible way.

We aim to predict whether a generic MILP instance will be solved before timing out, only relying on information from a first portion of the resolution process. More specifically, given problem P and a time-limit TL, we look at the partial resolution of P, up to a certain time τ, $0 < \tau < TL$, and ask whether P will be solved to proven optimality within TL. We summarize the partial resolution of P, and exploit Machine Learning (ML) tools to cast a prediction about it being solved or not before TL. Thus, the prediction we aim at is one that takes as input (a summary of) the evolution of a partial MILP run, up to time τ, and outputs a yes/no response, in the framework of binary classification. Note the inherent difference between our approach and the problem of directly predicting the "difficulty" of a MILP instance – e.g., in terms of tree-size [3,8] or runtime prediction, the latter being a common interest for both the optimization and the ML communities since the work of Knuth [20] (a more recent approach can be found in [17]).

The sequential nature of B&B makes it natural to interpret our question as a sequence classification task. However, the transformation of a stream of data from the MILP resolution process into a valid input for traditional classification algorithms cannot be performed with off-the-shelf techniques [33]. To this end, we design specific features to describe the development and behavior of a MILP run in a quantitative way, taking into account the complex interplay between the solver's components. The broad generality of the proposed features makes them apt to be re-used every time one needs to evaluate the B&B development of a general MILP, thus conferring even more impact to this contribution, especially given that applications of ML to discrete optimization have lately been flourishing as recently surveyed in [4]. For example, in the context of MILP, ML has been proposed to establish good solver's parametric configurations [16]; learn heuristics for B&B (see [28] for a survey); choose resolution options ([6,18,22]), and also predict solution-related outcomes ([12,25]). Our work represents a novel contribution in this thread of research: ML is employed to provide an accurate prediction on the resolution outcome of MILPs, which can readily be implemented within solvers to enable tailored optimization and enhance the comprehension of the resolution process, too often hard to unravel given the solver's complexity. In fact, despite the abundance of data and events in the MILP resolution framework, to the best of our knowledge no statistical analysis presently happens within the solver; in particular, information is not exploited in any structural way via ML algorithms to make decisions.

Applying to generic MILP problems and opening new opportunities on the solvers' side, our results affect a broad audience and assume greater methodological relevance for the discrete optimization community.

We end the introduction by stressing that discovering early in the process that the run will very likely not terminate with a proof of optimality is of fundamental value for MILP development, and opens promising scenarios for both developers and end-users. Indeed, on the one hand, MILP developers can adapt the resolution through algorithmic changes in the attempt of avoiding the issue, or can switch mode so as to try to improve the incumbent solution as much as possible giving up optimality. On the other hand, this can be achieved by an end-user too, although that would likely require restarting the run with a different parameter setting. Finally, note that the indicators we developed could, in turn, shed some light on the type of required algorithmic changes.

2 Background: Solving MILPs

As already mentioned, the resolution of MILPs is fundamentally based on the B&B paradigm. In its basic version, B&B sequentially partitions the solution space of (1) into sub-MILPs, which are mapped into nodes of a binary decision tree. At each node, the integrality requirements $x_i \in \mathbb{Z}$ for variables $i \in \mathcal{I}$ are dropped, and a *linear*, or *continuous, relaxation* (polynomially tractable) of the sub-problem is solved, providing a valid lower bound to the optimal solution value of the original MILP. When in the relaxed solution all variables $x_i, i \in \mathcal{I}$ take integer values, the solution is feasible for (1) as well and provides an upper bound of its optimal value. Otherwise, variables $x_i \notin \mathbb{Z}, i \in \mathcal{I}$ are *integer infeasible* (iinf) and among them one is selected for further branching: the tree is extended with two additional child nodes so that the current relaxed solution is removed from the sub-problems' feasible space; the new nodes also inherit from their parent an estimate of the objective function value. Global lower and upper bounds (called *best bound* and *incumbent*, respectively) are maintained throughout the resolution process and smartly used to prune unpromising regions of the feasible space, so that the resulting algorithm is only implicitly enumerating the exponentially many solutions of (1). The normalized difference between global bounds (known as *gap*) allows to measure at any point in time the quality of a solution and the progress of the optimization. For example, CPLEX implements the following (relative) gap measure:

$$\text{gap} = \frac{|\text{best bound} - \text{incumbent}|}{1e^{-10} + |\text{incumbent}|}. \tag{2}$$

A MILP is solved when the gap is fully closed, i.e., when it reaches 0, with upper and lower bounds coinciding (up to numerical tolerances). The branching and bounding operations are combined with other solver's building blocks – the cutting planes algorithm [14], presolving, primal heuristics – to form a very rich and interconnected resolution framework [26], in which single events and data become hard to disentangle.

The ability to identify the resolution phases of a MILP [5] and analyze the outputs of the B&B algorithm can help recognizing causes of performance issues, and explaining instance-specific trends [19]. In particular, many indicators interact in describing the progress of the MILP resolution process, and need to be taken into account when casting a prediction about the resolution outcome. To provide a simple example, we plot in Fig. 1 basic information from the resolution log of CPLEX, for an "Easy" instance of MIPLIB2010 [21]. We report the development of the global bounds and the gap, the number of *nodes left* (i.e., the leaves yet to be explored) and the *depth* of the nodes as the algorithm traverses the tree. The interconnection between these figures is, for this easy case, quite clear to observe: for example, an update of the incumbent value naturally reduces the gap, triggers a drop in the number of nodes left (due to pruning by bound), and possibly ends a (depth-first) dive in the tree traversal exploration, a common practice when looking for initial feasible solutions with primal heuristics.

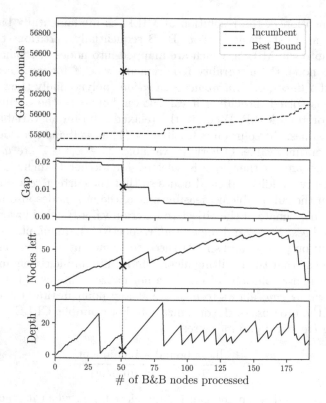

Fig. 1. Basic information from the CPLEX log from the resolution of MIPLIB2010 instance air04. Interpreting the evolution and interaction of these indicators enables a quantitative description of the optimization process.

3   Problem Formalization

We can re-phrase our question more formally by considering a MILP P, a time-limit TL, and a certain percentage ratio $\rho \in [0,1]$ yielding $\tau = \rho \cdot TL \in [0, TL]$. We solve problem P with time-limit TL and take into account the evolution of its resolution process up to time τ. We denote with t_{sol}^{P} the moment in which P is solved to proven optimality by the solver. We want to describe and evaluate the progress (in other words, the *"work done"*) in solving P, given that only a share of the total available time has passed; ultimately, we aim at casting a prediction on such a description. With respect to the defined parameters, we achieve 100% of work done at t_{sol}^{P}, and 100% of available time at TL. In practice, there is a discrepancy between t_{sol}^{P} and TL, the latter specified by a user, the former unknown and subject to variability.

Graphically, one could depict the advancement of the solver with a non-decreasing *"progress measure"*, describing the proportion of work done given the proportion of time passed (Fig. 2). Our classification question translates precisely into predicting whether the 100% of the work will be done before TL, i.e., whether $t_{sol}^{P} \leq TL$, only observing the resolution up to time τ. The function we aim to learn is thus the indicator function $\mathbf{1}_{\{t_{sol}^{P} \leq TL\}}$.

The task of feature design, on the other hand, aims at defining the progress measure used to represent the % of work done, given the triplet (TL, ρ, P). Instead of relying on a single feature to describe the optimization process (as could be done, e.g., using the gap), we try to capture the complexity of MILP resolution by considering heterogeneous measurements, and design a feature map Φ, describing the progress measure for (TL, ρ, P) with a vector in $\mathbb{R}^d$.

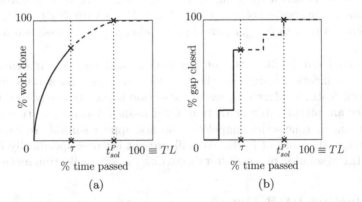

Fig. 2. (a) Graphical example of "progress measure" for a triplet (TL, ρ, P); we assume a smooth behavior for drawing purposes. The observed portion of the resolution (up to time τ) is drawn in solid. (b) If we were to measure the progress by looking at the % of gap closed only, we would draw a step-wise linear function.

3.1   Sequence Classification

The sequential character of B&B makes it natural to think about the partial resolution of P as a progressive stream of information and events. In the MILP context, it appears reasonable to discretize the time dimension by considering information being retrieved at every node of the B&B tree, starting from the root and up to the last one being processed before time τ (say η). In other words, one could describe the output of a MILP run with a multivariate time series $\mathcal{S}_{TL,\rho,P}$,

$$\mathcal{S}_{TL,\rho,P} = \big\{ (N^1, \langle v_1^1, \cdots, v_s^1 \rangle), \\ (N^2, \langle v_1^2, \cdots, v_s^2 \rangle), \\ \vdots \\ (N^\eta, \langle v_1^\eta, \cdots, v_s^\eta \rangle) \big\}, \tag{3}$$

a sequence of vectors $v^k \in \mathbb{R}^s$, each carrying information about the optimization state at node N^k, up to η.

Classifying $\mathcal{S}_{TL,\rho,P}$ depending on P's optimization outcome can be seen as a *(conventional) sequence classification* task. Sequence classification is typically employed in genomic applications, anomaly-detection and information retrieval (see, e.g., [11,24,32], respectively), and generally deals with learning a *sequence classifier* for data of sequential type. Few alternatives to tackle sequence classification can be found in the literature (see [33] for a brief survey). We opt for a feature-based approach: simply put, we transform the sequence $\mathcal{S}_{TL,\rho,P}$ into a single vector of numerical features $\Phi(TL, \rho, P) \in \mathbb{R}^d$, to which we will then apply traditional classification algorithms. In our setting, a data-point for the learning algorithm consists of a tuple $\big(\Phi(TL, \rho, P), y\big)$ with $\Phi(TL, \rho, P)$ describing the time series data $\mathcal{S}_{TL,\rho,P}$, and binary label $y \in \{0, 1\}$ assigned according to $\mathbf{1}_{\{t_{sol}^P \leq TL\}}$.

As pointed out in [33], one of the major challenges when dealing with sequence classification resides in the fact that sequence data does not come with explicit features. Moreover, feature selection is usually costly, and needs to account for an interpretable prediction. Off-the-shelf feature selection methods – like k-grams or time series shapelets – do not appear suitable to capture the special temporal nature of B&B. We will present features specifically designed for the MILP resolution process after discussing the data collection methodology.

4   Collecting B&B Data

As we said, the B&B framework produces a lot of heterogeneous information, whose combination can provide interesting insights about the optimization status of a MILP run. Extracting data from the resolution process is allowed by means of implementing custom callbacks in the solver's APIs, and comes with some computational overhead. From an application perspective, it seems reasonable that a user might be willing to spend some additional resources in the first part of the resolution process, say up to time τ, in order to get a prediction on the more lengthy horizon of TL. Nevertheless, especially in our setting, time is important: any appreciable overhead during the run could bias the yes/no response with respect to the fixed TL, so data collection has to be as cheap as possible.

In fact, the overhead we experienced comes from the computation of few indicators and the need to interface the solver through its API. For example, extracting the number of iinf variables at every branched node cannot be done with an API method directly, so that one needs to examine feasibility statuses for all variables. However, the same indicator would come almost for free if implemented internally, on the solver's side: the value of iinf at every node is available and systematically printed in the resolution log.

To comply with the need of collecting non-biased data – and certain that a data collection procedure implemented internally on the solver side would incur in much less overhead than that experienced by any user dealing with its interfaces – we devise a two-step proof-of-concept implementation. We use CPLEX 12.7.1 as solver, together with its Python API. Given (TL, ρ, P), we perform

1. *Label computation*: run P with time-limit TL, and determine a label for the run by checking if $t_{sol}^{P} \leq TL$. During the run record η^{P}, the number of nodes processed up to time τ.
2. *Data collection*: run again P (the deterministic run of Step 1 can be reproduced by setting the same random seed), and actively collect data during the optimization, up to η^{P} nodes.

Having detached data collection from label computation, we do not need to worry anymore about the overhead incurred in Step 2, nor about the integrity of the labeled data; the produced sequence $\mathcal{S}_{TL,\rho,P}$ records the real "work done" up to the sought fraction ρ of TL.

4.1 Producing Diversification

For fixed TL and ρ, a data-point corresponds to a single run of a problem P. The need of a reasonable amount of data for applying ML thus requires many MILP instances – definitely more than those currently part of MILP libraries (see Sect. 6). Instead of resorting to random problems generation, we try to create additional data from existing benchmark instances.

A first general diversification of data from the same problem P can be produced exploiting the so-called *performance variability* of MILPs [27]. Perturbations can be obtained simply by setting different random seeds in the solver, to obtain diverse runs of P. Other diversification schemes, specific to our setting, consist in varying the main parameters TL and ρ. In particular, one could (*i*) vary TL and keep ρ fixed, and/or (*ii*) vary ρ and keep TL fixed. Intuitively, approach (*i*) seems more promising at generating heterogeneous points: a change of

TL allows for a sensible re-scaling of τ as well, potentially producing data labeled differently, despite coming from the same problem P. We graphically describe this intuition in Fig. 3.

Having discussed how to produce and collect valuable MILP time series data, we now turn to the task of handling it, in order to craft a vector of features.

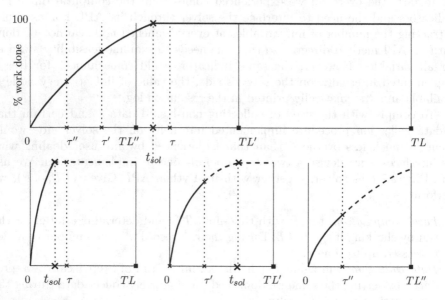

Fig. 3. Graphical example of approach (i) to obtain multiple data-points from fixed (ρ, P), varying TL. The run of P is represented at the top. Below, using additional TL', TL'' we get $\big(\Phi(TL, \rho, P), 1\big)$, $\big(\Phi(TL', \rho', P), 1\big)$ and $\big(\Phi(TL'', \rho'', P), 0\big)$. The observed portions of the resolution process are drawn in solid.

5   Feature Design

We undertake a feature-based approach for sequence classification, and transform MILP sequential data $\mathcal{S}_{TL,\rho,P}$ into a single vector of features $\Phi(TL, \rho, P) \in \mathbb{R}^d$, to be fed as input to traditional classification algorithms. As already mentioned, feature selection is not a straightforward process when dealing with serial data, especially if one wants to retain a certain degree of interpretability. We rely on MILP domain-knowledge to define features that shall encompass the optimization progress encoded in $\mathcal{S}_{TL,\rho,P}$.

In practice, we extract 25 raw numerical attributes from each callback call during Step 2 of our data collection procedure, i.e., each vector v^k of $\mathcal{S}_{TL,\rho,P}$ has dimension 25. Note, however, that the length η of the series varies considerably across instances and seeds, ranging between a few dozens and hundreds of thousands. At each branched node of the tree we collect information about the general state of the optimization (e.g., gap, value of incumbent and best bound, total number of processed nodes and count of simplex iterations performed),

Table 1. Description of the 37 features employed for learning experiments.

#	Group name	Features general description
7	Last observed global measures	Gap; ratio between best bound and incumbent; fraction of nodes left attaining max (resp. min) objective estimate; ratio between max (resp. min) estimate across nodes left and incumbent; primal-dual integral [1]
4	Nodes left and pruned, iterations count	Throughput of pruned nodes; ratio between nodes pruned and nodes left; last measure of nodes left over max observed one; throughput of simplex iterations
4	Node LP integer infeasibilities (iinf)	Max (resp. min, avg) number of observed iinf over $|\mathcal{I}|$; fraction of nodes with iinf below 5% quantile value
5	Incumbent	Throughput of incumbent updates (i.e., frequency); average improvement (resp. distance) of updates normalized by incumbent value (resp. total # of nodes); distance from last observed update over the average one; was an incumbent found before an integer feasible node (boolean)?
4	Best bound	Throughput of best bound updates (i.e., frequency); average improvement (resp. distance) of updates normalized by best bound value (resp. total # of nodes); distance from last observed update over the average one
3	Node LP objective	Fraction of nodes with objective above the 95% quantile value; differences in absolute value between quantile threshold and global bounds
4	Node LP fixed variables	Fraction of max (resp. min) observed # of fixed variables; fraction of nodes with # of fixed variables above 95% quantile value; distance from last observed peak over total # of nodes
6	Depth and tree traversal	Ratio between max observed depth and # of processed nodes; ratio between height of last full level (resp. waist) and maximal depth [8]; maximal and average length of backtracks; frequency of backtracks in the traversal

together with node-specific data (e.g., current node LP objective value, number of iinf in the LP solution, node depth). At few points in time, we extract information about the list of nodes left (e.g., its length, the maximum and minimum objective estimates, and the number of nodes attaining them). Data traditionally reported in the solver's log are included in these 25 attributes.

Let us point out a few remarks on the nature of the extracted B&B data, and on the guidelines that should be observed to transform them into MILP "progress measures".

1. Some pieces of raw information already describe the *global* optimization state, and can be considered in all respects as "progress measures" for the MILP resolution. An example in this sense is provided by the *gap* measure: the last datum collected about the gap refers to the entire resolution process up to that point, and can be used directly as feature in $\Phi(TL, \rho, P)$.

2. Some other information is instead *local*, referring to a particular node LP, and need to be embedded and interpreted within a broader and global context. For example, a single datum about the *depth* of a node is not informative of the tree evolution, but combined depth data can provide indications about the tree profile (e.g., in terms of maximal depth, width and full levels, see [8]), as well as describe dives and backtracks in the traversal.

3. Some traits are global (in the sense that they refer to the totality of the optimization process), but are not significant if taken individually. This is the case, for example, of data about the global bounds values, which present themselves as a crude sequence of decreasing (or increasing) scalar values. Measuring their development and changes, instead, can be more informative of the optimization progress.

4. Finally, the wide range of MILP benchmark instances requires features to be comparable across the dataset. For example, exact values linked to parameters $(c, A, b, |\mathcal{I}|)$ and solutions should be avoided. Global counters, e.g., the number of processed nodes, should be used to rescale other indicators, in order not to affect the learning process (and subsequent data normalizations) with data of different magnitudes.

With these guidelines in mind, by means of combining different raw indicators with each other and interpreting them from a development perspective, we design (and select) 37 features to represent the MILP progress. We describe the features set in Table 1.

Besides the canonical use of statistical functions (like max, min, average) to synthesize some serial information, and the use of *throughputs* measures (e.g., to infer the rates at which nodes are processed and pruned), we apply our domain-knowledge to summarize the optimization progress. For example, we tackle measures that can vary significantly even between consecutive nodes in the B&B tree, but for which we are interested in localizing extreme behaviors only, by employing quantile values as statistically meaningful thresholds. We use them to track peaks for values of node LP *objective*, number of *iinf* and number of *fixed variables*. Instead, for data that is updating throughout the optimization process (e.g., for *incumbent* and *best bound* values), we focus on interpreting their changes in time, deduce how often and how distant are updates happening, and what is their average improvement.

6   Experimental Results

Dataset Composition and Setup. We employ instances of MIPLIB2010 [21] and [30] for our experiments. An assessment of the distribution of solving times

Table 2. Dataset composition in terms of labels and original MILP libraries.

	Class 0	Class 1	Total (%)
Benchmark78	106	405	511 (52.7)
Challenge160	219	6	225 (23.2)
Mittelmann48	9	225	234 (24.1)
Total (%)	334 (34.4)	636 (65.6)	**970**

seemed necessary in order to produce a balanced and meaningful dataset. Evaluation runs with 10 different seeds on the MIPLIB2010 *Benchmark* set suggested the use of $TL \in \{3600, 2400, 1200\}$ seconds. A projection of the resulting labels distribution was performed, to select $\rho = 0.2$ (i.e., we stop the observation after 20% of TL).

To build our dataset, we collect B&B data from the following MILP problems:

– Benchmark78: 78 instances from MIPLIB2010 *Benchmark* set (problems belonging to *Infeasible* and *Primal* subsets are removed, since they do not appear meaningful for our question);
– Challenge160: 160 problems from MIPLIB2010 *Challenge* set (with *Infeasible* and *Primal* removed);
– Mittelmann48: 48 instances from Mittelmann *MILPlib* collection [30].

Problems in Benchmark78 and Mittelmann48 are solved with three different random seeds, while those in Challenge160 with a single one. As expected, Mittelmann48 runs are very short, with few cases of time-limiting problems. Counterbalancing this effect, the majority of instances in Challenge160 cannot be solved within 1 h time-limit; Benchmark78 run times are distributed more evenly. All MILP runs were performed on a cluster of 640 48-cores machines, each equipped with a 2.1 GHz Intel Platinum 8160F "Skylake" processor and 192 GB of RAM. Apart from time-limit specifications, we do not modify the solver's default setting; in particular, we leave in place CPLEX default presolve, cuts and primal heuristics.

The heterogeneity of the collected time series data makes necessary a thorough phase of data cleaning and scaling. We discard troublesome runs to get 1315 data-points, which then reduce to 970 after computing the hand-crafted features and performing basic data cleaning (data with missing values are removed). Note that a single MILP problem can generate up to 9 different data-points, given the variations in seeds and time-limits used. In the final dataset of 970 points, Class 1 (Class 0) represents the 65.6% (34.4%) of the total; a snapshot of the dataset composition is given in Table 2.

Train and Test Splits. In order to account for the different composition of MILP libraries and the role of performance variability, we define and try three different ways of splitting our data into training and test set.

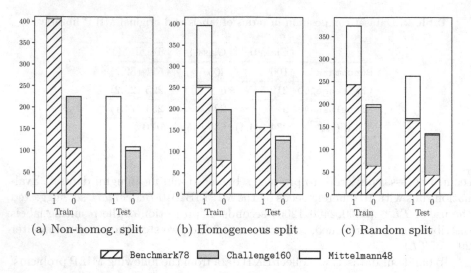

(a) Non-homog. split   (b) Homogeneous split   (c) Random split

Benchmark78   Challenge160   Mittelmann48

Fig. 4. Training and test set composition with respect to different labels and MILP libraries, reported for the three considered train-test splits.

1. *Non-homogeneous* split: data-points from `Benchmark78` are used for training, while those from `Mittelmann48` for test; data from `Challenge160` is divided between train and test, taking care of keeping together points arising from the same MILP instance.
2. *Homogeneous* split: both training and test sets are built using a share of each dataset. Again, points arising from the same instance are kept together.
3. *Random* split: data from all runs are mixed together and randomly split. In this case, points that originated from the same MILP instance can appear in both training and test sets.

Proportions between training and test set are roughly maintained around a 60%–40% repartition, with slight variations across splits. Figure 4 illustrates the datasets composition in more detail.

6.1   Learning Experiments

We train and test five different learning models, namely, Logistic Regression (LR), Support Vector Machines (SVM) with RBF kernel [9], Random Forest (RF) [7], Extremely Randomized Trees (ExT) [13], and Multi-Layer Perceptron (MLP) [15]. All algorithms are compared against a dummy classifier (dum) following a stratified strategy, i.e., predicting by respecting the class distribution. The learning phase is implemented entirely in Python with Scikit-learn [31], and run on a PC with Intel Core i5, 2.3 GHz and 8 GB of memory. Each feature is normalized to have a mean of 0 and a standard deviation of 1, and each experiment comprises a training phase with 3-fold cross validation to grid-search hyper-parameters, and a test phase on the neutral test set.

Table 3. Classification results for the three considered train-test split settings; measures are rounded to the second decimal. Best scores and classifiers are bold-faced.

	Accuracy	*Precision*	*Recall*	*F1-score*	*Accuracy*	*Precision*	*Recall*	*F1-score*	*Accuracy*	*Precision*	*Recall*	*F1-score*
dum	0.55	0.56	0.55	0.56	0.59	0.58	0.59	0.59	0.57	0.57	0.57	0.57
LR	0.94	0.94	0.94	0.94	0.90	0.91	0.90	0.90	0.93	0.93	0.93	0.93
SVM	0.94	0.94	0.94	0.94	0.91	0.91	0.91	0.91	0.94	0.94	0.94	0.94
RF	**0.96**	**0.96**	0.96	**0.96**	0.94	0.94	0.94	0.94	**0.94**	**0.95**	**0.94**	**0.94**
ExT	0.96	0.96	**0.96**	0.96	**0.95**	**0.95**	**0.95**	**0.95**	0.93	0.94	0.93	0.93
MLP	0.91	0.92	0.91	0.90	0.86	0.86	0.86	0.85	0.93	0.93	0.93	0.93
	(a) Non-homogeneous split				(b) Homogeneous split				(c) Random split			

Table 4. Confusion matrices (w/o normalization) for RF in different split settings. Note that support sizes are varying.

		Predicted					Predicted					Predicted		
		0	1	support			0	1	support			0	1	support
True	0	100	9	109	True	0	125	11	136	True	0	130	5	135
	1	3	222	225		1	12	228	240		1	18	244	262
	(a) Non-homogeneous				(b) Homogeneous				(c) Random			

Results. Table 3 reports the standard performance measures for binary classification: for all classifiers we compare accuracy, precision, recall and f1-score, the last three metrics averaged between classes and weighted by supports. Overall, RF and ExT are the best performing models, with SVM following close behind. We additionally report confusion matrices for RF in Table 4. The high accuracy scores obtained in all three train-test settings attest that there is indeed a statistical pattern to be learned during MILP resolution, and that the designed features are capturing it.

Taking a closer look at class-specific precision and recall scores, we note distinct behaviors with respect to different train-test splits. In particular, models in the *Non-homogeneous* case present a sensitivity (i.e., recall for Class 1) being higher than specificity, accompanied by high precision for Class 0. The trend is much less accentuated in the *Homogeneous* setting, and blurs completely (if not reverses itself) in the *Random* one. An explanation of these behaviors could be linked to the intrinsic difference in composition of the MILP libraries employed for the experiments. In fact, instances in `Benchmark78` do not exhibit clear-cut behaviors as those in `Challenge160` and `Mittelmann48`. Finally, the fact of *Random* being the setting in which MLP is best performing might be a sign of the model being able to recognize akin data-points arising from the same instance (now scattered in both training and test set), and thus linked to the presence of problems with low variability scores.

Table 5. Subset of features appearing in the top-10s for RF: scores are averaged among split cases; features marked with ∗ appear in the top-10 of each setting.

Rank	Score (avg)	Feature description
1	0.1856	∗ Throughput of pruned nodes (over total # of processed ones)
2	0.1839	∗ Ratio between pruned nodes and last measured number of nodes left
3	0.0805	∗ Last measured number of nodes left over maximal number of nodes left observed
4	0.0758	∗ Fraction of nodes left attaining max objective estimate
5	0.0632	∗ Fraction of nodes left attaining min objective estimate
6	0.0622	∗ Frequency of backtracks
7	0.0453	∗ Throughput of best bound updates
8	0.0324	∗ Last measured gap
9	0.0196	Ratio between last measured best bound and incumbent
10	0.0181	Maximal length of observed backtracks
11	0.0165	Difference in absolute value between objective 5% quantile threshold and best bound
12	0.0164	Distance from last observed best bound update over the average one

Feature Analysis. Our best performing methods, RF and ExT, have the advantage of interpretability. We employ feature scores returned by Scikit-learn, measuring the mean decrease in impurity [29], to provide a first evaluation of those factors that proved valuable for the predictions. We look at the sets of top-10 scoring features for RF, for each train-test split case, and note a very stable scoring pattern: 8 features appear in the top-10 of each setting, and a total of 12 different features covers the three top rankings. We report them in Table 5, where scores have been averaged among cases. In particular, throughputs and trends of nodes pruned, processed and left seem to be crucial for proper classification. Information on the proportions of nodes attaining maximum and minimum objective estimates within the list of nodes left is also valuable. Indeed, such estimates at the frontier of the B&B tree are somehow quantifying the amount of work to be done to close the upper and lower bounds in the remaining subtrees, and hence measuring the "difficulty" of what is yet to be explored. Together with the gap, few top-ranked features focus on dives and backtracks happened during the traversal, while few others on best bound updates. Note that, despite having provided the same set of features to capture updates of incumbent and best bound, only those relative to the latter are top-ranked by the algorithm. This is in line with the composition of MILP benchmarking libraries and the experience of MILP practitioners, who often witness slow B&B searches due to difficulty in improving the LP (dual) bound.

7 Conclusions and Outlook

We propose a learning approach to predict the outcome of a general MILP problem after only a share of the available computing time has passed. We summarize the sequential MILP resolution process with hand-crafted features, and successfully classify it with traditional learning models. In particular, our novel features can be applied to any type of MILP instance, and hence used in future application of ML for B&B studies, making this work of interest for a wide audience. Our positive results show that there is indeed a pattern to be learned across MILP instances, and represent (to the best of our knowledge) the first structural statistical use of the data provided by the solver throughout the resolution. The proposed framework could be readily implemented internally on the solver side, in order to strategically specialize the optimization process on the fly, before timing out, providing better options for the user. In other words, an early detection of a potential time out can trigger algorithmic changes that, in turn, could prevent such a time out to happen. The developed setting can be extended in a number of different directions. We plan to deepen data analysis – possibly augmenting our dataset – and frame the role of performance variability in the learning process. It would be interesting to consider other ways to tackle sequence classification, e.g., by following a pattern-based approach.

References

1. Achterberg, T., Berthold, T., Hendel, G.: Rounding and propagation heuristics for mixed integer programming. In: Klatte, D., Lüthi, H.J., Schmedders, K. (eds.) Operations Research Proceedings 2011, pp. 71–76. Springer, Heidelberg (2012). https://doi.org/10.1007/978-3-642-29210-1_12
2. Achterberg, T., Wunderling, R.: Mixed integer programming: analyzing 12 years of progress. In: Jünger, M., Reinelt, G. (eds.) Facets of Combinatorial Optimization, pp. 449–481. Springer, Heidelberg (2013). https://doi.org/10.1007/978-3-642-38189-8_18
3. Belov, G., Esler, S., Fernando, D., Bodic, P.L., Nemhauser, G.L.: Estimating the size of search trees by sampling with domain knowledge. In: Proceedings of the Twenty-Sixth International Joint Conference on Artificial Intelligence, IJCAI 2017, pp. 473–479 (2017). https://doi.org/10.24963/ijcai.2017/67
4. Bengio, Y., Lodi, A., Prouvost, A.: Machine learning for combinatorial optimization: a methodological tour d'horizon (2018). Preprint: arXiv:1811.06128
5. Berthold, T., Hendel, G., Koch, T.: From feasibility to improvement to proof: three phases of solving mixed-integer programs. Optim. Methods Softw. 33(3), 499–517 (2017). https://doi.org/10.1080/10556788.2017.1392519
6. Bonami, P., Lodi, A., Zarpellon, G.: Learning a classification of mixed-integer quadratic programming problems. In: van Hoeve, W.-J. (ed.) CPAIOR 2018. LNCS, vol. 10848, pp. 595–604. Springer, Cham (2018). https://doi.org/10.1007/978-3-319-93031-2_43
7. Breiman, L.: Random forests. Mach. Learn. 45(1), 5–32 (2001). https://doi.org/10.1023/A:1010933404324

8. Cornuéjols, G., Karamanov, M., Li, Y.: Early estimates of the size of branch-and-bound trees. INFORMS J. Comput. **18**(1), 86–96 (2006). https://doi.org/10.1287/ijoc.1040.0107
9. Cortes, C., Vapnik, V.: Support-vector networks. Mach. Learn. **20**(3), 273–297 (1995). https://doi.org/10.1023/A:1022627411411
10. CPLEX.   http://www-01.ibm.com/software/commerce/optimization/cplex-optimizer/index.html. Accessed 2018
11. Deshpande, M., Karypis, G.: Evaluation of techniques for classifying biological sequences. In: Chen, M.-S., Yu, P.S., Liu, B. (eds.) PAKDD 2002. LNCS (LNAI), vol. 2336, pp. 417–431. Springer, Heidelberg (2002). https://doi.org/10.1007/3-540-47887-6_41
12. Fischetti, M., Fraccaro, M.: Using OR + AI to predict the optimal production of offshore wind parks: a preliminary study. In: Sforza, A., Sterle, C. (eds.) Optimization and Decision Science: Methodologies and Applications. Springer Proceedings in Mathematics & Statistics, vol. 217, pp. 203–211. Springer, Cham (2017). https://doi.org/10.1007/978-3-319-67308-0_21
13. Geurts, P., Ernst, D., Wehenkel, L.: Extremely randomized trees. Mach. Learn. **63**(1), 3–42 (2006). https://doi.org/10.1007/s10994-006-6226-1
14. Gomory, R.: An algorithm for the mixed integer problem. Technical report RM-2597, The Rand Corporation (1960)
15. Goodfellow, I., Bengio, Y., Courville, A.: Deep Learning. MIT Press, Cambridge (2016). http://www.deeplearningbook.org
16. Hutter, F., Hoos, H.H., Leyton-Brown, K.: Automated configuration of mixed integer programming solvers. In: Lodi, A., Milano, M., Toth, P. (eds.) CPAIOR 2010. LNCS, vol. 6140, pp. 186–202. Springer, Heidelberg (2010). https://doi.org/10.1007/978-3-642-13520-0_23
17. Hutter, F., Xu, L., Hoos, H.H., Leyton-Brown, K.: Algorithm runtime prediction: methods & evaluation. Artif. Intell. **206**, 79–111 (2014). https://doi.org/10.1016/j.artint.2013.10.003
18. Khalil, E.B., Dilkina, B., Nemhauser, G., Ahmed, S., Shao, Y.: Learning to run heuristics in tree search. In: 26th International Joint Conference on Artificial Intelligence (IJCAI) (2017)
19. Klotz, E., Newman, A.M.: Practical guidelines for solving difficult mixed integer linear programs. Surv. Oper. Res. Manag. Sci. **18**(1), 18–32 (2013)
20. Knuth, D.E.: Estimating the efficiency of backtrack programs. Math. Comput. **29**(129), 122–136 (1975)
21. Koch, T., et al.: MIPLIB 2010. Math. Program. Comput. **3**(2), 103–163 (2011). https://doi.org/10.1007/s12532-011-0025-9
22. Kruber, M., Lübbecke, M.E., Parmentier, A.: Learning when to use a decomposition. In: Salvagnin, D., Lombardi, M. (eds.) CPAIOR 2017. LNCS, vol. 10335, pp. 202–210. Springer, Cham (2017). https://doi.org/10.1007/978-3-319-59776-8_16
23. Land, A., Doig, A.: An automatic method of solving discrete programming problems. Econometrica **28**, 497–520 (1960)
24. Lane, T., Brodley, C.E.: Temporal sequence learning and data reduction for anomaly detection. ACM Trans. Inf. Syst. Secur. **2**(3), 295–331 (1999). https://doi.org/10.1145/322510.322526
25. Larsen, E., Lachapelle, S., Bengio, Y., Frejinger, E., Lacoste-Julien, S., Lodi, A.: Predicting solution summaries to integer linear programs under imperfect information with machine learning (2018). Preprint: arXiv:1807.11876

26. Lodi, A.: Mixed integer programming computation. In: Jünger, M., et al. (eds.) 50 Years of Integer Programming 1958–2008, pp. 619–645. Springer, Heidelberg (2009). https://doi.org/10.1007/978-3-540-68279-0_16

27. Lodi, A., Tramontani, A.: Performance variability in mixed-integer programming, Chap. 1, pp. 1–12. INFORMS (2013). https://doi.org/10.1287/educ.2013.0112

28. Lodi, A., Zarpellon, G.: On learning and branching: a survey. TOP **25**(2), 207–236 (2017). https://doi.org/10.1007/s11750-017-0451-6

29. Louppe, G.: Understanding random forests: from theory to practice. Ph.D. thesis, October 2014. https://doi.org/10.13140/2.1.1570.5928

30. Mittelmann, H.D.: MILPlib (2018). http://plato.asu.edu/ftp/milp/. Accessed 2018

31. Pedregosa, F., et al.: Scikit-learn: machine learning in Python. J. Mach. Learn. Res. **12**, 2825–2830 (2011)

32. Sebastiani, F.: Machine learning in automated text categorization. ACM Comput. Surv. **34**(1), 1–47 (2002). https://doi.org/10.1145/505282.505283

33. Xing, Z., Pei, J., Keogh, E.: A brief survey on sequence classification. ACM SIGKDD Explor. Newsl. **12**(1), 40–48 (2010). https://doi.org/10.1145/1882471.1882478

An Improved Subsumption Testing Algorithm for the Optimal-Size Sorting Network Problem

Cristian Frăsinaru[(✉)] and Mădălina Răschip

"Alexandru Ioan Cuza" University, General Berthelot, 16, Iaşi, Romania
{acf,mionita}@info.uaic.ro,
http://www.uaic.ro

Abstract. In this paper a new method for checking the subsumption relation for the optimal-size sorting network problem is described. The new approach is based on creating a bipartite graph and modelling the subsumption test as the problem of enumerating all perfect matchings in this graph. Experiments showed significant improvements over the previous approaches when considering the number of subsumption checks and the time needed to find optimal-size sorting networks. We were able to generate all the complete sets of filters for comparator networks with 9 channels, confirming that the 25-comparators sorting network is optimal. The running time was reduced more than 10 times, compared to the state-of-the-art result described in [6].

Keywords: Comparator network · Optimal-size sorting network · Subsumption relation · Perfect matching

1 Introduction

Sorting networks are a special class of sorting algorithms with an active research area since the 1950's [2,3,10]. A sorting network is a comparison network which for every input sequence produces a monotonically increasing output. Since the sequence of comparators does not depend on the input, the network represents an oblivious sorting algorithm. Such networks are suitable in parallel implementations of sorting, being applied in graphics processing units [9] and multiprocessor computers [3].

Over time, the research was focused on finding the optimal sorting networks relative to their size or depth. When the size is considered, the network must have a minimal number of comparators, while for the second objective a minimal number of layers is required. In [1] a construction method for sorting network of size $O(n \log n)$ and depth $O(\log n)$ is given. This algorithm has good results in theory but it is inefficient in practice because of the large constants hidden in the big-O notation. On the other side, the simple algorithm from [3] which constructs networks of depth $O(\log^2 n)$ has good results for practical values of n.

© Springer Nature Switzerland AG 2019
L.-M. Rousseau and K. Stergiou (Eds.): CPAIOR 2019, LNCS 11494, pp. 292–303, 2019.
https://doi.org/10.1007/978-3-030-19212-9_19

Because optimal sorting networks for small number of inputs can be used to construct efficient larger networks, the research in the area focused in the last years on finding such small networks. Optimal-size and optimal-depth networks are known for $n \leq 8$ [10]. In [12] the optimal-depth sorting networks were provided for $n = 9$ and $n = 10$. The results were extended for $11 \leq n \leq 16$ in [4]. The approaches use search with pruning based on symmetries on the first layers. The last results for parallel sorting networks are for 17 to 20 inputs and are given in [5,8]. On the other side, the paper [6] proved the optimality in size for the case $n = 9$ and $n = 10$. The proof is based on exploiting symmetries in sorting networks and on encoding the problem as a satisfiability problem. The use of powerful modern SAT solvers to generate optimal sorting networks is also investigated in [11]. Other recent results can be found in [7], where a revised technique to generate, modulo symmetry, the set of saturated two-layer comparator networks is given. Finding the minimum number of comparators for $n > 10$ is still an open problem. In this paper, we consider the optimal-size sorting networks problem.

Heuristic approaches were also considered in the literature, for example approaches based on evolutionary algorithms [15] that are able to discover new minimal networks for up to 22 inputs, but these methods cannot prove their optimality.

One of the most important and expensive operation used in [6] is the subsumption testing. This paper presents a new better approach to implement this operation based on matchings in bipartite graphs. The results show that the new approach makes the problem more tractable by scaling it to larger inputs.

The paper is organized as follows. Section 2 describes the basic concepts needed to define the optimal-size sorting-network problem and a new model of the subsumption problem. Section 3 presents the problem of finding the minimal-size sorting network. Section 4 discusses the subsumption problem while Sect. 5 the subsumption testing. Section 6 presents the new way of subsumption testing by enumerating all perfect matchings. Section 7 describes the experiments made to evaluate the approach and presents the results.

2   Basic Concepts

A *comparator network* $C_{n,k}$ with n *channels* (also called *wires*) and *size* k is a sequence of *comparators* $c_1 = (i_1, j_1); \ldots; c_k = (i_k; j_k)$ where each comparator c_t specifies a pair of channels $1 \leq i_t < j_t \leq n$. We simply denote by C_n a comparator network with n channels, whenever the size of the network is not significant in a certain context.

Graphically, a comparator network may be represented as a Knuth diagram [10]. A channel is depicted as a horizontal line and a comparator as a vertical segment connecting two channels (Fig. 1).

An *input* to a comparator network C_n may be any sequence of n objects taken from a totally ordered set, for instance elements in $\mathbb{Z}^n$. Let $\bar{x} = (x_1, \ldots, x_n)$ be an input sequence. Each value x_i is assigned to the channel i and it will "traverse" the comparator network from left to right. Whenever the values on

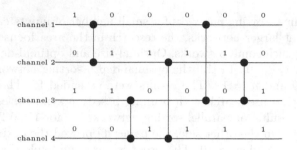

Fig. 1. The sorting network $C = (1,2); (3,4); (2,4); (1,3); (2,3)$, having 4 channels and 5 comparators, operating on the input sequence 1010. The output sequence is 0011.

two channels reach a comparator $c = (i,j)$ the following happens: if they are not in ascending order the comparator permutes the values (x_i, x_j), otherwise the values will pass through the comparator unmodified. Therefore, the *output* of a comparator network is always a permutation of the input. If $\overline{x}$ is an input sequence, we denote by $C(\overline{x})$ the output sequence of the network C.

A comparator network is called a *sorting network* if its output is sorted ascending for every possible input.

The *zero-one principle* [10] states that if a comparator network C_n sorts correctly all 2^n sequences of zero and one, then it is a sorting network. Hence, without loss of generality, from now on we consider only comparator networks with binary input sequences. In order to increase readability, whenever we represent a binary sequence we only write its bits; so 1010 is actually the sequence $(1,0,1,0)$.

The *output set* of a comparator network is $outputs(C) = \{C(\overline{x}) | \forall \overline{x} \in \{0,1\}^n\}$. Let $\overline{x}$ be a binary input sequence of length n. We make the following notations: $zeros(\overline{x}) = \{1 \leq i \leq n | x_i = 0\}$ and $ones(\overline{x}) = \{1 \leq i \leq n | x_i = 1\}$. The output set of a comparator network C_n can be partitioned into $n+1$ *clusters*, each cluster containing sequences in $outputs(C)$ having the same number of ones. We denote by $cluster(C,p)$ the cluster containing all sequences having p ones: $cluster(C,p) = \{\overline{x} \in outputs(C) \mid |ones(\overline{x})| = p\}$.

Consider the following simple network $C = (1,2); (3,4)$. The output clusters of C are: $cluster(C,0) = \{0000\}$, $cluster(C,1) = \{0001, 0100\}$, $cluster(C,2) = \{0011, 0101, 1100\}$, $cluster(C,3) = \{0111, 1101\}$, $cluster(C,4) = \{1111\}$.

The following proposition states some simple observations regarding the output set and its clusters.

Proposition 1. *Let C be a comparator network having n channels.*

(a) C is the empty network $\Leftrightarrow |outputs(C)| = 2^n$.
(b) C is a sorting network $\Leftrightarrow |outputs(C)| = n+1$ (each cluster contains exactly one element).
(c) $|cluster(C,p)| \leq \binom{n}{p}$, $1 \leq p \leq n-1$.
(d) $|cluster(C,0)| = |cluster(C,n)| = 1$.

We extend the *zeros* and *ones* notations to output clusters in the following manner. Let C be a comparator network. For all $0 \leq p \leq n$ we denote

$zeros(C,p) = \bigcup\{zeros(\overline{x})|\overline{x} \in cluster(C,p)\}$ and $ones(C,p) = \bigcup\{ones(\overline{x})|\overline{x} \in cluster(C,p)\}$. These sets contain all the positions between 1 and n for which there is at least one sequence in the cluster having a zero and a one at that position, respectively. Considering the clusters from the previous example, we have: $zeros(C,0) = zeros(C,1) = zeros(C,2) = \{1,2,3,4\}$, $zeros(C,3) = \{1,3\}$, $zeros(C,4) = \emptyset$, $ones(C,0) = \emptyset$, $ones(C,1) = \{2,4\}$, $ones(C,2) = ones(C,3) = ones(C,4) = \{1,2,3,4\}$.

We introduce the following equivalent representation of the $zeros$ and $ones$ sets, as a sequence of length n, where n is the number of channels of the network, and elements taken from the set $\{0,1\}$. Let Γ be a cluster:

- $\overline{zeros}(\Gamma) = (\gamma_1, \ldots, \gamma_n)$, where $\gamma_i = 0$ if $i \in zeros(\Gamma)$, otherwise $\gamma_i = 1$,
- $\overline{ones}(\Gamma) = (\gamma'_1, \ldots, \gamma'_n)$, where $\gamma'_i = 1$ if $i \in ones(\Gamma)$, otherwise $\gamma'_i = 0$.

In order to increase readability, we will depict 1 values in $\overline{zeros}$, respectively 0 values in $\overline{ones}$ with the symbol $-$. Considering again the previous example, we have: $\overline{zeros}(C,3) = (0-0-)$ and $\overline{ones}(C,1) = (-1-1)$.

If C is a comparator network on n channels and $1 \leq i < j \leq n$ we denote by $C;(i,j)$ the *concatenation* of C and (i,j), i.e. the network that has all the comparators of C and in addition a new comparator connecting channels i and j. The concatenation of two networks C and C' having the same number of channels is denoted by $C;C'$ and it is defined as the sequence of all comparators in C and C', first the ones in C and then the ones in C'. In this context, C represents a *prefix* of the network $C;C'$. Obviously, $size(C;C') = size(C) + size(C')$.

Let π be a permutation on $\{1, \ldots, n\}$. Applying π on a comparator network $C = (i_1,j_1); \ldots; (i_k,j_k)$ will produce the *generalized* network $\pi(C) = (\pi(i_1), \pi(j_1)); \ldots; (\pi(i_k), \pi(j_k))$. It is called generalized because it may contain comparators (i,j) with $i > j$, which does not conform to the actual definition of a standard comparator network. An important result in the context of analyzing sorting networks (exercise 5.3.4.16 in [10]) states that a generalized sorting network can always be *untangled* such that the result is a standard sorting network of the same size. The untangling algorithm is described in the previously mentioned exercise. Two networks C_a and C_b are called *equivalent* if there is a permutation π such that untangling $\pi(C_b)$ results in C_a.

Applying a permutation π on a binary sequence $\overline{x} = (x_1, \ldots, x_n)$ will permute the corresponding values: $\pi(\overline{x}) = (x_{\pi(1)}, \ldots, x_{\pi(n)})$. Applying π on a set of sequences S (either a cluster or the whole output set) will permute the values of all the sequences in the set: $\pi(S) = \{\pi(\overline{x})|\forall \overline{x} \in S\}$. For example, consider the permutation $\pi = (4,3,2,1)$ and the set of sequences $S = \{0011, 0101, 1100\}$. Then, $\pi(S) = \{1100, 1010, 0011\}$

3 Optimal-Size Sorting Networks

The *optimal size problem* regarding sorting networks is: "Given a positive integer n, what is the minimum number of comparators s_n needed to create a sorting network on n channels?".

Since even the problem of verifying whether a comparator network is a sorting network is known to be Co-$\mathcal{NP}$ complete [13], we cannot expect to design an algorithm that will easily answer the optimal size problem.

In order to prove that $s_n \leq k$, for some k, it is enough to find a sorting network of size k. On the other hand, to show that $s_n > k$ one should prove that no network on n channels having at most k comparators is a sorting network.

Let R_k^n denote the set of all comparator networks having n channels and k comparators. The naive approach to identify the sorting networks is by generating the whole set R_k^n, starting with the empty network and adding all possible comparators. In order to find a sorting network on n channels of size k, one could iterate through the set R_k^n and inspect the output set of each network. According to Proposition 1(b), if the size of the output is $n+1$ then we have found a sorting network. If no sorting network is found, we have established that $s_n > k$.

Unfortunately, the size of R_k^n grows rapidly since $|R_k^n| = (n(n-1)/2)^k$ and constructing the whole set R_k^n is impracticable even for small values of n and k.

We are actually interested in creating a set of networks N_k^n that does not include all possible networks but contains only "relevant" elements.

Definition 1. A complete set of filters [6] is a set N_k^n of comparator networks on n channels and of size k, satisfying the following properties:

(a) If $s_n = k$ then N_k^n contains at least one sorting network of size k.
(b) If $k < s_n = k'$ then $\exists C_{n,k'}^{opt}$, an optimal-size sorting network and $\exists C_{n,k} \in N_k^n$ such that C is a prefix of C^{opt}.

Since the existence of N_k^n is guaranteed by the fact that R_k^n is actually a complete set of filters, we are interested in creating such a set that is small enough (can be computed in a 'reasonable' amount of time).

4   Subsumption

In order to create a complete set of filters [6] introduces the relation of *subsumption*.

Definition 2. Let C_a and C_b be comparator networks on n channels. If there exists a permutation π on $\{1, \ldots, n\}$ such that $\pi(outputs(C_a)) \subseteq outputs(C_b)$ we say that C_a subsumes C_b, and we write $C_a \preceq C_b$ (or $C_a \leq_\pi C_b$ to indicate the permutation).

For example, consider the networks $C_a = (1,2); (2,3); (1,4)$ and $C_b = (1,2); (1,3); (2,4)$. Their output sets are:
$outputs(C_a) = \{\{0000\}, \{0001, 0010\}, \{0011, 0110\}, \{0111, 1011\}, \{1111\}\}$,
$outputs(C_b) = \{\{0000\}, \{0001, 0010\}, \{0011, 0101\}, \{0111, 1011\}, \{1111\}\}$.
It is easy to verify that $\pi = (1,2,4,3)$ has the property that $C_a \leq_\pi C_b$.

Proposition 2. Let C_a and C_b be comparator networks on n channels, having $|outputs(C_a)| = |outputs(C_b)|$. Then, $C_a \preceq C_b \Leftrightarrow C_b \preceq C_a$.

Proof. Assume that $C_a \leq_\pi C_b \Rightarrow \pi(outputs(C_a)) \subseteq outputs(C_b)$ and since $|outputs(C_a)| = |outputs(C_b)| \Rightarrow \pi(outputs(C_a)) = outputs(C_b)$. That means that π is actually mapping each sequence in $outputs(C_a)$ to a distinct sequence in $outputs(C_b)$. The inverse permutation π^{-1} is also a mapping, this time from $outputs(C_b)$ to $outputs(C_a)$, implying that $\pi^{-1}(outputs(C_b)) = outputs(C_a) \Rightarrow C_b \leq_{\pi^{-1}} C_a$.

The following result is the key to creating a complete set of filters:

Lemma 1. *Let C_a and C_b be comparator networks on n channels, both having the same size, and $C_a \preceq C_b$. Then, if there exists a sorting network $C_b; C$ of size k, there also exists a sorting network $C_a; C'$ of size k.*

The proof of the lemma is presented in [6] (Lemma 2) and [4] (Lemma 7).

The previous lemma "suggests" that when creating the set of networks R_k^n using the naive approach, and having the goal of creating actually a complete set of filters, we should not add two networks in this set if one of them subsumes the other.

The algorithm to generate N_k^n

Require: $n, k \in \mathbb{Z}^+$
Ensure: Returns N_k^n, a complete set of filters
  $N_0^n = \{C_{n,0}\}$ {Start with the empty network}
  for all $p = 1 \ldots k$ **do**
    $N_p^n = \emptyset$ {Generate N_p^n from N_{p-1}^n, adding all possible comparators}
    for all $C \in N_{p-1}^n$ **do**
      for all $i = 1 \ldots n - 1, j = i + 1 \ldots n$ **do**
        if the comparator (i, j) is redundant **then**
          continue
        end if
        $C^* = C; (i, j)$ {Create a new network C^*}
        if $\not\exists C' \in N_p^n$ such that $C' \preceq C^*$ **then**
          $N_p^n = N_p^n \cup C^*$
          Remove from N_p^n all the networks C'' such that $C^* \preceq C''$.
        end if
      end for
    end for
  end for
  return N_k^n

A comparator c is *redundant* relative to the network C if adding it at the end of C does not modify the output set: $outputs(C; c) = outputs(C)$. Testing if a comparator $c = (i, j)$ is redundant relative to a network C can be easily implemented by inspecting the values x_i and x_j in all the sequences $\overline{x} \in outputs(C)$. If $x_i \leq x_j$ for all the sequences then c is redundant.

The key aspect in implementing the algorithm above is the test for subsumption.

5   Subsumption Testing

Let C_a and C_b be comparator networks on n channels. According to Definition 2, in order to check if C_a subsumes C_b we must find a permutation π on $\{1, \ldots, n\}$ such that $\pi(outputs(C_a)) \subseteq outputs(C_b)$. If no such permutation exists then C_a does not subsume C_b.

In order to avoid iterating through all $n!$ permutations, in [6] several results are presented that identify situations when subsumption testing can be implemented efficiently. We enumerate them as the tests ST_1 to ST_4.

(ST_1) **Check the total size of the output**
If $|outputs(C_a)| > |outputs(C_b)|$ then C_a cannot subsume C_b.

(ST_2) **Check the size of corresponding clusters** (Lemma 4 in [6])
If there exists $0 \le p \le n$ such that $|cluster(C_a, p)| > |cluster(C_b, p)|$ then C_a cannot subsume C_b. When applying a permutation π on a sequence in $outputs(C_a)$, the number of bits set to 1 remains the same, only their positions change. So, if $\pi(outputs(C_a)) \subseteq outputs(C_b)$ then $\forall 0 \le p \le n$ $\pi(cluster(C_a), p) \subseteq cluster(C_b, p)$, which implies that $|cluster(C_a)| = |\pi(cluster(C_a), p)| \le |cluster(C_b, p)|$ for all $0 \le p \le n$.

(ST_3) **Check the ones and zeros** (Lemma 5 in [6])
Recall that *zeros* and *ones* represent the sets of positions that are set to 0, respectively to 1. If there exists $0 \le p \le n$ such that $|zeros(C_a, p)| > |zeros(C_b, p)|$ or $|ones(C_a, p)| > |ones(C_b, p)|$ then C_a cannot subsume C_b.
For example, consider the networks $C_a = (1, 2); (3, 4); (2, 4); (1, 5); (1, 3)$ and $C_b = (1, 2); (3, 4); (1, 3); (3, 5); (1, 3)$. $cluster(C_a, 2) = \{00011, 00110, 01010\}$, $cluster(C_b, 2) = \{00011, 01001, 01010\}$, $ones(C_a, 2) = \{2, 3, 4, 5\}$, $ones(C_b, 2) = \{2, 4, 5\}$, therefore $C_a \not\preceq C_b$.

(ST_4) **Check all permutations** (Lemma 6 in [6])
The final optimization presented in [6] states that if there exists a permutation π such that $\pi(outputs(C_a)) \subseteq outputs(C_b)$ then $\forall 0 \le p \le n$ $zeros(\pi(C_a, p)) \subseteq zeros(C_b, p)$ and $ones(\pi(C_a, p)) \subseteq ones(C_b, p)$. So, before checking the inclusion for the whole output sets, we should check the inclusion for the *zeros* and *ones* sets, which is computationally cheaper.

The tests (ST_1) to (ST_3) are very easy to check and are highly effective in reducing the search space. However, if none of them can be applied, we have to enumerate the whole set of $n!$ permutations, verify (ST_4) and eventually the definition of subsumption, for each one of them. In [6] the authors focused on $n = 9$ which means verifying $362, 880$ permutations for each subsumption test. They were successful in creating all sets of complete filters N_k^9 for $k = 1, \ldots, 25$ and actually proved that $s_9 = 25$. Using a powerful computer and running a parallel implementation of the algorithm on 288 threads, the time necessary for creating these sets was measured in days (more than five days only for N_{14}^9).

Moving from $9!$ to $10! = 3, 628, 800$ or $11! = 39, 916, 800$ does not seem feasible. We also have to take in consideration the size of the complete filter sets, for example $|N_{14}^9| = 914, 444$.

We present a new approach for testing subsumption, which greatly reduces the number of permutations which must be taken into consideration. Instead of enumerating all permutations we will enumerate all perfect matchings in a bipartite graph created for the networks C_a and C_b being tested.

6   Enumerating Perfect Matchings

Definition 3. *Let C_a and C_b be comparator networks on n channels. The subsumption graph $G(C_a, C_b)$ is defined as the bipartite graph $(A, B; E(G))$ with vertex set $V(G) = A \cup B$, where $A = B = \{1, \ldots, n\}$ and the edge set $E(G)$ defined as follows. Any edge $e \in E(G)$ is a 2-set $e = \{i, j\}$ with $i \in A$ and $j \in B$ (also written as $e = ij$) having the properties:*

- $i \in zeros(C_a, p) \Rightarrow j \in zeros(C_b, p), \forall 0 \leq p \leq n;$
- $i \in ones(C_a, p) \Rightarrow j \in ones(C_b, p), \forall 0 \leq p \leq n.$

So, the edges of the subsumption graph G represent a relationship between positions in the two output sets of C_a and C_b. An edge ij signifies that the position i (regarding the sequences in $outputs(C_a)$) and the position j (regarding C_b) are "compatible", meaning that a permutation π with the property $\pi(outputs(C_a)) \subseteq outputs(C_b)$ might have the mapping i to j as a part of it.

As an example, consider the following *zeros* and *ones* sequences, corresponding to $C_a = (1, 2); (3, 4); (2, 4); (2, 5)$ and $C_b = (1, 2); (3, 4); (1, 4); (2, 5)$.
$\overline{zeros}(C_a) = \{$`00000`,`00000`,`000-0`,`000-`,`000--`,`-----`$\}$,
$\overline{zeros}(C_b) = \{$`00000`,`00000`,`00000`,`000--`,`000--`,`-----`$\}$,
$\overline{ones}(C_a) = \{$`-----`,`---11`,`1-111`,`11111`,`11111`,`11111`$\}$,
$\overline{ones}(C_b) = \{$`-----`,`---11`,`-1111`,`11111`,`11111`,`11111`$\}$.
The subsumption graph $G(C_a, C_b)$ is pictured below:

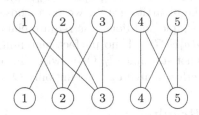

Fig. 2. The subsumption graph corresponding to the comparator networks $C_a = (1, 2); (3, 4); (2, 4); (2, 5)$ and $C_b = (1, 2); (3, 4); (1, 4); (2, 5)$

A *matching* M in the graph G is a set of independent edges (no two edges in the matching share a common node). If $ij \in M$ we say that i and j are *saturated*. A *perfect matching* is a matching that saturates all vertices of the graph.

Lemma 2. *Let C_a and C_b be comparator networks on n channels. If $C_a \leq_\pi C_b$ then $M = \{i\pi(i) : i = 1, \ldots, n\} \subseteq E(G)$ is a perfect matching in the subsumption graph $G(C_a, C_b)$.*

Proof. Suppose that $C_a \leq_\pi C_b$, $\pi(i) = j$ and $ij \notin E(G)$. That means that $\exists 0 \leq p \leq n$ such that $i \in zeros(C_a, p) \wedge j \notin zeros(C_b, p)$ or $i \in ones(C_a, p) \wedge j \notin ones(C_b, p)$. We will assume the first case. Let $\overline{x}$ a sequence in $cluster(C_a, p)$ such that $\overline{x}(i) = 0$. Since $\pi(outputs(C_a)) \subseteq outputs(C_b) \Rightarrow \pi(\overline{x}) \in cluster(C_b, p)$. But $\pi(i) = j$, therefore in $cluster(C_b, p)$ there is the sequence $\pi(\overline{x})$ having the bit at position j equal to 0, contradiction.

The previous lemma leads to the following result:

Corollary 1. *Let C_a and C_b be comparator networks on n channels. Then C_a subsumes C_b if and only if there exists a perfect matching π in the subsumption graph $G(C_a, C_b)$.*

The graph in Fig. 2 has only four perfect matchings: $(2, 1, 3, 4, 5)$, $(3, 1, 2, 4, 5)$, $(2, 1, 3, 5, 4)$, $(3, 1, 2, 5, 4)$. So, when testing subsumption, instead of verifying $5! = 120$ permutations it is enough to verify only 4 of them.

If two clusters are of the same size, then we can strengthen the previous result even more. If there is a permutation π such that $\pi(cluster(C_a, p)) = cluster(C_b, p)$ then $\pi^{-1}(cluster(C_b, p)) = cluster(C_a, p)$. Using the same reasoning, when creating the subsumption graph $C(G_a, C_b)$ we add the following two condition when defining an edge ij:

- $j \in zeros(C_b, p) \Rightarrow i \in zeros(C_a, p)$, $\forall 0 \leq p \leq n$ such that $|cluster(C_a, p)| = |cluster(C_b, p)|$,
- $j \in ones(C_b, p) \Rightarrow i \in ones(C_a, p)$, $\forall 0 \leq p \leq n$ such that $|cluster(C_a, p)| = |cluster(C_b, p)|$.

In order to enumerate all perfect matchings in a bipartite graph, we have implemented the algorithm described in [14]. The algorithm starts with finding a perfect matching in the subsumption graph $G(C_a, C_b)$. Taking into consideration the small size of the bipartite graph, we have chosen the Ford-Fulkerson algorithm which is very simple and does not require elaborate data structures. Its time complexity is $O(n|E(G)|)$. If no perfect matching exists, then we have established that C_a does not subsume C_b. Otherwise, the algorithm presented in [14] identifies all other perfect matchings, taking only $O(n)$ time per matching.

7   Experimental Results

We implemented both variants of subsumption testing:

- (1) enumerating all permutations and checking the inclusions described by (ST_4) before verifying the actual definition of subsumption;
- (2) verifying only the permutations that are actually perfect matchings in the subsumption graph, according to Corollary 1.

We made some simple experiments on a regular computer (Intel i7-4700HQ @2.40 GHz), using 8 concurrent threads. The programming platform was Java SE Development Kit 8.

Several suggestive results are presented in the table below:

(n,k)	$|N_k^n|$	total	sub	$perm_1$	$time_1$	$perm_2$	$time_2$
$(7,9)$	678	1,223,426	5,144	26,505,101	2.88	33,120	0.07
$(7,10)$	510	878,995	5,728	25,363,033	2.82	24,362	0.06
$(8,7)$	648	980,765	2,939	105,863,506	13.67	49,142	0.14
$(8,8)$	2088	9,117,107	9,381	738,053,686	94.50	283,614	0.49
$(8,9)$	5703	24,511,628	29,104	4,974,612,498	650.22	1,303,340	1.96

The columns of the table have the following significations:

- (n,k) - n is the number of channels, k is the number of comparators;
- $|N_k^n|$ - the size of the complete set of filters generated for the given n and k;
- total - the total number of subsumption checks;
- sub - the number of subsumptions that were identified;
- $perm_1$ - how many permutations were checked, using the variant (1);
- $time_1$ - the total time, measured in seconds, using the variant (1);
- $perm_2$ - how many permutations were checked, using the variant (2);
- $time_2$ - the total time, measured in seconds, using the variant (2);

As we can see from this results, using the variant (2) the number of permutations that were verified in order to establish subsumption is greatly reduced. Despite the fact that it is necessary to create the subsumption graph and to iterate through its set of perfect matchings, this leads to a much shorter time needed for the overall generation of the complete set of filters.

This new approach enabled us to reproduce the state-of-the-art result concerning optimal-size sorting networks, described in [6]. Using an Intel Xeon E5-2670 @ 2.60 GHz computer, with a total of 32 cores, we generated all the complete set of filters for $n = 9$. The results are presented in the table below.

k	1	2	3	4	5	6	7	8
$|N_k^9|$	1	3	7	20	59	208	807	3415
$time(s)$	0	0	0	0	0	0	0	0
k	9	10	11	12	13	14	15	16
$|N_k^9|$	14343	55991	188730	490322	854638	914444	607164	274212
$time(s)$	4	48	769	6688	25186	40896	24161	5511
k	17	18	19	20	21	22	23	24	25
$|N_k^9|$	94085	25786	5699	1107	250	73	27	8	1
$time(s)$	610	36	2	0	0	0	0	0	0

In [6] the necessary time required to compute $|N_{14}^9|$ using the generate-and-prune approach was estimated at more than 5 days of computation on 288 threads. Their tests were performed on a cluster with a total of 144 Intel E8400 cores clocked at 3 GHz. In our experiments, the same set was created in only 11 h, which is actually a significant improvement.

8   Conclusions

In this paper we have extended the work in [6], further investigating the relation of subsumption. In order to determine the minimal number of comparators needed to sort any input of a given length, a systematic BFS-like algorithm generates incrementally complete sets of filters, that is sets of comparator networks that have the potential to prefix an optimal-size sorting network. To make this approach feasible it is essential to avoid adding into these sets networks that subsume one another. Testing the subsumption is an expensive operation, invoked a huge number of times during the execution of the algorithm. We described a new approach to implement this test, based on enumerating perfect matchings in a bipartite graph, called the subsumption graph. Computer experiments have shown significant improvements, greatly reducing the number of invocations and the overall running time. The results show that, using appropriate hardware, it might be possible to approach in this manner the optimal-size problem for sorting networks with more than 10 channels.

Acknowledgments. We would like to thank Michael Codish for introducing us to this research topic and Cornelius Croitoru for his valuable comments. Furthermore, we thank Mihai Rotaru for providing us with the computational resources to run our experiments.

References

1. Ajtai, M., Komlós, J., Szemerédi, E.: An 0(n log n) sorting network. In: Proceedings of the Fifteenth Annual ACM Symposium on Theory of Computing, STOC 1983, pp. 1–9. ACM, New York (1983)
2. Baddar, S.W.A.H., Batcher, K.E.: Designing Sorting Networks: A New Paradigm. Springer, New York (2012). https://doi.org/10.1007/978-1-4614-1851-1
3. Batcher, K.E.: Sorting networks and their applications. In: Proceedings of the 30 April–2 May 1968, Spring Joint Computer Conference, pp. 307–314. ACM, New York (1968)
4. Bundala, D., Závodný, J.: Optimal sorting networks. In: Dediu, A.-H., Martín-Vide, C., Sierra-Rodríguez, J.-L., Truthe, B. (eds.) LATA 2014. LNCS, vol. 8370, pp. 236–247. Springer, Cham (2014). https://doi.org/10.1007/978-3-319-04921-2_19
5. Codish, M., Cruz-Filipe, L., Ehlers, T., Müller, M., Schneider-Kamp, P.: Sorting networks: to the end and back again. J. Comput. Syst. Sci. (2016)
6. Codish, M., Cruz-Filipe, L., Frank, M., Schneider-Kamp, P.: Twenty-five comparators is optimal when sorting nine inputs (and twenty-nine for ten). In: 2014 IEEE 26th International Conference on Tools with Artificial Intelligence (ICTAI), pp. 186–193. IEEE (2014)
7. Codish, M., Cruz-Filipe, L., Schneider-Kamp, P.: The quest for optimal sorting networks: Efficient generation of two-layer prefixes. In: 2014 16th International Symposium on Symbolic and Numeric Algorithms for Scientific Computing (SYNASC), pp. 359–366. IEEE (2014)

8. Ehlers, T., Müller, M.: New bounds on optimal sorting networks. In: Beckmann, A., Mitrana, V., Soskova, M. (eds.) CiE 2015. LNCS, vol. 9136, pp. 167–176. Springer, Cham (2015). https://doi.org/10.1007/978-3-319-20028-6_17

9. Kipfer, P., Westermann, R.: Improved GPU sorting. GPU Gems **2**, 733–746 (2005)

10. Knuth, D.E.: The Art of Computer Programming. Sorting and Searching, vol. 3, 2nd edn. Addison Wesley Longman Publishing Co. Inc., Redwood City (1998)

11. Morgenstern, A., Schneider, K.: Synthesis of parallel sorting networks using SAT solvers. In: MBMV, pp. 71–80 (2011)

12. Parberry, I.: A computer-assisted optimal depth lower bound for nine-inputsorting networks. Math. Syst. Theory **24**(1), 101–116 (1991). https://doi.org/10.1007/BF02090393

13. Parberry, I.: On the computational complexity of optimal sorting network verification. In: Aarts, E.H.L., van Leeuwen, J., Rem, M. (eds.) Parle' 91 Parallel Architectures and Languages Europe. LNCS, pp. 252–269. Springer, Heidelberg (1991). https://doi.org/10.1007/978-3-662-25209-3_18

14. Uno, T.: Algorithms for enumerating all perfect, maximum and maximal matchings in bipartite graphs. In: Leong, H.W., Imai, H., Jain, S. (eds.) ISAAC 1997. LNCS, vol. 1350, pp. 92–101. Springer, Heidelberg (1997). https://doi.org/10.1007/3-540-63890-3_11

15. Valsalam, V.K., Miikkulainen, R.: Using symmetry and evolutionary search to minimize sorting networks. J. Mach. Learn. Res. **14**, 303–331 (2013)

Investigating Constraint Programming for Real World Industrial Test Laboratory Scheduling

Tobias Geibinger$^{(\boxtimes)}$, Florian Mischek, and Nysret Musliu

Christian Doppler Laboratory for Artificial Intelligence and Optimization
for Planning and Scheduling, DBAI, TU Wien, Favoritenstraße 9-11,
1040 Vienna, Austria
{tgeibing,fmischek,musliu}@dbai.tuwien.ac.at

Abstract. In this paper we deal with a complex real world scheduling
problem closely related to the well-known Resource-Constrained Project
Scheduling Problem (RCPSP). The problem concerns industrial test laboratories in which a large number of tests has to be performed by qualified
personnel using specialised equipment, while respecting deadlines and
other constraints. We present different constraint programming models
and search strategies for this problem. Our approaches are evaluated
using CP solvers and a MIP solver on a set of generated instances of
different sizes. With our best approach we could find feasible and several
optimal solutions for instances that are generated based on real-world
test laboratory problems.

1 Introduction

Project scheduling includes various problems of high practical relevance. Such
problems arise in many areas and include different constraints and objectives.
Usually project scheduling problems require scheduling of a set of project activities over a period of time and assignment of resources to these activities. Typical
constraints include time windows for activities, precedence constraints between
the activities, assignment of appropriate resources etc. The aim is to find feasible schedules that optimize several criteria such as the minimization of total
completion time.

In this paper we investigate solving a real-world project scheduling problem
that arises in an industrial test laboratory of a large company. This problem,
Industrial Test Laboratory Scheduling (TLSP), which is an extension of the well
known Resource-Constrained Project Scheduling Problem (RCPSP), was originally described in [13,14]. It consists of a grouping stage, where smaller activities
(tasks) are joined into larger jobs, and a scheduling stage, where those jobs are
scheduled and have resources assigned to them. In this work, we deal with the
second stage and assume that a grouping of tasks into jobs is already provided.
Since we focus on the scheduling part, we denote the resulting problem TLSP-S.

L.-M. Rousseau and K. Stergiou (Eds.): CPAIOR 2019, LNCS 11494, pp. 304–319, 2019.
https://doi.org/10.1007/978-3-030-19212-9_20

The investigated problem has several features of previous project scheduling problems in the literature, but also includes some specific features imposed by the real-world situation, which have rarely been studied before. Among others, these include heterogeneous resources, with availability restrictions on the activities each unit of a resource can perform. While work using similar restrictions exists ([5,23]), most problem formulations either assume homogeneous, identical units of each resource or introduce additional activity modes for each feasible assignment, which quickly becomes impractical for higher resource requirements and multiple resources. Another specific feature of TLSP(-S) is that of linked activities, which require identical assignments on a subset of the resources. To the best of our knowledge, a similar concept appears only in [18], where modes should be identical over subsets of all activities. We also deal with several non-standard objectives instead of the usual makespan minimization, which arise from various business objectives of our industrial partner. Most notably, we try to minimize the total completion time of each project, i.e. the time between the start of the first and the end of the last job in the project.

In practice, exact solutions for this problem are desired especially in situations where it is necessary to check if a feasible solution exists at all. In the application that we consider, checking quickly if activities of additional projects can be added on top of an existing schedule is very important. In this paper we investigate exact methods for solving this problem. Although it is known from previous papers [21,23] that constraint programming techniques can give good results for similar project scheduling problems, it is an interesting question if Constraint Programming (CP) techniques can also solve TLSP-S that includes additional features and larger instances.

We provide a CP model for our problem by exploiting some previous ideas for a similar problem from [21,23] and extend it to model the additional features of TLSP-S. This includes, for example, the handling of the problem specific differences discussed above but also new redundant constraints as well as search procedures tailored to the problem. Using the MiniZinc [15] constraint programming language we experiment with various strategies involving the formulation of resource constraints, the reduction of the search space, and search procedures based on heuristics.

Our final experiments show that constraint programming techniques can reach very good results for realistic instances and outperform MIP solvers on the same model. Our results strengthen the conclusion of previous studies and show that CP technology can be applied successfully for solving large project scheduling problems.

The rest of the paper is organised as follows. In the next Section we give the related work. Section 3 introduces the problem that we investigate in this paper. A constraint model is given in Sect. 4. Experimental results are presented in Sect. 5 and the last section gives conclusions.

2   Literature Overview

The Resource-Constrained Project Scheduling Problem (RCPSP) has been investigated by numerous researchers over the last decades. For a comprehensive overview over publications dealing with this problem and its many variants, we refer to surveys e.g. by Brucker et al. [3], Hartmann and Briskorn [9], or Mika et al. [12].

Of particular interest for the problem treated in this work are various extensions to the classical RCPSP.

Multi-Mode RCPSP (MRCPSP) formulations allow for activities that can be scheduled in one of several modes. This variant has been extensively studied since 1977 [7], we refer to the surveys by Węglarz et al. [22] and Hartmann and Briskorn [9]. A good example of a CP-Model for the MRCPSP was given by Szeredi and Schutt [21].

Many formulations, including TLSP, make use of release dates, due dates, deadlines, or combinations of those. An example of this can be found in [6]. Further relevant extensions deal with multi-project formulations, including alternative objective functions (e.g. [17]). Usually, the objective in (variants of) RCPSP is the minimization of the total makespan [9]. However, also other objective values have been considered. Of particular relevance to TLSP are objectives based on total completion time and multi-objective formulations (both appear in e.g. [16]). Salewski et al. [18] include constraints that require several activities to be performed in the same mode. This is similar to the concept of linked jobs introduced in the TLSP.

RCPSP itself and most variants assume that individual units of each resource are identical and interchangeable. A problem closely related to TLSP-S is Multi-Skill RCPSP (MSPSP), first introduced by Bellenguez and Néron [2]. In this problem, each resource unit possesses certain skills, and an activity can only have those resources with the required skills assigned to it. This is similar to the availability restrictions on resources that appear in TLSP. Just like for our problem, they also deal with the problem that while availability restrictions could be modeled via additional activity modes corresponding to feasible resource assignments (e.g. in [1,17,20]), this is intractable due to the large number of modes that would have to be generated [2]. To the best of our knowledge, the best results for the MSPSP problem have been achieved by Young et al. [23], who use a CP model to solve the problem.

Bartels and Zimmermann [1] describe a problem for scheduling tests of experimental vehicles. It contains several constraints that also appear in similar form in TLSP-S, but includes a different resource model and minimises the number of experimental vehicles used.

3   Problem Description

As mentioned before, we deal with a variant of TLSP [14], where we assume that a grouping of tasks into jobs is already provided for each project, and focus on the scheduling part of the problem instead (TLSP-S). Thus, the goal is to find an assignment of a mode, time slot and resources to each (given) job, such that all constraints are fulfilled and the objective function is minimized.

In the following, we introduce the TLSP-S problem.

Each instance consists of a scheduling period of h discrete *time slots*. Further, it lists resources of different kinds:

- *Employees* $E = \{1, \ldots, |E|\}$ who are qualified for different types of jobs.
- A number of *workbenches* $B = \{1, \ldots, |B|\}$ with different facilities.
- Various auxiliary lab *equipment* groups $G_g = \{1, \ldots, |G_g|\}$, where g is the group index. These each represent a set of similar devices. The set of all equipment groups is denoted G^*.

Further we have given the set of *projects* labeled $P = \{1, \ldots, |P|\}$, and the set of *jobs* to be scheduled $J = \{1, \ldots, |J|\}$. For a project p, the jobs of this project are given as $J_p \subseteq J$.

Each job j has several properties[1]:

- A time window, given via a *release date* α_j and a *deadline* ω_j. In addition, it has a *due date* $\bar{\omega}_j$, which is similar to the deadline, except that exceeding it is only a soft constraint violation.
- A set of *available modes* $M_j \subseteq M$, where M is the set of all modes.
- A *duration* d_{mj} for each available mode $m \in M_j$.
- The resource requirements for the job:
  - The number of *required employees* r_m^{Em} depends on the mode $m \in M_j$. Each of these employees must be chosen from the set of *qualified employees* $E_j \subseteq E$. Additionally, there is also a set of *preferred employees* $E_j^{Pr} \subseteq E_j$.
  - The number of *required workbenches* $r_j^{Wb} \in \{0, 1\}$. If a workbench is required, it must be chosen from the *available workbenches* $B_j \subseteq B$.
  - For each equipment group $g \in G^*$, the job requires r_{gj}^{Eq} devices, which must be taken from the set of *available devices* $G_{gj} \subseteq G_g$ for the group.
- The *predecessors* $\mathcal{P}_j$ of the job, which must be completed before the job can start. Precedence relations will only occur between jobs of the same project.
- *Linked jobs* L_j of this job. All linked jobs must be performed by the same employee(s). As before, such links only occur between jobs of the same project.
- Optionally, the job may contain *initial assignments*.
  - An *initial mode* $\dot{m}_j$.
  - An *initial starting time slot* $\dot{s}_j$.
  - *Initial resource assignments*: For each employee $e \in E$, the boolean parameter $\dot{a}_{ej}^{Em}$ indicates whether e is initially assigned to j. Analogously, $\dot{a}_{bj}^{Wb}$ and $\dot{a}_{dj}^{Eq}$ perform the same function for each workbench $b \in B$ and each device $d \in G_g, g \in G^*$, respectively.

Some or all of these assignments may be present for any given job.

---

[1] In TLSP, these are derived from the tasks contained within a job. Since we assume the distribution of tasks into jobs to be fixed, they can be given directly as part of the input for TLSP-S.

Out of all jobs, a subset are *started jobs* $J^S \subseteq J$. A started job will always fulfill the following conditions:

- It must have a preassigned mode.
- Its start time must be set to 1.
- It must have initial resource assignments fulfilling all requirements.

The initial assignments of a started job must not be changed in the solution.

A complete description of all constraints of the original model can be found in [14]. The hard and soft constraints that we consider for the TLSP-S will be described in the next section, where we will introduce the CP model.

The aim for this problem is to find an assignment of a mode, time slot and resources to each (given) job, such that all hard constraints are fulfilled and the violation of soft constraints is minimized.

4   Constraint Programming Model

We developed our model using the solver-independent modeling language MiniZinc [15]. Using MiniZinc we can easily compare different solvers. Furthermore, previous studies have shown that CP gives very good results for similar project scheduling problems. Most notably, the approaches by Young et al. [23] and Szeredi et al. [21] for MSPSP and MRCPSP respectively. MiniZinc also enables the use of user defined search strategies, which were shown to be very effective for MSPSP [23]. For both scheduling problems, the LCG solver Chuffed [4] was able to achieve very good results.

In order to provide an additional comparison, we also modeled our problem with the IBM ILOG CP Optimizer [10]. This model uses a different formulation of constraints and decision variables than the MiniZinc model and is described in Subsect. 4.6.

In order to represent a solution for the scheduling problem we use the following decision variables. The start time variable s_j assigns a start time to each job j. Similarly, for each job j, mode variable m_j assigns it a mode. For resource assignments we need the following variables: For each job j, the variable a_{ej}^{Em} is set to 1 if employee e is assigned to j and 0 otherwise, the variable a_{bj}^{Wb} is 1 if j is performed on workbench b and 0 otherwise, and the variable a_{dj}^{Eq} is 1 if device d is used by j and 0 otherwise.

4.1   Basic Hard Constraints

The following constraints follow directly from the problem definition.

$$s_j \geq \alpha_j \ \wedge \ (s_j + d_{m_j j}) \leq \omega_j \qquad j \in J \qquad (1)$$

$$s_j \geq (s_k + d_{m_k k}) \qquad j \in J, \ k \in \mathcal{P}_j \qquad (2)$$

$$m_j \in M_j \qquad j \in J \tag{3}$$

$$a_{ej}^{Em} = 1 \;\rightarrow\; e \in E_j \qquad j \in J,\, e \in E \tag{4}$$

$$a_{bj}^{Wb} = 1 \;\rightarrow\; b \in B_j \qquad j \in J,\, b \in B \tag{5}$$

$$a_{dj}^{Eq} = 1 \;\rightarrow\; d \in G_{gj} \qquad j \in J,\, g \in G^*,\, d \in G_g \tag{6}$$

$$\sum_{e \in E} a_{ej}^{Em} = r_{m_j}^{Em} \qquad j \in J \tag{7}$$

$$\sum_{b \in B} a_{bj}^{Wb} = r_j^{Wb} \qquad j \in J \tag{8}$$

$$\sum_{d \in G_g} a_{dj}^{Eq} = r_{gj}^{Eq} \qquad j \in J,\, g \in G^* \tag{9}$$

$$a_{ej}^{Em} = a_{ek}^{Em} \qquad j \in J,\, k \in L_j,\, e \in E \tag{10}$$

$$s_j = 1 \;\wedge\; m_j = \dot{m}_j \;\wedge\; \qquad j \in J^S,\, e \in E,\, b \in B,$$
$$a_{ej}^{Em} = \dot{a}_{ej}^{Em} \;\wedge\; a_{bj}^{Wb} = \dot{a}_{bj}^{Wb} \;\wedge\; a_{dj}^{Eq} = \dot{a}_{dj}^{Eq} \qquad g \in G^*,\, d \in G_g \tag{11}$$

Constraint (1) makes sure that each job is executed in its time window, (2) enforces that the prerequisite jobs of a job are always completed before it starts. Constraints (3–6) ensure that assigned modes, employees, workbenches, and devices are available for the respective job. In order to make sure that each job has exactly as many resources as required, we have constraints (7–9). Furthermore, we need constraint (10) to make sure that linked jobs are assigned to the same employees and constraint (11) to fix the resource assignments of jobs which are already started.

The above set of constraints is however not enough to ensure a valid solution. Additionally, we have to consider constraints which enforce that no resource (employee, workbench, or equipment) is assigned to two or more jobs at the same time. Like it was the case with MSPSP [23], the constraints used for modeling those *unary resource requirements* have a tremendous impact on the practicability of the model and in the next subsection we will present different options for modeling such constraints.

4.2   Unary Resource Constraints

We will now present three different approaches for modeling unary resource constraints, each of which is designed with CP solvers in mind. Two of those three quickly proved to be impractical for our problem.

Time-Indexed Approach. The probably most straightforward way to model the non-overuse of any resource at any given time is captured by the following constraints.

$$\sum_{j \in J,\, s_j \leq t < (s_j + d_{m_j j})} a_{ej}^{Em} \leq 1 \qquad e \in E,\, 1 \leq t \leq h \tag{12}$$

$$\sum_{j\in J, s_j \leq t < (s_j + d_{m_{jj}})} a_{bj}^{Wb} \leq 1 \qquad b \in B,\ 1 \leq t \leq h \tag{13}$$

$$\sum_{j\in J, s_j \leq t < (s_j + d_{m_{jj}})} a_{dj}^{Eq} \leq 1 \qquad g \in G^*,\ d \in G_g,\ 1 \leq t \leq h \tag{14}$$

The number of constraints generated by MiniZinc based on (12–14) is of course directly dependent on the planning horizon h and the total number of resources. Because of the long compilation time and the high computer resource consumption, it quickly became immanent that for our larger instances the time-indexed approach is not efficient. This is of course not surprising since Young et al. [23] came to a similar conclusion for MSPSP. Hence, we discarded this option after some preliminary testing.

Overlap Constraint. For MSPSP, Young et al. [23] achieved their best results using a so-called *order constraint*. This constraint basically enforces that two activities cannot overlap in their execution when they use a common resource. During the initial modeling phase we tried a very similar approach. First, we introduced the new predicate `overlap`:

$$\texttt{overlap}(j,k) := s_k < (s_j + d_{m_{jj}}) \ \wedge \ (s_k + d_{m_kk}) > s_j$$

In MSPSP, resources are assigned to activities with respect to the needed skill of the activity. For the overlap constraint it is not important which skill requirement the resource contributes to, so Young et al. [23] had to introduce an auxiliary variable to express that a resource is used by an activity. We on the other hand assign the resources directly and thus can model our *overlap constraint* without any new variables.

$$\texttt{overlap}(j,k) \rightarrow (\bigwedge_{e \in E} (\neg a_{ej}^{Em} \vee \neg a_{ek}^{Em}) \ \wedge$$
$$\bigwedge_{b \in B} (\neg a_{bj}^{Wb} \vee \neg a_{bk}^{Wb}) \ \wedge$$
$$\bigwedge_{g \in G^*, d \in G_g} (\neg a_{dj}^{Eq} \vee \neg a_{dk}^{Eq})\ ) \qquad\qquad j, k \in J,\ j \neq k,$$
$$\alpha_k < \omega_j \wedge \omega_k > \alpha_j \tag{15}$$

Just like with the time-indexed approach, it turned out that the overlap constraint produced too many constraints and was thus impractical for larger instances. This is interesting because Young et al. had no such problems, but their biggest instances only had 60 resources and 42 activities, whereas we have instances with more than 300 resources and jobs, respectively. It should however be noted that Young et al. [23] reduced the number of generated constraints by considering only *unrelated* activity pairs, i.e. activities which do not depend on the execution of each other via precedence constraints (related activities can obviously never overlap). We on the other hand generate constraints for all pairs of jobs which are allowed to overlap based on their release dates and deadlines. Comparing only unrelated jobs requires the computation of the transitive closure of the job precedence relation and because our instances have a lot of unrelated jobs, we don't expect any significant improvement.

Cumulative Constraints. Another way to model the unary resource constraints is to use a global constraint like `cumulative`. The `cumulative` constraint takes as input the start times, durations and resource requirements of a list of jobs and ensures that their resource assignments never exceed a given bound. This is of course a perfect way to enforce non overload of any resource and both MSPSP and MRCPSP have efficient models which make use of `cumulative` in some way [21,23]. In order to enforce the non-overload of any resource we need three constraints (one for each resource type).

$$\texttt{cumulative}((s_j)_{j\in J}, (d_{m_{jj}})_{j\in J}, (a_{ej}^{Em})_{j\in J}, 1) \quad e \in E \tag{16}$$

$$\texttt{cumulative}((s_j)_{j\in J}, (d_{m_{jj}})_{j\in J}, (a_{bj}^{Wb})_{j\in J}, 1) \quad b \in B \tag{17}$$

$$\texttt{cumulative}((s_j)_{j\in J}, (d_{m_{jj}})_{j\in J}, (a_{dj}^{Eq})_{j\in J}, 1) \quad g \in G^*, d \in G_g \tag{18}$$

In difference to our first two modeling approaches, this one turned out to scale well. Since the others performed so poorly on large instances, the rest of our experiments were performed with the `cumulative` unary resource constraints.

4.3 Soft Constraints

There are several soft constraints in our problem definition [14]. Since we consider the job grouping fixed in TLSP-S, we can drop the first soft constraint regarding the number of jobs. In order to avoid confusion with the original formulation, we thus start the numbering of our soft constraints with two.

MiniZinc has no direct support for soft constraints, hence we define them as sums which should be minimised. Those sums are given as follows.

We want to prefer solutions where the assigned employees of a job are taken from the set of preferred employees:

$$s_2 = w_2 \cdot \sum_{j\in J} \sum_{e\in(E\setminus E_j^{Pr})} a_{ej}^{Em}$$

For each project, the total number of employees assigned to it should be minimised:

$$s_3 = w_3 \cdot \sum_{p\in P} \sum_{e\in E} ((\sum_{j\in J_p} a_{ej}^{Em}) > 0)$$

For each job, violating its due date should be avoided:

$$s_4 = w_4 \cdot \sum_{j \in J} max(s_j + d_{m_j j} - \bar{\omega}_j, 0)$$

Lastly, project durations should be as small as possible:

$$s_5 = w_5 \cdot \sum_{p \in P} (max_{j \in J_p}(s_j + d_{m_j j}) - min_{j \in J_p}(s_j))$$

The objective of the search is then given by $min \; \sum_{2 \leq i \leq 5} s_i$.

At the moment, the values of the weights w_i $(2 \leq i \leq 5)$ are being determined in correspondence with a real-world laboratory. Currently all these weights are set to 1.

4.4 Redundant Constraints

Finding good redundant constraints for our problem proved to be very hard since the search space is usually very big and at the beginning of the search there is little knowledge about the final duration of the jobs. To deal with this issue we introduced a relaxed `cumulative` constraint enforcing a global resource bound.

$$
\begin{aligned}
\texttt{cumulative}((s_j)_{j \in J}, \\
(min_{m \in M}(d_{m_j j}))_{j \in J}, \\
(min_{m \in M_j}(r_m^{Em}) + r_j^{Wb} + \sum_{g \in G^*} r_{gj}^{Eq})_{j \in J}, \\
|E| + |B| + \sum_{g \in G^*} |G_g|)
\end{aligned}
\tag{19}
$$

This enables the search to discard scheduling options which are impossible regardless of the chosen modes early on.

On top of that, we can also formulate more straightforward `cumulative` constraints which enforce the global resource bounds for each resource at any point in time.

$$\texttt{cumulative}((s_j)_{j \in J}, (d_{m_j j})_{j \in J}, (r_{m_j}^{Em})_{j \in J}, |E|) \tag{20}$$

$$\texttt{cumulative}((s_j)_{j \in J}, (d_{m_j j})_{j \in J}, (r_j^{Wb})_{j \in J}, |B|) \tag{21}$$

$$\texttt{cumulative}((s_j)_{j \in J}, (d_{m_j j})_{j \in J}, (r_{gj}^{Eq})_{j \in J}, |G|) \qquad g \in G^* \tag{22}$$

Given the large search space, trying to restrict the scope of the decision variables seems like a worthwhile idea. We achieve this by using *global cardinality constraints*. Those constraints allow us to give tight bounds for the total number of resources which should be used.

$$\texttt{global\_cardinality\_low\_up}((a_{ej}^{Em})_{e \in E, j \in J}, \ 1, \sum_{j \in J} min_{m \in M_j}(r_m^{Em}),$$

$$\sum_{j \in J} max_{m \in M_j}(r_m^{Em})) \quad (23)$$

$$\texttt{global\_cardinality\_low\_up}((a_{bj}^{Wb})_{b \in B, j \in J}, \ 1, \sum_{j \in J} r_j^{Wb}, \sum_{j \in J} r_j^{Wb}) \quad (24)$$

$$\texttt{global\_cardinality\_low\_up}((a_{dj}^{Eq})_{g \in G^*, d \in G_g, j \in J}, \ 1, \sum_{j \in J} \sum_{g \in G^*} r_{gj}^{Eq},$$

$$\sum_{j \in J} \sum_{g \in G^*} r_{gj}^{Eq}) \quad (25)$$

Constraint (23) enforces that no more employees can be assigned than the sum of the highest possible employee requirements and no less than the sum of the minimum requirements. The other two constraints analogously ensure that the number of assigned workbenches and equipment is tightly bounded by the cumulative requirement of all jobs.

4.5 Search Strategies

During initial testing it quickly became immanent that the default search strategy of Chuffed (or Gecode) was not even able to find feasible solutions for most instances. This was not surprising since Young et al. [23] already had a similar issue with MSPSP. However, they were able to improve their results drastically by employing a new MiniZinc search annotation called `priority_search` which is supported by Chuffed [8]. Based on their research we have experimented with four slightly different versions of `priority_search`:

  (i) `ps_startTimeFirst_aff`
  (ii) `ps_startTimeFirst_ff`
 (iii) `ps_modeFirst_aff`
 (iv) `ps_modeFirst_ff`

All four search strategies branch over the jobs and their resource assignments. The order of the branching is the same for all strategies and is determined by the smallest possible start times of the jobs in ascending order. For each branch, searches (i) and (ii) initially assign the smallest start time to the selected job followed by assigning it the mode which minimises the job duration. Search procedures (iii) and (iv) start with the mode assignment and then assign the start time. Once the start time and the mode have been assigned for the selected job, all of the search strategies make resource assignments for the job. Searches (i) and (iii) start by assigning those resources to the job which are available and have the biggest domain, whereas (ii) and (iv) start with assignments which are either unavailable or have only one value in their domain.

4.6   Alternative CP Model

We also modelled our problem with CP Optimizer [10]. In that model the decision variables are given by the following interval variables:

$$
\begin{aligned}
&\text{interval } a_j \subset [\alpha_j, \omega_j) && \forall j \in J \\
&\text{interval } b_p && \forall p \in P \\
&\text{interval } a_{ij} \text{ optional size } d_{ij} && \forall j \in J, i \in M_j \\
&\text{interval } a_{eij}^{EmM} \text{ optional} && \forall j \in J, i \in M_j, e \in E_j \\
&\text{interval } a_{ej}^{Em} \text{ optional} && \forall j \in J, e \in E_j \\
&\text{interval } a_{bj}^{Wb} \text{ optional} && \forall j \in J, b \in B_j \\
&\text{interval } a_{gdj}^{Eq} \text{ optional} && \forall j \in J, g \in G^*, d \in G_{gj}
\end{aligned}
$$

The intervals a_j represents the jobs and are constrained to the time windows of the respective job. The second set of intervals b_p are auxiliary variables representing the total duration of the projects. Those intervals enable an easy formulation of the project duration soft constraint. The next intervals a_{ij} are optional and the presence of such an interval indicates that job j is performed in mode i. For a job j several employee allocations are possible depending on its mode. The presence of an optional interval a_{eij}^{EmM} represents the allocation of employee e to perform job j in mode i. The last three sets of optional intervals are used to indicate resource allocation.

The constraints of the model ensure that for each job j only one interval a_{ij} is present and that the length of the main interval a_j matches the duration of the selected mode interval a_{ij} by using `alternative` constraints. Similarly for the resource intervals. The unary resource constraints described in Subsect. 4.2 are modelled with `noOverlap` constraints. Furthermore, the model ensures the given job precedences given in the problem instance by using `endBeforeStart` constraints and linked jobs are modelled by constraining the presence of the optional employee resource intervals. Lastly, the availability of resources is modelled by restricting the presence of their respective optional intervals and the project intervals b_p are constrained to span over all the job intervals belonging to the respective project. Redundant constraints similar to (20–22) are formulated using `pulse` constraints and the soft constraints are again formulated as sums.

5   Experiments and Comparison

We ran our experiments on a benchmark server with 224 GB RAM and two AMD Opteron 6272 Processors each with max. 2.1 GHz and 16 logical cores. Since all of the solvers we experimented with are single threaded, we usually ran two independent sets of benchmarks in parallel. We used MiniZinc 2.2.3 [15] with Chuffed 0.10.3 [4] and CPLEX 12.8.0 [11]. Furthermore, we also experimented with the ILOG CP Optimizer 12.8.0 [10] which was not run from MiniZinc but with the ILOG Java API. We have also tested Gecode [19] as an additional CP solver included in MiniZinc, but it quickly proved to be inferior to Chuffed even when run with multiple threads. Regarding comparison to other approaches in the literature, to the best of our knowledge no solutions exist yet for the problem we consider in this paper.

5.1   Instances

We use a total of 30 randomly generated instances (based on real-life situations) of different sizes for our experiments. A summary of the instances grouped by their size is given in Table 1. The instances all have three modes: a *single* mode requiring only one employee, a *shift* mode which requires two employees but has a reduced duration, and an *external* mode that requires no employees at all. In general, jobs can be done in single mode or optionally in shift mode. Some instances however also include jobs which can only be performed in external mode. Also, in the test instances the initial assignments are restricted to jobs which are already started or are fixed to their current value via availability restrictions and time windows.

While all of the instances were generated randomly, they are still modelled after real-world scenarios. Half of the instances are modelled very closely to a real-world laboratory, whereas the other half is more general and makes full use of the problem features. The details of how this generation works as well as the exact differences between the laboratory instances and general instances are given in [14]. Furthermore, our 30 instances are a selection from a total of 120 instances given in the report. We chose the first two instances of each size (scheduling horizon and number of projects) and two additional instances for the 3 smallest sizes. This selection was necessary, because of the long time it would have taken to experiment with all 120 instances. Those 120 instances as well as the 30 we selected for this paper can be found at https://www.dbai.tuwien. ac.at/staff/fmischek/TLSP/. Since those instances were generated with the full TLSP in mind and in TLSP-S we take the initial job grouping as fixed and unchangeable, the instances had to be converted. This is achieved by viewing the jobs as the smallest planning unit and assigning the job parameters – which are defined by the tasks contained in the job in TLSP – directly to the jobs.

Table 1. Instances summary

	No. of instances	h	$|P|$	$|J|$	$|E|$	$|B|$	$|G^*|$	$|\bigcup_{g \in G^*} G_g|$
1	8	89	5–10	7–37	7–13	7–13	3–6	6–107
2	12	175	15–60	29–212	12–46	12–46	3–6	16–271
3	6	521	20–60	71–260	6–18	6–18	3–6	18–218
4	4	783	60–90	247–401	13–19	13–19	3–5	16–284

5.2   Results

Table 2 shows the comparison of search procedures described in Sect. 4.5. The column "♯ sat" shows for how many instances the model-search combination found feasible solutions, whereas "♯ opt" contains the number of instances solved to optimality. Furthermore, the values in the column "cum. obj" show the cumulative objective value over all instances, and "avg. rt sat" is the average time it took to find the first feasible solution.

Table 2. Priority search experiments (Runtime 30 m)

Constraints	Search	♯ sat	♯ opt	cum. obj.	avg. rt sat
(1–11), (16–18), (19–25)	Default	13	8	–	–
(1–11), (16–18), (19–25)	ps_modeFirst_ff	**30**	**14**	46529	20.806 s
(1–11), (16–18), (19–25)	ps_modeFirst_aff	**30**	**14**	46530	20.878 s
(1–11), (16–18), (19–25)	ps_startTimeFirst_ff	**30**	**14**	45534	13.728 s
(1–11), (16–18), (19–25)	ps_startTimeFirst_aff	**30**	**14**	44202	**13.177 s**
(1–11), (16–18), (20–25)	ps_startTimeFirst_aff	**30**	**14**	**44103**	13.496 s

Each model was run using Chuffed with free search enabled. Free search alternates between user-defined and activity-based search on each restart. The time limit was set to 30 min for each instance. It can be easily seen that any version of priority_search is vastly superior to the default search of Chuffed. priority_search strategies solve more instances to optimality and also found feasible solution for every instance. It should be noted that the fourteen optimally solved instances are the same over all search configurations. They all have less than or equal to 20 projects (eight of those instances are from class 1, five are from class 2 and one is from class 3 (as described in Table 1). The search strategy ps_startTimeFirst_aff achieved the best cumulative objective value.

While initial experiments showed that redundant constraints (20–25) have a high impact on the search, Table 2 shows that while constraint (19) has no positive impact on the solution quality, it does improve the time to the first feasible solution. However, we decided to drop the constraint in our final experiment since we were mainly interested in solution quality and not in quickly found feasible solutions.

Table 3. CPLEX/CP optimizer comparison (Runtime 2 h)

Solver	Constraints	Search	♯ sat	♯ opt	♯ best
Chuffed	(1–11), (16–18), (20–25)	ps_startTimeFirst_aff	**30**	**15**	15
CPLEX	(1–11), (16–18), (19–25)	Default	10	2	4
CP optimizer	–	Default	**30**	4	**24**

Table 4. Detailed results for the best solutions found in the final experiment (all solutions are feasible for their instance)

Instance	h	$|P|$	$|J|$	$|E|$	$|B|$	$|G^*|$	$|\bigcup_{g \in G^*} G_g|$	s_2	s_3	s_4	s_5	Opt. gap	Solver
1	89	5	7	7	7	3	6	0	7	0	84	0.00%	Chuffed, CPO
2	89	5	8	7	7	3	95	0	9	0	56	0.00%	Chuffed, CPO
3	89	5	24	7	7	3	48	0	14	0	111	0.00%	Chuffed, CPO
4	89	5	14	7	7	3	49	1	12	0	78	0.00%	Chuffed, CPO
5	89	10	29	13	13	4	100	0	22	0	232	0.00%	Chuffed, CPO
6	89	10	18	13	13	6	107	0	20	0	124	0.00%	Chuffed, CPO
7	89	10	37	13	13	3	93	0	29	0	241	0.00%	Chuffed, CPO
8	89	10	29	13	13	3	98	0	27	0	254	0.00%	Chuffed, CPO
9	175	15	29	12	12	5	115	0	32	0	279	0.00%	Chuffed, CPO
10	175	15	53	12	12	3	98	0	38	0	329	0.00%	Chuffed
11	175	20	60	16	16	5	70	0	49	0	392	0.00%	Chuffed
12	175	20	84	16	16	4	16	0	53	0	429	4.98%	CPO
13	175	20	65	16	16	3	134	3	48	0	740	0.00%	Chuffed
14	175	20	62	16	16	3	132	0	50	0	544	0.00%	Chuffed
15	175	30	113	23	23	3	18	2	87	0	902	5.05%	CPO
16	175	30	105	23	23	3	201	1	92	0	105	20.04%	CPO
17	175	40	126	31	31	3	100	1	113	0	963	4.36%	CPO
18	175	40	138	31	31	3	271	2	128	0	1136	19.67%	CPO
19	175	60	208	46	46	6	219	7	213	0	1671	22.95%	CPO
20	175	60	212	46	46	3	397	13	213	0	1846	28.23%	CPO
21	521	20	76	6	6	5	42	0	55	0	548	0.00%	Chuffed
22	521	20	71	6	6	3	67	1	51	0	642	0.00%	Chuffed
23	521	40	196	12	12	4	18	7	124	0	1880	33.96%	CPO
24	521	40	187	12	12	3	146	3	123	0	1550	22.49%	CPO
25	521	60	260	18	18	6	148	15	232	2	2042	36.53%	CPO
26	521	60	239	18	18	3	218	9	199	2	2350	22.11%	CPO
27	783	60	270	13	13	4	16	8	204	0	1856	13.73%	CPO
28	783	60	247	13	14	3	196	15	213	0	1927	21.67%	CPO
29	783	90	384	19	19	5	139	21	363	6	2944	35.45%	CPO
30	783	90	401	19	19	3	284	28	353	0	4213	35.79%	CPO

Table 3 shows our final experiment, which is a comparison between Chuffed, the MIP solver CPLEX and the CP solver CP Optimizer. The time limit was set to 2 h for each instance. The column "♯ best" indicates the number of instances for which the solver found a solution with the best known objective for the specific instance. CP Optimizer was run with 8 threads and the parameter *FailureDirectedSearchEmphasis* was set to 4. CPLEX performed very poorly in comparison to Chuffed and CP Optimizer. Chuffed even found optimal solutions for one more instance with the longer time-limit (from class 3). The results for CP Optimizer are particularly interesting. While it only the manages to prove the optimality of the 4 smallest instances, it generally finds better solutions for instances which cannot be closed by Chuffed. Regarding CPLEX it should be noted that our model was developed with CP solvers in mind and thus might not be a perfect fit for MIP solvers and that CPLEX was only run in single-threaded mode.

Table 4 lists the best solutions found in the final experiment for all 30 test instances. The "Solver" column indicates which solvers found the listed solution or one with an equal objective value. The columns "s_i" ($i \in \{2, \ldots, 5\}$) show the objective values of the respective soft constraints. The total objective value of each solution is the sum of those values.

6   Conclusion

In this paper we have investigated different possibilities to model a complex real-world project scheduling problem. For some of the constraints, we first experimented with approaches which were already used in related project scheduling problems. To deal with this more complex problem and larger instances we introduced several extensions in modeling. We have evaluated our approach on a set of 30 benchmark instances. Using CP techniques we could find feasible solutions for all considered instances. Furthermore, optimal solutions for 15 instances could be provided for the first time.

For the future work, we plan to investigate exact techniques to solve both stages of the TLSP including grouping and scheduling simultaneously.

Acknowledgments. The financial support by the Austrian Federal Ministry for Digital and Economic Affairs and the National Foundation for Research, Technology and Development is gratefully acknowledged. We would also like to thank the anonymous reviewers for their feedback, in particular regarding CP-modelling.

References

1. Bartels, J.H., Zimmermann, J.: Scheduling tests in automotive R&D projects. Eur. J. Oper. Res. **193**(3), 805–819 (2009). https://doi.org/10.1016/j.ejor.2007.11.010
2. Bellenguez, O., Néron, E.: Lower bounds for the multi-skill project scheduling problem with hierarchical levels of skills. In: Burke, E., Trick, M. (eds.) PATAT 2004. LNCS, vol. 3616, pp. 229–243. Springer, Heidelberg (2005). https://doi.org/10.1007/11593577_14
3. Brucker, P., Drexl, A., Möhring, R., Neumann, K., Pesch, E.: Resource-constrained project scheduling: notation, classification, models, and methods. Eur. J. Oper. Res. **112**(1), 3–41 (1999). https://doi.org/10.1016/S0377-2217(98)00204-5
4. Chu, G.: Improving combinatorial optimization. Ph.D. thesis, University of Melbourne, Australia (2011). http://hdl.handle.net/11343/36679
5. Dauzère-Pérès, S., Roux, W., Lasserre, J.: Multi-resource shop scheduling with resource flexibility. Eur. J. Oper. Res. **107**(2), 289–305 (1998). https://doi.org/10.1016/S0377-2217(97)00341-X
6. Drezet, L.E., Billaut, J.C.: A project scheduling problem with labour constraints and time-dependent activities requirements. Int. J. Prod. Econ. **112**(1), 217–225 (2008). https://doi.org/10.1016/j.ijpe.2006.08.021. Special Section on Recent Developments in the Design, Control, Planning and Scheduling of Productive Systems

7. Elmaghraby, S.E.: Activity Networks: Project Planning and Control by Network Models. Wiley, New York (1977)
8. Feydy, T., Goldwaser, A., Schutt, A., Stuckey, P.J., Young, K.D.: Priority search with MiniZinc. In: ModRef 2017: The Sixteenth International Workshop on Constraint Modelling and Reformulation at CP 2017 (2017)
9. Hartmann, S., Briskorn, D.: A survey of variants and extensions of the resource-constrained project scheduling problem. Eur. J. Oper. Res. **207**(1), 1–14 (2010). https://doi.org/10.1016/j.ejor.2009.11.005
10. IBM, CPLEX: 12.8.0 IBM ILOG CPLEX optimization studio CP optimizer user's manual (2017). https://www.ibm.com/analytics/cplex-cp-optimizer
11. IBM, CPLEX: 12.8.0 IBM ILOG CPLEX optimization studio CPLEX user's manual (2017). https://www.ibm.com/analytics/cplex-optimizer
12. Mika, M., Waligóra, G., Węglarz, J.: Overview and state of the art. In: Schwindt, C., Zimmermann, J. (eds.) Handbook on Project Management and Scheduling. International Handbooks on Information Systems, vol. 1, pp. 445–490. Springer, Cham (2015). https://doi.org/10.1007/978-3-319-05443-8_21
13. Mischek, F., Musliu, N.: A local search framework for industrial test laboratory scheduling. In: Proceedings of the 12th International Conference on the Practice and Theory of Automated Timetabling (PATAT-2018), Vienna, Austria, 28–31 August 2018, pp. 465–467 (2018)
14. Mischek, F., Musliu, N.: The test laboratory scheduling problem. Technical report, Christian Doppler Laboratory for Artificial Intelligence and Optimization for Planning and Scheduling, TU Wien, CD-TR 2018/1 (2018)
15. Nethercote, N., Stuckey, P.J., Becket, R., Brand, S., Duck, G.J., Tack, G.: MiniZinc: towards a standard CP modelling language. In: Bessière, C. (ed.) CP 2007. LNCS, vol. 4741, pp. 529–543. Springer, Heidelberg (2007). https://doi.org/10.1007/978-3-540-74970-7_38
16. Nudtasomboon, N., Randhawa, S.U.: Resource-constrained project scheduling with renewable and non-renewable resources and time-resource tradeoffs. Comput. Ind. Eng. **32**(1), 227–242 (1997). https://doi.org/10.1016/S0360-8352(96)00212-4
17. Pritsker, A.A.B., Waiters, L.J., Wolfe, P.M.: Multiproject scheduling with limited resources: a zero-one programming approach. Manage. Sci. **16**(1), 93–108 (1969). https://doi.org/10.1287/mnsc.16.1.93
18. Salewski, F., Schirmer, A., Drexl, A.: Project scheduling under resource and mode identity constraints: model, complexity, methods, and application. Eur. J. Oper. Res. **102**(1), 88–110 (1997). https://doi.org/10.1016/S0377-2217(96)00219-6
19. Schulte, C., Lagerkvist, M., Tack, G.: Gecode 6.10 reference documentation (2018). https://www.gecode.org
20. Schwindt, C., Trautmann, N.: Batch scheduling in process industries: an application of resource-constrained project scheduling. OR-Spektrum **22**(4), 501–524 (2000)
21. Szeredi, R., Schutt, A.: Modelling and solving multi-mode resource-constrained project scheduling. In: Rueher, M. (ed.) CP 2016. LNCS, vol. 9892, pp. 483–492. Springer, Cham (2016). https://doi.org/10.1007/978-3-319-44953-1_31
22. Węglarz, J., Józefowska, J., Mika, M., Waligóra, G.: Project scheduling with finite or infinite number of activity processing modes–a survey. Eur. J. Oper. Re. **208**(3), 177–205 (2011). https://doi.org/10.1016/j.ejor.2010.03.037
23. Young, K.D., Feydy, T., Schutt, A.: Constraint programming applied to the multi-skill project scheduling problem. In: Beck, J.C. (ed.) CP 2017. LNCS, vol. 10416, pp. 308–317. Springer, Cham (2017). https://doi.org/10.1007/978-3-319-66158-2_20

An Approach to Robustness in the Stable Roommates Problem and Its Comparison with the Stable Marriage Problem

Begum Genc[1](✉), Mohamed Siala[2], Gilles Simonin[3], and Barry O'Sullivan[1]

[1] Insight, Centre for Data Analytics, University College Cork, Cork, Ireland
{begum.genc,barry.osullivan}@insight-centre.org
[2] LAAS-CNRS, Université de Toulouse, CNRS, INSA, Toulouse, France
mohamed.siala@laas.fr
[3] IMT Atlantique, LS2N, CNRS, UBL, 44307 Nantes, France
gilles.simonin@imt-atlantique.fr

Abstract. Recently a robustness notion for matching problems based on the concept of a (a, b)-supermatch is proposed for the Stable Marriage problem (SM). In this paper we extend this notion to another matching problem, namely the Stable Roommates problem (SR). We define a polynomial-time procedure based on the concept of reduced rotation poset to verify if a stable matching is a $(1, b)$-supermatch. Then, we adapt a local search and a genetic local search procedure to find the $(1, b)$-supermatch that minimises b in a given SR instance. Finally, we compare the two models and also create different SM and SR instances to present empirical results on the robustness of these instances.

1 Introduction

Robustness to change is an important property that has a variety of definitions in different settings [15]. There exist many robustness notions within the context of matching problems. These robustness notions mostly focus on handling uncertainty and erroneous data in the input [1–3,12]. Genc et al. introduced a novel notion of robustness for the Stable Marriage problem (SM) where the robustness of a solution refers to its capability to be repaired at a small bounded cost in case of an unforeseen event [4]. The notion of (a, b)-supermatches differs from the other robustness notions in this context since it specifies a degree of repairability. This property is often referred as fault-tolerance. The (a, b)-supermatch concept defines the notion of robustness for matching problems by using the fault-tolerance framework [8,9].

The SM is defined by a set of men and a set of women, each of which has a set of preferences over people of opposite sex. The task is to find a (monogamous) matching between men and women that is stable. A matching is said to be stable if there are no two pairs that are not matched to each other, but they prefer being together than being with their current partners. The robust variant of the problem is called *Robust Stable Marriage (RSM)* [4], in which the robustness of

© Springer Nature Switzerland AG 2019
L.-M. Rousseau and K. Stergiou (Eds.): CPAIOR 2019, LNCS 11494, pp. 320–336, 2019.
https://doi.org/10.1007/978-3-030-19212-9_21

a stable matching is measured by the minimum number of changes required to obtain another stable matching in the case of break-up of some pairs. If a pair appears in all the stable matchings, the pair is said to be fixed, otherwise, non-fixed. An (a, b)-supermatch is a stable matching such that if any a non-fixed agents (men/women) break-up, it is possible to find another stable matching by changing the partners of those a agents and also changing the partners of at most b others. The previous work on the RSM includes the proposal of the problem, a complexity study, a polynomial-time verification procedure for a given $(1, b)$-supermatch, and three different models (constraint programming, genetic algorithm, local search) to find the $(1, b)$-supermatch that minimises b for a given SM instance [4,6]. We investigate in this paper this robustness concept further on a generalised version of the SM, namely the Stable Roommates problem (SR). The Stable Roommates problem is a one-sided generalisation of SM, where any two agents regardless of their gender can be matched. We define the *Robust Stable Roommates problem (RSR)* analogous to the RSM. To the best of our knowledge, there is no previous research on finding the (a, b)-supermatches of the SR.

The motivation behind studying RSR is due to the large applicability of SR and the importance to handle the dynamism of the real world. Take the example of P2P networks where peers (computers for instance) are connected to each other for file sharing purposes [14]. Each peer has a preference list towards the other peers and a matching that respect stability is required. However, as the network evolves during time, peers continuously seek new partners. That is, if a peer that provides the file loses the connection, an alternative peer is needed for downloading a file. In this situation, we have to maintain stability with possibly the minimum changes to the current solution. An (a, b)-supermatch guarantees finding other peers to the broken ones at a small number of additional changes while preserving stability.

The paper is organised as follows: In Sect. 2, we give a formal background and introduce the robust stable roommates problem. Then, in Sect. 3, we show that one can verify in polynomial time if a given stable matching (in the SR context) is a $(1, b)$-supermatch. Next, we adapt a local search procedure and a hybrid (genetic local search) model for finding robust solutions in Sect. 4. Finally, we present in Sect. 5 our empirical study.

2   Background and Notation

The Stable Roommates problem (SR) consists of a set of $2 \times n$ agents, where each agent has a preference list in which he/she ranks all other agents in strict order of preference. In this context, given a set of people P, a *matching* corresponds to a partition of P into disjoint pairs (or partners). A matching is *stable* if it admits no blocking pairs. A pair $\{p_i, p_j\}$ blocks a matching if: p_i is unassigned or prefers p_j to his/her current partner, or p_j is unassigned or prefers p_i to his/her current partner. The solution to an SR instance is a stable matching. If such a solution does not exist, then the instance is *unsolvable*. A pair is *stable* if it appears in some stable matching. If a pair appears in all stable matchings, it is called a *fixed* pair. If a person p has at least two different partners among all stable matchings, p is said to be *non-fixed*. We measure the distance between any two stable matchings M, M' by the number of different pairs $d(M, M') = \mid M \setminus M' \mid$. The stable matching M' among all the stable matchings of the instance that has the minimum distance to M is said to be the *closest stable matching* to M.

Irving defines an $O(n^2)$ procedure to find a solution to SR or to report if none exists [10]. The procedure consists of two phases. Let us first define some notations to describe these phases. A *preference table* (denoted by T) is, for a given problem instance, a set of preference lists for which zero or more entries have been deleted. We use T_{init} to denote the initial preference table. During the two phases, some pairs are removed from T_{init}. We denote the preference list of a person p_i in a table T by $L_T(i)$. Let $f_T(p_i), s_T(p_i), l_T(p_i)$ denote the first, second and last entries of $L_T(i)$. The first phase is based on each person proposing to the first available person in their lists starting from T_{init} until every person has made a proposal that has been accepted, i.e. became semi-engaged. If a person p_i becomes semi-engaged to p_j, all pairs $\{p_j, p_k\}$ such that p_j prefers p_i to p_k are deleted from the table. The table obtained after applying the Phase 1 algorithm is called the *Phase-1 table* and is denoted by T_0.

The second phase of the algorithm is based on finding and eliminating *rotations* starting from T_0. A rotation ρ is a circular list denoted as $\rho = (x_0, y_0), (x_1, y_1), \ldots, (x_{r-1}, y_{r-1})$, where all $x_i, y_j \in P$. Each rotation has the property that $y_i = f_T(x_i)$ and $y_{i+1} = s_T(x_i)$ in a table T for all i, $0 \leq i \leq r - 1$, where i+1 is taken modulo r. The set of people $\{x_0, \ldots, x_{r-1}\}$ is called the *X-set* of ρ, denoted by $X(\rho)$. Similarly, $\{y_0, \ldots, y_{r-1}\}$ is called the *Y-set* of ρ, denoted by $Y(\rho)$. Additionally, given a set of rotations R, $X(R) = \cup_{\rho \in R} X(\rho)$. Similar for the Y-set. The *elimination* of a rotation ρ from a table T means for each pair $\{p_i, p_j\}$, where $p_i = x_m$ and $p_j = y_m$ and $(x_m, y_m) \in \rho$, the deletion of $\{p_i, p_j\}$ and all pairs $\{y_m, z\}$ such that y_m prefers x_{m-1} to z from T. In this case, ρ is said to be *exposed* on T and the table after eliminating ρ is denoted by T/ρ. If after Phase 1 or Phase 2, all lists in T contain exactly one entry, then T represents a stable matching. Note that sometimes we use (p_i, p_j) and (x_m, y_m) interchangeably. The notation (x_m, y_m) is used for denoting the position of the pair (p_i, p_j) in ρ. Lemma 4.4.1 from [7] states that $\{p_i, p_j\}$ is a stable non-fixed pair if and only if (p_i, p_j) or (p_j, p_i) is in a non-singular rotation.

There are two types of rotations: *singular* and *non-singular*. A rotation $\rho = (x_0, y_0), (x_1, y_1), \ldots, (x_{r-1}, y_{r-1})$ is called a non-singular rotation if $\bar{\rho} = (y_1, x_0), (y_2, x_1), \ldots, (y_0, x_{r-1})$ is also a rotation. In this case, ρ and $\bar{\rho}$ are called as *duals* of each other. If a rotation does not have a dual, then it is a singular rotation. We denote by T_S the table where all singular rotations are eliminated from T_0. A rotation ρ' is said to precede another rotation ρ (denoted by $\rho' \prec \rho$) if ρ' is eliminated for ρ to become exposed. In this case, we say ρ' is a *predecessor* of ρ and ρ is a *successor* of ρ'. A rotation ρ' is an *immediate predecessor* of ρ, and ρ is an *immediate successor* of ρ', if $\rho' \prec \rho$ and there does not exist a ρ^* such that $\rho' \prec \rho^* \prec \rho$. All predecessors and successors of a rotation, not necessarily immediate, are denoted by $N^-(\rho)$ and $N^+(\rho)$. The set of both singular and non-singular rotations under $\prec$ defines the *roommates rotation poset*. The set of non-singular rotations under $\prec$ defines the *reduced rotation poset* and is denoted by $\Pi = (\mathcal{V}, E)$. We refer to any two rotations as *incomparable* if none of them precede the other one, *comparable* otherwise. Let us illustrate these concepts on an SR instance $\mathcal{I}$. We use a sample instance of 10 people from page 180 in [7]. Figure 1 represents the T_S of $\mathcal{I}$. Figure 2 represents the reduced rotation poset of $\mathcal{I}$, where the pairs involved in the rotations are given next to their corresponding rotations for convenience.

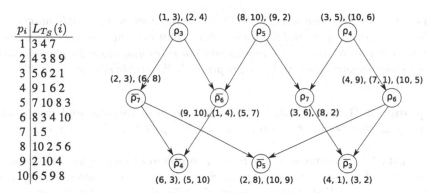

p_i	$LT_S(i)$
1	3 4 7
2	4 3 8 9
3	5 6 2 1
4	9 1 6 2
5	7 10 8 3
6	8 3 4 10
7	1 5
8	10 2 5 6
9	2 10 4
10	6 5 9 8

Fig. 1. The T_S for an SR instance $\mathcal{I}$ of size 10.

Fig. 2. Reduced rotation poset of $\mathcal{I}$ given in Fig. 1.

Table 1. A list of all the stable matchings and their corresponding complete closed subsets of $\mathcal{I}$.

$M_1 = \{(1,3),(2,4),(5,7),(6,8),(9,10)\}$	$S_1 = \{\bar\rho_3, \rho_4, \rho_5, \rho_6, \rho_7\}$
$M_2 = \{(1,7),(2,8),(3,5),(4,9),(6,10)\}$	$S_2 = \{\rho_3, \bar\rho_4, \rho_5, \bar\rho_6, \bar\rho_7\}$
$M_3 = \{(1,4),(2,9),(3,6),(5,7),(8,10)\}$	$S_3 = \{\rho_3, \rho_4, \bar\rho_5, \rho_6, \bar\rho_7\}$
$M_4 = \{(1,4),(2,3),(5,7),(6,8),(9,10)\}$	$S_4 = \{\rho_3, \rho_4, \rho_5, \rho_6, \rho_7\}$
$M_5 = \{(1,4),(2,8),(3,6),(5,7),(9,10)\}$	$S_5 = \{\rho_3, \rho_4, \rho_5, \rho_6, \bar\rho_7\}$
$M_6 = \{(1,7),(2,3),(4,9),(5,10),(6,8)\}$	$S_6 = \{\rho_3, \rho_4, \rho_5, \bar\rho_6, \rho_7\}$
$M_7 = \{(1,7),(2,8),(3,6),(4,9),(5,10)\}$	$S_7 = \{\rho_3, \rho_4, \rho_5, \bar\rho_6, \bar\rho_7\}$

A subset of the rotations in Π, containing one of each dual rotations and all their predecessors, is called a *complete closed subset*, denoted by S. There exists a 1-1 correspondence between the complete closed subsets of Π and the stable matchings of the underlying instance [7]. Any stable matching can be obtained by eliminating one of each dual rotations starting from T_S. A rotation ρ is said to *eliminate* $\{p_i, p_j\}$ if there exists a table T such that $\{p_i, p_j\} \in T$ and $\{p_i, p_j\} \notin T/\rho$. On the other hand, a rotation ρ is said to *produce* $\{p_i, p_j\}$ if there exists a table T such that $|L_T(i)| > 1$, $|L_T(j)| > 1$, $L_{T/\rho}(i)$ contains only p_j, and $L_{T/\rho}(j)$ contains only p_i. We use the term *flipping* ρ from S as the process of removing $\rho \in S$ from S and adding its dual $\bar\rho$ to S. A neighbour rotation

ρ is $\rho \notin S$ and either the rotation has no predecessors $(N^-(\rho) = \emptyset)$ or for all predecessors $\rho' \in N^-(S), \rho' \in S$. The set $\mathbf{N}(S)$ denotes the set of *neighbour rotations*. The set of all sink nodes of the graph induced by S is referred as the *sink rotations* of S, denoted as $\mathbf{L}(S)$. Table 1 presents all the 7 stable matchings of $\mathcal{I}$ given in Fig. 1 and their corresponding complete closed subsets.

Throughout the paper, we denote by M a given stable matching, and its corresponding complete closed subset by S. If there are any subscripts or superscripts for M such as M_i^*, then they are applied to the corresponding complete closed subset (i.e. S_i^*). Lemmas 1 and 2 are included here to be used in our proofs later.

Lemma 1 (Lemma 4.1.1 [7]). *Given an instance of the stable marriage problem involving n men and n women, there is an instance (in fact there are many instances) of the stable roommates problem involving those $2n$ persons such that the stable roommates matchings are precisely the stable matchings for the original SM instance.*

Lemma 2 (Lemma 4.3.7 [7]). *If ρ, σ are non-singular and π is a singular rotation, then: (1) $\rho \not\prec \bar{\rho}$; (2) $\rho \prec \sigma \iff \bar{\sigma} \prec \bar{\rho}$; (3) $\tau \prec \pi \implies \tau$ is singular.*

Robust Stable Roommates: We refer the problem of finding an (a, b)-supermatch to a given SR instance as the *Robust Stable Roommates problem (RSR)*. A stable matching of an RSR instance is called an (a, b)-*supermatch* if any a non-fixed pairs do not want to be partners anymore (i.e. *leave* the stable matching), it is possible to find another stable matching by changing the partners of the people involved in those a pairs and at most b other pairs.

Definition 1 ((a, b)-supermatch). *Given an SR instance $\mathcal{I}$, and two positive integers $a, b \in \mathbb{N}$, a stable matching M of $\mathcal{I}$ is said to be an (a, b)-supermatch if for any set $\Psi \subseteq M$ of non-fixed stable pairs, where $|\Psi| = a$, there exists a stable matching M' such that $M' \cap \Psi = \emptyset$ and $d(M, M') \le b + a$.*

The intractability result of the RSM is lifted to the RSR as the SR is a generalisation of the SM.

Theorem 1. *RSR is $\mathcal{NP}$-hard.*

Proof. The proof is straightforward as it is possible to create an SR instance I_{SR} from any given SM instance I_{SM} with the exact same stable matchings in polynomial-time by padding every other person of the same sex to the preference list of each person (see Lemma 1). Every (a, b)-supermatch in the I_{SM} is also an (a, b)-supermatch in the I_{SR} and vice versa. Hence, RSR is $\mathcal{NP}$-hard because RSM is $\mathcal{NP}$-hard [6]. $\square$

3   Verification of $(1, b)$-supermatch in Polynomial Time

We prove in this section that checking if a stable matching M is a $(1, b)$-supermatch can be done in polynomial time. Indeed, we show in Theorem 3 how to construct the closet matching to M if any non-fixed pair in M wants to leave.

In order to show our main result, we first prove in Theorem 2 that any non-fixed pair can be:(1) produced by a unique rotation and eliminated by another one; or (2) eliminated by two different rotations and produced by two others (see later Example 1). In the first case, we shall denote by ρ_e the elimination rotation and by ρ_p the production rotation. In the second case, we shall denote by ρ_{p1}, ρ_{p2} the two production rotations and by ρ_{e1}, ρ_{e2} the two elimination rotations.

We assume w.l.o.g that the input instance admits at least two stable matchings. For any non-fixed stable pair (p_i, p_j), there are two possible cases to consider:

Case 1: (A) $f_{T_S}(i) = p_j$ and $l_{T_S}(j) = p_i$, or (B) $l_{T_S}(i) = p_j$ and $f_{T_S}(j) = p_i$;
Case 2: Otherwise.

Case 1 is a special case indicating that if one of the persons in the pair is the other ones' most preferred person in T_S (respectively, the other one is the least preferred person in T_S). Note that, in both cases $L_{T_S}(i) > 1$ and $L_{T_S}(j) > 1$, because the pairs are non-fixed. Later, we refer to these cases for identifying scenarios. In Lemma 3, we show how to identify the elimination rotation(s) for a given pair regardless of its case.

Lemma 3. *A non-fixed stable pair $\{p_i, p_j\}$ is eliminated by a rotation ρ if and only if $(p_i, p_j) \in \rho$ or $(p_j, p_i) \in \rho$.*

Proof. $\rightarrow$ Let $\rho = (x_0, y_0), (x_1, y_1) \dots, (x_{|\rho|-1}, y_{|\rho|-1})$ be a rotation that eliminates $\{p_i, p_j\}$. Observe first that ρ is non-singular (otherwise $\{p_i, p_j\}$ is not stable). Recall that the elimination of ρ from a table T means for each pair $(x_m, y_m) \in \rho$, the deletion of $\{x_m, y_m\}$ and all pairs $\{y_m, z\}$ such that y_m prefers x_{m-1} to z from T. Table 2 gives an illustration of the preferences of x_m and y_m. The eliminating ρ moves x_m from y_m to y_{m+1} and deletes some $\{y_m, z\}$. In a similar way, eliminating $\bar{\rho}$ moves y_m from x_{m-1} to x_m and deletes some $\{x_m, z'\}$. Since every close complete subset contains either ρ or $\bar{\rho}$, then any pair $\{y_m, z\}$

Table 2. An illustration of the preferences

p	Preference lists
$\dots$	$\dots$
x_m	$\dots, y_m, z', y_{m+1} \dots$
$\dots$	$\dots$
y_m	$\dots, x_{m-1}, z, x_m, \dots$
$\dots$	$\dots$

and $\{x_m, z'\}$ cannot be part of any solution. Therefore, if ρ eliminates $\{p_i, p_j\}$ and $\{p_i, p_j\} \notin \rho$ then $\{p_i, p_j\}$ is not stable. This contradicts the fact that our pair $\{p_i, p_j\}$ is a non-fixed stable pair.

$\leftarrow$ By the definition of eliminating a rotation ρ from a table T, where $(p_i, p_j) \in \rho$, the elimination results in the deletion of p_j from p_i's list. Similarly, if $(p_j, p_i) \in \rho$ then it results in the deletion of p_i from p_j's list. $\square$

Lemma 4 identifies the production rotations.

Lemma 4. *If a non-fixed stable pair $\{p_i, p_j\}$ is eliminated by ρ_e, then $\{p_i, p_j\}$ is produced by the dual of it, $\rho_p = \bar{\rho}_e$.*

Proof. A rotation is said to produce $\{p_i, p_j\}$ if eliminating it from a table T reduces $L_{T/\rho}(i)$ to a single entry, namely to p_j and $L_{T/\rho}(j)$ to p_i. We prove the existence of the production rotations over the two cases (Case 1 and Case 2) identified above.

We have two sub-cases in Case 1. First case is when $f_{T_S}(i) = p_j, l_{T_S}(j) = p_i$. In order to reduce p_i's list to only p_j, we need a rotation that moves p_i from his/her second best choice up to the first choice. We refer to this operation as *limiting p_i from right*. Similarly, to reduce the p_j's list to only p_i, we need a rotation that moves p_j from his/her second least-preferred person to the least preferred person. We refer to this operation as *limiting p_j from left*. Referring back to Table 2 for notation, the production rotation ρ_p of the pair $\{p_i, p_j\} = (x_m, y_m)$ must contain the pair $(y_{m+1}, x_m) \in \rho_p$ to limit x_m from right. Additionally, it must contain (y_m, x_{m-1}) to limit y_m from left. To illustrate, the production rotation has the shape: $\rho_p = \ldots, (y_m, x_{m-1}), (y_{m+1}, x_m), \ldots$. Note that, each ordered pair can only appear in exactly one rotation. Observe that, the dual of ρ_p contains the pair (x_m, y_m) by definition of dual. By Lemma 3, we know that the rotation that contains (x_m, y_m) is the elimination rotation of the pair $\{p_i, p_j\}$. Therefore, $\rho_p = \bar{\rho}_e$ The proof for the second sub-case is similar, where $(y_m, x_m) \in \rho_e$.

For a pair $\{p_i, p_j\}$ of Case 2, each person has both more and less preferred people in their lists. Therefore, in order to produce a pair, their lists must be limited from both left and right. Let ρ_{p1} denote the rotation that limits p_i from left and p_j from right, and ρ_{p2} denote the rotation that limits p_i from right and p_j from left, respectively. Let the preference lists for the pair $\{p_i, p_j\}$ denoted by $L_{T_S}(i) = [\ldots, y_{m-1}, y_m, y_{m+1}]$ and $L_{T_S}(j) = [\ldots, x_{m-1}, x_m, x_{m+1}]$ where $\{p_i, p_j\} = (x_m, y_m)$. The pair (x_m, y_{m-1}) must be in ρ_{p1} to limit p_i from left and (x_{m+1}, y_m) be in ρ_{p1} to limit p_j from right. Additionally, the pair (y_{m+1}, x_m) must be in ρ_{p2} to limit p_i from right and (y_m, x_{m-1}) to limit p_j from left. Note that, the dual of ρ_{p1} contains (y_m, x_m), the dual of ρ_{p2} contains (x_m, y_m) by the definition of a dual rotation. By Lemma 3, we know these rotations are elimination rotations of the pair $\{p_i, p_j\}$.

Note that the two rotations ρ_{p1} and ρ_{p2} do not require one of them to be eliminated from the table first; they are incomparable. Therefore, depending on the order of elimination, both of them are production rotations. $\square$

We sum up the findings above for the non-fixed stable pairs. If a pair is of Case 1, then there exists only one elimination rotation for this pair and only one production rotation as the dual of the elimination one. Because the preference list needs to be limited in only one direction. However, for the pairs of Case 2, there exist two elimination rotations for this pair, and also two other production rotations. Observe that, for each non-fixed stable pair $\{p_i, p_j\}$ in a stable matching M, the corresponding complete closed subset of M contains all production rotations of $\{p_i, p_j\}$. It is important to note that, especially for the pairs of Case 2, including one production rotation in the complete closed subset and not the other one, results in producing other partners for that pair. Subsequently, Theorem 2 is an immediate result of Lemmas 3 and 4.

Theorem 2. *Let $\{p_i, p_j\}$ be a non-fixed stable pair. If $\{p_i, p_j\}$ is of **Case 1**, then there exists a unique elimination rotation ρ_e, where $(p_i, p_j) \in \rho_e$ or $(p_j, p_i) \in \rho_e$, and a unique production rotation ρ_p, where $\rho_p = \bar{\rho}_e$. Otherwise (**Case 2**), there exist two different elimination rotations ρ_{e1} and ρ_{e2}, where $(p_i, p_j) \in \rho_{e1}, (p_j, p_i) \in \rho_{e2}$ and two rotations $\rho_{p1} = \bar{\rho}_{e1}, \rho_{p2} = \bar{\rho}_{e2}$ that produce the pair.*

Let S_P denote the set of all the complete closed subsets for the underlying SR instance. Lemma 5 gives a characterisation for the complete closed subsets.

Lemma 5. *Let $S \in S_P$. For each sink rotation ρ of S, the set $S \setminus \{\rho\} \cup \{\bar{\rho}\} \in S_P$.*

Proof. By definition of closed subset, every predecessor $\rho' \in N^-(\rho)$ is in S. Since ρ is a sink rotation, any successor $\rho^* \in N^+(\rho)$ is not in S. Therefore, by definition of the complete closed subset, we have $\bar{\rho}^* \in S$ and $\bar{\rho}$ is not in S. Using Lemma 2, we know that $\bar{\rho}^* \prec \bar{\rho}$. Hence, all predecessors of $\bar{\rho}$ are already in S, making $\bar{\rho}$ a neighbour rotation and results in $S \setminus \{\rho\} \cup \{\bar{\rho}\} \in S_P$.   □

The distance between two stable matchings $d(M, M')$ is previously defined in Sect. 2 as the number of different pairs between M and M'. Observe that the distance can be calculated by also using their corresponding complete closed subsets. If $S \setminus S' = \{\rho\}$, it means $\rho \in S$ and $\bar{\rho} \in S'$. We know that, $X(\{\rho\}) = Y(\{\bar{\rho}\})$ and $Y(\{\rho\}) = X(\{\bar{\rho}\})$. Therefore, between M and M', only the people in ρ (or $\bar{\rho}$) have different partners. This can also be generalised to a set of rotations. Hence, the distance can also be denoted as $d(S, S') = |X(S \setminus S') \cup Y(S \setminus S')|/2$. Note that $d(S', S) = d(S, S')$.

Lemma 6 identifies the closest stable matching to a stable matching M, when a rotation from its corresponding complete closed subset is to be removed.

Lemma 6. *Given a stable matching M and its corresponding complete closed subset S, if $\rho \in S$ is a rotation to remove from S, the closest stable matching M' to M such that $\rho \notin S'$ is found by the formula[1]:*

$$C(S, \rho) = S' = (S \setminus (\{\rho\} \cup N^+(\rho))) \cup \{\bar{\rho}\} \cup \bigcup_{\rho^* \in N^+(\rho)} \bar{\rho}^* \tag{1}$$

---

[1] The parentheses are used to indicate priority.

Proof. The proof of the defined set S' being a complete closed subset is obvious by using Lemmas 2 and 5 as flipping a sink rotation of S yields in another complete closed subset. However, if ρ is not a sink rotation in S, we must flip all the successors of ρ to obtain a complete closed subset.

Let M^* denote the stable matching after flipping $\rho \in S$. Then, $d(M, M^*) = d(S, S^*) = |X(\{\rho\}) \cup Y(\{\rho\})|/2$. Now, let M^* denote the stable matching after flipping both $\rho, \sigma \in S$. Then, $d(M, M^*) = |X(\{\rho\}) \cup Y(\{\rho\}) \cup X(\{\sigma\}) \cup Y(\{\sigma\})|/2$. Observe that, flipping more rotations can only increase the distance between matchings. In Formula 1, the required number of flips is minimum. Therefore the function $C(S, \rho)$ returns the closest stable matching to M when $\rho \in S$ to be removed from S. □

Finally, Theorem 3 concludes how to find the closest stable matching M' to M if $\{p_i, p_j\} \in M$ wants to leave the M.

Theorem 3. *Given a stable matching M and a pair $\{p_i, p_j\}$ to leave M, the closest stable matching M' to M is identified by its corresponding S' using the Formula 1 as follows:*

1. *If Case 1, then $S' = C(S, \rho_p)$.*
2. *If Case 2, let M_1 and M_2 be the two stable matchings s.t. $S_1 = C(S, \rho_{p1})$ and $S_2 = C(S, \rho_{p2})$. Then $S' = S_1$ if $d(M, M_1) < d(M, M_2)$, otherwise $S' = M_2$.*

Proof. The proof is immediate from Theorem 2 and Lemma 6. □

In order to verify if a given M is a $(1, b)$-supermatch, all closest stable matchings to the given stable matching are found under the assumption that each non-fixed pair wants to leave the stable matching, one at a time. For each pair, its production rotation is identified and then Theorem 3 is applied to find the closest stable matching. Among all the closest stable matchings, the matching that results in the maximum distance to M sets the robustness of M, i.e. $b = d(M, M') - 1$, where 1 denotes the pair that wants to leave.

Example 1. [Computing robustness] Let us calculate the closest matching to M_6 given in Table 1. In Table 3, we identify the cases, and the production/elimination rotation(s) for assuming each pair leaves the M_6 at a time, and we apply Theorem 3 to find the robustness. The pair that has the maximum cost to be repaired sets the robustness value of the matching. Therefore, for this case, the robustness of M_6 is 3.

The production and elimination rotations of each pair can be identified in a preprocessing step. We show that checking if a stable matching is a $(1, b)$-supermatch can be performed in $O(n \times |\mathcal{V}|)$ time after the $O(n^3 log(n))$ preprocessing step for an instance where $2 \times n$ people are involved. The preprocessing step consists in identifying the rotations and building the reduced rotation poset $(O(n^3 log n))$ [7]; identifying all the predecessors and successors of each rotation ρ $(O(|\mathcal{V}|^2))$; and identifying elimination and production rotations for each pair

Table 3. Computing the closest matching to M_6

$\{p_i, p_j\}$	Case	ρ_p	ρ_e	$C(S, \rho)$	S	$d(M, M')$	S'	b
$\{p_1, p_7\}$	1	$\rho_p = \bar{\rho}_6$	$\rho_e = \rho_6$	$\{\rho_3, \rho_4, \rho_5, \rho_6, \rho_7\}$	S_4	4	S_4	3
$\{p_2, p_3\}$	2	$\rho_{p1} = \rho_7$	$\rho_{e1} = \bar{\rho}_7$	$\{\rho_3, \rho_4, \rho_5, \bar{\rho}_6, \bar{\rho}_7\}$	S_7	2	S_7	1
		$\rho_{p2} = \rho_3$	$\rho_{e2} = \bar{\rho}_3$	$\{\bar{\rho}_3, \rho_4, \rho_5, \rho_6, \rho_7\}$	S_1	4		
$\{p_4, p_9\}$	1	$\rho_p = \bar{\rho}_6$	$\rho_e = \rho_6$	$\{\rho_3, \rho_4, \rho_5, \rho_6, \rho_7\}$	S_4	4	S_4	3
$\{p_5, p_{10}\}$	2	$\rho_{p1} = \rho_4$	$\rho_{e1} = \bar{\rho}_4$	$\{\rho_3, \bar{\rho}_4, \rho_5, \bar{\rho}_6, \bar{\rho}_7\}$	S_2	3	S_2	2
		$\rho_{p2} = \bar{\rho}_6$	$\rho_{e2} = \rho_6$	$\{\bar{\rho}_3, \rho_4, \rho_5, \rho_6, \rho_7\}$	S_1	4		
$\{p_6, p_8\}$	1	$\rho_p = \rho_7$	$\rho_e = \bar{\rho}_7$	$\{\rho_3, \rho_4, \rho_5, \bar{\rho}_6, \bar{\rho}_7\}$	S_7	2	S_7	1

$\{p_i, p_j\}$ whenever applicable in $(O(n^2))$. Given a stable matching M, its corresponding complete closed subset S is found by finding and adding the production rotation(s) of each pair and their predecessors into S by starting from an empty set $(O(n \times |\mathcal{V}|))$. Conversely, given a closet complete S, M can be constructed by eliminating all the rotations in S from T_S by respecting their precedence order. The order is found by applying sorting $(O(|\mathcal{V}| \times log|\mathcal{V}|))$. The main algorithm is to compute for each pair in M, the closest stable matching M' by using Theorem 3. Observe that computing the distance between two stable matchings takes $O(n)$ time and flipping a rotation takes a constant time. Moreover, the worst case of finding the closest stable matching is to flip all the non-singular rotations in S, where the number of all non-singular rotations is $|\mathcal{V}|/2$. Therefore, this computation takes $O(|\mathcal{V}|)$ time.

4 Finding Robust Solutions to the SR

We consider in this section two meta-heuristic approaches to solve the problem of finding a $(1, b)$-supermatch to a given Stable Roommates instance that minimizes the value of b.

4.1 Local Search

Considering the structural similarities between the RSM and the RSR, we tailored the local search model (LS) for the RSM, as it is shown that the LS model produces near optimal solutions for RSM and is better than the proposed genetic algorithm [4]. In the generic LS model, there exists a neighbourhood N for the current solution. The algorithm works by searching the neighbourhood of the current solution, finding the best neighbour M_n in the neighbourhood and then proceeding the search by checking the neighbourhood of M_n. The aim is to find the stable matching that has the minimum b value. The search is restarted by a random stable matching at every few iterations to avoid getting stuck at a local optimum. The search continues until a termination criterion is met.

In our model, we have four termination criteria. The first one is a cut-off limit lim_{cutoff}, which "counts" the number of steps since the last best solution is found. The second one is the depth limit lim_{depth}, which indicates the depth of the neighbourhood search starting from a random stable matching. Another criterion is the optimality opt, which indicates if the algorithm has already found a solution with $b = 1$. Finally, we use a time limit lim_{time} for each instance.

The procedure starts by creating a random stable matching M_c as follows. We first mark all the non-singular rotations as available. Let A denote the set of rotations that are available. Then, we randomly select a rotation ρ from A and add it to the initially empty S_c. Subsequently, we remove ρ and $\bar\rho$ from A. We also add all predecessors ρ' of ρ that are not in S_c to S_c and remove ρ' and $\bar\rho'$ from A. This operation operates in a loop until $|S_c| = |\mathcal{V}|/2$. Once the complete closed subset S_c is found, its corresponding stable matching is computed by eliminating all rotations in S_c from T_S by respecting their precedence order.

After creating a random stable matching M_c, the neighbourhood N of M_c is found by checking all the sink rotations in S_c. By using Lemma 5, we know that flipping any sink rotation in M_c creates another stable matching M_n, which we refer as a *neighbour* of the M_c. The general procedure is the same as the one developed for the RSM [5]. In brief, the process starts by descending from the M_c by finding N of M_c. The next iteration descends from the neighbour of M_c that has the lowest b value. This loop is restarted every lim_{depth} iteration by a random M_c. The stable matching that has the minimum value of b as found during the search is returned as the solution.

The complexity of the LS procedure depends on the computation of the b values. Finding neighbours is based on the identification of the sink rotations of S_c, where there can be at most $|\mathcal{V}|/2$ sink rotations and then a constant cost for flipping each sink rotation. The best neighbour is identified after computing b values of $|N|$ stable matchings. This procedure takes $O(k \times n \times |\mathcal{V}| \times |N|)$, where k is the number of iterations and n is the number of non-fixed people.

4.2   Genetic Local Search (Hybrid)

Combining different search techniques to enhance the performance of a single model is proven to improve solution quality and the models [13,18]. Genc et al. propose three different models (constraint programming, local search and genetic algorithm) for finding $(1, b)$-supermatches to the RSM in [4]. The results indicate that genetic algorithm (GA) procedure has poor performance when compared to the LS. In this work, we consider combining the two metaheuristics: the genetic algorithm and the local search to provide a hybrid procedure. We denote this hybrid model as HB. The overview of the GA procedure we use in the HB model is the same as the one used for RSM (details can be found in [5]).

The procedure begins by initialising a population of random stable matchings. Then, the population is evolved by randomly selecting individuals from the population, applying crossover, searching for neighbours of the products of crossover, applying mutation. This process is repeated until some termination criteria is met (no improvement, time-limit exceeded, optimal solution found). The procedure below gives a pseudo-code of the evolution phase of HB.

```
1: procedure EVOLUTION()
2:     M₁ ← SELECTION()
3:     M₂ ← SELECTION()
4:     if M₁ ≠ M₂ then
5:         (M_c1, M_c2) ← CROSSOVER(M₁, M₂)
6:         N ←FINDNEIGHBOURS(M_c1)
7:         M_c1 ← BEST(N)
8:         N ←FINDNEIGHBOURS(M_c2)
9:         M_c2 ← BEST(N)
10:        REFINE(M_c1, M_c2)
11:        EVALUATION()
12:    M_fit ← GETFITTEST(P)
13:    M_m ← SELECTION()
14:    rand ← RANDOM(0, 1)
15:    if M_m ≠ M_fit and rand < p_m then
16:        MUTATION(M_m)
```

As can be seen from the procedure, the only LS enhancement to the GA algorithm is the search for the neighbours of the stable matchings after crossover (see Lines 6–9). Let M_{c1}, M_{c2} be the two stable matchings produced by the crossover. We update M_{c1} by its best neighbour after the neighbour search (same applies to the M_{c2}). Creating a random stable matching and finding neighbours are already discussed in Sect. 4.1.

If the original methods from LS and GA as described in [5], where the evolution phase is updated with the one here are used, we obtain the HB model for the RSM. In the RSR model, only the crossover and mutation operations are different than the original GA model defined for RSM. Instead of defining the crossover by adding rotations to the closed subset or removing them as we did for the RSM, we use the terminology *flip* for the RSR. Considering the Lemma 6, we define the crossover procedure for two stable matchings M_1, M_2 as follows. First, we find a random rotation $\rho_1 \in S_1$, and a random rotation $\rho_2 \in S_2$. If $\rho_1 \notin S_2$, then $\bar{\rho}_1 \in S_2$ due to the completeness property of the closed subsets in SR. Therefore, we flip $\bar{\rho}_1$ in S_2 and the duals of all of its predecessors $\rho' \in N^-(\rho)$ if ρ' is not included in S_2. We apply the same procedure to the other stable matching as well. Moreover, for the mutation operation, we select a random rotation ρ from the reduced rotation poset of the underlying instance and also a stable matching M. If $\rho \in S$, we flip ρ and all the required predecessors. If its dual $\bar{\rho} \in S$, then we flip $\bar{\rho}$ and the predecessors.

5 Experiments

In this section, we first compare the performances of the HB and the LS proposed for the RSR. Then, we investigate the robustness of different sets of RSM and RSR instances[2]. The code is implemented in Java, reusing the RSM

---

[2] Our datasets are publicly available at: github.com/begumgenc/rsmData.

experiments from [4]. All experiments are performed on a Dell M600 with 2.66 GHz processors under Linux, using three different randomisation seeds and fixing time limit $lim_{time} = 20$ min, number of iterations without improvement $lim_{cutoff} = 10000$, the number of stable matchings in LS that descend from a random stable matching $lim_{depth} = 50$. We use the population size for the HB as $|P| = 30$ and the mutation probability as $p_m = 0.7$. We use a high p_m as GA is suffering from getting stuck at local minima and randomisation helps with it. We discuss the size of the population for HB later.

HB v. LS for the RSR. Our first experiment is about the comparison of LS and HB models. Random SR instances have only a small number of stable matchings as we verify on the dataset RANDOM later [16]. For the comparison of HB and LS models, we look for instances that are likely to contain many stable matchings to gain more insight on their performances. For this purpose, we first created a dataset of purely random SM instances as each SM instance contains at least one stable matching, then we converted these instances to the SR. This conversion tackles the problem of random SR instances having only a few stable matchings, while preserving the randomness. Our SM dataset consists of 30 random instances of each size $n \in \{100 \times k \mid k \in \{1, \ldots, 10\}\}$. Note that, the resulting SR instances have size $2 \times n$.

Figures 3 and 4 provide detail on the comparison between the LS and the HB.[3] Figure 3 compares the average minimum b value found by the two models for each instance in the set. In the x-axis, the range shows the size of the instances such that all the instances that have x-values between $[0, 200]$ is of size 200, $[201-400]$ is of size 400, etc. We confirm by our experiments and also observe in Fig. 4 that for each instance that has size $200 \le n \le 600$, both models complete the search within the given time limit. Additionally, they either produce similar results (b values) or HB performs slightly better as can be observed in Fig. 3. The reason for exceeding the time limit in Fig. 4 is due to us not interrupting the construction of a stable matching. The construction of a stable matching consists of exposing all rotations in its complete closed subset in order starting from T_S. Then, the b value is computed. For large stable matchings, this computation becomes very costly. We can conclude that for small instances, both HB and LS perform well in terms of finding solutions with low b values. If the time is essential, HB model can be preferred over LS as it converges faster. Additionally, HB is able to find better solutions for larger instances.

Random RSM v. Random RSR. Next, we perform some tests for SM-SR comparison on our dataset RANDOM. Our dataset RANDOM consists of 30 purely randomly created SM and SR instances for each size $n \in \{100 \times k \mid k \in \{1, \ldots, 10\}\}$. Note that, for an SM instance of size n, there exists n men and n women in the problem. We have $2 \times n$ people in the corresponding SR instance. However, both have n pairs. All SR instances in RANDOM have at least two stable matchings. Considering the good performance of LS for small instances, we used the LS models for both RSM and RSR.

---

[3] The reader is referred to the online version for coloured version.

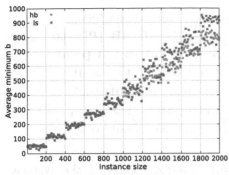

Fig. 3. Avg min b value found by LS and HB.

Fig. 4. Average time spent by LS and HB.

Table 4. Results on uniformly random instances for RSM.

n	$|\mathcal{V}|$	sm	np	b	ratio	t_{best}
100	22.02	47.9	75.12	48.27	0.64	0.02
200	41.59	116.9	166.19	115.34	0.69	0.10
300	60.22	182.4	263.94	193.08	0.73	0.37
400	74.51	244.1	356.58	265.98	0.74	0.77
500	91.47	322.5	456.00	350.16	0.76	2.18
600	103.82	394.9	551.10	425.51	0.77	3.67
700	117.08	449.6	646.69	505.61	0.78	5.89
800	131.81	527.6	749.98	595.64	0.79	9.09
900	146.34	585.5	848.32	679.82	0.80	14.60
1000	156.00	632.4	943.23	758.82	0.80	21.16

Table 5. Results on uniformly random instances for RSR.

n	$|\mathcal{V}|$	sm	np	b	ratio	t_{best}
100	3.91	3.78	17.91	5.31	0.3	0.003
200	3.87	3.94	26.76	8.52	0.32	0.003
300	4.36	4.56	35.53	11.22	0.32	0.017
400	4.71	5.92	37.64	10.93	0.29	0.048
500	4.29	4.81	37.62	11.70	0.31	0.066
600	4.16	4.48	42.44	14.47	0.34	0.130
700	4.58	5.50	48.71	16.02	0.33	0.312
800	4.93	5.99	55.02	17.39	0.32	0.498
900	4.82	7.07	57.64	18.50	0.32	0.662
1000	4.60	5.19	55.16	18.36	0.33	0.557

Tables 4 and 5 present a summary of the robustness of random RSM and RSR instances. The columns report for each size the average value of: the total number of pairs in the instance (**n**), the number of rotations in the rotation poset or the reduced rotation poset ($|\mathcal{V}|$), the number of different stable matchings found during the search of LS (**sm**), the number of non-fixed pairs (**np**), the b value of the solution found (**b**), the ratio $\frac{b}{np}$ (**ratio**), and the time spent until finding the best solution by LS in seconds (t_{best}).

Observe from the tables that the random RSM instances contain many more stable matchings than the random RSR instances of similar sizes. Recall that, the value of sm denotes only the number of a subset of the stable matchings found during the search. However, we can confirm the RSR instances not containing many stable matchings by looking at the number of rotations in their rotation posets. Note that, for the RSR instances, when 1000 pairs are included, the corresponding rotation posets, on the average, contain $|\mathcal{V}| \approx 5$ rotations. This is mainly caused by the large numbers of fixed-pairs in the random RSR instances. For instance, the average number of non-fixed pairs in the RSM instances of size 100 is $np = 75.12$. However, we observe in the large RSR instances that

Table 6. Summary of the results on large instances for RSM.

Instance			LS		HB, $|P| = 10$		HB, $|P| = 60$	
n	np	$|\mathcal{V}|$	b	t (min)	b	t (min)	b	t (min)
16	15.99	100.43	1.12	0.003	1.21	0.003	1.1	0.004
32	31.99	447.26	1.03	0.054	1.30	0.024	1.04	0.029
64	64	1889.95	1.685	3.158	1.74	0.824	1.28	0.916
128	128	7788.02	14.055	8.367	1.02	13.989	1.01	17.609

contain 1000 pairs that there are only 55.16 non-fixed pairs on the average, which is less than the smallest sized RSM instances that we tested. Note that, we measure the robustness ratio over the non-fixed pairs of the instances. It is desirable to obtain a smaller value for the ratio to indicate a better robustness for the instance. Because a smaller ratio indicates that a smaller proportion of the people that have alternative partners need to change their partners for a repair. Observe that, the ratio of the RSR instances is lower when compared to RSM. The ratio shows that the breakage of the pairs in the RSR instances are less costly to be repaired. Thus, we conclude that purely random RSR instances require a smaller proportion of the people to change their partners in the case of a breakage, when compared to the RSM.

Large RSM and RSR Instances. In this experiment, we search for instances with potentially many number of stable matchings and low b values. Therefore, we generated a dataset called MANY consisting of 100 SM instances for each size $n = \{16, 32, 64, 128\}$ using the family described by Irving and Leather [11], and then used in [17]. Note that, each SM instance in this set has a corresponding SR instance (see Lemma 1), where the corresponding SR instance has a reduced rotation poset of twice the size of the rotation poset of the SM instance. First, let us introduce this family of instances described by Irving and Leather. They prove that any instance in the original family contains at least 2^{n-1} stable matchings for an instance of size $n = 2^i$. They define this family over two matrices for the preferences of each gender, and the preference lists of these large instances are obtained recursively by appending the following matrices until the desired instance size is found. In our dataset MANY, we slightly modify each instance of this original family by first randomly selecting two random men m_i, m_j. Then, we modify m_i's preference list by swapping the positions of two randomly selected women within the list. We repeat the same for m_j. We also modify the preference lists of two random women in the same way. In other words, the original preference set between the original and the modified instances have a Hamming Distance of 8.

Table 6 reports for each size the average value of: the number of all men or women (n), the number of non-fixed men (np), the number of rotations in the rotation poset ($|\mathcal{V}|$). Additionally, it reports the average minimum b found by the model LS, HB where population size $|P| = 10$, and HB where population size $|P| = 60$ (b); followed by the total time spent in minutes for each of the three models (t (min)).

This dataset shows that the robustness of instances that have many stable matchings is very high (i.e the value of b is low). For each instance size, our best model for that size is able to find solutions whose average b values evaluate to a b value that is $opt = 1 < b < 2$. For instance, for size $n = 16$, the LS model finds solutions such that for the breakage of any man on the solution, on the average, 1.12 other men need to break-up from their current partners. Similarly, for size $n = 128$, HB models find that the solution is guaranteed to be repaired by only 1.02 additional men's break-up.

As one can observe from Table 6, we ran the HB model by using different sizes of population. Observe that, reducing the number of individuals in the population of HB (60 to 10) causes the algorithm to find slightly worse solutions (i.e. larger b). For instance, for size $n = 64$, the average minimum b is found as 1.74 by a population of size 10, and 1.28 by a population of size 60. This is due to having an increased chance of getting stuck at local minima for a smaller population. On the other hand, LS finds competitive values for b for sizes $16 \leq n \leq 64$. However, as we can see for $n = 128$, LS finds solutions that are far away from the optimal solution. We conclude that, an improvement for HB by changing population size is possible in exchange of obtaining slightly worse solutions. LS performs well for smaller instances.

Recall that, each SM instance in MANY has a corresponding SR instance that has exactly the same stable matchings. We do not run the RSR models on this dataset as they are much slower. However, this test provides an insight to some RSM and RSR instances that are repairable at low additional costs.

6   Conclusions

We study the notion of (a, b)-supermatch in the context of Stable Roommates problem. We propose a polynomial-time algorithm based on the reduced rotation poset to verify if a stable matching is a $(1, b)$-supermatch. Next, we use this procedure to design local search (LS) and hybrid genetic local search (HB) models to find robust solutions for the $(1, b)$ case (i.e., (1,b)-supermatch with (possibly) the minimum b). We empirically show that the HB model usually performs better than LS for RSR. Furthermore, we perform an RSM/RSR comparison and identify a family of instances that are rich in stable matchings and very robust.

Acknowledgement. This material is based upon works supported by the Science Foundation Ireland under Grant No. 12/RC/2289 which is co-funded under the European Regional Development Fund.

References

1. Ashlagi, I., Gonczarowski, Y.A.: Dating strategies are not obvious. CoRR abs/1511.00452 (2015)
2. Aziz, H., Biró, P., Gaspers, S., de Haan, R., Mattei, N., Rastegari., B.: Stable matching with uncertain linear preferences. CoRR abs/1607.02917 (2016)

3. Drummond, J., Boutilier, C.: Elicitation and approximately stable matching with partial preferences. In: Proceedings of the Twenty-Third International Joint Conference on Artificial Intelligence, IJCAI 2013, pp. 97–105. AAAI Press (2013)
4. Genc, B., Siala, M., O'Sullivan, B., Simonin, G.: Finding robust solutions to stable marriage. In: Proceedings of the Twenty-Sixth International Joint Conference on Artificial Intelligence, IJCAI 2017, Melbourne, Australia, 19–25 August 2017, pp. 631–637 (2017)
5. Genc, B., Siala, M., O'Sullivan, B., Simonin, G.: Finding robust solutions to stable marriage. CoRR abs/1705.09218 (2017)
6. Genc, B., Siala, M., Simonin, G., O'Sullivan, B.: Complexity study for the robust stable marriage problem. In: Theoretical Computer Science (2019)
7. Gusfield, D., Irving, R.W.: The Stable Marriage Problem: Structure and Algorithms. MIT Press, Cambridge (1989)
8. Hebrard, E.: Robust solutions for constraint satisfaction and optimisation under uncertainty. PhD thesis, University of New South Wales (2007)
9. Holland, A., O'Sullivan, B.: Super solutions for combinatorial auctions. In: Faltings, B.V., Petcu, A., Fages, F., Rossi, F. (eds.) CSCLP 2004. LNCS (LNAI), vol. 3419, pp. 187–200. Springer, Heidelberg (2005). https://doi.org/10.1007/11402763_14
10. Irving, R.W.: An efficient algorithm for the "stable roommates" problem. J. Algorithms 6(4), 577–595 (1985)
11. Irving, R.W., Leather, P.: The complexity of counting stable marriages. SIAM J. Comput. 15(3), 655–667 (1986)
12. Jacobovic, R.: Perturbation robust stable matching. CoRR abs/1612.08118 (2016)
13. Kolen, A., Pesch, E.: Genetic local search in combinatorial optimization. Discret. Appl. Math. 48(3), 273–284 (1994)
14. Lebedev, D., Mathieu, F., Viennot, L., Gai, A.-T., Reynier, J., De Montgolfier, F.: On using matching theory to understand P2P network design. In: INOC 2007, International Network Optimization Conference (2007)
15. Lussier, B., Chatila, R., Ingrand, F., Killijian, M.O., Powell, D.: On fault tolerance and robustness in autonomous systems. In: Proceedings of the Third IARP/IEEE-RAS/EURON Joint Workshop on Technical Challenge for Dependable Robots in Human Environments, Manchester, GB, 7–9 September 2004
16. Pittel, B.: The "stable roommates" problem with random preferences. Ann. Probab. 21, 1441–1477 (1993)
17. Siala, M., O'Sullivan, B.: Rotation-based formulation for stable matching. In: Beck, J.C. (ed.) CP 2017. LNCS, vol. 10416, pp. 262–277. Springer, Cham (2017). https://doi.org/10.1007/978-3-319-66158-2_17
18. Ulder, N.L.J., Aarts, E.H.L., Bandelt, H.-J., van Laarhoven, P.J.M., Pesch, E.: Genetic local search algorithms for the traveling salesman problem. In: Schwefel, H.-P., Männer, R. (eds.) PPSN 1990. LNCS, vol. 496, pp. 109–116. Springer, Heidelberg (1991). https://doi.org/10.1007/BFb0029740

Optimality Clue for Graph Coloring Problem

Alexandre Gondran[1]($\boxtimes$)(iD) and Laurent Moalic[2](iD)

[1] ENAC, French Civil Aviation University, Toulouse, France
alexandre.gondran@enac.fr
[2] UHA, University of Haute Alsace, Mulhouse, France
laurent.moalic@uha.fr

Abstract. In this paper, a new approach is presented to qualify or not, a solution found by a heuristic for a potential optimal solution. Our approach is based on the following observation: for a minimization problem, the number of admissible solutions decreases with the value of the objective function. Concerning the Graph Coloring Problem (GCP), we confirm this observation and present a new way in which to prove optimality. This proof is based on the counting of the number of different k-colorings and the number of independent sets of a given graph G.

Finding the exact solution for counting problems is difficult (#P-complete). However, we show that in using only randomized heuristics, it is possible to define an estimation of the upper bound of the number of k-colorings. This estimate has been calibrated on a broad benchmark of graph instances for which the exact number of optimal k-colorings is known.

Our approach, called *optimality clue*, constructs a sample of k-colorings from a given graph by running one randomized heuristic a number of times on the same graph instance. We use the evolutionary algorithm *HEAD* [26], which is one of the most efficient heuristics for GCP.

Optimality clue matches the standard definition of optimality on a wide number of instances for DIMACS and RBCII benchmarks where the optimality is known.

Keywords: Optimality · Metaheuristics · Near-optimal

1 Introduction

For any given integer $k \geq 1$, a k-coloring of a given graph $G = (V, E)$ assigns one of k distinct colors to each vertex $v \in V$ in the graph, so that no two adjacent vertices (linked by an edge $e \in E$) are given the same color. The Graph Coloring Problem (GCP) is to find, for a given graph G, the smallest k so that a k-coloring of G exists; This minimum k is called the *chromatic number* of G and is denoted by $\chi(G)$. GCP is NP-hard [19] for $k \geq 3$. The k-coloring problem (k-CP) is the associated decision problem. Concerning an optimization problem

© Springer Nature Switzerland AG 2019
L.-M. Rousseau and K. Stergiou (Eds.): CPAIOR 2019, LNCS 11494, pp. 337–354, 2019.
https://doi.org/10.1007/978-3-030-19212-9_22

which is NP-hard, there is no efficient exact polynomial-time algorithm with which to solve it, unless P = NP. For large size instances of a minimization NP-hard problem therefore, the exact algorithms must be aborted before they terminate. In this instance, exact algorithms such as branch and bound methods find both lower and upper bounds of the optimal value of the objective function. Heuristic approaches are then the only ways in which to find, in reasonably fast running-time, a "good" solution in terms of objective function value, i.e. an upper bound of the optimal value. Even if an admissible solution is found however, its distance from the optimal solution remains unknown, excepting approximation algorithms[1]. The optimality gap is the difference between the upper bound (found by a heuristic) and the lower bound (found by a partial exact method). Optimality is proven only when this gap is equal to zero. Unfortunately for large size instances of an NP-hard problem, this gap is often large. This is particularly true for challenging instances [15,26] in the GCP of the DIMACS benchmark [18]. This paper addresses the following question: What can be done in this situation? Is it possible to prove the optimality of a graph coloring problem instance using only heuristic algorithms?

The response is *Yes*, for a specific class of graphs: for example, efficient polynomial-time exact algorithms to find $\chi(G)$ for *interval graphs, chordal graphs, cographs* [27,31] exist. For certain graphs such as *1-perfect graphs*[2], for which the chromatic number $\chi(G)$ is equal to the size of a maximum clique $\gamma(G)$, it is possible to solve the problem of the Maximum Clique Problem (MCP), with another heuristic and to conclude with optimality if the size of the maximum clique found is equal to the smallest number of colors used for coloring G also found by a heuristic. In this specific case, the optimality gap (or duality gap between GCP and MCP) is zero.

The response to the question is however *No*, in general as follows; a heuristic finds approximate solutions (upper bound); although the coloring found may be optimal, it is not feasible to prove this potential optimality. The question therefore becomes: what can we do better than find an approximate solution using only a heuristic? Is it possible to define a form of optimality index for a graph coloring problem instance?

In this article, we show that a heuristic finds not only an upper bound of $\chi(G)$ but it is also able to count the number of different k-colorings (i.e. the number of admissible solutions having the same objective function value). Our approach is based on the fact that the number of different k-colorings decreases dramatically when the number of colors, k, decreases too. Indeed Fig. 1 presents a typical example of a random graph with 30 vertices, a density of 0.9 and $\chi(G) = 16$.

---

[1] Notice that it is still NP-hard to approximate $\chi(G)$ within $n^{1-\epsilon}$ for any $\epsilon > 0$ [35].

[2] A perfect graph is a graph for which the chromatic number of every induced subgraph is the same as the size of the largest clique of that subgraph. 1-perfect graphs are more general than perfect graphs. Polynomial-time exact algorithms with the aim of finding $\chi(G)$ for perfect graphs [13] exist, but are in reality slow in performance. Line graphs, chordal graphs, interval graphs or cographs are subclasses of perfect graphs.

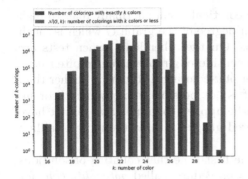

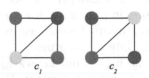

Fig. 1. Number of colorings with exactly k colors (blue bars) and number of total colorings with k colors or less (red bars), noted $\mathcal{N}(G, k)$ in function of $k = 16...30$, for a random graph with 30 vertices, density 0.9 and $\chi(G) = 16$. (Color figure online)

Fig. 2. Two 3-colorings c_1 and c_2 of the same graph with four vertices. These two colorings have to be considered as identical because $d(c_1, c_2) = 0$ with d the set-theoretic partition distance; we go from one to the other, simply through a permutation of color classes.

The number of colorings with exact k colors (blue bars) and the total colorings with k colors or less (red bars), noted $\mathcal{N}(G, k)$, are computed with precision for all values from $k = 16$ to $k = 30$. $\mathcal{N}(G, k)$ decreases exponentially when k decreases to $\chi(G)$. We prove a theorem which shows that when the number of k-colorings is lower than a given value (the number of independent sets of G^3), then we achieve the optimum: $\chi(G) = k$.

In this article, we try to apply the proposed theorem in order to prove optimality.

Brief Solutions Counting Review. The counting of NP-complete problems solutions has been widely studied for the boolean SATisfiability problems, known as #SAT, or for the Constraint Satisfaction Problem (CSP), known as #CSP; the k-coloring problem is a special case in CSP. These problems are known as #P-complete [33]. A recent survey on #CSPs is conducted in [17]. Even if a problem is not NP-hard, the problem of solution counting is often hard. Specific studies on counting solutions of k-CP are made in [7,16,25]. Because precise counting is in many cases a complex problem, statistical or approximate counting is often considered. Then, uniform sampling of all solutions is related to the problem of solution counting. Many papers examine uniform or near uniform sampling as in [11,12,34]. The objective is to count by sampling. Frieze and Vigoda [8] provided a survey on the use of Markov Chain Monte Carlo algorithms for the approximative counting of the number of k-colorings. The features of ergodicity or quasi-ergodicity of heuristics which guarantee uniform sampling

---

[3] An independent set is a subset of vertices of G, so every two distinct vertices in the independent set are not adjacent.

are discussed in detail in [6]. However, theoretical results normally refer to high value of $k \geq \Delta$ where Δ is the maximum degree of the graph G which is much higher than $\chi(G)$ for challenging graphs. On the other hand, when tests are performed with $k = \chi(G)$ as in [7], the considered graph instances often have more than 10^{20} k-colorings. If the number of k-colorings is too high (higher than the number of independent sets), then it is not possible to apply our theorem. Therefore, in practice, our approach can be applied to graphs that do not have too many optimal colorings; we considered graphs with at the most 1 million different *optimal* colorings.

To the best of our knowledge, it is the first time that counting of solutions are used to prove optimality. We define a procedure, called *optimality clue*, for application of the proposed theorem. First, we build a sample of k-colorings for a given graph G by running the same randomized heuristic algorithm many times (about 1,000). In this study, we use $HEAD^4$, our open-source memetic algorithm (i.e. hybridization of tabu search and evolutionary algorithm), which is very a efficient heuristic for solving GCP [26].

In this sample some colorings may appear several times and others only once. The number of different k-colorings within the sample is used to build up an estimate of the total number of colorings with k colors. This estimator has been calibrated on a wide benchmark of graph instances for which the number of optimal k-colorings is known with precision. Because we have no guarantee that the sampling is uniform, in general we can not then guarantee that our estimator is always correct.

Moreover, building a sample of k-colorings is time-consuming and the size of the sample should be "reasonable". Graphs for which our optimality clue can be calculated are therefore graphs without too many optimal k-colorings (i.e. about less than one million). Of course it is not possible to known a priori if a given graph has more or less than 1 million optimal colorings. Our approach for this reason provides a clue that coloring found by the heuristic may be optimal (positive conclusion), but in many cases we can not come to a conclusion.

This article is organized as follows. In Sect. 2, we present new optimality proof for GCP based on solution counting. Our general approach, called *optimality clue*, is defined in Sect. 3. In Sect. 4, we detail how we calculate the estimate of the number of k-colorings using benchmark graph instances. Numerical tests and experiments are presented in Sect. 5. Finally, we conclude in Sect. 6.

2   Proof of Optimality by Solutions Counting

Notice that there are different ways in which to count the k-colorings of a given graph G. When counting the number of different k-colorings, we have to take into account the permutations of the color classes. We consider one k-coloring not as the assigning of one color among k to each vertex but as a partition of the graph vertices into k independent sets. An Independent Set (IS) or stable set is

---

[4] Open-source code available at: github.com/graphcoloring/HEAD.

a set of G vertices, no two of which are adjacent. Two k-colorings c_1 and c_2 are considered identical if they correspond to the same G partition. The distance between two k-colorings taken into account is the *set-theoretic partition distance* used in [10,14,26], which is independent of the permutation of the color classes. In previous studies on solution counting of k-CP [7], authors counted the total number of k-colorings including every permutation such as in the example of Fig. 2; such a calculation of the number of different k-colorings is $k!$ times higher than the way we count. These methods are therefore inapplicable to our study. We write as $\Omega(G, k)$ the set of all k-colorings of graph G. A k-coloring can use exactly k colors or fewer, then $\Omega(G, k-1) \subset \Omega(G, k)$. The cardinal of $\Omega(G, k)$ is noted $\mathcal{N}(G, k) = |\Omega(G, k)|$.

Our approach is based on the following fact:

Lemma 1. *Let G be a graph and $k \geq 1$ an integer. If at least one k-coloring of G exists, then there exist at least $i(G) - k + 1$ different $(k+1)$-colorings of G:*

$$\mathcal{N}(G, k+1) \geq i(G) - k + 1,$$

where $i(G)$ is the number of independent sets of G.

Proof. Notice that a k-coloring of a graph $G = (V, E)$ is a partition of $|V|$ vertices into k IS. Indeed vertices colored with the same color inside a k-coloring are necessarily an IS. In other words, it is always possible to color all vertices of any IS with the same color. We note $IS(G) = \{U \subset V \mid \forall x, y \in U^2, \{x, y\} \notin E\}$ the set of all the IS of G, then $i(G) = |IS(G)|$.

Let C be an initial coloring of G with exactly k colors: $C = (V_1, V_2..., V_k)$, where V_i is the set of vertices colored with color i, for all $i = 1..k$. For each independent set of G, noted I, except for the k IS of C, it is possible to recolor all vertices of I with a new color (the $(k+1)$th color). We obtain in this way one different $(k+1)$-coloring, $C' = (V_1 \backslash I, V_2 \backslash I..., V_k \backslash I, I)$ for each different independent set, $I \in IS \backslash \cup_{i=1}^{k} \{V_i\}$, afterwards counting a total of at least $i(G) - k$ different colorings with exactly $(k+1)$ colors. Then, $\mathcal{N}(G, k+1) \geq i(G) - k + 1$ because we also have to count the initial k-coloring.

After, we obtain the following theorem:

Theorem 1. *Let G be a graph and $k \geq 1$ an integer. Let $\mathcal{N}(G, k)$ be the number of k-colorings of G and $i(G)$ the number of independent sets of G.*
If $i(G) - k > \mathcal{N}(G, k) > 0$, then $\chi(G) = k$.

Proof. $\chi(G) \leq k$ because $\mathcal{N}(G, k) > 0$. If $\chi(G) < k$, it means that at least one $(k-1)$-coloring exists (i.e. $\mathcal{N}(G, k-1) > 0$). If we add a new color, it is possible to consider this $(k-1)$-coloring and to recolor any independent set of G with the new color. We obtain in this way $i(G) - k$ different k-colorings (by Lemma 1). Therefore $i(G) - k \leq \mathcal{N}(G, k)$ which refutes initial assumption.

For example, the studied graph in Fig. 1 (30 vertices and density 0.9) has 38 different colorings with 16 colors: $\mathcal{N}(G, k = 16) = 38$; moreover this graph has

78 IS: $i(G) = 78$, The theorem is then applicable with $k = 16$ because: $i(G) - k = 78 - 16 = 62 > 38 = \mathcal{N}(G, k) > 0$. Thanks then to the theorem we can conclude that $\chi(G) = 16$. Moreover, for $k = 17$, $\mathcal{N}(G, k = 17) = 3121 > i(G) - k = 61$, so the theorem is not applicable.

Corollary 1. *Let G be a graph and $k \geq 1$ an integer. Let $\overline{\mathcal{N}}(G, k)$ be an upper bound of the number of $\mathcal{N}(G, k)$ and $\underline{i}(G)$ a lower bound of $i(G)$.*
  If $\mathcal{N}(G, k) > 0$ and $\underline{i}(G) - k > \overline{\mathcal{N}}(G, k)$, then $\chi(G) = k$.

3   Optimality Clue

We propose to apply in this paper the Corollary 1, so as to find an appropriate upper bound of the number of k-colorings of G, $\overline{\mathcal{N}}(G, k)$, and a lower bound of the number of independent sets of G, $\underline{i}(G)$.

3.1   IS Counting

Many algorithms [4,5,28,30] exist for the counting of all the maximal independent sets of a graph G (or similarly counting all the maximal cliques[5] in $\overline{G}$, the complement graph of G). By definition, the number of maximal IS, noted $i_{max}(G)$, is a lower bound of $i(G)$. Those algorithms are based on enumeration. Because the focus of this study is on graphs having less than 1 million optimal solutions, we can stop enumeration after finding 1 million IS. Generally, $i(G)$ is very high except for graphs with very high density. Real-life graphs often have low density, in which case $i(G)$ is very high. Moreover, a simple lower bound is given by [29]: $i(G) \geq 2^{\alpha(G)} + n - \alpha(G)$, where $\alpha(G)$ is the size of the largest independent set of G and n the number of vertices. Bollobás' book [2] (p. 283) also provides a statistical number of maximal cliques of size p for a random graph. We can conclude that: $i_{max}(G) \approx i_B(G) = \sum_{p=1}^{n} \binom{n}{p}(1 - d)^{\binom{p}{2}}$ with n the number of vertices and d the density of a random graph G.

In this study, we use Cliquer[6], an exact branch-and-bound algorithm developed by Patric Östergård [28] which enumerates all cliques not necessarily maximal (an IS is a clique in the complement graph).

It is more complex to evaluate $\overline{\mathcal{N}}(G, k)$ and Sect. 4 presents a way in which to build an *experimental* upper bound of $\mathcal{N}(G, k)$. We characterize this upper bound as *experimental* because it is based on experimental tests on benchmark graph instances. There is then no total guarantee that it is an upper bound.

---

[5] A maximal clique is a clique that cannot be extended by including an additional adjacent vertex. A maximum clique is a clique that has the largest size in a given graph; a maximum clique is therefore always maximal, but the converse is not so. Analogue definition for IS.

[6] Code available at: https://users.aalto.fi/~pat/cliquer.html. To count all IS of a graph, you just execute: ./cl <complement graph> -a -m 1 -M <k>.

3.2   Procedure

We define here the procedure of what we call *Optimality Clue* for graph coloring: let it be G a graph and $k > 0$ a positive integer, that we suspect to be the chromatic number of G. The proposed approach is based on the five following steps:

1. The building of a sample of $t = 1,000$ k-colorings of G: we run the algorithm *HEAD* as many times as needed to obtain t legal k-colorings. These solutions are *the solution samples*. The size of the sample is equal to t. In general we use case $t = 1,000$ when possible.
2. Count the number of different k-colorings inside the sample. This number is equal to p. Of course $0 \le p \le t$.
3. Estimate an upper bound of $\mathcal{N}(G, k)$ as $UB(p, t)$ (cf. Sect. 4); this upper bound is the function of t and p.
4. Compute $i(G)$, the number of IS, or at least a lower bound if $i(G) > 10^6$, with an exact algorithm (Cliquer).
5. If $i(G) > UB(p, t)$, then we conclude that solutions of the sample have a **clue so as to be optimal:**

$$\text{Chances are that } k \text{ is equal to } \chi(G)$$

Uniform Sample. If the sample is uniform[7], then statistical methods exist for the counting of solutions and the building of an upper bound with statistical guarantee, for example through the capture-recapture methods: The Peterson method [20], and Jolly-Seber method [1] which is commonly used in ecology to estimate an animal population size. However, this is not our case: we have no guarantee that our solution sample is uniform or quasi uniform. HEAD is a memetic algorithm that explores the space of non-legal k-colorings: a non-legal k-coloring is a coloring with at most k colors and one where two adjacent vertices (linked by an edge) may have the same color (these are known as conflicting edges). The objective of HEAD is to reduce the number of conflicting edges to zero, that is to get a legal k-coloring. HEAD is an evolutionary algorithm with a population size equal to two. In each generation, the two non-legal k-colorings crossover before performing a tabu search. The sample distribution depends on the fitness landscape properties [23,24][8] and there is no reason for this distribution to be uniform. A smooth landscape (respectively a rugged landscape) around a legal k-coloring will increase (resp. decrease) the probability of finding this k-coloring. Figure 3 represents the frequency of the 319 optimal 46-colorings of <r140_90.4> graph of the RCBII benchmark (140 vertices and density 0.9) in a sample of size 100,000 found by the HEAD heuristic. In this typical graph instance, the ratio between the least frequent and the most frequently found coloring is around a factor of 10^3 which corresponds to the same scale as similar studies [34].

---

[7] All k-colorings in the sample are uniformly drawn at random in $\Omega(G, k)$.

[8] The fitness landscape itself depends on the neighborhood used for both the tabu search and the crossover.

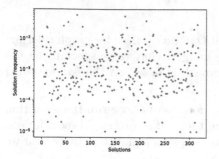

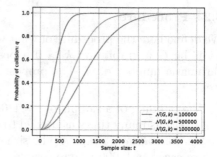

Fig. 3. Sampling of 46-colorings for the <r140_90.4> graph from RCBII benchmark (140 vertices and density 0.9).

Fig. 4. Collision probability q, given the sample size t, and the total number of k-colorings $\mathcal{N}(G, k)$.

Another approach is to take into account the ergodicity of an algorithm, which is its capability to explore all the search space. More precisely, an algorithm is ergodic if there is the possibility of reaching (probability not null) any k-coloring from any other k-coloring in a finite number of iterations. Random walks or Metropolis algorithms (with a positive temperature sufficiently high) are ergodic algorithms since there is always a finite probability of escaping from local minimum. However, those algorithms are very inefficient in practice for the finding of an optimal k-coloring in general.

Sample Size. The choice of t, the size of the sample, is very important for two reasons. Firstly, in practice, the building of a sample of k-colorings can be very time-consuming, and the size of the sample should be reasonable. We use $t = 1,000$ for most of the graph instances. However, the more challenging the graph instance, the longer HEAD takes to find one k-coloring. It is therefore not possible to build a sample of size 1,000 for all graphs, such as for the <DSJC500.5> graph of DIMACS (cf. Table 2).

The second reason is more theoretical. We have limited the maximum number of different optimal solutions to 1 million, in order for a graph to be considered by our approach. In fact, we choose 1 million because it equals to t^2 with $t = 1,000$. Indeed, if the sample is uniformly drawn at random in $\Omega(G, k)$, the probability q that at least two colorings of the sample are identical is equal to[9]: $q = 1 - \frac{\mathcal{N}!}{\mathcal{N}^t(\mathcal{N} - t)!} \simeq 1 - e^{-\frac{t(t-1)}{2\mathcal{N}}}$ then $\mathcal{N} \simeq -\frac{t(t-1)}{2ln(1-q)}$. We also call q the probability of collision. So, if $q = 0.5$ then $\mathcal{N} \simeq 720626$, if $q = 0.393$ then $\mathcal{N} \sim t^2 = 10^6$. Figure 4 represents the collision frequency, q, in function of the sample size, t, for different values of the $\Omega(G, k)$ size. When $\mathcal{N}(G, k) = 10^5$ and $t = 1,000$, it is almost impossible to miss a collision in the sample, but for $\mathcal{N} = 10^6$, there

---

[9] This problem is linked to the birthday problem that shows that in a room of just 23 people there's a 50-50 chance that two people have the same birthday. In our case, the number of days in a year is $\mathcal{N}$ and the number of people is the size t of the sample.

is around a 60% chance of missing a collision. However, it is not a disaster to miss a collision for our approach. Indeed, the consequence is that the clue of optimality may be not applicable but the risk of *false positive* is avoided. A false positive occurs if our Procedure 3.2 improperly indicates the optimality clue, when in reality the k-colorings are not optimal. Moreover, the collision frequency is higher for a non-uniform sample than for a uniform one.

4   Estimate of the Number of k-Colorings: $UB(G, k, p, t)$

4.1   Data Sets

In order to define an estimator or at least an upper bound of the number of k-colorings, we need to have a large number of graph instances for which we know the exact number of k-colorings. Fabio Furini et al. [9] have published an open-source and a very efficient version of the backtracking DSATUR algorithm [3] which returns the chromatic number of a given graph[10]. DSATUR is one of the best exact algorithms for GCP, particularly for graphs with high density. We suggest readers that interested in an overview of exact methods used for GCP read [15, 22].

We modified the DSATUR algorithm in order to count the total number of k-colorings. The pseudo code of the algorithm, CDSATUR, is presented in Algorithm 1. CDSATUR returns, for all values k, the exact value of $\mathcal{N}(G, k)$ taking into account the permutation of colors, especially $\mathcal{N}(G, k = \chi(G))$.

Algorithm 1. CDSATUR which returns the number of all k-colorings of G: $\mathcal{N}(G, k)$.

Data : $G = (V, E)$ a graph and k a positive integer.
$\mathcal{N} \leftarrow 0$
  $C[v] \leftarrow None$, $\forall v \in V$: C is the empty coloring.
  $l \leftarrow 0$: number of colors used by C.
  $CDSATUR(C, l)$
return $\mathcal{N}$
Procedure $CDSATUR(C', k')$:
  if *all the vertices of C' are colored* **then**
      if $k' \leq k$ **then**
          $\lfloor$ $\mathcal{N} \leftarrow \mathcal{N} + 1$
  else
      Select an uncolored vertex v of C'
      for *every feasible color* $i \in [1 \, ; k' + 1]$ **do**
          $C'' \leftarrow C'$, $C''[v] \leftarrow i$
          $k'' \leftarrow max(k', i)$: number of colors used in C''.
          if $k'' \leq k$ **then**
              $\lfloor$ $CDSATUR(C'', k'')$

---

[10] Code available at: lamsade.dauphine.fr/coloring/doku.php.

Fabio Furini et al. also published 2031 random GCP instances called RCBII[11] with vertices from 60 to 140 and density between 0.1 and 0.9. This wide variety of graphs is our reference dataset. We complete this dataset with easy DIMACS graphs [18] for which $\chi(G)$ and $\mathcal{N}(G, \chi)$ is computable with CDSATUR.

The 2031 graphs of the RCBII benchmark have the characteristics described in Table 1. We can see that $\chi(G)$ is recognized in all these graphs [9]. First we calculated $\mathcal{N}(G, \chi)$ with CDSATUR, with a time limit equal to 2400 s, a time sufficient for most of the graphs. There are only 210 graph instances of RBCII (on the 2031) in which CDSATUR does not have enough time to find $\mathcal{N}(G, \chi)$. These 210 graphs are used to test our approach (*test dataset*).

Among the graphs for which $\mathcal{N}(G, \chi)$ can be determined, we consider only those with less than 1 million optimal solutions: they form the *reference dataset* (959 graph instances). Finally, we can distinguish inside the reference dataset, 566 graph instances on the 959 verifying $i(G) > \mathcal{N}(G, \chi)$.

862 graphs remain on the 2031 of the RBCII benchmark with more than 1 million optimal solutions. We decided to test our approach on the graphs (called *control dataset*) to check whether or not the proposed algorithm can produce *false positives* or not.

Table 1. Distribution of 2031 RCBII graph instances

$\mathcal{N} > 10^6$		χ known $\mathcal{N} \leq 10^6$				$\mathcal{N}$?		Total #instances
862 (control dataset)		959 (reference dataset)				210 (test dataset)		2031
		$i(G) \leq \mathcal{N}$ 393		$i(G) > \mathcal{N}$ 566				
opt. clue	not opt. clue	opt. clue	not opt. clue	opt. clue	not opt. clue	opt. clue	not opt. clue	
0	862	0	393	449	117	39	171	

4.2 Analysis of Graph Instances

Before determining an upper bound of $\mathcal{N}(G, \chi)$, we investigate possible links between standard features in a graph such as its size (number of vertices), its density.

Graphs with the same size (number of vertices) and same density can have a number of optimal colorings which prove very different from each other A typical example is given in Fig. 5 where the distribution of 49 graph instances with 80 vertices and density 0.3 (<r80_30.*> of RBCII benchmark) is represented in function to the number of solutions $\mathcal{N}(G, \chi)$. Half of the graphs (25/49) have fewer than 100 000 optimal solutions while a third (18/49) have more than 1 million optimal solutions. There are no simple laws which characterize this distribution.

However, we can see that the lower the density, the higher the optimal solution number. Indeed, Fig. 6 presents the proportion of graphs with 70 vertices of the RBCII benchmark having more than 1 million colorings depending on graph

---

[11] Instances available in the same address.

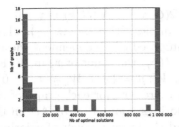

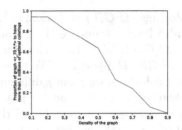

Fig. 5. Histogram characterizing the distribution of the 49 random graphs with 80 vertices and a density of 0.3 (<r80_30.*> for the RBCII benchmark) knowing the number of optimal colorings, $\mathcal{N}(G,\chi)$.

Fig. 6. Proportion of graphs with 70 vertices (<r70_*.*> of the RBCII benchmark) having more than 1 million optimal colorings given the density.

density. For a low density such as 0.1, nearly all graphs have more than 1 million optimal solutions, while no graph with high density (equal to 0.9) has more than 1 million optimal solutions.

In order to have a finer view of the link between the number of optimal colorings and the graph density, we generated 1,000 random graphs with 50 vertices and density d ($d = 0.1, 0.2, ..., $ or 0.9). Each line in Fig. 7 represents (for each density) the proportion of graphs having less than n optimal colorings with n between 10^2 and 10^6. The pink line in Fig. 7 shows for example that 50% of graphs (with 50 vertices and density $= 0.3$) have fewer than 10^5 optimal colorings. The plots are quite similar for graphs with 60 or 70 vertices. The graph size seems to have a slight influence on the number of optimal solutions.

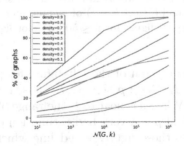

Fig. 7. Proportion of random graphs (with 50 vertices and a given density) having fewer than n optimal colorings with a n range between 1 hundred and 1 million. (Color figure online)

4.3 Upper Bound Function

We define in this section an upper bound of $\mathcal{N}(G,k)$ based on the 953 graphs of the reference dataset. Suppose we have, for a given graph G, a set of n different

k-colorings: $\Omega(G,k) = \{x_1, ..., x_n\}$, i.e. $n = |\Omega(G,k)| = \mathcal{N}(G,k)$ is unknown. We also have a sequence W of t independent samples: $W = (w_1, ..., w_t)$, where $w_k \in \Omega(G,k)$, $\forall k = 1...t$. This sample W is composed of t independent success runs of *HEAD* algorithm. We note $\forall j = 1...n$, $\#(x_j)$ the count of x_j in W. For these t colorings, we count p different colorings in W: $p = |\{x_j \in W, \#(x_j) > 0\}|$. So then, $\mathcal{N}(G,k) \geq p$ and $t \geq p \geq 1$. Figures 8 represents for each graph of the reference dataset, the number of **different** colorings p found by *HEAD* on the total of $t = 1,000$ success runs (in abscissa) and the exact number of colorings, $\mathcal{N}(G,k)$, calculated with CDSATUR (in ordinate). Each dot corresponds to one graph of the reference dataset. The immediate objective is to determine an as small as possible upper bound of $\mathcal{N}(G,k)$, UB. Indeed, in order to apply the Theorem 1, we must obtain $i(G) - k > UB$.

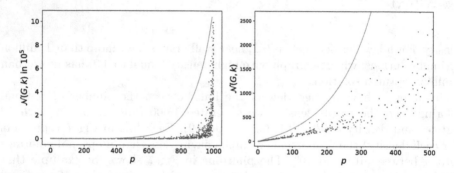

Fig. 8. Each blue dot corresponds to one of the 959 graph instances that have fewer than 1 million optimal colorings (reference dataset). The number of optimal colorings (calculated with CDSATUR algorithm) is in ordinate. The number p of different optimal solutions found by our *HEAD* algorithm after 1,000 success runs is in abscissa. The red line is an upper bound $UB(G,k,p,t)$ of the number of optimal colorings. The right figure is a zoom of the left figure for $p \leq 500$. (Color figure online)

Figure 8-right which is a zoom of the left figure for $p \leq 500$ shows that for $p \ll t$, p is near linear to $\mathcal{N}(G,k)$: $p \sim \mathcal{N}(G,k)$. p is then a good candidate for an estimator of $\mathcal{N}(G,k)$. When p is near to t, the range of $\mathcal{N}(G,k)$ values is very wide, close to $p^2 = 10^6$, and p is a very bad estimation of $\mathcal{N}(G,k)$. Notice that $\mathcal{N}(G,k) < p^2$. We add to those figures a red line which represents a possible upper bound of $\mathcal{N}(G,k)$ equal to:

$$UB(G,k,p,t) = \begin{cases} p + p^{\alpha \frac{t+p}{t}} & \text{if } p < t \times 0.99 \\ +\infty & \text{otherwise} \end{cases} \quad (1)$$

with $\alpha = 1.01$. Indeed, when $p \ll t$, $UB(G,k,p,t) \sim 2p$ and when p is close to t, $UB(G,k,p,t) \sim p^2$. Between these extreme values, the cloud of blue dots follows very approximately an exponential curve. $UB(G,k,p,t)$ was also built to be above all blue dots; i.e. it is a valid upper bound for all graphs of the reference dataset. Of course, there is no guarantee that this upper bound is still valid for all other graphs. So, our approach is never able to prove optimality in the strict sense. It can only provide a clue to it.

5   Experiments and Analysis

5.1   Tests

The upper bound UB was built based on the graphs of the *reference dataset*. In order now to test the *optimality clue* (procedure Sect. 3.2), we use this upper bound on graphs of the *test dataset* and the *control dataset* and for some graphs stemming from the DIMACS benchmark.

Results on the RCBII benchmark are presented in Table 1 in the last two lines. The first column concerns the 862 graphs with more than 1 million optimal solutions, corresponding to the *control dataset*. There is no *false positive*: Procedure 3.2 concludes that for all graphs there is no optimality clue. The two following columns concern the *reference dataset*. More precisely, the second column concerns graphs possessing fewer than 1 million optimal solutions which do not verify Theorem 1: the number of IS is lower than the number of optimal solutions. Of course, there are no *false positives* in this case, because UB was built to validate these graphs (*reference dataset*). The third column concerns the 566 graphs verifying Theorem 1. The *optimality clue* is proven for 449 of them because $i(G) > UB(G, k, p, t) > \mathcal{N}(G, k)$. The optimality clue is not shown on the 117 (= 566 − 449) other graphs because $UB(G, k, p, t) \geq i(G) > \mathcal{N}(G, k)$. $UB(G, k, p, t)$ is a too high an upper bound in this case. To prove the optimality clue on those graphs, we would have to increase the size of the solutions sample, t. The fourth column concerns the *test dataset* i.e. graphs for which the number of optimal solutions is unknown. We prove the *optimality clue* for nearly 20% of these graphs (39/210). There are three reasons why we did not prove the optimality clue for the other 171 (=210-39) graphs: (1) graph instances have more than 1 million solutions; (2) graph instances do not verify the Theorem 1; Nothing can be done for these two first reasons. (3) p is too close to t so the upper bound UB is too high. In order to obtain a more accurate upper bound, i.e. still valid but not too high, we have to increase the size of the sample or to choose a formula other than Eq. (1). Our approach therefore applies to about 20% of the random graphs in the RCBII benchmark. For control and reference datasets, we obtain more or less the same proportion: 25% (449/1821).

The results on selected DIMACS benchmark graphs are presented in Table 2. We only present graph instances for which the solutions sample generated by *HEAD*, are not all different (i.e. $p < t = 1,000$) and are susceptible to be optimal. The first column in Table 2 indicates the name of the graph instance. Columns 2–7 indicate for each graph, its number of vertices $|V|$, its density d, its chromatic number $\chi(G)$, when it is known, the number of colors k, used for the test ($k = \chi(G)$ if $\chi(G)$ is known), its number of independent sets $i(G)$, and the exact number of legal k-colorings $\mathcal{N}(G, k)$, when it is possible to calculate this with CDSATUR. Columns 8–10 then indicate the size of the solutions sample t,

Table 2. Results of *optimality clue* tests for graphs of DIMACS benchmark with $p < t$.

Instances	$|V|$	d	$\chi(G)$	k	$i(G)$	$N(G,k)$	t	p	$UB(G,k,p,t)$	Opt. clue	time (s)	$\|\chi(G)$	time(s)[15]
DSJC125.5	125	0.5	17	17	537,508	?	1,000	767	141,503	True	161	17	274
DSJC125.9	125	0.9	44	44	1,249	?	1,000	998	$+\infty$	False	28	44	7
DSJC250.9	250	0.9	72	72	6,555	?	1,000	889	423,733	False	1,963	72	11,094
flat1000_50_0	1,000	0.49	50	50	$>10^7$	?	1,000	1	2	True	25,694	50	3,331
flat1000_60_0	1,000	0.49	60	60	$>10^7$	?	1,000	1	2	True	44,315	60	29,996
le450_5a	450	0.06	5	5	$>10^7$	32	1,000	32	69	True	60	5	<0.1[21]
le450_5b	450	0.06	5	5	$>10^7$	1	1,000	1	2	True	138	5	<0.1[21]
le450_5c	450	0.1	5	5	$>10^7$	1	1,000	1	2	True	28	5	<0.1[21]
le450_5d	450	0.1	5	5	$>10^7$	8	1,000	8	16	True	20	5	<0.1[21]
le450_15c	450	0.17	15	15	$>10^7$	?	1,000	919	554,866	True		15	<0.1[21]
le450_15d	450	0.17	15	15	$>10^7$	?	1,000	579	26,041	True		15	<0.1[21]
myciel3	11	0.36	4	4	102	520	1,000	435	7,105	False	10	4	<0.1
queen5_5	25	0.53	5	5	461	2	1,000	2	4	True	9	5	<0.1[21]
queen6_6	36	0.46	7	7	2,634	20	1,000	20	42	True	10	7	<0.1
queen7_7	49	0.4	7	7	16,869	4	1,000	4	8	True	10	7	<0.1[21]
queen8_8	64	0.36	9	9	118,968	>154,068	1,000	993	$+\infty$	False	11	9	<1
r125.1c	125	0.97	46	46	787	?	1,000	977	934,514	False	5,962	46	<0.1[21]
DSJC250.5	250	0.5	?	28	24,791,612	?	1,000	999	$+\infty$	False	1,696	26	18
DSJC500.5	500	0.5	?	47	$>10^7$	?	1,000	281	32,731	True	out of time	43	439
				48		?	100,000	100,000	$+\infty$	False			
DSJC500.9	500	0.9	?	126	35,165	?	1,000	927	59,623	False	234,496	123	100
DSJC+300.1_8	300	0.1	?	8	$>10^7$	?	1,000	3	6	True	22,896	5	<0.1[21]
DSJC+300.5_31	300	0.5	?	31	$>10^7$	?	1,000	2	4	True	69,363	29	20
DSJC+400.5_39	400	0.5	?	39	$>10^7$	?	1,000	96	252	True	386,037	36	135

the number of different solutions in the sample p, the experimental upper bound of the number of k-colorings $UB(G, k, t, p)$. Columns 11 and 12 provide the result of the procedure of Sect. 3.2 for the *optimality clue* and the total computation time in seconds to generate all the samples. The last two columns indicate the lower bound of $\chi(G)$ found through the best known exact method [15] (or by IncMaxCLQ [21], which found the maximum clique) and the computation time of this method.

The first part of Table 2 corresponds to 17 graphs for which $\chi(G)$ is already known by other methods. We prove the *optimality clue* for 12 of them. The computation time of the optimality clue is higher than that for finding the lower bound with exact methods excepting two graphs. However the computation time of the optimality clue may be considerably reduced because the 1,000 runs of HEAD can be switched on to 2, 3... or 1,000 different processors or on to a cluster of computers.

The second part of Table 2 (below the horizontal line) corresponds to 6 graphs for which $\chi(G)$ is unknown. We have generated 3 new graph instances called <DSJC+*.*_k> with rules almost similar to those of <DSJC*.*> but for which a k-coloring is hidden so that very few other k-colorings can exist. For <DSJC500.5> which is one of the most challenging graphs in DIMACS, we have the *optimality clue* for $k = 47$[12].

Notice that when the optimality clue is proven for a given k, we check that the optimality clue for $k + 1$ can not be proven as well. For example, for <DSJC500.5>, we check the optimality clue for $k = 48$ has not been achieved: for $t = 100,000$ 48-colorings found by HEAD, all of them are different. If the randomized algorithm *HEAD* is biased, i.e. finds k-colorings always in the same

---

[12] For <DSJC500.5> the computation time is not reported because it takes several weeks and no accurate time has been recorded.

subset of $\Omega(G,k)$ (and therefore undervalues the upper bound), this bias does not reveal when we test the optimality clue for $k+1$.

Notice that the impact of the size of the sample has a great impact on the test. In order to have a not too high upper bound of $\mathcal{N}(G,k)$, we should have $p \ll t$. The size t of the solutions sample can be chosen therefore in function of the results of each graph instance. For example, we can extend the sample until all colorings of the sample are found at least twice (cf. Good-Turing estimator).

6   Conclusions and Perspectives

Based on Theorem 1, we propose a procedure, known as *optimality clue*, for determining if the global optimum is reached or not, by a heuristic method. This approach estimates an upper bound of the number of legal k-colorings by running a randomized heuristic several times. This process is contextual with the instance to solve. No general conclusion can be drawn on the heuristic itself, which is used for the building of solutions. This definition can be seen as an experimental criterion which evaluates the convergence of a randomized algorithm to the chromatic number. However, since it is not possible to be certain of the exactitude of the upper bound, it is also not feasible to prove optimality in the strict sense.

Our approach is nevertheless an alternative when exact methods are not applicable (high optimality gap). It is a new way in which to provide a criterion for the *proximity* of the optimality. The general idea is that the number of solutions with the same objective function value decreases when the objective function approaches the optimal value. *Optimality clue* matches with the standard definition of optimality in a large number of instances for DIMACS and RBCII benchmarks where the optimality is known. Furthermore, we proved the *optimality clue* for <DSJC500.5> graph of DIMACS with $k = 47$ colors which is a very challenging instance (only two algorithms are able to find 47-colorings [26, 32]). Tests on small random graphs (under 140 vertices) show that the *optimality clue* can be proven for 20% of them.

Finally, we defined quite a high upper bound in order to avoid *false positives*: graphs for which we prove the optimality clue for a given k, while $\chi(G) \neq k$.[13]

The proposed approach is based on a sampling of the legal k-colorings space, $\Omega(G,k)$. This sampling is built by running the *HEAD* algorithm a number of times; every successful run provides one element of the sample. Ideally, in order to obtain a representative sample, *HEAD* has to uniformly draw one k-coloring inside the legal k-colorings space. It is of course not possible to guarantee this feature in every case which is why we built an upper bound function (Eq. 1) of $\mathcal{N}(G,k) = |\Omega(G,k)|$. In order to improve our approach and to get closer to the ergodic objective, we plan to use more powerful counting models such as those presented in [6, 12] and to study the ergodic propriety of *HEAD*. Our work provides only an initial contribution to the study of optimality through

---

[13] In this context, we propose on our website a challenge to find a counterexample (false positive graph).

counting. Other methods of estimating the population should be tested, such as Good-Turing methods which estimate missing mass (i.e. missing k-colorings in the sample) or Peterson-type methods used to obtain statistical guarantees.

References

1. Baillargeon, S., Rivest, L.P.: Rcapture: loglinear models for capture-recapture in R. J. Stat. Softw. Art. **19**(5), 1–31 (2007)
2. Bollobás, B.: Random Graphs. Cambridge Studies in Advanced Mathematics, 2 edn. Cambridge University Press (2001). https://doi.org/10.1017/CBO9780511814068
3. Brélaz, D.: New methods to color the vertices of a graph. Commun. ACM **22**(4), 251–256 (1979)
4. Bron, C., Kerbosch, J.: Algorithm 457: finding all cliques of an undirected graph. Commun. ACM **16**(9), 575–577 (1973). https://doi.org/10.1145/362342.362367
5. Carraghan, R., Pardalos, P.M.: An exact algorithm for the maximum clique problem. Oper. Res. Lett. **9**(6), 375–382 (1990). https://doi.org/10.1016/0167-6377(90)90057-C
6. Ermon, S., Gomes, C.P., Selman, B.: Uniform solution sampling using a constraint solver as an oracle. In: de Freitas, N., Murphy, K.P. (eds.) Proceedings of the Twenty-Eighth Conference on Uncertainty in Artificial Intelligence, Catalina Island, CA, USA, 14–18 August 2012, pp. 255–264. AUAI Press (2012)
7. Favier, A., de Givry, S., Jégou, P.: Solution counting for CSP and SAT with large tree-width. Control Syst. Comput. **2**, 4–13 (2011)
8. Frieze, A., Vigoda, E.: A survey on the use of Markov chains to randomly sample colourings. Oxford University Press, Oxford (2007). Chap. 4. https://doi.org/10.1093/acprof:oso/9780198571278.003.0004
9. Furini, F., Gabrel, V., Ternier, I.: An improved DSATUR-based branch-and-bound algorithm for the vertex coloring problem. Networks **69**(1), 124–141 (2017). https://doi.org/10.1002/net.21716
10. Galinier, P., Hao, J.K.: Hybrid evolutionary algorithms for graph coloring. J. Comb. Optim. **3**(4), 379–397 (1999). https://doi.org/10.1023/A:1009823419804
11. Gomes, C.P., Hoffmann, J., Sabharwal, A., Selman, B.: From sampling to model counting. In: IJCA Proceedings IJCAI 2007, pp. 2293–2299. IJCAI (2007)
12. Gomes, C.P., Sabharwal, A., Selman, B.: Model counting. In: Biere, A., Heule, M., van Maaren, H., Walsh, T. (eds.) Handbook of Satisfiability, Frontiers in Artificial Intelligence and Applications, vol. 185, pp. 633–654. IOS Press (2009). https://doi.org/10.3233/978-1-58603-929-5-633
13. Grötschel, M., Lovász, L., Schrijver, A.: Polynomial algorithms for perfect graphs. In: Berge, C., Chvátal, V. (eds.) Topics on Perfect Graphs, North-Holland Mathematics Studies, vol. 88, pp. 325–356. North-Holland (1984). https://doi.org/10.1016/S0304-0208(08)72943-8
14. Gusfield, D.: Partition-distance: a problem and class of perfect graphs arising in clustering. Inform. Process. Lett. **82**(3), 159–164 (2002)
15. Held, S., Cook, W., Sewell, E.: Maximum-weight stable sets and safe lower bounds for graph coloring. Math. Program. Comput. **4**(4), 363–381 (2012). https://doi.org/10.1007/s12532-012-0042-3
16. Jerrum, M.: A very simple algorithm for estimating the number of k-colorings of a low-degree graph. Random Struct. Algorithms **7**(2), 157–165 (1995). https://doi.org/10.1002/rsa.3240070205

17. Jerrum, M.: Counting constraint satisfaction problems. In: Krokhin, A., Zivny, S. (eds.) The Constraint Satisfaction Problem: Complexity and Approximability, Dagstuhl Follow-Ups, vol. 7, pp. 205–231. Schloss Dagstuhl-Leibniz-Zentrum fuer Informatik, Dagstuhl, Germany (2017)

18. Johnson, D.S., Trick, M. (eds.): Cliques, Coloring, and Satisfiability: Second DIMACS Implementation Challenge, 1993, DIMACS Series in Discrete Mathematics and Theoretical Computer Science, vol. 26. American Mathematical Society, Providence (1996)

19. Karp, R.: Reducibility among combinatorial problems. In: Miller, R.E., Thatcher, J.W. (eds.) Complexity of Computer Computations, pp. 85–103. Plenum Press, New York (1972)

20. Krebs, C.J.: Ecology, 6th edn. Pearson, London (2009)

21. Li, C., Fang, Z., Xu, K.: Combining MaxSAT reasoning and incremental upper bound for the maximum clique problem. In: 2013 IEEE 25th International Conference on Tools with Artificial Intelligence, pp. 939–946, November 2013. https://doi.org/10.1109/ICTAI.2013.143

22. Malaguti, E., Toth, P.: A survey on vertex coloring problems. Int. Trans. Oper. Res. **17**, 1–34 (2009)

23. Marmion, M.É., Jourdan, L., Dhaenens, C.: Fitness landscape analysis and metaheuristics efficiency. J. Math. Model. Algorithms Oper. Res. **12**(1), 3–26 (2013). https://doi.org/10.1007/s10852-012-9177-5

24. Merz, P.: Memetic algorithms for combinatorial optimization problems: fitness landscapes and effective search strategies. Ph.D. thesis, Department of Electrical Engineering and Computer Science, University of Siegen, Germany (2000)

25. Miracle, S., Randall, D.: Algorithms to approximately count and sample conforming colorings of graphs. Discret. Appl. Math. **210**(Suppl. C), 133–149 (2016). lAGOS 2013: Seventh Latin-American Algorithms, Graphs, and Optimization Symposium, Playa del Carmen, México – 2013

26. Moalic, L., Gondran, A.: Variations on memetic algorithms for graph. J. Heuristics **24**(1), 1–24 (2018). https://doi.org/10.1007/s10732-017-9354-9

27. Orlin, J., Bonuccelli, M., Bovet, D.: An $O(n^2)$ algorithm for coloring proper circular arc graphs. SIAM J. Algebraic Discret. Methods **2**(2), 88–93 (1981). https://doi.org/10.1137/0602012

28. Östergård, P.R.: A fast algorithm for the maximum clique problem. Discret. Appl. Math. **120**(1), 197–207 (2002). https://doi.org/10.1016/S0166-218X(01)00290-6. Special Issue devoted to the 6th Twente Workshop on Graphs and Combinatorial Optimization

29. Pedersen, A.S.P., Vestergaard, P.D.: Bounds on the number of vertex independent sets in a graph. Taiwan. J. Math. **10**(6), 1575–1587 (2006)

30. Samotij, W.: Counting independent sets in graphs. Eur. J. Comb. **48**, 5–18 (2015). https://doi.org/10.1016/j.ejc.2015.02.005

31. Shih, W.K., Hsu, W.L.: An $O(n^{1.5})$ algorithm to color proper circular arcs. Discret. Appl. Math. **25**(3), 321–323 (1989). https://doi.org/10.1016/0166-218X(89)90011-5

32. Titiloye, O., Crispin, A.: Parameter tuning patterns for random graph coloring with quantum annealing. PLoS ONE **7**(11), e50060 (2012). https://doi.org/10.1371/journal.pone.0050060

33. Valiant, L.G.: The complexity of enumeration and reliability problems. SIAM J. Comput. **8**(3), 410–421 (1979). https://doi.org/10.1137/0208032

34. Wei, W., Selman, B.: A new approach to model counting. In: Bacchus, F., Walsh, T. (eds.) SAT 2005. LNCS, vol. 3569, pp. 324–339. Springer, Heidelberg (2005). https://doi.org/10.1007/11499107_24
35. Zuckerman, D.: Linear degree extractors and the inapproximability of max clique and chromatic number. Theory Comput. **3**(6), 103–128 (2007). https://doi.org/10.4086/toc.2007.v003a006. http://www.theoryofcomputing.org/articles/v003a006

Computing Wasserstein Barycenters via Linear Programming

Gennaro Auricchio[1], Federico Bassetti[2], Stefano Gualandi[1(✉)],
and Marco Veneroni[1]

[1] Dipartimento di Matematica "F. Casorati",
Università degli Studi di Pavia, Pavia, Italy
stefano.gualandi@unipv.it
[2] Dipartimento di Matematica, Politecnico di Milano, Milan, Italy

Abstract. This paper presents a family of generative Linear Programming models that permit to compute the exact Wasserstein Barycenter of a large set of two-dimensional images. Wasserstein Barycenters were recently introduced to mathematically generalize the concept of averaging a set of points, to the concept of averaging a set of clouds of points, such as, for instance, two-dimensional images. In Machine Learning terms, the Wasserstein Barycenter problem is a generative constrained optimization problem, since the values of the decision variables of the optimal solution give a new image that represents the *"average"* of the input images. Unfortunately, in the recent literature, Linear Programming is repeatedly described as an inefficient method to compute Wasserstein Barycenters. In this paper, we aim at disproving such claim. Our family of Linear Programming models rely on different types of Kantorovich-Wasserstein distances used to compute a barycenter, and they are efficiently solved with a modern commercial Linear Programming solver. We numerically show the strength of the proposed models by computing and plotting the barycenters of all digits included in the classical MNIST dataset.

Keywords: Wasserstein Barycenter ·
Kantorovich-Wasserstein distance · Linear Programming ·
Constrained optimization

1 Introduction

In several Machine Learning problems, a fundamental step is the computation of a similarity measure between a pair of objects. These objects very often correspond to uncertain measure quantities, which are represented as probability density functions or discrete N-dimensional histograms. The Kantorovich-Wasserstein distance is a mathematical metric which permits to compute the distance between probability density functions or N-dimensional histograms by solving a constrained optimization problem [11, 17]. Intuitively, the distance

L.-M. Rousseau and K. Stergiou (Eds.): CPAIOR 2019, LNCS 11494, pp. 355–363, 2019.
https://doi.org/10.1007/978-3-030-19212-9_23

Fig. 1. Left: Euclidean average of 100 images of digit "1". Right: Wasserstein Barycenter of the same set of images.

between two (discrete) measures is equal to the total cost of "transporting" all the mass of the first (discrete) distribution into the second. The cost for transporting a unit of mass from a location of the first distribution to a location of the second distribution is called the *ground distance*. In case of N-dimensional histograms, the locations are associated to the centers of the bins of the histograms, and the ground distance can be, for example, any of the standard norms: ℓ_1, ℓ_2, or ℓ_∞. Compared to other methods used in probability and information theory to measure the similarity between probability distributions, the Kantorovich-Wasserstein distance is more robust to noise of the input data.

A very interesting problem, which uses the Kantorovich-Wasserstein distance as a building block, is the **Wasserstein Barycenter problem** [2]: Given a set of m discrete measures γ_k, with $k = 1, \ldots, m$, defined on a space $X \subseteq \mathbb{R}^N$, we have to find a discrete measure y that has the minimal overall Kantorovich-Wasserstein *distance* to all γ_k measures. The new discrete measure y is called the Wasserstein Barycenter, since it generalizes the idea of "averaging": Given k vectors of $\mathbb{R}^n$, the usual average computed as $\left(\frac{1}{k} \sum_{i=1}^{k} x_i \right)$ is the minimizer of the problem $\arg\min_x \sum_i \|x - x_i\|_2^2$, where $\|\cdot\|_2$ is the Euclidean norm. Note that the Wasserstein Barycenter problem is a "generative" problem, since its optimal solution is a new discrete measure on the same space X of the input measures γ_k. In case of images, which are a very specific type of histograms, the barycenter is a new image. For instance, given 100 images of the digit "1", Fig. 1 shows on the left the corresponding Euclidean average, while on the right it shows the Wasserstein Barycenter of the same set of images, computed with the ℓ_2 ground distance.

The state-of-the-art approach to compute Wasserstein Barycenters relies on entropic regularized formulations of the constrained optimization problem [9], which are solved with derivations of the Sinkhorn's algorithm [8,13,14]. The main advantages of this class of algorithms are two: (i) they are very easy to understand and to implement, and (ii) they can be implemented to run in parallel on multiple Graphics Processing Units (GPU). However, in the regularized formulation of Optimal Transport there is a crucial parameter that has to be tuned manually. As shown in [5], these methods can be numerically unstable and can provide solutions which are very far from the optimal value. Other approaches to compute the Wasserstein Barycenters include stochastic algorithms [7] and gradient descent based algorithms [15]. Unfortunately, Linear Programming approaches to compute Wasserstein Barycenters [4] are repeatedly considered to be extremely inefficient on larger instances [7,15].

Our Contribution. In this paper, we propose a new class of generative Linear Programming models to solve the Wasserstein Barycenter problem. Building on the results recently presented in [6], and depending on the norm used as ground distance in the pairwise Kantorovich-Wasserstein distances, we exploit the geometric structure of the problem to reduce the size of the linear programs. For the ℓ_1 and ℓ_∞ ground distances, we provide exact formulations that can compute the barycenter of up to 3 200 images of size 28×28. For the ℓ_2 ground distance, we provide an approximation scheme that permits to approximate the optimal barycenter within a guaranteed percentage error.

The outline of this paper is as follows. Section 2 formally presents the definition of the Kantorovich-Wasserstein distance and of the Wasserstein Barycenter problem. Section 3 introduces our generative Linear Programming models based on uncapacitated minimum cost flow formulations. In Sect. 4, we discuss preliminary computational results on the computation of the Wasserstein Barycenters of all the digits included into the well-known MNIST dataset [1].

2 Background

In this section, we present the basic formulation of the Kantorovich-Wasserstein distance and we introduce the corresponding Wasserstein Barycenter problem. In the following, for the sake of clarity, we tailor our exposition to the case of discrete measures represented as N-dimensional histograms. In particular, the grey scale images of the MNIST dataset used in Fig. 1 can be viewed as 2D histograms.

2.1 Kantorovich-Wasserstein Distances

Let μ and ν be two probability measures defined on a space $X \subseteq \mathbb{R}^N$. We can think of μ and ν as two vectors in the unitary simplex $\mathcal{S}^n$, that is, vectors in $\mathbb{R}^n_+$ such that $\sum_{i=1}^n \mu_i = 1$ and $\sum_{i=1}^n \nu_i = 1$. In addition, we have a cost function $c : X \times X \to \mathbb{R}_+$ that gives the cost c_{ij} of transporting a unit of mass from position i to j, where both $i, j \in X$. The computation of the Kantorovich-Wasserstein functional [16] between the two given discrete measures μ and ν is equivalent to solve the following Linear Program:

$$\mathcal{W}_c(\mu, \nu) = \min \ \sum_{i=1}^n \sum_{j=1}^n c_{ij} x_{ij} \qquad (1)$$

$$\text{s.t.} \ \sum_{j=1}^n x_{ij} = \mu_i \qquad\qquad i = 1, \ldots, n \qquad (2)$$

$$\sum_{i=1}^{n} x_{ij} = \nu_j \qquad\qquad j = 1,\ldots,n \qquad (3)$$

$$x_{ij} \geq 0, \qquad\qquad i = 1,\ldots,n, j = 1,\ldots,n. \qquad (4)$$

Note that this problem is clearly a special case of the Transportation Problem [10,12] and, hence, it can be formulated and solved as an uncapacitated minimum cost flow problem in a bipartite graph with $2n$ nodes and n^2 arcs [3].

Whenever the costs in (1) are defined as $c_{ij} = d_{ij}^p$, where d_{ij} is the distance between positions i and j in X, we define the **Kantorovich-Wasserstein distance of order** p as

$$W_p(\mu, \nu) := W_{d^p}(\mu, \nu)^{\min\{\frac{1}{p}, p\}}. \qquad (5)$$

In this paper, we are interested in the Kantorovich-Wasserstein distance of order 1, that is, $p = 1$ and the distance d_{ij} is measured with the ℓ_1, ℓ_2 and ℓ_∞ norms. Recall that i and j are points in $\mathbb{R}^N$.

2.2   Wasserstein Barycenters

In the discrete Wasserstein Barycenter problem, we have to find a new discrete measure $y^* \in \mathcal{S}^n$ that minimizes the sum of the distances to a given set γ of m discrete measures γ_k:

$$\mathcal{B}(\gamma, c) := \min_{y \in \mathcal{S}^n} \sum_{k=1}^{m} W_p(\gamma_k, y). \qquad (6)$$

If we denote by γ_{ik} the i-th element of vector γ_k, the discrete Wasserstein Barycenter problem is equivalent to the following Linear Program [4]:

$$\mathcal{B}(\gamma, c) = \min_{y} \ \sum_{k=1}^{m}\sum_{i=1}^{n}\sum_{j=1}^{n} c_{ij} x_{ijk} \qquad (7)$$

$$\text{s.t.} \ \sum_{j=1}^{n} x_{ijk} = \gamma_{ik} \qquad\qquad i = 1,\ldots,n, \ k = 1,\ldots,m \qquad (8)$$

$$\sum_{i=1}^{n} x_{ijk} = y_j \qquad\qquad j = 1,\ldots,n, \ k = 1,\ldots,m \qquad (9)$$

$$\sum_{i=1}^{n} y_j = 1 \qquad\qquad (10)$$

$$x_{ijk} \geq 0, \qquad\qquad i, j = 1,\ldots,n, \ k = 1,\ldots,m \qquad (11)$$

$$y_j \geq 0, \qquad\qquad j = 1,\ldots,n. \qquad (12)$$

The size of problem (7)–(12) depends on the size of each discrete measure γ_k, that is equal to n, and on the number m of given discrete measures: there are $n^2 m$

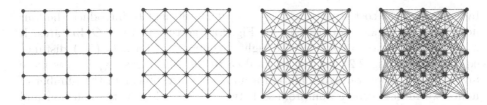

Fig. 2. The first two networks are for ground distances ℓ_1 and ℓ_∞. The third and fourth networks are for ℓ_2, and, respectively $L = 2$ and $L = 3$.

variables and $2nm$ constraints. Indeed, the size of this Linear Program in practice can become very large. Note that we have m bipartite graphs with n^2 arcs each, and hence, its solution raises very interesting computational challenges.

The strong structure of problem (7)–(12) is more evident if we introduce in the model the dummy variables y_{ik}, and we replace constraints (9) with the following pairs of constraints:

$$\sum_{i=1,\ldots,n} x_{ijk} = y_{jk} \qquad\qquad j = 1,\ldots,n, \ \ k = 1,\ldots,m \qquad (13)$$

$$y_{jk} = y_i \qquad\qquad j = 1,\ldots,n, \ \ k = 1,\ldots,m. \qquad (14)$$

At this point, if we relax constraints (14), we get two classes of independent subproblems: An optimization subproblem for each discrete measure γ_k, and a single minimization problem defined on the variables y_i. Each subproblem in the first class considers the values of y_{ik} as given, and computes the Kantorovich-Wasserstein distances $W_p(\gamma_k, y_k)$, where y_k is the vector of n values y_{ik}. The second subproblem looks for a vector y belonging to the probability simplex $\mathcal{S}^n$.

In the following section, using the results recently introduced in [6], we show how we can reduce the size of problem (7)–(12) by exploiting the structure of the cost c_{ij}.

3   Network Flow Formulations

Given two discrete measures μ and ν, when the exponent p in (5) is equal to 1, we are computing the so-called Kantorovich-Wasserstein distance of order 1, denoted by $W_1(\mu, \nu)$. In this case, we can formulate the problem of computing $W_1(\mu, \nu)$ using a complete flow network $K = (N, E, c, b)$ defined as follows: For each position $i \in X$ where we have defined the two measures μ_i and ν_i, we introduce a node in N. The node balance b_i is set equal to the difference between the two measures at position i, that is, $b_i = \mu_i - \nu_i$. The nodes with $b_i > 0$ are the supply nodes, the nodes where $b_i < 0$ are the demand nodes, and all the others are the transit nodes. For each ordered pair of positions i, j, i.e. $(i, j) \in X \times X$, with $i \neq j$, we introduce an arc (i, j) in E with cost $c_{ij} = d_{ij}$.

As shown in [6], it is enough to consider a flow problem on a smaller network. In particular the number of arcs added to the network can be significantly limited by exploiting the cost structure. If $c_{ij} = \ell_1$ it is sufficient to introduce the four neighbors at distance equal to 1 (see Fig. 2.1), getting in total $O(4n)$ arcs. If $c_{ij} = \ell_\infty$, we have to consider the eight neighbors within a $\ell_\infty(i,j)$ distance equal to 1 (see Fig. 2.2), getting in total $O(8n)$ arcs. The case $c_{ij} = \ell_2$ is more involved, and if we want to maintain a number of arcs linear in the number of nodes we must accept a compromise between the quality of the solution and the size of the arc set, which depends on a parameter L. Figure 2.3 shows the network obtained with $L = 2$ and Fig. 2.4 with $L = 3$.

In all the three cases just considered, the discrete Wasserstein Barycenter problem can be reformulated using a flow network $G = (N, A, c, b)$, with $A \subseteq E$, as follows:

$$\mathcal{B}_G(\gamma, c) = \min \sum_{k=1}^{m} \sum_{(i,j) \in A} c_{ij} x_{ijk} \tag{15}$$

$$\text{s.t.} \sum_{(i,j) \in A} x_{ijk} - \sum_{(j,i) \in A} x_{ijk} = \gamma_{ik} - y_i \qquad \forall i \in N, k = 1, \dots, m, \tag{16}$$

$$\sum_{i \in N} y_i = 1 \tag{17}$$

$$x_{ijk} \geq 0, y_i \geq 0 \qquad \forall (i,j) \in A, k = 1, \dots, m. \tag{18}$$

The size of this LP directly depends on the size of the arc set A, which in turn depends on the choice of ℓ_1, ℓ_∞, or ℓ_2 for c_{ij}. In the first two cases, there is a network G for which $\mathcal{B}_G(\gamma, \ell_1)$ and $\mathcal{B}_G(\gamma, \ell_\infty)$ are exact formulations which use an order of magnitude less of variables x_{ijk} with respect to problem (7)–(12).

Proposition 1. *If $d_{ij} = \ell_1(i,j)$ or $d_{ij} = \ell_\infty(i,j)$, then there is a graph G with $O(n)$ arcs such that $\mathcal{B}_G(\gamma, c) = \mathcal{B}(\gamma, c)$ and hence any optimal solution y^* to the LP problem (15)–(18) is a W_1-barycenter of $\gamma_1, \dots, \gamma_m$.*

The third case, regarding $\mathcal{B}_G(\gamma, \ell_2)$, is more involved and it is described in the next Theorem, which can be proved by using the results in [6].

Theorem 1. *Let $d_{ij} = \ell_2(i,j)$ and y^* be a solution to the LP problem (15)–(18). Then:*

(i) *There is a graph G with $O(\frac{6}{\pi^2} n^2)$ arcs such that $\mathcal{B}_G(\gamma, c) = \mathcal{B}(\gamma, c)$. In particular, y^* is a W_1-barycenter.*
(ii) *For every $1 \leq L < \sqrt{n} - 1$, there is a graph G with $O(n)$ arcs such that*

$$(1 - \Gamma_G)\mathcal{B}_G(\gamma, c) \leq \mathcal{B}(\gamma, c) \leq \mathcal{B}_G(\gamma, c). \tag{19}$$

with

$$1 - \frac{\sqrt{1 + 4L^2}}{L + \sqrt{1 + L^2}} \leq \Gamma_G \leq 0.26 \left(1 - \frac{L}{\sqrt{1 + L^2}} \right), \qquad as\ L \to +\infty. \tag{20}$$

Table 1. Comparison of runtime (in seconds) in computing the barycenter of 50 images using the bipartite [4] and our flow models. For ℓ_1 and ℓ_∞, we report the runtime and the relative speedup (s.up). For ℓ_2 with $L = 5$, in addition, we report the percentage gap to the optimal solution.

Digits	ℓ_1			ℓ_∞			ℓ_2, L=5			
	Bipartite	Flow	s.up	Bipartite	Flow	s.up	Bipartite	Flow	s.up	Gap
0	1154.2	7.5	153.2	1114.9	12.0	93.1	1402.8	204.1	6.9	0.02%
1	370.9	6.7	55.3	410.8	11.1	36.9	437.5	222.1	2.0	0.01%
2	902.6	7.6	118.5	1019.9	11.8	86.2	1066.5	210.7	5.1	0.03%
3	791.6	7.7	103.4	917.0	12.5	73.6	1049.8	206.6	5.1	0.03%
4	632.7	7.6	83.4	711.3	11.6	61.2	749.7	215.0	3.5	0.03%
5	690.6	7.8	88.7	835.3	11.4	73.4	815.0	207.6	3.9	0.03%
6	684.0	7.6	89.9	744.9	11.5	64.8	801.3	202.7	4.0	0.04%
7	603.9	7.8	77.8	667.8	12.2	54.7	690.9	214.5	3.2	0.03%
8	791.9	7.6	104.1	924.5	13.0	71.1	1108.5	230.7	4.8	0.05%
9	800.8	7.6	106.0	964.7	12.9	74.7	1008.0	213.3	4.7	0.03%

In this case, y^* is an approximate W_1-barycenter and asymptotically

$$\Gamma_G = \frac{1}{8L^2} - \frac{11}{128L^4} + O\Big(\frac{1}{L^6}\Big).$$

4 Computational Results

This section reports our preliminary computational results on the solution of the Wasserstein Barycenter problem using the Linear Programming models presented in the previous sections. As benchmark, we used the set of images contained in the MNIST dataset [1].

All models are solved using the commercial solver Gurobi v8.1 on a Dell Workstation with a Intel Xeon W-2155 CPU with 10 physical cores at 3.3 GHz, and 32 GB of RAM. In particular, we set the following parameters: (i) we force the solver to use the Barrier algorithm, which can run in parallel, and which, specifically on the larger instances, is much faster than both the primal and the dual simplex algorithms. (ii) We disabled the crossover operator, since we do not need a basic solution. Note that disabling the final crossover yields a significant speedup on the larger instances. (iii) We let the solver decide how many physical cores to use. All the other parameters were left at their default values.

Table 1 shows the results of our test on computing the Wasserstein Barycenter of 50 different images of the same digits using as ground distances the ℓ_1, ℓ_∞, and ℓ_2 norms. We compare the running time in seconds of the bipartite model (7)–(12) with the running time of our flow formulation (15)–(18). For the ℓ_1 and ℓ_∞ norms, we report the runtime along with the relative speedup (column

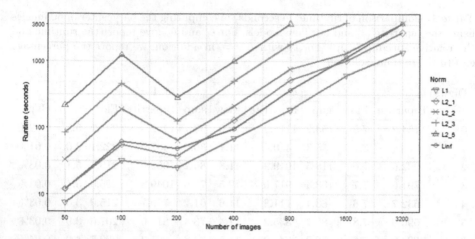

Fig. 3. Comparison of runtime (in seconds) as a function of the number of images used to compute the Wasserstein Barycenters. For the ℓ_2 norms, we used $L = \{1, 2, 3, 5\}$.

Fig. 4. The first row shows the barycenters of 3200 images of each digit, computed using the ℓ_1 ground distance. The second row shows the barycenters of 800 images, using the ℓ_2 ground distance with $L = 5$. The ℓ_2 norm generates slightly sharper images.

"s.up") obtained with our flow model: in both cases, we observe a speedup of two or three orders of magnitude. For the ℓ_2 norm, we used the flow network that depends on the parameter L defined in (20), using the value $L = 5$, in order to obtain a good tradeoff between computational speed and solution quality. The last column of the table reports the observed percentage gap to the optimal solution, which, indeed, is always very small.

Figure 3 shows how the runtime (in seconds) of the different proposed LP models scales with the number of images given as input. The runtime refers to the average over all the 10 different digits, and the plot is given in a log-log scale. The plot confirms what we would expect: on the larger problem with 3600 input images, the problem that uses the ℓ_1 norm is the fastest, since it has the smaller flow network G (these results are shown in the first row of Fig. 4). In addition, for the ℓ_2 norm, we remark that with the parameter $L = 5$, the solver run out of memory already with 1600 images in input, while for $L = 3$ it run out of memory with 3200 images. Again, this is a consequence of the number of arcs in the flow network, which determines the number of variables in the flow formulation of the Wasserstein Barycenter problem (15)–(18). However, already using "only" 800 images, the ℓ_2 norm with $L = 5$ produces sharp barycenters.

For the sake of reproducibility, our source code is available at the following public repository: https://github.com/stegua/barycenters-cpaior2019. We have published all the optimal solutions, that is, the Wasserstein Barycenters found in our computations.

References

1. MNIST dataset. http://yann.lecun.com/exdb/mnist/. Accessed 29 Nov 2018
2. Agueh, M., Carlier, G.: Barycenters in the Wasserstein space. SIAM J. Math. Anal. **43**(2), 904–924 (2011)
3. Ahuja, R.K., Magnanti, T.L., Orlin, J.B.: Network Flows: Theory, Algorithms, and Applications. Alfred P. Sloan School of Management, Massachusetts Institute of Technology, Cambridge, Massachusetts (1988)
4. Anderes, E., Borgwardt, S., Miller, J.: Discrete Wasserstein Barycenters: optimal transport for discrete data. Math. Methods Oper. Res. **84**(2), 389–409 (2016)
5. Auricchio, G., Bassetti, F., Gualandi, S., Veneroni, M.: Computing Kantorovich-Wasserstein distances on d-dimensional histograms using (d+1)-partite graphs. In: Advances in Neural Information Processing Systems (2018)
6. Bassetti, F., Gualandi, S., Veneroni, M.: On the computation of Kantorovich-Wasserstein distances between 2D-histograms by uncapacitated minimum cost flows (2018). arXiv preprint: arXiv:1804.00445
7. Claici, S., Chien, E., Solomon, J.: Stochastic Wasserstein Barycenters (2018). arXiv preprint: arXiv:1802.05757
8. Cuturi, M.: Sinkhorn distances: lightspeed computation of optimal transport. In: Advances in Neural Information Processing Systems, pp. 2292–2300 (2013)
9. Cuturi, M., Doucet, A.: Fast computation of Wasserstein Barycenters. In: International Conference on Machine Learning, pp. 685–693 (2014)
10. Flood, M.M.: On the Hitchcock distribution problem. Pac. J. Math. **3**(2), 369–386 (1953)
11. Santambrogio, F.: Optimal Transport for Applied Mathematicians, pp. 99–102. Birkäuser, New York (2015)
12. Schrijver, A.: On the history of the transportation and maximum flow problems. Math. Program. **91**(3), 437–445 (2002)
13. Sinkhorn, R., Knopp, P.: Concerning nonnegative matrices and doubly stochastic matrices. Pac. J. Math. **21**(2), 343–348 (1967)
14. Solomon, J., et al.: Convolutional Wasserstein distances: efficient optimal transportation on geometric domains. ACM Trans. Graph. (TOG) **34**(4), 66 (2015)
15. Staib, M., Claici, S., Solomon, J.M., Jegelka, S.: Parallel streaming Wasserstein Barycenters. In: Advances in Neural Information Processing Systems, pp. 2647–2658 (2017)
16. Vershik, A.M.: Long history of the Monge-Kantorovich transportation problem. Math. Intell. **35**(4), 1–9 (2013)
17. Villani, C.: Optimal Transport: Old and New, vol. 338. Springer, Heidelberg (2008). https://doi.org/10.1007/978-3-540-71050-9

Repairing Learned Controllers
with Convex Optimization: A Case Study

Dario Guidotti[1(✉)], Francesco Leofante[1], Claudio Castellini[2],
and Armando Tacchella[1]

[1] DIBRIS, Università degli Studi di Genova,
Via all'Opera Pia, 13, 16145 Genoa, Italy
{dario.guidotti,francesco.leofante}@edu.unige.it,
armando.tacchella@unige.it
[2] Institute of Robotics and Mechatronics, German Aerospace Center,
Münchener Straße, 20, Oberpfaffenhofen, 82234 Weßling, Germany
claudio.castellini@dlr.de

Abstract. Despite the increasing popularity of Machine Learning methods, their usage in safety-critical applications is sometimes limited by the impossibility of providing formal guarantees on their behaviour. In this work we focus on one such application, where Kernel Ridge Regression with Random Fourier Features is used to learn controllers for a prosthetic hand. Due to the non-linearity of the activation function used, these controllers sometimes fail in correctly identifying users' intention. Under specific circumstances muscular activation levels may be misinterpreted by the method, resulting in the prosthetic hand not behaving as intended. To alleviate this problem, we propose a novel method to verify the presence of this kind of intent detection mismatch and to repair controllers leveraging off-the-shelf LP technology without using additional data. We demonstrate the feasibility of our approach using datasets gathered from human participants.

1 Introduction

In the last few years Machine Learning techniques proved to be successful in many domains of application such as image classification [1] or speech recognition [2], with some architectures even claiming to match the cognitive abilities of humans [3]. Despite this popularity, the usage of Machine Learning (ML) methods in safety-critical applications is still sometimes limited by the absence of effective methods to provide formal guarantees on their behavior. In this paper we focus on one such safety-critical application, where Kernel Ridge Regression with Random Fourier Features [4,5] is used to learn controllers for prosthetic hands. In this framework, multi-fingered, self-powered prosthetic hands [6] are controlled using signals generated from a certain number of surface electromyography [7] sensors. Ensuring the correct behavior of the controller is critical to the safety of the amputee wearing the robotic artifact.

© Springer Nature Switzerland AG 2019
L.-M. Rousseau and K. Stergiou (Eds.): CPAIOR 2019, LNCS 11494, pp. 364–373, 2019.
https://doi.org/10.1007/978-3-030-19212-9_24

Although several approaches have been proposed to build ML models enforcing reliable myocontrol [8,9], reliability is still an issue. In particular we tackle a problem of *intent mismatch*: whenever a subject increases her muscle activation, the prosthesis should in turn increase the applied force. However, if the ML model is a non-linear one, there is no guarantee that this will happen. To reduce the chance of intent mismatch, we propose an approach that couples a standard ML-based myocontrol system with a Linear Programming (LP) solver to *automatically repair* and improve the learned model *without requiring additional data*. This last point is crucial as gathering more data from the subject to amend the ML model is not desirable: to gather relevant data, the subject would need to apply a large amount of force leading to muscle strain, fatigue and frustration.

By leveraging LP technology, we can represent the ML model, as well as the property of interest, as a set of arithmetic constraints and establish algorithmically whether a controller satisfies the given property. If the property is not satisfied, a LP solver is used to iteratively repair the controller until the resulting model is mathematically guaranteed to be safe.

To demonstrate the feasibility of our approach, we compare results obtained with a standard ML-based myocontroller against a myocontroller that was repaired using our methodology on datasets gathered from human subjects at the German Aerospace Center. Remarkably, we show not only that LP-based repair is effective, but it also comes at a reasonably low computational price.

2   Preliminaries

Kernel Ridge Regression with Random Fourier Features. KRR-RFF with Tikhonov regularization [4] has been demonstrated multiple times in literature to match most of the requirements of myocontrol [8,9]. Ridge Regression (RR) builds a linear model of the input space data $\mathbf{x}$ of the form $f(\mathbf{x}) = \mathbf{w}^T \mathbf{x}$, where $\mathbf{w}$ is the vector of weights computed as a result of the training phase. KRR-RFF modifies standard RR by introducing a feature map $\Phi : \mathbb{R}^d \to \mathbb{R}^D$ that maps a sample $\mathbf{x} \in \mathbb{R}^d$ onto a D-dimensional space as shown in Eq. 1

$$f(\mathbf{x}) = \mathbf{w}^T \Phi(\mathbf{x})$$
$$\Phi(\mathbf{x}) = \sqrt{2}\cos(\mathbf{\Omega}\mathbf{x} + \beta) \tag{1}$$

where $\mathbf{\Omega}$ is a $D \times d$ matrix and β is a D-dimensional vector, whose values are drawn from a normal distribution and a uniform distribution from 0 to 2π, respectively – we refer the reader to, e.g., [4] for more details.

This method can be seen as a finite-(D-)dimensional approximation to a RBF-kernel Least-Squares SVM [10], therefore it can be made arbitrarily accurate by tuning D; nevertheless, since its kernel is finite-dimensional and can be explicitly written, it enjoys most useful properties of Ridge Regression such as, e.g., space-boundedness (making it ideal for online learning) and extremely fast training and testing. On top of this, using a rank-1 update method such as, e.g., the Sherman-Morrison formula, it can be made incremental, paving the way to interactive myocontrol.

Myocontrol and Intent Detection. Natural, simultaneous and proportional myocontrol is an instance of (multi-variate) regression. Let $S = (X, Y)$ be a dataset composed by a set of observation $X = \{\mathbf{x}_i \mid \mathbf{x}_i \in \mathbb{R}^d, i = 1, \ldots, N\}$ and a set of corresponding target values $Y = \{\mathbf{y}_i \in \mathbb{R}^m \mid \forall \mathbf{x}_i \in X\}$. Each observation consists of d features evaluated from a set of sensors and denotes the muscular activation corresponding to an action (e.g., wrist flexion, power grasp, etc.); each associated target value, in turn, is a vector of m motor activation values (currents, torques, ...) for a prosthetic device and corresponds to the desired action as enacted by the device itself. In practice S is built by gathering, for each desired action, an adequate number of observations recorded while the subject is stimulated to perform it; each such observation is then coupled with the target value enforcing the action by the prosthetic device. For instance, the subject is asked to power grasp ("make a fist"); once the experimenter verifies that the signals have reached a stable pattern which is also sufficiently distinct from the baseline, a representative amount of observations is recorded and associated to (synthetic) target values denoting maximal activation of all fingers. At the end of the data gathering phase, S consists of one or more observation clusters for each action considered, coupled with adequate target values – see, e.g., [11] for more details.

KRR-RFF builds an approximant function $f(\mathbf{x}) : \mathbb{R}^d \to \mathbb{R}^m$ which best fits S and offers the best generalization power on so-far unseen data. The approximant f can be seen as an *intent detector*: whenever the wearer's muscles are activated to enforce a specific action, the prosthesis should behave accordingly. If properly built out of S, f will smoothly and timely activate every motor of the prosthesis whenever required; minimal and maximal muscular activations, as gathered from the subject, will correspond to minimal and maximal motor activations; under plausible assumptions, intermediate activation values too will be correctly predicted in a monotonically-increasing fashion [4,8,11]. Finally, increasing the muscle activation beyond the maximal values obtained during data gathering should correspond to coordinated increased activation of the motors of the prosthesis, but this cannot be guaranteed by learning techniques only.

Linear Programming. Linear Programming solves linear problems over a set of *decision variables*, a set of *linear constraints* over these variables and a *convex objective function* that is linear in the decision variables [12].

Let $x_1, \ldots, x_n$ be a set of decision variables, a general linear program can be written as

$$\text{minimize} \sum_{i=1} c_i x_i$$

$$\text{subject to} \sum_{i=1} a_{ij} x_i \geq b_j, j \in [1, m]$$

$$x_i \in \mathbb{R}, i \in [1, n]$$

Values c_i, a_{ij} and b_j are constants that are specified during problem formulation. General approaches to solving LPs include methods such as *simplex* or the *ellipsoid* – for further details see, e.g., [12].

Verification and Repair of Learned Models. Several approaches have been proposed to verify different classes of ML models automatically. Even though such approaches might differ in several aspects, most of them consider trained models, i.e., they do not intermingle with the learning process, and they seek a transformation from the ML model to a decision or optimization problem in some constraint system – see, e.g., [13] for a recent account on the subject. For instance, in [14] Boolean satisfiability solvers are proposed to verify robustness of Binarized Neural Networks. *Satisfiability Modulo Theories* engines are leveraged to verify neural networks in, e.g., [15,16] and Support Vector Machines in [17]. MILP-based approaches have been proposed to verify neural networks by, e.g., [18,19], while a combination of SAT and LP techniques is used by [20] for the same purpose. A different approach is taken in [21], where abstract interpretation is used to certify safety and robustness of Deep Neural Networks with ReLU and max pooling layers. Repair has received less attention compared to verification, with [15] and [17] being the only contributions in this direction. However, the repair they propose involves retraining the ML model with data generated by solvers.

3   Verification and Repair

In order to enhance the reliability of the controllers we consider in this work, we propose an automated procedure that iteratively brings the controller to a condition where intent detection mismatch does not occur anymore. In the following, we describe the assumptions on which our approach rests, show how our ideas can be formalized into an executable algorithm and provide experimental evidence to support the approach we propose.

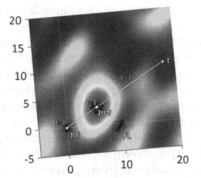

Fig. 1. The heat map representing the value of the approximant for power grasping f_{PW}, learned using readings from two sensors.

3.1   A Linear Model of Intent Detection

The data gathering procedure produces a cluster of samples for each action of interest. An example is shown in Fig. 1, where observations are gathered for three actions: rest (RE), power grasp (PW) and wrist flexion (FL). For a specific action, we assume that values of the input space lying on the line joining the resting cluster and the cluster corresponding to the action of interest denote an execution of the action with increasing strength (see Fig. 1). This assumption is justified by physiological reasons: muscle activation remains roughly coordinated as the action is performed with more force (see, e.g., [22–26]).

More precisely, for a generic action a, let $\mathbf{x}_a$ denote the coordinates corresponding to the average of points belonging to the cluster corresponding to a.

Performing an action a with increasing strength can be seen as moving along the line $\mathbf{x}_{RE} + (\mathbf{x}_a - \mathbf{x}_{RE})t_a$, where $t_a \in \mathbb{R}_{\geq 0}$[1] is a parameter proportional to the muscle activation value. By construction, $t_a = 1$ corresponds to the maximum activation value registered during the data gathering phase. Figure 1 shows an example for $a = PW$: for $t_{PW} = 0$, the subject is at rest; as t_{PW} increases the subject starts power-grasping moving along $\mathbf{x}_{RE} + (\mathbf{x}_{PW} - \mathbf{x}_{RE})t_{PW}$. Under this assumption, absence of intent mismatch for action a can be encoded as follows:

$$\forall t_a.(\mathbf{x} = \mathbf{x}_{RE} + (\mathbf{x}_a - \mathbf{x}_{RE})t_a \wedge t_a > 1) \implies f_a(\mathbf{x}) \geq a_{max} \tag{2}$$

where $f_a(\mathbf{x})$ is the value of the approximant function for a applied to the input $\mathbf{x}$ and a_{max} is the full activation value (e.g., the value corresponding to the full closure of the hand when the action is PW). Therefore, a mismatch happens whenever the value of muscular activation t_a is greater than 1 (i.e., the maximum value observed during data gathering) and the corresponding motor activation value predicted by the approximant f is less than a_{max}.

3.2   The Algorithm

Our repair procedure LP-REPAIR leverages training data and our prior knowledge about the problem in order to modify the parameters of the machine learning model used in the controller. In particular, LP-REPAIR analyses the initial model in order to determine if there exists an unsafe point, i.e., an instance of intent detection mismatch. Such point is considered to create additional constraints for an optimization problem which, once solved, gives as solution a new set of parameters for the machine learning model. This constraint generation approach is justified by the fact that, in the feature space, the model is a linear function of the weights as per Eq. (1). The resulting model is guaranteed to fix activation insofar the unsafe point is concerned. The procedure proceeds iteratively verifying safety of the modified controller and, possibly, adding a new constraint (corresponding to a new unsafe point) to the problem; when the controller is deemed safe, the procedure ends.

Algorithm 1 shows the pseudocode of LP-REPAIR: X and Y are, respectively, the inputs and outputs training data of our controller, w is the vector of the weights of the machine learning model and A is a set of actions of interest. **safetyCheck**(...) uses a set of points, in the original input space, with uniform distances along the straight lines of interest in order to find the most unsafe point for each action, i.e., the point whose output has the highest difference from the expected value. This function returns a boolean variable $safe_a$ (which is false if an unsafe point has been found) and an unsafe point $unsafePoint_a$.

The procedure **Opt**(...) defines and solves the optimization problem presented in Fig. 2, where w_i is the i-th component of the vector of the weights, and ϵ_i is the variation on the above mentioned component. D is the dimensionality of the feature space, N is the number of samples in the dataset, and η corresponds to

---

[1] In practice, the value of t_a is upper bounded by operating range of EMG sensors.

Algorithm 1. Repair procedure

1: **procedure** LP-REPAIR(X, Y, w, A)
2:     $safe \leftarrow False$
3:     $\hat{w} \leftarrow w$
4:     $unsafeSet \leftarrow \emptyset$
5:     **while** not $safe$ **do**
6:         $safe \leftarrow True$
7:         **for each action** $a \in A$ **do**
8:             $(safe_a, unsafePoint_a) \leftarrow$ **safetyCheck**$(X, Y, \hat{w}, a)$
9:             **if** not $safe_a$ **then**
10:               $safe \leftarrow False$
11:               $unsafeSet \leftarrow (unsafeSet \cup unsafePoint_a)$
12:         $\hat{w} \leftarrow$ **Opt**$(X, Y, \hat{w}, unsafeSet)$
13:     **return** $\hat{w}$

---

the sampling rate which determines the subset of the training set we consider in the cost function. x_i^{rest} is the i-th component of the data-point corresponding to the center of the rest cluster, and $x_i^{act_a}$ is the i-th component of the data-point corresponding to the center of the cluster associated to the activation of a certain action a. Finally, $x_i^{unsafe_j}$ denotes the i-th component of the j-th unsafe point.

In more detail, Eq. 3a defines the cost function to be minimized, where: *(i)* the first member corresponds to the modification of the parameters of the controllers (w_i), *(ii)* the second member tries to minimize the error on a subset of the training set, and we do this in order to guarantee that the prediction performances are not degraded and *(iii)* the third member is a tolerance on the error δ. Equations 3b and 3c encode the constraints to guarantee the correct output value respectively for the center of the rest cluster and the centers of the activation clusters. Finally, Eq. 3d presents the constraints we use to force the output values of the unsafe points to be correct, where k is the total number of unsafe points found. It is important

$$\min \sum_{i=1}^{D} |\epsilon_i| + \sum_{l=1}^{floor(N/\eta)} \left| Y^{l \cdot \eta} - \sum_{i=1}^{D} (w_i + \epsilon_i) X_i^{l \cdot \eta} \right| + \delta \tag{3a}$$

$$\sum_{i=1}^{D} (w_i + \epsilon_i) x_i^{rest} = 0 \tag{3b}$$

$$a_{max} - \delta \le \sum_{i=1}^{D} (w_i + \epsilon_i) x_i^{act_a} \le a_{max} + \delta \quad \forall a \in A \tag{3c}$$

$$\sum_{i=1}^{D} (w_i + \epsilon_i) x_i^{unsafe_j} \ge a_{max}, \quad \forall j = 1, ..., k \tag{3d}$$

Fig. 2. LP model used for repair.

to highlight that, since all the points we consider in the problem are given, i.e., data-points are not variables, we can apply the feature mapping beforehand and formulate the problem of repairing a non-linear classifier without introducing non-linear constraints. For this reason, all the data-points in Eq. 3 are data-points in the feature space, not in the input space.

3.3   Experimental Results

A prototypical implementation of our approach was tested[2] on 18 datasets, each consisting of 320 samples, gathered from human subjects. The actions of interests are rest, power grasp, wrist flexion and wrist extension. In all our tests our procedure managed to terminate successfully, e.g. it managed to bring the controller to a safe configuration. In Fig. 3a we show an example of how the output of the controller for power-grasping is modified by LP-Repair. As it can be seen, at the last iteration, the output is greater than one[3] for all admissible values of t above a_{max}.

In Fig. 3b we show the quantities of interest in our analysis: the sampling rate (S.R.) is the rate at which we pick samples from the training data to use in the cost function. D is the parameter which determines the dimension of the feature space. MSE-o (resp. MSE-r) is the mean square error computed on the training set before (resp. after) the repair process. Time corresponds to the CPU time needed for the repair process (in seconds). MSE and time values are both computed as means on 18 datasets. We do not display the last three rows of the Table when the sampling rate is 100 because the repair procedure did not complete successfully, i.e., none of the 18 datasets yields a successful repair due to conditioning problems reported by the LP engine. We have chosen to compare MSEs before and after repair in order to verify that our repair process does not degrade the performances of the controller substantially. As it can be seen from Fig. 3b, the time complexity of the problem grows monotonically with the sampling rate and the dimension of the feature space, but on average, the runtime of the repair procedure is always less than 3 CPU minutes. In particular, for values of D which are relevant from an engineering point of view, i.e., those that yield MSE errors of less than 0.1%, we notice that the repair procedure is feasible for all the sampling rates considered, and that the final accuracy, albeit decreased, is still viable for practical applications. This is more evident as D increases because as the representativity increases the repair process can modify the controller without decreasing too much the accuracy of the prediction.

---

[2] We implemented our procedure using Python version 2.7. and the libraries *sklearn* and *cvxopt* for learning and optimization respectively. The default solver of *cvxopt*, i.e., *conelp*, was used – see [27] for more details. All the experiments are capped at 10 min of CPU time and 4 GBs of memory; experiments ran on a Ubuntu 18.04 machine equipped with a quad-core i5 Intel CPU running at 2.60 GHz.

[3] Notice that for power-grasping a_{max} is equal to one.

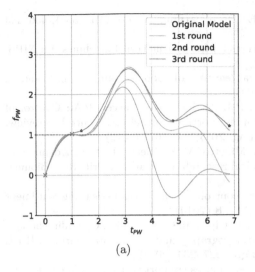

S. R.	D	MSE-o	MSE-r	Time
10	10	0.0140	0.4370	6.967s
10	50	0.0020	0.0040	18.456s
10	100	0.0012	0.0025	39.752s
10	150	0.0008	0.0019	68.984s
10	200	0.0007	0.0019	103.650s
10	250	0.0006	0.0015	138.009s
10	300	0.0005	0.0012	179.034s
50	10	0.0146	0.5177	4.880s
50	50	0.0021	0.0282	14.988s
50	100	0.0011	0.0151	26.941s
50	150	0.0009	0.0216	41.134s
50	200	0.0007	0.0195	61.431s
50	250	0.0006	0.0134	83.657s
50	300	0.0005	0.0157	110.448s
100	10	0.0157	0.9239	4.041s
100	50	0.0019	0.0648	13.461s
100	100	0.0011	0.0517	25.440s
100	150	0.0009	0.0349	36.183s

(a)   (b)

Fig. 3. Results for power-grasping.

4   Conclusion

Our paper provides empirical evidence that convex optimization techniques can be used to *repair* machine learned controllers for prosthetic hands insofar as detection mismatch is concerned and controllers are learned using KRR-RFF. The key factors of our successful evaluation are (*i*) a physiology-rooted modeling of intent detection mismatch along regions that conjoin data clusters corresponding to actions to those corresponding to rest conditions and (*ii*) a formalization of the repair problem based on feature space rather than input space. The former allows us to mathematically define intent detection mismatch, while the latter allows us to solve the problem using a linear program, rather than a non-linear one. The main feature of our method is that it repairs the controller *without requiring additional data* to be gathered. This is very important in applications involving human subjects, where additional data acquisition can be time-consuming, expensive or just plain impossible. While our method is effective on a specific case study, we expect that our findings can help approach the problem in different contexts where KRR-RFF is viable, as well as provide guidance to the repair of controllers learned with different methods: our future work includes furthering research along these directions.

References

1. Taigman, Y., Yang, M., Ranzato, M., Wolf, L.: DeepFace: closing the gap to human-level performance in face verification. In: CVPR, pp. 1701–1708 (2014)
2. Yu, D., Hinton, G.E., Morgan, N., Chien, J.-T., Sagayama, S.: Introduction to the special section on deep learning for speech and language processing. IEEE Trans. Audio Speech Lang. Process. **20**(1), 4–6 (2012)

3. LeCun, Y., Bengio, Y., Hinton, G.E.: Deep learning. Nature **521**(7553), 436–444 (2015)
4. Rahimi, A., Recht, B.: Random features for large-scale kernel machines. In: NIPS, pp. 1177–1184 (2008)
5. Gijsberts, A., Metta, G.: Incremental learning of robot dynamics using random features. In: ICRA, pp. 951–956 (2011)
6. Fougner, A., Stavdahl, Ø., Kyberd, P.J., Losier, Y.G., Parker, P.A.: Control of upper limb prostheses: terminology and proportional myoelectric control - a review. IEEE Trans. Neural Syst. Rehabil. Eng. **20**(5), 663–677 (2012)
7. Merletti, R., Botter, A., Cescon, C., Minetto, M.A., Vieira, T.M.M.: Advances in surface EMG: recent progress in clinical research applications. Crit. Rev. Biomed. Eng. **38**(4), 347–379 (2011)
8. Gijsberts, A., et al.: Stable myoelectric control of a hand prosthesis using non-linear incremental learning. Front. Neurorobot. **8** (2014)
9. Strazzulla, I., Nowak, M., Controzzi, M., Cipriani, C., Castellini, C.: Online bimanual manipulation using surface electromyography and incremental learning. IEEE Trans. Neural Syst. Rehabil. Eng. **25**(3), 227–234 (2017)
10. Gestel, T.V., et al.: Benchmarking least squares support vector machine classifiers. Mach. Learn. **54**(1), 5–32 (2004)
11. Patel, G., Nowak, M., Castellini, C.: Exploiting knowledge composition to improve real-life hand prosthetic control. IEEE Trans. Neural Syst. Rehabil. Eng. **25**(7), 967–975 (2017)
12. Schrijver, A.: Theory of Linear and Integer Programming. Wiley-Interscience Series in Discrete Mathematics and Optimization. Wiley, Hoboken (1999)
13. Leofante, F., Narodytska, N., Pulina, L., Tacchella, A.: Automated verification of neural networks: advances, challenges and perspectives. arXiv preprint arXiv:1805.09938 (2018)
14. Narodytska, N., Kasiviswanathan, S.P., Ryzhyk, L., Sagiv, M., Walsh, T.: Verifying properties of binarized deep neural networks. In: AAAI, pp. 6615–6624 (2018)
15. Pulina, L., Tacchella, A.: An abstraction-refinement approach to verification of artificial neural networks. In: Touili, T., Cook, B., Jackson, P. (eds.) CAV 2010. LNCS, vol. 6174, pp. 243–257. Springer, Heidelberg (2010). https://doi.org/10.1007/978-3-642-14295-6_24
16. Katz, G., Barrett, C., Dill, D.L., Julian, K., Kochenderfer, M.J.: Reluplex: an efficient SMT solver for verifying deep neural networks. In: Majumdar, R., Kunčak, V. (eds.) CAV 2017. LNCS, vol. 10426, pp. 97–117. Springer, Cham (2017). https://doi.org/10.1007/978-3-319-63387-9_5
17. Leofante, F., Tacchella, A.: Learning in physical domains: mating safety requirements and costly sampling. In: Adorni, G., Cagnoni, S., Gori, M., Maratea, M. (eds.) AI*IA 2016. LNCS (LNAI), vol. 10037, pp. 539–552. Springer, Cham (2016). https://doi.org/10.1007/978-3-319-49130-1_39
18. Cheng, C.-H., Nührenberg, G., Ruess, H.: Maximum resilience of artificial neural networks. In: D'Souza, D., Narayan Kumar, K. (eds.) ATVA 2017. LNCS, vol. 10482, pp. 251–268. Springer, Cham (2017). https://doi.org/10.1007/978-3-319-68167-2_18
19. Fischetti, M., Jo, J.: Deep neural networks and mixed integer linear optimization. Constraints **23**(3), 296–309 (2018)
20. Ehlers, R.: Formal verification of piece-wise linear feed-forward neural networks. In: D'Souza, D., Narayan Kumar, K. (eds.) ATVA 2017. LNCS, vol. 10482, pp. 269–286. Springer, Cham (2017). https://doi.org/10.1007/978-3-319-68167-2_19

21. Gehr, T., Mirman, M., Drachsler-Cohen, D., Tsankov, P., Chaudhuri, S., Vechev, M.: AI2: safety and robustness certification of neural networks with abstract interpretation. In: 2018 IEEE Symposium on Security and Privacy (SP) (2018)
22. Valero-Cuevas, F.J.: Predictive modulation of muscle coordination pattern magnitude scales fingertip force magnitude over the voluntary range. J. Neurophysiol. **83**(3), 1469–1479 (2000)
23. Poston, B., Danna-Dos Santos, A., Jesunathadas, M., Hamm, T.M., Santello, M.: Force-independent distribution of correlated neural inputs to hand muscles during three-digit grasping. J. Neurophysiol. **104**(2), 1141–1154 (2010)
24. de Rugy, A., Loeb, G.E., Carroll, T.J.: Muscle coordination is habitual rather than optimal. J. Neurosci. **32**(21), 7384–7391 (2012)
25. He, J., Zhang, D., Sheng, X., Li, S., Zhu, X.: Invariant surface emg feature against varying contraction level for myoelectric control based on muscle coordination. IEEE J. Biomed. Heal. Inform. **19**(3), 874–882 (2015)
26. Al-Timemy, A.H., Khushaba, R.N., Bugmann, G., Escudero, J.: Improving the performance against force variation of emg controlled multifunctional upper-limb prostheses for transradial amputees. IEEE Trans. Neural Syst. Rehabil. Eng. **24**(6), 650–661 (2016)
27. Vandenberghe, L.: The CVXOPT linear and quadratic cone program solvers (2010)

A Hybrid Approach for Exact Coloring
of Massive Graphs

Emmanuel Hebrard[1(✉)] and George Katsirelos[2]

[1] LAAS-CNRS, Université de Toulouse, CNRS, Toulouse, France
hebrard@laas.fr
[2] MIAT, UR-875, INRA, Toulouse, France
gkatsi@gmail.com

Abstract. The graph coloring problem appears in numerous applications, yet many state-of-the-art methods are hardly applicable to real world, very large, networks. The most efficient approaches for massive graphs rely on "peeling" the graph of its low-degree vertices and focus on the maximum k-core where k is some lower bound on the chromatic number of the graph. However, unless the graphs are extremely sparse, the cores can be very large, and lower and upper bounds are often obtained using greedy heuristics.

In this paper, we introduce a combined approach using local search to find good quality solutions on massive graphs as well as locate small subgraphs with potentially large chromatic number. The subgraphs can be used to compute good lower bounds, which makes it possible to solve optimally extremely large graphs, even when they have large k-cores.

1 Introduction

The *Vertex Coloring Problem* (VCP) asks for the minimum number of colors that can take the vertices of a graph G so that no two adjacent vertices share a color. This number $\chi(G)$ is called the *chromatic number* of the graph.

The VCP has numerous applications. For instance, when allocating frequencies, devices on nearby locations should work on different frequencies to avoid interference. The chromatic number of this distance-induced graph is thus the minimum span of required frequencies [1,23]. In compilers, finding an optimal register allocation is a coloring problem on an interference graph of value live ranges [6]. In timetabling, assigning time slots to lectures so that no two classes attended by a common subset of student happen in parallel is a VCP [8].

The best performing approaches to the VCP often do not scale to extremely large graphs such as, for instance, social networks. In fact, on networks with several million nodes, even local search methods are seldom used and the best approaches rely on scale reduction and greedy heuristics both for lower and upper bounds [16,28]. Indeed, the main technique used for reducing the graph consists

G. Katsirelos—The second author was partially supported by the french "Agence nationale de la Recherche", project DEMOGRAPH, reference ANR-16-C40-0028.

L.-M. Rousseau and K. Stergiou (Eds.): CPAIOR 2019, LNCS 11494, pp. 374–390, 2019.
https://doi.org/10.1007/978-3-030-19212-9_25

in removing vertices of degree lower than some lower bound on the chromatic number. This technique might be very effective on sparse graphs especially when a maximum or a maximal clique provides a good lower bound. Several real-world extremely large sparse graphs can be efficiently tackled, even via complete algorithms, after such preprocessing. However, even relatively sparse graphs can have a large core of vertices whose degree within the core is higher than the chromatic number. In this case, there are not many practical techniques for upper bounds and most proposed approaches rely on greedy\heuristics, in particular Brelaz' DSATUR [5]. Likewise, in this context there is virtually no method for computing a lower bound other than finding a large clique in the graph. As a result, there is little hope to optimally solve an instance with a large core, and whose chromatic number is strictly larger than the size of its largest clique.

In this paper we consider two datasets of very large graphs. The first, dimacs10, contains 30 graphs from the 10th DIMACS challenge [3]. It consists of two subclasses, one of graphs with heavy-tailed distribution of degrees and the other quasi-regular graphs. The second, snap, contains 75 graphs from the Stanford Large Network Dataset Collection [15]. These graphs correspond to social, citation, collaboration, communication, road or internet networks. They range from tens of thousands to several million vertices and all have extremely low density.

Whereas about half of these graphs are easy or even trivial for the state-of-the-art approaches, the rest remain too large and hard to color even after preprocessing. By combining several methods including local search, heuristics and complete algorithms, we can solve a significant proportion to optimality (close to 40%) of these hardest instances, even if they contain hundreds of thousands of vertices after preprocessing and even if their chromatic number is larger than their clique number. We survey the related work in Sect. 2, describe our main contribution, a method to obtain good lower bounds on very large graphs in Sect. 3, an effective local search approach to obtain good upper bounds in Sect. 4, and a way to combine these in Sect. 5. We report on an experimental comparison with the state of the art in Sect. 6.

2   Related Work

Heuristic methods are very relevant since they easily scale to very large inputs. In particular, the DSATUR heuristic proposed by Brelaz [5] is instrumental in the state-of-the-art method on the datasets we consider, FastColor [16]. The DSATUR heuristic builds a coloring C mapping vertices to colors. It iteratively choses a vertex from a set U initially containing all vertices V of the graph. The chosen vertex v is the one with maximum *saturation degree* $\delta^{sat}(v)$ defined as the number of colors among its neighbors $N(v)$, i.e., $\delta^{sat}(v) = |\{C(u) \mid u \in (N(v) \setminus U)\}|$. In case of a tie, the vertex with maximum degree $|N(v)|$ is selected. Then it sets $C(v)$ to the smallest possible color $\min(\mathbb{N} \setminus \{C(u) \mid u \in (N(v) \setminus U)\})$.

DSATUR-based branch and bound algorithms [9,26] are among the best complete methods, alongside column generation approaches [18,20] and SAT-based models and hybrid algorithms [11,25,27,30]. However, none of these scale to graphs with more than a few thousands vertices.

2.1 Local Search

Local search and meta-heuristics have long been applied to graph coloring (e.g. [12]), and with great success. All the best known colorings on the commonly used dataset from the second dimacs challenge [13] were obtained by such methods[1].

In principle, local search approaches seem very well suited for coloring large graphs, and indeed most algorithms scale very well to relatively large graphs. However, surprisingly, we could not find a report of a local search or a meta-heuristic approach applied to the large graphs of the snap and dimacs10 datasets, or on graphs of similar magnitude.

When the number of vertices grows really large, then one must be very careful about the implementation details. As a matter of fact, several off-the-shelf algorithms we tried used data structures with a space complexity quadratic in the number of vertices, and are de facto irrelevant. Another critical point is the size of the neighborhood. For instance, the most common tabu scheme considers all the (non-tabu) moves of any node sharing a color with a neighbor, to a different color. Typical methods evaluate every such move and choose the one that decreases the most the number of conflicts. The number of such moves to consider in a graph with millions of vertices can be prohibitive, especially when starting from low quality initial solutions. The state-of-the-art memetic algorithm HEAD [21] uses a similar tabu search, and although we made superficial changes to make it capable of loading massive graphs in memory, it performed poorly on those. After a non-exhaustive review of the literature and of the available software, our belief is that these methods *could* be adapted to extremely large and sparse graphs, but it would require non-trivial implementation work.

Blöchliger and Zufferey's local search algorithm [4] appears to be relatively promising in this context. The idea is to try to complete a partial coloring, i.e., a partition of the vertices into of k disjoint independent sets $\{C_1, \ldots, C_k\}$ plus an extra set U of "uncolored" vertices. A move consists in swapping a node $v \in U$ with the vertices $N(v) \cap C_i$ for some color $i \in \{1, \ldots, k\}$. A move (v, i) minimizing $|N(v) \cap C_i|$ is randomly chosen. In order to escape local minima, after each move (v, i), the moves (u, i) for $u \in N(v)$ are added to a tabu list so that v will stay with color i for a given number of iterations. When the set U becomes empty, a k-coloring is obtained and the process can continue by randomly eliminating one color i, that is, setting $U = C_i$ and removing C_i from the partition.

2.2 Independent Set Extraction

Whereas sequence-based coloring heuristics (such as DSATUR) explore the vertices and insert them into the smallest possible color class (or independent set), Leighton's RLF heuristic [14] extracts one maximal independent set (or color

---

[1] http://www.info.univ-angers.fr/~porumbel/graphs/.

class) at a time. This technique has been shown to be more effective than DSATUR on some graphs, however it has a higher computational cost.

Recent effective methods for large graphs rely on this principle. For instance, Hao and Wu [10] recently proposed a method which iteratively extracts maximal independent sets until the graphs contains no more than a given number of vertices. Then, any algorithm can be used on the residual graph to produce a k-coloring which can be trivially extended to a $k + p$-coloring of the whole graph if p independent sets have been extracted. Moreover, the authors show that it may be effective to iteratively expand the residual graph by re-inserting the vertices of some independent set extracted in the first phase and run again the coloring method on the larger residual graph. This method, however, was not tested on graphs larger than a few thousand vertices.

2.3   Peeling-Based Approaches

The so-called "peeling" procedure is an efficient scale reduction technique introduced by Abello *et al.* [2] for the maximum clique problem. Since vertices of $(k + 1)$-cliques have each at least k neighbors, one can ignore vertices of degree $k - 1$ or less. As observed in [28], this procedure corresponds to restricting search to the maximum χ^{low}-core of G where χ^{low} is some lower bound on $\omega(G)$:

Definition 1 (k-Core and denegeracy). *A subset $S \subseteq V$ is called a k-core of the the graph $G = (V, E)$ if the minimum degree of any vertex in the subgraph of G induced by S is k. The maximum value of k for which G has a non-empty k-core is called the* degeneracy *of G.*

As observed by Verma *et al.* [28], the peeling technique can also be used for graph coloring, since low-degree vertices can be colored greedily.

Theorem 1 (Verma *et al.* 2015). *G is k-colorable if and only if the maximum k-core of G is k-colorable.*

Indeed, starting from a k-coloring of the maximum k-core of G, one can explore the vertices of G that do not belong to the core and add them back in the inverse of the *degeneracy order*, so that any vertex is preceded by at most $k - 1$ of its neighbors, and hence can be colored without introducing a $k + 1$th color. The other direction is trivial as the maximum k-core is a subgraph of G.

This preprocessing technique can be extremely effective on very sparse graphs, and computing a lower bound of the chromatic number is relatively easy: computing the clique number of a graph is NP-hard, but in practice it is much easier than computing its chromatic number. However, the χ^{low}-core might be too large, and therefore a second use of the peeling technique was proposed in [28]. The idea is to find a coloring of the maximum $(\chi^{up} - 1)$-core of G where χ^{up} is an *upper bound* on $\chi(G)$. The maximum $(\chi^{up} - 1)$-core has several good properties: it is often small, its chromatic number is a lower bound on $\chi(G)$, and if there exists such a k-coloring with $k < \chi^{up}$, then it can be extended, in the worst case, to a $(\chi^{up} - 1)$-coloring of G.

Therefore, Verma *et al.* proposed the following method: Starting from the bounds $\chi^{low} \leq \chi(G) \leq \chi^{up}$, the algorithm solves the maximum $(\chi^{up} - 1)$-core of G to optimality, and extends the corresponding k-coloring greedily following the inverse degeneracy order to a k'-coloring. Then it sets χ^{low} to $\max(\chi^{low}, k)$ and χ^{up} to k'. The algorithm converges since since χ^{low} cannot decrease and χ^{up} is guaranteed to decrease at each step.

Unfortunately, some graphs simply do not have small k-cores, even for k larger than their chromatic number, so this method is limited to extremely sparse graphs. Moreover, notice that the core must be solved to optimality in order to extract relevant information from the iteration and converge.

The algorithm `FastColor` proposed by Lin *et al.* [16] also uses peeling, but in a slightly different way. A k-*bounded independent set* is an independent set whose vertices all have a degree strictly smaller than k. Their method iteratively finds a maximal clique using a very effective sampling-based heuristic; removes a χ^{low}-bounded independent set where χ^{low} is the size of the clique from the graph; and computes an upper bound using the DSATUR heuristic.

This method is very effective, outperforming the approach of Verma *et al.* on graphs with large cores. However, notice that the vertices in a χ^{low}-bounded independent set cannot be in a χ^{low}-core since their degree is strictly thess than χ^{low}, and therefore this variant of peeling is less effective than Verma's. The two main components are the method to find a clique and the DSATUR heuristic to find upper bounds. The former essentially samples a set of vertices to be expanded to a maximal clique. When extending a clique, a number p of neighbors are probed and the one that maximizes the size of the residual candidate set of vertices to expand the clique is chosen. Several runs are performed with the parameter p growing exponentially at every run. However, it cannot prove a lower bound greater than the clique number. The runs of DSATUR are randomized and augmented with the *recolor* technique [24]: when a new color class i is created for a vertex v, if there exist two color classes C_j, C_k with $j < k$ and a vertex u such that $N(v) \cap C_j = \{u\}$ and $N(u) \cap C_k = \emptyset$, then v and u can be recolored to j and k respectively, thus leaving the color i free.

3   Iterated Dsatur

The overwhelmingly most common lower bound technique is to find a large clique. Several other lower bounds have been used. For instance, two extra lower bounds were proposed in [9]: the Lovász Theta number [17] and a second lower bound based on a mapping between coloring and independent sets on a reformulation of the graph [7]. Another lower bound based on finding embedded Mycielskian graphs [22] was proposed in [11]. Moreover, the bounds obtained by linear relaxation of either the standard model or the set covering problem from the branch & price approach are very strong. However, it is difficult to make any of these methods scale up to graphs with millions of vertices.

Many graphs of the `dimacs10` and `snap` datasets have a chromatic number equal to their clique number. Moreover, finding a maximum clique turns out to

---

Algorithm 1. Iterated Dsatur

Algorithm: I-Dsatur

Data: Graph G, Initial order O^0, color assignment C^0, bounds χ^{low}, χ^{up}

Result: $\chi(G)$

$i \leftarrow 0$

while $\chi^{low} < \chi^{up}$ **do**

$\quad p \leftarrow 1 + \max\{j \mid C^i(o_k^i) \leq \chi^{low} \; \forall k < j\}$

$\quad O^{i+1} \leftarrow \{o_1^i, \ldots o_p^i\}$

$\quad i \leftarrow i + 1$

$\quad C_{core} = \texttt{ExactColoring}(G_{O^i})$

$\quad$ **if** $\max(C_{core}) > \chi^{low}$ **then**

$\quad\quad \chi^{low} \leftarrow \max(C_{core})$

$\quad\quad C^i \leftarrow C^{i-1}$

$\quad$ **else**

$\quad\quad C^i \leftarrow C_{core}$

$\quad\quad (O^i, C^i) \leftarrow \texttt{Dsatur}(O^i, C^i)$

$\quad\quad$ **if** $\max(C^i) < \chi^{up}$ **then**

$\quad\quad\quad \chi^{up} \leftarrow \max(C')$

return (χ^{low}) // $= \chi(G)$

---

be much easier in practice than solving the VCP. Therefore, it is often possible to find a maximum clique and they often provide a good lower bound.

In this section, we introduce a method to solve the VCP that scales up to very large graphs. Moreover, it may compute non-trivial lower bounds, that is, larger than the clique number. As a consequence, this method can produce optimality proofs, even when $\omega(G) < \chi(G)$. The principle is to iteratively compute a coloring with DSATUR, and optimize its prefix up to the first occurrence of the color $\chi^{low} + 1$. If there exists a χ^{low}-coloring of the prefix, then the next iteration of DSATUR will follow the optimized prefix, whose length will thus increase. Otherwise, the lower bound can be incremented.

Algorithm 1 uses a variant of DSATUR which takes a total order O of a subset of the vertices and a coloring C for these vertices. It assigns first vertices in the given order and coloring, then colors the rest of the vertices using the standard DSATUR heuristic. It returns the coloring C as well as the total order $O = \langle o_1, \ldots, o_{|V|} \rangle$ that it followed. In the following, we write $\max(C)$ for the maximum color used, and $C(v)$ for the color of v.

Algorithm 1 proceeds as follows. Given initial bounds χ^{low} and χ^{up}, as well as a coloring and ordering that witness the upper bound, we extract the *core graph*, which is the subgraph G_{O^1} of G induced by the vertices $\{o_1, \ldots, o_p\}$ where p is the maximum index for which all vertices $o_1, \ldots, o_{p-1}$ are assigned colors in $[1, \chi^{low}]$. In other words, p is the index of the first vertex that is assigned a color greater than the current lower bound χ^{low}. The order of these p vertices is fixed for all subsequent runs of DSATUR. We then compute $\chi(G_{O^1})$, using any exact coloring algorithm. In our implementation this is the satisfiability-based algorithm from [11]. If $\chi(G_{O^1}) > \chi^{low}$ then we can update $\chi^{low} = \chi(G_{O^1})$. This is because G_{O^1} is an induced subgraph of G, so $\chi(G_{O^1})$ is a lower bound on $\chi(G)$. On the other hand, if $\chi(G_{O^1}) \leq \chi^{low}$, we fix the first p vertices to their order and color them as in the optimal coloring of G_{O^1} and use them as the starting point for a run of Dsatur[2]. In either case, we proceed to the next iteration.

Algorithm 1 converges because at every iteration a growing subset of the vertices are included in the core. Indeed, if $\chi(G_{O^i}) > \chi^{low}$, then the lower bound is increased, which means that $G_{O^{i+1}}$ is larger. If $\chi(G_{O^i}) \leq \chi^{low}$, then the next run of Dsatur is constrained to assign at least o_p to a color in $[1, \chi^{low}]$, so the core graph at the next iteration contains at least one more vertex. In the extreme, the algorithm will terminate when $G_{O^i} = G$.

4   Local Search for Massive Graphs

As far as we know, the best upper bound for the datasets we consider were obtained using either Brelaz' heuristic [16], or by greedily extending the optimal solution of a k-core [28]. Therefore, whether local search can help remains to be seen. In this section we describe the modifications we made to Blöchliger and Zufferey's tabu-search algorithm in order to adapt it to extremely large graphs.

Initialization. A first very modest, but significant, addition is a method to efficiently initialize the solution of the local search. The algorithm described in [4] is given an integer k and tries to find a k-coloring. Since our method produces colorings during preprocessing (from the computation of the degeneracy ordering and from DSATUR) it is immediate to initialize the solution with such a coloring whereby the vertices of any one color class are considered "uncolored". However, we observed that it was important to choose a small color class, as they can be extremely unbalanced and chosing randomly could lead to a prohibitively large neighborhood to explore in the initial steps.

Chained Flat Moves. Recall that a move consists in swapping a node v from the set U of uncolored vertices with its neighbors $N(v) \cap C_i$ in some color class i. When $N(v) \cap C_i = \emptyset$ this is an *improving move* as we have one less uncolored node. Now we call a move (v, i) such that $|N(v) \cap C_i| = \{u\}$ a *flat move*. We know that no strictly improving move was possible, so if there is an improving or a flat move involving u it is likely to be selected next. Therefore, in the event of a flat move we greedily follow chains of flat moves from the previous vertex until reaching an improving move, or until no flat or improving move is possible for that vertex. This technique does not change the neighborhood, but explores it in a more greedy way and is often beneficial. Moreover, we observed that it was relatively easy to assess if such moves were effective, by counting how many of them lead to an improving move, and by checking their length.

---

[2] Dsatur denotes our implementation of the DSATUR heuristic.

Algorithm 2. Local Search

Algorithm: TabuSearch

Data: Graph $G = (V, E)$, Coloring C, Parameters I, t
Result: A coloring of G

$best \leftarrow C, k \leftarrow 0$
foreach $v \in V', 1 \leq i \leq \max(C)$ **do** $T_v^i = 0$
while $k \leq I$ **do**

> 1 $c \leftarrow \arg\min_i(|C_i|)$
> $U \leftarrow C_c$
>   **while** $i \leq I$ *and* $C_i \neq \emptyset$ **do**
>> $v, i \leftarrow \arg\min_{u \in U, j \neq c | T_u^j \leq k}(|N(u) \cap C_j|)$
>> 2   **if** $|N(v) \cap C_i| = 1$ **then**
>>> **repeat**
>>>> $C(v) \leftarrow i$
>>>> $v' \leftarrow v, i' \leftarrow i$
>>>> $v, i \leftarrow \arg\min_{u \in C_{i'}, j \notin \{c, i'\} | T_u^j \leq k}(|N(u) \cap C_j|)$
>>>
>>> **until** $|N(v) \cap C_i| = 1$
>>> **if** $|N(v) \cap C_i| > 1$ **then**
>>>> $C(v) \leftarrow c$
>>>> $T_v^{i'} \leftarrow k + t$
>>
>> 3   **else**
>>> $C(v) \leftarrow i$
>>> **foreach** $u \in N(v) \cap C_i$ **do**
>>>> $C(u) \leftarrow c$
>>>> $T_u^i \leftarrow k + t$
>>
>> $k \leftarrow k + 1$
>
> **if** $U = \emptyset$ **then** $best \leftarrow C$
> **return** $best$

Algorithm 2 is a pseudo-code of our implementation of Blöchliger and Zufferey's tabu search. We denote C_i the set of vertices of color i, that is $C_i = \{v \mid C(v) = i\}$. The outer loop and the color selection in line 1 are not in the original implementation, as well as the random path of flat moves corresponding to the lines between 2 and 3. Notice that ties are broken randomly in every "arg min" operator. Moreover, the management of the tabu list (T_v^i) as well as of the iteration limit, and the choice of applying a random path move is more complex than the pseudo-code shows. We set the parameters as follows.

Tabu List. Here we used a relatively straightforward scheme which is in fact a simplified version of what is done in the original code. Every 10000 iterations, the tabu tenure parameter t is decremented, unless it is null or the delta between the lowest and largest size for U (the set of "uncolored" vertices) is lower than or equal to 1 since the last update of the tabu tenure. In both of the latter cases, t is increased by its initial value (the initial value was 10 in all our experiments).

Iteration Limit. In order to dynamically adapt the number of iterations to the progress made by the tabu search, we used the following policy: Let k be the current number of iterations and I the current limit. When the limit is reached within the outer loop, we check if there was any progress on the upper bound χ^{up} since the last limit update. If there was some progress, then we increase the limit by the current number of iterations ($I = I + k$). Now, let δ be the value of $I - k$ at the start of the inner loop. When the limit is reached within the inner loop, we check if there was any progress on the number of uncolored vertices ($|U|$) since the last limit update. If there was some progress, then we increase the limit by δ, otherwise we increase it by $\delta/2$. We used an initial limit of 250000.

Limit on Chains of Flat Moves. In some cases it is possible to explore very long paths of flat moves hence slowing down the algorithm. We introduce a parameter p (originally set to 1) controlling the probablity $1/p$ of preferring such moves. Then we simply check the average length l of these moves and their frequency f and adjust p in consequence. In practice, we double p when $l \times f \geq 20$ and decrement it when it is strictly greater than 1 and $l \times f \leq 3$.

5   Overall Approach

Our approach combines the peeling preprocessing from Sect. 2, the tabu search described in Sect. 4 and the iterated DSATUR scheme described in Sect. 3.

The principle we use for choosing the exact sequence of techniques is to apply first those that have the greatest effect for the least computational cost. Therefore, we first call `DegeneracyOrder` to compute not only the degeneracy of the graph, but also the smallest-last ordering [19] O, which is the order in which vertices are processed by the degeneracy algorithm and the array D, which contains the degrees of the vertices during the elimination procedure. The actual degeneracy D is only implicitly contained there as the maximum value in the array, and $D+1$ is an upper bound on the chromatic number. We also compute a lower bound by finding a clique. Using this lower bound and the order O, we can compute the peeled graph H by removing the vertices whose degree D during the degeneracy computation is at most k.

Although finding the maximum clique is NP-hard, it turns out to be much easier than coloring in the dataset we used, so we solve the problem exactly rather than use a heuristic. It also has a great effect on the rest of the algorithm, as a better initial lower bound results in greater scale reduction from peeling and hence improves all heuristics used further on.

After peeling, we first improve the upper bounds using the DSATUR heuristic (`Dsatur`) and then local search. Finally, we switch to iterated DSATUR (`I-Dsatur`), which is exact and hence the most computationally expensive phase.

Algorithm 3. Graph Coloring

Algorithm: LS+I-Dsatur

Data: Graph $G = (V, E)$, Parameters I, t

Result: The chromatic number of G

```
/* Preprocessing phase                                              */
```
1 $(O, D) \leftarrow \texttt{DegeneracyOrder}(G)$

$\chi^{up} \leftarrow \max(D) + 1$

$\chi^{low} \leftarrow |\texttt{FindClique}(G)|$

$H \leftarrow$ subgraph of G induced by $\{o_k, \ldots, o_{|V|}\}$ with

$k = \max\{i \mid j \geq i \text{ or } D(j) < \chi^{low}\}$

$(O, C) \leftarrow \texttt{Dsatur}(H)$

$\chi^{up} \leftarrow \max(\chi^{up}, \max(C))$

```
/* Local search phase                                               */
```
$C \leftarrow \texttt{TabuSearch}(H, C, I, t)$

$\chi^{up} \leftarrow \min(\chi^{up}, \max(C))$

foreach $v \in V'$ **do** $\delta^{sat}(v) \leftarrow |\{C(u) \mid u \in N(v)\}|$

2 $O = \{o_1, \ldots, o_{|V'|}\}$ with $i < j \implies \delta^{sat}(o_i) \geq \delta^{sat}(o_j)$

```
/* Iterated DSATUR phase                                            */
```
$(O, C') \leftarrow \texttt{Dsatur}(H, O, C)$

return $\texttt{I-Dsatur}(H, O, C', \chi^{low}, \chi^{up})$

One complication is that the iterated DSATUR phase is initialized with the current best solution. If this solution was found by the local search algorithm, there is no ordering that I-Dsatur can use to extract a core. We can produce a relevant ordering from the local search solution simply by sorting the vertices by saturation degree *within the local search coloring*[3] as shown in line 2. However, this coloring may not use the smallest colors for the first vertices in the order, therefore, we apply the following transformation:

We run Dsatur following the ordering O. When processing node v, we check if the color $C(v)$ assigned by the tabu search to v has already been mapped to some color, if not, we map it to the minimum color c that v can take and assign c to v. We do the same if the color $C(v)$ happens to be already mapped to c. Otherwise, we switch to the standard DSATUR from that point on.

The resulting coloring is similar (at least in the prefix) to the LS solution, however it is in a form that might have been produced by DSATUR.

---

[3] Ties broken by overall degree.

6   Experimental Results

Our implementaton uses dOmega [29] for finding the initial maximum clique, and MiniCSP[4] as the underlying CDCL CSP solver during the I-*Dsatur* phase.[5]

We compare it to the state of the art: the FastColor approach [16]. Unfortunately, we could not compare with the approach described in [28] since the coloring part of this code is now lost.[6] However, this latter approach is dominated by FastColor on instances with large cores, hence the hardest.

Every method was run 20 times with different random seeds and with a time limit of one hour and a memory limit of 10 GB. The memory limit was an issue only for dOmega which exceeded the memory limit on 3 instances. We raised the limit to 50 GB in these three cases. We used 4 cluster nodes, each with 35 Intel Xeon CPU E5-2695 v4 2.10 GHz cores running Linux Ubuntu 16.04.4.

Table 1. CPU time (easy dimacs10 instances)

	$|V|/|E|$	(scaled)	FastColor			LS+I-Dsatur		
			CPU time (ms)			CPU time (ms)		
			min	*avg*	*max*	*min*	*avg*	*max*
as-22july06	23K/48K	144/2758	**13**	**18**	**23**	2666	6083	9700
caidaRouterLevel	192K/609K	2861/56K	**229**	**432**	**694**	430	2785	29066
citationCiteseer	268K/1157K	2779/33K	489	1131	3143	**404**	**552**	**661**
cnr-2000	326K/2739K	0/0	1997	2360	2649	**375**	**426**	**548**
coAuthorsCiteseer	227K/814K	0/0	**107**	**189**	383	215	300	**367**
coAuthorsDBLP	299K/978K	0/0	**130**	**301**	**564**	321	434	592
coPapersCiteseer	100K/498K	0/0	**25**	**49**	**93**	73	96	147
coPapersDBLP	540K/15M	0/0	**1175**	**1439**	**1903**	1769	2091	2541
cond-mat-2005	40K/176K	0/0	**19**	41	**74**	23	**40**	**54**
eu-2005	333K/3949K	2128/106K	3383	3912	4844	**542**	**690**	**824**
in-2004	163K/2602K	0/0	721	1726	2042	**206**	**263**	**331**
rgg-n-2-17-s0	131K/729K	0/0	**108**	235	319	155	**217**	**281**
rgg-n-2-19-s0	524K/3270K	0/0	**615**	1678	2888	843	**1233**	**1702**
rgg-n-2-20-s0	1049K/6892K	59/637	**1486**	3131	7094	1953	**2962**	**4056**
rgg-n-2-21-s0	2097K/14M	0/0	5386	10664	15991	**4476**	**6329**	**8262**
rgg-n-2-22-s0	4194K/30M	0/0	**9673**	24810	45292	10642	**14192**	**17222**
rgg-n-2-23-s0	8389K/64M	0/0	**17501**	56107	92511	24693	**30174**	**36390**
rgg-n-2-24-s0	17M/133M	0/0	**33786**	137946	439554	56001	**63153**	**89313**
belgium_osm	1441K/1550K	5/8	**229**	**342**	**905**	1061	1398	1665
ecology1	1000K/1998K	1000K/1998K	**500**	**907**	4008	1288	1568	1816
luxembourg_osm	115K/120K	0/0	**9**	**19**	**46**	38	65	85
preferentialAttachment	100K/500K	0/0	266	1199	3146	**136**	**187**	**242**
Average CPU time			**3538**	11302	28553	4923	**6147**	**9358**

The first two columns of Tables 1, 2, 3 and 5 give the size of the graph (number of vertices/edges) before and after scale reduction. In all these tables, bold font

---

[4] Sources available at: https://bitbucket.org/gkatsi/minicsp.

[5] Sources available at: https://bitbucket.org/gkatsi/gc-cdcl/src/master/.

[6] Personnal communication with the authors.

Table 2. CPU time (easy `snap` instances)

	$|V|/|E|$	(scaled)	FastColor CPU time (ms)			LS+I-Dsatur CPU time (ms)		
			min	*avg*	*max*	*min*	*avg*	*max*
as-skitter	1696K/11M	4410/318K	**9168**	**12775**	**15628**	24464	36847	75037
ca-AstroPh	19K/198K	0/0	**12**	32	68	18	**27**	**32**
ca-CondMat	23K/93K	0/0	**10**	**18**	**28**	9	19	34
ca-GrQc	5246/14K	0/0	1	3	8	0	**2**	**3**
ca-HepPh	12K/118K	0/0	39	44	83	**8**	**13**	**18**
ca-HepTh	9880/26K	0/0	2	**4**	**6**	**1**	5	8
cit-HepPh	35K/421K	8491/188K	110	5095	40103	5827	34997	194368
athletes_edges	14K/87K	42/793	**6**	**14**	28	10	17	**26**
com-amazon.ungraph	335K/926K	0/0	**174**	**290**	**532**	423	555	704
com-dblp.ungraph	317K/1050K	0/0	**157**	**302**	722	415	510	**715**
com-lj.ungraph	3925K/34M	383/73K	17624	49986	74537	**17069**	**23097**	**29383**
company_edges	14K/52K	0/0	**5**	**8**	**12**	6	10	13
government_edges	7057/89K	856/26K	**6**	**25**	**54**	48	67	77
new_sites_edges	28K/206K	36/615	**16**	47	74	25	**37**	**47**
politician_edges	5908/42K	527/11K	**25**	**46**	**74**	5012	5596	6467
public_figure_edges	12K/67K	544/16K	**13**	**48**	73	40	54	**66**
tvshow_edges	3892/17K	0/0	1	2	5	**0**	**2**	**3**
wiki-topcats	1788K/25M	106K/5163K	**20920**	**50524**	**85890**	74802	91791	103097
loc-gowalla_edges	197K/950K	3420/121K	**181**	**556**	920	509	624	**775**
loc-gowalla_totalCheckins	5669K/6442K	841K/1630K	**5260**	**7083**	**13508**	5833	6842	8193
Amazon0302	262K/900K	0/0	**175**	**366**	**681**	323	515	776
Amazon0312	401K/2350K	0/0	**417**	**671**	**1023**	899	1109	1811
Amazon0505	410K/2439K	0/0	**442**	**694**	**1185**	943	1185	1789
Amazon0601	403K/2443K	0/0	**415**	**705**	**1176**	715	1124	1649
roadNet-CA	1965K/2767K	0/0	**507**	**835**	**1902**	1916	2491	3264
roadNet-PA	1088K/1542K	0/0	**269**	**520**	**1303**	948	1323	2089
roadNet-TX	1380K/1922K	0/0	**304**	**498**	**1023**	1356	1584	1973
soc-sign-epinions	132K/711K	251/21K	352	1131	1673	**300**	**361**	**417**
HU_edges	48K/223K	0/0	**22**	144	405	56	69	86
RO_edges	42K/126K	147/722	**18**	59	101	22	**46**	**71**
soc-LiveJournal1	4847K/43M	474/106K	79695	107923	129442	**22337**	**29664**	**35301**
soc-pokec-relationships	1633K/22M	262K/8307K	**8430**	54346	149444	22600	**27607**	**31308**
twitter_combined	81K/1342K	699/48K	1252	1836	3548	**319**	**422**	**487**
web-BerkStan	685K/6649K	392/41K	4893	5350	6718	**969**	**1209**	**1907**
web-Google	876K/4322K	48/1121	**692**	**1662**	4137	1328	1863	**3197**
web-NotreDame	326K/1090K	1367/108K	**122**	**182**	**259**	302	385	514
web-Stanford	282K/1993K	1252/72K	940	1457	1794	**470**	**606**	**746**
wiki-RfA	38K/94K	7286/65K	**48**	**56**	**71**	100	119	137
Average CPU time			**4019**	**8035**	**14164**	5011	**7179**	**13331**

is used to highlight the (strictly) best outcomes. In Tables 1 and 2 we report the CPU time in milliseconds for the "easy" instances of the `dimacs10` and `snap` sets, respectively. We say that an instance is easy when both `I-Dsatur` and `FastColor` solved to optimality. We give the minimum, maximum and average CPU time – parsing excluded – across the 20 random runs on the same instance.

Table 3. Lower and upper bounds (hard `dimacs10` instances)

	$|V|/|E|$	(scaled)	FastColor				LS+I-Dsatur			
			χ^{low}		χ^{up}		χ^{low}		χ^{up}	
			max	avg	min	avg	max	avg	min	avg
kron_g500-logn16	55K/2456K	6885/1495K	136	136.00	151	**152.42**	^{1}136	136.00	^{3}145	153.40
333SP	3713K/11M	2261K/6759K	4	4.00	5	5.00	^{1}4	4.00	^{0}5	5.00
G_n_pin_pout	100K/501K	100K/501K	4	**4.00**	6	6.00	^{3}4	3.95	^{2}5	**5.00**
audikw1	944K/38M	936K/38M	36	36.00	40	40.89	^{1}36	36.00	^{2}39	**39.30**
cage15	5155K/47M	5134K/47M	6	6.00	12	12.00	^{1}6	6.00	^{2}11	**11.00**
ldoor	952K/23M	952K/23M	21	21.00	32	32.75	^{3}23	**21.65**	^{2}28	**29.85**
smallworld	100K/500K	100K/500K	6	6.00	7	7.00	^{1*}6	6.00	^{2}6	**6.00**
wave	156K/1059K	156K/1058K	6	6.00	8	8.00	^{3}7	**6.05**	^{1}8	8.00

Table 4. Summary (hard `dimacs10` instances)

Method	χ^{low}		χ^{up}		Opt.	CPU
	avg	avg (G)	avg	avg (G)	avg	avg
LS+I-Dsatur	27.456	11.752	32.194	14.857	0.125	635682
FastColor	27.182	11.680	32.792	15.814	0.000	346630

Tables 3 and 5 show the lower (χ^{low}) and upper bounds (χ^{up}) found by I-Dsatur and FastColor on the rest of the dataset ("hard" instances). Both for the lower and upper bound, we give the best and average value across the 20 random runs on the same instance. We use an asterisk (*) to denote that the maximum lower bound found over the 20 runs is as high as the minimum upper bound, signifying that the method is able close the instance. Moreover, for the results of I-Dsatur, we denote via a superscript in which phase of the approach the best outcome was found. A value of 0 stands for the computation of the degeneracy ordering, 1 for the preprocessing phase, 2 for the local search and 3 for the iterated DSATUR phase.

Finally, Tables 4 and 6 give a summary view for hard instances, of respectively the `dimacs10` and `snap` datasets, with the arithmetic and geometric mean bounds; overall ratio of optimality; and overall mean CPU time.

We first observe that for many of these graphs (see Tables 1 and 2) finding an optimal coloring is easy. One reason is that their clique and chromatic numbers are equal. However, this is also the case for some graphs classified here as "hard". Whereas we use a complete maximum clique algorithm in our approach, FastColor does not and yet it finds a maximum clique in all the "easy" graphs and in most of the "hard" ones. Moreover, both solvers were able to quickly find a maximum clique and an optimal coloring. In particular, many easy graphs are solved during the preprocessing phase, the maximum ($\chi^{low} - 1$)-core being very small. Those graphs are therefore trivial both for FastColor and for our approach, which are in fact similar on those. There is a slight advantage to our method in terms of average run time, both for easy `dimacs10` and easy `snap`

Table 5. Lower and upper bounds (hard **snap** instances)

	$|V|/|E|$	(scaled)	FastColor				LS+I-Dsatur			
			χ^{low}		χ^{up}		χ^{low}		χ^{up}	
			max	avg	min	avg	max	avg	min	avg
cit-HepTh	28K/352K	6819/188K	*23	**23.00**	23	**23.68**	^{3*}23	22.25	^{3}23	24.00
artist_edges	51K/819K	18K/591K	18	18.00	**19**	**19.94**	^{1}18	18.00	^{3}20	20.15
com-orkut.ungraph	3072K/117M	742K/57M	50	49.44	75	77.83	^{1}51	**51.00**	^{1}73	**73.00**
com-youtube.ungraph	1135K/2988K	27K/708K	17	17.00	23	**23.00**	^{3}18	18.00	^{1}24	24.00
email-Eu-core	986/16K	527/13K	18	18.00	19	19.00	^{3*}19	19.00	^{3}19	19.00
email-Enron	37K/184K	2707/76K	20	**20.00**	23	**23.47**	^{3}20	19.05	^{3}24	24.00
email-EuAll	265K/364K	1570/40K	16	16.00	18	18.00	^{3}18	18.00	^{3}18	18.00
p2p-Gnutella04	11K/40K	6899/35K	4	4.00	5	5.00	^{3}4	4.00	^{1}5	5.00
p2p-Gnutella05	8850/32K	4994/25K	4	4.00	5	5.00	^{1}4	4.00	^{1}5	5.00
p2p-Gnutella06	8717/32K	5548/27K	4	4.00	5	5.00	^{3}4	4.00	^{1}5	5.00
p2p-Gnutella08	6301/21K	2541/13K	5	5.00	6	6.00	^{3*}6	**6.00**	^{1}6	6.00
p2p-Gnutella09	8114/26K	3835/19K	5	5.00	6	6.00	^{3*}6	**6.00**	^{1}6	6.00
p2p-Gnutella24	27K/65K	11K/46K	4	**4.00**	5	5.00	^{3}4	3.80	^{1}5	5.00
p2p-Gnutella25	23K/55K	7892/33K	4	4.00	5	5.00	^{1}4	4.00	^{1}5	5.00
p2p-Gnutella30	37K/88K	12K/53K	4	4.00	5	5.00	^{1}4	4.00	^{1}5	5.00
p2p-Gnutella31	63K/148K	20K/87K	4	4.00	5	5.00	^{1}4	4.00	^{1}5	5.00
soc-sign-Slashdot081106	77K/469K	4760/164K	26	26.00	29	29.00	^{3*}29	**28.90**	^{3}29	29.00
soc-sign-Slashdot090216	82K/498K	4654/163K	27	27.00	29	**29.00**	^{3*}29	**28.95**	^{3}29	29.05
soc-sign-Slashdot090221	82K/500K	4703/165K	27	27.00	29	**29.00**	^{3*}29	**28.75**	^{3}29	29.30
soc-sign-bitcoinalpha	3783/14K	400/5352	10	10.00	12	12.00	^{3*}12	**12.00**	^{2}12	12.00
soc-sign-bitcoinotc	5881/21K	513/7516	11	11.00	12	12.00	^{3*}12	**12.00**	^{3}12	12.00
HR_edges	55K/498K	20K/299K	12	12.00	13	13.00	^{1}12	12.00	^{2}13	13.00
Wiki-Vote	7115/101K	2262/83K	17	17.00	22	**22.00**	^{3}19	17.55	^{3}22	22.85
facebook_combined	4039/88K	480/29K	69	69.00	70	70.00	^{3*}70	**70.00**	^{3}70	70.00
gplus_combined	108K/12M	13K/6831K	325	324.05	327	327.84	^{3*}326	**324.40**	^{3}326	**327.40**
soc-Epinions1	76K/406K	4782/205K	23	23.00	**28**	**28.00**	^{1}23	23.00	^{1}29	29.00
CollegeMsg	1899/14K	911/12K	7	7.00	9	**9.00**	^{3*}9	**8.30**	^{3}9	9.05
sx-askubuntu	157K/456K	1834/59K	23	23.00	25	**25.00**	^{3}24	24.00	^{3}25	25.10
sx-mathoverflow	25K/188K	1584/80K	30	30.00	**35**	**35.95**	^{3}32	31.90	^{3}36	36.45
sx-stackoverflow	2584K/28M	111K/11M	55	55.00	**66**	**66.16**	^{1}55	55.00	^{1}67	67.00
sx-superuser	192K/715K	2868/118K	29	29.00	30	30.00	^{3*}30	**30.00**	^{3}30	30.00
wiki-talk-temporal	1094K/2788K	12K/643K	25	25.00	46	**46.00**	^{3}27	25.95	^{3}46	46.25
wiki-Talk	2394K/4660K	15K/771K	26	26.00	48	**48.35**	^{3}29	28.30	^{3}48	48.80
wiki-Vote	7120/101K	2262/83K	17	17.00	22	**22.00**	^{3}19	17.80	^{3}22	22.70

Table 6. Summary (hard **snap** instances)

Method	χ^{low}		χ^{up}		Opt.	CPU
	avg	avg (G)	avg	avg (G)	avg	avg
LS+I-Dsatur	28.938	15.893	32.591	18.480	0.332	209093
FastColor	27.784	15.049	32.137	18.203	0.009	178857

instances, which can presumably be attributed to our peeling method being more efficient than the independent set extraction in `FastColor`.

Of the hard `dimacs10` instances in Table 3, all but `kron_g500-logn16` are quasi-regular, i.e., every vertex has roughly the same degree. These graphs do not have small cores, hence the peeling phase is irrelevant. We can see that on these graphs, the tabu search algorithm significantly outperforms DSATUR and therefore our approach dominates `FastColor` for the upper bound. For instance, on `1door`, `LS+I-Dsatur` finds a 29.85-coloring on average whereas the best coloring found by `FastColor` has 32 colors. On the instance `kron_g500-logn16`, the tabu search performs poorly and is on average dominated by `FastColor`. In one run, however, the iterated DSATUR algorithm is able to find a much better coloring using 6 fewer colors than the best one found by `FastColor`. The aggregated results given in Table 4 show that `LS+I-Dsatur` outperforms `FastColor` both for the lower and upper bounds on this dataset.

The iterated DSATUR algorithm is also able to improve the lower bound of 2 instances out of 8 (`1door` and `G_n_pin_pout`). However, for the latter, `FastColor` produces the same lower bound (4) which is larger than the maximum clique found by `dOmega`. We do not know how to explain this.

On hard instances of the `snap` dataset (Table 5), the picture is very different with in particular the tabu search being almost useless. The best coloring found by our method was obtained during the local search phase only once, for the instance `HR_edges`. In all other cases the best coloring was produced either during preprocessing via DSATUR, or during the iterated DSATUR phase. Overall, as shown in Table 6, this is slightly less efficient for the upper bound than `FastColor` which repeatedly uses DSATUR and eventually finds better colorings in several instances whilst `LS+I-Dsatur` is best only on four instances.

The iterated DSATUR phase, however, is very effective with respect to the lower bound. It improves on the maximum clique found by `dOmega` in 25 out of 34 instances, and it matches the best upper bound for 14 instances. Here again, on three instances (`cit-HepTh`, `email-Enron` and `p2p-Gnutella24`) `FastColor` outputs a lower bound greater than that found by `dOmega`. Overall, our approach can close 14 of the hard instances, for 10 of which[7], the optimal coloring was not previously known, as far as we know. `FastColor` can only close one of them.

7   Conclusions

We have presented a new algorithm for exactly computing the chromatic number of large real world graphs. This scheme combines a novel local search component that performs well on massive graphs and gives improved upper bounds as well as an iterative reduction method that produces much smaller graphs than previous state of the art scale reduction methods. This scheme involves extracting more information than simply a coloring from the DSATUR greedy coloring heuristic and iteratively solving reduced instances with a complete, branch-and-bound

---

[7] `email-Eu-core`,   `email-EuAll`,   `Gnutella08/09`,   `bitcoinalpha`,   `bitcoinotc`, `facebook`, `gplus`, `CollegeMsg` and `sx-superuser`.

solver, in such a way that lower bounds produced for the reduced graphs are also lower bounds of the original graph. Combined with the fact that we achieve more significant reduction than the current state of the art means that we can find non-trivial lower bounds even when peeling-based reduction cannot reduce the graph to fewer than hundreds of thousands of vertices. Indeed, in our experimental evaluation on a set of massive graphs, this method is able to produce both better lower and upper bounds than existing solvers and proves optimality on several (almost 75%) of them.

We expect that finding a method to extract cores from other heuristics, such as our local search procedure will further improve performance.

References

1. Aardal, K.I., Hoesel, S.P.M.V., Koster, A.M.C.A., Mannino, C., Sassano, A.: Models and solution techniques for frequency assignment problems. Ann. Oper. Res. **153**(1), 79–129 (2007)
2. Abello, J., Pardalos, P., Resende, M.G.C.: On maximum clique problems in very large graphs. In: Abello, J.M., Vitter, J.S. (eds.) External Memory Algorithms, pp. 119–130. American Mathematical Society, Boston (1999)
3. Bader, D.A., Meyerhenke, H., Sanders, P., Wagner, D.: Graph partitioning and graph clustering (2012). http://www.cc.gatech.edu/dimacs10/
4. Blöchliger, I., Zufferey, N.: A graph coloring heuristic using partial solutions and a reactive tabu scheme. Comput. Oper. Res. **35**(3), 960–975 (2008)
5. Brélaz, D.: New methods to color the vertices of a graph. Commun. ACM **22**(4), 251–256 (1979)
6. Chaitin, G.J., Auslander, M.A., Chandra, A.K., Cocke, J., Hopkins, M.E., Markstein, P.W.: Register allocation via coloring. Comput. Lang. **6**(1), 47–57 (1981)
7. Cornaz, D., Jost, V.: A one-to-one correspondence between colorings and stable sets. Oper. Res. Lett. **36**(6), 673–676 (2008)
8. de Werra, D.: An introduction to timetabling. Eur. J. Oper. Res. **19**(2), 151–162 (1985)
9. Furini, F., Gabrel, V., Ternier, I.-C.: Lower bounding techniques for DSATUR-based branch and bound. Electron. Notes Discrete Math. **52**, 149–156 (2016). INOC 2015–7th International Network Optimization Conference
10. Hao, J.-K., Qinghua, W.: Improving the extraction and expansion method for large graph coloring. Discrete Appl. Math. **160**(16–17), 2397–2407 (2012)
11. Hebrard, E., Katsirelos, G.: Clause learning and new bounds for graph coloring. In: Hooker, J. (ed.) CP 2018. LNCS, vol. 11008, pp. 179–194. Springer, Cham (2018). https://doi.org/10.1007/978-3-319-98334-9_12
12. Hertz, A., de Werra, D.: Using tabu search techniques for graph coloring. Computing **39**(4), 345–351 (1987)
13. Johnson, D.J., Trick, M.A. (eds.): Cliques, Coloring, and Satisfiability: Second DIMACS Implementation Challenge, Workshop, October 11–13, 1993. American Mathematical Society, Boston (1996)
14. Leighton, F.T.: A graph coloring algorithm for large scheduling problems. J. Res. Natl. Bur. Stand. **84**, 489–506 (1979)
15. Leskovec, J., Krevl, A.: SNAP datasets: Stanford large network dataset collection (2014). http://snap.stanford.edu/data

16. Lin, J., Cai, S., Luo, C., Su, K.: A reduction based method for coloring very large graphs. In: Proceedings of the Twenty-Sixth International Joint Conference on Artificial Intelligence, IJCAI 2017, pp. 517–523 (2017)
17. Lovász, L.: On the Shannon capacity of a graph. IEEE Trans. Inf. Theor. **25**(1), 1–7 (2006)
18. Malaguti, E., Monaci, M., Toth, P.: An exact approach for the vertex coloring problem. Discrete Optim. **8**(2), 174–190 (2011)
19. Matula, D.W., Beck, L.L.: Smallest-last ordering and clustering and graph coloring algorithms. J. ACM **30**(3), 417–427 (1983)
20. Mehrotra, A., Trick, M.A.: A column generation approach for graph coloring. INFORMS J. Comput. **8**, 344–354 (1995)
21. Moalic, L., Gondran, A.: Variations on memetic algorithms for graph coloring problems. CoRR, abs/1401.2184 (2014)
22. Mycielski, J.: Sur le coloriage des graphes. Colloq. Math. **3**, 161–162 (1955)
23. Park, T., Lee, C.Y.: Application of the graph coloring algorithm to the frequency assignment problem. J. Oper. Res. Soc. Jpn. **39**(2), 258–265 (1996)
24. Rossi, R.A., Ahmed, N.K.: Coloring large complex networks. CoRR, abs/1403.3448 (2014)
25. Schaafsma, B., Heule, M.J.H., van Maaren, H.: Dynamic symmetry breaking by simulating zykov contraction. In: Kullmann, O. (ed.) SAT 2009. LNCS, vol. 5584, pp. 223–236. Springer, Heidelberg (2009). https://doi.org/10.1007/978-3-642-02777-2_22
26. Segundo, P.S.: A new DSATUR-based algorithm for exact vertex coloring. Comput. Oper. Res. **39**(7), 1724–1733 (2012)
27. Van Gelder, A.: Another look at graph coloring via propositional satisfiability. Discrete Appl. Math. **156**(2), 230–243 (2008)
28. Verma, A., Buchanan, A., Butenko, S.: Solving the maximum clique and vertex coloring problems on very large sparse networks. INFORMS J. Comput. **27**(1), 164–177 (2015)
29. Walteros, J.L., Buchanan, A.: Why is maximum clique often easy in practice? Optimization Online (2018). http://www.optimization-online.org/DB_HTML/2018/07/6710.html
30. Zhou, Z., Li, C.-M., Huang, C., Ruchu, X.: An exact algorithm with learning for the graph coloring problem. Comput. Oper. Res. **51**, 282–301 (2014)

Modelling and Solving the Minimum Shift Design Problem

Lucas Kletzander[✉] and Nysret Musliu

Christian Doppler Laboratory for Artificial Intelligence and Optimization
for Planning and Scheduling, DBAI, TU Wien, Vienna, Austria
{lkletzan,musliu}@dbai.tuwien.ac.at

Abstract. When demand for employees varies throughout the day, the minimum shift design (MSD) problem aims at placing a minimum number of shifts that cover the demand with minimum overstaffing and understaffing. This paper investigates different constraint models for the problem, using a direct representation, a counting representation and a network flow based model and applies both constraint programming (CP) and mixed integer programming (MIP) solvers. The results show that the model based on network flow clearly outperforms the other models. While a CP solver finds some optimal results, with MIP solvers it can for the first time provide optimal solutions to all existing benchmark instances in short computational time.

Keywords: Shift design · Constraint programming ·
Mixed integer programming · Network flow

1 Introduction

In many professions there is demand to work in different shifts. However, there are several steps in the process of generating schedules for employees. When demand fluctuates throughout the day, a decision has to be made regarding the placement of various shifts along the day to cover the demand as well as possible. This could be done in an integrated fashion, performing the assignment of employees and the placement of shifts at the same time, or the steps could be considered separately, first determining the shifts and the required number of employees per shift before assigning concrete employees in the next step.

The minimum shift design (MSD) problem aims at this first step of designing the required shifts while minimizing the amount of understaffing and overstaffing regarding the given demand and in addition keeping the total number of different shifts to a minimum as well, as schedules with fewer shifts are deemed to be easier to read and manage. As the problem has high practical importance, there has been work in literature as well as implementations in commercial software like the Operating Hours Assistant (OPA) [21] from XIMES Corp.

Although previous techniques in the literature give good results in practice, solving of some challenging instances to optimality is still an open question.

© Springer Nature Switzerland AG 2019
L.-M. Rousseau and K. Stergiou (Eds.): CPAIOR 2019, LNCS 11494, pp. 391–408, 2019.
https://doi.org/10.1007/978-3-030-19212-9_26

This paper presents new models for the problem built in the solver independent modelling language MiniZinc [28]. The first uses a direct representation of the problem and serves for comparison. Another model is built based on counting remaining shift times. The final model turns some of the constraints into a network flow constraint to model the assignment of shifts including understaffing and overstaffing.

All models are then evaluated using both the lazy clause generation solver Chuffed [10] and the MIP solvers Gurobi [22] and CPLEX [14] on the standard benchmark datasets that range from instances allowing a perfect cover to hard instances where optima were not known so far. The results show that the model based on network flow clearly outperforms the other models and that MIP solvers, especially Gurobi, perform very well using the new model. With this model, for the first time we can solve all benchmark instances to optimality within short computational time.

2 Related Work

The need to schedule employees in different shifts in order to cover different kinds of demand arise in many professions like healthcare, transportation, production or call centers. As results influence the lives of many people, large amounts of research exist on these topics and are described in several reviews [7,9,16,20].

The MSD problem is described in [26] where a tabu search approach with several neighbourhood relations is proposed. This work also considers a forth objective that is not used in the following works, setting bounds for the average number of duties per week. Di Gaspero et al. [17] provide an improved local search technique and analyse the complexity of the problem. They describe the relation of the problem to a min-cost max-flow problem, therefore raising the connection to the usage of network flow. More precisely, the problem could be represented as a min-cost max-flow problem as long as the minimization of shifts is not taken into account. They use this relation to provide a hybrid method combining min-cost max-flow with local search.

Recent approaches include a formulation in answer set programming [1] that is able to provide several optimal results, while it struggles with instances that do not admit an exact cover of the demand. Good results have been achieved using MIP in [24], providing optimal solutions for most instances with Gurobi and CPLEX using a runtime of two hours.

A related problem is shift scheduling [3,4,8,15,31], which deals with one or more days, potentially including activities, skills or breaks. In contrast, MSD deals with the minimization of the number of shifts across several days including cyclicity. Recently, regular and context-free languages [11–13,23,29] as well as column generation techniques [25,30] have been applied to shift scheduling successfully. In [25,30], networks are used where paths correspond to feasible assignments of activities.

Further the scheduling of breaks within shifts [5,6,32] or the combined problem of shift and break scheduling [2,18,19] are considered by several authors.

In [2] results for the first MSD benchmark dataset are given as well, however, applying weights for under- and overstaffing per timeslot instead of per minute.

3   Problem Description

The MSD problem aims at generating a set of shifts and determining the required number of employees for each shift on each day in the planning period. The specification is mainly based on [26]. The problem input is defined as follows.

- n consecutive timeslots $[a_1, a_2), [a_2, a_3), \ldots, [a_n, a_{n+1})$, each slot $[a_i, a_{i+1})$ with the same slotlength sl in minutes and with a requirement for workers R_i indicating the optimal number of employees required at that timeslot. A full day consists of a whole number of slots, therefore, sl divides $24 \cdot 60$. The time from a_1 to a_{n+1} is called the planning period and consist of d days, where d is integer.
- s shift types $t_1, \ldots, t_s$, each of them associated with a minimum and maximum start time $smin_s$ and $smax_s$, given in timeslots of the current day, as well as minimum and maximum length $lmin_s$ and $lmax_s$, given in timeslots. Note that shifts can extend to the following day. We consider a cyclic schedule where shifts extending beyond a_{n+1} continue from a_1. For simplicity, we assume that shift types never overlap, i.e, no shift can be of two types at the same time. If necessary, new shift types can be introduced to resolve overlap.

For illustration purpose throughout the paper consider the following example TOY with $d = 1$ and $sl = 180$. The two shift types are given in Table 1. Corresponding values are provided both in timeslots and hour:minute notation. The demand is given as $R = [1, 1, 4, 3, 5, 5, 2, 3]$.

Table 1. Shift types for TOY

Shift type	$smin_s$	time	$smax_s$	time	$lmin_s$	time	$lmax_s$	time
t_1	2	6:00	2	6:00	2	6:00	4	12:00
t_2	4	12:00	7	21:00	2	6:00	4	12:00

In order to provide a feasible solution, shifts are only allowed within the time windows specified by the shift types. Among feasible solutions, the following criteria have to be minimized.

- T_1: Sum of excesses of workers across the whole planning period (per minute).
- T_2: Sum of shortages of workers across the whole planning period (per minute).
- T_3: Number of shifts.

$$T = \sum_{i=1}^{3} \alpha_i \cdot T_i \tag{1}$$

This results in a multi-criteria optimization problem where the individual objectives are combined according to given weights α_i. Equation (1) provides the combined objective T.

A solution is given as a set of feasible shifts and their corresponding numbers of assigned employees for each day. Note that a shift is considered in use if there is any day in the planning period where employees are assigned to the shift.

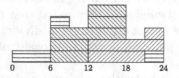

shift start	shift end	employees
6:00	18:00	3
12:00	24:00	2
21:00	9:00	1

Fig. 1. Optimal solution for TOY

For TOY, assuming $\alpha_1 = \alpha_2 = 1$ and $\alpha_3 = 180$, an optimal solution with no understaffing or overstaffing and three used shifts is easily found and given in Fig. 1. Remember that usually instances consider multiple days.

4   Constraint Models

In this section we describe three different ways to provide a constraint model for the problem. This includes a direct model DIRECT, a model using a counting representation COUNT and a model based on a network flow constraint NETWORK. For all models, we define the number of intervals per day $int = \frac{24 \cdot 60}{sl}$, and the sets $\mathbf{I} = \{1, \ldots, int\}$, $\mathbf{D} = \{1, \ldots, d\}$, $\mathbf{S} = \{1, \ldots, s\}$ and $\mathbf{N} = \{1, \ldots, n\}$.

4.1   Direct Model

The direct model DIRECT represents a solution directly by using decision variables that model the number of employees assigned to each possible shift on each day. More precisely, for each combination of a shift start time and shift length of any shift type, the number of employees is represented for each day. This can be done by a 3-dimensional matrix E of size $int \times \max(lmax_s) \times d$. Here, the first dimension represents all possible shift start times in the day, the second all possible lengths and the third all days. All elements that are not allowed by any shift type can be set to 0 immediately. To reduce the size of the matrix, the following sets are defined.

$$\mathbf{St} = \{\min(smin), \ldots, \max(smax)\} \tag{2}$$
$$\mathbf{L} = \{\min(lmin), \ldots, \max(lmax)\} \tag{3}$$

Equation (2) defines the set of indices in use for the first dimension of E from the earliest minimum start time to the latest maximum start time, (3) defines the set of indices for the second dimension from the smallest minimum length to the largest maximum length. Next, we deal with infeasible shifts as follows.

$$Check_{i,j} = (\exists k \in \mathbf{S} : i \geq smin_k \wedge i \leq smax_k \wedge j \geq lmin_k \wedge j \leq lmax_k) \quad (4)$$

$$\neg Check_{i,j} \rightarrow (\forall k \in \mathbf{D} : E_{i,j,k} = 0) \qquad \forall i \in \mathbf{St}, j \in \mathbf{L} \quad (5)$$

For convenience, Eq. (4) defines a predicate that is true if the shift starting at i with length j is a valid shift for any type k. In Eq. (5), for any combination of a start time i and a length j, if there is no shift type allowing this combination, then for all days the number of employees is set to 0. Note that this only depends on input parameters and is already used during compilation of the model.

In order to track the number of employees working at any timeslot i, we introduce the workforce array W of length n.

$$W_{(i-1)\cdot int+j} = \sum_{\substack{k \in \mathbf{St} \\ k < j}} \sum_{\substack{\ell \in \mathbf{L} \\ k+\ell \geq j}} E_{k,\ell,i} + \sum_{\substack{k \in \mathbf{St} \\ k-int < j}} \sum_{\substack{\ell \in \mathbf{L} \\ k+\ell-int \geq j}} E_{k,\ell,i-1}$$

$$\forall i \in \mathbf{D}, j \in \mathbf{I} \quad (6)$$

Equation (6) sums up the shifts active at timeslot j on day i. The first sum spans all shifts that start before j and end not earlier than j. The second sum deals with shifts that are active after midnight. Those are still assigned to day $i - 1$, but contribute to the timeslot on day i if start and end, corrected by int, match the requirements.

Now the optimization goals can be defined in terms of the model.

$$T_1 = sl \cdot \sum_{i \in \mathbf{N}} \max\{W_i - R_i; 0\} \quad (7)$$

$$T_2 = sl \cdot \sum_{i \in \mathbf{N}} \max\{R_i - W_i; 0\} \quad (8)$$

$$T_3 = \sum_{i \in \mathbf{St}} \sum_{j \in \mathbf{L}} (\exists k \in \mathbf{D} : E_{i,j,k} > 0) \quad (9)$$

Equations (7) and (8) hold the sums corresponding to workforce over respectively under the demand. The factor sl is used to count overstaffing and understaffing per minute. Equation (9) counts all shifts that have an employee assigned on at least one day.

4.2   Counting Model

The counting model COUNT uses a similar idea to the residual lengths in [1] to keep track of the remaining lengths of shifts for each timeslot instead of just storing which shifts are in use. For this purpose, the decision variables in E are replaced with a 2-dimensional counting matrix C of size $n \times |\mathbf{L}|$. Now, each

Table 2. Counting representation for TOY (transposed)

j	i							
	1	2	3	4	5	6	7	8
2	0	1	0	0	3	0	2	0
3	1	0	0	3	0	2	0	0
4	0	0	3	0	2	0	0	1

element $C_{i,j}$, with $i \in \mathbf{N}$ and $j \in \mathbf{L}$, states that in timeslot i there are $C_{i,j}$ active shifts of remaining length j. Note that index i is considered cyclic. For illustration, Table 2 shows the matrix C for the optimal solution for TOY.

We can now define the counting behaviour as follows.

$$Prev_{i,j} = \begin{cases} 0 & \text{if } j = \max(lmax) \\ C_{i-1,j+1} & \text{otherwise} \end{cases} \tag{10}$$

$$\begin{cases} C_{i,j} \geq Prev_{i,j} & \text{if } Check_{(i-1) \bmod int, j} \vee Check_{(i-1) \bmod int + int, j} \\ C_{i,j} = Prev_{i,j} & \text{otherwise} \end{cases}$$
$$\forall i \in \mathbf{N}, j \in \mathbf{L} \tag{11}$$

Equation (10) defines the previous element in the matrix along the secondary diagonal. Shifts counted by $C_{i,j}$ either had a remaining length of $j+1$ in timeslot $i - 1$ or are newly introduced in timeslot i. This is captured in Eq. (11), where this inequality is expressed whenever it is possible to start a new shift at $C_{i,j}$. Note that the existence of shifts is checked both for the current day and shifts overlapping from the previous day. The second line deals with infeasible shifts, as in this case values are propagated without change along the diagonal, corresponding to Eq. (5) in the direct model.

Now we have an easier time defining the workforce W.

$$W_i = \sum_{j \in \mathbf{L}} C_{i,j} + \sum_{j \in \{1,\dots,\min(lmin)-1\}} C_{i-j,\min(lmin)} \qquad \forall i \in \mathbf{N} \tag{12}$$

In Eq. (12) we first sum up the row of C. Note that we could extend the lengths in C to start from index 1, however, below $\min(lmin)$ we would only have direct propagation along the diagonal as no new shifts can start in this region. Omitting this region makes C smaller and we can simply continue the sum along the first column of C to get shifts with remaining length $< \min(lmin)$.

Equations (7) and (8) still define T_1 and T_2, only T_3 needs to be redefined.

$$T_3 = \sum_{i \in \mathbf{I}} \sum_{j \in \mathbf{L}} (\exists k \in \mathbf{D} : C_{(k-1) \cdot int + i, j} > Prev_{(k-1) \cdot int + i, j}) \tag{13}$$

Equation (13) now uses the fact that new shifts introduce a difference along the diagonal, while overall still counting all shifts that have an employee assignment on at least one day.

4.3 Network Flow Model

There is a known relation between the MSD problem and the min-cost max-flow problem [17]. In fact, without taking the minimization of shifts into account, the problem can be formulated using a flow network. However, we want to combine utilization of network flow with the minimization of shifts by formulating parts of the problem using a global constraint for network flow, while additional constraints are enforced on the flow variables to represent the whole problem in our model NETWORK. In the following the definition of the network graph is presented, first for exact cover and then extending it for over- and understaffing. The whole network will later be illustrated using TOY in Fig. 2.

- $\mathbf{V} = \mathbf{N}$: Vertices correspond to timeslots.
- $\mathbf{A} = \{(i,j) \mid i,j \in \mathbf{N}, \ Check_{(i-1) \bmod int, j-i} \vee Check_{(i-1) \bmod int+int, j-i}\}$: Arcs correspond to feasible shifts as defined by $Check$.
- $B_i = R_i - R_{i-1}$: The balance, i.e., the flow that is consumed or produced per node corresponds to the difference in demand from a timeslot to the previous one.
- $F_{(i,j)}$: The decision variables in this model represent the flow along the arcs, corresponding to the number of employees working the represented shift.

Note that this network does not constitute a usual flow network as it is cyclic. Due to cyclicity of the whole problem it rather forms a ring-like structure. Also it does not have a single source and target node. However, the network flow constraint in MiniZinc can deal with the given network as it essentially models the flow conservation constraints taking into account the balance.

Note that so far we only model the demand changes, not the total demand. This has to be done using additional constraints.

$$Span_{i,j,k} = j \leq i < k \vee (j > k \wedge (i \geq j \vee i < k)) \tag{14}$$

$$R_i = \sum_{\substack{(j,k) \in \mathbf{A} \\ Span_{i,j,k}}} F_{(i,j)} \qquad \forall i \in \mathbf{N} \tag{15}$$

Equation (14) is a predicate defined to check whether arc (j, k) spans across timeslot i. This usually indicated by $j \leq i < k$, but due to cyclicity the second case needs to be considered as well. Equation (15) sums up the flows of all arcs that span a given timeslot i, which has to match the demand for each timeslot.

$$T_3 = \sum_{\substack{(i,j) \in \mathbf{A} \\ Check_{i-1,j-i}}} (\exists k \in \mathbf{D} : F_{((k-1)\cdot int+i, (k-1)\cdot int+j)} > 0) \tag{16}$$

Equation (16) defines the number of shifts in the usual way, counting all shifts that have at least one day with positive flow. This time the sum can be represented going through all arcs assigned to the first day in the planning period (via the $Check$ condition) and then adding the required number of days to both start and end node.

However, so far the model only represents exact cover without the possibility for overstaffing or understaffing. To include these in the network itself, we introduce new arcs and switch to a weighted network.

- $\mathbf{A}_{OC} = \{(i, i+1) \mid i \in \mathbf{N}\}$; $\mathbf{A}_{UC} = \{(i, i+1) \mid i \in \mathbf{N}\}$; $\mathbf{A}' = \mathbf{A} \cup \mathbf{A}_{OC} \cup \mathbf{A}_{UC}$:
  New arcs are defined from each node to its immediate successor, both for undercover and overcover. The total set of arcs $\mathbf{A}'$ is the union of all arc sets.

- $weight_a = \begin{cases} 0 & \text{if } a \in \mathbf{A} \\ -\alpha_1 & \text{if } a \in \mathbf{A}_{OC} \\ \alpha_2 & \text{if } a \in \mathbf{A}_{UC} \end{cases}$

The weights are 0 for all shift arcs. Overcover arcs are defined to hold a certain overflow capacity that should stay on the overcover arcs, therefore, we use negative weights. If overcover is needed on shift arcs, the corresponding capacity is removed from the negatively weighted overcover arcs, resulting in higher cost. As for any demand peak $\max(R)$ is an upper bound for the number of assigned employees and for a slot of maximum overcover shifts from both sides (earlier demand peaks and later demand peaks) might overlap, an upper bound of $2 \cdot \max(R)$ provides enough overcover capacity for the worst case. Undercover arcs are used if demand cannot be fulfilled using shift arcs, as the additional demand uses the undercover arcs and increases the cost.

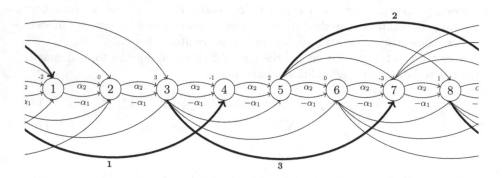

Fig. 2. Full network flow representation for TOY

Figure 2 shows the network flow representation for TOY including overcover and undercover arcs. The main flow corresponding to the optimal solution is highlighted using bold arcs. As the cover is exact, all undercover arcs with weight α_2 have flow 0, all overcover arcs with weight $-\alpha_1$ have flow $2 \cdot \max(R) = 10$.

The flows for the new arcs, denoted by F^{OC} and F^{UC} and the overcover capacity also need to be included in the demand sum.

$$R_i + 2 \cdot \max(R) = \sum_{\substack{(j,k) \in \mathbf{A} \\ Span_{i,j,k}}} F_{(j,k)} + F^{OC}_{(i,i+1)} + F^{UC}_{(i,i+1)} \qquad \forall i \in \mathbf{N} \qquad (17)$$

Equation (17) sums up all relevant arc flows for each timeslot i.

The global constraint used for weighted network flow includes summing up the total network cost, represented by $cost$. Finally, the combined objective T needs to be defined.

$$T = (cost + 2 \cdot \max(R) \cdot \alpha_1 \cdot n) \cdot sl + \alpha_3 \cdot T_3 \tag{18}$$

Equation (18) corrects the network cost by the corresponding value for the negatively weighted overcover capacity and takes the slotlength into account.

5   Evaluation

This section presents the evaluation of the different models on the standard benchmark datasets[1]. The instances are organized in four datasets. The first three use a set of four shift types with different start times across the day and a duration of 7 to 9 h. The slotlength is either 15, 30 or 60 min, the length of the planning period is 7 days. The first two datasets are designed in a way that a solution with exact cover exists. However, dataset 3 is constructed to make the existence of such exact covers very unlikely. Dataset 4 contains a real-life instance using three shift types and two modified instances from dataset 3. All experiments were executed on an Intel Core i7-7500 CPU with 2.7 GHz and 16 GB RAM.

The models were implemented using the solver-independent modelling language MiniZinc 2.2.2 [28]. It allows to directly specify the constraint models including a range of global constraints like network flow and compiles them into a format called FlatZinc, which is understood by a wide range of solvers. This allows to compare the performance of different solvers. Regarding CP, the lazy clause generation solver Chuffed [10] was used in version 0.10.3. For MIP, we compare Gurobi [22] version 8.0.1 and CPLEX [14] version 12.8.

5.1   Search Efficiency with Chuffed

In all models the goal is to minimize T. However, while this goal will directly be used with the MIP solvers in the comparison, Chuffed struggles within a large search space filled with a lot of feasible solutions that are far off the optimum. E.g., an empty schedule is always a feasible solution with zero shifts and no overstaffing, but a huge amount of understaffing.

Typically, good solutions will be found when the actual workforce is somewhere around the demand. Therefore, we define an array WD for the workforce difference.

$$WD_i = |R_i - W_i| \tag{19}$$

Equation (19) defines the distance as the absolute difference between workforce and demand. For Chuffed, we can now use free search, alternating between

---

[1] http://www.dbai.tuwien.ac.at/proj/Rota/benchmarks.html.

the internal search and the strategy to first assign the smallest possible values to the workforce distance. As preliminary evaluations of our models showed, the variable selection strategy first_fail, assigning the variable with the smallest domain first, together with the value selection strategy indomain_min, assigning the smallest value in the domain first, used on *WD*, provided the best results.

In a similar way, this is also done for the network flow model, assigning smallest possible flow values to the undercover arcs, largest possible flow values to the overcover arcs and finally also to the regular flow arcs. The reason for using the maximum for regular flow arcs is to go towards the minimization of the number of shifts by rather assigning more employees to the same shift than splitting assignments across many shifts.

5.2   Results

We executed all three models with all three solvers. As NETWORK performed best for all solvers, we report results for NETWORK on all datasets. The comparison to the other models is presented on selected datasets.

Dataset 1. The first dataset contains instances which admit solutions with exact cover of the demand. Table 3 provides the results for this set of instances. The summary shows solved (S), optimal (O) and proven optimal (P) solutions. We compare all three models for Chuffed to highlight the differences in performance for this solver, further the results for the MIP solvers using NETWORK.

For Chuffed we can see a clear ranking of the models in terms of performance. For DIRECT, 11 optimal solutions can be found and proved within the timeout of 3600 s, while for 17 instances no feasible solution is found at all. The results for COUNT are already much better, giving 18 proven optimal solutions within the timeout, while only 7 instances do not reach any feasible solution within the runtime. However, best results are achieved with NETWORK, finding feasible solutions for all instances (even though the solution for instance 23 is far off), and for 23 instances the optimum can be proved within the timeout. This relation between the different models using Chuffed also holds for the following datasets, where actually apart from NETWORK hardly any feasible solutions are found within the timeout.

Using either Gurobi or CPLEX, all instances from dataset 1 can be solved within less than 14 s per instance. While optimal solutions for these instances were already known, the best approaches so far still needed up to 56 s for some instances [24], resulting in a major runtime improvement using NETWORK. Note that all three models reach optimal solutions for all instances in this dataset, however, the other models need more runtime for several instances. We will present this comparison in detail for dataset 3, as these instances are more challenging and therefore show more differences between the models.

A major factor contributing to runtime needs to be discussed here as well. In fact, Gurobi never spends more than 4 s on any instance of dataset 1. However, the conversion from MiniZinc to FlatZinc also takes some time, in case of

Table 3. Results for dataset 1

Instance	Chuffed						Gurobi		CPLEX	
	DIRECT		COUNT		NETWORK		NETWORK		NETWORK	
	Result	Time	Result	Time	Result	Time	Result	Time	Result	Time
1	480	13.80	480	6.44	480	1.83	480	0.60	480	0.50
2	-	3600	300	3600	300	102.26	300	2.56	300	2.82
3	600	295.13	600	1732.00	600	2.02	600	0.60	600	0.50
4	-	3600	-	3600	570	3600	450	2.62	450	4.12
5	480	28.21	480	8.48	480	1.05	480	0.60	480	0.50
6	420	3.47	420	3.94	420	1.17	420	0.77	420	0.53
7	-	3600	270	624.17	270	21.90	270	2.32	270	1.93
8	-	3600	-	3600	150	1462.39	150	11.60	150	9.28
9	-	3600	150	3600	150	1262.67	150	10.57	150	9.17
10	-	3600	330	3600	330	42.60	330	2.37	330	2.07
11	30	2.89	30	4.09	30	10.27	30	10.83	30	8.78
12	-	3600	90	45.72	90	46.20	90	10.32	90	8.78
13	-	3600	105	374.69	105	64.36	105	10.32	105	9.05
14	-	3600	-	3600	285	3600	195	13.68	195	12.64
15	180	0.44	180	0.77	180	0.78	180	0.57	180	0.48
16	-	3600	-	3600	225	3600	225	12.88	225	12.28
17	-	3600	-	3600	630	3600	540	3.94	540	3.45
18	720	3600	720	916.80	720	7.58	720	0.83	720	0.54
19	-	3600	180	3600	180	3600	180	12.44	180	9.44
20	540	27.98	540	17.20	540	1.26	540	0.72	540	0.52
21	-	3600	120	2102.38	120	118.82	120	11.61	120	9.05
22	75	1372.65	75	18.95	75	21.70	75	10.99	75	9.03
23	-	3600	-	3600	655335	3600	150	10.48	150	9.22
24	480	18.06	480	22.36	480	1.19	480	0.81	480	0.65
25	-	3600	-	3600	720	3600	480	5.17	480	13.03
26	-	3600	600	599.08	600	44.47	600	1.16	600	0.92
27	480	68.06	480	40.58	480	1.15	480	0.68	480	0.52
28	5550	3600	270	996.53	270	17.83	270	2.38	270	1.90
29	-	3600	360	3600	360	396.86	360	2.21	360	1.85
30	75	927.67	75	16.27	75	14.81	75	10.31	75	9.00
S/O/P	13/12/11		23/23/18		30/25/23		30/30/30		30/30/30	

the network model combined with Gurobi actually more than the solver itself. The scaling factor that decides the needed compilation time is clearly the time granularity, as it also determines the number of possible shifts when shift types are otherwise unmodified. Table 4 shows average compilation times for different models, independent of the solver.

Table 4. Approximate compilation times in seconds

Model	$sl = 60$	$sl = 30$	$sl = 15$
DIRECT	0.2	0.5	2.4
COUNT	0.6	1.4	3.6
NETWORK	0.5	1.8	10

We can see that DIRECT compiles fastest, while NETWORK needs the most time. This actually allows some faster results for easy instances, where compilation constitutes most of the runtime, using DIRECT or COUNT. The compilation times show a quadratic correlation to the inverse of the slotlength sl.

We also evaluated the results for dataset 1 applying α_1 and α_2 per timeslot instead of per minute to compare to [2], where we can reach all optimal solutions using an average of only 17.2 instead of 108 s per instance. They do not present comparable results for other datasets.

Dataset 2. The second dataset consists of further instances with exact cover solutions. However, it requires increasing numbers of shifts for later instances. This makes the dataset harder than set 1, otherwise the results are quite similar. The results for NETWORK are presented in Table 5.

Regarding Chuffed, 5 instances do not reach a feasible solution, rather towards the larger instances. 16 optimal solution can be proved. Gurobi and CPLEX both find the optimal solutions for all instances. Note that previous solution methods were not able to find the optimum for instance 27.

Chuffed only finds one feasible solution using DIRECT and 11 feasible (5 proven optimal) solutions using COUNT. Gurobi can still find and prove all optimal solutions using DIRECT, however, using significantly more runtime on several instances, and all but two with COUNT. CPLEX finds all but one with DIRECT and all but two with COUNT.

Datasets 3 and 4. Next the results for dataset 3 are presented. As these instances are constructed such that it is very unlikely an exact cover exists, this dataset is the most challenging for solvers. The results, including the detailed comparison of different models using Gurobi, are presented in Table 6.

The results show that Chuffed can find 24 feasible solutions for this dataset, however, no optimal solutions are found and the gap to the optimal solutions is rather large, never less than 25%. Using the other models, Chuffed does not find feasible solutions before the timeout.

Regarding Gurobi, again all instances can be solved to optimality using NETWORK. While previous work was not able to solve all instances using a timeout of 7200 s, the new model can provide a solution within 1 min for all instances except instance 23, which can still be solved in less than 8 min. This constitutes a major speedup compared to previous work and, for the first time, allows to optimally solve all benchmark instances, even within short runtime.

Table 5. Results for dataset 2 (NETWORK)

Instance	Chuffed		Gurobi		CPLEX	
	Result	Time	Result	Time	Result	Time
1	720	34.44	720	0.67	720	0.76
2	720	11.25	720	0.92	720	0.55
3	360	323.19	360	2.46	360	2.56
4	360	778.56	360	2.33	360	1.90
5	720	30.42	720	1.08	720	1.03
6	360	460.39	360	2.90	360	1.86
7	720	4.19	720	1.83	720	0.68
8	315	3600	180	13.66	180	13.38
9	360	174.22	360	2.49	360	1.98
10	660	5.98	660	1.29	660	1.67
11	840	3600	480	7.17	480	20.87
12	900	3109.58	900	0.99	900	1.10
13	900	572.12	900	2.03	900	1.90
14	840	45.50	840	1.09	840	0.85
15	480	3600	480	8.52	480	17.06
16	240	3600	240	10.83	240	9.01
17	960	2889.60	960	0.90	960	0.86
18	840	178.92	840	1.99	840	0.94
19	-	3600	240	10.96	240	9.50
20	960	423.70	960	1.15	960	0.91
21	690	3600	600	2.78	600	2.62
22	1080	3600	1080	1.47	1080	2.13
23	-	3600	300	34.65	300	31.72
24	780	3600	600	3.97	600	6.90
25	-	3600	600	3.96	600	7.31
26	1020	1337.78	1020	1.76	1020	2.69
27	-	3600	300	25.87	300	323.70
28	-	3600	300	11.76	300	17.52
29	1140	3600	1140	1.78	1140	2.26
30	1140	3600	1020	2.08	1020	2.60
S/O/P	25/20/16		30/30/30		30/30/30	

In comparison to the other models, DIRECT is able so prove all optimal results but one, however, for several instances using significantly longer runtime. While for some instances COUNT is faster than DIRECT, in the average it is worse, also running into timeout on three instances. Note that for some instances both

Table 6. Results for dataset 3

Instance	Chuffed		Gurobi						CPLEX	
	NETWORK		DIRECT		COUNT		NETWORK		NETWORK	
	Result	Time	Result	Time	Result	Time	Result	Time	Result	Time
1	-	3600	2385	5.02	2385	5.34	2385	10.80	2385	9.07
2	10500	3600	7590	23.52	7590	149.61	7590	3.66	7590	11.43
3	20700	3600	9540	4.46	9540	5.54	9540	3.28	9540	2.73
4	10410	3600	6540	101.84	6540	32.80	6540	5.87	6540	66.58
5	12240	3600	9720	2.31	9720	3.68	9720	0.87	9720	7.40
6	2940	3600	2070	5.74	2070	4.91	2070	10.41	2070	8.99
7	434595	3600	6075	6.25	6075	16.96	6075	10.67	6075	9.34
8	11820	3600	8580	2.71	8580	3.73	8580	2.52	8580	2.32
9	147660	3600	6000	16.49	6000	34.65	6000	11.22	6000	12.29
10	4920	3600	2940	4.94	2940	7.79	2940	2.69	2940	7.71
11	19500	3600	5190	74.73	5190	137.62	5190	6.80	5190	28.91
12	-	3600	4110	122.88	4110	841.17	4110	14.44	4110	47.79
13	-	3600	4605	1274.54	4605	3600	4605	35.84	4605	205.46
14	13800	3600	9600	7.42	9600	13.11	9600	1.93	9600	2.09
15	21840	3600	11250	49.97	11250	106.73	11250	7.36	11250	20.35
16	14940	3600	10620	8.40	10620	3.08	10620	2.16	10620	6.06
17	1198110	3600	4680	21.23	4680	204.38	4680	13.78	4680	14.34
18	9210	3600	6540	24.05	6540	57.99	6540	4.27	6540	15.60
19	-	3600	4890	2489.77	4890	3600	4890	39.28	4890	143.03
20	-	3600	8910	38.28	8910	46.49	8910	4.61	8910	18.34
21	-	3600	5910	3600	5910	3600	5910	444.50	5910	2558.94
22	20460	3600	12600	23.05	12600	89.91	12600	4.16	12600	32.28
23	15000	3600	8280	13.54	8280	12.78	8280	1.87	8280	6.02
24	17520	3600	10260	4.28	10260	9.69	10260	1.29	10260	3.20
25	17040	3600	13020	6.72	13020	5.84	13020	1.76	13020	1.22
26	24330	3600	12780	39.23	12780	94.53	12780	4.11	12780	25.31
27	22620	3600	10020	6.06	10020	3.92	10020	1.52	10020	0.93
28	15000	3600	10440	7.36	10440	11.07	10440	1.83	10440	8.30
29	14190	3600	6510	109.94	6510	332.04	6510	54.56	6510	102.83
30	17760	3600	13320	2.63	13320	5.79	13320	1.07	13320	0.80
S/O/P	24/0/0		30/30/29		30/30/27		30/30/30		30/30/30	

DIRECT and COUNT are faster than NETWORK. This is the result of the longer compilation time as already mentioned before. On such easy instances DIRECT and COUNT are better, however, on more difficult instances they cannot keep up with NETWORK.

The results obtained by CPLEX are similar to the previous datasets. Again, CPLEX can prove optimal solutions for all instances, on some instances slightly faster than Gurobi, but on several instances significantly slower. Again, DIRECT and COUNT are mostly slower and can not prove optimality for a few instances (DIRECT 1, COUNT 2) similar to the comparison using Gurobi.

Dataset 4 consist of one real-life instance and two modified versions of instance 5 from dataset 3. The results are similar to those of dataset 3 and are presented in Table 7. The results from Gurobi and CPLEX are again optimal.

Table 7. Results for dataset 4 (NETWORK)

Instance	Chuffed		Gurobi		CPLEX	
	Result	Time	Result	Time	Result	Time
1	42090	3600	18420	0.569	18420	0.768
2	12540	3600	9720	3.375	9720	64.194
3	24660	3600	18780	0.871	18780	4.027
S/O/P	3/0/0		3/3/3		3/3/3	

Summary. To sum up, our new model based on network flow NETWORK clearly outperforms the other models across different datasets and solvers. The CP solver Chuffed can provide optimal solutions for several instances where exact cover solutions exist, but struggles with instances admitting no such solution. However, compared to [1], it can improve the results for many instances across all datasets when working only with CP-related methods.

The MIP solvers Gurobi and CPLEX can both provide optimal solutions for all instances. Further, without timeout this is only possible using NETWORK, showing its superiority compared to the other models. The only drawback is longer compilation time, resulting in more runtime for easy instances, however, more difficult instances greatly benefit from the new model. In direct comparison, while most results are close, CPLEX can provide slightly faster results on several instances, but Gurobi can provide significantly faster results on some difficult instances.

6 Conclusion

In this paper we presented three different constraint models for the minimum shift design problem. Those models were implemented in MiniZinc and evaluated on the standard benchmark datasets using three different solvers, the lazy clause generation solver Chuffed and the MIP solvers Gurobi and CPLEX.

The results show that our new model NETWORK, using a network flow based formulation, clearly outperforms the other models. Further, while Chuffed can provide optimal solutions for several instances in the first two datasets, all optimal solutions can be found using Gurobi or CPLEX. This conclusion is similar to [27], which also applies global constraints in MiniZinc and Gurobi, however, in a different way to a different problem, using a regular constraint for rotating workforce scheduling. Comparing Gurobi and CPLEX, most results are close, but Gurobi provides significantly shorter runtimes on several difficult instances.

This is the first time in literature that all available instances can be solved to optimality within short runtime, using less than 1 min for all instances but one, which can be solved in less than 8 min, highlighting the strength of our approach. Across different instances from all datasets, runtimes can be reduced significantly in comparison to previous work. Therefore, our approach constitutes the new state-of-the-art for minimum shift design.

Future work might involve the integration of additional constraints into the problem or trying to apply a network flow based model to break scheduling or combined shift and break scheduling problems.

Acknowledgements. The financial support by the Austrian Federal Ministry for Digital and Economic Affairs and the National Foundation for Research, Technology and Development is gratefully acknowledged.

References

1. Abseher, M., Gebser, M., Musliu, N., Schaub, T., Woltran, S.: Shift design with answer set programming. Fundamenta Informaticae **147**(1), 1–25 (2016). https://doi.org/10.3233/FI-2016-1396
2. Akkermans, A., Post, G., Uetz, M.: Solving the shifts and breaks design problem using integer linear programming. In: PATAT 2018: Proceedings of the 12th International Conference of the Practice and Theory of Automated Timetabling, pp. 137–152 (2018)
3. Aykin, T.: A comparative evaluation of modelling approaches to the labour shift scheduling problem. Eur. J. Oper. Res. **125**, 381–397 (2000)
4. Bechtold, S., Jacobs, L.: Implicit modelling of flexible break assignments in optimal shift scheduling. Manage. Sci. **36**(11), 1339–1351 (1990)
5. Beer, A., Gaertner, J., Musliu, N., Schafhauser, W., Slany, W.: Scheduling breaks in shift plans for call centers. In: Proceedings of the 7th International Conference on the Practice and Theory of Automated Timetabling, pp. 1–17 (2008)
6. Beer, A., Gärtner, J., Musliu, N., Schafhauser, W., Slany, W.: An AI-based break-scheduling system for supervisory personnel. IEEE Intell. Syst. **25**(2), 60–73 (2010). https://doi.org/10.1109/MIS.2010.40
7. Van den Bergh, J., Beliën, J., De Bruecker, P., Demeulemeester, E., De Boeck, L.: Personnel scheduling: a literature review. Eur. J. Oper. Res. **226**(3), 367–385 (2013). https://doi.org/10.1016/j.ejor.2012.11.029
8. Bhulai, S., Koole, G., Pot, A.: Simple methods for shift scheduling in multiskill call centers. Manufact. Serv. Oper. Manage. **10**(3), 411–420 (2008)
9. Burke, E.K., De Causmaecker, P., Berghe, G.V., Van Landeghem, H.: The state of the art of nurse rostering. J. Sched. **7**(6), 441–499 (2004). https://doi.org/10.1023/B:JOSH.0000046076.75950.0b
10. Chu, G., Stuckey, P.J., Schutt, A., Ehlers, T., Gange, G., Francis, K.: Chuffed, a lazy clause generation solver (2018). https://github.com/chuffed/chuffed
11. Côté, M.C., Gendron, B., Quimper, C.G., Rousseau, L.M.: Formal languages for integer programming modeling of shift scheduling problems. Constraints **16**(1), 55–76 (2011)
12. Côté, M.C., Gendron, B., Rousseau, L.M.: Grammar-based integer programming models for multiactivity shift scheduling. Manage. Sci. **57**(1), 151–163 (2011)

13. Côté, M.C., Gendron, B., Rousseau, L.M.: Grammar-based column generation for personalized multi-activity shift scheduling. INFORMS J. Comput. **25**(3), 461–474 (2013). https://doi.org/10.1287/ijoc.1120.0514

14. CPLEX, I.I.: V12. 1: User's manual for cplex. International Business Machines Corporation **46**(53), 157 (2009)

15. Dantzig, G.B.: A comment on Eddie's traffic delays at toll booths. Oper. Res. **2**, 339–341 (1954)

16. De Bruecker, P., Van den Bergh, J., Beliën, J., Demeulemeester, E.: Workforce planning incorporating skills: state of the art. Eur. J. Oper. Res. **243**(1), 1–16 (2015). https://doi.org/10.1016/j.ejor.2014.10.038

17. Di Gaspero, L., Gärtner, J., Kortsarz, G., Musliu, N., Schaerf, A., Slany, W.: The minimum shift design problem. Ann. Oper. Res. **155**(1), 79–105 (2007). https://doi.org/10.1007/s10479-007-0221-1

18. Di Gaspero, L., Gärtner, J., Musliu, N., Schaerf, A., Schafhauser, W., Slany, W.: A hybrid LS-CP solver for the shifts and breaks design problem. In: Blesa, M.J., Blum, C., Raidl, G., Roli, A., Sampels, M. (eds.) HM 2010. LNCS, vol. 6373, pp. 46–61. Springer, Heidelberg (2010). https://doi.org/10.1007/978-3-642-16054-7_4

19. Di Gaspero, L., Gärtner, J., Musliu, N., Schaerf, A., Schafhauser, W., Slany, W.: Automated shift design and break scheduling. In: Uyar, A.S., Ozcan, E., Urquhart, N. (eds.) Automated Scheduling and Planning. SCI, vol. 505, pp. 109–127. Springer, Heidelberg (2013). https://doi.org/10.1007/978-3-642-39304-4_5

20. Ernst, A., Jiang, H., Krishnamoorthy, M., Sier, D.: Staff scheduling and rostering: a review of applications, methods and models. Eur. J. Oper. Res. **153**(1), 3–27 (2004). https://doi.org/10.1016/S0377-2217(03)00095-X

21. Gärtner, J., Musliu, N., Slany, W.: Rota: a research project on algorithms for workforce scheduling and shift design optimization. AI Commun. **14**(2), 83–92 (2001)

22. Gurobi Optimization, L.: Gurobi optimizer reference manual (2018). http://www.gurobi.com

23. Hernández-Leandro, N.A., Boyer, V., Salazar-Aguilar, M.A., Rousseau, L.M.: A matheuristic based on Lagrangian relaxation for the multi-activity shift scheduling problem. Eur. J. Oper. Res. **272**(3), 859–867 (2019). https://doi.org/10.1016/j.ejor.2018.07.010

24. Kocabas, D.: Exact Methods for Shift Design and Break Scheduling. Master's thesis, Technische Universität Wien (2015)

25. Lequy, Q., Bouchard, M., Desaulniers, G., Soumis, F., Tachefine, B.: Assigning multiple activities to work shifts. J. Sched. **15**(2), 239–251 (2012). https://doi.org/10.1007/s10951-010-0179-8

26. Musliu, N., Schaerf, A., Slany, W.: Local search for shift design. Eur. J. Oper. Res. **153**(1), 51–64 (2004). https://doi.org/10.1016/S0377-2217(03)00098-5

27. Musliu, N., Schutt, A., Stuckey, P.J.: Solver independent rotating workforce scheduling. In: van Hoeve, W.-J. (ed.) CPAIOR 2018. LNCS, vol. 10848, pp. 429–445. Springer, Cham (2018). https://doi.org/10.1007/978-3-319-93031-2_31

28. Nethercote, N., Stuckey, P.J., Becket, R., Brand, S., Duck, G.J., Tack, G.: MiniZinc: towards a standard CP modelling language. In: Bessière, C. (ed.) CP 2007. LNCS, vol. 4741, pp. 529–543. Springer, Heidelberg (2007). https://doi.org/10.1007/978-3-540-74970-7_38

29. Quimper, C.G., Rousseau, L.M.: A large neighbourhood search approach to the multi-activity shift scheduling problem. J. Heuristics **16**(3), 373–391 (2010)

30. Restrepo, M.I., Lozano, L., Medaglia, A.L.: Constrained network-based column generation for the multi-activity shift scheduling problem. Int. J. Prod. Econ. **140**(1), 466–472 (2012). https://doi.org/10.1016/j.ijpe.2012.06.030
31. Thompson, G.: Improved implicit modeling of the labor shift scheduling problem. Manage. Sci. **41**(4), 595–607 (1995)
32. Widl, M., Musliu, N.: The break scheduling problem: complexity results and practical algorithms. Memetic Comput. **6**(2), 97–112 (2014). https://doi.org/10.1007/s12293-014-0131-0

A Computational Comparison
of Optimization Methods for the Golomb
Ruler Problem

Burak Kocuk[1]([envelope])[ID] and Willem-Jan van Hoeve[2][ID]

[1] Sabancı University, 34956 Istanbul, Turkey
burakkocuk@sabanciuniv.edu
[2] Carnegie Mellon University, Pittsburgh, PA 15213, USA
vanhoeve@andrew.cmu.edu

Abstract. The Golomb ruler problem is defined as follows: Given a positive integer n, locate n marks on a ruler such that the distance between any two distinct pair of marks are different from each other and the total length of the ruler is minimized. The Golomb ruler problem has applications in information theory, astronomy and communications, and it can be seen as a challenge for combinatorial optimization algorithms. Although constructing high quality rulers is well-studied, proving optimality is a far more challenging task. In this paper, we provide a computational comparison of different optimization paradigms, each using a different model (linear integer, constraint programming and quadratic integer) to certify that a given Golomb ruler is optimal. We propose several enhancements to improve the computational performance of each method by exploring bound tightening, valid inequalities, cutting planes and branching strategies. We conclude that a certain quadratic integer programming model solved through a Benders decomposition and strengthened by two types of valid inequalities performs the best in terms of solution time for small-sized Golomb ruler problem instances. On the other hand, a constraint programming model improved by range reduction and a particular branching strategy could have more potential to solve larger size instances due to its promising parallelization features.

Keywords: Golomb ruler · Integer programming · Constraint programming

1 Introduction

For a given positive integer n, let us denote the positions of n marks on a ruler as $x_1, x_2, \ldots, x_n$. Without loss of generality, we assume that the position of the first mark is zero, i.e. $x_1 = 0$, and the locations are ordered, i.e. $x_1 \leq x_2 \leq \cdots \leq x_n$. A *Golomb ruler* satisfies the property that the pairwise distances between distinct marks are all different, in other words, $x_j - x_i \neq x_k - x_l$ unless $i = l$ and $j = k$. The *optimal* Golomb ruler is the one with the smallest length, that is, a

© Springer Nature Switzerland AG 2019
L.-M. Rousseau and K. Stergiou (Eds.): CPAIOR 2019, LNCS 11494, pp. 409–425, 2019.
https://doi.org/10.1007/978-3-030-19212-9_27

Golomb ruler with the minimum x_n. The Golomb ruler problem has interesting applications in several fields [3], including information theory [14], astronomy, and communications [1,2,4,12].

In general, constructing a Golomb ruler with a given number of marks is an easy task, and there are many heuristic methods that provide high quality rulers. For instance, previous literature on heuristics has focused on affine and projective plane constructions [7,15], genetic algorithms [19], and local search [6,13] while exact methods based on constraint programming [8,17] or hybrid methods [16] exist as well. Although not proven to be NP-hard yet, solving the Golomb ruler problem exactly proved to be notoriously difficult. For instance, the optimal rulers for $n = 24, 25, 26, 27$ have been proven by a parallel search with thousands of workstations coordinated by the website `distributed.net`, and it took approximately 4, 4, 1, and 5 years to complete, respectively. Currently, a search for the 28-mark problem is under way for more than 4 years.

As summarized above, most of the effort to prove that a given Golomb ruler is an optimal one is devoted to explicit enumeration techniques. However, such brute force approaches seem to be the only viable option since it is very difficult to establish strong valid lower bounds for the Golomb ruler problem.

At this point, we would like to state the main purpose of this paper, which is to *certify the optimality of a given Golomb ruler through optimization methods*. Most optimization algorithms inherently solve relaxations and hence, naturally provide lower bounds for minimization type problems. Therefore, it is worth focusing on optimization models to better understand the structure of the Golomb ruler problem, and hopefully, propose efficient methods which we can use to solve the Golomb ruler problem instances.

In this paper, we consider three classes of optimization problems to carry out the aforementioned analysis. In particular, we formulate the Golomb ruler problem as linear integer programming, constraint programming and quadratic integer programming problems. Some of these models exist in the literature while and the others are introduced, to the best of our knowledge, by us. Since the performance of the basic models is not satisfactory to solve instances with more than 10 marks, we propose several enhancements to improve the scalability of each method by means of bound tightening, valid inequalities, cutting planes and effective search strategies. Our computational experiments show that linear integer programming models scale up to 13 marks given a budget of 8 h while constraint programming models can solve up to 13-mark instances in about an hour and 14-mark problem in about 10 h. Quadratic integer programming models, on the other hand, are able to solve 14-mark instance in about four hours, all using a modest personal computer. As a comparison, it took 2.8 h for the constraint programming model in [8] to find an optimal ruler for the 13-mark instance and another 11.8 h to prove its optimality. The lean implementation of the search method in [16] reduced the respective computational effort to 0.6 and 1.3 h, albeit at the expense of a significantly larger search tree.

The rest of the paper is organized in three sections, which respectively cover the linear integer programming, constraint programming and quadratic integer programming models in detail. Each section contains a formulation, enhancements and computational experiments subsections together with some discussions and comparisons of these different optimization paradigms.

2 Linear Integer Programming Models

2.1 Two Formulations

In this section, we present two known linear integer programming formulations for the Golomb ruler problem with n marks. One of these formulations is *exact* and the other one is only a *relaxation* but can be made exact by the use of additional features as detailed later. For both of the models, we assume that an upper bound, say L, on the optimal length is known (such a bound can be obtained as the length of any feasible Golomb ruler with n marks).

"$d + e$" **Formulation.** We first present an exact linear integer programming model for the Golomb ruler problem [9]. In this formulation, there are two sets of decision variables: Let e_{ijv} be one if the distance between marks i and j is v, and zero otherwise. Also, we define d_{ij} as the distance between marks i and j. Then, the optimization problem is given as follows:

$$\min_{d,e} \sum_{i=1}^{n-1} d_{i,i+1} \tag{1a}$$

$$\text{s.t.} \sum_{v=1}^{L} e_{ijv} = 1 \qquad 1 \le i < j \le n \tag{1b}$$

$$\sum_{i<j} e_{ijv} \le 1 \qquad v = 1,\ldots,L \tag{1c}$$

$$\sum_{v=1}^{L} v e_{ijv} = d_{ij} \qquad 1 \le i < j \le n \tag{1d}$$

$$\sum_{k=i}^{j-1} d_{k,k+1} = d_{ij} \qquad 2 \le i+1 < j \le n \tag{1e}$$

$$e_{ijv} \in \{0,1\}, d_{ij} \in \mathbb{Z}_+ \qquad 1 \le i < j \le n, v = 1,\ldots,L. \tag{1f}$$

Here, the objective function (1a) minimizes the length of the Golomb ruler (alternatively, it can be simply given as $d_{1,n}$). Constraint (1b) assigns a distance between $1,\ldots,L$ to every pair of marks i and j while constraint (1c) ensures that each distance between $1,\ldots,L$ is assigned at most once. Constraint (1d) is simply a definition of the d variables in terms of the e variables. Finally, constraint (1e) is an identity guaranteeing that the distance between the marks i and j is the sum of the basic distances between consecutive marks. We note that d variables can be projected out by substituting the definition given in (1d) into constraint (1e) although the resulting lower dimensional model does not seem to be more advantageous empirically in terms of computational efficiency.

"d" Formulation. Now, we present a relaxation of the Golomb ruler problem by eliminating the e variables from the "$d + e$" Formulation. The resulting optimization problem is defined as follows:

$$\min_d \sum_{i=1}^{n-1} d_{i,i+1} \tag{2a}$$

s.t. (1e)

$$\sum_{(i,j)\in R} d_{ij} \geq \frac{1}{2}|R|(|R|+1) \qquad R \subseteq \{(k,l) \in \mathbb{Z}^2 : 1 \leq k < l \leq n\}. \tag{2b}$$

The constraint set (2b), called "Subset Sum Inequalities", was introduced in [9] to strengthen the integer programming (IP) formulation, and defines the facets of the convex hull of the all-different constraint [20]. In practice, the model (2) can be solved with a constraint generation scheme in which the subset sum inequalities are gradually added. We also note that the complexity of the separation of these inequalities is polynomial-time as it requires sorting $\binom{n}{2}$ many numbers.

As opposed to the "$d + e$" Formulation, the "d" Formulation is not an exact representation of the Golomb ruler problem. However, it is known that the optimal value of the "d" Formulation is equal to the linear programming (LP) relaxation value of "$d + e$" Formulation [11].

We note that even if the "d" Formulation is solved as an IP, it is still not an exact formulation since it does not guarantee that $d_{ij} \neq d_{kl}$ for $i \neq k$ and $j \neq l$, i.e., the uniqueness of the distances. However, this observation leads to a natural way to make the "d" Formulation exact: We can solve the problem (2) as an LP with two callbacks: Firstly, we add lazy constraint callbacks to ensure that subset sum inequalities (2b) are satisfied. Secondly, we add a branch callback such that the missing constraint $d_{ij} \neq d_{kl}$ is enforced by the solver as we go down the branch-and-bound tree. In particular, once $d_{ij} = d_{kl}$ for $i \neq k$ and $j \neq l$, we can create a dichotomy as $d_{ij} \leq d_{kl} - 1$ and $d_{ij} \geq d_{kl} + 1$.

2.2   Enhancements

We propose several enhancements to the models presented above based on bound tightening and branching strategies. We also carried out a preliminary polyhedral study of the Golomb ruler problem with the hope of obtaining strong valid inequalities. Although we have discovered several families of valid inequalities, they have not helped solving the problem more efficiently. Therefore, we leave this line of research as future work for further inquiry.

In the sequel, let G_m denote the length of the optimal Golomb ruler of order m, $m = 1, \ldots, n - 1$. We assume that all the G_m values are known for $m < n$ when we are trying to solve the n-mark problem.

Bound Tightening. Bound tightening is a widely used strategy in optimization algorithms to reduce the range of the decision variables in order to save computational effort. It can also be used as a way to strengthen relaxations and improve the performance of search methods (see [8] for an application to the Golomb ruler problem). Our bound tightening procedure starts with the following simple observation: If the difference between marks i and j is small (large), then d_{ij} cannot be too large (small). In particular, we can infer the following initial bounds on d_{ij} variables:

$$G_{j-i+1} =: \underline{d}_{ij} \le d_{ij} \le \overline{d}_{ij} := L - G_i - G_{n-j+1}. \tag{3}$$

After this initialization step, we can further improve the bounds $\underline{d}_{ij}$ and $\overline{d}_{ij}$ by solving bounding LPs. In particular, we can minimize/maximize d_{ij} variables over a suitable relaxation (for instance, over the feasible region of the LP relaxation of (1)) to try to improve the bounds iteratively. Once new bounds $\underline{d}_{ij}$ and $\overline{d}_{ij}$ are obtained after rounding up and down the minimum and maximum values, we repeat this procedure until the fixed point is reached, that is, none of the bounds are improved further. We will refer to this procedure as *LP Bounding*.

As an additional mechanism to tighten the variable bounds on the d variables, we extend the LP Bounding approach in the following sense. Now, we optimize the d variables over the feasible region of (1) (not its LP relaxation as in the LP Bounding approach) with a limited computational budget. We will refer to this procedure as *IP Bounding*. Although this approach requires an additional non-trivial effort, it pays off in terms of reducing the solution time of the "$d + e$" formulation.

We also use the bounds $\underline{d}_{ij}$ and $\overline{d}_{ij}$ to fix some of the binary variables e_{ijv} to zero as follows:

$$e_{ijv} = 0 \text{ if } v < \underline{d}_{ij} \text{ or } v > \overline{d}_{ij}.$$

This procedure reduces the total range of the d variables and the number of e variables considerably although it does not improve the LP relaxation bound.

Branching Strategies. The choice of branching strategies may significantly affect the computational performance of the mixed-integer programming solvers. Branching decisions made by the solvers can be altered by either explicitly choosing the variables to be branched on through branch callbacks, or implicitly by assigning priorities to the integer variables. We experimented with both of these choices by exploiting the structure of the Golomb ruler problem.

In terms of imposing explicit branching decisions, we experimented with two strategies which can be applied to both "$d + e$" and "d" Formulations. The first strategy, which we will call as "Left Branching", is described as follows: We first branch on the variable d_{12} by creating $\overline{d}_{12} - \underline{d}_{12} + 1$ many child nodes, each taking an integer from the interval $[\underline{d}_{ij}, \overline{d}_{ij}]$. Then, we proceed by solving the node relaxations. Whenever we have to make another branching decision, we decide on the next variable still undecided from the left of the ruler (for instance, the second variable would be d_{23}). Such an algorithm is based on the intuition that the classical dichotomy branching is probably ineffective for the Golomb ruler problem since assigning particular values to d_{ij} variables can help detecting infeasibility of more branches than simply branching on d_{ij} by assigning intervals, or traditional binary branching on the e_{iju} variables. The shortcoming of this strategy is of course the increased number of child nodes for each level of the branch-and-bound tree.

Table 1. Results of the "$d+e$" Formulation with LP and IP Bounding strategies. Here, BTT, OT, TT and #BBN respectively represent the bound tightening, optimization, total time (in seconds unless otherwise stated), and the number of branch-and-bound nodes.

n	LP Bounding				IP Bounding after LP Bounding			
	BTT	OT	TT	#BBN	BTT	OT	TT	#BBN
9	0.10	0.36	0.46	0	41.35	0.60	42.05	615
10	0.11	1.42	1.53	333	37.69	0.68	38.48	337
11	0.12	199.87	199.99	155,421	20.55	140.38	161.05	138,820
12	0.19	297.62	297.81	153,244	68.27	208.39	276.85	126,405
13	0.26	32,435.04	32,435.30	13,949,679	2,874.62	24,993.13	27,868.01	13,363,776
14	0.32	>10 h	-	-	3,721.56	>10 h	-	-

The second strategy for imposing explicit branching decisions is called "Difference Branching", and implemented with the inclusion of a branch callback function. Suppose that in a certain node in the branch-and-bound tree, we have two variables d_{ij} and d_{kl}, $i \neq k$ and $j \neq l$, such that $|d_{ij} - d_{kl}| \leq 1$. Then, we can branch on constraints $d_{ij} \leq d_{kl} - 1$ and $d_{ij} \geq d_{kl} + 1$ as this is a valid partitioning of the feasible region.

In terms of imposing implicit branching decisions, we experimented with different priority assignment strategies. The most successful strategy seems to be the one that assigns higher priorities to e_{iju} variables with smaller u indices. Here, the intuition is that if smaller distances are decided first, then we can either find feasible solutions or detect infeasibility faster.

2.3 Computations

We first report the results of our computational experiments with the "$d + e$" formulation in Table 1 for $n = 9, \ldots, 14$. We compare the following two settings (we note that the bound tightening techniques are only applied at the root node assuming that $L = G_n$):

- LP Bounding: Bound tightening is applied over the LP Relaxation of the "$d + e$" formulation in parallel for five rounds.
- IP Bounding after LP Bounding: Bound tightening is applied over the "$d+e$" formulation with a budget of 1 s for $n \leq 12$ and 1 min for $n \geq 13$ in parallel for five rounds.

CPLEX 12.8 is used as the mixed-integer linear programming (MILP) solver on a 64-bit computer with Intel Core i7 CPU 2.60GHz processor and 16 GB RAM. Since our aim to prove the optimality of the n-mark ruler of length G_n, we set $L = G_n - 1$ while solving the MILPs.

We note that the LP Bounding scheme is quite cheap to obtain better variable bounds than the initial bounds derived in (3). On the other hand, IP Bounding requires an additional nontrivial effort to further improve those bounds. We observe that this additional computational effort can be justified when $n \geq 12$ as the reduction in the optimization step overweighs the increase in the bound tightening step and the 13-mark instance can be solved in less than 8 h in total. Due to the sharp increase in the CPU time necessary, we were not able to solve the Golomb ruler problem with 14 marks in less than 10 h.

Table 2. Results of the "d" Formulation with different variable bounding and branching strategies.

n	LP Bounding				IP Bounding after LP Bounding			
	Diff. Branching		Left Branching		Diff. Branching		Left Branching	
	TT	#BBN	TT	#BBN	TT	#BBN	TT	#BBN
9	1.45	3,277	3.05	5310	42.59	3,257	44.45	5,338
10	0.55	805	2.00	3248	38.21	785	39.74	3,297
11	9,915.67	4,256,165	1,920.96	1,651,695	9,785.72	4,285,299	1,862.69	1,655,038

We report the results of our experiments with the "d" formulation in Table 2 for $n = 9, 10, 11$. We compare the Difference (Diff.) Branching and Left Branching strategies as introduced in Sect. 2.2 in combination with LP Bounding and IP Bounding. We observe that the Left Branching becomes significantly better than the Difference Branching approach as the number of marks increases. Since we were able to solve only very small Golomb ruler problem instances with the "d" Formulation in comparison to the "$d + e$" Formulation, we have not pursued this direction further. Nevertheless, the Left Branching strategy has proved to be quite effective and is utilized multiple times in this paper.

3 Constraint Programming Model

In the previous section, we presented two linear integer programming models to solve the Golomb ruler problem. Although several enhancement of these models are introduced and the computational effort is reduced significantly, we were not able prove the optimality of a given 14-mark ruler in a reasonable amount of time. In this section, we switch our attention to constraint programming models which prove to be more successful for the Golomb ruler problem.

3.1   Formulation

A constraint programming model of the Golomb ruler problem can be formulated as follows [16,18]:

$$\min_{d} d_{1,n} \tag{4a}$$

$$\text{s.t. alldiff}(\{d_{ij} : 1 \leq i < j \leq n\}) \tag{4b}$$

$$d_{ij} + d_{jk} = d_{ik} \qquad\qquad 1 \leq i < j < k \leq n \tag{4c}$$

$$d_{ij} \in \{\underline{d}_{ij}, \ldots, \overline{d}_{ij}\} \qquad\qquad 1 \leq i < j \leq n. \tag{4d}$$

Here, constraint (4b) ensures that each distance d_{ij} are different from each other. Constraint (4c) guarantees that the distances respect the "triangle" constraint, that is, the distance between marks i, k is the sum of the distances between marks i, j and j, k, where j is strictly between i and k. Finally, constraint (4d) specifies the ranges of the decision variables.

3.2   Enhancements

The constraint programming model (4) can easily solve small instances of the Golomb ruler problem, e.g., $n \leq 10$, but runs into slow convergence issues for even slightly larger instances. Similar to the integer programming models considered in the previous sections, we propose some enhancements to improve the scalability of the constraint programming model. These enhancements utilize bound tightening, table constraints and search strategies.

Bound Tightening. The bound tightening procedures proposed in Sect. 2.2 based on LPs and IPs are quite effective in reducing the ranges of the d variables. We now discuss another similar procedure based on constraint programming techniques (see [10] for a related method called "shaving"). The proposed idea is quite simple: We fix a d_{ij} variable to its current lower or upper bound and solve the feasibility version of the constraint programming model (4) for a limited amount of time. If the infeasibility of this restricted model can be proven, this implies that we can tighten the range of the d_{ij} variable by excluding the value that we have fixed. We will refer to this procedure as *CP Bounding*. This approach is implemented in an iterative fashion with limited computational budget and proved to be helpful to further reduce the range of the decision variables.

Table Constraints. Table constraints can be crucial to speed up constraint programming solvers. By exploiting the specific structure of the Golomb ruler problem, we also define certain "forbidden assignments". The construction is as follows: Consider the subruler with marks numbered as $\{i, \ldots, i+4\}$. We first enumerate the set S_j of all triplets $(d_{j,j+1}, d_{j+1,j+2}, d_{j+2,j+3})$ that constitute a feasible subruler with respect to the variable bounds and all-different constraints, for $j = i$ and $j = i + 1$. Now, suppose that for some $(\overline{d}_{i,i+1}, \overline{d}_{i+1,i+2},$

$\bar{d}_{i+2,i+3}) \in S_i$, there does not exists any $\bar{d}_{i+3,i+4}$ such that $(\bar{d}_{i+1,i+2}, \bar{d}_{i+2,i+3},$ $\bar{d}_{i+3,i+4})$ belongs to the set S_{i+1}. In this case, we can declare the triplet $(\bar{d}_{i,i+1},$ $\bar{d}_{i+1,i+2}, \bar{d}_{i+2,i+3})$ as "forbidden". We can repeat this procedure for a few rounds across different subrulers to identify more forbidden assignments.

Table 3. Results with LP Bounding, IP Bounding and CP Bounding strategies.

n	LP Bounding		IP Bounding after LP		CP Bounding after LP and IP		
	OT	TT	OT	TT	BTT	OT	TT
9	2.82	2.92	0.27	41.71	1.03	0.06	42.54
10	11.95	12.06	0.28	38.08	1.17	0.07	39.04
11	140.30	140.42	19.63	40.30	11.85	0.54	33.07
12	554.71	554.90	23.39	91.85	27.77	1.52	97.75
13	11,615.19	11,615.45	1,780.54	4,655.42	1,665.97	209.87	4,750.71
14	>10 h	-	31,839.14	35,561.02	3,872.62	29,795.25	37,389.74

Search Strategies. Search strategies are extremely important in constraint programming as they significantly alter the performance of the solvers. Inspired by the Left Branching for the linear integer programming model and its adaptation to the quadratic integer model, we have decided to employ a variable selection rule based on lexicographical ordering. We also set the search phase parameter to depth first search as our aim is to prove the optimality of a given ruler efficiently. Finally, we experiment with different value selection strategies and decide to use the one based on the smallest impact.

3.3 Computations

We report the results of our computational experiments with the constraint programming formulation in Table 3 for $n = 9, \ldots, 14$. In addition to the "LP Bounding" and "IP Bounding after LP Bounding" settings introduced in Sect. 2.3, we also experimented with the following version:

- CP Bounding after LP and IP Bounding: Bound tightening is applied over the constraint programming formulation with a budget of 1 s for $n \leq 12$ and 1 min for $n \geq 13$ in parallel for five rounds. This includes the generation of forbidden assignments based on the table constraints.

CPLEX CP Optimizer is used as the constraint programming solver with the default settings unless otherwise stated.

We now summarize our observations: Firstly, a comparison with Tables 1 and 3 indicates that the constraint programming formulation takes less time than the "$d + e$" formulation under the same version of bounding for $n \geq 11$. This allows us to solve the 14-mark problem with the constraint programming approach with IP Bounding in 10 h, which was not possible with the "$d + e$" formulation. Secondly, the overhead of the CP Bounding approach is quite large, hence, the reduction in the optimization time compared to the IP Bounding may not be fully justified always. However, we believe that further inquiry along this direction should be pursued. For instance, we extended the CP Bounding approach for the 14-mark problem by allowing 10 min of budget for each subruler length. This increases the total bounding time to 18,251 s but reduces the optimization time to 16,851 s. Although the total time remains more or less unchanged, this additional experiment shows that a carefully executed bounding mechanism may have a potential to be efficient overall.

4   Quadratic Integer Programming Models

So far, we presented classical linear integer and constraint programming formulations for the Golomb ruler problem. In this section, we focus on a less-explored approach based on quadratic integer programming.

4.1   Two Formulations

In this section, we discuss two possible quadratic integer programming formulations for the Golomb ruler problem with n marks, one based on an *optimization* model and the other based on a *feasibility* version. To the best of our knowledge, such formulations have not been proposed before in the literature.

Let us define a single set of binary variables y_l, which takes value one if there is a mark at location l and zero otherwise, $l = 1, \ldots, L$. Here, L is again an upper bound on the length of a shortest Golomb ruler with n marks.

Optimality Version. We first present an alternative integer programming formulation of the Golomb ruler problem as follows:

$$\min_{y} \ \max_{l=1,\ldots,L} \ l \times y_l \tag{5a}$$

$$\text{s.t.} \ \sum_{l=0}^{L-v} y_l y_{l+v} \leq 1 \qquad\qquad v = 1, \ldots, L \tag{5b}$$

$$y_l \in \{0, 1\} \qquad\qquad l = 1, \ldots, L. \tag{5c}$$

Here, the objective (5a) minimizes the position of the last mark on a ruler of length L, which corresponds to the length of an optimal ruler. We note that the objective function can be easily linearized using an auxiliary variable and enforcing L additional constraints. Constraint (5b) guarantees that each distance v is used at most once in a feasible solution. Observe that the model (5) can be reformulated as a quadratically constrained program, which contains two types of nonconvexities, one due to the bilinear inequalities (5b), and another due to the integrality of the y variables.

Convexification techniques can be utilized to solve or approximate the non-convex problem (5). We note that this formulation admits a straightforward semidefinite programming (SDP) relaxation given as follows:

$$\min_{y,z,Y} z \tag{6a}$$

$$\text{s.t. } l \times y_l \leq z \qquad\qquad l = 1, \ldots, L \tag{6b}$$

$$\sum_{l=0}^{L-v} Y_{l,l+v} \leq 1 \qquad\qquad v = 1, \ldots, L \tag{6c}$$

$$0 \leq y_l \leq 1 \qquad\qquad l = 1, \ldots, L \tag{6d}$$

$$\begin{bmatrix} 1 & y^T \\ y & Y \end{bmatrix} \succeq 0. \tag{6e}$$

Unfortunately, the dual bound obtained from solving the SDP relaxation (6) is extremely weak (for instance, the bound obtained for the 10-mark instance is only 15.14 while the length of the optimal Golomb ruler is 55). Therefore, we have not pursued this line of research direction further.

Feasibility Version. Now, we consider a "complementary" version of the problem defined as follows: Given the length of a ruler L, maximize the number of marks that can be located onto such a ruler that satisfies the Golomb ruler requirements. This version of the problem can be formulated as follows:

$$n_L := \max_y \sum_{l=0}^{L} y_l \tag{7a}$$

$$\text{s.t. } (5b) - (5c). \tag{7b}$$

Note that the formulation (7) can be seen as the feasibility version of the model (5) in the following sense: If $n_L = n$ but $n_{L-1} = n - 1$, then we can certify that $G_n = L$. Hence, in order to obtain the length of a shortest Golomb ruler with n marks, that is G_n, we can first solve problem (7) with $L = G_{n-1}$, and then increase the value of L until we can locate all of the n marks. Such a procedure gives an indirect way of solving the Golomb ruler problem.

Problem (7) is again a nonconvex, quadratically constrained integer program. Below, we propose two linearization methods that can be used to solve problem (7) via an appropriate branch-and-bound method.

Linearization via SDP The problem (7) can be reformulated as a mixed-integer SDP as follows:

$$\max_{y,Y} \sum_{l=0}^{L} y_l$$

$$\text{s.t. } (6c), (6e), (5c). \tag{8a}$$

Since the problems in this class are not typically supported by commercial solvers, we implemented our own branch-and-bound algorithm. In this algorithm, we solve the SDP relaxation of the model (8), which replaces the binary restriction (5c) with its continuous relaxation (6d), at each node of the tree. Our algorithm decides which y variables to choose for branching, which is discussed in more detail in Sect. 4.2.

Linearization via LP. The problem (7) can be also reformulated as a mixed-integer LP as follows:

$$\max_{y,Y} \sum_{l=0}^{L} y_l \tag{9a}$$

$$\text{s.t. } (6c), (5c)$$

$$y_l + y_k - 1 \leq Y_{lk} \qquad l, k = 1, \ldots, L \tag{9b}$$

$$0 \leq Y_{lk} \leq y_l \qquad l, k = 1, \ldots, L. \tag{9c}$$

Here, constraints (9b)–(9c) correspond to the McCormick envelopes for the equation $Y_{lk} = y_l y_k$. In general, solving problem (9) directly as an MILP is quite expensive, partly due to the fact that its LP relaxation is highly degenerate. Therefore, we adopt a Benders decomposition approach, whose problem specific details are presented in Sect. 4.2.

4.2   Enhancements

We again propose some enhancements to speed up the solution procedure of the feasibility version of the quadratic formulation of the Golomb ruler problem. In particular, we develop two types of valid inequalities, Benders decomposition for the linearized model (9) and branching strategies. Improved variable bounds obtained via the bound tightening procedure presented in Sect. 2.2 are also used whenever applicable.

Valid Inequalities. In this section, we present two families of valid inequalities, which we refer to as "Golomb" and "Clique" inequalities. Below, we present their precise formulations together with the intuition behind them.

Golomb Inequalities. Since the number of marks that can be placed onto any subruler of length t is upper bounded by n_t, the following inequalities are valid and are added to the root node relaxation:

$$\sum_{j=l}^{l+\min\{G_{i+1},L\}+1} y_j \leq i \quad i = 2, \ldots, n_L; \; l = 0, \ldots, L - (\min\{G_{i+1}, L\} + 1). \tag{10}$$

More inequalities of this kind can be obtained as follows: Instead of summing the consecutive y variables, we can consider any subset of these variables whose indices are separated by exactly the same integer c, $c = 2, \ldots, \lfloor L/2 \rfloor$, such as $y_j, y_{j+c}, y_{j+2c}, \ldots$.

Clique Inequalities In order to better explain the construction of the clique inequalities, it is more suitable to present the Golomb ruler problem as a special *maximum cardinality clique problem* defined as follows: Let us consider a complete graph $G = (V, E)$, where the set of vertices is $V := \{0, \dots, L\}$. We partition the edge set E into L subsets E_l defined as $E_l := \{(i, i+l) : i = 0, \dots, L-l\}$ for $l = 1, \dots, L$. Then, in order to solve the Golomb ruler problem, we search for a largest clique $G' = (V', E')$ in this graph such that at most one edge from each subset E_l belongs to E', that is, $|E_l \cap E'| \leq 1$.

Motivated by the above construction, let us introduce the clique inequalities, which are easily implementable in a cutting plane framework. Consider a fractional solution $\tilde{y}$, and construct two sets $\mathcal{L}_1 := \{l : \tilde{y}_l = 1\}$ and $\mathcal{L}_f := \{l : \tilde{y}_l \in (0, 1)\}$. Let us define the distances induced by the solution $\tilde{y}$ as $\mathcal{D} := \{|k - l| : k, l \in \mathcal{L}_1\}$. We will now construct an auxiliary graph $\tilde{G} = (\tilde{V}, \tilde{E})$, where $\tilde{V} := \mathcal{L}_f$ and $\tilde{E} := \{(k, l) \in \tilde{V} \times \tilde{V} : |k - l| \in \mathcal{D}\}$. We also associate node weights $\tilde{y}_l$ for each $l \in \tilde{E}$. Then, each maximal clique $\tilde{C}$ in the graph $\tilde{G}$ whose weight is more than 1 gives rise to a cutting plane of the following form:

$$\sum_{l \in \tilde{C}} y_l \leq 1. \tag{11}$$

We note that the set of all maximal cliques in a graph can be found by the Bron-Kerbosch algorithm [5] in reasonable time for such small graphs, and the inequalities (11) can be added as local user cuts in a branch-and-cut algorithm.

Benders Decomposition. Since we observe that solving problem (9) directly is not computationally efficient, we employ a Benders decomposition technique instead. In this approach, we solve the following master problem

$$\max_{y} \sum_{l=0}^{L} y_l \tag{12a}$$

s.t. (5c)

$$\sum_{l \in C} |C_u(l)| y_l \leq \frac{|C|}{2} + 1 \qquad C \subseteq \{0, \dots, L\}, u = 1, \dots, L, \tag{12b}$$

where constraints (12b) are added in lazy fashion until feasibility is proven. Here, $C_u(l) := \{k : |k - l| = u\}$. This is achieved through the separation procedure (feasibility check) described as follows: Given a binary vector $\tilde{y}$, we first define the set of marks as $M := \{l : \tilde{y}_l = 1\}$. For this candidate ruler to be a Golomb ruler, the distances between each pair of mark should be distinct. Therefore, the cardinality of the set $M_u(l) := \{k \in M : |k - l| = u\}$ should be 1 for $l \in M$ and $u = 1, \dots, L$. Otherwise, we detect infeasibility and can add the following cut:

$$\sum_{l \in M} |M_u(l)| y_l \leq \frac{|M|}{2} + 1. \tag{13}$$

In other words, we expand the constraint set (12b) by the inclusion of the set M_u. We note that our approach allows to add multiple cuts of the form (13) corresponding to different u values for a given solution.

What we described so far amounts to a classical implementation of the Benders decomposition technique in which a "multi-tree" approach is employed, that is, at each iteration, we solve the master problem (12) as an MILP. Therefore, multiple branch-and-bound trees are created. An alternative approach would be to use a single branch-and-bound tree, and add the Benders feasibility cuts (13) via lazy constraint callbacks. Such an approach is commonly referred to as a "one-tree" approach, branch-and-cut or "branch-and-check" and works much better than its multi-tree counterpart for model (12).

Table 4. Results of the "y" Formulation (linearization via LP) with and without Golomb and Clique cuts using one-tree Benders decomposition. Total time and branch-and-nodes for each mark n and different ruler lengths L are reported.

n	L	without Golomb Cuts				with Golomb Cuts			
		w/o Clique Cuts		with Clique Cuts		w/o Clique Cuts		with Clique Cuts	
		TT	#BBN	TT	#BBN	TT	#BBN	TT	#BBN
9	35–43	7.91	17,753	2.63	278	5.73	12,630	3.49	597
10	45–54	32.24	81,036	11.45	354	17.73	59,770	5.80	846
11	56–71	3,850.04	1,712,947	730.56	824	619.7	1,384,860	35.49	2,101
12	73–84	>10 h	-	5,576.83	674	4,119.82	4,710,139	57.80	1,661
13	86–105	>10 h	-	>10 h	-	>10 h	-	1,689.52	4,869
14	107–126	>10 h	-	>10 h	-	>10 h	-	12,960.17	7,776

Branching Strategy. The branching decisions are extremely important for both the SDP and LP based branch-and-bound algorithms. Following the intuition from Left Branching idea from linear integer programming models as mentioned in Sect. 2.2, we propose a similar scheme that decides the next mark from the left of the ruler. In particular, suppose that the first m marks from the left are located at the positions $\ell_1, \ldots, \ell_m$. Then, the location of the mark $m + 1$ can be chosen from the set

$$\left\{ v : \underline{d}_{1,m+1} \leq v \leq \overline{d}_{1,m+1}, \ v - \ell_k \notin \{\ell_j - \ell_i : 1 \leq i < j \leq m\} \ \forall k = 1, \ldots, m \right\}.$$

Therefore, we again prefer to create multiple child nodes rather than the more traditional dichotomous branching.

4.3   Computations

We report the results of our computational experiments with the quadratic integer programming formulation with LP linearization in Table 4 for $n = 9, \ldots, 14$. Since quadratic integer programming formulations are based on feasibility version of the Golomb ruler problem, we solve a sequence of models with increasing ruler length to certify that a given ruler is optimal (see the explanation at the end of Sect. 4.1). For instance, to prove that $G_9 = 44$, we solve problem (9) with $L = 35, \ldots, 43$ and certify that $n_L = 8$ (here, we assume that G_8 is known as 34). We report the computational results with and without the Golomb and Cliques cuts giving rise to four different settings.

We observe that both types of cuts are quite effective to solve the subproblems from different perspectives: Golomb cuts are especially helpful in reducing the computational time more directly whereas Clique cuts significantly lowers the number of branch-and-bound nodes and indirectly reduces the total time. The reason that these two cuts behave differently is that Golomb cuts are added from scratch and their number is limited whereas Clique cuts are added on the fly at each node of the tree and their number can be quite large. We believe that the separation of Clique cuts can be made more efficient and selective so that the total computational effort can be further improved.

Finally, we report the results of our computational experiments with the quadratic integer programming formulation with SDP linearization in Table 5 for $n = 9, 10, 11$ with and without Golomb cuts (MOSEK 8.1 is used as the SDP solver). Although the total number of branch-and-bound nodes is reduced by solving the SDP relaxation of the model (8) at each node, the total time increases quite significantly which prevents this line of research to be practical. However, we point out that our implementation is quite naïve and perhaps the value of stronger relaxations provided by the SDP relaxations can be made useful.

Table 5. Results of the "y" Formulation (linearization via SDP) with and without Golomb cuts.

n	L	w/o Golomb Cuts		with Golomb Cuts	
		TT	#BBN	TT	#BBN
9	35–43	21.38	227	22.58	209
10	45–54	33.03	231	40.77	200
11	56–71	6,776.11	25,591	6,607.07	24,737

5 Concluding Remarks

In this paper, we provided a comprehensive comparison of computational methods to solve the Golomb ruler problem using optimization techniques. In particular, we analyzed three formulations based on linear integer programming, constraint programming and quadratic integer programming, and proposed several enhancements based on valid inequalities, variable bounding and branching strategies. According to our experiments with a budget of 10 h, integer linear programming models can solve up to 13-mark instances whereas constraint programming and quadratic integer programming formulations can scale up to 14-mark instance, with the latter being faster. We observed that proposed enhancements significantly alter the solution procedures and provide substantial savings in terms of computational effort.

Although the methods in this paper can solve relatively small-size instances of the Golomb ruler problem, we think that there are some promising research directions which might utilize them more effectively. As an example, if a large number of processors is available, then bound tightening subproblems can be parallelized asynchronously so that they can exchange information whenever a new bound is improved. Since the availability of tight variable bounds seems to accelerate the constraint programming solver, this can potentially enable us to solve larger instances. Another potential line of research would be to make the cut generation procedure for the quadratic integer programming model more efficient and selective so that the overhead associated with solving large MILPs is reduced while keeping the strength of the relaxations intact.

References

1. Babcock, W.C.: Intermodulation interference in radio systems. Bell Labs Tech. J. **32**(1), 63–73 (1953)
2. Biraud, F., Blum, E.J., Ribes, J.C.: On optimum synthetic linear arrays with application to radioastronomy. IEEE Trans. Antennas Propag. **22**, 108–109 (1974)
3. Bloom, G.S., Golomb, S.W.: Applications of numbered undirected graphs. Proc. IEEE **65**(4), 562–570 (1977)
4. Blum, E.J., Ribes, J.C., Biraud, F.: Some new possibilities of optimum synthetic linear arrays for radioastronomy. Astron. Astrophys. **41**, 409–411 (1975)
5. Bron, C., Kerbosch, J.: Algorithm 457: finding all cliques of an undirected graph. Commun. ACM **16**(9), 575–577 (1973)
6. Dotú, I., Van Hentenryck, P.: A simple hybrid evolutionary algorithm for finding Golomb rulers. In: The 2005 IEEE Congress on Evolutionary Computation, vol. 3, pp. 2018–2023. IEEE (2005)
7. Drakakis, K., Gow, R., O'Carroll, L.: On some properties of Costas arrays generated via finite fields. In: 40th Annual Conference on Information Sciences and Systems, pp. 801–805. IEEE (2006)
8. Galinier, P., Jaumard, B., Morales, R., Pesant, G.: A constraint-based approach to the Golomb ruler problem. Montréal: Centre for Research on Transportation = Centre de recherche sur les transports (CRT) (2003)
9. Lorentzen, R., Nilsen, R.: Application of linear programming to the optimal difference triangle set problem. IEEE Trans. Inf. Theor. **37**(5), 1486–1488 (1991)
10. Martin, P., Shmoys, D.B.: A new approach to computing optimal schedules for the job-shop scheduling problem. In: Cunningham, W.H., McCormick, S.T., Queyranne, M. (eds.) IPCO 1996. LNCS, vol. 1084, pp. 389–403. Springer, Heidelberg (1996). https://doi.org/10.1007/3-540-61310-2_29
11. Meyer, C., Jaumard, B.: Equivalence of some LP-based lower bounds for the Golomb ruler problem. Discrete Appl. Math. **154**(1), 120–144 (2006)
12. Oshiga, O., Abreu, G.: Design of orthogonal Golomb rulers with applications in wireless localization. In: 2014 48th Asilomar Conference on Signals, Systems and Computers, pp. 1497–1501, November 2014

13. Prestwich, S.: Trading completeness for scalability: hybrid search for cliques and rulers. In: Proceedings of the Third International Workshop on Integration of AI and OR Techniques in Constraint Programming for Combinatorial Optimization Problems, pp. 159–174 (2001)
14. Robinson, J., Bernstein, A.: A class of binary recurrent codes with limited error propagation. IEEE Trans. Inf. Theor. **13**(1), 106–113 (1967)
15. Singer, J.: A theorem in finite projective geometry and some applications to number theory. Trans. Am. Math. Soc. **43**(3), 377–385 (1938)
16. Slusky, M.R., van Hoeve, W.-J.: A Lagrangian relaxation for golomb rulers. In: Gomes, C., Sellmann, M. (eds.) CPAIOR 2013. LNCS, vol. 7874, pp. 251–267. Springer, Heidelberg (2013). https://doi.org/10.1007/978-3-642-38171-3_17
17. Smith, B.M., Stergiou, K., Walsh, T.: Modelling the Golomb ruler problem. Research Report Series-University of Leeeds School of Computer Studies LU SCS RR (1999)
18. Smith, B.M., Stergiou, K., Walsh, T.: Using auxiliary variables and implied constraints to model non-binary problems. In: Proceedings of AAAI/IAAI, pp. 182–187 (2000)
19. Soliday, S.W., Homaifar, A., Lebby, G.L.: Genetic algorithm approach to the search for Golomb rulers. In: ICGA, pp. 528–535 (1995)
20. Williams, H.P., Yan, H.: Representations of the all-different predicate of constraint satisfaction in integer programming. INFORMS J. Comput. **13**(2), 96–103 (2001)

A New CP-Approach for a Parallel Machine Scheduling Problem with Time Constraints on Machine Qualifications

Arnaud Malapert[1] and Margaux Nattaf[2(✉)]

[1] Université Côte d'Azur, CNRS, I3S, Sophia Antipolis, France
`arnaud.malapert@unice.fr`
[2] Univ. Grenoble Alpes, CNRS, Grenoble INP, G-SCOP, 38000 Grenoble, France
`margaux.nattaf@grenoble-inp.fr`

Abstract. This paper considers the scheduling of job families on parallel machines with time constraints on machine qualifications. In this problem, each job belongs to a family and a family can only be executed on a subset of qualified machines. In addition, machines can lose their qualifications during the schedule. Indeed, if no job of a family is scheduled on a machine during a given amount of time, the machine loses its qualification for this family. The goal is to minimize the sum of job completion times, i.e. the flow time, while maximizing the number of qualifications at the end of the schedule. The paper presents a new Constraint Programming (CP) model taking more advantages of the CP feature to model machine disqualifications. This model is compared with two existing models: an Integer Linear Programming (ILP) model and a Constraint Programming model. The experiments show that the new CP model outperforms the other model when the priority is given to the number of disqualifications objective. Furthermore, it is competitive with the other model when the flow time objective is prioritized.

Keywords: Parallel machine scheduling · Time constraint ·
Machine qualifications · Integer Linear Programming ·
Constraint Programming

1 Introduction

Process industries, and specially semiconductor industries, need to be more and more competitive and they are looking for strategies to improve their productivity, decrease their costs and enhance quality. In this context, companies must pay constant attention to manufacturing processes, establish better and more intelligent controls at various steps of the fabrication process and develop new scheduling techniques. One way of doing it is to integrate scheduling and process control [16]. This paper considers such a problem: the integration of constraints coming from process control into a scheduling problem.

Grenoble INP—Institute of Engineering Univ. Grenoble Alpes.

L.-M. Rousseau and K. Stergiou (Eds.): CPAIOR 2019, LNCS 11494, pp. 426–442, 2019.
https://doi.org/10.1007/978-3-030-19212-9_28

Semiconductor fabrication plants (or fabs) have characteristics that make scheduling a very complex issue [8]. Typical ones include a very large number of jobs/machines, multiple job/machine types, hundreds of processing steps, re-entrant flows, frequent breakdowns... Scheduling all jobs in a fab is so complex that jobs are scheduled in each workshop separately. In this paper, the focus is on the photolithography workshop, which is generally a bottleneck area. In this area, scheduling can be seen as a scheduling problem on non-identical parallel machines with job family setups (also called s-batching in [8]).

Fabrication processes of semiconductors are very precise and require a high level of accuracy. Reliable equipments are required and valid recipe parameters should be provided. Advanced Process Control (APC) systems ensure that each process is done following predefined specifications and that each equipment is reliable to process different product types. APC is usually associated with the combination of Statistical Process Control, Fault Detection and Classification, Run to Run (R2R) control, and more recently Virtual Metrology [9]. The main interest of this paper is to consider, in scheduling decisions, constraints induced by R2R controllers. As shown in the survey paper of [14], R2R control is becoming critical in high-mix semiconductor manufacturing processes.

R2R controller uses data from past process runs to adjust settings for the next run as presented for example in [10] and [4]. Note that a R2R controller is associated to one machine and one job family. In order to keep the R2R parameters updated and valid, a R2R control loop should regularly get data. Hence, as presented in [12,13], an additional constraint is defined on the scheduling problem to impose that the execution of two jobs of the same family lies within a given time interval on the same (qualified) machine. The value of the time threshold depends on several criteria such as the process type (critical or not), the equipment type, the stability of the control loop, etc. If this time constraint is not satisfied, a qualification run is required for the machine to be able to process again the job family on the machine. This procedure ensures that the machine works within a specified tolerance and is usually time-consuming. In this paper, we assume that qualification procedures are not scheduled either because the scheduling horizon is not sufficiently long or because qualification procedures have to be manually performed and/or validated by process engineers. Therefore, maintaining machine qualifications as long as possible is crucial. More precisely, it is important to have as many remaining machine qualifications as possible at the end of the schedule, so that future jobs can also be scheduled.

To our knowledge, there are few articles dealing with scheduling decisions while integrating R2R constraints. [2] and [7] study related problems, except that they allow qualification procedures to be performed, the number or the type of machines is different and the threshold is expressed in number of jobs instead of in time. The scheduling problem addressed in this paper has been introduced in [13], where two Integer Linear Programs (ILP) and two constructive heuristics are proposed. More recently, [11,12] develop a new ILP, modelling problem constraints in a better way. The paper also presents one Constraint Programming (CP) model, as well as two improvement procedures of existing heuristics.

In this paper, a new CP model, which takes more advantages of the CP features, is presented. The main idea of this model is to exploit the fact that once a machine is disqualified, it is until the end of the schedule. The consequence of this is that it is possible to model machine disqualifications more accurately. Then, the performance of this model is compared with the two exact solution methods described in [11]. Both CP models use the CP Optimizer (CPO) framework [3]. Indeed, CPO allows to model elegantly and to propagate the precedent constraints efficiently and optional jobs.

The paper is organized as follows. Section 2 gives a formal description of the problem. Section 3 presents the ILP and CP models of [11]. Section 4 describes the new CP model and finally, Sect. 5 provides a detailed comparison of the performance of each model.

2   Problem Description

Formally, the problem takes as input a set of jobs, $\mathcal{N} = \{1, \ldots, N\}$, a set of families $\mathcal{F} = \{1, \ldots, F\}$ and a set of machines, $\mathcal{M} = \{1, \ldots, M\}$. Each job j belongs to a family and the family associated with j is denoted by $f(j)$. For each family f, only a subset of the machines $\mathcal{M}_f \subseteq \mathcal{M}$, is able to process a job of f. A machine m is said to be qualified to process a family f if $m \in \mathcal{M}_f$. Each family f is associated with the following parameters:

- n_f denotes the number of jobs in the family. Note that $\sum_{f \in \mathcal{F}} n_f = N$.
- p_f corresponds to the processing time of jobs in f.
- s_f is the setup time required to switch the production from a job belonging to a family $f' \neq f$ to the execution of a job of f. Note that this setup time is independent of f'. In addition, no setup time is required between the execution of two jobs of the same family.
- γ_f is the threshold value for the time interval between the execution of two jobs of f on the same machine. Note that this time interval is computed on a start-to-start basis, i.e. the threshold is counted from the start of a job of family f to the start of the next job of f on machine m. Then, if there is a time interval $]t, t + \gamma_f]$ without any job of f on a machine, the machine lose its qualification for f.

The objective is to minimize both the sum of job completion times, i.e. the flow time, and the number of qualification looses or disqualifications. Note that the interest of minimizing the number of disqualifications comes from the fact that, even if the time horizon considered is relatively small, the problem is solved in a rolling horizon. Hence, it is interesting to preserve machine qualifications for future jobs. In addition, it is relevant to consider that a machine cannot lose its qualification for a family after the end of the schedule. Thus, this assumption is made in the remaining of the paper. This problem, introduced in [13], is called the scheduling Problem with Time Constraints (PTC). An example of PTC together with two feasible solutions is now presented.

Example 1. Consider the following instance with $N = 10$, $M = 2$ and $F = 3$:

f	n_f	p_f	s_f	γ_f	$\mathcal{M}_f$
1	3	9	1	25	$\{2\}$
2	3	6	1	26	$\{1, 2\}$
3	4	1	1	21	$\{1, 2\}$

Figure 1 shows two feasible solutions. The first solution, described by Fig. 1a, is optimal in terms of flow time. For this solution, the flow time is equal to $1 + 2 + 9 + 15 + 21 + 1 + 2 + 12 + 21 + 30 = 114$ and the number of qualification losses is 3. Indeed, machine 1 (m_1) loses its qualification for f_3 at time 22 since there is no job of f_3 starting in interval $]1, 22]$ which is of size $\gamma_3 = 21$. The same goes for m_2 and f_3 at time 22 and for m_2 and f_2 at time 26.

The second solution, described by Fig. 1b, is optimal in terms of number of disqualifications. Indeed, in this solution, none of the machines loses their qualifications. However, the flow time is equal to $1 + 2 + 9 + 17 + 19 + 9 + 18 + 20 + 27 + 37 = 159$. This shows that the flow time and the number of qualification losses are two conflicting criteria. Indeed, to maintain machine qualifications, one needs to regularly change the job family executed on machines. This results in many setup time and then to a large flow time value.

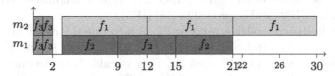

(a) An optimal solution for the flow time objective

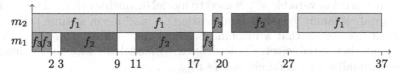

(b) An optimal solution for qualification losses

Fig. 1. Two solution examples for PTC.

Note also that disqualifications may occur after the last job on the machine. For example, in Fig. 1a, m_1 become disqualified for f_3 at time 22 whereas the last job scheduled on m_1 finishes at time 21. However, no disqualifications can occur after the makespan C_{max}.

Remark 1 (Bi-objective optimization). In [13], PTC is studied using a weighted sum of the flow time and number of disqualifications. The weight associated to the flow time is α and is always equal to 1. The weight associated with the number of disqualifications is β and is set to 1 when the priority is given to the flow time and to $N \cdot T$ when the priority is given to the number of disqualifications.

In this paper, we use instead the lexicographical order for the CP models where the minimization of the disqualifications is prioritized.

Remark 2 (Bound on the makespan). In the remaining of the paper, the following upper and lower bound on the makespan are defined. The upper bound used is the same as in [13], i.e. $T = \overline{C_{max}} = \sum_{f \in \mathcal{F}} n_f \cdot (p_f + s_f)$. A trivial lower bound is $\underline{C_{max}} = \lceil (\sum_{f \in \mathcal{F}} n_f * p_f)/M \rceil$.

3   Existing Models

This section describes the two exact methods developed in [11]. First, the ILP is described and then, the CP model.

3.1   ILP Model

The ILP model in [11] is an improvement of two existing models introduced in [13]. The first ILP model of [13] relies on a job-based formulation. Indeed, in this model, a variable $x_{j,t}^m$ is defined for each job j, each machine m and each time t. This variable is then equal to 1 if and only if job j starts at time t on machine m. However, in a solution, there is no need to know which job starts at which time on which machine. Indeed, only the family of the job is important. Hence, a family-based model is developed in [13] (IP2) and improved in [11] (IP3).

In the family-based model, a variable $x_{f,t}^m$ is introduced for each family $f \in \mathcal{F}$, each machine $m \in \mathcal{M}_f$ and each time $t \in \mathcal{T} = \{0, \ldots, T-1\}$, with T the upper bound on the makespan (see Remark 2). This variable is set to one if and only if one job of f starts at time t on machine m. Therefore, the number of binary variables is reduced compared to the job-based model.

Similarly, a set of variable $y_{f,t}^m$ is used to model disqualifications. This variable is set to 1 only if family f lose its qualification on machine m at time t. However, in (IP2), it may occur that a machine becomes disqualified after C_{max}. Thus, in (IP3), another variable set Y_f^m is defined to model the fact that a machine becomes disqualified for a family before C_{max}.

$$\text{min. } \alpha \cdot \sum_{f \in \mathcal{F}} C_f + \beta \cdot \sum_{f \in \mathcal{F}} \sum_{m \in \mathcal{M}} Y_f^m \tag{1}$$

$$\sum_{m \in \mathcal{M}_f} \sum_{t=0}^{T-p_f} x_{f,t}^m = n_f \qquad\qquad \forall f \in \mathcal{F} \tag{2}$$

$$\sum_{m \in \mathcal{M}_f} \sum_{t=0}^{T-p_f} (t + p_f) \cdot x_{f,t}^m \leq C_f \qquad\qquad \forall f \in \mathcal{F} \tag{3}$$

$$n_f \cdot x_{f',t}^m + \sum_{\tau=t-p_f-s_{f'}+1}^{t} x_{f,\tau}^m \leq n_f \qquad\qquad \forall f \neq f' \in \mathcal{F}^2,$$
$$\qquad\qquad\qquad\qquad\qquad\qquad \forall m \in \mathcal{M}_f \cap \mathcal{M}_{f'}, \ \forall t \in \mathcal{T} \tag{4}$$

$$y_{f,t}^m + \sum_{\tau=t-p_f+1}^{t} x_{f,\tau}^m \leq 1 \qquad\qquad \forall f \in \mathcal{F}, \ \forall t \in \mathcal{T}, \ \forall m \in \mathcal{M}_f \tag{5}$$

$$y_{f,t}^m + \sum_{\tau=t-\gamma_f+1}^{t} x_{f,\tau}^m \geq 1 \qquad\qquad \forall f \in \mathcal{F},\ \forall t \geq \gamma_f \in \mathcal{T},\ \forall m \in \mathcal{M}_f \qquad (6)$$

$$y_{f,t-1}^m \leq y_{f,t}^m \qquad\qquad \forall f \in \mathcal{F},\ \forall t \in \mathcal{T},\ \forall m \in \mathcal{M}_f \qquad (7)$$

$$\frac{1}{M \cdot (T-t)} \sum_{f' \in \mathcal{F}} \sum_{\tau=t-p_{f'}}^{T-1} \sum_{m' \in \mathcal{M}_{f'}} x_{f',\tau}^{m'} + y_{f,t-1}^m - 1 \leq Y_f^m$$

$$\forall t \in \mathcal{T},\ \forall f \in \mathcal{F},\ \forall m \in \mathcal{M}_f \qquad (8)$$

$$x_{f,t}^m \in \{0,1\} \qquad\qquad \forall t \in \mathcal{T}, \forall f \in \mathcal{F},\ \forall m \in \mathcal{M}_f \qquad (9)$$

$$y_{f,t}^m \in \{0,1\} \qquad\qquad \forall t \in \mathcal{T}, \forall f \in \mathcal{F},\ \forall m \in \mathcal{M}_f \qquad (10)$$

$$Y_f^m \in \{0,1\} \qquad\qquad \forall f \in \mathcal{F},\ \forall m \in \mathcal{M}_f \qquad (11)$$

The objective of the model is described by (1). It is expressed as the weighted sum of the flow times and the number of disqualifications. Constraints (2) ensure that all jobs are executed. Constraints (3) is used to compute the completion time of family f, i.e. the sum of completion time of all jobs of f. Constraints (4) ensure that jobs of f and jobs of f' do not overlap and that the setup times are satisfied. Constraints (5) are used to model both the fact that the execution of two jobs of the same family cannot occur simultaneously and the fact that a machine has to be qualified to process a job. Constraints (6) make sure that if no jobs of family f start on m during an interval $]t - \gamma_f, t]$, then m becomes disqualified for f at time t. Constraints (7) maintain the disqualification of the machine once it becomes disqualified. Finally, Constraints (8) ensure that it is no longer necessary to maintain a qualification on a machine if there is no job which starts on any machine in the remainder of the horizon, i.e. $\frac{1}{M \cdot (T-t)} \sum_{f' \in \mathcal{F}} \sum_{\tau=t-p_{f'}}^{T-1} \sum_{m' \in \mathcal{M}_{f'}} x_{f',\tau}^{m'} = 0$.

The number of variables of the model is $F \cdot M \cdot (2T+1)$ and the number of constraints is at most $2F + T \cdot M \cdot (4F + F^2)$.

3.2   CP Model

In this section, the CP model defined in [11] is described. The first part of the model concerns the modelling of a classical parallel machine scheduling problem (PMSP) with setup time and the second part deals with the modelling of the disqualifications. Attention will be given to this part of the model.

The Parallel Machine Scheduling Problem with Setup Time. The PMSP with setup time can be modeled using optional (or not) interval variables introduced by [5,6]. An (optional) interval variable J is associated with four variables: a start time, $st(J)$; a duration, $d(J)$; an end time, $et(J)$ and a binary execution status $x(J)$, equal to 1 if and only if the interval variable is present in the final solution. If the job J is executed, it behaves as a classical job that is executed on its time interval, otherwise it is not considered by any constraint.

In the considered scheduling problem, a job j of family f can be scheduled on any machine belonging to $\mathcal{M}_f$. Therefore, a set of optional interval variables $altJ_{j,m}$ is associated with each job j and each machine belonging to $\mathcal{M}_{f(j)}$. The domain of such variables is $dom(altJ_{j,m}) = \{[st,et] \mid [st,et] \subseteq [0,T),\ st + p_{f(j)} = et\}$. Furthermore, a non-optional interval variable, $jobs_j$ is associated with each job j. Its domain is $dom(jobs_j) = \{[st,et] \mid [st,et] \subseteq [0,T),\ st + p_{f(j)} = et\}$.

To model the PMSP with setup time, the following two sets of global constraints is used [15].

Alternative Constraints. Introduced in [5], this constraint models an exclusive alternative between a bunch of jobs.

$$alternative\left(jobs_j, \{altJ_{j,m}|m \in \mathcal{M}_{f(j)}\}\right) \qquad \forall j \in \mathcal{N} \qquad (12)$$

It means that when $jobs_j$ is executed, then exactly one of the $altJ_{j,m}$ jobs must be executed, i.e. the one corresponding to the machine m on which the job is scheduled. Furthermore, the start date and the end date of $jobs_j$ must be synchronised with the start and end date of the $altJ_{j,m}$ jobs. However, if $jobs_j$ is not executed, none of the other jobs can be executed. In our model, $jobs_j$ is a mandatory job. This constraint models the fact that each job must be executed on one and only one machine.

No-Overlap Constraints. An important constraint is that jobs cannot be executed simultaneously on the same machine. It is a unary resource constraint. Each machine can then be used by only one job at a time. To model this feature, we use noOverlap constraints. This constraint ensures that the executions of several interval variables do not overlap. It can also handle the setup time. Let S be the matrix of setup times of the problem, i.e. $(S_{f',f}) = \begin{cases} 0 & \text{if } f = f' \\ s_f & \text{otherwise} \end{cases}$. Then, the following noOverlap constraint makes sure that, for all pairs of jobs (i,j) s.t. $m \in \mathcal{M}_i \cap \mathcal{M}_j$, either the start of $altJ_{j,m}$ occurs after the end of $altJ_{i,m}$ plus $s_{f(j)}$ or the opposite:

$$noOverlap\left(\{altJ_{j,m}|\forall j \text{ s.t. } m \in \mathcal{M}_{f(j)}\}, S\right) \qquad \forall m \in \mathcal{M} \qquad (13)$$

The exact semantic of this constraint is presented in [6].

Additional Ordering Constraints. The authors of [11] add a non-mandatory set of constraints to the model. Indeed, the model is correct without these constraints but adding them remove many symmetry in the model. The constraint order the start of jobs belonging to the same family.

$$startBeforeStart(jobs_j, jobs'_j) \qquad \forall j, j' \in \mathcal{N}, \ j > j', f(j') = f(j) \qquad (14)$$

Modelling of the Number of Disqualifications. In the model of [11], disqualifications are modelled as optional interval variables. The variable will be present in the final solution if and only if, the machine became disqualified for the family. The start time of the variable corresponds to the time at which the machine becomes disqualified. Therefore, a set of optional interval variable, $disq_{f,m}$, of length 0 is defined for each family f and each machine m such that $m \in \mathcal{M}_f$. The domain of these variables is $dom(disq_{f,m}) = \{[st, et) \,|\, [st, et) \subseteq [\gamma_f, T), \ st = et\}$. In addition, the model will use a C_{max} interval variable of length 0 modelling the end time of the last job executed on all machine, i.e. the end of the schedule. Its domain is $dom(C_{max}) = \{[st, et) \,|\, [st, et) \subseteq [0, T), \ st = et\}$.

Then, the constraints used to model machine disqualifications are stated below. The first set of constraints model the fact that each job has to be executed before C_{max}.

$$endBeforeStart(jobs_j, C_{max}) \qquad \forall j \in \mathcal{N} \qquad (15)$$

Another set of constraints ensures that no job of a family f is scheduled on m if m is disqualified for f, i.e. after the $disq_{f,m}$ job.

$$startBeforeStart(altJ_{j,m}, disq_{f(j),m}, \gamma_{f(j)}) \qquad \forall j \in \mathcal{N}, \; \forall m \in \mathcal{M}_{f(j)} \qquad (16)$$

Finally, the following constraints sets enforce a machine to become disqualified if no job of family f is scheduled on m during an interval of size γ_f. Indeed, the first set state that if a job of f is scheduled on m, either there is another job of f scheduled on m less than γ_f units of time later, or the machine become disqualified, or the end of the scheduled (C_{max}) is reached.

$$x(altJ_{j,m}) \Rightarrow \bigvee_{\substack{j' \neq j \\ f(j)=f(j')}} \left(st(altJ_{j',m}) \leq t_{j,m} \right) \vee \left(st(disq_{f(j),m}) = t_{j,m} \right) \vee \left(C_{max} \leq t_{j,m} \right)$$

$$\forall j \in \mathcal{N}, \; \forall m \in \mathcal{M}_{f(j)} \qquad (17)$$

with $t_{j,m} = st(altJ_{j,m}) + \gamma_{f(j)}$. The second set of constraints ensures that if no job of f is scheduled on m, then m becomes disqualified for f.

$$\bigvee_{\substack{j \in \mathcal{N} \\ f(j)=f(j')}} \left(st(altJ_{j,m}) \leq \gamma_f \right) \vee \left(st(disq_{f(j),m}) = \gamma_f \right) \vee \left(st(C_{max}) \leq \gamma_f \right)$$

$$\forall f \in \mathcal{F}, \forall m \in \mathcal{M}_f \qquad (18)$$

Objective Functions. The objective is to minimize both the flow time and the number of disqualifications. In this CP model, the flow time can be expressed as $flowTime = \sum_{j \in \mathcal{N}} et(jobs_j)$ and the number of disqualifications as $nbDisq = \sum_{f \in \mathcal{F}} \sum_{m \in \mathcal{M}} x(disq_{f,m})$.

Model Size. The number of variables of the model is at most $N \cdot (M+1) + M \cdot F + 1$ and the number of constraints is at most $N^2 + 2N + M \cdot (1 + 2N + F)$.

4   New CP Model

This section presents a new CP model that can be used to solve PTC. As said earlier, PTC can be decomposed into two sub-problems: a PMSP with setup time and a *machine qualifications problem*. The model described in this section uses the same idea as in [11] to formulate the first sub-problem of PTC. However, to model the machine qualification sub-problem a novel approach is developed modelling qualifications as resource.

The first part of this section described the difference of modelling of the PMSP between the model of Sect. 3.2 and the model of this section. The second part is dedicated to the machine qualification sub-problem.

In the model, without loss of generality, the two following assumptions are made. First, it is assumed that jobs of the same family have consecutive index in $\mathcal{N}$. More precisely, with n_f the number of jobs in family f then jobs with index in $\{1, \ldots, n_1\}$ belong to family 1, jobs with index in $\{n_1 + 1, \ldots, n_1 + n_2\}$ are jobs of family 2, etc. The second assumption made in the model is that it is equivalent to consider the threshold either on an end-to-end basis or on a start-to-start basis. Indeed, if a job of family f starts at time t on m, another job of f has to start before $t + \gamma_f$. This is equivalent to: if a job of family f ends at time $t + p_f$ on m, another job of f has to end before $t + p_f + \gamma_f$. Therefore, the model considers the threshold on an end-to-end basis. The motivation for this second assumption will be given later in the section.

The Parallel Machine Scheduling Problem with Setup Time. As for the model of Sect. 3.2, the parallel machine scheduling problem with setup time is modeled using interval variables $jobs_j$, $\forall j \in \mathcal{N}$, and optional interval variables $altJ_{j,m}$. The constraints used are the same and, therefore, are not described in this section.

Cumulative Constraints. The model is also reinforced by considering the set of machines as a cumulative resource of capacity M. Indeed, each job consumes one unit of resource (one machine) during its execution and the total capacity of the resource (total number of machines available) is M. This is expressed using the global constraint *cumulative* [1].

$$cumulative(\{(jobs_j, 1) \mid \forall j \in \mathcal{N}\}, M) \tag{19}$$

Makespan Modelling. As for the previous model, the makespan of the scheduling is needed to model machine disqualifications. The constraints presented in this section concern the link between the makespan and the PMSP. A constraint linking the makespan with the number of disqualifications will be presented later in the paper.

Unlike the previous model, the makespan is modeled here as an interval variable starting at time 0 and spanning the execution of all jobs. This is modelled using *span* constraints. Introduced in [6], this constraint states that an executed job must span over a set of other executed jobs by synchronising its start date with the earliest start date of other executed jobs and its end date with the latest end date. It is expressed by the following constraints:

$$span(C_{max}, \{jobs_j \mid \forall j \in \mathcal{N}\}) \tag{20}$$
$$st(C_{max}) = 0 \tag{21}$$

In addition, the size of the interval has to be between the upper and the lower bound on the makespan defined in Remark 2.

Machine Qualifications Problem. In this section, the model for the machine qualifications problem is described. The main idea of the model is that, each time a job of family f is scheduled on a machine m, a *qualification interval* of size γ_f will occur right after. This interval "models" the fact that machine m remains qualified for family f until, at least, the end of the interval. To model this feature, optional interval variable are used. Indeed, for each job j and each machine $m \in \mathcal{M}_{f(j)}$, an optional interval variable, $qual_{j,m}$, of size $\gamma_{f(j)}$ and taking its value in $\{0, \ldots, T + \max_f \gamma_f\}$. is created. Then, a variable $qual_{j,m}$ will be present in the solution only if $altJ_{j,m}$ is present and will start at the end of $altJ_{j,m}$. This is expressed by the following set of constraints.

$$x(altJ_{j,m}) = x(qual_{j,m}) \qquad \forall j \in \mathcal{N}, \ \forall m \in \mathcal{M}_{f(j)} \qquad (22)$$

$$endAtStart(altJ_{j,m}, qual_{j,m}) \qquad \forall j \in \mathcal{N}, \ \forall m \in \mathcal{M}_{f(j)} \qquad (23)$$

Hence, a job of f can only be scheduled on m during a qualification interval of f on m. This is modeled using cumulative functions. A cumulative function $Q_{f,m}$ counts, at each time t, the number of qualification intervals for (f, m) in which t is. If the number of qualification intervals for (f, m) is greater than 1, then a job of f can be scheduled on m. Otherwise, the number of interval is zero and m is disqualified for f. $Q_{f,m}$ is expressed as:

$$Q_{f,m} = pulse(0, \gamma_f + p_f, 1) + \sum_{\substack{j \in \mathcal{N} \\ f(j)=f}} \sum_{m \in \mathcal{M}_{f(j)}} pulse(qual_{j,m}, 1)$$

Indeed, at the beginning of the scheduled, the machine is qualified from time 0 to $\gamma_f + p_f$. In addition, each time an interval variable $qual_{j,m}$ is scheduled, $Q_{f(j),m}$ increases by one. Then, when a job of f is scheduled on m, $Q_{f,m}$ has to be greater than one and one can show that $Q_{f,m}$ is always smaller than $n_f + 1$.

$$alwaysIn(Q_{f(j),m}, altJ_{j,m}, 1, n_{f(j)} + 1) \qquad \forall j \in \mathcal{N}, \ \forall m \in \mathcal{M}_{f(j)} \qquad (24)$$

Example 2 (Example of cumulative function). Considering the instance of Example 1. The cumulative function Q_{f_3,m_1} corresponding to Fig. 1b is described by Fig. 2.

Each time a job of f_3 ends, the value of the function Q_{f_3,m_1} increases by one and decreases when the qualification interval ends. While the value of Q_{f_3,m_1} is greater than one, it is possible to schedule jobs of f_3 on m_1. Here, Q_{f_3,m_1} is always greater than one for $t \in [0, C_{max})$ meaning that m_1 remains qualified for f_3 at the end of the schedule.

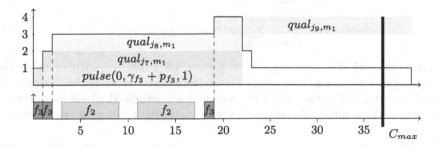

Fig. 2. Example of cumulative function to model qualifications.

Machine Disqualifications. Another dummy optional interval variable set, $endQ_{f,m}$, is introduced to check if a machine has been disqualified for a family during the schedule. The variable is present in the final solution only if the machine is still qualified at the end of the schedule. In this case, the variable starts at time 0, ends at time $C_{max} + p_f$ and the function $Q_{f,t}$ has to be greater than one during the whole execution of job $endQ_{f,m}$ (otherwise, the machine has been disqualified).

$$st(endQ_{f,m}) = 0 \qquad\qquad \forall f \in \mathcal{F},\ \forall m \in \mathcal{M}_f \qquad (25)$$

$$endAtEnd(C_{max}, endQ_{f,m}, p_f) \qquad \forall f \in \mathcal{F},\ \forall m \in \mathcal{M}_f \qquad (26)$$

$$alwaysIn(Q_{f,m}, endQ_{f,m}, 1, n_f + 1) \qquad \forall f \in \mathcal{F},\ \forall m \in \mathcal{M}_f \qquad (27)$$

Thresholds were considered on an end-to-end basis so that the constraints *alwaysIn* can be used.

Ordering Constraints. The following sets of constraint (partially) order variables in the solution. These (partial) ordering is used to break symmetries in the model. Recalling that it is assumed that jobs of the same family have consecutive index in $\mathcal{N}$. Then constraints (28) state that $jobs_{j-1}$ has to start before $jobs_j$ and constraints (29) that the maximum time lag between these jobs is $\gamma_{f(j)}$. Constraints (30) order jobs that cannot be executed in parallel. Indeed, job j can overlap at most $M_{f(j)} - 1$. Hence, job $j - M_{f(j)}$ cannot overlap job j and has to end before. Constraints (31) ensure that the qualification interval corresponding to job j is separated from the qualification interval of $j - 1$ by at least the duration of the job. Finally, constraints (32) model the fact that, on a machine m, jobs of a same family are ordered, i.e. smaller index scheduled first.

$$startBeforeStart(jobs_{j-1}, jobs_j) \qquad \forall j \in \mathcal{N}\ s.t.\ f(j) = f(j-1) \quad (28)$$

$$startBeforeSart(jobs_j, jobs_{j-1}, -\gamma_{f(j)}) \qquad \forall j \in \mathcal{N}\ s.t.\ f(j) = f(j-1) \quad (29)$$

$$endBeforeStart(jobs_{j-M_{f(j)}}, jobs_j) \qquad \forall j \in \mathcal{N}\ s.t.\ f(j) = f(j - M_{f(j)}) \quad (30)$$

$$startBeforeStart(qual_{j-1,m}, qual_{j,m}, p_{f(j)})$$
$$\forall m \in \mathcal{M}_{f(j)},\ \forall j \in \mathcal{N}\ s.t.\ f(j) = f(j-1) \quad (31)$$

$$endBeforeStart(altJ_{i,m}, altJ_{j,m}) \quad \forall m \in \mathcal{M}_{f(j)},\ \forall i < j \in \mathcal{N}\ s.t.f(i) = f(j) \quad (32)$$

Objective Functions. The objective function is modeled using two integer variables: $flowTime \in \{\underline{C_{max}}, \ldots, \overline{C_{max}}\}$ and $qualified \in \{1, \ldots, \sum_{f \in \mathcal{F}} M_f\}$. The expressions of these variables are given below:

$$flowTime = \sum_{j \in \mathcal{N}} et(jobs_j) \qquad (33)$$

$$qualified = \sum_{f \in \mathcal{F}} \sum_{m \in \mathcal{M}_{f(j)}} x(endQ_{f,m}) \qquad (34)$$

Then, the objective is expressed as a sum, i.e. $(flow - qual)$, or using the lexicographical order, e.g. $lex(-qual, flow)$. Note that, in this model, the number of machine qualified at the end of the schedule is maximized which is equivalent to minimize the number of machine becoming disqualified during the schedule.

Model Size. The number of variables of the model is at most $N \cdot (2M + 1) + M \cdot F + 3$ and the number of constraints is at most $N^2 \cdot M + 4N + M \cdot (1 + 4N + 3F) + 6$.

5   Experiments

This section starts with the presentation of the instances used in the experiments (Sect. 5.1). Then, the general framework of the experiments is described in Sect. 5.2. Finally, the three model presented in the paper are used to solve the instances and the results are compared and analysed (Sect. 5.3).

5.1   Instance Generation

The benchmark instances used to perform our experiments are extracted from [11]. In this paper, 19 instance sets are generated with different number of jobs (N), machines (M), family (F) and qualification schemes. Each of the instance sets is a group of 30 instances and are generated as follows.

In each generated instances, each family can be executed by at least one machine and each machine is qualified to process at least one job family. Furthermore, since short thresholds may lead to very quick machine disqualifications, the time thresholds of job families are chosen sufficiently large compared to their associated processing times, i.e. $\max_{f\in\mathcal{F}} p_f \leq \min_{f\in\mathcal{F}} \gamma_f$. Then, to ensure diversity, each set of instances contains 10 instances with small threshold (corresponding to duration needed to process one to two jobs of another family than f), 10 with medium threshold (two to three jobs) and 10 with large threshold (three to four jobs). In addition, setup times are not chosen too large so that the risk of disqualifying a machine due to a setup time insertion is "acceptable", i.e. $\max_{f\in\mathcal{F}} s_f \leq \min_{f\in\mathcal{F}} p_f$.

Table 1 presents the parameters of the different instance sets. In the first row, the different number of jobs N is given, the number of machines M is described by the second row and number of families F is detailed in the third row. Note that each triplet (n, m, f) corresponds to 30 instances. Among those instances, at least $99,5\%$ are feasible. Indeed, experiments in [11] show that only one 60-job instance and two 70-job instances have an unknown status. For all other instances, at least one of the algorithms presented in [11] is able to find a feasible solution.

Table 1. Instance characteristics

N	\multicolumn{6}{c|}{20}	\multicolumn{6}{c|}{30}	40	50	60	\multicolumn{2}{c}{70}

N	20						30						40	50	60	70	
M	3		4				3				4	5	3			4	
F	4	5	2	3	4	5	2	3	4	5	4	5	3	3	4	5	5	4	5

The instances generated are relatively small compared to industrial instances. However, due to the complexity of the problem, it is important to first analyse and compare the results of the three models described in this paper. Finding good solutions for industrial instances is a real challenge and is an important research direction for future work.

5.2   Framework

The experiment framework is defined so the following questions are addressed:

Question 1. Which model is the best at finding a feasible solution, proving the optimality or finding good upper bounds (especially when solving large instances)?

Question 2. Does the performance of a model change depending on the objective function or on the time limit?

The models are implemented using IBM ILOG CPLEX Optimization Studio 12.8 [3]. That is CPLEX for the ILP model and CP Optimizer for CP models. All the experiments were led on a computer running on Ubuntu 16.04.5 with 32 GB of RAM and one Intel Core i7-3930K 3.20 GHz processors (6 cores). Furthermore, two time limits are used in the experiments: 30 and 600 s.

Two heuristics are used to find solutions which are used as a basis for the models. These heuristics are called *Scheduling Centric Heuristic* and *Qualification Centric Heuristic* [11]. The goal of the first heuristic is to minimize the flow time while the second one tries to minimize the number of disqualifications.

In the following of the section ILP model, CP_O model and CP_N model denotes respectively the ILP model of section 3.1, the previous CP model described in Sect. 3.2 and the new CP model detailed in Sect. 4. Furthermore, to describe the performance of the different models, the following indicators are used in the table of Sect. 5.3: *%sol.* gives the percentage of instances for which feasible solution is found; *%opt.* shows the percentage of instances for which the optimality is proven; *%vbs* provides the percentage of instances for which the model is the virtual best solver, i.e. has found the best solution compared to others; *#dis.* gives the average number of disqualified machines and finally, *obj.* is used to show the average of the sum of the flow time and the number of disqualified machines.

In addition, a bold value in the table means that the corresponding indicator has the best values among its row, i.e. compared to other model.

5.3   Comparison of the Three Models

This section aims at comparing the results of the three models. First, the results are described for the *tight time limit*, i.e. 30 s. Then, the results with the 600-seconds time limits are given.

30-seconds Time Limit

Minimizing the Number of Disqualifications Over the Flow Time. Table 2 gives indicators for the three models solved using the $lex(-qual, flow)$ objective with 30-seconds time limit.

Table 2 shows that the ILP model finds less feasible solutions than the CP models. Furthermore, the ILP model does not scale well for large instances.

Table 2. Lexicographic minimization of the disqualified machines and the flow time within 30 s.

N	ILP model				CP$_O$ model				CP$_N$ model			
	%sol.	%opt.	%vbs	#dis.	%sol.	%opt.	%vbs	#dis.	%sol.	%opt.	%vbs	#dis.
20	100	54.4	55.6	1.1	100	69.4	86.1	**0.6**	100	**82.2**	**90.6**	0.6
30	97.2	21.7	23.3	3.1	**99.4**	51.1	59.4	1.4	98.9	**56.7**	**71.1**	1.2
40	100	23.3	26.7	0.9	100	63.3	63.3	0.6	100	**83.3**	**90**	**0.2**
50	100	0	6.7	2.9	100	33.3	36.7	1.4	100	**56.7**	**73.3**	**0.8**
60	88.3	0	0	7.5	90	8.3	33.3	3.4	90	**21.7**	**56.7**	**2.8**
70	86.7	0	0	9.5	**91.1**	4.4	41.1	5	91.1	**15.6**	**52.2**	**4.1**

Indeed, the ILP model is never the VBS and its average number of disqualified machines is very high for the largest instances compared to the CP models.

On the other hand, the CP$_N$ model obtains better results than the CP$_O$ model. Indeed, the percentage of proof of optimality is higher with CP$_N$ model. The model is also more often the VBS regardless of the instance size. Furthermore, the difference between the average numbers of qualified machines of both model increases with the instance size. This shows that the CP$_N$ model scales better than the CP$_O$ model.

Minimizing the Flow Time Over the Number of Disqualifications. Table 3 gives indicators for the three models solved using the $(flow - qual)$ objective with 30-seconds time limit.

Table 3. Weighted sum minimization of the flow time and number of disqualified machines within 30 s.

N	ILP model				CP$_O$ model				CP$_N$ model			
	%sol.	%opt.	%vbs	obj.	%sol.	%opt.	%vbs	obj.	%sol.	%opt.	%vbs	obj.
20	100	**96.7**	**97.8**	**334.7**	100	0	87.8	334.8	100	65.6	90	**334.7**
30	97.8	**69.4**	**72.2**	782.5	**99.4**	0	71.1	770	98.9	24.4	57.8	**766.8**
40	100	**90**	90	1536	100	0	93.3	1530	100	60	100	**1529**
50	100	**60**	70	2265	100	0	76.7	2159	100	10	73.3	**2151**
60	88.3	**5**	8.3	3228	90	0	50	**2792**	90	0	36.7	2805
70	86.7	**4.4**	5.6	4256	90	0	**52.2**	3583	91.1	0	43.3	**3562**

Table 3 shows that the ILP model is more competitive when the priority is given to the number of disqualifications. Indeed, despite the fact that it finds a few less feasible solutions than the CP Models, it is better at proving the optimality of its solution. However, the ILP model does not scale well as shown by the high objective values for the largest instances.

On the other hand, the CP$_O$ model is the most efficient for finding good upper bounds, but completely fails at proving the optimality of its solution. The

CP_N model proves optimality less often than the ILP model. However, it is only slightly dominated by the CP_O model in terms of being the VBS. However, the CP_N model still have the lowest objective values.

600-seconds Time Limit. Table 4 gives indicators for the three models solved using both lexicographic and weighted sum minimization with 30-seconds and 600-seconds time limit. Only challenging instances with 60 jobs are used to save computation time.

For the lexicographic minimization, the CP_N model confirms its predominance. For all three models, the percentages of solved instances remain constant, the percentages of optimality proof only slightly increase, and the average numbers of disqualified machines significantly decrease.

For the weighted sum minimization, the ILP model becomes the best model. The percentages of solved instances and optimality proof significantly improve and the model often becomes the VBS. Nevertheless, the CP_N model model has the best average objective.

Most of the time, the low improvements of the number of solved instances or optimality proofs suggest that the solvers is subject to thrashing and therefore cannot diversify the search.

Table 4. Weighted sum and leximin minimization over instances of 60 jobs within 600 s.

t	ILP model				CP_O model				CP_N model			
	%sol.	%opt.	%vbs	#dis.	%sol.	%opt.	%vbs	#dis.	%sol.	%opt.	%vbs	#dis.
30s	88.3	0	0	7.5	**90**	8.3	33.3	3.4	**90**	**21.7**	**56.7**	**2.8**
600s	**90**	0	3.3	4.2	**90**	11.7	28.3	2.9	**90**	**23.3**	**61.7**	**2.2**
t	%sol.	%opt.	%vbs	obj.	%sol.	%opt.	%vbs	obj.	%sol.	%opt.	%vbs	obj.
30s	88.3	**5**	8.3	3228	**90**	0	**50**	**2792**	**90**	0	36.7	2805
600s	**98.3**	**55**	**75**	2873	90	0	33.3	2755	90	0	33.3	**2744**

6 Conclusions and Further Work

A parallel machine scheduling problem was studied where some Advanced Process Control constraints are integrated: minimal time constraints between jobs of the same family to be processed on a qualified machine to avoid losing the qualification. Two criteria to minimize are considered: the sum of completion times and the number of disqualifications.

For this problem, a new CP model was proposed. This model improves the modelling of machine disqualifications. Indeed, when the number of disqualifications is prioritized, this model is better than the existing methods (ILP model and CP_O model) in terms of objective value and in terms of optimality proof. However, when the flow time is prioritized, the performance of the model is less impressive. In this case, the CP_O model tends to have better performance for small-time limit and the ILP model performs better in case of larger time limit.

Experiment results show that a good CP model needs to make some improvements on the modelling and/or the solving of the parallel machine scheduling problem with the flow time objective. Interesting research directions include the improvement of variable bounds, especially the makespan. It also includes the study of good relaxations of the problem to enhance the performance of constraint programming models.

Another relevant research perspective consists in scheduling jobs on a longer time horizon, where lost qualifications could be automatically recovered after a given qualification procedure. Qualification procedures, requiring time on machines, would then also be scheduled.

References

1. Beldiceanu, N., Carlsson, M., Rampon, J.X.: Global constraint catalog (revision a), January 2012
2. Cai, Y., Kutanoglu, E., Hasenbein, J., Qin, J.: Single-machine scheduling with advanced process control constraints. J. Sched. **15**(2), 165–179 (2012). https://doi.org/10.1007/s10951-010-0215-8
3. IBM: IBM ILOG CPLEX Optimization Studio (2019). https://www.ibm.com/products/ilog-cplex-optimization-studio
4. Jedidi, N., Sallagoity, P., Roussy, A., Dauzère-Pérès, S.: Feedforward run-to-run control for reduced parametric transistor variation in CMOS logic 0.13 μm technology. IEEE Trans. Semicond. Manufact. **24**(2), 273–279 (2011)
5. Laborie, P., Rogerie, J.: Reasoning with conditional time-intervals. In: Proceedings of the Twenty-First International Florida Artificial Intelligence Research Society Conference, Coconut Grove, Florida, USA, 15–17 May 2008, pp. 555–560 (2008). http://www.aaai.org/Library/FLAIRS/2008/flairs08-126.php
6. Laborie, P., Rogerie, J., Shaw, P., Vilím, P.: Reasoning with conditional time-intervals. Part II: an algebraical model for resources. In: Proceedings of the Twenty-Second International Florida Artificial Intelligence Research Society Conference, Sanibel Island, Florida, USA, 19–21 May 2009 (2009). http://aaai.org/ocs/index.php/FLAIRS/2009/paper/view/60
7. Li, L., Qiao, F.: The impact of the qual-run requirements of APC on the scheduling performance in semiconductor manufacturing. In: Proceedings of 2008 IEEE International Conference on Automation Science and Engineering(CASE), pp. 242–246 (2008)
8. Moench, L., Fowler, J.W., Dauzère-Pérès, S., Mason, S.J., Rose, O.: A survey of problems, solution techniques, and future challenges in scheduling semiconductor manufacturing operations. J. Sched. 1–17 (2011). https://doi.org/10.1007/s10951-010-0222-9
9. Moyne, J., del Castillo, E., Hurwitz, A.M.: Run-to-Run Control in Semiconductor Manufacturing, 1st edn. CRC Press, Boca Raton (2000)
10. Musacchio, J., Rangan, S., Spanos, C., Poolla, K.: On the utility of run to run control in semiconductor manufacturing. In: Proceedings of 1997 IEEE International Symposium on Semiconductor Manufacturing Conference, pp. 9–12 (1997)

11. Nattaf, M., Dauzère-Pérès, S., Yugma, C., Wu, C.H.: Parallel machine scheduling with time constraints on machine qualifications, Manuscript submitted for publication

12. Nattaf, M., Obeid, A., Dauzère-Pérès, S., Yugma, C.: Méthodes de résolution pour l'ordonnancement de familles de tâches sur machines parallèles et avec contraintes de temps. In: 19ème édition du congrès annuel de la Société Française de Recherche Opérationnelle et d'Aide à la Décision, ROADEF2018

13. Obeid, A., Dauzère-Pérès, S., Yugma, C.: Scheduling job families on non-identical parallel machines with time constraints. Ann. Oper. Res. **213**(1), 221–234 (2014). https://doi.org/10.1007/s10479-012-1107-4

14. Tan, F., Pan, T., Li, Z., Chen, S.: Survey on run-to-run control algorithms in high-mix semiconductor manufacturing processes. IEEE Trans. Ind. Inform. **11**(6), 1435–1444 (2015)

15. Wolf, A.: Constraint-based task scheduling with sequence dependent setup times, time windows and breaks. GI Jahrestagung **154**, 3205–3219 (2009)

16. Yugma, C., Blue, J., Dauzère-Pérès, S., Obeid, A.: Integration of scheduling and advanced process control in semiconductor manufacturing: review and outlook. J. Sched. **18**(2), 195–205 (2015). https://doi.org/10.1007/s10951-014-0381-1

Efficient Solution Methods for the Cumulative-Interference Channel Assignment Problem Using Integer Optimization and Constraint Programming

Paul J. Nicholas[1,2](✉) and Karla L. Hoffman[2]

[1] Johns Hopkins University Applied Physics Laboratory, Laurel, MD 20723, USA
paul.nicholas@jhuapl.edu
[2] George Mason University, Fairfax, VA 22030, USA
khoffman@gmu.edu

Abstract. Interest in the channel assignment problem (CAP) has been growing rapidly with both the spread of wireless data networks and the increasing scarcity of electromagnetic (EM) spectrum. The ability to efficiently reuse available EM channels is heavily dependent on co-channel interference, i.e., interference occurring between two radios using the same channel but not communicating on the same network. The vast majority of CAP research considers only the interference between any pair of radios, but many radio systems – including the mobile ad-hoc networks we consider – are sensitive to the effects of cumulative interference. In previous work, we describe the vast computational challenges of considering cumulative interference within a CAP. We present a new method to solve this problem via heuristics, integer optimization, and constraint programming techniques. We apply our methods to realistic data sets from a large U.S. Marine Corps operational scenario and provide detailed performance results. To our knowledge, we are the first to describe algorithms for solving realistic, large-scale cumulative-interference minimum-order and minimum-cost channel assignment problems to global or near-global optimality.

Keywords: Constraint programming · Integer optimization ·
Channel assignment problem · Electromagnetic spectrum

1 Introduction

Interest in the *channel assignment problem* (*CAP*) has been growing rapidly with both the spread of wireless data networks and the increasing scarcity of

Partial support for this research was provided by the Office of Naval Research grant N00014-15-1-2176.

© Springer Nature Switzerland AG 2019
L.-M. Rousseau and K. Stergiou (Eds.): CPAIOR 2019, LNCS 11494, pp. 443–460, 2019.
https://doi.org/10.1007/978-3-030-19212-9_29

electromagnetic (EM) spectrum [1, 13, 35]. Efficient channel allocation schemes leverage *channel reuse*, where a channel is a contiguous block of EM spectrum. The ability to reuse a channel is dependent on (among other things) *co-channel interference*, i.e., interference occurring between two radios assigned the same channel but that do not wish to communicate. The vast majority of research on the CAP considers only pairwise interference constraints [1] due to the computational challenges of explicitly representing cumulative interference, and the ease with which the problem can be represented as a *graph-coloring problem* [4, 9, 25, 27, 38]. This seemingly simple problem is *NP-complete* [6], and yet the realistic cumulative interference constraints we model are much more difficult [26–28].

We consider the challenge of a *spectrum manager* who must determine an efficient channel allocation scheme to support radio communications over a certain period of time for mobile units operating on rough terrain. We specifically consider the use of wideband *mobile ad-hoc network* (*MANET*) radios fielded by the United States Marine Corps (USMC), but our approach generalizes to other military services and any EM transceiver system requiring a discrete channel assignment. The spectrum manager knows the capabilities of each radio and their starting locations, and has a rough understanding of their future locations within the *operating area*. Using this information and terrain elevation data, the spectrum manager must determine the minimum number of channels required to support communications with an acceptable level of co-channel interference. Further, since each radio requires manual assignment, the spectrum manager is responsible for the reallocation of channels whenever the situation changes, and therefore desires to minimize the number of channel changes over time.

Due to the computational difficulties of exactly solving the CAP, heuristics are often used to solve the problem [1, 23]. While heuristics may provide usable solutions in reasonable amounts of time, we feel that optimality bounds are important for understanding the goodness of a particular solution, especially since spectrum is increasingly crowded and scarce, and communications may be critical to the success of a military operation.

Dunkin et al. [9] describe the challenge of using cumulative interference constraints, and instead use simple binary and tertiary constraints (i.e., groups of three interfering radios) using a *constraint satisfaction* approach. Daniels et al. [7] formulate an integer CAP that considers cumulative interference and establish the NP-hardness of the problem. Fischetti et al. [11] use pre-processing and *branch-and-cut* to solve their cumulative interference CAP, but their problem sizes are much smaller than those studied here and they consider relatively few sources of interference, i.e., they have a relatively small number of constraints.

We use integer and constraint programming methods to develop more efficient methods of channel allocation. Our first problem minimizes the number of required channels (i.e., *minimum order*), subject to cumulative co-channel interference constraints for any given instance in time, and the second problem minimizes the number of channel assignment changes over time (i.e., *minimum cost*). We use realistic radio performance data from large-scale, high-fidelity simulations of U.S. Marine

Corps operational scenarios (the data are available to the research community at
[12]). To our knowledge, we are the first to solve to global- or near-global optimality the minimum-order and minimum-cost channel assignment problems for large,
realistic datasets while also considering the effects of cumulative co-channel interference and the costs of manual channel changes. We believe that the tools created
for this application are likely to be appropriate for other complex graph-coloring
problems, as well.

This paper is organized as follows. Section 2 provides an overview of our
model of MANET operations. Section 3 describes our formulation and computational results in solving the minimum-order channel assignment problem, and
Sect. 4 does the same for the minimum-cost CAP. Section 5 provides conclusions
and suggestions for future research.

2 Model of MANET Communications

We create a network model to simulate the key aspects of a MANET formed by
tactical wideband radios at a given moment in time (i.e., time step). Let $r \in R$
(alias s) represent each MANET radio. Each radio is permanently assigned to
a MANET *unit* $u \in U$, indicated by the set of *logical arcs* $(r, u) \in L$. In a
military scenario, a unit may represent a tactical military organization, such as
an infantry company or battalion headquarters. Let the set of *nodes* N (indexed
by n) consist of both radios R and units U, i.e., $n \in N = R \cup U$. Let a channel
$c \in C$ be a contiguous range of EM frequencies, where C is the set of available
orthogonal (i.e., non-interfering) channels. Each unit u and the radios assigned
to it require a channel assignment.

Let $(r, s) \in W$ indicate the set of arcs representing wireless transmissions
between all radios $r, s \in R$. A unit $u \in U$ forms a separate MANET among its
assigned radios using the available wireless arcs $(r, s) \in W : (r, u) \in L, (s, u) \in L$.
Figure 1 shows two separate units (indicated in blue and green) and their
assigned radios. The solid lines indicate bidirectional wireless arcs $(r, s) \in W$
between radios. Any radio (e.g., radio r in Fig. 1) communicates with its *network control radio* (e.g., radio s) via these arcs (a radio may route through other
radios in the same unit to reach the network control radio). All radios are subject to co-channel interference from any other radios assigned to different units
but operating on the same channel and geographically close enough to cause
interference; this is indicated by dashed gray arrows directed to r (other lines
withheld for clarity).

To calculate both co-channel interference and the strength of desired wireless
transmissions between intra-unit radios, we calculate the *received signal strength*
(RSS) along all wireless arcs $(r, s) \in W$ in dBm (power ratio in decibels relative
to milliwatts). We instantiate our scenarios in Systems Toolkit (STK) [3] and
then use Python and the Terrain Integrated Rough Earth Model (TIREM) of
Alion Science & Technology Corporation [2] to calculate path loss considering
the technical specifications of each radio and the effects of terrain, atmospheric
absorption, etc.

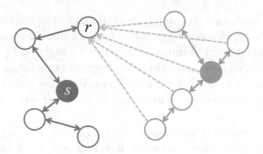

Fig. 1. Simple example of two units (indicated in blue and green) with network control radios (solid circles) and other radios (open circles). Wireless arcs are indicated by arrows. The radios within each unit must be capable of bi-directional communication with their unit's network control radio via direct communication or routing through other radios in the same network. All radios are subject to co-channel interference (dashed arrows) from other radios assigned to different units but operating on the same channel. (Color figure online)

For each radio $s \in R$, we follow [1] and pre-calculate the maximum allowable interference in watts $max\_interference_s^c$. This calculation is based on the RSS between radios and each particular radio's required *signal-to-interference ratio* (*SIR*), a measure of signal quality [33]. Any co-channel interference above this level severs the shortest path and thus disconnects the radio from its assigned network control radio. Among radios not assigned to the same unit but operating on the same channel, the RSS represents co-channel interference. The magnitude of co-channel interference along all arcs $(r, s) \in W$ for each available channel $c \in C$ is pre-calculated in watts, and is indicated by $interference_{rs}^c$.

We use realistic datasets generated from high-fidelity simulations of U.S. Marine Corps operations. We find the largest scenario, depicting a Marine Expeditionary Force (MEF) of 60,000 Marines conducting a large amphibious operation and based on Integrated Security Construct B [8], to be the most computationally interesting. We generate separate datasets at 20 different *time steps* (i.e., discrete moments) within the scenario (each containing the locations of 118 units comprising 1887 total radios). See [26] for full details of our scenarios.

3   Minimum-Order Channel Assignment Problem (MO-CAP)

3.1   Problem Formulation

The MO-CAP aims to minimize the total number of channels required to support MANET operations at a given moment in time.

Variables:

$$X_n^c = \begin{cases} 1, & \text{if node } n \text{ uses channel } c \\ 0, & \text{otherwise} \end{cases} \qquad \forall n \in N, c \in C \qquad (1)$$

$$Y^c = \begin{cases} 1, & \text{if channel } c \text{ is used} \\ 0, & \text{otherwise} \end{cases} \qquad \forall c \in C \qquad (2)$$

Constraints:

We enforce the definition of Y^c via:

$$X_u^c \le Y^c \quad \forall u \in U, c \in C. \qquad (3)$$

Each radio is assigned the same channel as its associated unit:

$$X_r^c = X_u^c \quad \forall c \in C, (r, u) \in L. \qquad (4)$$

To ensure each unit u is assigned one and only one channel, we add the constraints:

$$\sum_{c \in C} X_u^c = 1 \quad \forall u \in U. \qquad (5)$$

Two radios from different units are subject to interference if they are both assigned to the same channel. This assignment will only be allowed if the received $interference_{rs}^c$ between these two radios is less than the precalculated allowable total interference, $max\_interference_s^c$. One way of representing this pairwise interference is:

$$interference_{rs}^c X_r^c X_s^c \le max\_interference_s^c \quad \forall (r, s) \in W, c \in C. \qquad (6)$$

To model the total aggregate interference that a radio receives, we follow the lead of [19], and assume the cumulative effects of jamming sources on the same channel are additive (in watts). That is, a radio $s \in R$ may be unable to use a channel $c \in C$ because the total sum of interference exceeds the threshold $max\_interference_s^c$, even if the interference received from any single radio is less than the threshold. Summing along all arcs yields:

$$\sum_{r:(r,s) \in W} interference_{rs}^c X_r^c X_s^c \le max\_interference_s^c \quad \forall s \in R, c \in C. \qquad (7)$$

To linearize these constraints, we introduce the binary variable Z_{rs}^c where:

$$Z_{rs}^c = \begin{cases} 1, & \text{if } X_r^c = X_s^c = 1 \\ 0, & \text{otherwise} \end{cases} \qquad \forall r, s \in R, c \in C \qquad (8)$$

which is enforced via:

$$Z_{rs}^c \ge X_r^c + X_s^c - 1 \qquad \forall r, s \in R, c \in C \qquad (9)$$
$$Z_{rs}^c \le X_r^c \qquad \forall r, s \in R, c \in C \qquad (10)$$
$$Z_{rs}^c \le X_s^c \qquad \forall r, s \in R, c \in C. \qquad (11)$$

Our cumulative co-channel interference constraints are thus represented:

$$\sum_{r:(r,s)\in W} interference_{rs}^c Z_{rs}^c \leq max\_interference_s^c \quad \forall s \in R, c \in C. \qquad (12)$$

Given the results of radio propagation simulation in a military scenario, we pre-calculate the $max\_interference_s^c$ values (using the method described above), and fix the assignment of radios to their respective units (indicated by arcs $(r, u) \in L$).

Since the goal is to minimize the total number of channels required, our objective function is:

$$\min \sum_{c \in C} Y^c. \qquad (13)$$

3.2   Computational Challenges

The MO-CAP is relatively easy to understand and describe. However, it suffers from several serious computational difficulties when the full problem is simply provided to a commercial solver (e.g., CPLEX or Gurobi) with our realistic datasets. First, commercial solvers may be sensitive to vast differences in input parameters. In our simulated datasets, our interference values vary by 24 orders of magnitude, and are generally quite small. Also, non-integral input data may result in highly *fractionalized* LP solutions, as the solver will attempt to pack the most units (including fractions of units) onto the same channel.

Another computational problem (also observed by [32]) is that of *symmetry*, which occurs when channel assignments may be changed among units with no corresponding change in the objective function value [24]. The very *near* symmetry that is characteristic of our datasets (as opposed to *exact* symmetry) results from some units being located near each other, and is especially difficult for solvers to detect and mitigate [31].

Some of these computational problems could be avoided if we considered only pairwise interference constraints, as IP and constraint satisfaction solvers reformulate these pairwise constraints into clique constraints and then handle these structures very efficiently. Unfortunately, these constraints alone do not adequately represent our real-world problem, and will cause at least a few radios to be disconnected from their respective MANETs.

A simple "brute force" IP method (i.e., using CPLEX to solve the full problem as-is, without providing any initial solution or conducting preprocessing) fails to obtain useful answers to the Marine Corps scenario, even after two weeks of computation on a cluster of 14 high-performance desktop computers. In an attempt to improve the solution process, we create a simple greedy heuristic that iteratively "packs" units onto channels until the channel is "full," and then starts with the next channel. We provide the heuristic solution as an initial feasible solution to CPLEX and attempt to solve the problem for a single time step. We find that after 60 h of runtime, CPLEX improves upon the initial feasible solution, but the obtained solution has an optimality gap of 77%. This indicates that our heuristic may not be finding very good solutions and/or the lower bounding technique is not very effective, and that we require more sophisticated methods if we are going to solve realistic instances of this problem with certifiably-good solutions.

3.3   Integer Programming Solution Method and Results

Rather than simply "throw" the computationally-challenging cumulative interference constraints (12) at a solver, we preprocess the constraints to create simplified and more computationally tractable *packing constraints*. For example, suppose two specific nodes r and s (not assigned to the same unit) are not both allowed to be assigned to channel c because to do so would violate the associated interference constraint. This may be represented as:

$$X_r^c + X_s^c \leq 1. \tag{14}$$

We use Python and the `mpmath` library [18], which allows the use of arbitrary-precision floating point mathematics, to identify unacceptable pairs of radios and handle the extremely small interference values present in our realistic data sets.

To generalize for larger n-tuples of units above pairs (triplets, quadruplets, etc.), let $S \subset U$ be a subset of units that cannot all be assigned to the same channel c. We can represent such a restriction of assignments as:

$$\sum_{r \in S} X_r^c \leq |S| - 1. \tag{15}$$

Preprocessing all such unacceptable combinations and adding them as constraints would effectively replace the cumulative co-channel interference constraints (12). However, identifying all combinations would be computationally prohibitive (as they grow exponentially with both the number of units and available channels) and unnecessary, as many combinations will be redundant and/or represent negligible levels of co-channel interference.

Instead, we dynamically add these higher-order constraints to the formulation only as needed via *lazy constraints*, which are constraints that are checked for violation whenever an integer solution to the current formulation is found. They are added on an as-needed basis [17]. This approach avoids the problem of very small numbers in CPLEX, as we can process the constraints outside of the solver (e.g., in Python), and then add the much-simplified packing constraints (15) dynamically. Also, since the solver is no longer required to calculate cumulative interference at each radio, the formulation no longer requires the index $r \in R$. That is, we are now concerned only with the cumulative interference received at each unit. By removing the index $r \in R$, we reduce the number of decision variables in the problem by an order of magnitude.

After building an initial problem instance with pairwise constraints using Python and Pyomo [15], we send the problem to CPLEX via the Python API and indicate to the solver that we wish to initiate lazy constraint *callbacks*. Upon

Table 1. MO-CAP results by time step using pairwise and lazy constraints. "Time" indicates the time at which the displayed solution and optimality gap is obtained, during a total runtime of 9,000 s.

Time Step	Number lazy constraints	Highest- order lazy constraint	Solution value	Time (s)	Gap	Improvement over heuristic
1	49	5	46	1356.53	2.17%	9.80%
2	25	5	37	1333.92	0%	22.92%
3	87	6	36	4432.53	5.56%	21.74%
4	62	5	34	7828.09	5.88%	27.66%
5	9	5	33	678.23	0%	23.26%
6	104	6	36	4086.04	2.78%	29.41%
7	67	5	37	1737.45	0%	24.49%
8	57	5	31	8614.79	6.45%	26.19%
9	21	8	32	271.19	0%	25.58%
10	0	0	34	248.16	0%	30.61%
11	121	11	33	5997.82	3.03%	26.67%
12	29	5	36	927.38	2.78%	16.28%
13	104	6	32	2510.22	3.12%	25.58%
14	69	6	31	1780.48	3.23%	27.91%
15	147	6	38	1669.23	2.63%	22.45%
16	119	8	36	4194.86	5.56%	23.40%
17	128	6	37	3092.38	2.70%	24.49%
18	8	4	31	245.20	0%	22.50%
19	99	5	30	1673.56	3.33%	25.00%
20	47	5	37	1268.30	0%	24.49%
Aver	**67.6**	**5.6**	**34.9**	**2697.32**	**2.46%**	**24.02%**

finding an integer solution that is feasible with the current constraints, the solver runs our lazy constraint callback code. The code checks the feasibility of the current solution in the full problem, i.e., it checks if the solution satisfies each of the constraints (12). This can be calculated in polynomial time, specifically $\mathcal{O}\left(|R|^2|C|\right)$. If infeasibility exists, we add the lowest-order constraint (15) to the constraint set to prevent the same units from being assigned the same channel again. CPLEX then continues the search process with these new constraints added into the formulation. The process repeats until optimality is achieved or a time limit is reached.

Table 1 displays results for each time step in the Marine Corps scenario, including the number of lazy constraints (and the order of the highest-order lazy constraint), and solution results. Each time step is run for 9,000 s, or until optimality is obtained. The times in Table 1 indicate the time when the displayed solution value and optimality gap is obtained; those time steps with a non-zero optimality gap fail to converge within 9,000 s. Our results are obtained using a Dell Mobile Precision 6800 laptop with 32 GB of RAM and an Intel Core i7-4940MX processor running at 3.1 GHz. We use IBM ILOG CPLEX version 12.6.2 and Python 2.7.

The lazy constraint approach to solving the MO-CAP yields results far superior to our previous methods. The solutions are on average 24% lower than the heuristic, and each solution has an associated optimality gap. On seven time steps, optimality is achieved. Even for those for which optimality is not proven, the method finds solutions within one or two channels of optimality.

We next improve on our lazy constraint method by adding the constraints specifying the *maximum clique*, which is the largest *complete sub-graph* formed from among the pairwise interference constraints. We use the `NetworkX` Python library [14] to find the maximum clique, which relies on the algorithm of [5] as adapted by [36]. Let $M \subset U$ be the subset of units in the maximum clique. The maximum clique constraint takes the form:

$$\sum_{u \in M} X_u^c \le 1 \quad \forall c \in C. \tag{16}$$

That is, only one unit in the clique may be assigned any given channel. Adding this constraint forces the lower bound up significantly and allows the optimization engine to search in a much smaller feasible region. After we add the maximum clique, we then add all remaining pairwise constraints that are not included in this clique constraint. The lazy constraint method is used again to generate any higher-order interference constraints.

The results of this method are displayed in Table 2, where bolded values indicate an improvement over the previously-described technique. Again, each time step is run for 9,000 s, or until optimality is obtained, and "Time" indicates solver time when the displayed solution value and optimality gap is obtained. Overall, inclusion of the maximum clique reduces average runtime to obtain solutions within one channel of optimality. For the problem associated with time step 3, this method obtains a solution that requires one less channel than that identified without use of the maximum clique. On eight time steps, this method reduces the known optimality gap, and on 12 time steps, the method obtains the provably-optimal solution (five more time steps than the previous method). It is interesting to note that the size of the maximum clique (which itself provides a lower bound on the number of required channels) is within one of the best-known solution for each time step. This is indicative of the power of the maximum clique constraint. We note that there is a clique constraint generator within CPLEX, but this procedure does not find this very strong clique; the overall solution times obtained when the maximum clique constraint is removed and CPLEX clique generator is turned on to aggressive yields results no better than those obtained with default parameters for CPLEX.

We also note that we do not obtain shorter solutions times or better bounds when we provide CPLEX with our feasible solution obtained using our greedy heuristic. This indicates that the solutions found with the heuristic are of little use to CPLEX.

3.4 Constraint Programming Solution Method and Results

We reformulate MO-CAP as a constraint programming (CP) problem in an attempt to quickly find lower bounds to the problem. We use the Optimization Programming Language (OPL) to formulate the problem using integer variables, where each variable $w_u \in C$ indicates the channel number that unit $u \in U$ is assigned, and the domain of each variable is equal to the number of available channels $|C|$. (We originally formulate this problem using binary variables, but find that the CP solver is much less efficient in determining feasibility using binary variables for this particular problem.)

Table 2. MO-CAP results by time step using pairwise and lazy constraints, and a maximum clique constraint. Bold values indicate an improvement over the previous method. "Time" indicates the time at which the displayed solution and optimality gap is obtained, during a total runtime of 9,000 s.

Time Step	Max Clique Size	Number lazy constraints	Highest-order lazy constraint	Sol'n value	Time (s)	Gap	Improvement over heuristic
1	46	45	6	46	552.68	**0%**	9.80%
2	37	4	4	37	273.40	0%	22.92%
3	34	143	7	**35**	4338.21	0%	**23.91%**
4	33	85	6	34	3831.19	**2.94%**	27.66%
5	33	10	3	33	1010.00	0%	23.26%
6	35	95	6	36	3128.34	2.78%	29.41%
7	37	13	6	37	266.68	0%	24.49%
8	30	45	5	31	4415.48	**3.23%**	26.19%
9	32	2	4	32	226.37	0%	25.58%
10	34	6	4	34	323.08	0%	30.61%
11	33	42	8	33	856.69	**0%**	26.67%
12	35	30	5	36	1577.96	2.78%	16.28%
13	31	131	6	32	3172.95	3.12%	25.58%
14	30	214	9	31	2702.16	3.23%	27.91%
15	38	105	6	38	1047.00	**0%**	22.45%
16	35	16	5	36	600.91	**2.78%**	23.40%
17	36	89	5	37	1495.15	2.70%	24.49%
18	31	13	4	31	322.90	0%	22.50%
19	30	74	6	30	1653.29	0%	25.00%
20	37	33	4	37	1387.28	0%	24.49%
Aver	**34.4**	**59.8**	**5.5**	**34.8**	**1659.09**	1.18%	**24.13%**

We add all pairwise constraints to the problem by indicating that two given units u and v are not allowed to be assigned the same channel, for all pairs $(u, v) \in P$. We solve the problem using IBM ILOG CPLEX CP Optimizer [17].

We begin with a small (infeasible) number of available channels $|C|$ (i.e., the domain of each $w_u \in C$), and iteratively increase $|C|$ until the solver either determines that the problem is feasible, or it cannot resolve the problem within 12 h. At each $|C|$, we have a *relaxation* of the original MO-CAP. If a problem is infeasible with the given number of channels, then we have established that the original MO-CAP (with all constraints) is also infeasible. This indicates that at least $|C| + 1$ channels are required, establishing a MO-CAP lower bound. If the lower bound equals the upper bound (obtained using CPLEX), we have obtained an optimal solution.

Table 3. MO-CAP results by time step using constraint programming. "Optimal solution?" indicates whether the obtained value proves the optimality of a solution, and bolded values indicate new lower bounds (i.e., not found in the previous analyses.

Time step	Infeasible	Optimal solution?
1	45	Yes
2	36	Yes
3	33	
4	32	
5	32	Yes
6	34	
7	36	Yes
8	29	
9	31	Yes
10	33	Yes
11	32	Yes
12	34	
13	31	**Yes**
14	29	
15	37	Yes
16	34	
17	36	**Yes**
18	30	Yes
19	29	Yes
20	36	Yes

The results are displayed in Table 3, where "Infeasible" indicates the largest value at which the solver detects infeasibility, i.e., at least one more channel is required for the problem to be feasible. "Optimal solution?" indicates whether the obtained value proves the optimality of a solution, where bolded values indicate new lower bounds (i.e., not found in the previous analyses). While the solver does not find the exact lower bound at each time step, it does establish two new exact lower bounds (for the problems associated with time steps 13 and 17). When infeasibility is detected by the solver, it is detected extremely quickly (less than a tenth of a second in each case).

On the other hand, the constraint programming procedure was not capable of proving optimality. In general, one can quickly determine that one needs at least k channels (because one can establish that $k-1$ channels are not feasible). If the CP solver cannot establish whether k channels are infeasible within a few seconds, one is likely to find that the solver will not establish the satisfiability of these constraints within 12 h. In order to improve the solver's capabilities to prove optimality, we try adding *symmetry-breaking* constraints (following [34]) but that does not alter the performance result. Next, we try adding all triplet constraints and the known maximum clique constraint (via CP `allDifferent` constraints), as well as adding constraints iteratively, to no avail.

Thus, we conclude that the CP approach is very efficient at finding infeasibilities (and thus establishing lower bounds), but is incapable of finding feasible solutions close to or at the actual lower bound. We conclude that a very good approach to obtaining optimal or near-optimal solutions to the problem is to integrate CP and IP in a complementary fashion ([16]), where IP is used to search for good solutions and establish upper bounds, and CP is used to quickly tighten lower bounds. Similar techniques have been used for large-scale spectrum auctions [20] and scheduling [30, 37].

4 Minimum-Cost Channel Assignment Problem over Time (MC-CAP-T)

4.1 Problem Formulation

Given the number of channels needed at a moment in time (established using MO-CAP), a spectrum manager may now wish to reduce the total number of times a radio must change channels. Excessive channel changes waste the time of radio operators and require coordination and synchronization among potentially many dispersed units, which may be difficult to achieve in battlefield conditions.

The *minimum-cost channel assignment problem over time* (*MC-CAP-T*) aims to minimize the cost incurred by channel changes over time, given the number of channels required at each time step. Let the index $t \in T$ represent each time step, and let $g \in G$ (alias h) be a *group* of units that must be assigned the same channel at a given time step. Groups are obtained at each time step from the MO-CAP. A *naïve* approach would simply assign channel numbers to the groups as they appear in order. In practice, this produces surprisingly bad solutions as group membership (i.e., the units assigned to each group) may change significantly from time step to time step, and thus an excessively large cost is incurred if one simply dictates that group 1 is always assigned channel 1, etc. We instead use a *decomposition* approach that takes the solutions from each MO-CAP time step and minimizes the "distance" (i.e., number of channel changes) from one time step to the next. We obtain globally-optimal solutions to this problem in polynomial time using our decomposition approach.

We wish to associate each group g at time t to a group h at time $t+1$ at least cost. Let the binary variable Y_{gh}^t indicate if group g at t is associated with group h at $t+1$, and let $(g,h,t) \in A$ be the arcs representing the set of possible associations between g and h. One could simplify this formulation further by dropping the $t \in T$ index, but we retain the notation to aid in describing our solution approach. At each time step, each group g at t must be assigned a group h at $t+1$, and vice versa, which is enforced via the assignment constraints:

$$\sum_{h \in G} Y_{gh}^t = 1 \quad \forall\, (g, \cdot, t) \in A \tag{17}$$

$$\sum_{g \in G} Y_{gh}^t = 1 \quad \forall\, (\cdot, h, t) \in A. \tag{18}$$

The cost of associating a group g at time t to a group h at time $t+1$, $cost_{gh}^t$, is a function of the difference in unit membership between g and h. Specifically, if $radios_u$ is the number of radios assigned to unit u,

$$cost_{gh}^t = \sum_{u \in h \backslash g} radios_u \quad \forall\, (g, h, t) \in A. \tag{19}$$

That is, the cost from g to h is the number of radios from units that are in group h but not in group g. This method of calculating costs prevents double-counting when a unit moves from an existing channel to a new channel. Our objective function minimizes the sum total costs of associating each group g at t with group h at $t+1$:

$$\min_{Y} \sum_{(g,h,t) \in A} cost_{gh}^t Y_{gh}^t. \tag{20}$$

Note this cost function assumes all units and radios have the same importance, but that need not be the case: one could associate scalar weights with each radio or to an entire unit to model its relative importance.

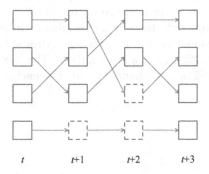

Fig. 2. Example of the association of groups (blue boxes) at each time step. Virtual groups (comprising no units) are represented by dashed boxes. (Color figure online)

4.2   Decomposition Solution Method and Results

To solve the MC-CAP-T, we use a decomposition approach based on the key insight that the actual channel number (or color, or any other label) is arbitrary. We also observe that the cost of changing assignment of a group from t to $t+1$ depends only on the unit membership of each group at t and $t+1$; i.e., the costs can be decomposed by time step. These properties allow us to decompose the problem by time step while maintaining global convergence.

Figure 2 provides a visual representation of the process of associating groups at each time step, where for each time step the column of squares on the left represents groups g and on the right groups h. The number of groups (and their unit membership) is determined by the solutions from the MO-CAP, so some time steps may have more or fewer groups than others. For those time steps with fewer groups than the maximum, we create *virtual groups* (indicated in Fig. 2 by dashed boxes), representing a placeholder group with no assigned units. In this sense, a group represents both a collection of units to be assigned the same channel, and a placeholder for the channel itself, i.e., $|G|$ is equal to the maximum number of available channels across all time steps in the scenario.

At each time step, each group g must be associated with a group h, indicated by gray lines between groups in Fig. 2. When a real group g (i.e., comprising units) is associated with a virtual group h, no cost is incurred because the units in g are assigned to other groups (not in h) at $t+1$. When a virtual group g is associated with a real group at h, the cost equals the number of radios in h, since, according to (18), each unit in h was previously assigned a different group.

We implement our solution in Python. We first calculate all of the $cost_{gh}^t$ values for each possible $(g, h, t) \in A$, and then solve a classic integer *assignment problem* at each time step using a variation of the Hungarian (or Munkres) algorithm [21], which solves to optimality in $\mathcal{O}\left(n^3\right)$ time. Global convergence is maintained because at each time step, the cost of channel changes depends only on the assignments at t and $t+1$. The actual assigned channel (i.e., its number) is arbitrary, since all channels provide the same performance and each group must have a channel. Thus this formulation exhibits *optimal substructure* that allows us to efficiently solve each time step to optimality and then combine our results to solve the entire problem to optimality.

Note that in this approach, there is no variable or index representing a particular channel; the association Y_{gh}^t implies one. After solving the problem, the *paths* created by associating each g with an h at the next time step represent discrete channels. By assigning a channel number to each of these paths (i.e., the gray lines in Fig. 2), we effectively solve the MC-CAP-T. The following pseudo-code describes our algorithm for solving the problem:

Algorithm 1. *MC-CAP-T*

Input: MO-CAP solutions at each time step

Output: $X_u^{ct}, \forall u \in U, c \in C, t \in T$ (unit channel assignments for all time steps)

begin

    Calculate $cost_{gt}^t, \forall (g, h, t) \in A$

    $channel \leftarrow 1$

    for $g \in G : t = 1$

        $\Gamma_g \leftarrow channel$ // Assign channels to groups during first time step

        $channel \leftarrow channel + 1$

    next;

    for $t = 1, 2, \ldots, t - 1$

        Solve the MC-CAP-T for t using Hungarian / Munkres algorithm

        Store Y_{gh}^t values

        for $g, h \in (g, h, t)$

            if $Y_{gh}^t = 1$

                $\Gamma_h \leftarrow \Gamma_g$ // Assign channels to groups for time step t

            endif;

        next;

    next;

    for $g \in G$

        for $u \in g$

            $X_u^{\Gamma_g^t} \leftarrow 1$ // Assign channels to units

        next;

    next;

end;

We solve for each time step in the Marine Corps scenario. The naïve method requires a total of 33,340 channel changes, whereas our decomposition method (which solves to optimality in less than 53 s) requires 21,915 channel changes, a reduction of 34%. Figure 3 is a method of visualizing the results of this comparison. For both the naïve and decomposition methods, a row represents a unit, where reddish units are larger (comprising up to 25 radios each) and greenish units are smaller, each column represents a time step, and a blank entry indicates that no channel change is required for that unit at that time step. This visualization provides a qualitative sense of how much better the decomposition method (which provides an exact solution) is at reducing channel changes, especially for larger (and thus more penalizing) units.

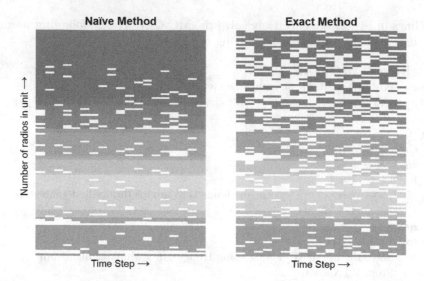

Fig. 3. Results of MC-CAP-T, where each row represents a unit, and each column represents a time step. White indicates that the channel assignment remains the same (i.e., no cost), and color indicates that a different channel is assigned at the next time step. Red indicates larger units (more radios); green indicates smaller units. (Color figure online)

5   Conclusions and Future Research

We present new integer optimization and constraint programming methodologies that solve large, realistic instances of the minimum-order and minimum-cost channel assignment problems to global or near-global optimality in reasonable amounts of time. Our approach can be used to support military spectrum managers who must quickly make spectrum allocation decisions in congested EM environments. Our ongoing and future research explores robustness and resiliency in the presence of an adversary determined to jam portions of the EM spectrum (see, e.g., [10, 22, 29, 39]).

References

1. Aardal, K.I., Van Hoesel, S.P., Koster, A.M., Mannino, C., Sassano, A.: Models and solution techniques for frequency assignment problems. Ann. Oper. Res. **153**(1), 79–129 (2007)
2. Alion Science and Technology Corporation: TIREM RF Modeling (2017). https://www.alionscience.com/terrain-integrated-rough-earth-model-tirem/
3. Analytical Graphics Inc.: Engineering Tools (2018). http://www.agi.com/products/engineering-tools
4. Berry, L.: The potential contribution of optimum frequency assignment to efficient use of the spectrum. In: IEEE International Symposium on Electromagnetic Compatibility, pp. 409–412. IEEE (1990)

5. Bron, C., Kerbosch, J.: Algorithm 457: finding all cliques of an undirected graph. Commun. ACM **16**(9), 575–577 (1973)
6. Cuppini, M.: A genetic algorithm for channel assignment problems. Eur. Trans. Telecommun. **5**(2), 285–294 (1994)
7. Daniels, K., Chandra, K., Liu, S., Widhani, S.: Dynamic channel assignment with cumulative co-channel interference. ACM SIGMOBILE Mob. Comput. Commun. Rev. **8**(4), 4–18 (2004)
8. Department of Defense: Integrated Security Construct-B. Multi-Service Force Deployment, scenario 3 (2013)
9. Dunkin, N., Bater, J., Jeavons, P., Cohen, D.: Towards high order constraint representations for the frequency assignment problem. University of London, Egham, Surrey, UK, Technical report (1998)
10. El-Bardan, R., Brahma, S., Varshney, P.K.: Power control with jammer location uncertainty: a game theoretic perspective. In: 48th Annual Conference on Information Sciences and Systems, pp. 1–6. IEEE (2014)
11. Fischetti, M., Lepschy, C., Minerva, G., Romanin-Jacur, G., Toto, E.: Frequency assignment in mobile radio systems using branch-and-cut techniques. Eur. J. Oper. Res. **123**(2), 241–255 (2000)
12. Github: CAP datasets (2019). https://github.com/nickelpickle1/cap_dataset/
13. Goldstein, P.: Pentagon strikes deal with broadcasters, clearing way for 1755–1780 MHz auction. Fierce Wireless, February 2013. http://www.fiercewireless.com
14. Hagberg, A.A., Schult, D.A., Swart, P.J.: Exploring network structure, dynamics, and function using NetworkX. In: Proceedings of the 7th Python in Science Conference (SciPy 2008), pp. 11–15. Pasadena, CA USA, August 2008
15. Hart, W.E., Laird, C., Watson, J.P., Woodruff, D.L.: Pyomo-Optimization modeling in Python, vol. 67. Springer Science & Business Media, New York (2012). https://doi.org/10.1007/978-1-4614-3226-5
16. Hooker, J.N., Ottosson, G.: Logic-based Benders decomposition. Math. Program. **96**(1), 33–60 (2003)
17. IBM: IBM CPLEX Optimization Studio (2018). http://www-01.ibm.com/software/commerce/optimization/CPLEX-optimizer/
18. Johannson, F., et al.: mpmath: A Python library for arbitrary-precision floating-point arithmetic (version 0.18) (2013). http://mpmath.org
19. Katzela, I., Naghshineh, M.: Channel assignment schemes for cellular mobile telecommunication systems: a comprehensive survey. IEEE J. Pers. Commun. **3**(3), 10–31 (1996)
20. Kiddoo, J., et al.: Operations research enables auction to repurpose television spectrum for next-generation wireless technologies. INFORMS J. Appl. Anal. (2018, submitted)
21. Kuhn, H.W.: The Hungarian method for the assignment problem. Naval Res. Logistics Q. **2**(1–2), 83–97 (1955)
22. London, J.P.: The new wave of warfare-Battling to dominate the electromagnetic spectrum. J. Electron. Defense (JED) **38**(9), 68–76 (2015)
23. Mannino, C., Sassano, A.: An enumerative algorithm for the frequency assignment problem. Discrete Appl. Math. **129**(1), 155–169 (2003)
24. Margot, F.: Symmetry in integer linear programming. In: Jünger, M., et al. (eds.) 50 Years of Integer Programming 1958–2008, pp. 647–686. Springer, Heidelberg (2010). https://doi.org/10.1007/978-3-540-68279-0_17
25. Metzger, B.: Spectrum management technique. In: 38th National ORSA Meeting (1970)

26. Nicholas, P.J.: Optimal spectrum allocation to support tactical mobile ad-hoc networks. Ph.D. thesis, George Mason University (2016)
27. Nicholas, P.J., Hoffman, K.L.: Computational challenges of dynamic channel assignment for military MANET. In: Proceedings of the Military Communications Conference (MILCOM), pp. 1150–1157. IEEE (2015)
28. Nicholas, P.J., Hoffman, K.L.: Optimal channel assignment for military MANET using integer optimization and constraint programming. In: Proceedings of the Military Communications Conference (MILCOM), pp. 1114–1120. IEEE (2016)
29. Nicholas, P.J., Hoffman, K.L.: Analysis of spectrum allocation to support mobile ad-hoc networks in contested environments. In: Proceedings of the Military Communications Conference (MILCOM), pp. 145–150. IEEE (2018)
30. O'Neil, R.J., Hoffman, K.: Integer models for the asymmetric traveling salesman problem with pickup and delivery (2018). optimization-online.org
31. Ostrowski, J., Linderoth, J., Rossi, F., Smriglio, S.: Orbital branching. Math. Program. **126**(1), 147–178 (2011)
32. Palpant, M., Oliva, C., Artigues, C., Michelon, P., Didi Biha, M.: Models and methods for frequency assignment with cumulative interference constraints. Int. Trans. Oper. Res. **15**(3), 307–324 (2008)
33. Poisel, R.: Modern Communications Jamming: Principles and Techniques. Artech House (2011)
34. Ramani, A., Aloul, F.A., Markov, I.L., Sakallah, K.A.: Breaking instance-independent symmetries in exact graph coloring. In: Proceedings of the Design, Automation and Test in Europe Conference and Exhibition, vol. 1, pp. 324–329. IEEE (2004)
35. Selyukh, A.: In switch, U.S. military offers to share airwaves with industry, Thomson Reuters (2013)
36. Tomita, E., Tanaka, A., Takahashi, H.: The worst-case time complexity for generating all maximal cliques and computational experiments. Theor. Comput. Sci. **363**(1), 28–42 (2006)
37. Trick, M.: Sports scheduling. In: Van Hentenryck, P., Milano, M. (eds.) Hybrid Optimization: The Ten Years of CPAIOR, pp. 489–508. Springer Science & Business Media, New York (2010)
38. Wang, S.W., Rappaport, S.S.: Signal-to-interference calculations for balanced channel assignment patterns in cellular communications systems. IEEE Trans. Commun. **37**(10), 1077–1087 (1989)
39. Wu, Y., Wang, B., Liu, K.R., Clancy, T.C.: Anti-jamming games in multi-channel cognitive radio networks. IEEE J. Sel. Areas Commun. **30**(1), 4–15 (2012)

Heat Exchanger Circuitry Design by Decision Diagrams

Nikolaos Ploskas[1], Christopher Laughman[2], Arvind U. Raghunathan[2],
and Nikolaos V. Sahinidis[3]($\boxtimes$)

[1] Department of Informatics and Telecommunications Engineering,
University of Western Macedonia, 50100 Kozani, Greece
nploskas@uowm.gr
[2] Mitsubishi Electric Research Laboratories, Cambridge, MA 02139, USA
{laughman,raghunathan}@merl.com
[3] Department of Chemical Engineering,
Carnegie Mellon University, Pittsburgh, PA 15213, USA
sahinidis@cmu.edu

Abstract. The interconnection pattern between the tubes of a tube-fin heat exchanger, also referred to as its circuitry, has a significant impact on its performance. We can improve the performance of a heat exchanger by identifying optimized circuitry designs. This task is difficult because the number of possible circuitries is very large, and because the dependence of the heat exchanger performance on the input (i.e., a given circuitry) is highly discontinuous and nonlinear. In this paper, we propose a novel decision diagram formulation and present computational results using the mixed integer programming solver CPLEX. The results show that the proposed approach has a favorable scaling with respect to number of tubes in the heat exchanger size and produces configurations with 9% higher heat capacity, on average, than the baseline configuration.

Keywords: Optimization · Decision diagram ·
Heat exchanger design · Refrigerant circuitry · Heat capacity

1 Introduction

Heat exchanger performance is important in many systems, ranging from the heating and air-conditioning systems that are widely used in residential and commercial applications, to plant operation for process industries. A variety of shapes and configurations can be used for the constituent components of the heat exchanger, depending on its application [2]. The most common configuration in heating and air-conditioning is the crossflow fin-and-tube type. In this type, a refrigerant flows through a set of pipes and moist air flows across a possibly enhanced surface on the other side of the pipe, allowing thermal energy to be transferred between the air and the refrigerant.

© Springer Nature Switzerland AG 2019
L.-M. Rousseau and K. Stergiou (Eds.): CPAIOR 2019, LNCS 11494, pp. 461–471, 2019.
https://doi.org/10.1007/978-3-030-19212-9_30

Heat exchanger performance can be improved according to a number of different metrics; these typically include maximization of heating or cooling capacity, size reduction, component material reduction, manufacturing cost reduction, reduction of pumping power, or a combination of these metrics. While the concept of many of these metrics is reasonably straightforward (e.g., size reduction and manufacturing cost reduction), the heat capacity is influenced by various parameters (like the geometry of the heat exchanger and the inlet conditions) and the dependence of the heat exchanger performance on these parameters tends to be highly discontinuous and nonlinear.

The *circuitry* determines the sequence of tubes through which the refrigerant flows and has a significant influence on the thermal performance of the heat exchanger. As heat exchangers for contemporary air-source heat pumps often have between 60 and 200 tubes, design engineers are faced with a very large number of potential circuitry choices that must be evaluated to identify a suitable design that meets performance and manufacturing specifications. Current design processes typically involve the manual choice of the configuration based upon expert knowledge and the results of an enumerated set of simulations. This task is inherently challenging, and does not guarantee that a manually found configuration will be optimal.

Systematic optimization of heat exchangers has been a long-standing research topic [3,4]. Circuitry optimization is a particularly challenging task because: (i) the search space is enormous, making exhaustive search algorithms impractical for large numbers of tubes, and (ii) there is a highly discontinuous and nonlinear relationship between the circuitry design and the heat exchanger performance. Many researchers [5–9] have studied the effect of improving the refrigerant circuitry, and have concluded that circuitry optimization is often more convenient and less expensive than optimizing the geometry of the fins and tubes. Moreover, it has also been found that the optimal circuitry design for a specific heat exchanger is different from that of other heat exchangers [10].

A variety of methods have thus been proposed to tackle the circuitry optimization problem [11–17]. These methods generally require either a significant amount of time to find the optimal circuitry design or generate a circuitry that is difficult to manufacture. In [1], we presented a binary constrained formulation for the heat exchanger circuitry optimization problem that generates circuitry designs without requiring extensive domain knowledge. Derivative-free optimization algorithms were applied to optimize heat exchanger performance and constraint programming methods were used to verify the results for small heat exchangers.

In this paper, we extend our work in [1] by providing a novel *relaxed* decision diagram formulation for the heat exchanger circuitry optimization problem. Decision diagrams have played a variety of roles in discrete optimization [18–31]. In a number of applications the decision diagram formulation has vastly outperformed existing formulations [20,23,27,29]. Our new formulation produces smaller optimization instances and is able to find optimized circuitry configurations on heat exchangers with 128 tubes. In contrast, the approach in [1] could only optimize coils up to 36 tubes.

The remainder of this paper is organized as follows. In Sect. 2, we present circuitry design principles of a heat exchanger. Section 3 describes the proposed formulation for optimizing the performance of heat exchangers. Section 4 presents the computational experiments on finding the best circuitry arrangements on various heat exchangers and also provides a discussion on the advantages of the formulation. Conclusions from the research are presented in Sect. 5.

2   Heat Exchanger Circuitry

In this paper, we assume that all geometric and inlet parameters are predefined. As described in the introduction, the main problem of interest is to determine the circuitry configuration that optimizes the heat exchanger performance. This configuration, which is typically realized during the manufacturing process, includes both the circuitry design and the identification of the inlet and outlet tubes. Figure 1(a) is an illustration of the circuitry for a representative heat exchanger. The manufacturing process for fin-tube heat exchangers typically proceeds by first stacking layers of aluminum fins together that contain preformed holes, and then press-fitting copper tubes into each set of aligned holes. The copper tubes are often pre-bent into a U shape before insertion, so that two holes are filled at one time. After all of the tubes are inserted into the set of aluminum fins, the heat exchanger is flipped over and the other ends of the copper tubes are connected in the desired circuitry pattern. Figures 1(b) and (c) illustrate circuitry configurations for a heat exchanger of eight tubes. A crossed sign inside a circle indicates that the refrigerant flows into the page, while a dotted sign indicates that the refrigerant flows out of the page. There are two types of connections: (i) a connection at the far end of the tubes, and (ii) a connection at the front end of the tubes. Therefore, a dotted line between two tubes represents a connection

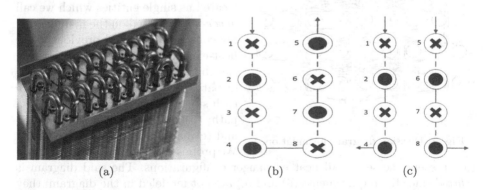

(a)   (b)   (c)

Fig. 1. (a) Illustration of heat exchanger (Image licensed from S. S. Popov/ Shutterstock.com). (b) and (c) are examples of valid circuitry configurations with one and two circuits, respectively.

on the far end (pre-bent tube), while a solid line represents a connection on the front end of the tubes. In this example, the pairs of tubes 1–2, 3–4, 5–6 and 7–8 are the pre-connected tubes (i.e. tubes with bends on the far end of the coil). In Fig. 1(b), the inlet stream is connected to tube 1 and outlet stream is connected to tube 5. In Fig. 1(c), tubes 1 and 5 are connected to inlet streams, while tubes 4 and 8 are connected to outlet streams. A given *circuit* is a set of pipes through which the refrigerant flows from inlet to outlet. Figures 1(b) and (c) depict circuitry configuration with one and two circuits, respectively.

A set of realistic manufacturing constraints are imposed on the connections of the tubes: (i) adjacent pairs of tubes in each column, starting with the bottom tube, are always connected (this constraint is imposed by the manufacturing process since one set of bends on the far end are applied to the tubes before they are inserted into the fins), (ii) the connections on the far end cannot be across rows unless they are at the edge of the coil, (iii) plugged tubes, i.e., tubes without connections, are not allowed, (iv) inlets and outlets must always be located at the near end, and (v) merges and splits are not allowed.

3   Decision Diagram Formulation

In [1], we proposed a new approach for formulating the refrigerant circuitry design problem. We formulated the problem as a binary constrained optimization problem with a black-box objective function and we applied derivative-free optimization algorithms to solve this problem. Each connection was represented using a binary variable and cycles were excluded by adding inequality constraints. As a result, the constraint matrix was dense. In this paper, we propose a new formulation based on decision diagrams.

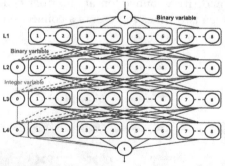

Fig. 2. Decision diagram formulation.

The main idea is that pre-connected tubes (i.e., tubes with bends on the far end of the coil) are treated as single entities which we call *super-nodes*. Based on the manufacturing constraint outlined previously, the heat-exchanger circuitry configuration can be defined as: (a) a collection of paths involving super-nodes where each super-node occurs only once in a path; (b) paths cover all super-nodes; and (c) paths are super-node disjoint. We propose a *relaxed* decision diagram to represent the set of all heat exchanger configurations. The said diagram is *relaxed* since the requirements (a) and (c) are not modeled in the diagram; they are not ignored though, rather will be enforced by additional constraints later on. Figure 2 shows such a relaxed decision diagram for a heat exchanger with eight tubes.

Let us assume that we have n tubes. The number of layers in the decision diagram is equal to $N = \frac{n}{2}$. The layers are indexed sequentially and every layer consists of the set of super-nodes. In addition, a 0-node is introduced into layers with index 2 and above. The 0-node represents the end of a circuit. Arcs are drawn between the nodes (collection of super-nodes and 0-node) of two successive layers. Root and terminal nodes are introduced that respectively connect to the first and last layers in the diagram. In this representation, a path from the root to terminal can repeat super-nodes (refer to Fig. 2). For example, the path (r, 1–2, 3–4, 0, 0, t) is a path satisfying (a), while the path (r, 1–2, 3–4, 1–2, 3–4, t) is a path that does not satisfy (a). However, we impose additional constraints that ensure that we identify configurations satisfying the requirements (a)–(c). The constraints ensure that the identified path is indeed a *circuit*.

Before presenting the mixed integer programming model derived from this decision diagram formulation, we introduce the following notation:

- N: the number of layers in the decision diagram
- L_i: represents the i-th layer in the decision diagram, where $i = 1, \ldots, N$
- s: super-nodes (not including 0-node)
- S: set of super-nodes
- r, t: the root and terminal nodes in the decision diagram
- (s, i) or $(0, i)$: node in layer i of decision diagram
- a: arcs in the decision diagram
- $\text{head}(a)$ ($\text{tail}(a)$): starting (ending) node of the arc in the decision diagram
- $A_{s,i}^{in}$ ($A_{s,i}^{out}$): set of input arcs to (output arcs from) super-node s in L_i
- $A_{0,i}^{in}$ ($A_{0,i}^{out}$): set of input arcs to (output arcs from) 0 in L_i
- $x_a \in \{0, 1\}$ for $a \in A(x) := \bigcup_{i=1}^{N} \bigcup_{s \in S} \left(A_{s,i}^{in} \cup A_{s,i}^{out} \right)$: binary variables encoding flow on the arcs between $s, s' \in S$ and flow on arcs between $s \in S$ and 0
- $z_a \in \{0, 1, \ldots\}$ for $a \in A(z) := \bigcup_{i=2}^{N} A_{0,i}^{out}$: integer variables encoding flow on the arcs between 0 in successive layers
- C_{lb}: the minimum number of circuits
- C_{ub}: the maximum number of circuits

Therefore, the mixed integer programming model derived from the decision diagram formulation can be expressed as:

$$\max \quad Q(x, z) \left(\text{ or } \tfrac{Q(x,z)}{\Delta P(x,z)} \right) \tag{1}$$

$$\text{s.t.} \quad \sum_{a \in A_{s,i}^{in}} x_a = \sum_{a \in A_{s,i}^{out}} x_a, \qquad \forall s \in S, i \in \{1, \ldots, N\} \tag{2}$$

$$\sum_{a \in A_{0,i}^{in}: \text{tail}(a) \in S} x_a + \sum_{a \in A_{0,i}^{in}: \text{tail}(a) = 0} z_a = \sum_{a \in A_{0,i}^{out}} z_a, \forall i = 2, \ldots, N \tag{3}$$

$$\sum_{i=1}^{N} \sum_{a \in A_{s,i}^{in}} x_a = 1, \qquad \forall s \in S \tag{4}$$

$$C_{lb} \leq \sum_{a \in \bigcup_{s \in S} A_{s,1}^{in}} x_a \leq C_{ub}, \tag{5}$$

$$x_a \in \{0, 1\}, \quad a \in A(x), \quad z_{a'} \in \mathbb{Z}, \qquad \forall a' \in A(z). \tag{6}$$

where $Q(x, z)$ is the heat capacity related to the solution vectors x and z, and ΔP is the pressure difference across the heat exchanger.

Two targets of the refrigerant circuitry optimization are considered in this work: (i) maximization of the heat capacity $(Q(x, z))$, and (ii) maximization of the ratio of the heat capacity to the pressure difference across the heat exchanger $(Q(x, z)/\Delta P(x, z))$. Constraint (2) is the flow balance for the super-nodes in all different levels, while constraint (3) is the flow balance for the 0-nodes. Note that $\{a \in A_{0,2}^{in} \mid tail(a) = 0\} = \emptyset$. Constraint (4) is imposed for each super-node s and invalidates any repetition of super-nodes, so there can be no cycles. Constraint (5) sets a limit on the number of circuits in the circuitry configuration.

The total number of variables in the decision diagram formulation is equal to $|S|^3 - |S|^2 + 3|S| - 1$ and the total number of constraints is equal to $|S|^2 + 2|S| + 2$. Table 1 shows the superiority of the proposed formulation compared to the one proposed in [1]. The formulation in [1] is memory bound, i.e., it requires an exponentially increasing amount of memory for only a constant increase in problem size. Moreover, the constraint matrix in [1] is dense, while the constraint matrix in the proposed formulation is sparse. For example, the constraint matrix for a heat exchanger with 40 tubes of the formulation in [1] needs $\sim 6\,GB$ of memory, while the constraint matrix of the proposed formulation needs only $\sim 300\,KB$.

Table 1. Reduction in the problem size of the proposed decision diagram formulation compared to the formulation proposed in [1]

# of tubes	Problem size of the formulation proposed in [1]	Problem size of the proposed formulation	Reduction in problem size (%)
16	263 × 120	471 × 82	−22%
24	4,107×276	1,619×170	76%
32	65,551×496	3,887×290	97%
40	1,048,595×780	7,659×442	100%
128	-	258,239×4,226	-

We provide a brief discussion on the advantages of the decision diagram based representation. The width and the depth (number of layers) of the diagram grows linearly in the number of tubes. This can be quite prohibitive in that it leads to a very dense formulation for heat exchangers with a large number of tubes. However, operational and manufacturing constraints help to alleviate this complexity. From practical operational considerations, it is not desirable to have long circuits since they incur large pressure drops and increased costs (pump power) to flow the refrigerant. Hence, from pressure drop considerations it is desirable to limit the depth of the diagram and this can be easily accomplished by truncating the diagram. From manufacturing considerations it is not desirable to allow connections between all pairs of tubes. This can also be accomplished by eliminating arcs between pairs of super-nodes for which a connection is forbidden. This results in a much sparser diagram and an associated integer program which is smaller in size. We will explore these aspects in a future work.

4   Computational Results

In order to validate the proposed model, we performed a computational study with the aim of optimizing the heat capacity $(Q(x, z))$ and the ratio of the heat capacity to the pressure difference $(Q(x, z)/\Delta P(x, z))$ across the heat exchanger. The analytical form of $Q(x, z)$ and $Q(x, z)/\Delta P(x, z)$ as a function of x, z is typically not available and hence, the optimization problem in (1)–(6) cannot be solved by mixed integer programming solvers such as CPLEX. The quantities $Q(x, z)$ and $Q(x, z)/\Delta P(x, z)$ can only be obtained by specifying a particular circuitry configuration defined by x, z as input to a heat exchanger simulation program such as CoilDesigner [32]. CoilDesigner is a steady-state simulation and design tool for air to refrigerant heat exchangers, to simulate the performance of different refrigerant circuitry designs.

We replaced the objective in (1) by a constant and applied the mixed integer programming solver CPLEX on (2)–(6) to produce 2,500 feasible circuitry configurations. In order to achieve that, we used the appropriate CPLEX parameters to create a diverse solution pool of 2,500 feasible solutions for this problem (parameters: PopulateLim, SolnPoolCapacity, and SolnPoolReplace). Then, we evaluated all the feasible configurations using CoilDesigner. We created a test suite with seven circuitry architectures with a varying number of tubes. The structural parameters and work conditions for all instances are the same; the only difference between the test cases is in the number of tubes per row, ranging from 2 to 64 that correspond in heat exchangers with 4 to 128 tubes.

Tables 2 and 3 present the results of the optimization of the two objective functions, $Q(x, z)$ and $Q(x, z)/\Delta P(x, z)$, respectively. We compare the optimized results generated by the proposed approach with the results in [1] and the results obtained using a baseline configuration, which includes two circuits: one that connects all tubes in the first column of the coil and one that connects all tubes in the second column of the coil. The inlet tubes of the baseline configuration are the tubes in the first row and the outlet tubes are the tubes in the last row. The baseline configuration is a heat exchanger design that is typically used in practice today. The results show that the proposed approach can generate optimized configurations in a short amount of time. On average, the proposed approach produces configurations with 4% and 9% higher heat capacity than the approach in [1] and the baseline configuration, respectively. In addition, the proposed approach produces configurations with 90% and $8,826\%$ higher ratio of the heat capacity to the pressure difference than the approach in [1] and the baseline configuration, respectively. It is worth noting that the formulation in [1] can only be used to solve problems with up to 36 coils, while the present formulation can be used to solve much larger problems. Therefore, the current approach not only produces better results than the approach in [1] but it can also solve much larger problems. In addition, the current approach needs an order of magnitude less execution time than the approach proposed in [1].

The limit of function evaluations (2,500) prevents the proposed approach from finding even better results for very large coils. When optimizing $Q(x, z)$, the best circuitry configurations include a small number of long circuits and the circuits usually contain at least one connection between tubes across columns. On the other hand, when optimizing $Q(x, z)/\Delta P(x, z)$, the best circuitry configurations include many circuits that are not very long and most of the connections on these circuits are between adjacent tubes. This was expected since longer circuits incur more pressure drop.

Table 2. Computational results for Q optimization

# of tubes	Baseline Q	Optimized Q	Optimized Q in [1]	Improvement over baseline (%)	Improvement over [1] (%)
4	1,388	1,754	1,754	26%	0%
8	1,884	2,189	2,017	16%	9%
16	2,179	2,391	2,230	10%	7%
24	2,249	2,353	2,294	5%	3%
32	2,234	2,269	2,244	2%	1%
40	2,154	2,255	-	5%	-
128	9,694	9,790	-	1%	-

Table 3. Computational results for $Q/\Delta P$ optimization

# of tubes	Baseline $\frac{Q}{\Delta P}$	Optimized $\frac{Q}{\Delta P}$	Optimized $\frac{Q}{\Delta P}$ in [1]	Improvement over baseline (%)	Improvement over [1] (%)
4	3,727	3,727	3,727	0%	0%
8	2,640	14,664	12,464	455%	18%
16	1,668	51,219	32,985	2,971%	55%
24	1,162	110,289	47,865	9,391%	130%
32	854	156,181	45,292	18,188%	245%
40	632	193,654	-	30,541%	-
128	448	1,518	-	239%	-

5   Conclusions

The performance of a heat exchanger can be significantly improved by optimizing its circuitry configuration. Design engineers currently select the circuitry design based on their domain knowledge and some simulations. However, the design of an optimized circuitry is difficult and needs a systematic approach to be used. In this paper, we extended our work in [1] and proposed a novel decision diagram formulation for the circuitry optimization problem. The generated mixed integer programming problem is much smaller than the problem derived from the formulation in [1] and leads us to optimize coils with a very large number of tubes. We applied CPLEX to generate feasible configurations for seven different heat exchangers and we evaluated them using CoilDesigner. The results show that the proposed formulation can improve the baseline configuration by 9% for the heat capacity and by 8,826% for the ratio of the heat capacity to the pressure difference than the baseline configuration. Finally, the proposed approach produces on average 4% higher heat capacity and 90% higher ratio of the heat capacity to the pressure difference than the approach in [1].

References

1. Ploskas, N., Laughman, C., Raghunathan, A.U., Sahinidis, N.V.: Optimization of circuitry arrangements for heat exchangers using derivative-free optimization. Chem. Eng. Res. Des. **131**, 16–28 (2018). https://doi.org/10.1016/j.cherd.2017.05.015
2. Hewitt, G.F., Shires, G.L., Bott, T.R.: Process Heat Transfer, vol. 113. CRC Press, Boca Raton (1994)
3. Fax, D.H., Mills, R.R.: Generalized optimal heat exchanger design. ASME Trans. **79**, 653–661 (1957)
4. Hedderich, C.P., Kelleher, M.D., Vanderplaats, G.N.: Design and optimization of air-cooled heat exchangers. J. Heat Transfer **104**, 683–690 (1982)
5. Liang, S.Y., Wong, T.N., Nathan, G.K.: Study on refrigerant circuitry of condenser coils with exergy destruction analysis. Appl. Therm. Eng. **20**, 559–577 (2000). https://doi.org/10.1016/s1359-4311(99)00043-5
6. Wang, C.C., Jang, J.Y., Lai, C.C., Chang, Y.J.: Effect of circuit arrangement on the performance of air-cooled condensers. Int. J. Refrig. **22**, 275–282 (1999). https://doi.org/10.1016/s0140-7007(98)00065-6
7. Yun, J.Y., Lee, K.S.: Influence of design parameters on the heat transfer and flow friction characteristics of the heat exchanger with slit fins. Int. J. Heat Mass Transf. **43**, 2529–2539 (2000). https://doi.org/10.1016/s0017-9310(99)00342-7
8. Liang, S.Y., Wong, T.N., Nathan, G.K.: Numerical and experimental studies of refrigerant circuitry of evaporator coils. Int. J. Refrig. **24**, 823–833 (2001). https://doi.org/10.1016/s0140-7007(00)00050-5
9. Matos, R.S., Laursen, T.A., Vargas, J.V.C., Bejan, A.: Three-dimensional optimization of staggered finned circular and elliptic tubes in forced convection. Int. J. Therm. Sci. **43**, 477–487 (2004). https://doi.org/10.1016/j.ijthermalsci.2003.10.003
10. Domanski, P.A., Yashar, D., Kim, M.: Performance of a finned-tube evaporator optimized for different refrigerants and its effect on system efficiency. Int. J. Refrig. **28**, 820–827 (2005). https://doi.org/10.1016/j.ijrefrig.2005.02.003
11. Domanski, P.A., Yashar, D., Kaufman, K.A., Michalski, R.S.: An optimized design of finned-tube evaporators using the learnable evolution model. HVAC&R Res. **10**, 201–211 (2004). https://doi.org/10.1080/10789669.2004.10391099

12. Domanski, P.A., Yashar, D.: Optimization of finned-tube condensers using an intelligent system. Int. J. Refrig. **30**, 482–488 (2007). https://doi.org/10.1016/j.ijrefrig.2006.08.013

13. Wu, Z., Ding, G., Wang, K., Fukaya, M.: Application of a genetic algorithm to optimize the refrigerant circuit of fin-and-tube heat exchangers for maximum heat transfer or shortest tube. Int. J. Therm. Sci. **47**, 985–997 (2008). https://doi.org/10.1016/j.ijthermalsci.2007.08.005

14. Bendaoud, A.L., Ouzzane, M., Aidoun, Z., Galanis, N.: A new modeling procedure for circuit design and performance prediction of evaporator coils using CO2 as refrigerant. Appl. Energy **87**, 2974–2983 (2010). https://doi.org/10.1016/j.apenergy.2010.04.015

15. Lee, W.J., Kim, H.J., Jeong, J.H.: Method for determining the optimum number of circuits for a fin-tube condenser in a heat pump. Int. J. Heat Mass Transf. **98**, 462–471 (2016). https://doi.org/10.1016/j.ijheatmasstransfer.2016.02.094

16. Yashar, D.A., Lee, S., Domanski, P.A.: Rooftop air-conditioning unit performance improvement using refrigerant circuitry optimization. Appl. Therm. Eng. **83**, 81–87 (2015). https://doi.org/10.1016/j.applthermaleng.2015.03.012

17. Cen, J., Hu, J., Jiang, F.: An automatic refrigerant circuit generation method for finned-tube heat exchangers. Can. J. Chem. Eng. (2018). https://doi.org/10.1002/cjce.23150

18. Andersen, H.R., Hadzic, T., Hooker, J.N., Tiedemann, P.: A constraint store based on multivalued decision diagrams. In: Bessière, C. (ed.) CP 2007. LNCS, vol. 4741, pp. 118–132. Springer, Heidelberg (2007). https://doi.org/10.1007/978-3-540-74970-7_11

19. Behle, M.: Binary decision diagrams and integer programming. Ph.D. thesis, Saarland University (2007)

20. Bergman, D., Cire, A.A., van Hoeve, W.J., Hooker, J.N.: Discrete optimization with decision diagrams. INFORMS J. Comput. **28**(1), 47–66 (2016). https://doi.org/10.1287/ijoc.2015.0648

21. Bergman, D., Cire, A.A., van Hoeve, W.J., Hooker, J.N.: Decision Diagrams for Optimization. Artificial Intelligence: Foundations, Theory, and Algorithms, 1st edn. Springer, Cham (2016). https://doi.org/10.1007/978-3-319-42849-9

22. Bergman, D., Cire, A.A.: Discrete nonlinear optimization by state-space decompositions. Manage. Sci. **64**(10), 4700–4720 (2017). https://doi.org/10.1287/mnsc.2017.2849

23. Cire, A.A., van Hoeve, W.J.: Multivalued decision diagrams for sequencing problems. Oper. Res. **61**(6), 1411–1428 (2013). https://doi.org/10.1287/opre.2013.1221

24. Davarnia, D., van Hoeve, W.J.: Outer approximation for integer nonlinear programs via decision diagrams (2018). http://www.optimization-online.org/DB_HTML/2018/03/6512.html

25. Haus, U.U., Michini, C., Laumanns, M.: Scenario aggregation using binary decision diagrams for stochastic programs with endogenous uncertainty. CoRR abs/1701.04055, https://arxiv.org/abs/1701.04055 (2017)

26. Hooker, J.N.: Decision diagrams and dynamic programming. In: Gomes, C., Sellmann, M. (eds.) CPAIOR 2013. LNCS, vol. 7874, pp. 94–110. Springer, Heidelberg (2013). https://doi.org/10.1007/978-3-642-38171-3_7

27. Lozano, L., Smith, J.C.: A binary decision diagram based algorithm for solving a class of binary two-stage stochastic programs. Math. Program. (2018). https://doi.org/10.1007/s10107-018-1315-z

28. Morrison, D.R., Sewell, E.C., Jacobson, S.H.: Solving the pricing problem in a branch-and-price algorithm for graph coloring using zero-suppressed binary decision diagrams. INFORMS J. Comput. **28**(1), 67–82 (2016). https://doi.org/10.1287/ijoc.2015.0667
29. Raghunathan, A.U., Bergman, D., Hooker, J.N., Serra, T., Kobori, S.: Seamless multimodal transportation scheduling. CoRR abs/1807.09676 https://arxiv.org/abs/1807.09676 (2018)
30. Serra, T., Hooker, J.N.: Compact representation of near-optimal integer programming solutions (2017). http://www.optimization-online.org/DB_HTML/2017/09/6234.html
31. Tjandraatmadja C., van Hoeve, W.J.: Target cuts from relaxed decision diagrams. INFORMS J. Comput. (2018, to appear)
32. Jiang, H., Aute, V., Radermacher, R.: CoilDesigner: a general-purpose simulation and design tool for air-to-refrigerant heat exchangers. Int. J. Refrig. **29**, 601–610 (2006). https://doi.org/10.1016/j.ijrefrig.2005.09.019

Column Generation for Real-Time Ride-Sharing Operations

Connor Riley[1], Antoine Legrain[2], and Pascal Van Hentenryck[1(✉)]

[1] Georgia Institute of Technology, Atlanta, GA 30332, USA
ctriley@gatech.edu, pvh@isye.gatech.edu
[2] Polytechnique Montréal, Montreal, QC H3T 1J4, Canada
antoine.legrain@polymtl.ca

Abstract. This paper considers real-time dispatching for large-scale ride-sharing services over a rolling horizon. It presents RTDARS which relies on a column-generation algorithm to minimize wait times while guaranteeing short travel times and service for each customer. Experiments using historic taxi trips in New York City for instances with up to 30,000 requests per hour indicate that the algorithm scales well and provides a principled and effective way to support large-scale ride-sharing services in dense cities.

Keywords: Real-time dial-a-ride · Large-scale optimization

1 Introduction

In the past decade, commercial ride-hailing services such as Didi, Uber, and Lyft have decreased reliance on personal vehicles and provided new mobility options for various population segments. More recently, ride-sharing has been introduced as an option for customers using these services. Ride-sharing has the potential for significant positive impact since it can reduce the number of cars on the roads and thus congestion, decrease greenhouse emissions, and make mobility accessible to new population segments by decreasing trip prices. However, the algorithms used by commercial ride-sharing services rarely use state-of-the-art techniques, which reduces the potential positive impact. Recent research by Alonso-Mora et al. [1] has shown the benefits of more sophisticated algorithms. Their algorithm uses shareability graphs and cliques to generate all possible routes and a MIP model to select the routes. They impose significant constraints on waiting times (e.g., 420 s), which reduces the potential riders to consider for each route at the cost of rejecting customers.

This paper considers large-scale ride-sharing services where *customers are always guaranteed a ride*, in contrast to prior work. The Real-Time Dial-A-Ride System (RTDARS) divides the days into short time periods called epochs, batches requests in a given epoch, and then schedules customers to minimize

© Springer Nature Switzerland AG 2019
L.-M. Rousseau and K. Stergiou (Eds.): CPAIOR 2019, LNCS 11494, pp. 472–487, 2019.
https://doi.org/10.1007/978-3-030-19212-9_31

average waiting times. RTDARS makes a number of modeling and solving contributions. At the modeling level, RTDARS has the following innovations:

1. RTDARS follows a Lagrangian approach, relaxing the constraint that all customers must be served in the static optimization problem of each epoch. Instead, RTDARS associates a penalty with each rider, representing the cost of not serving the customer.
2. To balance the minimization of average waiting times and ensure that the waiting time of every customer is reasonable, RTDARS increases the penalty of an unserved customer in the next epoch, making it increasingly harder not to serve the waiting rider.
3. RTDARS exploits a key property of the resulting formulation to reduce the search space explored for each epoch.
4. To favor ride-sharing, RTDARS uses the concept of virtual stops used in the RITMO project [12] and being adopted by ride-sourcing services.

RTDARS solves the static optimization problem for each epoch with a column-generation algorithm based on the three-index MIP formulation [6]. The main innovation here is the pricing problem which is organized as a series of waves, first considering all the insertions of a single customer, before incrementally adding more customers.

RTDARS was evaluated on historic taxi trips from the New York City Taxi and Limousine Commission [8], which contains large-scale instances with more than 30,000 requests an hour. The results show that RTDARS can provide service guarantees while improving the state-of-the-art results. For instance, for a fleet of 2,000 vehicles of capacity 4, RTDARS obtains an average wait of 2.2 min and an average deviation from the shortest path of 0.62 min. The results also show that large-occupancy vehicles (e.g., 8-passenger vehicles) provide additional benefits in terms of waiting times with negligible increases in in-vehicle time. RTDARS is also shown to generate a small fraction of the potential columns, explaining its efficiency. The Lagrangian modeling also helps in reducing computation times significantly.

The rest of this paper is organized as follows. Section 2 presents the related work in more detail. Section 3 describes the real-time setting. Section 4 specifies the static problem and gives the MIP formulation. Section 5 describes the column generation. Section 6 specifies the real-time operations. Section 7 presents the experimental results and Sect. 8 concludes the paper.

2   Related Work

Dial-a-ride problems have been a popular topic in operations research for a long time. Cordeau and Laporte [6] provided a comprehensive review of many of the popular formulations and the starting point of RTDARS's column generation is their three-index formulation. Constraint programming and large neighborhood search were also proposed for dial-a-ride problems (e.g., [4,7]). Progress in communication technologies and the emergence of ride-sourcing and ride-sharing services have stimulated further research in this area. Rolling horizons are often used to batch requests and were used in taxi pooling previously [10,11]. In addition, stochastic scenarios along with waiting and reallocation strategies have been previously explored in [2,3]. Bertsimas, Jaillet, and Martin [5] explored the taxi routing problem (without ride-sharing) and introduced a "backbone" algorithm which increases the sparsity of the problem by computing a set of candidate paths that are likely to be optimal. Alonso-Mora et al. proposed an anytime algorithm which uses cliques to generate vehicle paths combined with a vehicle rebalancing step to move vehicles towards demand [1]. Their "results show that 2,000 vehicles (15% of the taxi fleet) of capacity 10 or 3,000 of capacity 4 can serve 98% of the demand within a mean waiting time of 2.8 min and mean trip delay of 3.5 min." [1]. Both [1] and [5] use hard time windows to reject riders when they cannot serve them quickly enough (e.g., 420 s in the aforementioned results). This decision significantly reduces the search space as only close riders can be served by a vehicle. In contrast, RTDARS provides service guarantees for all riders, while still reducing the search space through a Lagrangian reformulation. The results show that RTDARS is capable of providing these guarantees while improving prior results in terms of average waiting times. Indeed, for 2,000 vehicles of capacity 4, RTDARS provides an average waiting time of 2.2 min with a standard deviation of 1.24 and a mean trip deviation of 0.62 min (standard deviation 1.13). For 3,000 vehicles of capacity 4, the average waiting time is further reduced to 1.81 min with a standard deviation of 1.03 and an average trip deviation of 0.23 min.

3   Overview of the Approach

RTDARS divides time into epochs, e.g., time periods of 30 s. During an epoch, RTDARS performs two tasks: It batches incoming requests and it solves the epoch optimization problem for all unserved customers from prior epochs. The epoch optimization takes, as inputs, these unserved customers and their penalties, as well as the *first* stop of each vehicle after the start of the epoch: Vehicle schedules prior to this stop are committed since, for safety reasons, RTDARS does not allow a vehicle to be re-routed once it has departed for its next customer. These first stops are called *departing stops* in this paper. All customers served before and up to the departing stops of the vehicles are considered served. All others, even if they were assigned a vehicle in the prior epoch optimization, are considered unserved.

Once the epoch is completed, a new schedule and a new set of requests are available. The schedule commits the vehicle routes for the entire next epoch and determines their next departing stops. The customer penalties are also updated to make it increasingly harder not to serve them. RTDARS then moves to the next epoch.

4   The Static Problem

This section defines and presents the static (generalized) dial-a-ride problem solved for each epoch. its objective is to schedule a set of requests on a given set of vehicles while ensuring that no customer deviates too much from their shortest trip time.

The inputs consist primarily of the vehicle and request data. The set of vehicles is denoted by V and each vehicle $v \in V$ is associated with a tuple $(u_0^v, w_0^v, I_v, T_v^B, T_v^E, Q_v)$, where u_0^v is the time the vehicle arrives at its *departing stop* for the epoch, w_0^v is the number of passengers currently in the vehicle, I_v is the set of dropoff requests for on-board passengers, T_v^B is the vehicle start time, T_v^E is the vehicle end time, and Q_v is the capacity of the vehicle. In other words, a vehicle v can only insert new requests after time u_0^v and it must serve the dropoffs in I_v. The request data is given in terms of a complete graph $\mathcal{G} = (\mathcal{N}, \mathcal{A})$, which contains the nodes for each possible pickup and delivery. There are five types of nodes: the pickup nodes $P = \{1, \ldots n\}$, their associated dropoff nodes $D = \{n+1, \ldots 2n\}$, the dropoff nodes $I = \cup_{v \in V} I_v$ of the passengers inside the vehicles, the source 0, and the sink s (the last node in terms of indices). Each node i is associated with a number of people q_i to pick up ($q_i > 0$) or drop off ($q_i < 0$) and the time $\Delta_i \geq 0$ it takes to perform them. If $i \in P$, then the corresponding delivery node is $n+i$ and $q_i = -q_{n+i}$. Also, q_i and Δ_i are zero for the source and the sink. Each node $i \in P$ is associated with a request, which is a tuple of the form (e_i, o_i, d_i, q_i) where e_i is the earliest possible pickup time, o_i is the pickup location, d_i is the dropoff location, and q_i is the number of passengers. Every request i in I is associated with the time u_i^P on which the request was picked up. Every request $i \in P \cup I$ is associated with the shortest time t_i from the request origin to its destination. Finally, the input contains a matrix $(t_{i,j})_{(i,j) \in \mathcal{A}}$ of travel times from any node i to any node j satisfying the triangle inequality, the constants α and β which constrain the deviation from the shortest path, and the penalty p_i of not serving the request $i \in P$.

A MIP model for the static problem is presented in Fig. 1. The MIP variables are as follows: u_i^v represents the time at which vehicle v arrives at node i, w_i^v the number of people in vehicle v when v leaves node i, x_{ij}^v denotes whether edge (i, j) is used by vehicle v, and z_i captures whether request $i \in P$ is served. Objective (1a) balances the minimization of wait times for every pickups with the penalties incurred by unserved riders. Note that the wait times for riders in I are not included in the objective because these riders are already in vehicles: only the constraints on their deviations must be satisfied. Constraints (1b) ensure that only one vehicle serves each request and that, if the request is not served, z_i is set to 1 to activate the penalty in the objective. Constraints (1c) are flow balance constraints. Constraints (1d) and (1e) are flow constraints for the source and the sink. Constraints (1f) ensure that every request is dropped off by the same vehicle that picks it up. Constraints (1g) ensure that every passenger currently in a vehicle is dropped off. Constraints (1h) define the arrival times at the nodes. Constraints (1i) and (1j) ensure that the vehicle is operational during its working hours. Constraints (1k) ensure that each rider is picked up no

$$\min \quad \sum_{i \in P} \sum_{v \in V} (u_i^v - e_i) + \sum_{i \in P} p_i z_i \tag{1a}$$

subject to

$$\left(\sum_{v \in V} \sum_{j \in \mathcal{N}} x_{ij}^v \right) + z_i = 1 \qquad\qquad \forall i \in P \tag{1b}$$

$$\sum_{j \in \mathcal{N}} x_{ij}^v = \sum_{j \in \mathcal{N}} x_{ji}^v \qquad\qquad \forall i \in \mathcal{N} \setminus \{0, s\}, \forall v \in V \tag{1c}$$

$$\sum_{j \in \mathcal{N}} x_{0j}^v = 1 \qquad\qquad \forall v \in V \tag{1d}$$

$$\sum_{j \in \mathcal{N}} x_{j,s}^v = 1 \qquad\qquad \forall v \in V \tag{1e}$$

$$\sum_{j \in \mathcal{N}} x_{ij}^v - \sum_{j \in \mathcal{N}} x_{n+i,j}^v = 0 \qquad\qquad \forall i \in P, \forall v \in V \tag{1f}$$

$$\sum_{i \in \mathcal{N}} x_{ij}^v = 1 \qquad\qquad \forall j \in I_v, \forall v \in V \tag{1g}$$

$$u_j^v \geq (u_i^v + \Delta_i + t_{ij}) x_{ij}^v \qquad\qquad \forall i, j \in \mathcal{N}, \forall v \in V \tag{1h}$$

$$u_0^v \geq T_v^B \qquad\qquad \forall v \in V \tag{1i}$$

$$u_s^v \leq T_v^E \qquad\qquad \forall v \in V \tag{1j}$$

$$u_i^v \geq e_i \qquad\qquad \forall i \in P, v \in V \tag{1k}$$

$$t_i \leq u_{n+i}^v - (u_i^v + \Delta_i) \leq \max\{\alpha t_i, \beta + t_i\} \qquad\qquad \forall i \in P, \forall v \in V \tag{1l}$$

$$t_i \leq u_i^v - (u_i^P + \Delta_i) \leq \max\{\alpha t_i, \beta + t_i\} \qquad\qquad \forall i \in I_v, \forall v \in V \tag{1m}$$

$$w_j^v \geq (w_i^v + q_j) x_{ij}^v \qquad\qquad \forall i, j \in \mathcal{N}, \forall v \in V \tag{1n}$$

$$0 \leq w_i^v \leq Q_v \qquad\qquad \forall i \in \mathcal{N}, \forall v \in V \tag{1o}$$

$$x_{ij}^v \in \{0, 1\} \qquad\qquad \forall i, j \in \mathcal{N}, \forall v \in V \tag{1p}$$

Fig. 1. The static formulation of the dial-a-ride problem.

earlier than its lower bound. Constraints (1l) ensure that the travel time of each served passenger does not deviate too much from the shortest path between its origin and destination. Passengers are allowed to spend either $\alpha * t_i$ (a percentage of the shortest path), or $\beta + t_i$ (a constant deviation time from the shortest path) traveling in the vehicle, whichever is larger. Constraints (1m) do the same for passengers already in a vehicle. Constraints (1n) define the vehicle capacities. Lastly, constraints (1o) ensure that the vehicle capacities are not exceeded. Constraints (1h) and (1n) can be linearized using a Big M formulation.

The following theorem provides a way to prune the search space significantly. It shows that, in an optimal solution, a rider cannot be picked up by a vehicle v if the smallest possible wait time incurred using v is greater than her penalty.

Theorem 1. *A feasible solution where rider l is assigned to vehicle v such that $u_0^v + t_{0,l} - e_l > p_l$ is suboptimal.*

Proof. Suppose that there exists a feasible solution *(I)* that serves a passenger l such that $u_0^v + t_{0,l} - e_l > p_l$. Let r be the route of vehicle v (i.e., a sequence of edges in $\mathcal{A}$). Removing the pickup and dropoff of rider l from route r produces a new feasible route $\hat{r}$ since the deviation time cannot increase by the triangular inequality and the number of riders in v decreases. Solution *(II)* is derived from solution *(I)* by replacing the route r by route $\hat{r}$ and fixing z_l to 1. Using $\hat{u}$ and $\hat{z}$ to denote the variables of solution *(II)*, the cost $C_{(II)}$ of solution *(II)* is:

$$C_{(II)} = \sum_{i \in P\setminus\{l\}} \sum_{v \in V} (\hat{u}_i^v - e_i) + \sum_{i \in P\setminus\{l\}} p_i \hat{z}_i + p_l \tag{2a}$$

$$< \sum_{i \in P\setminus\{l\}} \sum_{v \in V} (\hat{u}_i^v - e_i) + \sum_{i \in P\setminus\{l\}} p_i \hat{z}_i + u_0^v + t_{0,l} - e_l \tag{2b}$$

$$\leq \sum_{i \in P\setminus\{l\}} \sum_{v \in V} (u_i^v - e_i) + \sum_{i \in P\setminus\{l\}} p_i \hat{z}_i + u_l^v - e_l \tag{2c}$$

$$= \sum_{i \in P} \sum_{v \in V} (u_i^v - e_i) + \sum_{i \in P} p_i z_i = C_{(I)} \tag{2d}$$

Equality (2a) is just the definition of the objective of solution *(II)*. Inequality (2b) is induced by the hypothesis. Inequality (2c) is induced by the triangular inequality on the travel times. Inequality (2d) just factors the equation to get the objective of solution *(I)*. Solution *(I)* is thus suboptimal.   □

5   The Column-Generation Algorithm

This section presents the column-generation algorithm, starting with the master problem before presenting the pricing subproblem, and the specifics of the column-generation process. Upon completion of the column generation, RTDARS solves a final MIP that imposes integrality constraints on the master problem variables.

The Master Problem The restricted master problem, RMP, (presented in Fig. 2) selects a route for each vehicle. In order for a route to be assigned to a vehicle, the route must contain dropoffs for every current passenger of that vehicle. The set of routes is denoted by R and its subset of routes that can be assigned to vehicle v is denoted R_v. The variables in the master problem are the following: $y_r \in [0, 1]$ is set to 1 if potential route r is selected for use and variable $z_i \in [0, 1]$ is set to 1 if request i is not served by any of the selected routes. The constants are as follows: c_r is the sum of the wait time incurred by customers served by route r, p_i is the cost of not scheduling request i for this period, and $a_i^r = 1$ if request i is served by route r. The objective minimizes the waiting times incurred by all customers on each route and the penalties for the customers not

$$\min \quad \sum_{r \in R} c_r y_r + \sum_{i \in P} p_i z_i \tag{3a}$$

$$\text{subject to} \tag{3b}$$

$$\left(\sum_{r \in R} y_r a_i^r \right) + z_i = 1 \qquad \forall i \in P \qquad (\pi_i) \tag{3c}$$

$$\sum_{r \in R_v} y_r = 1 \qquad \forall v \in V \qquad (\sigma_v) \tag{3d}$$

$$z_i \in \mathbb{N} \qquad \forall i \in P \tag{3e}$$

$$y_r \in \{0,1\} \qquad \forall r \in R \tag{3f}$$

Fig. 2. The master problem formulation.

scheduled during the current period. Constraints (3c) ensure that z_i is set to 1 if request i is not served by any of the selected routes and constraints (3d) ensure that only one route is selected per vehicle. The dual variables associated with each constraint are specified in between parentheses next to the constraint in the model.

The Pricing Problem. The routes for each vehicle v are generated via a pricing problem depicted in Fig. 3. The pricing problem (4) is defined for a given vehicle v. Theorem 1 makes it possible to remove some passengers from the set P to obtain the subset P_v and thus a new graph $\mathcal{G}_v = (\mathcal{N}_v, \mathcal{A}_v)$. The pricing problem minimizes the reduced cost of the route being generated. Constraints (4b)–(4o) correspond to constraints (1c)–(1p) in the static problem.

The Column Generation. In traditional column generation for dial-a-ride problems, the pricing problem is formulated as a resource-constrained shortest-path problem and solved using dynamic programming. However, the minimization of waiting times, i.e., $\sum_{i \in P}(u_i - e_i)$, is particularly challenging, as it cannot be formulated as a classical resource-constrained shortest-path problem. One option is to discretize time and use time-expanded graphs. However, this raises significant computational challenges for large instances. As a result, this paper solves the pricing problem through an anytime algorithm that takes into account the real-time constraints RTDARS operates under.

The column-generation algorithm is specified in Algorithm 1: *It generates multiple columns with disjoint sets of customers.* In the algorithm, function Pricing(v, R) solves the pricing problem for a vehicle v and a set R of requests, while Route(v, R) returns the optimal route for a vehicle v and a set of request R. Lines 1–5 is the high-level column-generation procedure: It alternates the generation of columns and the solving of the master problem with the generated columns until no more columns can be generated. It proceeds in waves, first generating columns with one customer before progressively increasing the number of considered requests. Procedure GenerateColumn (lines 6–12) generates

$$\min \quad \sum_{i \in P_v} (u_i - e_i) - \sum_{i \in P_v} \sum_{j \in \mathcal{N}_v} x_{ij}\pi_i - \sigma_v \tag{4a}$$

subject to

$$\sum_{j \in \mathcal{N}_v} x_{ij} = \sum_{j \in \mathcal{N}_v} x_{ji} \qquad\qquad \forall i \in \mathcal{N}_v \setminus \{0, s\} \tag{4b}$$

$$\sum_{j \in \mathcal{N}_v} x_{0j} = 1 \tag{4c}$$

$$\sum_{j \in \mathcal{N}_v} x_{js} = 1 \tag{4d}$$

$$\sum_{j \in \mathcal{N}_v} x_{ij} - \sum_{j \in \mathcal{N}_v} x_{n+i,j} = 0 \qquad\qquad \forall i \in P_v \tag{4e}$$

$$\sum_{i \in \mathcal{N}_v} x_{ij} = 1 \qquad\qquad \forall j \in I_v \tag{4f}$$

$$u_j \geq (u_i + \Delta_i + t_{ij})x_{ij} \qquad\qquad \forall i, j \in \mathcal{N}_v \tag{4g}$$

$$u_0 \geq T_v^B \tag{4h}$$

$$u_s \leq T_v^E \tag{4i}$$

$$u_i \geq e_i \qquad\qquad \forall i \in P_v \tag{4j}$$

$$t_i \leq u_{n+i} - (u_i + \Delta_i) \leq \max\{\alpha t_i, \beta + t_i\} \qquad\qquad \forall i \in P_v \tag{4k}$$

$$t_i \leq u_i - (u_i^P + \Delta_i) \leq \max\{\alpha t_i, \beta + t_i\} \qquad\qquad \forall i \in I_v \tag{4l}$$

$$w_j \geq (w_i + q_j)x_{ij} \qquad\qquad \forall i, j \in \mathcal{N}_v \tag{4m}$$

$$0 \leq w_i \leq Q_v \qquad\qquad \forall i \in \mathcal{N}_v \tag{4n}$$

$$x_{ij} \in \{0, 1\} \qquad\qquad \forall i, j \in \mathcal{N}_v \tag{4o}$$

Fig. 3. The pricing problem formulation for vehicle v.

columns by increasing number of requests. Procedure GENERATESIZEDCOLUMN (lines 13–18) generates columns of size k, where k is the number of requests in the column. It first computes Q, a set in which each element is a k-sized set of possible requests. It then considers the various vehicles ranked in decreasing order of their dual values σ_v. Line 15 computes the sets of requests with the smallest pricing objective value. If the pricing objective is negative (line 16), all set of requests which contains a request covered by R_v are removed from Q to ensure that RTDARS generates a set of non-overlapping columns at each iteration (line 17). Finally, line 18 returns the routes for each vehicle with negative reduced costs.

6   The Real-Time Problem

RTDARS divides the time horizon into epochs of length ℓ, i.e., $[0, \ell), [\ell, 2\ell), [2\ell, 3\ell), \ldots$ and epoct τ corresponds to the time interval $[\tau\ell, (\tau+1)\ell)$.

Algorithm 1. COLUMNGENERATION

---

1 **while** *true* **do**
2     $\mathcal{C} \leftarrow$ GENERATECOLUMNS()
3   **if** $\mathcal{C} = \emptyset$ **then**
4   break;
5   Solve RMP after adding $\mathcal{C}$

   Function GENERATECOLUMNS():
6     $k \leftarrow 1$
7   **while** $k \leq |P|$ **do**
8         $\mathcal{C} \leftarrow$ GENERATESIZECOLUMNS(k)
9   **if** $\mathcal{C} \neq \emptyset$ **then**
10   **return** $\mathcal{C}$
11   **else**
12            $k{+}{+}$

   Function GENERATESIZEDCOLUMNS(k):
13    $Q \leftarrow \{R \subseteq P \mid |R| = k\}$
14   **forall** $v \in |V|$ *ordered by decreasing σ_v*
15        $R_v \leftarrow \arg\min_{R \subset Q}$ PRICING(v, R)
16   **if** PRICING(v, R_v)$\} < 0$ **then**
17            $Q \leftarrow \{R \subseteq Q \mid R \cap R_v = \emptyset\}$
18   **return** $\{$ROUTE(v, R_v)$|v \in V$ & PRICING(v, R_v) $< 0\}$

During period τ, RTDARS batches the incoming requests into a set P_τ, which is considered in the next epoch. It also optimizes the static problem using the requests accumulated in $P_{\tau-1}$ and those requests not yet committed to in the epochs $\tau - 1$ and before. The optimization is performed over the interval $[(\tau + 1)\ell, \infty)$.

It remains to specify how to compute the inputs to the optimization problem, i.e., the departing stops and times for each vehicle and the various set of requests to serve. To determine the starting stop for a vehicle v, the optimization in epoch τ uses the solution $\phi_{\tau-1}$ to the static problem in epoch $\tau - 1$ and considers the first stop s_v in $\phi_{\tau-1}$ in the interval $[(\tau + 1)\ell, \infty)$ if it exists. This stop becomes the starting stop u_0^v of the vehicle and its earliest time is given by the earliest departure time of vehicle v in $\phi_{\tau-1}$. If vehicle v is idle at stop s_v in $\phi_{\tau-1}$ and not scheduled on $[(\tau + 1)\ell, \infty)$, then the departing stop is s_v and the earliest departing time is $(\tau + 1)\ell$. Consider now the sets P, D, and I_v ($v \in V$) for period τ. For a vehicle v, all the requests before its departing stop s_v are said to be *committed* and are not reconsidered. The set I_v are the dropoffs of the requests that have been picked up before s_v but not yet dropped off. The set P corresponds to the requests that have not been picked up by any vehicle v before s_v, as well as the requests batched in $P_{\tau-1}$. The set D simply contains the dropoffs associated with P.

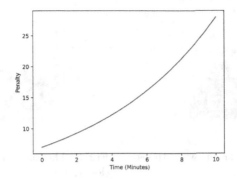

Fig. 4. The penalty function for unserved customers.

Finally, since the static problem may not schedule all the requests, it is important to update the penalty of unserved requests to ensure that they will not be delayed too long. The penalty for an unserved request c in period τ is given by $p_c = \delta 2^{(\tau \ell - e_c)/(10\ell)}$ and it increases exponentially over time as shown in Fig. 4. The δ parameter incentivizes the schedule of the request in its first available period. Figure 4 displays the function for $\delta = 420$ s and $\ell = 30$ s: It ensures that the penalty doubles every ten periods (in the example, every five minutes).

Observe that the static model schedules all the requests which have not been committed to any vehicle. This gives a lot of flexibility to the real-time system at the cost of more complex pricing subproblems.

7   Experimental Results

Instance Description. RTDARS was evaluated on the yellow trip data provided by the New York City Taxi and Limousine Commission [8]. This data provides *pickup and dropoff locations*, which were used to match trips to the closest virtual stops, *starting times*, which were used as the request time, and the *number of passengers*. This section reports results on a representative set of 24 instances, 1 h per day for two weekdays per month from July 2015 through June 2016. To capture the true difficulty of the problem, rush hours (7–8am) were selected. The instances have an average of 21,326 customers and range from 6,678 customers to 28,484 customers. Individual requests with more customers than the capacity of the vehicles were split into several trips. An additional test was performed on the largest instance with 32,869 customers.

Virtual Stops. The evaluation assumes a dial-a-ride system using the concept of virtual stops proposed in the RITMO system [12] (Uber and Lyft are now considering similar concepts). Virtual stops are locations where vehicles can pick up and drop off customers without impeding traffic. They also ensure that customers are ready to pick up and make ride-sharing more efficient since they

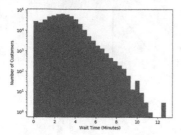
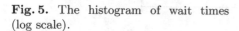
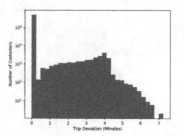

Fig. 5. The histogram of wait times (log scale).

Fig. 6. The histogram of trip deviations (log scale).

decrease the number of stops. To implement virtual stops, Manhattan was over-layed with a grid with cells of 200 squared meters and every cell had a virtual stop. The trip times were precomputed by querying OpenStreetMap for travel times between each virtual stop [9]. All customers at a virtual stop are grouped and can be picked up together.

Algorithmic Setting. Both the final master problem and the restricted master problem are solved using Gurobi 8.1. Empty vehicles are initially evenly dis-tributed over the virtual stops. The pricing problem uses parallel computing to implement line 15 of Algorithm 1, exploring potential requests simultaneously. To meet real-time constraints, the implementation greedily extends the "opti-mal" routes of size k to obtain routes of size $k+1$. Unless otherwise specified, all experiments are performed with the following default parameters: 2,000 vehicles of capacity 5, $\alpha = 1.5$, $\beta = 240$ s, and $\delta = 420$ s. The impact of these parameters is also studied.

Wait Times. Figure 5 reports the distribution of the waiting for all customers across all instances. The results demonstrate the performance of RTDARS: The average waiting time is about 2.58 min with a standard deviation of 1.31. On the instance with 32,869 customers, the average waiting time is 5.42 min.

Trip Deviation. Figure 6 depicts a histogram of trip deviations incurred because of ride-sharing. The results indicate that riders have an average trip deviation of 0.34 min with a standard deviation of 0.74. In percentage, this represents a deviation of about 12%. On the instance with 32,869 customers, the average trip deviation is 2.23 min, which shows the small overhead induced by ride-sharing.

The Impact of the Fleet Size. Figure 7 studies the impact of the fleet size on the waiting times and trip deviation. The plot reports the average waiting times for various numbers of riders, where capacity is 4, $\alpha = 1$, $\beta = 840$ s, and $\delta = 420$ s to facilitate comparisons to [1]. The results show that, even with 1,500 vehicles, the average waiting time remains below 6 min and the average deviation time below 40 s. Since RTDARS is guaranteed to serve all the requests,

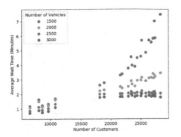
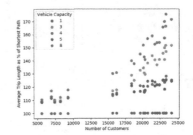

(a) The Impact on the Average Wait Times.

(b) The Impact on the Average Trip Deviations.

Fig. 7. The impact of the fleet size on the average wait times and average deviations on all instances.

these results demonstrate the potential of column generation and ride-sharing for large-scale real-time dial-a-ride platforms. Adopting RTDARS has the potential to substantially reduce traffic in large cities, while still guaranteeing service within reasonable times. Recall that the approach in [1] does not serve about 2% of the requests.

The Impact of Vehicle Capacity. Figure 8 studies the impact of the vehicle capacity (i.e., how many passengers a vehicle can carry) on the average waiting times and trip deviation. The parameters are set to 2,000 vehicles, $\alpha = 1$, $\beta = 840$ s, and $\delta = 420$ s to facilitate comparisons to [1]. The results on waiting times show that moving to vehicles of capacity 8 further reduces the average waiting times, especially on the large instances. On the other hand, moving from a capacity 5 to 3 does not affect the results too much. The results on deviations are more difficult to interpret. Obviously moving to a capacity 8 further increases the deviation (although it remains below one minute). However, moving to vehicles of capacity 3 also increases the deviation, which is not intuitive. This may be a consequence of myopic decisions that cannot be corrected easily given the tight capacity.

The Impact of the Penalty. The penalty p_i in the model is an exponential function of the current waiting time of customer i. Constant δ controls the initial penalty: If it is too small, the penalty for not scheduling a request for the first few periods is low, which causes an increase in wait times, as can be observed in Fig. 9. Once δ is large enough, the average wait times converge to the same values.

Final Vehicle Assignments. As a result of re-optimization, the vehicle to which a rider is assigned can change. Figure 10 reports the amount of time until riders receive their final vehicle assignment (the vehicle which actually picks them up). Not surprisingly, this histogram closely follows the waiting time distribution. The majority of riders receive this assignment quickly. However, it takes some

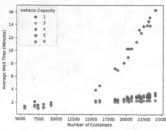
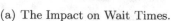
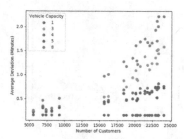

(a) The Impact on Wait Times.   (b) The Impact on Trip Deviations.

Fig. 8. The impact of the vehicle capacity on the average wait times and the average trip deviations on all instances.

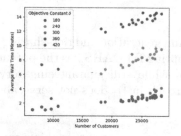

Fig. 9. The impact of the penalty on average wait times.

Fig. 10. Times until final vehicle assignments.

riders over 10 min to receive their final vehicle assignment, which shows that RTDARS takes advantage of the ability to re-assign riders to vehicles which will result in better overall assignments.

The Impact of Column Generation. Figure 11 depicts the impact of column generation and reports the number of columns in the final MIP as all possible columns of sizes 1 and 2 to be conservative. The results show that the algorithm only explores a small percentage of all potential columns, demonstrating the benefits of a column-generation approach.

The Impact of Pruning. Figure 12 shows the impact of Theorem 1, which provides a way to prune the number of requests considered at each step of the algorithm. The figures report the total optimization time for all time periods of each instance. Each optimization must be performed in less than 30 s, but the graph reports the total optimization time over the entire hour. As the results indicate, the pruning benefits become substantial as the instance sizes grow. The results show that the pruning significantly reduces the computational time. They also show that RTDARS should be able to handle even larger instances since, after exploiting Theorem 1, RTDARS uses only about a sixth of the available time. This creates opportunities to exploit stochastic information.

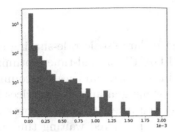
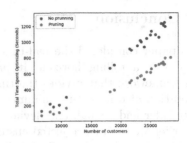

Fig. 11. The number of generated columns as a percentage of possible combinations of requests/vehicles. The x-Axis value are scaled by 10^{-3}.

Fig. 12. Optimization times with and without pruning.

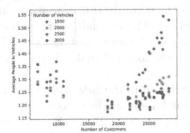
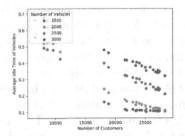

(a) The Impact on Average Vehicle Utilization.

(b) The Impact on Average Vehicle Idle Time.

Fig. 13. The impact of the fleet size on the average vehicle utilization and idle time on all instances.

The Impact of Ride Sharing. Figure 13 reports the average number of people in each vehicle at all times for each instance. The results show a significant amount of ride sharing, although single trips and idle time remain a significant portion of the rides, especially when the fleet is oversized. Lastly, Fig. 8 shows that wait times are reduced by a factor of 4 when moving from single-rider trips to ride-sharing for large instances while the trip deviation only increases to at most 2 min for vehicles of capacity 8, thus demonstrating the value of ride sharing.

Comparison with Prior Work. The results of [1] "show that 2,000 vehicles (15% of the taxi fleet) of capacity 10 or 3,000 of capacity 4 can serve 98% of the demand within a mean waiting time of 2.8 min and mean trip delay of 3.5 min." RTDARS relaxes the hard time-windows present in [1] and improves on these results, yielding an average wait time of 2.2 min with only 2,000 vehicles, while guaranteeing service for all riders.

8   Conclusion

This paper considered the real-time dispatching of large-scale ride-sharing services over a rolling horizon. It presented RTDARS, a real-time optimization framework that divides the time horizon into epochs and uses a column-generation algorithm that minimizes wait times while guaranteeing services for every rider and a small trip deviation compared to a direct trip. This contrasts to earlier work which rejected customers when the predicted waiting time was considered too long (e.g., 7 min). This assumption reduced the search space at the cost of rejecting a significant number of requests.

The column-generation algorithm of RTDARS is derived from a three-index formulation [6] which is adapted for use in real-time dial-a-ride applications. In addition, to ensure that all riders are served in reasonable times, the paper proposed an optimization model that balances the minimization of waiting times with penalties for riders that are not scheduled yet. These penalties are increased after each epoch to make it increasingly harder not to serve waiting riders. The paper also presented a key property of the formulation that makes it possible to reduce the search space significantly.

RTDARS was evaluated on historic taxi trips from the New York City Taxi and Limousine Commission [8], which contains large-scale instances with more than 30,000 requests an hour. The results indicated that RTDARS enables a real-time dial-a-ride service to provide service guarantees (every rider is served in reasonable time) while improving average waiting times and average trip deviations compared to prior work. The results also showed that larger occupancy vehicles bring benefits and that the fleet size can be further reduced while preserving very reasonable waiting times.

Substantial work remains to be done to understand the strengths and limitations of the approach. The current implementation is myopic and heavily driven by the dual costs to generate the columns. Different pricing implementation, including the use of constraint programming to replace our dedicated search algorithm, and the inclusion of stochastic information are natural directions for future research.

Acknowlegments. This research was partly supported by Didi Chuxing Technology Co. and Department of Energy Research Grant 7F-30154. We would like to thank the reviewers for their detailed comments and suggestions which dramatically improved the paper, and the program chairs for a rebuttal period that was long enough to run many experiments.

References

1. Alonso-Mora, J., Samaranayake, S., Wallar, A., Frazzoli, E., Rus, D.: On-demand high-capacity ride-sharing via dynamic trip-vehicle assignment. Proc. Natl. Acad. Sci. **114**(3), 462–467 (2017)
2. Bent, R., Van Hentenryck, P.: Waiting and relocation strategies in online stochastic vehicle routing. In: Proceedings of the 20th International Joint Conference on Artifical Intelligence, IJCAI 2007, pp. 1816–1821. Morgan Kaufmann Publishers Inc., San Francisco (2007). http://dl.acm.org/citation.cfm?id=1625275.1625569
3. Bent, R.W., Van Hentenryck, P.: Scenario-based planning for partially dynamic vehicle routing with stochastic customers. Oper. Res. **52**(6), 977–987 (2004)
4. Berbeglia, G., Cordeau, J.F., Laporte, G.: A hybrid tabu search and constraint programming algorithm for the dynamic dial-a-ride problem. INFORMS J. Comput. **24**(3), 343–355 (2012)
5. Bertsimas, D., Jaillet, P., Martin, S.: Online vehicle routing: the edge of optimization in large-scale applications (2018)
6. Cordeau, J.F., Laporte, G.: The dial-a-ride problem: models and algorithms. Ann. Oper. Res. **153**(1), 29–46 (2007)
7. Jain, S., Van Hentenryck, P.: Large neighborhood search for dial-a-ride problems. In: Lee, J. (ed.) CP 2011. LNCS, vol. 6876, pp. 400–413. Springer, Heidelberg (2011). https://doi.org/10.1007/978-3-642-23786-7_31
8. NYC: Nyc taxi & limousine commission - trip record data. http://www.nyc.gov/html/tlc/html/about/trip_record_data.shtml
9. OpenStreetMap contributors: Planet dump (2017). https://planet.osm.org. https://www.openstreetmap.org
10. Ota, M., Vo, H., Silva, C., Freire, J.: A scalable approach for data-driven taxi ride-sharing simulation. In: 2015 IEEE International Conference on Big Data (Big Data), pp. 888–897, October 2015
11. Ota, M., Vo, H., Silva, C., Freire, J.: Stars: simulating taxi ride sharing at scale. IEEE Trans. Big Data **3**(3), 349–361 (2017)
12. RITMO app introduces on-demand mass transit at U-M, with plans to expand. Concentrate (2018). www.secondwavemedia.com/concentrate/innovationnews/ritmorollout0443.aspx

Some Experiments with Submodular Function Maximization via Integer Programming

Domenico Salvagnin[✉]

Department of Information Engineering (DEI), University of Padova, Padova, Italy
domenico.salvagnin@unipd.it

Abstract. Submodular function maximization is a classic problem in optimization, with many real world applications, like sensor coverage, location problems and feature selection, among others. Back in the 80's, Nemhauser and Wolsey proposed an integer programming formulation for the general submodular function maximization. Being the number of constraints in the formulation exponential in the size of the ground set, a constraint generation technique was proposed. Since then, the method was not developed further. Given the renewed interest in recent years in submodular function maximization, the constraint generation method has been used as reference to evaluate both exact and heuristic approaches. However, the outcome of those experiments was that the method is utterly slow in practice, even for small instances. In this paper we propose several algorithmic enhancements to the constraint generation method. Preliminary computational results show that a proper implementation, while still not scalable to big instances, can be significantly faster than the obvious implementation by the book. A comparison with direct mixed integer linear programming formulations on some classes of models that admit one also show that the submodular framework, in its generality, is clearly slower than dedicated formulations, so it should be used only when those approaches are not viable.

1 Introduction

Let N be a finite set of n elements, called the ground set. A set function $f : 2^N \to \mathbb{R}$ is called *submodular* if it satisfies the following property:

$$f(S) + f(T) \geq f(S \cup T) + f(S \cap T) \quad \forall S, T \subseteq N \tag{1}$$

Equivalently [17], a set function f is submodular if and only of it satisfies the *diminishing returns* property:

$$f(S \cup j) \geq f(T \cup j) \quad \forall S \subseteq T \subseteq N, j \in N - T \tag{2}$$

In other words, the later we add an element j to a set, the smaller its effect is on the objective. Many real-world optimization problems give rise to submodular

© Springer Nature Switzerland AG 2019
L.-M. Rousseau and K. Stergiou (Eds.): CPAIOR 2019, LNCS 11494, pp. 488–501, 2019.
https://doi.org/10.1007/978-3-030-19212-9_32

functions like, e.g., location problems and sensor coverage [14]. In addition, many problems in computer science and machine learning, such as exemplar clustering [8], influence spread [13], image denoising [5], and feature selection [11], can be formulated as submodular maximization problems.

A submodular function is *nondecreasing* (or *monotone*) if $f(S) \leq f(T)$ $\forall S \subseteq T \subseteq N$, and $f(\emptyset) = 0$. Clearly, maximizing a nondecreasing submodular function is trivial in the absence of further constraints. In this paper we will consider the cardinality constrained version:

$$\max f(S)$$
$$S \subseteq N$$
$$|S| \leq k$$

for a given $0 < k < n$. We further assume the standard *value oracle* model, i.e., the submodular function f is known only through a black box oracle that is able to compute the value $f(S)$ for an arbitrary $S \subseteq N$. It is well-known that the simple greedy algorithm [18], which adds at each iteration the element j with the largest marginal benefit w.r.t. the current set until its cardinality is k, achieves an approximation ratio of $(1 - 1/e)$. The greedy algorithm uses $O(nk)$ functions evaluations in the worst case, although this can be improved in practice exploiting definition (2), i.e., the fact that as we add elements to the set, the marginal benefits of the remaining elements to consider can only decrease, see [16]. We will see in the following that the role of the greedy algorithm is quite prominent even in exact methods.

The structure of the paper is as follows. In Sect. 2 we will describe the MIP model of Nemhauser and Wolsey, which is the basis of all our MIP methods for general submodular function maximization. Then in Sect. 3 we describe a modern implementation of the basic constraint generation model based on such model, and introduce some algorithmic enhancements to the method, in terms of primal heuristics and cut separation. Computational results are given in Sect. 4. Finally, conclusions and future directions of research are drawn in Sect. 5.

2   The General IP Model

In [17], Nemhauser and Wolsey introduced a mixed-integer-programming formulation for the general submodular function maximization problem. Let us denote with $\Delta_j(S)$ the marginal value of adding an element j to S, i.e., $\Delta_j(S) = f(S \cup j) - f(S)$. The formulation is based on the following general property of submodular functions:

$$f(T) \leq f(S) + \sum_{j \in T-S} \Delta_j(S) - \sum_{j \in S-T} \Delta_j(S \cup T - j) \quad \forall S, T \subseteq N \quad (3)$$

If f is monotone, (3) can be further simplified to

$$f(T) \leq f(S) + \sum_{j \in T-S} \Delta_j(S) \quad \forall S, T \subseteq N \quad (4)$$

which immediately suggests the MIP formulation:

$$P = \max \quad z \tag{5}$$

$$z \leq f(S) + \sum_{j \notin S} \Delta_j(S) x_j \quad \forall S \subseteq N \tag{6}$$

$$\sum_{j \in N} x_j \leq k \tag{7}$$

$$x_j \in \{0, 1\} \quad \forall j \in N \tag{8}$$

where variables x_j basically encode the indicator function $1(S)$, i.e., $x_j = 1$ if and only if $j \in S$, and constraints (6) encode (4) as linear inequalities. Note that the model does not exploit any knowledge that cannot be obtained under the value oracle model. Model P has an exponential number of constraints (one for each subset of N, plus the cardinality constraint): thus a constraint generation approach was proposed in [17], where constraints (6) are added iteratively.

Given a solution (x^*, z^*) with x^* integer, it is trivial to solve the separation problem over the family of constraints (6): x^* uniquely defines a subset S^* and we just need to evaluate $f(S^*)$. If $z^* \leq f(S^*)$, then (x^*, z^*) cannot be cut. Otherwise, we just need to construct the cut corresponding to S^* at the cost of additional $n - k$ function evaluations, and this is guaranteed to be violated by (x^*, z^*) by the amount $z^* - f(S^*)$.

It is important to emphasize that formulation P does not give the convex hull of the points $(1(S), f(S))$, so even if we would add all constraints (6), we would not be able to solve the model as a linear program. Surprisingly, constraints (6) can be strengthened by ignoring the fact that the function is nondecreasing, and reverting to the general expression (3). Given that we do not know T in advance, we relax the coefficients $\Delta_j(S \cup T \cup -j)$ to $\Delta_j(N - j)$, and obtain the inequality:

$$z \leq [f(S) - \sum_{j \in S} \Delta_j(N - j)] + \sum_{j \in S} \Delta_j(N - j) x_j + \sum_{j \notin S} \Delta_j(S) x_j \quad \forall S \subseteq N \tag{9}$$

It is easy to show that (9) dominates (6): if $T \supseteq S$ the two expressions coincide, otherwise the right hand side expression of (9) is strictly better. Note that cuts (9) are a little denser than their (6) counterparts. On the other hand, they are not more expensive to compute, as the coefficients $\Delta_j(N - j)$ do not depend on S and can thus be computed once and for all at the beginning. Unfortunately, these cuts are still not enough to obtain a convex hull formulation, as can be easily proven by constructing small counterexamples. Still, we will use cuts (9) in place of (6) in the rest of the paper, as preliminary computational results showed that the resulting formulation gives a stronger dual bound.

3   A Modern Implementation

The constraint generation method based on model P is somehow reminiscent of Benders' decomposition [4]. The model P that is solved iteratively acts as a

Benders' master, while constraints (6)—or (9)—are the analogous of Benders' optimality cuts. While the analogy is only superficial—in the submodular case there is no variable splitting and there is no (LP) duality theory to exploit to derive a cut—some of the algorithmic improvements proposed for Benders' decomposition over the years easily carry over. In particular, we do not need to solve a MIP to proven optimality at each iteration, but we can separate constraints (9) on the fly each time an integer solution is found by the branch-and-cut, exploiting modern MIP solvers support for so-called *lazy constraints*. In other words, only a single enumeration tree is needed.

Then, we can improve the overall method in at least two directions: (i) use a more sophisticated primal heuristic to find a tighter primal bound at the beginning and (ii) separate constraints (9) not only at integer solutions, but also at fractional ones, in order to improve the dual bound more quickly.

3.1 GRASP Heuristic

The simple greedy algorithm [18] is known to perform very well in practice, a behaviour that was confirmed in our computational evaluation. Still, it yields the true optimal solution only on the smaller models. A first improvement can be obtained by adding a local search phase at the end of the greedy phase. In order to do so, we need to define a neighborhood structure on the solutions. In the following, we assume to have at hand a subset S of cardinality k. The easiest choice is to consider an *exchange* neighborhood, i.e., consider the set of subsets T that can be obtained by dropping an element from S and adding an element not in S. More formally:

$$\mathcal{N}(S) = \{T \subseteq N : T = S \cup i - j, \forall i \in S, \forall j \notin S\}$$

Each such neighborhood can be explored at the cost of additional $O(nk)$ evaluations. Then the local search phase consists in iteratively exploring the neighborhood of the current set S, and updating it until it is locally optimal.

We can extend the greedy plus local search combination into the full-blown meta-heuristic scheme called GRASP [9]. The main idea is to introduce a randomized component into the greedy procedure, where at each step, instead of picking the element with the largest marginal gain, we pick randomly among the best C (say) candidates, the so-called *restricted candidate list*. Then, each solution found this way is improved by local search, and the process is repeated until some iteration/resource limit is reached.

Interestingly, in our setting we can derive an additional benefit from this more sophisticated heuristic than just a hopefully better primal bound. For each locally optimal solution found by GRASP, we can construct a cut (9), so that we can warm-start the MIP P with an initial pool of cuts.

3.2 Separating Fractional Solutions

While separating integer solution is trivial, separating fractional solutions is far more challenging, as we cannot directly map the point to cut to a subset of N.

Ideally, we would like to solve, for a given (x^*, z^*), the separation problem:

$$\max \quad z^* - [f(S) - \sum_{j \in S} \Delta_j(N - j)] - \sum_{j \in S} \Delta_j(N - j)x_j^* - \sum_{j \notin S} \Delta_j(S)x_j^*$$

$$S \subseteq N$$

$$|S| \leq k$$

but this is basically as hard as the original problem. As such, we resort to heuristic separation algorithms. In the following, we will describe two such procedures, one based on the greedy algorithm and one based on the Lovász extension of f [15].

Let F_1 (resp. F_0) be the set of variables fixed to 1 (resp. 0) at the current node of the branch-and-cut tree. In the following, we will denote a node by the pair (F_1, F_0). Clearly, we can modify the greedy algorithm to take those local domains into account. Let S' be the greedy set computed in this way, i.e., $T = F_1 \cup \{x_{j_1}, \ldots, x_{j_p}\}$ for some p, with the variables added by the greedy algorithm exactly in this order. Then for each $0 \leq q \leq p$ we can consider the set $T_q = F_1 \cup \{x_{j_1}, \ldots, x_{j_q}\}$ (the current set after q greedy steps—note that $T_0 = F_1$), construct the corresponding cut and check whether it is violated by the current fractional solution. So we test each T_q in sequence, and keep adding elements as long as the violation increases.

There is a nice connection between the cuts that can be obtained this way and the modular heuristic h_{mod} used in A^* search approaches for submodular function maximization. We recall that A^* search picks the next node to explore as the one maximizing $f(F_1) + h(F_1, F_0)$, where $h(\cdot)$ is a so-called *(admissible) heuristic* function that bounds the objective value of any node in the current subtree—in the mathematical programming terminology, any such function would yield a valid dual bound. The admissible heuristic h_{mod}, proposed in [6], is computed as:

$$h_{mod}(F_1, F_0) = \sum_{j \in T - F_1} \Delta_j(F_1)$$

where T is the greedy solution computed at node (F_1, F_0). It is easy to see that any fractional solution with $z^* > f(F_1) + h_{mod}(F_1, F_0)$ would be violated by a cut computed by the procedure above. Indeed, the fractional solution is already violated by the cut computed from F_1, and we enlarge the set only if it gives an improvement w.r.t. violation. As a corollary, if the separation procedure above is applied at all nodes of the branch-and-bound tree, we have that the LP relaxation computed at node (F_1, F_0) cannot be worst than $f(F_1) + h_{mod}(F_1, F_0)$, and thus the LP bound dominates the h_{mod} bound, albeit at the cost of solving LPs.

The greedy solution T computed at node (F_1, F_0) is completely oblivious to the fractional solution (x^*, z^*), a fact that can make the discovery of violated inequalities quite rare. A different approach consists in using the so-called Lovász extension $\hat{f}$ of f, i.e., the extension of function f to the unit cube $[0, 1]^n$ [15]. The extension is defined as follows. Let $x^* \in \mathbb{R}^n_+$ be an arbitrary fractional vector in the unit cube. Then we can express x^* as a convex combination of $n + 1$

vertices $x^0, x^1, \ldots, x^n$ of the unit cube, with the additional property that those vertices form a nested sequence, i.e., $x^i \subseteq x^{i+1}$ (with a small abuse of notation, we identify the integer vertices with the subsets of N of which they are the indicator vectors). Given the coefficients $\lambda_0, \ldots, \lambda_n$ of the convex combination, we can then define $\hat{f}(x^*)$ as

$$\hat{f}(x^*) = \sum_{i=0}^{n} \lambda_i f(S_i)$$

where S_i is the set associated to vertex x^i. Instrumentally, an efficient procedure is known to compute, given an arbitrary point x^*, both the sequence $S_0, \ldots, S_n$, and the corresponding multipliers. Intuitively, the sequence is obtained by starting from the empty set, and adding elements according to a permutation of the indices that sorts the values x_j^* in non-increasing order, while the multipliers are obtained as differences of pairs of consecutive coefficients in the sorted sequence, see [15] for more details about the procedure. Once we have computed the sequence (and the corresponding multipliers), we can easily perform two operations:

1. compute $\hat{f}(x^*)$. If $z^* \leq \hat{f}(x^*)$, then we have a proof that no violated cut of the form (9) exists, as (x^*, z^*) belongs to the convex hull of the feasible solutions of P.
2. We can construct a cut for each set S_i in the sequence, and check whether it is violated.

Unfortunately, even this machinery does not give an exact separation procedure, as it can happen that $z^* > \hat{f}(x^*)$, but no cut obtained from the sets in the sequence is violated, and this does not rule out the existence of violated inequalities associated with other subsets of N. The reason is that, intuitively, the Lovász construction gives only one among many possible ways of constructing x^* as a convex combination of vertices of the unit cube.

4   Computational Results

We implemented all the methods under comparison in C++, using IBM ILOG CPLEX 12.8 [10] as MIP solver. In the following, we will denote by base the basic implementation of the constraint generation method, by bc its counterpart where constraints (9) are separated on the fly as lazy constraints, and by bc+ the improved version that also uses GRASP and separation of fractional solutions in the tree. All codes were run on a cluster of 24 identical machines, each equipped with an Intel Xeon CPU E3-1220 V2 CPU running at 3.10 GHz, and 16 GB of RAM. All codes take full advantage of multi-threading. Each method was run on each instance with a time limit of 1 h. The parameters used by our code are as follows:

- the GRASP heuristic is run with a limit of 100 iterations. Its restricted candidate list has size $\max(5, n/4)$.

- CPLEX is called with defaults parameters for base, as we use it as a pure black box there. On the other hand, for methods bc and bc+ we disable dual and nonlinear reductions, as needed to guarantee correctness because of lazy constraints, and we set the variable selection strategy to *strong branching* [2,3], as this resulted in an improved performance in preliminary tests.
- We separate fractional solutions at the root and at all nodes whose relaxation is within 1% of the best bound node. In addition, the generated cuts are added to the current node only if their violation is at least 0.1% of the best bound. The rationale is that in our case violation is a measure of how much z^* is currently overestimated, hence we can directly compare this value against the objective value. Finally, all separated cuts are added as local cuts—although they would legitimately be globally valid—to ensure a more aggressive purging.

4.1 Benchmark Sets

We considered three classes of problems that give rise to submodular functions, namely *location*, *weighted coverage* and *bipartite inference*. We will now briefly describe each of them.

Location [12,19]. We are given a ground set of N locations, a set M of clients, and a non-negative profit g_{ij} if client i is served by location j, for all possible pairs. Each client gets the profit from the best opened location, and we want to maximize the overall profit. This corresponds to the submodular function:

$$f(S) = \sum_{i \in M} \max_{j \in S} g_{ij}$$

Weighted Coverage [14,19]. We are given a ground set of N sensors, and a set M of possible targets. Each target i has an associated non-negative weight of $w_i \geq 0$. Each sensor j covers the subset of targets $M_j \subseteq M$. We want to maximize the total weight of the covered targets. This corresponds to the submodular function:

$$f(S) = \sum_{i \in \bigcup_{j \in S} M_j} w_i$$

Bipartite Inference[19]. We are given a ground set of N items, and a set M of targets. We are also given a bipartite graph $G = (N, M, A)$, where the set of arcs A encodes which targets can be *influenced* by which items. Finally, we get an activation probability p_j for each item j. Given a subset $S \subseteq N$, the graph structure and the activation probabilities p_j, we can compute the activation probability $p_S(i)$ of each target $i \in M$ as:

$$p_S(i) = 1 - \prod_{j \in S:(j,i) \in A} (1 - p_j)$$

We want to maximize the submodular function $f(S) = \sum_{i \in M} p_S(i)$.

We generated random instances for all the classes above, using the following rules:

- $n \in \{20, 50, 100, 200\}$
- $m = \mu n$ with $\mu \in \{2, 5, 10\}$
- $k \in \begin{cases} \{5, 10\} & \text{if } n = 20 \\ \{10, 20\} & \text{if } n \in \{50, 100\} \\ \{20, 50\} & \text{if } n = 200 \end{cases}$

where $n = |N|$ and $m = |M|$. For each combination $(n, m.k)$ we generated 5 random models, obtaining in total 360 instances. For location instances, g_{ij} are randomly picked in the interval $[0, 1]$. For weighted coverage instances, w_i are randomly picked in the interval $[0, 1]$, and a target is covered by a sensor with probability 0.07. Finally, for bipartite inference instances, p_j are again randomly picked in the interval $[0, 1]$, while each arc in the bipartite graph exists with probability 0.07.

4.2   Results

We first compared the three methods base, bc and bc+ on the whole testbed of 360 models. Aggregated results are reported in Table 1. The structure of the table is as follows. Instances are divided in different subsets, based on the *difficulty* of the models. To avoid any bias in the analysis, the level of difficulty is defined by taking into account all methods under comparison. The subclasses "[l, 3600}" ($l = 0, 1, 10, 100$), contain the subset of models for which at least one of the methods took at least l seconds to solve and that were solved to optimality within the time limit by at least one of the methods. The subset "all" contains all models. For each subset of models, we report three performance indicators for each of the compared methods: #S reports the number of instances solved to optimality, #T the number of instances for which the method hit the time limit, and the shifted geometric mean [1] of the solution time, with a shift of one second. Note that for all methods except the reference method (base in our case), we report the ratio w.r.t. to the reference of the runtime (columns "tQ") rather than the value itself. Ratios $t < 1$ indicate a speedup factor of $1/t$.

According to the table, bc+ outperforms bc by a large margin, both in terms of number of instances solved and average runtime, and in turn the same is true for bc with respect to the baseline method base, again according to both criteria. If we compare bc+ to base directly, we see that bc+ can solve 62 more models and is on average 4× faster. If we restrict to the set of models that at least one method can solve (193 models), then the speedup is even more impressive, up to 20×. In addition, the speedup further increases as we consider harder models.

More detailed results are given in Table 2, where we aggregate only over the 5 different models generated for each parameter combination. For each problem class, and for each combination of (n, m, k), we report, for all methods under comparison, the number of instances solved (out of 5), the shifted geometric mean of runtime, and the final relative gap at the end of the solve—the latter

Table 1. Aggregated results over whole testbed.

Instances	base			bc			bc+		
	#S	#T	time (s)	#S	#T	tQ	#S	#T	tQ
All	121	239	349.42	153	207	0.42	193	167	0.24
[0, 3600}	121	72	45.67	153	40	0.19	193	0	0.05
[1, 3600}	39	72	739.86	71	40	0.07	111	0	0.01
[10, 3600}	24	72	1703.62	56	40	0.05	96	0	0.01
[100, 3600}	17	72	2340.09	49	40	0.05	89	0	0.00

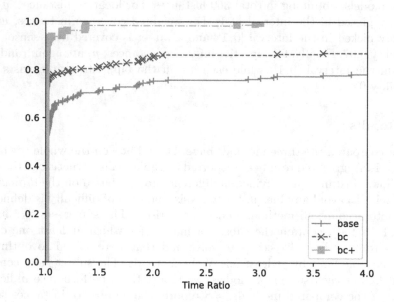

Fig. 1. Performance profile over whole testbed.

being a significant measure for the case in which we frequently hit the time limit. Note that we also report intermediate aggregate results by problem class. According to the table, most parameter combinations either end up being in the easy or unsolvable class, with only a few combinations in between. Still, the more sophisticated methods—and bc+ in particular—do not exhibit any slowdown on the easy models, while being up to two orders of magnitude faster on the medium models, most of which the base cannot even solve. As for the subsets on which all methods hit the time limit, the average final gap is reduced by approximately a factor of 2.

In Fig. 1 we report the performance profile [7] of the three methods on the whole testbed, which largely confirms the finding of Table 1.

4.3   Comparison with a Direct MIP Model

Both location and weighted coverage problems can be formulated directly as MIP models. For the location case, the model reads:

$$\max \sum_{i \in M} \sum_{j \in N} w_{ij} y_{ij}$$

$$\sum_{j \in N} y_{ij} \leq 1 \quad \forall i \in M$$

$$y_{ij} \leq x_j \quad \forall i \in M, j \in N$$

$$\sum_{j \in N} x_j \leq k$$

$$x_j \in \{0,1\} \quad \forall j \in N$$

$$y_{ij} \in \{0,1\} \quad \forall i \in M, j \in N$$

Binary variables x_j encode which locations are open—as in the Nemhauser and Wolsey model—while binary variables y_{ij} encode which location is serving a given client. Note that variables y_{ij} could be relaxed to continuous without affecting the model, as for x fixed the resulting matrix is totally unimodular.

Similarly, for weighted coverage models, the model reads:

$$\max \sum_{i \in M} w_i y_i$$

$$\sum_{j : i \in M_j} x_j \geq y_i \quad \forall i \in M$$

$$\sum_{j \in N} x_j \leq k$$

$$x_j \in \{0,1\} \quad \forall j \in N$$

$$y_i \in \{0,1\} \quad \forall i \in M$$

Here, binary variables x_j encode which sensors are deployed, while binary variables y_i encode which targets are covered. Again, variables y_i could be relaxed to continuous without affecting the model.

It is thus interesting to compare our specialized methods against a direct application of a black box MIP solver, which of course requires much less effort. We present the comparison for the subset of locations models—similar results can be obtained for the class of weighted coverage models—and we compare the best of our specialized methods, bc+, against CPLEX defaults (CPLEX) and the automatic Benders' decomposition of CPLEX (Benders)—Benders being a viable option assuming variables y_{ij} are relaxed to continuous. Aggregated results are given in Table 3, whose structure is identical to Table 1, and the corresponding performance profile is given in Fig. 2. The comparison allows us to put the performance of bc+ into perspective: while quite faster than what we started from (base), it is still no match for a black box MIP solver like CPLEX,

Table 2. Detailed results over all subsets of models. Note that a small gap is sometimes reported even for instances solved to proven optimality: this is a consequence of the default relative optimality gap in the solver.

Class	n	m	k	#solved			time(s)			gap		
				base	bc	bc+	base	bc	bc+	base	bc	bc+
loc	20	40	5	5	5	5	1.08	0.05	0.03	0.00%	0.00%	0.00%
			10	5	5	5	0.03	0.01	0.02	0.00%	0.00%	0.00%
		100	5	5	5	5	75.03	0.62	0.17	0.00%	0.01%	0.00%
			10	5	5	5	0.66	0.03	0.06	0.00%	0.01%	0.01%
		200	5	5	5	5	346.59	1.21	0.33	0.00%	0.01%	0.00%
			10	5	5	5	26.36	0.51	0.31	0.00%	0.01%	0.00%
	50	100	10	0	0	5	t.l.	t.l.	21.22	0.76%	0.34%	0.01%
			20	5	5	5	61.56	2.78	1.20	0.00%	0.01%	0.01%
		250	10	0	0	2	t.l.	t.l.	3077.35	2.19%	1.93%	0.47%
			20	0	0	3	t.l.	t.l.	1137.97	0.44%	0.33%	0.05%
		500	10	0	0	0	t.l.	t.l.	t.l.	2.82%	2.74%	1.88%
			20	0	0	0	t.l.	t.l.	t.l.	0.86%	0.73%	0.41%
	100	200	10	0	0	0	t.l.	t.l.	t.l.	3.02%	2.46%	1.56%
			20	0	0	0	t.l.	t.l.	t.l.	1.18%	0.81%	0.49%
		500	10	0	0	0	t.l.	t.l.	t.l.	4.06%	3.92%	3.17%
			20	0	0	0	t.l.	t.l.	t.l.	1.79%	1.58%	1.24%
		1000	10	0	0	0	t.l.	t.l.	t.l.	4.74%	4.74%	4.04%
			20	0	0	0	t.l.	t.l.	t.l.	2.31%	2.10%	1.77%
	200	400	20	0	0	0	t.l.	t.l.	t.l.	2.11%	1.79%	1.41%
			50	0	0	0	t.l.	t.l.	t.l.	0.39%	0.37%	0.23%
		1000	20	0	0	0	t.l.	t.l.	t.l.	2.71%	2.50%	2.09%
			50	0	0	0	t.l.	t.l.	t.l.	0.65%	0.64%	0.48%
		2000	20	0	0	0	t.l.	t.l.	t.l.	3.10%	2.96%	2.61%
			50	0	0	0	t.l.	t.l.	t.l.	0.83%	0.84%	0.69%
				35	35	45	725.32	375.15	269.63	1.42%	1.28%	0.94%
wcov	20	40	5	5	5	5	0.02	0.01	0.01	0.00%	0.00%	0.00%
			10	5	5	5	0.01	0.00	0.01	0.00%	0.00%	0.00%
		100	5	5	5	5	0.47	0.02	0.02	0.00%	0.00%	0.00%
			10	5	5	5	0.05	0.01	0.01	0.00%	0.00%	0.00%
		200	5	5	5	5	3.05	0.06	0.03	0.00%	0.00%	0.00%
			10	5	5	5	0.10	0.01	0.02	0.00%	0.00%	0.00%
	50	100	10	3	5	5	1442.22	27.44	0.67	0.56%	0.01%	0.01%
			20	5	5	5	0.87	0.23	0.07	0.00%	0.01%	0.00%
		250	10	0	0	5	t.l.	t.l.	36.44	6.06%	3.48%	0.01%
			20	3	5	5	837.59	39.35	1.88	0.31%	0.01%	0.00%
		500	10	0	0	1	t.l.	t.l.	2787.92	13.20%	11.08%	3.13%
			20	0	0	3	t.l.	t.l.	1503.03	4.48%	2.94%	0.65%
	100	200	10	0	0	0	t.l.	t.l.	t.l.	13.26%	10.25%	4.83%
			20	0	0	5	t.l.	t.l.	937.31	4.89%	1.51%	0.01%
		500	10	0	0	0	t.l.	t.l.	t.l.	20.36%	17.70%	10.03%
			20	0	0	0	t.l.	t.l.	t.l.	11.72%	9.64%	6.69%
		1000	10	0	0	0	t.l.	t.l.	t.l.	26.59%	23.34%	13.37%
			20	0	0	0	t.l.	t.l.	t.l.	17.37%	14.85%	11.97%
	200	400	20	0	0	0	t.l.	t.l.	t.l.	11.01%	8.80%	5.16%
			50	5	5	5	0.04	0.01	0.16	0.00%	0.00%	0.00%
		1000	20	0	0	0	t.l.	t.l.	t.l.	18.75%	16.66%	13.39%
			50	5	5	5	0.10	0.02	0.39	0.00%	0.00%	0.00%
		2000	20	0	0	0	t.l.	t.l.	t.l.	24.00%	21.96%	19.53%
			50	2	3	5	145.45	26.04	2.06	0.05%	0.05%	0.00%
				53	58	74	147.31	92.69	50.40	7.19%	5.93%	3.70%

Table 2. (*continued*)

Class	n	m	k	#solved			time(s)			gap		
				base	bc	bc+	base	bc	bc+	base	bc	bc+
binf	20	40	5	5	5	5	0.01	0.01	0.01	0.00%	0.00%	0.00%
			10	5	5	5	0.01	0.00	0.01	0.00%	0.00%	0.00%
		100	5	5	5	5	0.02	0.01	0.01	0.00%	0.00%	0.00%
			10	5	5	5	0.00	0.00	0.01	0.00%	0.00%	0.00%
		200	5	4	5	5	4.47	0.01	0.01	0.00%	0.00%	0.00%
			10	5	5	5	0.01	0.00	0.01	0.00%	0.00%	0.00%
	50	100	10	2	5	5	318.98	0.27	0.05	2.21%	0.01%	0.00%
			20	1	5	5	753.23	0.08	0.06	0.32%	0.01%	0.00%
		250	10	0	5	5	*t.l.*	17.41	0.23	1.35%	0.01%	0.01%
			20	0	5	5	*t.l.*	1.91	0.21	5.41%	0.01%	0.00%
		500	10	0	5	5	*t.l.*	88.92	0.82	2.59%	0.01%	0.01%
			20	1	5	5	886.66	7.15	0.69	0.33%	0.01%	0.00%
	100	200	10	0	0	5	*t.l.*	*t.l.*	14.12	8.62%	5.26%	0.01%
			20	0	0	2	*t.l.*	*t.l.*	2546.50	6.00%	4.15%	1.15%
		500	10	0	0	3	*t.l.*	*t.l.*	517.44	16.68%	11.26%	1.41%
			20	0	0	0	*t.l.*	*t.l.*	*t.l.*	11.52%	8.76%	6.25%
		1000	10	0	0	4	*t.l.*	*t.l.*	837.36	16.31%	11.51%	0.92%
			20	0	0	0	*t.l.*	*t.l.*	*t.l.*	12.41%	9.25%	7.01%
	200	400	20	0	0	0	*t.l.*	*t.l.*	*t.l.*	16.78%	12.95%	10.87%
			50	0	0	0	*t.l.*	*t.l.*	*t.l.*	2.84%	2.65%	2.20%
		1000	20	0	0	0	*t.l.*	*t.l.*	*t.l.*	21.51%	17.90%	15.67%
			50	0	0	0	*t.l.*	*t.l.*	*t.l.*	3.93%	3.81%	3.26%
		2000	20	0	0	0	*t.l.*	*t.l.*	*t.l.*	23.63%	20.07%	17.75%
			50	0	0	0	*t.l.*	*t.l.*	*t.l.*	4.76%	4.63%	4.16%
				33	60	74	398.46	93.58	42.83	6.55%	4.68%	2.94%
All				121	153	193	349.42	148.37	83.79	5.05%	3.96%	2.53%

which is able to solve 50% more models, and is overall twice as fast (and up to 10× faster as the models get harder). The automatic Benders' decomposition is even faster, solving additional models and being quite faster than CPLEX.

In hindsight, this is not unexpected: being able to express the full model as a mixed integer program allows for a lot of sophisticated techniques to be employed that can take advantage of a global view on the problem. Unsurprisingly, encoding the structure of f directly into the model results in being a better option than writing a model that can access f only through a black box oracle. On the other hand, a direct MIP formulation is not always a viable option, e.g., in the bipartite inference case, and this justifies the effort to make the general submodular framework more efficient. In addition, there is clearly still room for further research and improvements.

Table 3. Aggregated results on location models, comparing bc+ to CPLEX defaults and CPLEX automatic Benders.

	bc+			CPLEX			Benders		
Instances	#S	#T	time (s)	#S	#T	tQ	#S	#T	tQ
All	45	75	269.63	66	54	0.48	73	47	0.39
[0, 3600}	45	28	50.13	66	7	0.28	73	0	0.20
[1, 3600}	16	28	623.30	37	7	0.13	44	0	0.07
[10, 3600}	10	28	1551.18	31	7	0.10	38	0	0.05
[100, 3600}	5	28	2952.66	26	7	0.10	33	0	0.05

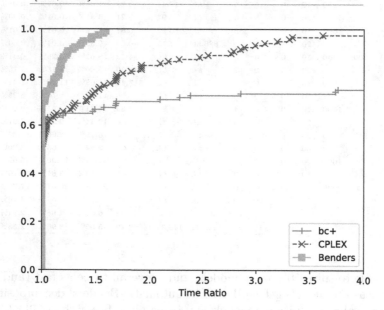

Fig. 2. Performance profile on location models, comparing bc+ to CPLEX defaults and CPLEX automatic Benders.

5   Conclusions and Future Research

In the present paper, we presented a modern implementation of the MIP model of Nemhauser and Wolsey for submodular function maximization, based on lazy constraint generation. We also developed some algorithmic improvements, namely a GRASP heuristic and two (heuristic) procedures for separating submodular cuts from fractional solutions. A computational evaluation on three classes of submodular functions showed that the developed methods significantly improve over the basic constraint generation model by the book. A comparison with a direct MIP formulation for one class of models also showed that, when available, this is usually a preferable option, being not only far easier to implement but also quite faster than the MIP-based framework based on the general model. Still, the general framework is in its computational infancy, and further research is needed in many areas. Among others, more effective (and possibly exact) separation procedures over the family of cuts (9)—or even (6)—and custom branching rules based on f could make the method more efficient in practice.

References

1. Achterberg, T.: Constraint integer programming. Ph.D. thesis, Technische Universität Berlin (2007)
2. Applegate, D.L., Bixby, R.E., Chvátal, V., Cook, W.J.: Finding cuts in the TSP (A preliminary report). Technical report 95–05, DIMACS (1995)
3. Applegate, D.L., Bixby, R.E., Chvátal, V., Cook, W.J.: The Traveling Salesman Problem: A Computational Study. Princeton University Press, Princeton (2007)
4. Benders, J.F.: Partitioning procedures for solving mixed-variables programming problems. Numerische Mathematik 4, 238–252 (1962)
5. Chambolle, A.: Total variation minimization and a class of binary MRF models. In: Rangarajan, A., Vemuri, B., Yuille, A.L. (eds.) EMMCVPR 2005. LNCS, vol. 3757, pp. 136–152. Springer, Heidelberg (2005). https://doi.org/10.1007/11585978_10
6. Chen, W., Chen, Y., Weinberger, K.Q.: Filtered search for submodular maximization with controllable approximation bounds. In: AISTATS, vol. 38, pp. 156–164 (2015)
7. Dolan, E., Moré, J.J.: Benchmarking optimization software with performance profiles. Math. Program. 91, 201–213 (2002)
8. Dueck, D., Frey, B.J.: Non-metric affinity propagation for unsupervised image categorization. In: ICCV, pp. 1–8. IEEE Computer Society (2007)
9. Feo, T.A., Resende, M.G.C.: A probabilistic heuristic for a computationally difficult set covering problem. Oper. Res. Lett. 8, 67–71 (1989)
10. IBM. ILOG CPLEX 12.8 User's Manual (2018)
11. Jović, A., Brkić, K., Bogunović, N.: A review of feature selection methods with applications. In: MIPRO, pp. 1200–1205 (2015)
12. Kawahara, Y., Nagano, K., Tsuda, K., Bilmes, J.A.: Submodularity cuts and applications. In: Bengio, Y., Schuurmans, D., Lafferty, J.D., Williams, C.K.I., Culotta, A. (eds.) NIPS, pp. 916–924 (2009)
13. Kempe, D., Kleinberg, J., Tardos, E.: Maximizing the spread of influence in a social network. In: KDD, pp. 137–146 (2003)
14. Krause, A., Golovin, D.: Submodular Function Maximization, pp. 71–104. Cambridge University Press, Cambridge (2014)
15. Lovász, L.: Submodular functions and convexity. In: Bachem, A., Grötschel, M., Korte, B. (eds.) Mathematical Programming - The State of the Art, pp. 235–257. Springer, Heidelberg (1983). https://doi.org/10.1007/978-3-642-68874-4_10
16. Minoux, M.: Accelerated greedy algorithms for maximizing submodular set functions. In: Stoer, J. (ed.) Optimization Techniques. LNCIS, vol. 7, pp. 234–243. Springer, Heidelberg (1978). https://doi.org/10.1007/BFb0006528
17. Nemhauser, G.L., Wolsey, L.A.: Maximizing submodular set functions: formulations and analysis of algorithms. In: Hansen, P. (ed.) Studies on Graphs and Discrete Programming, pp. 279–301 (1981)
18. Nemhauser, G.L., Wolsey, L.A., Fisher, M.L.: An analysis of approximations for maximizing submodular set functions - I. Math. Program. 14(1), 265–294 (1978)
19. Sakaue, S., Ishihata, M.: Accelerated best-first search with upper-bound computation for submodular function maximization. In: McIlraith, S.A., Weinberger, K.Q. (eds.) AAAI, pp. 1413–1421 (2018)

Metric Hybrid Factored Planning in Nonlinear Domains with Constraint Generation

Buser Say[(✉)] and Scott Sanner

Department of Mechanical and Industrial Engineering,
University of Toronto, Toronto, Canada
{bsay,ssanner}@mie.utoronto.ca

Abstract. We introduce a novel planner SCIPPlan for metric hybrid factored planning in nonlinear domains with general metric objectives, transcendental functions such as exponentials, and instantaneous continuous actions. Our key contribution is to leverage the spatial branch-and-bound solver of SCIP inside a nonlinear constraint generation framework where we iteratively check relaxed plans for temporal feasibility using a domain simulator, and repair the source of the infeasibility through a novel nonlinear constraint generation methodology. We experimentally evaluate SCIPPlan on a variety of domains, showing it is competitive with, or outperforms, ENHSP in terms of run time and makespan and handles general metric objectives. SCIPPlan is also competitive with a general metric-optimizing unconstrained Tensorflow-based planner (TF-Plan) in nonlinear domains with exponential transition functions and metric objectives. Overall, this work demonstrates the potential of combining nonlinear optimizers with constraint generation for planning in expressive metric nonlinear hybrid domains.

Keywords: Metric hybrid planning · Nonlinear optimization · Constraint generation

1 Introduction

Metric optimization is at the core of many real-world nonlinear hybrid [6] planning domains where the *quality* of the plan matters. Most nonlinear hybrid planners in the literature either ignore metric specifications [3,4,12], or leverage heuristics to guide their search for finding a plan quickly [9,13] with the notable exceptions COLIN [5] and ENHSP [17], which can handle metric optimization for a subset of PDDL+ [6] domains.

In this paper, we leverage the nonlinear constrained optimization solver SCIP [10] to present SCIPPlan for solving metric hybrid factored [2] nonlinear planning problems by decomposing the original problem into a master problem and a subproblem. In the master problem, we relax the original problem

© Springer Nature Switzerland AG 2019
L.-M. Rousseau and K. Stergiou (Eds.): CPAIOR 2019, LNCS 11494, pp. 502–518, 2019.
https://doi.org/10.1007/978-3-030-19212-9_33

to a system of sequential function updates[1], which allows us to handle arbitrary nonlinear functions (such as polynomial, exponential, logarithmic etc.) in the transition and metric objectives as well as instantaneous continuous action inputs that are beyond the expressivity of existing hybrid planners.

In the nonlinear hybrid planning literature, the time at which a conditional expression (e.g., a mode switch condition) is satisfied is known as a *zero-crossing* and when the dynamics of the planning problem are piecewise linear, one can use the TM-LPSAT compilation to find valid plans that avoid zero-crossings [18]. When the continuous change can be described more generally as polynomials, one can use the SMTPlan [4] compilation of the hybrid planning problem to avoid zero-crossings between two consecutive decision points (i.e. happenings). However, in general, problem dynamics can include arbitrary nonlinear change and only ENHSP [17] approaches the expressive dynamics of SCIPPlan.

In SCIPPlan, the candidate solution found by solving the master problem can include zero-crossings of general transcendental nonlinear conditions between two consecutive decisions which can either (i) violate the global constraints of the original problem, or (ii) contain mode switches that are not accounted for by the master problem. To identify and repair the source of zero-crossings, we use the simulate-and-validate approach [7] in the subproblem where domain simulators are used to simulate the candidate plan, and if the candidate plan is found to be infeasible, temporal constraints associated with zero-crossings are generated and added back to the master problem. SCIPPlan iteratively solves the master problem and the subproblem until a valid plan is found.

Experimentally, we show that SCIPPlan outperforms the state-of-the-art metric nonlinear hybrid planner ENHSP in almost all problem instances with respect to makespan and run time performance. We further experiment with the capabilities of SCIPPlan beyond the expressiveness limitations of ENHSP in the optimization of general metrics on a subset of modified domains, and verify its competitiveness versus an unconstrained Tensorflow-based planner (TF-Plan) [19] on a nonlinear metric domain with exponential transitions.

2   Preliminaries

In this section, we present the preliminary definitions, notation and solution methodologies required to define and solve the metric hybrid factored planning problem.

3   Metric Discrete Time Factored Planning: Π

Before we dive into the notationally heavy details of general nonlinear hybrid factored planning, we begin with a straightforward mixed-integer nonlinear program (MINLP) compilation of a *discrete time* factored nonlinear planning domain. A discrete time metric factored planning problem is a tuple $\Pi = \langle S, A, C, T, I, G, Q \rangle$ where

---

[1] Relaxation refers to the omission of temporal constraints from the master problem.

- $S = \{S^d, S^c\}$ is a set of discrete S^d and continuous S^c domains with state variables/assignments denoted $s \in S$,
- $A = \{A^d, A^c\}$ is a set of discrete A^d and continuous A^c domains with action variables/assignments denoted $a \in A$,
- $C : S \times A \rightarrow \{true, false\}$ is a function that returns true if action $a \in A$ and state $s \in S$ pair satisfies constraints that represent global constraints on state and action variables,
- $T : S \times A \rightarrow S$ denotes the state transition function between discrete time steps t and $t+1$, $T(s^t, a^t) = s^{t+1}$ if $C(s^t, a^t) = true$, and is undefined otherwise, and
- $Q : S \times A \rightarrow \mathbb{R}$ is the metric reward function to optimize.

In addition, I represents the initial state constraint $s^1 = \bar{s}^1$ and $G : S \rightarrow \{true, false\}$ represents goal state constraints. Given a finite planning horizon of H decision stages, a solution $\pi = \langle \bar{a}^1, \dots, \bar{a}^H \rangle$ (i.e. plan) to Π is a fixed value assignment to actions $a^t = \bar{a}^t$ that induces an assignment to state variables s^t satisfying the initial state I, transition T, goal G, and global C constraints for all $t \in \{1, \dots, H\}$. Our objective in solving metric planning problem Π is to find the action sequence π that maximizes the sum of rewards over the time horizon by optimizing the following model:

$$\max_{\pi = \langle a^1, \dots, a^H \rangle} \sum_{t=1}^{H} Q(s^{t+1}, a^t) \tag{1}$$

$$\text{subject to } I : s^1 = \bar{s}^1$$
$$G(s^{H+1})$$
$$T(s^t, a^t) = s^{t+1} \quad \forall t \in \{1, \dots, H\}$$
$$C(s^t, a^t) \quad \forall t \in \{1, \dots, H\}$$

Note that Π is a standard discrete-time model that does not consider the (potentially) changing values of states between pairs of consecutive time steps $t, t+1 \in \{1, \dots, H\}$. Under this simplifying assumption, there is no need to consider zero-crossing constraints that will become critical for relaxing time to be continuous in our subsequent hybrid generalization of the above framework. Before we present the hybrid generalization, however, we discuss the compilation and solution of the above problem followed by an example.

3.1   High-Level Syntax and SCIP MINLP Compilation

In order to compile the optimization formulation of (1) into a Mixed-Integer Nonlinear Programming (MINLP) formulation that can be solved via an off-the-shelf MINLP solver (e.g., SCIP [10]), we need (i) a high-level syntax such as the RDDL language [15] for specifying all constraints and functions and (ii) a compilation that can translate any formula in this syntax into the MINLP language. For example, piecewise functions induced by if-then-else constructs require use of the big-M trick to encode conditional constraints, while boolean

Table 1. (Left column) Grammar to recursively generate expression syntax of the RDDL language [15] extending [14] to nonlinear expressions in the last four rows. E_1 and E_2 belong to the same language as E and are acyclic. (Middle column) Conditions on grammar rule application. (Right column) MINLP compilation of the grammar rule: every RDDL expression E is represented by an MINLP variable v_E that evaluates to the value of that expression ($\{0, 1\}$ if boolean). M is a large constant.

Expression	Condition	Constraints
$E \to k$	k is a constant	$v_E = k$
$E_b \to \top$ (or $\bot$)		$v_{E^b} = 1$ (or 0)
$E_b \to p$	p is state or action variable	$v_E = v_p$
$E \to \wedge_{i=1}^n E_b^i \equiv \forall_i E_b^i$	E_b^i is a boolean expression	$n v_E \leq \sum_{i=1}^n v_{E_b^i} \leq n - 1 + v_E$
$E \to \vee_{i=1}^n E_b^i \equiv \exists_i E_b^i$	E_b^i is a boolean expression	$v_E \leq \sum_{i=1}^n v_{E_b^i} \leq n v_E$
$E \to \neg E_b$	E_b is a boolean expression	$v_E = 1 - E_b,\ v_E, v_{E_b^i} \in \{0, 1\}$
$E \to k E_1$	k is a constant	$v_E = k v_{E_1}$
$E \to E_1\ op\ E_2$	$op \in \{+, -\}$	$v_E = v_{E_1}\ op\ v_{E_2}$
$E_b \to E_1 \geq E_2$	E_b is a boolean expression	$M v_{E_b} - M \leq v_{E_1} - v_{E_2}$ $\leq M v_{E_b}, v_{E_b} \in \{0, 1\}$
$E \to$ if E_b then E_1 else E_2	E_b is a boolean expression	$v_{E_1} + M v_{E_b} - M \leq v_E$ $\leq M + v_{E_1} - M v_{E_b},$ $v_{E_2} - M v_{E_b} \leq v_E$ $\leq v_{E_2} + M v_{E_b}, v_{E_b} \in \{0, 1\}$
$E \to E_1\ op\ E_2$	$op \in \{\times, \div\}$	$v_E = v_{E_1}\ op\ v_{E_2}$
$E \to \exp(E_1)$		$v_E = e^{v_{E_1}}$
$E \to \log(E_1)$	E_1 is a positive expression	$v_E = log(v_{E_1})$
$E \to \mathrm{abs}(E_1)$		$v_E = |v_{E_1}|$

expressions in constraints and if-then-else conditions require special encodings as arithmetic expressions over integers.

In Table 1, we provide a grammar for the (nonlinear) expression syntax of the ground RDDL language and a compilation of each grammar rule to the SCIP MINLP format assuming each sub-expression has been recursively compiled.

3.2 Spatial Branch-and-Bound

To solve the compiled MINLP, SCIP uses Spatial Branch-and-Bound (SBB) [11] – an algorithm based on the *divide-and-conquer* strategy for solving MINLPs in the form of min $f(\mathbf{x})$ subject to $\mathbf{g}(\mathbf{x}) \leq 0$ where function $f(x)$ and function vector $\mathbf{g}(\mathbf{x})$ contain nonlinear expressions, and the decision variable vector $\mathbf{x}$ can have continuous and/or discrete domains. The SBB algorithm uses tree search where branching decisions are made on candidate solutions $\bar{\mathbf{x}}$, and the optimal value of the objective function $f(\bar{x})$ is bounded at each search node until a preset optimality gap is reached.

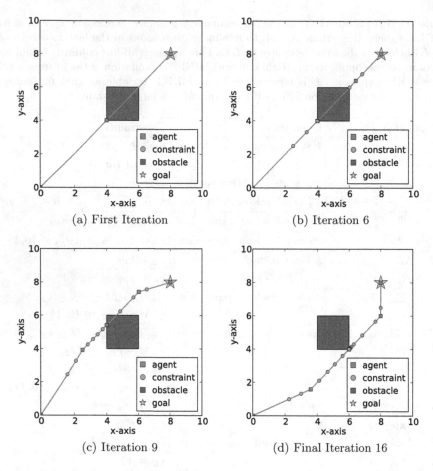

(a) First Iteration

(b) Iteration 6

(c) Iteration 9

(d) Final Iteration 16

Fig. 1. Visualization of iterative plan generation of SCIPPlan for the example *hybrid* navigation domain. In the first six iterations, the plan steps π (in red) generated to reach the goal (in orange) pass through the obstacle (in blue), violating zero-crossing constraint c_3 that is detected during plan simulation. At each iteration, additional zero-crossing constraints are generated *symbolically* at the midpoints of each violation interval (in green) to eliminate these zero-crossings from the solution space of the master problem. By iteration 9, SCIPPlan starts to converge to a valid plan and by iteration 16, SCIPPlan returns a valid plan. Note that sometimes there are overlaps of the position of the agent between time steps (i.e., the agent does not move). (Color figure online)

3.3   Illustrative Example

To illustrate how the MINLP compilation and solution works for the metric factored planning problem, we consider the following simple navigation domain with (i) three continuous action variables $(a_x, a_y, \Delta) \in A^c$ that move the agent a_x and a_y in respective x and y directions for duration Δ, (ii) two continuous state variables $(s_x, s_y) \in S^c$ representing agent location, (iii) and three constraints in C:

$$c_1 : 0 \leq s_x^t + a_x^t \Delta^t, s_y^t + a_y^t \Delta^t \leq 10 \quad \forall t \in \{1, \ldots, H\},$$

$$c_2 : -1 \leq a_x^t, a_y^t \leq 1 \quad \forall t \in \{1, \ldots, H\}, \text{and}$$

$$c_3 : 4 \geq s_x^t + a_x^t \Delta^t \vee 6 \leq s_x^t + a_x^t \Delta^t$$

$$\vee \, 4 \geq s_y^t + a_y^t \Delta^t \vee 6 \leq s_y^t + a_y^t \Delta^t \quad \forall t \in \{1, \ldots, H\}.$$

Here, constraints c_1, c_2 denote bounds on the domains of state variables s_x and s_y, and constraint c_3 represents an obstacle located in the middle of the maze. Initial and goal constraints are compiled as follows for $H = 4$:

$$I : s_x^1, s_y^1 = 0, \quad G : s_x^{H+1}, s_y^{H+1} = 8,$$

Given the transition function

$$T : s_x^{t+1} = s_x^t + a_x^t \Delta^t, \quad s_y^{t+1} = s_y^t + a_y^t \Delta^t \quad \forall t \in \{1, \ldots, H\}$$

and reward metric $Q(s^{t+1}, a^t) = -\Delta^t$ (minimize total time, a.k.a. makespan), the SSB solver can return a plan π as visualized by Fig. 1(a). The plan π passes through the obstacle since the discrete time formalization only checks constraints at the start and end points of each decision stage; we remedy this with a hybrid extension in the next section.

4   Metric Hybrid Factored Planning: Π^δ

In this section, we define the metric hybrid factored planning problem Π^δ by building on the notation, definitions and the solution methodology presented for the metric factored planning problem Π. But before we define Π^δ, first we need to distinguish one continuous action variable as the control duration $\Delta \in A^c$ to specify the duration of time step $t \in \{1, \ldots, H\}$ such that $0 \leq \Delta^2$. Similarly, we update the notation we use for the global constraint function $C(s^t, a^t, \Delta^t)$ and the state transition function $T(s^t, a^t, \Delta^t)$ to explicitly specify the duration Δ^t of time step $t \in \{1, \ldots, H\}$. Finally, we need to distinguish the set of boolean expressions that appear in if-else conditions of the state transition function T as transition modes M, that is,

$$T(s^t, a^t, \Delta^t) = \text{if } E_b^1(s^t, a^t, \Delta^t) \text{ then } E_1(s^t, a^t, \Delta^t)$$

$$\cdots$$

$$\text{elif } E_b^n(s^t, a^t, \Delta^t) \text{ then } E_n(s^t, a^t, \Delta^t)$$

$$\text{else } E_{n+1}(s^t, a^t, \Delta^t)$$

---

[2] In this work, we focus on hybrid planning problems where duration Δ is completely controlled by the planner. When there are exogenous events or processes that can change the total duration of a time step, we need to define a continuous state variable $\Delta' \in S^c$ as a function of s, a, Δ such that $f(s, a, \Delta) = \Delta'$ and transfer zero-crossing definitions onto Δ'. In this work, we assume $\Delta = \Delta'$ and omit Δ' for notational simplicity.

where $E_b^1(\bar{s}^t, \bar{a}^t, \Delta^t), \ldots, E_b^n(\bar{s}^t, \bar{a}^t, \Delta^t) \in \mathsf{M}$. We denote $\mathsf{M}^\delta : S \times A \times \Delta \to P(\mathsf{M})$ as a function that returns the set of transition modes evaluating to true for given values of state $\bar{s}^t$ and action $\bar{a}^t$ variables and duration $\bar{\Delta}^t$ for $t \in \{1, \ldots, H\}$ where notation $P(S)$ denotes the power set of S.

Definition 1. *(Zero-Crossing Certificate): Given the values of state $\bar{s}^t$ and action $\bar{a}^t$ variables and duration $0 < \bar{\Delta}^t$ for $t \in \{1, \ldots, H\}$, we say $x^t \in (0, \bar{\Delta}^t)$ is a zero-crossing certificate for time step t if at least one of the following holds:*

1. *Global Constraint Violation:* $C(\bar{s}^t, \bar{a}^t, x^t) = false$,
2. *Mode[3] Switch:* $\mathsf{M}^\delta(\bar{s}^t, \bar{a}^t, x^t) \neq \mathsf{M}^\delta(\bar{s}^t, \bar{a}^t, \bar{\Delta}^t)$.

Given the definition of the zero-crossing certificate, the metric hybrid factored planning problem is a tuple $\Pi^\delta = \langle S, A, C, C^\delta, T, I, G, Q \rangle$ where $C^\delta : S \times A \times \Delta \to \{true, false\}$ is a function defined as $C^\delta(\bar{s}^t, \bar{a}^t, \Delta^t) = true$ if and only if there does not exist $x^t \in (0, \Delta^t)$ that is a zero-crossing certificate. Given a planning horizon H, a plan $\pi^\delta = \langle \bar{a}^1, \bar{\Delta}^1 \ldots, \bar{a}^H, \bar{\Delta}^H \rangle$ to Π^δ is a plan π to Π where $C^\delta(\bar{s}^t, \bar{a}^t, x^t) = true$ for all $x^t \in (0, \bar{\Delta}^t)$ and $t \in \{1, \ldots, H\}$. Note that the definition Π^δ extends deterministic RDDL [15] to continuous time and allows instantaneous continuous actions $A^c \subseteq A$ that are not functions of time. Unlike the PDDL+ [6] formalism, we do not assume that the effects of instantaneous actions are realized ϵ time after their execution.

5   Solving Π^δ with Constraint Generation

In this section, we introduce our novel SCIP-based planner (SCIPPlan) to plan in metric hybrid planning problems with nonlinear dynamics. But before we outline SCIPPlan, we first need to define the zero-crossing interval.

Definition 2. *(Zero-Crossing Interval): Given the values of state $\bar{s}^t$ and action $\bar{a}^t$ variables, and the duration $0 < \bar{\Delta}^t$ of time step $t \in \{1, \ldots, H\}$, a zero-crossing is an interval $|_L x_1^t, x_2^t |_R$ where $|_L \in \{[, (\}$ and $|_R \in \{],)\}$ if and only if:*

1. *Non-empty: $0 < x_1^t \leq x_2^t < \bar{\Delta}^t$ such that if $x_1^t = x_2^t$ then $|_L x_1^t, x_2^t |_R$ is not an open interval (x_1^t, x_2^t), and*
2. *Uninterrupted and Contiguous: $\forall x \in |_L x_1^t, x_2^t |_R$ where x is a zero-crossing certificate.*

The novelty of SCIPPlan is that it decomposes Π^δ into a master problem $\mathcal{M}(\Pi, H)$ and a subproblem $\mathcal{S}(\pi, \epsilon)$, where $\mathcal{M}(\Pi, H)$ solves the metric factored planning problem Π for a given horizon H using a SBB solver, and $\mathcal{S}(\pi, \epsilon)$ checks whether π is also a plan for Π^δ using a domain simulator with respect to a time discretization parameter ϵ. If π is not a plan for Π^δ, $\mathcal{S}(\pi, \epsilon)$ returns the first zero-crossing interval $|_L x_1^t, x_2^t |_R$ with minimum time step $t \in \{1, \ldots, H\}$, and a temporal constraint is added back to the master problem $\mathcal{M}(\Pi, H)$ to update either function C or T, depending on whether the zero-crossing is due to a global constraint violation or a mode switch, respectively.

---

[3] The concept of a mode is analogous to its counterpart in the field of Hybrid Automata [8].

5.1   Master Problem

The master problem $\mathcal{M}(\Pi, H)$ solves the metric factored planning problem Π for a given horizon H, using the compilation presented for Π in the MINLP formulation of (1) assisted by complex expression compilation of Π to MINLP form provided in Table 1. We optimize the MINLP in (1) using the SCIP SBB solver [10].

5.2   Subproblem

Given a plan π for Π and a discretization parameter ϵ, the subproblem $\mathcal{S}(\pi, \epsilon)$ uses a domain simulator to check for a zero-crossing certificate by simulating the state transition T sequentially $\lfloor \frac{\bar{\Delta}^t}{\epsilon} \rfloor$ times for all time steps $t \in \{1, \dots, H\}$ such that $T(\bar{s}^t, \bar{a}^t, \epsilon) \dots T(\bar{s}^t, \bar{a}^t, \epsilon \lfloor \frac{\bar{\Delta}^t}{\epsilon} \rfloor)$. If a zero-crossing certificate is found, $\mathcal{S}(\pi, \epsilon)$ returns (i) the first zero-crossing interval $\lfloor_L x_1^t, x_2^t\rfloor_R$ with minimum time step $t \in \{1, \dots, H\}$ such that there does not exist another zero-crossing certificate x^t ? x_1^t found by $\mathcal{S}(\pi, \epsilon)$ where the relational operator ? is $<$ (i.e., greater) if $|_L = [$ (i.e., minimum bound of the interval is closed) and $\leq$ (i.e., greater or equal to) otherwise, and (ii) the set of compilation constraints g^t that cause the zero-crossing interval $\lfloor_L x_1^t, x_2^t\rfloor_R$.

Precisely, the zero-crossings due to (i) global constraint violation can be mapped to a set of compilation constraints representing the global constraint function C (as presented in the Illustrative Example Revisited section). Zero-crossings due to (ii) mode switch can be mapped to a set of boolean expressions E_b^t and to their respective compilation boolean decision variables $v_{E_b}^t$—these evaluate to different values within zero-crossing interval $\lfloor_L x_1^t, x_2^t\rfloor_R$ compared to the end of control duration $\bar{\Delta}^t$ at time step $t \in \{1, \dots, H\}$.

5.3   Temporal Constraint Generation

Given the interval $\lfloor_L x_1^t, x_2^t\rfloor_R$ identified by the domain simulator for a time step $t \in \{1, \dots, H\}$ and the respective set of constraints g^t, SCIPPlan generates a nonlinear constraint

$$g^t(k\Delta^t) \leq 0, \quad k = \frac{x_2^t + x_1^t}{2\bar{\Delta}^t}, \tag{2}$$

where Constraint 2 symbolically substitutes all Δ^t with $k\Delta^t$. Note that $k \in [0, 1]$ is a constant coefficient representing the ratio of the mid-point of the zero-crossing interval $\lfloor_L x_1^t, x_2^t\rfloor_R$ to the complete duration at time step t. There are four benefits of our constraint generation methodology: (i) We generate a symbolic[4] constraint ensuring the zero-crossing violation of the current plan is enforced, while generalizing as a valid constraint for all other plans. (ii) Instantiation of zero-crossing constraints at the violation midpoint is intended to induce a

---

[4] Symbolic refers to the fact that Constraint (2) is a function of decision variables (i.e., s^t, a^t, Δ^t) whose values are decided at optimization time.

binary search refinement in the constraint generation process. (iii) SCIPPlan only generates temporal constraints as needed, thus substantially reducing MINLP size. (iv) Constraint (2) only perturbs g^t by changing its coefficients and not adding any additional decision variables, thus allowing SBB solvers to reuse information between iterations (e.g., warm start features). Given the descriptions of $\mathcal{M}(\Pi, H)$, $\mathcal{S}(\pi, \epsilon)$ and Constraint (2), SCIPPlan is outlined by Algorithm 1.

Algorithm 1. SCIPPlan

1: $H \leftarrow 1$, $\pi \leftarrow \emptyset$, $x_1^t, x_2^t, g^t \leftarrow \emptyset$, $\epsilon \leftarrow$ small numerical constant
2: **while** π is $\emptyset$ **do**
3:     $\pi \leftarrow \mathcal{M}(\Pi, H)$
4:     **if** π is $\emptyset$ **then**
5:         $H \leftarrow H + 1$.
6:     **else** $|_L x_1^t, x_2^t|_R, g^t \leftarrow \mathcal{S}(\pi, \epsilon)$
7:         **if** $|_L x_1^t, x_2^t|_R, g^t$ are $\emptyset$ **then**
8:             **return** π
9:         **else** $\mathcal{M}(\Pi, H) \leftarrow g^t(k\Delta^t) \leq 0$ where $k = \frac{x_2^t + x_1^t}{2\Delta^t}$

5.4   Illustrative Example Revisited

We have previously ended the illustrative example where the master problem $\mathcal{M}(\Pi, H)$ (i.e., the SSB solver) returned the plan $\pi = \langle \bar{a}_x^1 = 0, \bar{a}_y^1 = 0, \bar{\Delta}^1 = 0, \bar{a}_x^2 = 1, \bar{a}_y^2 = 1, \bar{\Delta}^2 = 4, \bar{a}_x^3 = 0, \bar{a}_y^3 = 0, \bar{\Delta}^3 = 0, \bar{a}_x^4 = 1, \bar{a}_y^4 = 1, \bar{\Delta}^4 = 4 \rangle$ as visualized by Fig. 1(a). The subproblem $\mathcal{S}(\pi, \epsilon)$ will detect the zero-crossing interval by simulating the transition function T for all time steps $t \in \{1, \dots, H\}$ and detect the first violation of constraint c_3 which occurs within the interval $[0, 2]$ over duration $\bar{\Delta}^4 = 4$. Given the identified zero-crossing interval $[0, 2]$ for time step $t = 4$ and the violated constraint c_3, the following constraint (i.e., checking for the obstacle at the midpoint of the zero-crossing)

$$g_1^4 : 4 \geq s_x^4 + (0.25)a_x^4 \Delta^4 \vee 6 \leq s_x^4 + (0.25)a_x^4 \Delta^4$$
$$\vee \, 4 \geq s_y^4 + (0.25)a_y^4 \Delta^4 \vee 6 \leq s_y^4 + (0.25)a_y^4 \Delta^4$$

will be added to the master problem. As visualized by Fig. 1(b–d), the master problem would then be re-solved and further constraints will be generated if needed. Once no zero-crossings are detected in a solution, that plan would be returned as the final plan π^δ in Fig. 1(d).

6   Experimental Results

In this section, we test the computational efficacy of SCIPPlan on three metric hybrid factored planning problems Π^δ, namely HVAC [1],

ComplexPouring [17], [3], NavigationJail, against ENHSP [17], and on one metric factored planning problem Π, namely NavigationMud [16], against TF-Plan [19][5] with respect to run time and solution quality. Unless otherwise stated, all domains minimize total time (i.e., makespan) $Q(s^{t+1}, a^t) = -\Delta^t$.

6.1 Domain Descriptions

In this section, we describe the benchmark domains in detail. The domains were chosen to test the capabilities of SCIPPlan on metric optimization, handling nonlinear transitions and concurrency.

Heating, Ventilation and Air Conditioning is the problem of heating different rooms $r \in R$ of a building upto a desired temperature by sending heated air b_r. The temperature of a room h_r^{t+1} is bilinear function of its current temperature h_r^t, the volume of heated air sent to the room b_r, the temperature of the adjacent rooms $h_{r'}^t$ and the duration Δ^t of the control input at time step t where $r' \in Adj(r) \subset R$ denotes the set of adjacent rooms to room r. The dynamics of the domain are described as follows:

$$h_r^{t+1} = h_r^t + \frac{\Delta^t}{C_r}(b_r + \sum_{r' \in Adj(r)} \frac{h_{r'}^t - h_r^t}{W_{r,r'}}) \tag{3}$$

for all $r \in R, t \in \{1, \ldots, H\}$ where C_r and $W_{r,r'}$ are parameters denoting the heat capacity of room r and the heat resistance of the wall between r and r', respectively. Moreover, the initial and the goal constraints are described as $h_r^1 = H_r^{init}$ and $h_r^{H+1} = H_r^{goal}$ for all rooms $r \in R$ where the parameters H_r^{init} and H_r^{goal} denote the initial and goal temperatures of the rooms, respectively.

ComplexPouring is the problem of filling buckets $b \in B$ upto a desired volume with the water that is initially stored in the tanks $u \in U$. The volume of a bucket v_b^{t+1} (or a tank) is a nonlinear function of its current volume v_b^t (or v_u^t), volume of water poured in (and out) from (and to) other tanks and Δ^t at time step t. The dynamics of the domain are described as follows:

$$v_b^{t+1} = v_b^t + \overrightarrow{v}_b^t - \overleftarrow{v}_b^t \qquad\qquad \forall b \in B \cup U \tag{4}$$

$$\overrightarrow{v}_b^t = \sum_{u \in U} \Delta^t p_{u,b}^t (2R_u \sqrt{v_u^t} - R_u^2) \qquad\qquad \forall b \in B \cup U \tag{5}$$

$$\overleftarrow{v}_b^t = \sum_{u \in B \cup U} \Delta^t p_{b,u}^t (2R_b \sqrt{v_b^t} - R_b^2) \qquad\qquad \forall b \in U \tag{6}$$

$$0 \leq v_b^t + \overrightarrow{v}_b^t - \overleftarrow{v}_b^t \leq V_b^{max\,t} \qquad\qquad \forall b \in B \cup U \tag{7}$$

for all $t \in \{1, \ldots, H\}$ where $p_{b,u}^t \in \{0, 1\}$ is a binary decision variable denoting whether tank b pours into bucket (or tank) u at time step t, and R_b and $V_b^{max\,t}$ are

---

[5] We note that TF-Plan does not handle (i) discrete variables, (ii) global or goal constraints, or (iii) support dynamic time discretization, but can handle exponential transitions and complex metric objectives (e.g., NavigationMud).

parameters denoting the flow rate and capacity of bucket (or tank) b, respectively. Further, the initial and the goal constraints are described as $v_b^1 = V_b^{init}$ for all buckets and tanks $b \in B \cup U$ and $v_b^{H+1} \geq V_b^{goal}$ for all buckets $b \in B$ where the parameters V_b^{init} and V_b^{goal} denote the initial and goal volumes of tanks and buckets, respectively.

NavigationJail is a two-dimensional $d \in \{x, y\} = D$ path- finding domain that is designed to test the ability of planners to handle instantaneous events. The location of the agent l_d^{t+1} is a nonlinear function (i.e., cubic polynomial) of its current location l_d^t, speed v_d^t, acceleration a_d^t and Δ^t at time step t. Moreover, the agent can be instantaneously relocated to its initial position L_d^{init} for all dimensions $d \in D$ and set its speed to 0 if it travels through a two-dimensional jail area that is located in the middle of the maze with the corner points J_d^{min}, J_d^{max} for all $d \in D$. The system dynamics of the domain is described as follows:

$$l'^t_d = l_d^t + v_d^t \Delta^t + 0.5 a_d^t (\Delta^t)^2 \qquad \forall d \in D \qquad (8)$$

$$v'^t_d = v_d^t + a_d^t \Delta^t \qquad \forall d \in D \qquad (9)$$

$$\textbf{if} \quad \forall d \in D \quad J_d^{min} \leq l'^t_d \leq J_d^{max} \qquad (10)$$

$$\textbf{then} \quad l_d^{t+1} = L_d^{init}, \, v_d^{t+1} = 0 \qquad \forall d \in D \qquad (11)$$

$$\textbf{else} \quad l_d^{t+1} = l'^t_d, \, v_d^{t+1} = v'^t_d \qquad \forall d \in D \qquad (12)$$

$$L_d^{min} \leq l'^t_d \leq L_d^{max}, \quad A_d^{min} \leq a_d^t \leq A_d^{max} \qquad \forall d \in D \qquad (13)$$

for all $t \in \{1, \ldots, H\}$ where (L_d^{min}, L_d^{max}) and (A_d^{min}, A_d^{max}) are the minimum and the maximum boundaries of the maze and the control input for dimension $d \in D$, respectively. The goal of the domain is to find a path from the initial location L_d^{init} to the goal location L_d^{goal} for all dimensions $d \in D$. The initial and the goal constraints are described as $l_d^1 = L_d^{init}$, $v_d^1 = 0$, and $l_d^{H+1} = L_d^{goal}$ for all dimensions $d \in D$, respectively.

NavigationMud is a two-dimensional $d \in \{x, y\} = D$ domain that is designed to test the ability of planners to handle transcendental functions with general optimization metrics. The location of the agent l_d^{t+1} is a nonlinear function (i.e., exponential) of its current location l_d^t and positional displacement action p_d^t at time step t due to higher slippage in the center of the maze. The system dynamics of the domain is described as follows:

$$l_d^{t+1} = l_d^t - 0.99 + p_d^t \frac{2.0}{1.0 + e^{-2y^t}} \qquad \forall d \in D \qquad (14)$$

$$y^t = \sqrt{\sum_{d \in D} (l_d^t - \frac{L_d^{max} - L_d^{min}}{2.0})^2} \qquad (15)$$

$$L_d^{min} \leq l_d^t \leq L_d^{max}, \quad P_d^{min} \leq p_d^t \leq P_d^{max} \qquad \forall d \in D \qquad (16)$$

for all $t \in \{1, \ldots, H\}$ where (P_d^{min}, P_d^{max}) are the minimum and the maximum boundaries of the positional displacement for dimension $d \in D$, respectively.

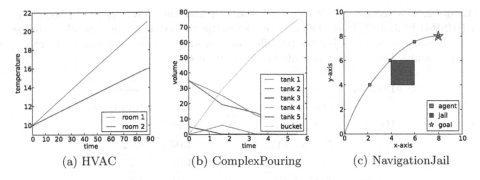

(a) HVAC   (b) ComplexPouring   (c) NavigationJail

Fig. 2. Visualization of example plans generated by SCIPPlan. The inspection of plan traces show from left to right: linear, piecewise linear, and nonlinear state transitions as a function of time. As observed in Table 2, we remark that the nonlinear domain (right) requires significantly more compute time than the linear (left) and piecewise linear (middle) domains.

The objective of the domain is to find a path from the initial location L_d^{init} that is described by the constraint $l_d^1 = L_d^{init}$ for all dimensions $d \in D$ with the minimum total Manhattan distance $\sum_{t \in \{1,\dots,H\}} \sum_{d \in D} |l_d^{t+1} - L_d^{goal}|$ from the goal location L_d^{goal} over all time steps t.

6.2 Implementation Details

SCIPPlan is a compilation-based planner that consists of the constraints compiled from RDDL [15] using the syntax presented in Table 1, RDDLsim domain simulator [15] and the dynamically generated temporal constraints (2). At every iteration, SCIPPlan only adds the set of constraints that correspond to the first zero-crossing interval, or terminates if the plan is valid with respect to the discretization parameter ϵ. In SCIPPlan, we modeled the actions b_r^t and a_d^t from HVAC and NavigationJail domains as decision variables with continuous domains. In PDDL+, we incremented and decremented the actions as a function of time with some constant rate z. Further in the NavigationJail domain, we have modeled the if-else-then statements (10–12) using events in PDDL+ (as opposed to using global constraints) since going into the jail location can still lead to feasible plans. In the HVAC and NavigationJail domains, we tested ENHSP with relaxed goal settings where the respective equality goal constraints were relaxed to the following constraints:

$$H_r^{goal} - z \leq h_r^{H+1} \leq H_r^{goal} + z \qquad \forall r \in R \qquad (17)$$
$$L_d^{goal} - z \leq l_d^{H+1} \leq L_d^{goal} + z \qquad \forall d \in D \qquad (18)$$

due to the continuous domains of state $s \in S^c$ and action $a \in A^c$ variables, and the fact that ENHSP identified these domains to be infeasible with equality constraints. We tested SCIPPlan under different optimality gap parameters g for

Table 2. Comparison of plan quality produced and run times by SCIPPlan (SP), ENHSP (EP) and TF-Plan (TF) with respect to the given domain metrics. We optimize both Makespan metric objectives (middle four columns) and General metric objectives (last column). Lower values indicate better solution quality.

Domain	Makespan				General
	Quality		Run time		Run time
HVAC	SP^0	$EP^{0.1}$	SP^0	$EP^{0.1}$	SP^0
(2,R)	**88.00**	145.00	**≤0.01**	1.02	0.02
(2,D)	**88.00**	145.00	**≤0.01**	1.02	0.19
Pouring	SP^0	EP	SP^0	EP	SP^0
(3,1)	**4.30**	11.00	0.10	0.32	**0.01**
(5,1)	**5.51**	19.00	1.38	**0.41**	0.87
(4,2)	**7.67**	22.00	0.93	**0.37**	0.58
(9,2)	**1.69**	10.00	0.90	0.37	**0.08**
NJail	$SP^{0.05}$	$EP^{0.1}$	$SP^{0.05}$	$EP^{0.1}$	-
(−1.0,1.0)	**13.59**	-	**281.75**	≥1800	-
(−0.5,0.5)	**13.63**	-	**60.94**	≥1800	-
(−0.2,0.2)	**13.35**	-	**59.29**	≥1800	-
NMud	$SP^{0.05}$	TF	$SP^{0.05}$	TF	-
(−1.0,1.0)	**64.25**	65.23	**15.46**	30.00	-
(−0.5,0.5)	140.35	**136.55**	232.84	240.00	-
(−0.2,0.2)	800.00	**360.38**	1800.00	**960.00**	-

the underlying SBB solver. For both parameter settings z and g, we will use the notation **SP**^{x} to report results for SCIPPlan under the optimality gap setting $g = x$, and **EP**^{x} for ENHSP under the rate setting $z = x$. Finally, when the total makespan is not minimized, in SCIPPlan we constrained the total makespan by a large constant such that $\sum_{t \in \{1,...,H\}} \Delta^t \le M$.

6.3   Comparison of the Solution Quality and Run Time Performance

In Table 2, we compare the quality of plans produced and the run times of SCIPPlan, ENHSP and TF-Plan with respect to the chosen optimization metric under the best performing parameter settings. From left to right, the first column of Table 2 specifies the domains and problem instances solved. The second and third columns present the optimal makespan found by the respective planners. The fourth and fifth columns present the computational effort that is required to produce the metrics presented in the second and third columns. The sixth column presents the running time (seconds) that is required to optimize the general metric variants of the original domains.

6.4   Computational Performance

In this section, we investigate the efficiency of using SCIPPlan for solving metric hybrid factored planning problems in nonlinear domains. We ran the experiments on MacBookPro with 2.8 GHz Intel Core i7 16 GB memory. We optimized the nonlinear encodings using SCIP 4.0.0 [10] with 1 thread, and 30 min total time limit per domain instance.

Comparison of Solution Qualities. The detailed inspection of the columns associated with solution quality shows that SCIPPlan can successfully find high quality plans in almost all the instances with optimality gap parameter $g \leq 0.05$, except the largest domain NavigationMud $(-0.2, 0.2)$. In contrast, we observe that in HVAC and ComplexPouring domains, ENHSP can find plans with on average 60% lower quality compared to SCIPPlan. Moreover in NavigationJail domain, neither $EP^{0.1}$ nor $EP^{0.01}$ found feasible plans within time limit. In NavigationMud domain, we tested the scalability of SCIPPlan against TF-Plan. We found that SCIPPlan is competitive with TF-Plan with respect to the solution quality of the plans found in the small and medium size instances, whereas the large instance NavigationMud(-0.2,0.2) is hard to optimize (i.e., the plan does not contain actions other than no-ops) for SCIPPlan. We note that unlike TF-Plan, we currently do not leverage parallel computing, which is one of the main strengths of Tensorflow to handle large scale optimization problems. In Fig. 2, we visualize the plan traces to get a better understanding of what makes a domain hard in terms of plan computability. The inspection of plan traces shows from left to right: linear, piecewise linear and nonlinear state transitions. Together with the computational results presented in Table 2, we confirm that domains with nonlinear state transitions (e.g., NavigationJail) are significantly computationally harder compared to linear (e.g., HVAC) and piecewise linear (e.g., ComplexPouring) domains.

Comparison of Run Time Performances. The inspection of the last three columns shows that SCIPPlan finds high quality plans with little computational effort in HVAC and ComplexPouring domains, whereas it takes on average 135 s and 125 s to find high quality plans for NavigationJail and NavigationMud domains, with the exception of the largest instance NavigationMud $(-0.2, 0.2)$ for horizon $H = 50$. The benefit of spending computational resources to provide stronger optimality guarantees is justified in Fig. 3, where we plot the increase in plan quality as a function of optimality gap parameter g. Figure 3 shows that spending more computational resources can significantly improve the quality of the plan found as the instances get harder to solve.

6.5   General Metric Specifications

Finally, we test SCIPPlan on general metrics of interest in HVAC and ComplexPouring domains and measure the effect on run time. In the HVAC

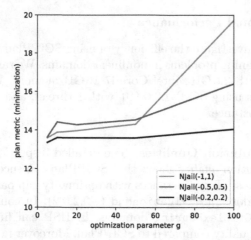

Fig. 3. The increase in plan quality (lower is better for minimization) as a function of optimality gap parameter g for SCIPPlan on NavigationJail domain.

domain, we modify the metric to minimize the total cost $\sum_{t \in \{1,...,H\}} \sum_{r \in R} c_r b_r^t$ of heating all rooms $r \in R$ of a building for all time steps where the parameter c_r denotes the unit cost of heating room $r \in R$. Similarly in ComplexPouring domain, we minimize the total number of times we pour from one tank to the bucket (or other tanks) across all time steps such that $\sum_{t \in \{1,...,H\}} \sum_{u \in U} \sum_{b \in B \cup U} p_{u,b}^t$. The results presented in the last column of Table 2 show that the performance of SCIPPlan is on average the same for general metric optimization and makespan optimization. As demonstrated in NavigationMud and modified HVAC and ComplexPouring domains, SCIPPlan finds high quality plans with respect to general metric specifications.

7   Conclusion

In this paper, we presented a novel SCIP-based planner (SCIPPlan) that can plan in metric hybrid factored planning domains with nonlinear transcendental functions such as exponentials and instantaneous continuous actions. In SCIPPlan, we leveraged the spatial branch-and-bound solver of SCIP inside a nonlinear constraint generation framework where candidate plans are iteratively checked for temporal infeasibility using a domain simulator, and the sources of infeasibilities are repaired through a novel nonlinear constraint generation algorithm. Experimentally, we have shown that SCIPPlan can plan effectively on a variety of domains and outperformed ENHSP in terms of plan quality and run time performance. We have further shown that SCIPPlan is competitive with the Tensorflow-based planner (TF-Plan) in highly nonlinear domains with exponential transitions and general metric specifications.

References

1. Agarwal, Y., Balaji, B., Gupta, R., Lyles, J., Wei, M., Weng, T.: Occupancy-driven energy management for smart building automation. In: ACM Workshop on Embedded Sensing Systems for Energy-Efficiency in Building, pp. 1–6 (2010)
2. Boutilier, C., Dean, T., Hanks, S.: Decision-theoretic planning: structural assumptions and computational leverage. JAIR **11**(1), 1–94 (1999). http://dl.acm.org/citation.cfm?id=3013545.3013546
3. Bryce, D., Gao, S., Musliner, D., Goldman, R.: SMT-based nonlinear PDDL+ planning. In: 29th AAAI, pp. 3247–3253 (2015). http://dl.acm.org/citation.cfm?id=2888116.2888168
4. Cashmore, M., Fox, M., Long, D., Magazzeni, D.: A compilation of the full PDDL+ language into SMT. In: ICAPS, pp. 79–87 (2016). http://dl.acm.org/citation.cfm?id=3038594.3038605
5. Coles, A.J., Coles, A.I., Fox, M., Long, D.: COLIN: planning with continuous linear numeric change. JAIR **44**, 1–96 (2012)
6. Fox, M., Long, D.: Modelling mixed discrete-continuous domains for planning. JAIR **27**(1), 235–297 (2006). http://dl.acm.org/citation.cfm?id=1622572.1622580
7. Fox, M., Long, D., Magazzeni, D.: Plan-based policies for efficient multiple battery load management. CoRR abs/1401.5859 (2014). http://arxiv.org/abs/1401.5859
8. Henzinger, T.A., Kopke, P.W., Puri, A., Varaiya, P.: What's decidable about hybrid automata? In: Proceedings of the Twenty-Seventh Annual ACM Symposium on Theory of Computing, pp. 373–382. ACM, New York (1995). https://doi.org/10.1145/225058.225162, http://doi.acm.org/10.1145/225058.225162
9. Löhr, J., Eyerich, P., Keller, T., Nebel, B.: A planning based framework for controlling hybrid systems. In: ICAPS, pp. 164–171 (2012). http://www.aaai.org/ocs/index.php/ICAPS/ICAPS12/paper/view/4708
10. Maher, S.J., et al.: The SCIP optimization suite 4.0. Technical report 17-12, ZIB, Takustr. 7, 14195 Berlin (2017)
11. Mitten, L.G.M.: Branch-and-bound methods: general formulation and properties. Oper. Res. **18**(1), 24–34 (1970). http://www.jstor.org/stable/168660
12. Penna, G.D., Magazzeni, D., Mercorio, F., Intrigila, B.: UPMurphi: a tool for universal planning on PDDL+ problems. In: ICAPS, pp. 106–113 (2009). http://dl.acm.org/citation.cfm?id=3037223.3037238
13. Piotrowski, W.M., Fox, M., Long, D., Magazzeni, D., Mercorio, F.: Heuristic planning for hybrid systems. In: AAAI, pp. 4254–4255 (2016). http://www.aaai.org/ocs/index.php/AAAI/AAAI16/paper/view/12394
14. Raghavan, A., Sanner, S., Tadepalli, P., Fern, A., Khardon, R.: Hindsight optimization for hybrid state and action MDPs. In: Proceedings of the Thirty-First AAAI Conference on Artificial Intelligence (AAAI 2017), San Francisco, USA (2017)
15. Sanner, S.: Relational dynamic influence diagram language (RDDL): Language description (2010)
16. Say, B., Wu, G., Zhou, Y.Q., Sanner, S.: Nonlinear hybrid planning with deep net learned transition models and mixed-integer linear programming. In: Proceedings of the Twenty-Sixth International Joint Conference on Artificial Intelligence, IJCAI 2017, pp. 750–756 (2017). https://doi.org/10.24963/ijcai.2017/104
17. Scala, E., Haslum, P., Thiébaux, S., Ramírez, M.: Interval-based relaxation for general numeric planning. In: ECAI, pp. 655–663 (2016). https://doi.org/10.3233/978-1-61499-672-9-655

18. Shin, J.A., Davis, E.: Processes and continuous change in a sat-based planner. Artif. Intell. **166**(1–2), 194–253 (2005). https://doi.org/10.1016/j.artint.2005.04.001
19. Wu, G., Say, B., Sanner, S.: Scalable planning with tensorflow for hybrid nonlinear domains. In: Proceedings of the Thirty First Annual Conference on Advances in Neural Information Processing Systems (NIPS 2017), Long Beach, CA (2017)

Last-Mile Scheduling Under Uncertainty

Thiago Serra[1], Arvind U. Raghunathan[1(✉)], David Bergman[2], John Hooker[3], and Shingo Kobori[4]

[1] Mitsubishi Electric Research Labs, Cambridge, MA, USA
{tserra,raghunathan}@merl.com
[2] University of Connecticut, Storrs, CT, USA
david.bergman@uconn.edu
[3] Carnegie Mellon University, Pittsburgh, PA, USA
jh38@andrew.cmu.edu
[4] Advanced Technology R&D Center, Mitsubishi Electric Corporation, Hyogo, Japan
Kobori.Shingo@cj.MitsubishiElectric.co.jp

Abstract. Shared mobility is revolutionizing urban transportation and has sparked interest in optimizing the joint schedule of passengers using public transit and last-mile services. Scheduling systems must anticipate future requests and provision flexibility in order to be adopted in practice. In this work, we consider a two-stage stochastic programming formulation for scheduling a set of known passengers and uncertain passengers that are realized from a finite set of scenarios. We present an optimization approach based on decision diagrams. We obtain, in minutes, schedules for 1,000 known passengers that are robust and optimized with respect to scenarios involving up to 100 additional uncertain passengers.

Keywords: Decision diagrams · Scheduling · Stochastic programming

1 Introduction

The transportation industry is transforming due to recently introduced mechanisms for shared mobility [9,16]. Transportation systems are a key element of integrative smart city operations, leading to a host of complex optimization problems [12,19–21]. Of critical importance is the joint scheduling of mass transportation systems with last-mile vehicles, which when scheduled in unison can lead to significant operational improvements [15,17,18,22].

This paper studies the joint scheduling of passengers on mass transit systems and last-mile vehicles under uncertainty. Passengers arrive by train to a central terminal and board limited-capacity pods called *commuter vehicles* (CVs) which are automated or otherwise operated, where some passengers are known and other passengers are uncertain (and thus may or may not request service). The goal is to assign passengers to trains and then to group passengers traveling together on a CV so as to minimize a combination of total travel time and

© Springer Nature Switzerland AG 2019
L.-M. Rousseau and K. Stergiou (Eds.): CPAIOR 2019, LNCS 11494, pp. 519–528, 2019.
https://doi.org/10.1007/978-3-030-19212-9_34

number of CV trips.[1] This objective models a tradeoff between quality of service (passenger travel time) and operational costs related to fuel consumption and maintenance requirements (number of trips). In the absence of uncertainty, this problem is known as the *integrated last-mile transportation problem* (ILMTP).

The uncertain setting is applicable to systems where a central scheduler takes requests from passengers and assigns them to trains and CV trips, while building in flexibility for passengers that might request transportation services but are yet to submit an associated request. This leads to a significant increase in problem complexity with respect to previous work [15,17,18], but also makes for a more realistic setting where the initial scheduling of passengers must account for additional demand from late requests that also needs to be accommodated.

Optimization under uncertainty, or *stochastic optimization*, defines a broad class of challenging problems [8]. A relatively recent and popular technique for handling uncertainty is robust optimization [2], where an uncertainty set is defined and worst-case operational decisions are employed. It is well known that this can lead to highly conservative solutions, since unlikely outcomes can drive decisions. A more classical approach is simulation-based optimization, where algorithms such as sample average approximation methods [11] are employed, which consider a finite set of possible realizations described as a sample of scenarios which are optimized over in order to maximize the expected value over the sampling.

This paper presents a two-stage stochastic programming formulation for the ILMTP under additional passenger uncertainty (ILMTP-APU). In addition to a set of *known* passengers, we model *uncertain passengers* through a finite set of scenarios. The first-stage decision is the scheduling of known passengers and the second-stage decision schedules the additional passengers in each scenario. Our approach relies on decision-diagram (DD)-based optimization (DDO) [1,3,6], and more specifically on decompositions based on DDs [4,5,7,13] and is inspired by the model presented in [17]. Specifically, a DD is built for known passengers going to each building and separately for unknown passengers in each scenario and for each building. The DDs are then integrated through channeling constraints that can be optimized with an integer programming (IP) formulation. This results in a large model. However, due to the tightness of the formulation, we obtain a reliable approach for optimally solving the problem. The resulting solutions are significantly better than what could be obtained by solving the problem for the known passengers to optimality and then scheduling the unknown passengers with the remaining capacity when a scenario is realized.

This paper adds to the recent literature on DDO for stochastic optimization, where BDDs have been used for determining decision-dependent scenario probabilities [10] and, more closely related to the current study, a study of a class of two-stage stochastic IP problems [14]. Our proposed approach differs from that of [14] in that we model both the first-stage and second-stage decisions using DDs and link them through assignment constraints. This provides an additional

---

[1] We assume a single destination per CV trip [17,18] and only a few destinations [15], which is operationally favorable since destination batching leads to efficiency.

mechanism by which decision making under uncertainty can be tackled through DDs. The main contributions of this paper are therefore (1) an extension of the ILMTP to incorporate uncertain passenger arrivals; (2) structural results on families of optimal solutions; and (3) the development of a novel DDO stochastic programming modeling framework for solving the ILMTP-APU based on these structural results. An experimental evaluation on synthetic instances shows great promise. In particular, the solutions obtained are far superior to a basic heuristic extension of the work in [17] and the algorithm scales to 100 uncertain passengers per scenario when 1,000 confirmed passengers are scheduled.

2   Problem Description

We first describe the elements of the problem, including the mass transit system, last-mile vehicles, destinations, passenger requests, and associated parameters.

Mass Transit System: We assume that the mass transit is a train system. Let $\mathbf{T_0}$ be the *terminal station* that links a mass transit system with a last-mile service system. The mass transit system is described by a set of *trips*, denoted by $\mathcal{C}$, with $n_c := |\mathcal{C}|$. Each trip originates at a station in set $\mathcal{S}$ and ends at $\mathbf{T_0}$. The trips are regular in the sense that the train stops at all stations in $\mathcal{S}$ sequentially, with $\mathbf{T_0}$ as the last stop of each trip. A trip c leaves station $s \in \mathcal{S}$ at time $\tilde{t}(c, s)$ and arrives to $\mathbf{T_0}$ at $\tilde{t}(c)$.

Destinations: Let $\mathcal{D}$ be the set of destinations where the CVs make stops, with $K := |\mathcal{D}|$, where we assume $\mathbf{T_0} \in \mathcal{D}$. For each destination $d \in \mathcal{D}$, let $\tau(d)$ be the total time it takes a CV to depart $\mathbf{T_0}$, travel to d (denoted by $\tau^1(d)$), stop at d for passengers to disembark (denoted by $\tau^2(d)$), and return to $\mathbf{T_0}$ (denoted by $\tau^3(d)$). Therefore, $\tau(d) = \tau^1(d) + \tau^2(d) + \tau^3(d)$. Let $\mathcal{T} := \{1, \ldots, t^{\max}\}$ be an index set of the operation times of both systems. We assume that the time required to board passengers into the CVs is incorporated in $\tau^1(d)$. For simplicity, the boarding time is independent of the number of passengers.

Last-Mile System: Let V be the set of CVs, with $m := |V|$. Denote by v^{cap} the number of passengers that can be assigned to a single CV trip. Each CV trip consists of a set of passengers boarding the CV, traveling from $\mathbf{T_0}$ to a destination $d \in \mathcal{D}$, and then returning back to $\mathbf{T_0}$. Therefore, passengers sharing a common CV trip must request transportation to a common building. We also assume that each CV must be back at the terminal by time $t^{\max}$.

Known Passengers: Let $\mathcal{J}$ be the set of known passengers. Each passenger $j \in \mathcal{J}$ requests transport from a station $s(j) \in \mathcal{S}$ to $\mathbf{T_0}$, and then by CV to destination $d(j) \in \mathcal{D}$, to arrive at time $t^r(j)$. The set of passengers that request service to destination d is denoted by $\mathcal{J}(d)$. Let $n := |\mathcal{J}|$ and $n_d := |\mathcal{J}(d)|$. Each passenger $j \in \mathcal{J}$ must arrive to $d(j)$ between $t^r(j) - T_w$ and $t^r(j) + T_w$.

Uncertain Passengers: We assume a finite set of scenarios, denoted by $\mathcal{Q}$, representing different realizations of the uncertain passengers. Let $\widehat{\mathcal{J}}(q)$ be the set of uncertain passengers in scenario $q \in \mathcal{Q}$. Each passenger $j \in \widehat{\mathcal{J}}(q)$ requests transport from a station $s(j) \in \mathcal{S}$ to $\mathbf{T}_0$, and then by CV to destination $d(j) \in \mathcal{D}$, to arrive at time $t^r(j)$. The set of passengers that request service to destination d is denoted by $\widehat{\mathcal{J}}(q,d)$. Let $\widehat{n}_q := |\widehat{\mathcal{J}}(q)|$ and $\widehat{n}_{q,d} := \left|\widehat{\mathcal{J}}(q,d)\right|$.

Problem Statement: The ILMTP-APU is the problem of assigning train trips and CVs to each known passenger so that the uncertain passengers in any of the scenarios $\mathcal{Q}$ can be feasibly scheduled and the expected value of a convex combination of the total transit time and the number of CV trips utilized is minimized. A solution therefore consists of:

– A partition $\mathbf{g} = \{g_1, \ldots, g_\gamma\}$ of $\mathcal{J}$, where each group g_l is associated with a departure time $t_l^{\mathbf{g}}$ indicating the time the CV carrying the passengers in g_l departs $\mathbf{T}_0$, satisfying all request time and operational constraints. For passenger $j \in \mathcal{J}$, let $\mathbf{g}(j)$ be the group in $\mathbf{g}$ where j belongs.

– For each scenario $q \in \mathcal{Q}$, a partition $\widehat{\mathbf{g}}(q) = \{\widehat{g}_{q,1}, \ldots, \widehat{g}_{q,\widehat{\gamma}(q)}\}$ of $\widehat{\mathcal{J}}(q)$, where each group $\widehat{g}_{q,k}$ is associated with a departure time $t_k^{\widehat{\mathbf{g}}(q)}$ and an indicator function $\sigma(q,k) \in \{1, \ldots, \gamma\} \cup \{\emptyset\}$. $\sigma(q,k) \neq \emptyset$ indicates that uncertain passenger group $\widehat{g}_{q,k}$ shares the last-mile trip with known passenger group $g_{\sigma(q,k)}$. In other words, groups leave from the terminal at the same time (i.e. $t_{\sigma(q,k)}^{\mathbf{g}} = t_k^{\widehat{\mathbf{g}}(q)}$) and the combination of confirmed passenger group and all such shared passenger groups in a scenario does not exceed the CV capacity, i.e. $|g_l| + \sum_{k \in \{1, \ldots, \widehat{\gamma}(q): \sigma(q,k)=l\}} |\widehat{g}_{q,k}| \leq v^{\mathrm{cap}}$ for each $l \in \{1, \ldots, \gamma\}$ and $q \in \mathcal{Q}$.

3 Structure of Optimal Solutions

The deterministic version of ILMTP has optimal solutions with a structure that is particularly helpful for defining compact models. For each destination, passengers can be sorted by their desired arrival times and then grouped sequentially. This structure is valid to minimize passenger average waiting time [18] as well as the number of CV trips [17] hence leading to the compact DD-based model in [17]. For ILMTP-APU, however, a more elaborate structure is required.

For example, let us suppose that for a particular time range there is a single CV of capacity 4 available, 4 known passengers, and just 1 unknown passenger in 1 out of 10 scenarios. Furthermore, let us assume that a first trip with the CV incurs no wait time whereas a second trip would impose a wait time of w on any passenger involved, and that the desired arrival time of the uncertain passenger falls strictly in the middle of those of the known passengers. If we sort and group all passengers regardless of their categories, then at least 2 trips will always be necessary and at least 1 known passenger has to wait w. But if we define a second trip only for the unknown passenger, then the average cost of the solution is reduced to a tenth because the second trip and the corresponding wait time w only materialize in 1 out of 10 scenarios. Since uncertain passengers have a smaller impact on the objective function, it is intuitive that they might be delayed with respect to known passengers if the schedule remains feasible. We formalize this two-tier structure using the groups from the previous section.

Proposition 1. *When ILMTP-APU is feasible, there is an optimal solution where the groups of passengers for each category – known or uncertain from a scenario q – are grouped sequentially by their desired arrival times.*

Proof. Let us assume, by contradiction, that there is an instance for which the statement does not hold for a group of known passengers involving a passenger j with destination d in group g_1 and another group $g_2 \supseteq \{j-1, j_+\}$ for some $j_+ > j$. Furthermore, among the optimal solutions for such an instance, let us choose the optimal solution for which the first index d of the destination where such a grouping of known passengers does not exist is maximized; and among those solutions the one that maximizes the index j of the first passenger for which there is a group defined by passengers before and after j is maximized.

The key to obtaining a contradiction is the fact that the length of the time window for arrival is identical for all passengers each of whom have access to the same public transit service. Let us denote by $g_A \in \{g_1, g_2\}$ the group with earliest arrival time, say t_A; let us denote the other group by g_B, for which the departure time t_B is such that $t_B \geq t_A$; and let us denote the indices of passengers on either group as $\{j_1, j_2, \ldots, j_{|g_A|+|g_B|}\}$, where $t^r_{j_i} \leq t^r_{j_{i+1}}$ for $i = 1, \ldots, |g_A| + |g_B| - 1$. Note that all of these passengers have either t_A or t_B in their time windows. Since at least $|g_A|$ passengers can arrive at time t_A, it follows that the time window of the first $|g_A|$ passengers includes t_A. If not, then some of these first passengers would contain t_B in their time window instead, implying that some among the remaining $|g_B|$ passengers have t_A in their time window, and thus that the passengers are not sorted by desired arrival times. Thus, $t_A \in [t^r_{j_{|g_A|}} - \omega, t^r_{j_1} + \omega]$. Similarly, at least $|g_B|$ passengers can arrive at time t_B, and thus $t_B \in [t^r_{j_{|g_A|+|g_B|}} - \omega, t^r_{j_{|g_A|+1}} + \omega]$. Hence, we can replace the passengers of group g_A with $\{j_1, \ldots, j_{|g_A|}\}$ and those of group g_B with $\{j_{|g_A|+1}, \ldots, j_{|g_A|+|g_B|}\}$ while preserving their arrival times, size, and consequently with no change to the feasibility or optimality of the solution. However, this exchange implies that up to destination d and passenger j all passengers are grouped by sorted arrival times, hence contradicting the choice of d and j.

Without loss of generality, we can apply the same argument for uncertain passengers by also choosing the maximum index of a scenario q for which the groupings are not continuous and finding the same contradiction. $\Box$

The following result, which is independent from Proposition 1, is also helpful to simplify the modeling of ILMTP-APU.

Proposition 2. *When ILMTP-APU is feasible, there is an optimal solution where at most one group of uncertain passengers for each scenario is assigned to each group of known passengers.*

Proof. If multiple groups are assigned, they have the same arrival times in any optimal solution and thus can be combined without loss of generality. $\Box$

4 Decision Diagram for Single Building

We use a DD to represent the groups from Proposition 1 for each destination. A DD is a set of paths between a universal source node (root) and a universal sink node (terminal), each containing a common number of arcs. The i-th arc corresponds to a decision regarding the i-th passenger sorted by desired arrival time: either passenger i is the last in a group and a departure time is chosen, or subsequent passengers join the group. The endpoints of each arc are nodes representing how many passengers have been accumulated to define a group. Arcs either increment this number by 1 (up to CV capacity) or set it back to 0.

Following similar notation as in [17], we define for each destination $d \in \mathcal{D}$ a DD $\mathsf{D}^d = (\mathsf{N}^d, \mathsf{A}^d)$ for the known passengers. N^d is partitioned into $n_d + 1$ ordered layers $\mathsf{L}_1^d, \mathsf{L}_2^d, \ldots, \mathsf{L}_{n_d+1}^d$ where $n_d = |\mathcal{J}(d)|$. Layer $\mathsf{L}_1^d = \{\mathbf{r}^d\}$ and layer $\mathsf{L}_{n_d+1}^d = \{\mathbf{t}^d\}$ consist of one node each; the *root* and *terminal*, respectively. Each arc $a \in \mathsf{A}^d$ is directed from its *arc-root* $\psi(a)$ to its *arc-terminal* $\omega(a)$. If $\psi(a) \in \mathsf{L}_i^d$, then $\omega(a) \in \mathsf{L}_{i+1}^d$. It is assumed that every maximal arc-directed path connects $\mathbf{r}^d$ to $\mathbf{t}^d$. Similarly, for each scenario $q \in \mathcal{Q}$, we define a DD $\mathsf{D}^{d,q} = (\mathsf{N}^{d,q}, \mathsf{A}^{d,q})$ by using the corresponding upper index q for disambiguation.

The arcs between layers of the diagram correspond to the passengers that request transportation to the destination. Each node u is associated with a *state* $\mathsf{s}(\mathsf{u})$ corresponding to the number of immediately preceding passengers that is grouped with the next passenger. Each arc a is associated with a label $\phi(a) \in \{0, 1\}$ on whether passenger $\psi(a)$ is not the last one in the group, in which case $\phi(a) = 0$ and $\mathsf{s}(\omega(a)) = \mathsf{s}(\psi(a)) + 1 \leq v^{\mathrm{cap}}$, or else $\phi(a) = 1$. There can be multiple arcs between the same nodes in the latter case, each arc a corresponding to a different CV start time $t^0(a)$. Accordingly, each arc a such that $\phi(a) = 1$ has a corresponding wait time $W(a)$ for all passengers in the group and a label $\chi(a, t) \in \{0, 1\}$ to indicate that a CV would be active at time t (i.e. $t^0(a) \leq t \leq t^0(a) + \tau(d)$) if arc a is chosen. Hence, $\phi(a) = 0$ implies $\chi(a, t) = 0$.

5 IP Formulation

We define a formulation by which we group the passengers using a path from each DD. Some of the groups of known and uncertain passengers are combined, and we aim for a feasible schedule of the resulting groups using the CV fleet.

We introduce binary variable $x_a \in \{0, 1\} \forall a \in \mathsf{A}^d, d \in \mathcal{D}$, to denote the choice of the particular arc in the DD for known passengers. Similarly, we introduce binary variable $y_a^q \in \{0, 1\} \forall a \in \mathsf{A}^{d,q}, d \in \mathcal{D}$ to denote the choice of the particular arc in the DD for uncertain passengers in scenario $q \in \mathcal{Q}$. Let $\widetilde{\mathsf{A}}^{d,q} = \{(a_1, a_2) \in \mathsf{A}^d \times \mathsf{A}^{d,q} \mid \phi(a_1) = 1, \phi(a_2) = 1, t^0(a_1) = t^0(a_2) \text{ and } \mathsf{s}(\psi(a_1)) + \mathsf{s}(\psi(a_2)) + 2 \leq v^{\mathrm{cap}}\}$. The set $\widetilde{\mathsf{A}}^{d,q}$ denotes the set of feasible pairs of known passenger group and uncertain passenger group of scenario q, i.e. identical start time on the CV and the capacity constraint is satisfied. Let $z_{a_1,a_2}^q \in \{0, 1\} \forall (a_1, a_2) \in \mathsf{A}^{d,q}$ denote the decision of pairing the group of known passengers represented by arc a_1 and group of uncertain passengers represented by arc a_2.

The objective function can be expressed as

$$f(\alpha) = \sum_{d \in D} \sum_{a \in A^d} [\alpha W(a) + (1 - \alpha)] x_a$$

$$+ \frac{1}{|Q|} \sum_{d \in D} \sum_{q \in Q} \left(\sum_{a \in A^{d,q}} [\alpha W(a) + (1 - \alpha)] y_a^q - (1 - \alpha) \sum_{(a_1,a_2) \in \widetilde{A}^{d,q}} z_{a_1,a_2}^q \right)$$

The following constraint imposes that only one group of uncertain passengers is paired with a group of known passengers if the latter is selected:

$$\sum_{a_2 : (a_1,a_2) \in \widetilde{A}^{d,q}} z_{a_1,a_2}^q \leq y_{a_1} \; \forall d \in D, q \in Q, a_1 \in A^d : \phi(a_1) = 1. \qquad (1a)$$

The fleet size constraint can be modeled for all $t \in T, q \in Q$ as

$$\sum_{d \in D} \left(\sum_{a \in A^d : \chi(a,t)=1} x_a + \sum_{a \in A^{d,q} : \chi(a,t)=1} y_a^q \right) - \sum_{d \in D} \sum_{(a_1,a_2) \in \widetilde{A}^{dq}} z_{a_1,a_2}^q \leq m. \qquad (1b)$$

The IP model for the ILMTP-APU is

min $f(\alpha)$

s.t.   Network flow constraints for D^d $\forall d \in D$ (2a)

Network flow constraints for $D^{d,q}$ $\forall d \in D, q \in Q$ (2b)

Eq. (1a), (1b)

$x_a \in \{0,1\}$ $\forall d \in D, a \in A^d$ (2c)

$y_a^q \in \{0,1\}$ $\forall d \in D, q \in Q, a \in A^d$ (2d)

$z_{a_1,a_2}^q \in \{0,1\}$ $\forall d \in D, q \in Q, (a_1, a_2) \in \widetilde{A}^{d,q}.$ (2e)

The network flow constraints in (2a)–(2b) guarantee that a path is taken on each decision diagram, which corresponds to the groupings of known passengers and uncertain passengers for each scenario.

6   Experimental Results

We ran experiments to test our approach on a machine with an Intel(R) Core(TM) i7-4770 CPU @ 3.40 GHz and 16 GB RAM. The code is in Python 2.7.6 and the ILPs are solved using Gurobi 7.5.1. We generated instances with 1000 passengers, 60 CVs with capacity 5, 10 destinations, time windows of 10 min, time units of 30 s, and 10 scenarios each containing 50 or 100 passengers. The instances are similar to those in [17], but smaller due to problem complexity.

For benchmarking, we also tested the following heuristic H: (1) solve the problem optimally for known passengers using the algorithm in [17]; and (2) for each scenario, solve the resulting MIP formulation to maximize the number of uncertain passengers that can be scheduled with the remaining capacity of the trips already scheduled and the remaining availability of the CVs for more trips.

Table 1 summarizes results for $\alpha = 0.5$. Each instance $P_{k,u,i}$ corresponds to the i-th instance with k known passengers and u unknown passengers on each of the 10 scenarios. We report the values for the first stage (known passengers) and for the second stage (uncertain passengers) as well as runtimes. The first- and second-stage values correspond to the first and second terms of $f(\alpha)$. If the second stage is infeasible, we report the percentage of scheduled passengers.

Table 1. Comparison of solution obtained with DDO and with heuristic H.

Instance	Heuristic H			DDO approach		
	1st Stage	2nd Stage	Time (s)	1st Stage	2nd Stage	Time (s)
$P_{1000,50,1}$	21,940.0	(41.6%)	23.8	22,991.5	1,254.2	630.4
$P_{1000,50,2}$	22,434,0	(33.4%)	25.2	23,410.5	1,257.9	535.1
$P_{1000,50,3}$	22,409.5	(35.8%)	25.2	23,403.5	1,271.1	532.2
$P_{1000,50,4}$	22,410.0	(38.2%)	25.6	23,397.0	1,302.2	643.5
$P_{1000,50,5}$	22,099.0	(37.4%)	25.4	23,152.5	1,282.0	1019.9
$P_{1000,50,6}$	22,552.0	(37.6%)	28.7	23,826.0	1,239.8	539.8
$P_{1000,50,7}$	22,397.0	(34.8%)	24.3	23,621.5	1,293.3	646.1
$P_{1000,50,8}$	22,002.0	(42.2%)	25.9	23,003.0	1,264.2	564.6
$P_{1000,50,9}$	22,745.5	(34.8%)	26.0	23,855.0	1,276.0	704.5
$P_{1000,50,10}$	22,206.0	(40.0%)	25.4	23,186.0	1,262.3	526.8
$P_{1000,100,1}$	22,389.5	(31.6%)	24.8	24,078.5	2,503.4	1,171.3
$P_{1000,100,2}$	21,992.5	(40.4%)	25.5	23,718.5	2,510.7	2,142.1
$P_{1000,100,3}$	22,181.0	(31.6%)	39.4	23,814.5	2,523.1	1,330.6
$P_{1000,100,4}$	22,409.5	(31.3%)	23.8	23,923.5	2,577.0	1,535.2
$P_{1000,100,5}$	22,447.5	(40.3%)	29.5	24,162.0	2,496.3	896.0
$P_{1000,100,6}$	22,725.5	(31.4%)	25.2	24,277.0	2,530.4	1,112.4
$P_{1000,100,7}$	22,236.5	(36.9%)	26.5	24,286.0	2,537.8	2,776.1
$P_{1000,100,8}$	22,584.0	(33.0%)	27.3	24,177.5	2,560.4	1,544.1
$P_{1000,100,9}$	22,541.0	(35.6%)	24.9	24,106.5	2,477.9	1,982.5
$P_{1000,100,10}$	22,552.0	(37.6%)	29.0	24,274.0	2,525.3	1,965.6

We note that ignoring the second stage leads to infeasible problems in all cases for heuristic H, and only a portion of the uncertain passengers could be scheduled. Interestingly, the proportion of uncertain passengers that are scheduled when the optimal solution of the known passengers alone is fixed remains approximately the same for 50 and 100 uncertain passengers per scenario. Hence, reducing the number of uncertain passengers does not make heuristic H more suitable.

For $\alpha = 0$, we found that the optimal solution of the deterministic case, which is used by H, has the same value as the first stage of DDO. In that case, DDO found solutions that are robust for the scenarios considered while also optimal for the known passengers. In contrast, for $\alpha > 0$ there is a difference between first stage values for both approaches, which is due to minimizing travel times.

7 Conclusion

We considered a two-stage optimization problem of last-mile passenger scheduling subject to a finite set of scenarios representing uncertain additional demand. Our approach based on decision diagram optimization produces solutions that, despite an increase in runtimes, are feasibly robust with respect to all scenarios while minimizing the expected number of last-mile trips necessary to satisfy the demand across all scenarios. The results show the potential of using decision diagrams to solve such challenging problems of scheduling under uncertainty.

References

1. Andersen, H.R., Hadzic, T., Hooker, J.N., Tiedemann, P.: A constraint store based on multivalued decision diagrams. In: Bessière, C. (ed.) CP 2007. LNCS, vol. 4741, pp. 118–132. Springer, Heidelberg (2007). https://doi.org/10.1007/978-3-540-74970-7_11. http://dl.acm.org/citation.cfm?id=1771668.1771682
2. Ben-Tal, A., Nemirovski, A.: Robust optimization - methodology and applications. Math. Program. **92**(3), 453–480 (2002). https://doi.org/10.1007/s101070100286
3. Bergman, D., Cire, A., van Hoeve, W., Hooker, J.: Decision Diagrams for Optimization. Springer, Cham (2016). https://doi.org/10.1007/978-3-319-42849-9
4. Bergman, D., Cire, A.A.: Decomposition based on decision diagrams. In: Quimper, C.-G. (ed.) CPAIOR 2016. LNCS, vol. 9676, pp. 45–54. Springer, Cham (2016). https://doi.org/10.1007/978-3-319-33954-2_4
5. Bergman, D., Cire, A.A.: Discrete nonlinear optimization by state-space decompositions. Manage. Sci. **64**(10), 4700–4720 (2018). https://doi.org/10.1287/mnsc.2017.2849
6. Bergman, D., van Hoeve, W.-J., Hooker, J.N.: Manipulating MDD relaxations for combinatorial optimization. In: Achterberg, T., Beck, J.C. (eds.) CPAIOR 2011. LNCS, vol. 6697, pp. 20–35. Springer, Heidelberg (2011). https://doi.org/10.1007/978-3-642-21311-3_5
7. Bergman, D., Lozano, L.: Decision diagram decomposition for quadratically constrained binary optimization. Optimization Online e-prints, October 2018
8. Birge, J.R., Louveaux, F.: Introduction to Stochastic Programming. Springer, New York (2011). https://doi.org/10.1007/978-1-4614-0237-4
9. Grosse-Ophoff, A., Hausler, S., Heineke, K., Möller, T.: How shared mobility will change the automotive industry (2017). https://www.mckinsey.com/industries/automotive-and-assembly/our-insights/how-shared-mobility-will-change-the-automotive-industry
10. Haus, U.U., Michini, C., Laumanns, M.: Scenario aggregation using binary decision diagrams for stochastic programs with endogenous uncertainty (2017)

11. Kleywegt, A., Shapiro, A., Homem-de Mello, T.: The sample average approximation method for stochastic discrete optimization. SIAM J. Optim. **12**(2), 479–502 (2002). https://doi.org/10.1137/S1052623499363220
12. Liu, Z., Jiang, X., Cheng, W.: Solving in the last mile problem: ensure the success of public bicycle system in Beijing. Procedia Soc. Behav. Sci. **43**, 73–78 (2012)
13. Lozano, L., Bergman, D., Smith, J.C.: On the consistent path problem. Optimization Online e-prints, April 2018
14. Lozano, L., Smith, J.C.: A binary decision diagram based algorithm for solving a class of binary two-stage stochastic programs. Math. Program. (2018). https://doi.org/10.1007/s10107-018-1315-z
15. Mahéo, A., Kilby, P., Hentenryck, P.V.: Benders decomposition for the design of a hub and shuttle public transit system. Transp. Sci. **53**(1), 77–88 (2018)
16. McCoy, K., Andrew, J., Glynn, R., Lyons, W.: Integrating shared mobility into multimodal transportation planning: improving regional performance to meet public goals. Technical report, Office of the Assistant Secretary of Transportation for Research and Technology, U.S. Department of Transportation (2018)
17. Raghunathan, A.U., Bergman, D., Hooker, J., Serra, T., Kobori, S.: Seamless multimodal transportation scheduling. arXiv:1807.09676 (2018)
18. Raghunathan, A.U., Bergman, D., Hooker, J.N., Serra, T., Kobori, S.: The integrated last-mile transportation problem (ILMTP). In: Proceedings of the Twenty-Eighth International Conference on Automated Planning and Scheduling, ICAPS 2018, Delft, The Netherlands, 24–29 June 2018, pp. 388–398 (2018). https://aaai.org/ocs/index.php/ICAPS/ICAPS18/paper/view/17720
19. Shaheen, S.: U.S. carsharing & station car policy considerations: monitoring growth, trends & overall impacts. Technical report, Institute of transportation studies, Working paper series, Institute of Transportation Studies, UC Davis (2004)
20. Shen, Y., Zhang, H., Zhao, J.: Simulating the first mile service to access train stations by shared autonomous vehicle. In: Transportation Research Board 96th Annual Meeting (2017)
21. Thien, N.D.: Fair cost sharing auction mechanisms in last mile ridesharing. Ph.D. thesis, Singapore Management University (2013)
22. Wang, H.: Routing and scheduling for a last-mile transportation problem. Transp. Sci. **53**(1), 1–17 (2017). https://doi.org/10.1287/trsc.2017.0753. http://pubsonline.informs.org/doi/abs/10.1287/trsc.2017.0753

Building Optimal Steiner Trees
on Supercomputers by Using
up to 43,000 Cores

Yuji Shinano[2]([envelope]) [iD], Daniel Rehfeldt[1] [iD], and Thorsten Koch[2] [iD]

[1] TU Berlin, Str. des 17. Juni 135, 10623 Berlin, Germany
rehfeldt@zib.de
[2] Zuse Institute Berlin, Takustr. 7, 14195 Berlin, Germany
{shinano,koch}@zib.de

Abstract. SCIP-JACK is a customized, branch-and-cut based solver for Steiner tree and related problems. ug [SCIP-JACK, MPI] extends SCIP-JACK to a massively parallel solver by using the Ubiquity Generator (UG) framework. ug [SCIP-JACK, MPI] was the only solver that could run on a distributed environment at the (latest) 11th DIMACS Challenge in 2014. Furthermore, it could solve three well-known open instances and updated 14 best known solutions to instances from the benchmark library STEINLIB. After the DIMACS Challenge, SCIP-JACK has been considerably improved. However, the improvements were not reflected on ug [SCIP-JACK, MPI]. This paper describes an updated version of ug [SCIP-JACK, MPI], especially branching on constrains and a customized racing ramp-up. Furthermore, the different stages of the solution process on a supercomputer are described in detail. We also show the latest results on open instances from the STEINLIB.

Keywords: Steiner tree problem · Branch-and-cut ·
Parallel computing · SCIP · UG

1 Introduction

The *Steiner tree problem in graphs* (SPG) is one of the fundamental $\mathcal{NP}$-hard optimization problems [5]. Given an undirected connected graph $G = (V, E)$, costs $c : E \rightarrow \mathbb{Q}_{\geq 0}$ and a set $T \subseteq V$ of *terminals*, the problem is to find a tree $S \subseteq G$ of minimum cost that includes T. The 2014 DIMACS Challenge, dedicated to Steiner tree problems, marked a revival of research on the SPG and related problems. SCIP-JACK [2], which is a customized SCIP solver for SPG and related problems, was initially developed to attend the DIMACS Challenge. SCIP-JACK was by far the most versatile solver participating in the Challenge, being able to solve the SPG and 10 related problems. After the DIMACS Challenge, the performance of SCIP-JACK has continuously improved, both for SPG [13] and related problems [11,12,14]. The improvements were for instance marked by

© Springer Nature Switzerland AG 2019
L.-M. Rousseau and K. Stergiou (Eds.): CPAIOR 2019, LNCS 11494, pp. 529–539, 2019.
https://doi.org/10.1007/978-3-030-19212-9_35

SCIP-JACK being the most successful solver at the PACE 2018 Challenge [1] dedicated to fixed-parameter tractable (FPT) SPG instances (although SCIP-JACK does not include any FPT specific algorithms).

ug [SCIP-JACK, MPI] is an extension of SCIP-JACK to a massively parallelized solver by using the *Ubiquity Generator (UG) framework* [16], a software package to parallelize branch-and-bound (B&B) based solvers. ug [SCIP-JACK, MPI] was the only solver which could run on a distributed environment at the 11th DIMACS Challenge. Moreover, it solved three open instances and updated 14 best known solutions to instances from the STEINLIB [7]. However, no detailed statistics on the solving process have been published yet. After the DIMACS Challenge, solving new open instances from the STEINLIB by ug [SCIP-JACK, MPI] looked hopeless for all open instances—judging from their run-time log files—and there have been no new result published prior to this paper. For the results presented throughout this paper, we used the ug [SCIP-JACK, MPI] code included in the SCIP Optimization Suite 6.0 [3].

ug [SCIP-JACK, MPI] was not implemented from scratch by using UG, but it was parallelized by using the *ug* [SCIP-*, MPI]-library, which is a software library to parallelize SCIP applications. SCIP is a plugin based software framework [3] and by adding new user-plugins it can be extended to create a customized solver like SCIP-JACK. The ug [SCIP-*, MPI]-library allows its users to include these user-plugins to PARASCIP by adding a small amount of glue-code (typically 100 – 200 lines). Usually, if a solver performance parallelized by UG is improved, this is directly reflected in the performance of its parallel extension. Since the SCIP-JACK performance has improved tremendously after the DIMACS Challenge, see Sect. 3, one would expect the same of ug [SCIP-JACK, MPI]. However, several idiosyncrasies of SCIP-JACK required to develop new features of the ug [SCIP-*, MPI]-library, in order to also obtain the performance improvements in its massively parallel extension.

In the following, we briefly describe UG, and go on to introduce the newly added features of the ug [SCIP-*, MPI]-library that aim to improve the performance of ug [SCIP-JACK, MPI]. Finally, first results obtained with the new features will be presented.

2   Key Features of UG and ug [SCIP-*, MPI]-Library

A uniqueness of UG is that it is a software framework to parallelize an existing state-of-the-art B&B based solvers. We call the B&B based solver parallelized by UG *base solver*. In UG, the base solver is encapsulated in an abstracted `ParaSolver`. The `ParaSolver` accesses the base solver via its API. At run-time on a supercomputer, there are `ParaSolver` processes, which solve subproblems, and there is a special process called `LoadCoordinator (LC)`, which makes all decisions about load balancing among the `ParaSolvers`. To realize the load balancing, message passing based protocols are defined between the LC and the `ParaSolvers`. The LC also has a base solver environment, which does presolving (all `ParaSolver`s solve the presolved instance internally) and converts the solution to the presolved problem to a solution to the original one. Key features of UG are:

Ramp-up. Ramp-up is the phase until all solvers have become active. In *Normal ramp-up*, only one `ParaSolver` receives the root node, and it distributes one of the branched nodes to the other solvers via the LC. All the `ParaSolver`s do the same when they receive a node. The transferred node (subproblem) data contains only the difference between the subproblem and the (presolved) instance data. *Racing ramp-up* exploits the performance variability commonly observed in MIP solving [6]. An instance is solved multiple times by `ParaSolver`s in parallel, each time with a different parameter setting. If the instance has not been solved to optimality once a predefined termination criterion, e.g., a time limit, is reached, the most promising branch-and-bound tree is distributed among the `ParaSolver`s and the default solving procedure is initiated. The effectiveness of racing ramp-up is described in [17,20].

Dynamic load balancing. UG provides a Supervisor-Worker load coordination scheme [10]. In the Master-Worker paradigm, all B&B search tree data is managed by the Master. In contrast to the Master-Worker paradigm, the idea of Supervisor-Worker is that the Supervisor functions only to make decisions about the load balancing, but does not actually store the data associated with the B&B search tree. In UG, the Supervisor is the LC and the Workers are the `ParaSolver`s. The B&B search tree data is managed by the `ParaSolver`s. The terminal nodes (subproblems) of the B&B search tree in the `ParaSolver`s are sent on demand to the LC; a set of subproblems in the LC works as a buffer to ensure subproblems are available to idle `ParaSolver`s as needed.
Load balancing is accomplished mainly by switching the collection mode in the `ParaSolver`. Turning collecting mode on results in additional "high quality" subproblems being sent to the LC, which can then be distributed to the `ParaSolver`s. The method of selecting which `ParaSolver` to collect from is crucial and is controlled very carefully. Some additional keys to avoid having the Supervisor becoming a communication bottleneck are:
 – Frequency of status updates can be controlled depending on the number of `ParaSolver`s.
 – The maximum number of `ParaSolver`s in collection mode is capped and the `ParaSolver`s are chosen dynamically.
A detailed description of the dynamic load balancing is presented in [18,20].

Checkpointing and restrating mechanism. By the dynamic load balancing of UG, B&B nodes in a sub-tree can be transferred recursively to the other solvers. Therefore, at each checkpoint, only essential B&B nodes, i.e., subtree roots whose ancestor node is not available on a run-time system, are saved. The number of such nodes is extremely small compared to the number of open nodes; thus the checkpointing is very light weight. However, a huge search tree has to be regenerated at restart. This regeneration might look redundant and inefficient. However, for MIP solvers, this procedure has been shown to be notably efficient [17], since dual bounds of the checkpoint nodes are calculated more precisely and the B&B tree is regenerated based on these values at restart—the regenerated B&B tree can thus be different than that of the previous run.

Deterministic mode for debugging. One of the most difficult parts of software development is debugging. Before running a parallel solver instantiated by UG, extensive debugging for a set of instances with different number of solvers is usually needed. Without having a deterministic mode, this would be extremely inefficient.

ParaSCIP (=ug [SCIP, MPI]) is an instantiated parallel solver that uses UG, in which SCIP is the base solver. Since SCIP is plugin-based, it is natural to make a ug [SCIP-*, MPI]-library in which user plugins are installed automatically by providing a small amount of glue code. ug [SCIP-Jack, MPI] is realized by using such a library and is distributed as a UG application. The Steiner tree application directory of SCIP Optimization Suite 6.0 contains only one source file stp_plugins.cpp and it has only 173 lines of glue code without empty and comment lines.

3   Improvements of SCIP-Jack After the DIMACS Challenge

SCIP-Jack has seen a large number of improvements after the 11th DIMACS Challenge, both for SPG and related problems. These developments include new primal and dual heuristics [2,14], reduction techniques [15], and various technical improvements such as a fast maximum-flow implementation [8] (used for separation). Of particular relevance for massive parallelization is the subsequently described improvement for domain propagation: During the solving process it is usually possible to fix many (binary) edge variables of the IP formulation to 0 or 1—for instance by using reduced cost arguments [2] or branching information. These fixings can be directly translated into edge deletions and contractions in the underlying graph, which can allow for further eliminations by the powerful graph reduction techniques of SCIP-Jack. However, as already observed by other authors [9], such graph reductions can change the graph in a complex way, which cannot be easily translated into variable fixings in the IP formulation. However, we have devised a simple mapping that given an original instance $P = (V, E, T, c)$ and reduced instance $P' = (V', E', T', c')$ allows to map P' to a problem P'' such that P'' can be obtained from P' by deletion of edges only. First, note that the reduction techniques of SCIP-Jack provide a mapping $p : E' \rightarrow \mathscr{P}(E)$ such that for each (optimal) solution $S' \subseteq E'$ to P', set $\bigcup_{e \in S'} p(e)$ is an (optimal) solution to P. With this information one obtains:

Proposition 1. *Let* $P = (V, E, T, c)$ *be an SPG and* (V', E', T', c') *be an instance obtained by using the reduction techniques of* SCIP-Jack. *Each solution* S'' *to the SPG* $P'' = (V'', E'', T'', c'')$ *defined by*

$$E'' := \bigcup_{e \in E'} p(e),$$

$$V'' := \{v \in V \mid \exists (v,w) \in E'', w \in V\},$$

$$T'' := \{t \in T \mid \exists (t,w) \in E'', w \in V\},$$

$$c'' := c|_{E''},$$

is a solution to P. Furthermore, if S'' is an optimal solution to P'', it is an optimal solution to P.

One readily acknowledges, that $V'' \subseteq V$ and $E'' \subseteq E$. Note, however, that usually $|V''| > |V'|$ and $|E''| > |E'|$, so we first apply only techniques that can be directly translated into variable fixings (such as deletion of edges) and apply the corresponding fixings to the IP; only afterward we perform more complex reductions (and use Proposition 1 to apply further fixings).

4   New Features of ug [SCIP-JACK, MPI]

In this section, we describe new general features added to ug [SCIP-*, MPI]-library, and also specialized new features added to ug [SCIP-JACK, MPI].

4.1   Branching on Constraints

After the DIMACS Challenge, instead of branching on variables, which in the case of Steiner tree problem correspond to edges, default SCIP-JACK uses vertex branching [4]. During the B&B process, SCIP-JACK selects a non-terminal vertex of the Steiner tree problem graph to be rendered a terminal in one B&B child node and to be excluded in the other child. These two operations are modeled in the underlying IP formulation by including one additional constraint. This procedure could not be used in previous versions of ug [SCIP-JACK,MPI], since branching on constrains was only possible in SCIP, but not in the ug [SCIP-*, MPI]-library. Therefore, a new feature for transferring branching constrains has been added to the ug [SCIP-*, MPI]-library. The new feature allows ug [SCIP-JACK,MPI] to use the vertex branching.

4.2   Callback to Initialize a Transferred Subproblem

A distinguishing feature of UG solvers is that it can naturally realize *layered presolving*, in which B&B tree nodes are transferred to the other ParaSolvers recursively and additional presolving is performed on the subproblems. The effectiveness of the layered presolving is documented in [19,20]. When using ug [SCIP-*, MPI]-library, MIP presolving realized by SCIP can work without any additional coding. However, SCIP-JACK performs presolving on the underlying graph before it formulates the subproblem as an IP. In order to realize the graph based presolving, a callback to initialize the transferred subproblem has been added to the ug [SCIP-*, MPI]-library. To retain previous graph based branching decisions, ug [SCIP-JACK, MIP] transfers the branching history together with a subproblem, enabling SCIP-JACK to change the underlying graph (by adding terminals and deleting vertices). Additionally, whenever a subproblem has been transferred, SCIP-JACK performs aggressive reduction routines to reduce the (modified) problem further and translates the reductions into variable fixings by means of Proposition 1.

4.3   Customized Racing

The latest ug [SCIP-*, MPI]-library includes *customized racing*, which allows the user to specify their own parameter settings for racing. If the number of UG solvers exceeds the number of provided parameter sets, then the customized parameter settings are combined with random number seeds. While the latest release version of ParaSCIP does not use customized racing by default, it is applied in ug [SCIP-JACK,MPI]. For this article we used 30 parameter settings, where we varied: the aggressiveness of the primal heuristics, the aggressiveness of domain propagation, the branching rule (LP-based [2] or based on primal solution [9]), and various parameter for the cut selection.

5   Updated Computational Results for Open Instances

For solving open instances of the PUC test set from STEINLIB as of 1st of November 2018, we used two supercomputers. One is an ISM (Institute of Statistical Mathematics) supercomputer which is a HPE SGI 8600 with 384 compute nodes, each node has two Intel Xeon Gold 6154 3.0 GHz CPUs (18 cores × 2) sharing 384 GB of memory, and an Infiniband (Enhanced Hypercube) interconnect. The other is HLRN III which is a Cray XC40 with 1872 compute nodes, each node with two 12-core Intel Xeon IvyBridge/Haswell CPUs sharing 64 GiB of RAM, and with an Aries interconnect. The interval time of checkpointing was set to 1,800 s. The maximum number of ParaSolvers in collection mode was capped at 500.

5.1   hc9p (Solved)

This instance was solved by five restarted runs and by using up to 24,576 cores. The initial primal solution was found by ug [SCIP-JACK, MPI] at the DIMACS Challenge. All computations were used to prove its optimality. The racing termination criteria was a node limit of 50, that is: once the number of open B&B nodes in a ParaSolver with the largest dual bound surpasses 50, racing is terminated. Table 1 shows the supercomputer used, the computing time in seconds (racing time is shown within parentheses), the idle time ratio for all ParaSolvers, the number of transferred B&B nodes to ParaSolvers, primal and dual bounds, gap, the number of B&B nodes generated, and the number of open B&B nodes for each run. The initial values are shown in the upper row and the final values of those are shown in the lower row for each run.

The final dual bound in the previous run is sometimes slightly different from that of the initial one in the following run. This means that the dual bound in the previous run was updated after the final checkpoint. The number of open B&B nodes decreases a lot at restart, since the checkpointing mechanism only saves essential sub-tree roots. For example, run 1.1 ends up with 1,257,112 open B&B nodes, but run 1.2 starts with 15 open ones. This means that only 15 B&B sub-tree roots existed at the end of run 1.1 and the other sub-tree roots were descendants of one of the 15 B&B nodes.

The number of transferred B&B nodes can be considered as an indicator of how frequently `ParaSolvers` became idle and also how frequently layered presolving was applied. It is natural that at larger scale we can expect more layered presolving. Actually, the number of transferred B&B nodes of run 1.1 with 72 cores was only 738 nodes in a one week long execution. It was increased by using 2,304 cores to 979,695 in another one week execution. In the following bigger jobs it was drastically increased.

Figure 1 shows the evolution of the computation for the final run 1.5. The number of B&B open nodes continuously increases and decreases during the computation and it looks sometimes difficult to make all `ParaSolvers` active. However, dynamic load balancing recovered the situation well and all the `ParaSolvers` were active during almost the entire computing time. The idle time ratio was only 1.5%. The number of checkpoint nodes also changed a lot during the computation.

We can obtain the idle time ratio for all `ParaSolvers` only if ug [SCIP-JACK, MPI] finishes its computation and cannot get it if the program is canceled by the system in case the time-limit is hit. After racing ramp-up, all `ParaSolver` statistics are collected. Therefore, by using its partial data, an upper bound on the idle time ratio is calculated. The lack of data is complemented by the maximum idle time ratio in the case of racing ramp-up, and complemented by the idle time ratio of run 1.5. Table 1 also shows the upper bounds of the idle time ratio. The idle time ratios for all runs are notably small, which indicates that the supercomputers are used efficiently.

5.2   hc11p (Updated the Best Known Solution)

During the new developments in ug [SCIP-JACK, MPI], the best known solution to the `hc11p` instance could be updated (with objective value 119,492 compared to 119,689 at the DIMACS Challenge). The first additional run 1 on the ISM supercomputer generated 11 new incumbent solutions, with the best objective value being 119,297. Afterwards we just solved it from scratch with the best solution in racing ramp-up (run 2.1) again, since it can be used for presolving, propagation, and heuristics. The racing termination criteria for run 1 was the same as that for `hc9p`, but the node limit 100 was used for run 2.1. The restarted job was conducted from the checkpoint file of run 2.1, since run 2.1 could not improve the incumbent solution. Run 1 consumed 12,095 cores-hours (=(72 ×

Table 1. Statistics for solving `hc9p` on supercomputers

Run	Computer	Cores	Time (sec.)	Idle (%)	Trans.	Primal bound (upper bound)	Dual bound (lower bound)	Gap (%)	Nodes	Open nodes
1.1	ISM	72	604,796 (317)	<0.3	738	30,242.0000	29,879.3721	1.21	0	0
						30,242.0000	30,058.9366	0.61	110, 012, 624	1, 257, 112
1.2	ISM	2, 304	604,794	<1.5	979, 695	30,242.0000	30,058.7930	0.61	0	15
						30,242.0000	30,102.7556	0.46	3, 758, 532, 600	723, 167
1.3	HLRN III	24, 576	86,336	<1.7	8, 811, 512	30,242.0000	30,102.6645	0.46	0	35
						30,242.0000	30,116.3592	0.42	2, 402, 406, 311	575, 678
1.4	HLRN III	12, 288	43,199	<1.5	1, 709, 027	30,242.0000	30,115.3331	0.42	0	3, 709
						30,242.0000	30,120.4801	0.40	664, 909, 985	602, 323
1.5	HLRN III	12, 288	118,259	1.5	9, 158, 920	30,242.0000	30,120.4801	0.40	0	285
						30,242.0000	30,242.0000	0.00	1, 677, 724, 126	0

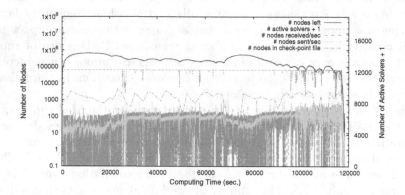

Fig. 1. Evolution of computation for solving `hc9p` by using 12,288 cores (Run 1.5)

Table 2. Statistics for solving `hc11p` on supercomputers

Run	Computer	Cores	Time (sec.)	Idle (%)	Trans.	Primal bound (upper bound)	Dual bound (lower bound)	Gap (%)	Nodes	Open nodes
1	ISM	72	604,799 (2,558)	<0.3	71	119,492.0000	117,388.8528	1.79	0	0
						119,297.0000	117,496.5470	1.53	4,314,198	1,109,629
2.1	HLRN III	12,288	43,149 (7,164)	<0.5	31,304	119,297.0000	117,388.7971	1.63	0	0
						119,297.0000	117,426.2226	1.59	28,491,470	5,433,482
2.2	HLRN III	43,000	86,354	<4.9	86,152	119,297.0000	117,426.2226	1.59	0	103
						119,297.0000	117,468.8459	1.56	267,513,609	40,499,188

604799)/3600) and it reached a 1.53(%) gap. Runs 2.1 and 2.2 consumed 118,582 cores-hours (=((1288 × 43149) + (4300 × 86354))/3600) reached a 1.56(%) gap. To improve the gap with the same amount of computing resources, initial longer run at small scale look more promising than large scale runs with short computing time (Table 2).

The numbers of transferred B&B nodes were very small compared to those for `hc9p`. This shows a fundamental hardness of `hc11p` compared to that of `hc9p`. Figure 2 shows the evolution of computation for run 2.2—the largest scale used with `ug` [SCIP-JACK, MPI] so far. The restart is always normal ramp-up from the nodes in checkpoint file. In the normal ramp-up, all `ParaSolvers` send

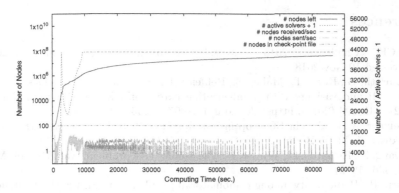

Fig. 2. Evolution of computation for solving `hc11p` by using 43,000 cores (Run 2.2)

one of two branched nodes to the other `ParaSolvers` via LC. This lasts until all `ParaSolvers` have become active. SCIP-JACK does presolving and adds cutting planes aggressively at its root node, making ramp-up difficult. Additionally, once in ramp-up the LC's internal mode changes to collection mode. In this mode, only a restricted number of `ParaSolvers` can be in collection mode. Therefore, the number of active `ParaSolvers` decreases after the first peak. However, once the LC has collected enough nodes again, the quality of the nodes in the LC is very good and less and less dynamic load balancing is needed. Figure 2 shows this behavior. Taking into account this difficulty of ramp-up, the idle time ratio of run 2.2 is less than 4.9%. The number of checkpoint nodes stays the same and the open B&B nodes keep increasing. Thus further improvements of SCIP-JACK or much larger runs are needed to solve `hc11p`.

6   Concluding Remarks

We have extended ug [SCIP-JACK, MPI] to immediately obtain the benefits of any SCIP-JACK improvements, allowing us to solve one previously unsolved benchmark instance to optimality. We also showed that ug [SCIP-JACK, MPI] can run on up to 43,000 cores efficiently in terms of computing resources usage. Therefore, when SCIP-JACK has been further improved (as planned for the near future) we expect to solve additional open instances. Also, the techniques presented in this paper work on other problems related to the SPG that can be handled by SCIP-JACK.

Acknowledgements. The authors would like to thank Utz-Uwe Haus for his help in tracking down a particularly insistent bug. This work has been supported by the Research Campus MODAL *Mathematical Optimization and Data Analysis Laboratories* funded by the Federal Ministry of Education and Research (BMBF Grant 05M14ZAM). This work was also supported by the North-German Supercomputing Alliance (HLRN). Supported by BMWi project BEAM-ME (fund number 03ET4023DE).

References

1. PACE Challenge 2018. https://pacechallenge.wordpress.com/pace-2018/. Accessed 10 Nov 2018
2. Gamrath, G., Koch, T., Maher, S., Rehfeldt, D., Shinano, Y.: SCIP-Jack—a solver for STP and variants with parallelization extensions. Math. Program. Comput. **9**(2), 231–296 (2017). https://doi.org/10.1007/s12532-016-0114-x
3. Gleixner, A., et al.: The SCIP optimization suite 6.0. Technical report, 18–26, ZIB, Takustr. 7, 14195 Berlin (2018)
4. Hwang, F., Richards, D., Winter, P.: The Steiner tree problem. Ann. Discret. Math. **53**, 336 (1992)
5. Karp, R.: Reducibility among combinatorial problems. In: Miller, R., Thatcher, J. (eds.) Complexity of Computer Computations, pp. 85–103. Plenum Press (1972)
6. Koch, T., et al.: MIPLIB 2010. Math. Program. Comput. **3**, 103–163 (2011)
7. Koch, T., Martin, A., Voß, S.: SteinLib: an updated library on Steiner tree problems in graphs. In: Du, D.Z., Cheng, X. (eds.) Steiner Trees in Industries, pp. 285–325. Kluwer (2001)
8. Maher, S.J., et al.: The SCIP optimization suite 4.0. Technical report 17–12, ZIB, Takustr. 7, 14195 Berlin (2017)
9. Polzin, T.: Algorithms for the Steiner problem in networks. Ph.D. thesis, Saarland University (2004). http://scidok.sulb.uni-saarland.de/volltexte/2004/218/index.html
10. Ralphs, T., Shinano, Y., Berthold, T., Koch, T.: Parallel solvers for mixed integer linear optimization. Handbook of Parallel Constraint Reasoning, pp. 283–336. Springer, Cham (2018). https://doi.org/10.1007/978-3-319-63516-3_8
11. Rehfeldt, D., Koch, T.: Combining NP-hard reduction techniques and strong heuristics in an exact algorithm for the maximum-weight connected subgraph problem. SIAM J. Optim. **29**(1), 369–398 (2019). https://doi.org/10.1137/17M1145963
12. Rehfeldt, D., Koch, T.: Reduction-based exact solution of prize-collecting Steiner tree problems. Technical report 18–55, ZIB, Takustr. 7, 14195 Berlin (2018)
13. Rehfeldt, D., Koch, T.: SCIP-Jack—a solver for STP and variants with parallelization extensions: an update. In: Kliewer, N., Ehmke, J.F., Borndörfer, R. (eds.) Operations Research Proceedings 2017. ORP, pp. 191–196. Springer, Cham (2018). https://doi.org/10.1007/978-3-319-89920-6_27
14. Rehfeldt, D., Koch, T.: Transformations for the prize-collecting Steiner tree problem and the maximum-weight connected subgraph problem to SAP. J. Comput. Math. **36**(3), 459–468 (2018)
15. Rehfeldt, D., Koch, T., Maher, S.J.: Reduction techniques for the prize collecting Steiner tree problem and the maximum-weight connected subgraph problem. Networks **73**(2), 206–233 (2019). https://doi.org/10.1002/net.2185. https://onlinelibrary.wiley.com/doi/abs/10.1002/net.21857
16. Shinano, Y.: The ubiquity generator framework: 7 years of progress in parallelizing branch-and-bound. In: Kliewer, N., Ehmke, J.F., Borndörfer, R. (eds.) Operations Research Proceedings 2017. ORP, pp. 143–149. Springer, Cham (2018). https://doi.org/10.1007/978-3-319-89920-6_20
17. Shinano, Y., Achterberg, T., Berthold, T., Heinz, S., Koch, T., Winkler, M.: Solving hard MIPLIP2003 problems with ParaSCIP on supercomputers: an update. In: IEEE (ed.) IPDPSW 2014 Proceedings of the 2014 IEEE, International Parallel & Distributed Processing Symposium Workshops, pp. 1552–1561 (2014). https://doi.org/10.1109/IPDPSW.2014.174

18. Shinano, Y., Achterberg, T., Berthold, T., Heinz, S., Koch, T., Winkler, M.: Solving open mip instances with ParaSCIP on supercomputers using up to 80,000 cores. In: Proceedings of 30th IEEE International Parallel & Distributed Processing Symposium (2016). https://doi.org/10.1109/IPDPS.2016.56
19. Shinano, Y., Berthold, T., Heinz, S.: A first implementation of ParaXpress: combining internal and external parallelization to solve MIPs on supercomputers. In: Greuel, G.-M., Koch, T., Paule, P., Sommese, A. (eds.) ICMS 2016. LNCS, vol. 9725, pp. 308–316. Springer, Cham (2016). https://doi.org/10.1007/978-3-319-42432-3_38
20. Shinano, Y., Heinz, S., Vigerske, S., Winkler, M.: FiberSCIP - a shared memory parallelization of SCIP. INFORMS J. Comput. **30**(1), 11–30 (2018). https://doi.org/10.1287/ijoc.2017.0762

Deep Inverse Optimization

Yingcong Tan[1][(✉)], Andrew Delong[2], and Daria Terekhov[1]

[1] Department of Mechanical, Industrial and Aerospace Engineering,
Concordia University, Montreal, Canada
`t_yingco@encs.concordia.ca, daria.terekhov@concordia.ca`
[2] Department of Electrical and Computer Engineering,
University of Toronto, Toronto, Canada
`andrew.delong@gmail.com`

Abstract. Given a set of observations generated by an optimization process, the goal of inverse optimization is to determine likely parameters of that process. We cast inverse optimization as a form of deep learning. Our method, called *deep inverse optimization*, is to unroll an iterative optimization process and then use backpropagation to learn parameters that generate the observations. We demonstrate that by backpropagating through the interior point algorithm we can learn the coefficients determining the cost vector and the constraints, independently or jointly, for both non-parametric and parametric linear programs, starting from one or multiple observations. With this approach, inverse optimization can leverage concepts and algorithms from deep learning.

Keywords: Inverse optimization · Deep learning · Interior point

1 Introduction

The potential for synergy between optimization and machine learning is well-recognized [6], with recent examples including [9,20,30]. Our work uses machine learning for *inverse optimization* (IO). In inverse optimization, we observe one or more decisions from an unknown optimization process, and the goal is to 'learn' an optimization model that is consistent with the observations. Aspects of the unknown optimization process that we may wish to learn include terms in the objective function or constraints on the decision variables.

An early example of IO is the inverse shortest path problem, used to learn the unobservable transmission times of seismic waves which are known to follow a shortest path [42]. Other applications include determining the tolls that would enforce a desired traffic flow [11], imputing the relative importance of treatment objectives from clinically-approved radiotherapy plans [13,31] in order to automate clinicians' decision-making, and predicting the behaviour of price-responsive customers [36].

© Springer Nature Switzerland AG 2019
L.-M. Rousseau and K. Stergiou (Eds.): CPAIOR 2019, LNCS 11494, pp. 540–556, 2019.
https://doi.org/10.1007/978-3-030-19212-9_36

We illustrate our framework in the context of parametric linear optimization. Specifically, consider the parametric linear program $\mathrm{PLP}(\mathbf{u}, \mathbf{w})$:

$$\begin{aligned}
\underset{\mathbf{x}}{\text{minimize}} \quad & \mathbf{c}(\mathbf{u}, \mathbf{w})'\mathbf{x} \\
\text{subject to} \quad & \mathbf{A}(\mathbf{u}, \mathbf{w})\mathbf{x} \leq \mathbf{b}(\mathbf{u}, \mathbf{w}),
\end{aligned} \tag{1}$$

where $\mathbf{x} \in \mathbb{R}^d$, $\mathbf{c}(\mathbf{u}, \mathbf{w}) \in \mathbb{R}^d$, $\mathbf{A}(\mathbf{u}, \mathbf{w}) \in \mathbb{R}^{m \times d}$ and $\mathbf{b}(\mathbf{u}, \mathbf{w}) \in \mathbb{R}^m$. The 'feature' vector $\mathbf{u}$ represents conditions (e.g., time, prices, weather) under which we may want to instantiate and solve the linear program. The 'weight' vector $\mathbf{w}$ represents parameters relating the features to the optimization model instance.

In inverse optimization, for a set of features $\{\mathbf{u}^1, \mathbf{u}^2, \ldots, \mathbf{u}^N\}$ we observe the corresponding decisions of some unknown optimization process. Call these decisions $\{\mathbf{x}_{\text{tru}}^1, \mathbf{x}_{\text{tru}}^2, \ldots, \mathbf{x}_{\text{tru}}^N\}$, as they are generated by the 'true' underlying process. Fundamentally, IO problems are learning problems: the goal of IO is to learn weights $\mathbf{w}$ such that, for each $n \in \{1, \ldots, N\}$, there exists an optimal solution of $\mathrm{PLP}(\mathbf{u}^n, \mathbf{w})$ that is consistent with the corresponding observed decision $\mathbf{x}_{\text{tru}}^n$. The learned model can then be applied to predict decisions under new conditions $\mathbf{u}$ that were not seen at training time.

In this paper, we cast inverse optimization as a form of deep learning. Our method, called *deep inverse optimization*, is to 'unroll' an iterative optimization process and then use backpropagation to learn model parameters $\mathbf{w}$ that generate the observations, i.e., training targets. Specifically, we use a deep learning framework to trace computations across the iterations of an optimization loop, resulting in a chain of dependent variables (a dynamically unrolled loop) which are then automatically differentiated with respect to a loss function so as to compute a gradient for $\mathbf{w}$.

Figure 1 shows the actual result of applying our deep IO method to three inverse optimization learning tasks. The top panel illustrates the *non-parametric*, single-point variant of model (1)—the case when exactly one $\mathbf{x}_{\text{tru}}$ is given—a classical problem in IO (see [1,14]). In Fig. 1(i), only $\mathbf{c}$ needs to be learned: starting from an initial cost vector $\mathbf{c}_{\text{ini}}$, our method finds $\mathbf{c}_{\text{lrn}}$ which makes $\mathbf{x}_{\text{tru}}$ an optimal solution of the LP by minimizing $\|\mathbf{x}_{\text{tru}} - \mathbf{x}_{\text{lrn}}\|^2$. In Fig. 1(ii), starting from $\mathbf{c}_{\text{ini}}$, $\mathbf{A}_{\text{ini}}$ and $\mathbf{b}_{\text{ini}}$, our approach finds $\mathbf{c}_{\text{lrn}}$, $\mathbf{A}_{\text{lrn}}$ and $\mathbf{b}_{\text{lrn}}$ which make $\mathbf{x}_{\text{tru}}$ an optimal solution of the learned LP through minimizing $\|\mathbf{x}_{\text{tru}} - \mathbf{x}_{\text{lrn}}\|^2$.

Figure 1(iii) shows learning $\mathbf{w} = [w_0, w_1]$ for the *parametric* problem instance

$$\begin{aligned}
\underset{\mathbf{x}}{\text{minimize}} \quad & \cos(w_0 + w_1 u)x_1 + \sin(w_0 + w_1 u)x_2 \\
\text{subject to} \quad & -x_1 \leq 0.2 w_0 u, \\
& -x_2 \leq -0.2 w_1 u, \\
& w_0 x_1 + (1 + \tfrac{1}{3} w_1 u)x_2 \leq w_0 + 0.1u.
\end{aligned} \tag{2}$$

The left panel of Fig. 1(iii) shows the true $\mathrm{PLP}(u, \mathbf{w}_{\text{tru}})$ with $\mathbf{w}_{\text{tru}} = [1.0, 1.0]$, along with four observations denoted as $\mathbf{x}(u^n, \mathbf{w}_{\text{tru}})$ corresponding to u values $\{-1.5, -0.5, 0.5, 1.5\}$. Starting from $\mathbf{w}_{\text{ini}} = [0.2, 0.4]$ with a loss (mean squared error) of 0.45, our method is able to find $\mathbf{w}_{\text{lrn}} = [1.0, 1.0]$ with a loss of zero,

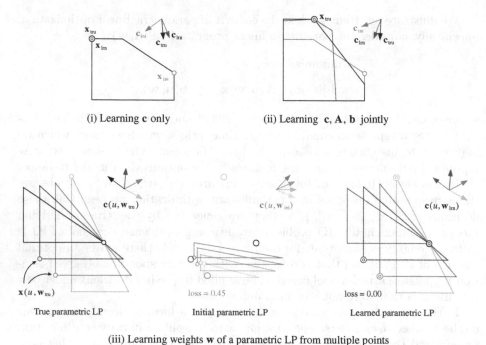

(i) Learning $\mathbf{c}$ only   (ii) Learning $\mathbf{c}, \mathbf{A}, \mathbf{b}$ jointly

True parametric LP   Initial parametric LP   Learned parametric LP

(iii) Learning weights $\mathbf{w}$ of a parametric LP from multiple points

Fig. 1. Three IO learning tasks in non-parametric and parametric linear programs.

thereby making the observed $\mathbf{x}_{\mathrm{tru}}^n$ optimal solutions of (2). In this case, the learned PLP will predict the same decisions as the true PLP when evaluated on new values of u. In other words, the learned model generalizes well.

The contributions of this paper are as follows. We propose a general framework for inverse optimization based on deep learning. This framework is applicable to learning coefficients of the objective function and constraints, individually or jointly; minimizing a general loss function; learning from a single or multiple observations; and solving both non-parametric and parametric problems. As a proof of concept, we demonstrate that our method obtains effectively zero loss on many randomly generated linear programs for all three types of learning tasks shown in Fig. 1, and always improves the loss significantly. Such a numerical study on randomly generated non-parametric and parametric linear programs with multiple learnable parameters has not previously been published for any IO method in the literature. Finally, to the best of our knowledge, we are the first to use unrolling and backpropagation for constrained inverse optimization.

We explain how our approach differs from methods in inverse optimization and machine learning in Sect. 2. We present our deep IO framework in Sect. 3 and our experimental results in Sect. 4. Section 5 discusses both the generality and the limitations of our work, and Sect. 6 concludes the paper.

2   Related Work

The goal of our paper is to develop a general-purpose IO approach that is applicable to problems for which theoretical guarantees or efficient exact optimization approaches are difficult or impossible to develop. Naturally, such a general-purpose approach will not be the method of choice for all classes of IO problems. In particular, for non-parametric linear programs, closed-form solutions for learning the c vector (Fig. 1(i)) and for learning the constraint coefficients have been derived by Chan et al. [14, 16] and Chan and Kaw [15], respectively. However, learning objective and constraint coefficients jointly (Fig. 1(ii)) has, to date, received little attention. To the best of our knowledge, this task has been investigated only by Troutt et al. [43, 44], who referred to it as linear system identification, using a maximum likelihood approach. However, their approach was limited to two dimensions [44] or required the coefficients to be non-negative [43].

In the parametric optimization setting, Keshavarz et al. [25] develop an optimization model that encodes KKT optimality conditions for imputing objective function coefficients of a convex optimization problem. Aswani et al. [3] focus on the same problem under the assumption of noisy measurements, developing a bilevel formulation and two algorithms which are shown to maintain statistical consistency. Saez-Gallego and Morales [36] address the case of learning c and b jointly in a parametric setting where the b vector is assumed to be an affine function of a regressor. The general case of learning the weights of a parametric linear optimization problem (1) where c, A and b are functions of u (Fig. 1(iii)) has not been addressed in the literature.

Recent work in machine learning [4, 5, 18] views IO through the lens of online learning, where the optimization model is incrementally updated based on new observations. Our approach may be applicable in online settings, but in the current paper we consider problems with a fixed training set.

It is worth noting that there are conceptual parallels between inverse optimization and *constraint acquisition* [7], including recent variants that incorporate machine learning [27]. In constraint acquisition, the goal is to allow non-expert users to specify constraint sets in the *constraint programming* formalism using an example-based approach.

Methodologically, our unrolling strategy is similar to McLaurin et al. [28] who directly optimize the hyperparameters of a neural network training procedure with gradient descent. Conceptually, the closest papers to our work are by Amos and Kolter [2] and Donti, Amos and Kolter [19]. However, these papers are written independently of the inverse optimization literature. Amos and Kolter [2] present the OptNet framework, which integrates a quadratic optimization layer in a deep neural network. The gradients for updating the coefficients of the optimization problem are derived through implicit differentiation. This approach involves taking matrix differentials of the KKT conditions for the optimization problem in question, while our strategy is based on allowing a deep learning framework to unroll an existing optimization procedure. Their method has efficiency advantages, while our unrolling approach is easily applicable, including to processes for which the KKT conditions may not hold or are difficult to implicitly differentiate. We include a more in-depth discussion in Sect. 5.

3   Deep Learning Framework for Inverse Optimization

Inverse optimization can be viewed as an approach to machine learning specialized to the case when the observed data is coming from an optimization process. Given this perspective on IO, and motivated by the success of deep learning for a variety of learning tasks in recent years (see [26]), this paper develops a deep learning framework for inverse optimization.

Deep learning is a set of techniques for training the parameters of a sequence of transformations (layers) that have been composed (chained) together. The more intermediate layers of computation, the 'deeper' the architecture. We refer the reader to the textbook by Goodfellow, Bengio and Courville [21] for details. The features of the intermediate layers can be trained/learned through backpropagation [35], an automatic differentiation technique that can efficiently compute a direction in which to update the weights of the model. Importantly, current machine learning libraries such as PyTorch provide built-in backpropagation capabilities [33], making this technique much more accessible and flexible than in the past.

Our deep IO framework cycles through three steps: (1) instantiate a forward optimization problem with the current weights $\mathbf{w}$, (2) solve the problem with a standard algorithm while tracing its execution, and (3) automatically compute an update to improve $\mathbf{w}$ by backpropagating through the traced steps.

Algorithm 1. Deep inverse optimization framework.

Input: initial weights $\mathbf{w}_{\text{ini}}$; training targets $\{(\mathbf{u}^n, \mathbf{x}_{\text{tru}}^n)\}_{n=1}^{N}$.
Output: learned weights $\mathbf{w}_{\text{lrn}}$

1: $\mathbf{w} \leftarrow \mathbf{w}_{\text{ini}}$
2: **for** s in $1 \mathbin{..} \texttt{max\_steps}$ **do**
3:      $\Delta\mathbf{w} \leftarrow \mathbf{0}$
4:      **for** n in $1 \mathbin{..} N$ **do**   ▷ For each training example
5:          $\mathbf{x} \leftarrow \mathbf{FO}(\mathbf{u}^n, \mathbf{w})$ ▷ Run forward optimizer to completion
6:          $\ell \leftarrow \mathcal{L}(\mathbf{x}, \mathbf{x}_{\text{tru}}^n)$ ▷ Compute loss w.r.t. target
7:          $\frac{\partial \ell}{\partial \mathbf{w}} \leftarrow \texttt{backprop}(\ell)$ ▷ Backpropagate gradient to weights
8:          $\Delta\mathbf{w} \leftarrow \Delta\mathbf{w} + \frac{1}{N}\frac{\partial \ell}{\partial \mathbf{w}}$ ▷ Accumulate average gradient
9:      **end for**
10:     $\Delta\mathbf{w} \leftarrow \alpha \odot \Delta\mathbf{w}$ ▷ Scale gradient component-wise
11:     $\beta \leftarrow \texttt{line\_search}(\mathbf{w}, \Delta\mathbf{w})$ ▷ Find safe step size
12:     $\mathbf{w} \leftarrow \mathbf{w} - \beta\Delta\mathbf{w}$ ▷ Update weights
13: **end for**
14: Return $\mathbf{w}$

Our approach, shown in Algorithm 1, takes the pairs $\{(\mathbf{u}^n, \mathbf{x}_{\text{tru}}^n)\}_{n=1}^{N}$ as input, and initializes $\mathbf{w}$ to $\mathbf{w}_{\text{ini}}$. For each n, the forward optimization problem (**FO**) is solved with the current weights (line 5), and the loss between the resulting

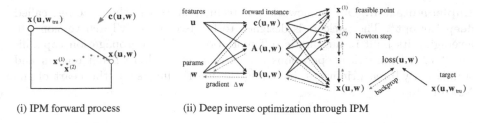

(i) IPM forward process　　　　(ii) Deep inverse optimization through IPM

Fig. 2. Illustration of the deep inverse optimization framework.

optimal solution $\mathbf{x}$ and $\mathbf{x}_{\mathrm{tru}}$ is computed (line 6). The gradient of the loss function with respect to $\mathbf{w}$ is computed by backpropagation through the layers of the forward process (line 7). The gradient is optionally scaled by component-wise product ($\odot$) with a vector $\boldsymbol{\alpha}$ that controls the relative learning rates (line 10). Line search then determines a safe step size β that precludes an increase in the overall loss and prevents the forward problem from becoming unbounded or infeasible (line 11). Finally, the weights are updated (line 12). This process repeats until `max_steps` iterations are complete.

Importantly, our framework is applicable in principle to any differentiable forward optimization procedure. Gradients are automatically computable even with non-linear constraints or non-linear objectives, as long as they can be expressed through standard differentiable primitives. For our experiments we implement the barrier interior point method (IPM) as described by Boyd and Vandenberghe [10] for our forward solver. The IPM forward process is illustrated in Fig. 2(i): the central path taken by IPM is illustrated for the current $\mathbf{u}$ and $\mathbf{w}$, which define both the current feasible region and the current $\mathbf{c}(\mathbf{u}, \mathbf{w})$. As shown in Fig. 2(ii), backpropagation starts from the loss between IPM solution $\mathbf{x}(\mathbf{u}, \mathbf{w})$ and the target $\mathbf{x}(\mathbf{u}, \mathbf{w}_{\mathrm{tru}})$ and proceeds backward to the initial state $\mathbf{x}^{(1)}$ of IPM. The key to backpropagating through each Newton step of IPM is to differentiate a matrix inverse operation, which PyTorch now does automatically. In practice, backpropagating all the way to $\mathbf{x}^{(1)}$ may not be necessary for computing sufficiently accurate gradients; see Sect. 5.

The framework requires setting three main hyperparameters: $\mathbf{w}_{\mathrm{ini}}$, the initial weight vector; `max_steps`, the total number of steps allotted to the training; and $\boldsymbol{\alpha}$, the learning rates for the different components of $\mathbf{w}$. The number of additional hyperparameters depends on the forward optimization process.

4　Experimental Results

In this section, we demonstrate the application of our framework on randomly-generated linear programs for the three types of problems shown in Fig. 1: learning $\mathbf{c}$ in the non-parametric case; learning $\mathbf{c}$, $\mathbf{A}$ and $\mathbf{b}$ together in the non-parametric case; and learning $\mathbf{w}$ in the parametric case.

Implementation. Our framework is implemented as a Python package called `deep_inv_opt`[1]. The package is designed to be used with PyTorch version 1.0, leveraging its built-in automatic differentiation and backpropagation capabilities [33]. All numerical operations are carried out with PyTorch tensors and standard PyTorch primitives, including the matrix inverse at the heart of the Newton step.

Hyperparameters. We limit learning to `max_steps` = 200 in all experiments. Four additional hyperparameters are set in each experiment:

- ϵ, which controls the precision and termination of IPM;
- $t^{(0)}$: the initial value of the barrier IPM sharpness parameter t;
- μ: the factor by which t is increased along the IPM central path;
- α: the vector of per-parameter learning rates, which in some experiments is broken down into α_c and α_{Ab}.

In all experiments, the ϵ hyperparameter is either a constant 10^{-5} or decays exponentially from 0.1 to 10^{-5} during learning.

Benchmark Methodology. To the best of our knowledge, there are no well-established benchmarks in the IO literature. Thus, we develop an IO benchmark comprising random instances with varying dimension and number of constraints. We generate a set of feasible regions having d dimensions and m constraints by sampling at least d points with components from $\mathcal{N}(0,1)$ and computing their convex hull via the *scipy.spatial.convexhull* package [34]. We refer to these as 'baseline' feasible regions. We generate 50 baseline feasible regions for each of the following six problem sizes: $d = 2$ with $m \in \{4, 8, 16\}$, and $d = 10$ with $m \in \{20, 36, 80\}$. The baseline regions and training/testing targets in our experiments can all be generated by scripts in the accompanying code repository. Though we observe that our method works for equality constraints, our experiments focus on inequality constraints, and we leave a systematic evaluation of equality constraints to future work.

4.1   Experiments on Non-parametric Linear Programs

We first demonstrate the performance of our method for learning $\mathbf{c}$ only, and learning $\mathbf{c}$, $\mathbf{A}$ and $\mathbf{b}$ jointly, on the single-point variant of model (1), i.e., when a single optimal target $\mathbf{x}_{tru}$ is given, a classical assumption in IO [1]. We use two loss functions, absolute duality gap (ADG) and squared error (SE), defined as follows:

$$\text{ADG} = \mathbf{c}'_{lrn}|\mathbf{x}_{tru} - \mathbf{x}_{lrn}|, \tag{3}$$

$$\text{SE} = \|\mathbf{x}_{tru} - \mathbf{x}_{lrn}\|_2^2. \tag{4}$$

Both ADG and SE have been used in IO [13,14,16], and SE is a standard metric in machine learning.

---

[1] Available at https://github.com/tankconcordia/deep_inv_opt.

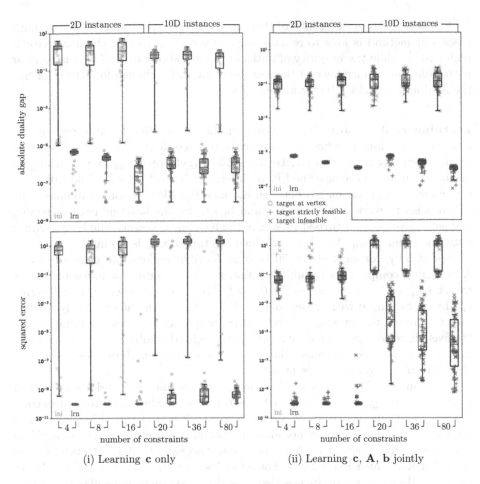

Fig. 3. Experimental results for non-parametric IO problems.

Learning c Only. For each of the 50 baseline feasible regions, we randomly select one vertex of the convex hull to be the training target x_{tru}. We set A and b to match the baseline feasible region, and we generate a random c_{ini} by drawing from $\mathcal{N}(0,1)$. The goal is to find a c_{lrn} for which x_{tru} is a solution.

We implement a randomized grid search by sampling 20 random combinations of the following three hyperparameter sets: $t^{(0)} \in \{0.5, 1, 5, 10\}$, $\mu \in \{1.5, 2, 5, 10, 20\}$, and $\alpha_c \in \{1, 10, 100, 1000\}$. These hyperparameter sets were chosen based on intuition from preliminary experiments. As in other applications of deep learning, it is not clear which hyperparameters will work best for a particular problem instance. For each instance we run our algorithm with the same 20 hyperparameter combinations, reporting the best final error values. Note that in this experiment a loss of zero is achievable by a closed-form expression [14], so the success of our method can be evaluated in absolute terms by the fraction of instances that achieve zero loss.

Figure 3(i) shows the results of this experiment for ADG and SE loss. In both cases, our method is able to reliably learn $\mathbf{c}$: for all instances, the final error is under 10^{-4}, while the majority of initial errors are above 10^{-1}. There is no clear pattern in the performance of the method as m and d change for ADG; for SE, the final loss is slightly larger for higher d.

Learning c, A, b Jointly. For each of the 50 baseline feasible regions, we generate a random baseline $\mathbf{c}$ vector to form a baseline LP. We then generate an either strictly feasible or strictly infeasible target point $\mathbf{x}_{\mathrm{tru}}$ by perturbing an optimal solution to the baseline LP. We interpret these strictly feasible/infeasible targets as being a mismatch between the baseline LP and some unknown true LP we wish to recover. We set $\mathbf{A}_{\mathrm{ini}}$ and $\mathbf{b}_{\mathrm{ini}}$ to be the baseline feasible region, and set $\mathbf{c}_{\mathrm{ini}}$ to be a perturbed version of the baseline $\mathbf{c}$ vector. The IO algorithm must then find a $\mathbf{c}_{\mathrm{lrn}}, \mathbf{A}_{\mathrm{lrn}}, \mathbf{b}_{\mathrm{lrn}}$ for which the target $\mathbf{x}_{\mathrm{tru}}$ is optimal.

Specifically, for each of the 50 baseline feasible regions, we generate a $\mathbf{c} \sim \mathcal{N}(0, 1)$ and compute its optimal solution $\mathbf{x}^*$. To generate an infeasible target we set $\mathbf{x}_{\mathrm{tru}} = \mathbf{x}^* + \eta$ where $\eta \sim U[-0.2, 0.2]$. We similarly generate a challenging $\mathbf{c}_{\mathrm{ini}}$ by corrupting $\mathbf{c}$ with noise from $U[-0.2, 0.2]$. To generate a strictly feasible target near $\mathbf{x}^*$, we set $\mathbf{x}_{\mathrm{tru}} = 0.9\mathbf{x}^* + 0.1\mathbf{x}'$ where $\mathbf{x}'$ is a random point within the feasible region, generated by a Dirichlet-weighted combination of all vertices; this method was used because adding noise to a vertex in 10 dimensions almost always results in an infeasible target.

In summary, the IO task involves both a misspecified $\mathbf{c}_{\mathrm{ini}}$ and a misspecified feasible region $\mathbf{A}_{\mathrm{ini}}$ and $\mathbf{b}_{\mathrm{ini}}$ relative to the target $\mathbf{x}_{\mathrm{tru}}$. The goal is to demonstrate the ability of our algorithm to alter the constraints and the objective so that the feasible/infeasible target becomes an optimum. For each of the six problem sizes, we randomly split the 50 instances into two subsets, one with feasible and the other with infeasible targets. For ADG loss we set $\epsilon = 10^{-5}$ and for SE we use the ϵ decay strategy. In practice, this decay strategy is similar to putting emphasis on learning $\mathbf{c}$ in the initial iterations and ending with emphasis on constraint learning.

The values of hyperparameters $\boldsymbol{\alpha_c}$ and $\boldsymbol{\alpha_{Ab}}$ are independently selected from $\{0.1, 1, 10\}$ and concatenated into one learning rate vector $\boldsymbol{\alpha}$. We generate 20 different hyperparameter combinations from the same hyperparameter sets as described above. We run our algorithm on each instance with all hyperparameter combinations and record the value of the best trial. The minimum achievable loss in this experiment is again zero, so the success of our method can be evaluated in absolute terms by the fraction of instances achieving zero loss.

Figure 3(ii) shows the results of this experiment for ADG and SE loss. In both cases, our method is able to learn model parameters that result in median loss of under 10^{-4}. For ADG, our method performs equally well for all problem sizes, and there is not much difference in the final loss for feasible and infeasible targets. For SE, however, the final loss is larger for higher d but decreases as m increases. Furthermore, there is a visible difference in performance of the method on feasible and infeasible points for 10-dimensional instances: learning from infeasible targets becomes a more difficult task.

4.2   Experiments on Parametric Linear Programs

Several aspects of the experiment for parametric LPs are different from the non-parametric case. First, we train by minimizing $\text{MSE}(\mathbf{w})$, defined as

$$\text{MSE}(\mathbf{w}) = \frac{1}{N} \sum_{n=1}^{N} \|\mathbf{x}(\mathbf{u}^n, \mathbf{w}_{\text{tru}}) - \mathbf{x}(\mathbf{u}^n, \mathbf{w})\|_2^2. \tag{5}$$

For the parametric experiments, we chose to train and evaluate using the SE loss instead of ADG for reasons discussed in Sect. 5. In the parametric case, we also assess how well the learned PLP generalizes, by evaluating its $\text{MSE}(\mathbf{w}_{\text{lrn}})$ on a held-out test set.

To generate parametric problem instances, we again started from the baseline feasible regions. To generate a true PLP, we used six weights to define linear functions of u for all elements of $\mathbf{c}$, all elements of $\mathbf{b}$, and one random element in each row of $\mathbf{A}$. For example, for 2-dimensional problems with four constraints, our instances have the following form:

$$
\begin{aligned}
\underset{\mathbf{x}}{\text{minimize}} \quad & (c_1 + w_1 + w_2 u)x_1 + (c_2 + w_1 + w_2 u)x_2 \\
\text{subject to} \quad & \begin{bmatrix} a_{11} & a_{12} + w_3 + w_4 u \\ a_{21} & a_{22} + w_3 + w_4 u \\ a_{31} + w_3 + w_4 u & a_{32} \\ a_{41} & a_{42} + w_3 + w_4 u \end{bmatrix} \leq \begin{bmatrix} b_1 + w_5 + w_6 u \\ b_2 + w_5 + w_6 u \\ b_3 + w_5 + w_6 u \\ b_4 + w_5 + w_6 u \end{bmatrix}.
\end{aligned} \tag{6}
$$

Specifically, the "true PLP" instances are generated by setting $w_1, w_3, w_5 = 0$ and $w_2, w_4, w_6 \sim \mathcal{N}(0, 0.2)$. This ensures that when $u = 0$ the true PLP feasible region matches the baseline feasible region. For each true PLP, we find a range $[u_{min}, u_{max}] \subseteq [-1, 1]$ over which the resulting PLP remains bounded and feasible. To find this 'safe' range we evaluate u at increasingly large values and try to solve the corresponding LP, expanding $[u_{min}, u_{max}]$ if successful. For each true PLP, we generate 20 equally spaced training points spanning $[u_{min}, u_{max}]$. We also sample 20 test points u sampled uniformly from $[u_{min}, u_{max}]$. We then initialize learning from a corrupted PLP by setting $\mathbf{w}_{\text{ini}} = \mathbf{w}_{\text{tru}} + \eta$ where each element of $\eta \sim U[-0.2, 0.2]$.

Hyperparameters are sampled from $t^{(0)} \in \{0.5, 1, 5, 10\}$, $\mu \in \{1.5, 2, 5, 10, 20\}$ and $\alpha_{\mathbf{Ab}} \in \{1, 10\}$, and $\alpha_{\mathbf{c}}$ is then chosen to be a factor of $\{0.01, 1, 100\}$ times $\alpha_{\mathbf{Ab}}$, i.e., a relative learning rate. The range of these values was based on preliminary experiments. Here, $\alpha_{\mathbf{c}}$ and $\alpha_{\mathbf{Ab}}$ control the learning rate of parameters within $\mathbf{w}$ that determine $\mathbf{c}$ and $(\mathbf{A}, \mathbf{b})$, respectively. In total, we generate 20 different hyperparameter combinations. We run our algorithm on each instance with all hyperparameter combinations and record the best final error value. A constant value of $\epsilon = 10^{-5}$ is used. In these experiments, the minimum achievable loss is again zero.

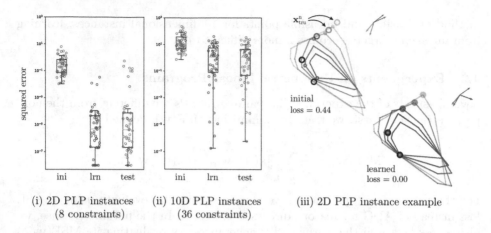

(i) 2D PLP instances   (ii) 10D PLP instances   (iii) 2D PLP instance example
   (8 constraints)   (36 constraints)

Fig. 4. Experimental results for parametric IO problems.

We demonstrate the performance of our method on learning parametric LPs of the form shown in (6) with $d = 2$, $m = 8$, and $d = 10$, $m = 36$. In Fig. 4, we report two metrics evaluated on the training set, namely $\mathrm{MSE}(\mathbf{w}_{\mathrm{ini}})$ and $\mathrm{MSE}(\mathbf{w}_{\mathrm{lrn}})$, and one metric for the test set, $\mathrm{MSE}(\mathbf{w}_{\mathrm{lrn}})$. Figure 4(iii) shows an example of an instance with $d = 2$, $m = 8$ from the training set. We see that, overall, our deep learning method works well on 2-dimensional problems with the training and testing error both being much smaller than the initial error. In the vast majority of cases the test error is also comparable to training error, though there are a few cases where it is worse, which indicates a failure to generalize well. For 10D instances, the algorithm significantly improves $\mathrm{MSE}(\mathbf{w}_{\mathrm{lrn}})$ over the initialization $\mathrm{MSE}(\mathbf{w}_{\mathrm{ini}})$, but in most cases fails to drive the loss to zero, either due to local minima or slow convergence. Again, performance on the test set is similar to that on training set.

5   Discussion

The conceptual message that we wish to reinforce is that inverse optimization can be viewed as a form of deep learning, and that unrolling gives easy access to the gradients of any parameter used directly or indirectly in the forward optimization process. There are many aspects of this view that merit further exploration. What kind of forward optimization processes can be inversely optimized this way? Which ideas and algorithms from the deep learning community will help? Are there characteristics of IO that make gradient-based learning more challenging than in deep learning at large? Conclusive answers are beyond the scope of this paper, but we discuss these and other questions below.

Relation to Neural Networks. Deep neural networks often have millions of trainable weights and are very flexible in what kinds of input-output relations they can learn, thus requiring very large training sets. The optimization models we consider have comparatively few trainable weights because they represent a strong prior over how features **u** determine decisions **x**. As such, they require less training data than a typical neural network, which is why we can train our parametric instances on only 20 training points and not observe over-fitting.

Generality and Applicability. As a proof of concept, this paper uses linear programs as the forward problems with barrier IPM as the optimization process. In principle, the framework is applicable to any forward process for which automatic differentiation can be applied. This observation does not mean that ours is the best approach for a specialized IO problem, such as learning **c** from a single point [14] or multiple points within the same feasible region [16], but it provides a new strategy.

The practical message of our paper is that, when faced with novel classes or novel parameterizations of IO problems, the unrolling strategy provides convenient access to a suite of general-purpose gradient-based algorithms for solving the IO problem at hand. This strategy is made especially easy by deep learning libraries that support dynamic 'computation graphs' such as PyTorch. Researchers working within this framework can rapidly apply IO to many differentiable forward optimization processes, without having to derive the algorithm for each case. Automatic differentiation and backpropagation have enabled a new level of productivity for deep learning research, and they may do the same for inverse optimization research. Applying deep inverse optimization does not require expertise in deep learning itself.

We chose IPM as a forward process because the inner Newton step is differentiable and because we expected the gradient to temperature parameter t to have a stabilizing effect on the gradient. For non-differentiable optimization processes, it may still be possible to develop differentiable versions. In deep learning, many advances have been made by developing differentiable versions of traditionally discrete operations, such as memory addressing [22] or sampling from a discrete distribution [29]. We believe the scope of differentiable forward optimization processes may similarly be expanded over time.

Finally, it may be possible to develop hybrid approaches, combining gradient-based learning with closed-form solutions, combinatorial algorithms, coordinate descent schemes, or techniques from black-box optimization.

Limitations and Possible Improvements. Deep IO inherits the limitations of most gradient-based methods. If learning is initialized to the right "basin of attraction", it can proceed to a global optimum. Even then, the choice of learning algorithm may be crucial. When implemented within a steepest descent framework, as we have here, the learning procedure can get trapped in local minima or exhibit very slow convergence. Such effects are why most instances in Fig. 4(ii) failed to achieve zero loss.

In deep learning with neural networks, poor local minima become exponentially rare as the dimension of the learning problem increases [17,39]. A typical strategy for training neural networks is therefore to over-parameterize (use a high search dimension) and then use regularization to avoid over-fitting to the data. In deep IO, natural parameterizations of the forward process may not permit an increase in dimension, or there may not be enough observations for regularization to compensate, so local minima remain a potential obstacle. We believe training and regularization methods specialized to low-dimensional learning problems such as those from Sahoo et al. [37] may be applicable here.

We expect that other techniques from deep learning, and from gradient-based optimization in general, will translate to deep IO. For example, learning algorithms with second-order aspects such as momentum [41] and L-BFGS [12] are readily available in deep learning frameworks. Deep learning 'tricks' may also help deep IO. For example, we observe that, when c is normal to a constraint, the gradient with respect to c can suddenly become very large. We stabilized this behaviour with line search, but a similar 'exploding gradient' phenomenon exists when training deep recurrent networks, and gradient clipping [32] is a popular way to stabilize training. A detailed investigation of applicable deep learning techniques is outside the scope of this paper.

Deep IO may be more successful when the loss with respect to the forward process can be annealed or 'smoothed' in a manner akin to graduated non-convexity [8]. Our ϵ-decay strategy is an example of this approach, as discussed below.

Loss Function and Metric of Success. One advantage of the deep inverse optimization approach is that it can accommodate various loss functions, or combinations of loss functions, without special development or analysis. For example one could substitute other p-norms, or losses that are robust to outliers, and the gradient will be automatically available. This flexibility may be valuable. Special loss functions have been important in machine learning, especially for structured output problems [23]. The decision variables of optimization processes are likewise a form of structured output.

In this study we chose two classical loss functions: absolute duality gap and squared error. The behaviour of our algorithm varied depending on the loss function used. Looking at Fig. 3(ii) it appears that deep IO performs better with ADG loss than with SE loss when learning c, A, b jointly. However, this performance is due to the theoretical property that ADG can be zero despite the observed target point being infeasible [14]. With ADG, all the IO solver needs to do is adjust c, A, b so that $x_{lrn} - x_{tru}$ is orthogonal to c, which in no way requires the learned model to be capable of generating x_{tru} as an optimum. In other words, ADG is meaningful mainly when the true feasible region is known, as in Fig. 3(i). When there is limited knowledge about the true feasible region, SE may be a more meaningful loss function because it prioritizes optimization models that can directly generate the observations x_{tru}^n. That is why we used SE for our parametric experiments (Fig. 4). However, SE penalizes *any* difference between the predicted and observed decision variables, even if those differences

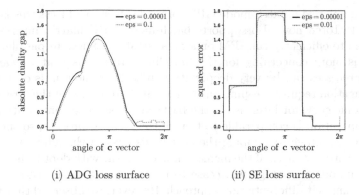

(i) ADG loss surface   (ii) SE loss surface

Fig. 5. Loss surfaces for the feasible region and target shown in Fig. 1 (i).

do not affect optimality. In short, ADG and SE both have conceptual drawbacks, and it may be beneficial to develop new or hybrid loss metrics.

In practice, minimizing the SE loss also appears to be more challenging for steepest descent. To get a sense for the characteristics of ADG versus SE from the point of view of varying $\mathbf{c}$, consider Fig. 5, which depicts the loss for the IO problem in Fig. 1 (i) using both high precision ($\epsilon = 10^{-5}$) and low precision ($\epsilon = 0.1, 0.01$) for IPM. Because the ADG loss is directly dependent on $\mathbf{c}$, the loss varies smoothly even as the corresponding optimum $\mathbf{x}^*$ stays fixed. The SE loss, in contrast, is piece-wise constant; an instantaneous perturbation of $\mathbf{c}$ will almost never change the SE loss in the limit of $\epsilon \to 0$. Note that the gradients derived by implicit differentiation [2] indicate $\frac{\partial \ell}{\partial \mathbf{c}} = \mathbf{0}$ everywhere in the linear case, which would mean $\mathbf{c}$ cannot be learned by gradient descent. With IPM one can learn $\mathbf{c}$ nonetheless because the barrier sharpness parameter t smooths the loss, especially at low values. The precision parameter ϵ limits the maximal sharpness during forward optimization, and so the gradient $\frac{\partial \ell}{\partial \mathbf{c}}$ is not zero in practice, especially when ϵ is weak. Notice that the SE loss surface in Fig. 5 becomes qualitatively smoother for weak ϵ, whereas ADG is not fundamentally changed. Also, when $\mathbf{c}$ is normal to a constraint (when the optimal point is about to transition from one point to another) the gradient $\frac{\partial \ell}{\partial \mathbf{c}}$ explodes even when the problem is smoothed.

Computational Efficiency. Our paper is conceptual and focuses on flexibility and the likelihood of success, rather than computational efficiency. Many applications of IO are not real-time, and so we expect methods with running times on the order of seconds or minutes to be of practical use. Researchers may also consider applying gradient-free solvers [24,38] to their IO problem instances. Still, we believe the gradient-based framework can be both flexible and fast.

Deep learning frameworks are GPU accelerated and scale well with the size of an individual forward problem, so large instances are not a concern. A bigger issue for GPUs is solving many small or moderate instances efficiently. Amos and

Kolter [2] developed a batch-mode GPU forward solver to address this issue. We note that PyTorch now also supports batch-mode GPU matrix inverse, which can be used to efficiently run IPM on several small instances in parallel.

What is more concerning for the unrolling strategy is that forward optimization processes can be very deep, with hundreds or thousands of iterations. Backpropagation requires keeping all the intermediate values of the forward pass resident in memory, for later use in the backward pass. The computational cost of backpropagation is comparable to that of the forward process, so there is no asymptotic advantage to skipping the backwards pass. Although memory usage was small in our instances, if the memory usage is linear with depth, then at some depth the unrolling strategy will cease to be practical compared to Amos and Kolter's [2] implicit differentiation approach. However, we observed that for IPM most of the gradient contribution comes from the final few Newton steps before termination. In other words, gradient contributions diminish as backpropagation returns 'deeper' along the central path. This means the gradient can be well-approximated in practice with *truncated backpropagation through time* (see [40] for review), which uses a small constant pool of memory regardless of the number of forward steps that were run (i.e., regardless of depth).

The unrolling approach is convenient and practical, especially during the development and exploration phase of IO research. Once an IO model is proven to work, its implementation can be made more efficient through a number of strategies, including deriving the implicit gradients [2] or by asymptotically faster learning algorithms being developed in the deep learning community.

6   Conclusion

We developed a deep learning framework for inverse optimization based on backpropagation through an iterative forward optimization process. We illustrate the potential of this framework via an implementation where the forward process is the interior point barrier method. Our results on linear non-parametric and parametric problems show promising performance. To the best of our knowledge, this paper is the first to explicitly connect deep learning and inverse optimization.

References

1. Ahuja, R.K., Orlin, J.B.: Inverse optimization. Oper. Res. **49**(5), 771–783 (2001)
2. Amos, B., Kolter, J.Z.: OptNet: differentiable optimization as a layer in neural networks. In: Proceedings of the 34th International Conference on Machine Learning, PMLR 70, pp. 136–145 (2017)
3. Aswani, A., Shen, Z.J., Siddiq, A.: Inverse optimization with noisy data. Oper. Res. **63**(3) (2018)
4. Bärmann, A., Martin, A., Pokutta, S., Schneider, O.: An online-learning approach to inverse optimization. arXiv preprint arXiv:1810.12997 (2018)
5. Bärmann, A., Pokutta, S., Schneider, O.: Emulating the expert: inverse optimization through online learning. In: International Conference on Machine Learning, pp. 400–410 (2017)

6. Bengio, Y., Lodi, A., Prouvost, A.: Machine learning for combinatorial optimization: a methodological tour d'Horizon. arXiv preprint arXiv:1811.06128 (2018)
7. Bessiere, C., Koriche, F., Lazaar, N., O'Sullivan, B.: Constraint acquisition. Artif. Intell. **244**, 315–342 (2017)
8. Blake, A., Zisserman, A.: Visual Reconstruction. MIT Press, Cambridge (1987)
9. Bonami, P., Lodi, A., Zarpellon, G.: Learning a classification of mixed-integer quadratic programming problems. In: van Hoeve, W.-J. (ed.) CPAIOR 2018. LNCS, vol. 10848, pp. 595–604. Springer, Cham (2018). https://doi.org/10.1007/978-3-319-93031-2_43
10. Boyd, S., Vandenberghe, L.: Convex Optimization. Cambridge University Press, New York (2004)
11. Burton, D., Toint, P.L.: On an instance of the inverse shortest paths problem. Math. Programm. **53**(1–3), 45–61 (1992)
12. Byrd, R.H., Lu, P., Nocedal, J., Zhu, C.: A limited memory algorithm for bound constrained optimization. SIAM J. Sci. Comput. **16**(5), 1190–1208 (1995)
13. Chan, T.C.Y., Craig, T., Lee, T., Sharpe, M.B.: Generalized inverse multi-objective optimization with application to cancer therapy. Oper. Res. **62**(3), 680–695 (2014)
14. Chan, T.C.Y., Lee, T., Terekhov, D.: Goodness of fit in inverse optimization. Manag. Sci. (2018)
15. Chan, T.C.Y., Kaw, N.: Inverse optimization for the recovery of constraint parameters. arXiv preprint arXiv:1811.00726 (2018)
16. Chan, T.C.Y., Lee, T., Mahmood, R., Terekhov, D.: Multiple observations and goodness of fit in generalized inverse optimization. arXiv preprint arXiv:1804.04576 (2018)
17. Dauphin, Y.N., Pascanu, R., Gulcehre, C., Cho, K., Ganguli, S., Bengio, Y.: Identifying and attacking the saddle point problem in high-dimensional non-convex optimization. In: Advances in Neural Information Processing Systems, pp. 2933–2941 (2014)
18. Dong, C., Chen, Y., Zeng, B.: Generalized inverse optimization through online learning. In: Advances in Neural Information Processing Systems, pp. 86–95 (2018)
19. Donti, P., Amos, B., Kolter, J.Z.: Task-based end-to-end model learning in stochastic optimization. In: Advances in Neural Information Processing Systems, pp. 5484–5494 (2017)
20. Fischetti, M., Jo, J.: Deep neural networks and mixed integer linear optimization. Constraints, 1–14 (2018)
21. Goodfellow, I., Bengio, Y., Courville, A.: Deep Learning. MIT Press (2016). http://www.deeplearningbook.org
22. Graves, A., et al.: Hybrid computing using a neural network with dynamic external memory. Nature **538**(7626), 471 (2016)
23. Hazan, T., Keshet, J., McAllester, D.A.: Direct loss minimization for structured prediction. In: Advances in Neural Information Processing Systems, pp. 1594–1602 (2010)
24. Hutter, F., Hoos, H.H., Leyton-Brown, K.: Sequential model-based optimization for general algorithm configuration. In: Coello, C.A.C. (ed.) LION 2011. LNCS, vol. 6683, pp. 507–523. Springer, Heidelberg (2011). https://doi.org/10.1007/978-3-642-25566-3_40
25. Keshavarz, A., Wang, Y., Boyd, S.: Imputing a convex objective function. In: 2011 IEEE International Symposium on Intelligent Control, pp. 613–619. IEEE (2011)
26. LeCun, Y., Bengio, Y., Hinton, G.: Deep learning. Nature **521**, 436–444 (2015)
27. Lombardi, M., Milano, M.: Boosting combinatorial problem modeling with machine learning. arXiv preprint arXiv:1807.05517 (2018)

28. Maclaurin, D., Duvenaud, D., Adams, R.: Gradient-based hyperparameter optimization through reversible learning. In: International Conference on Machine Learning, pp. 2113–2122 (2015)
29. Maddison, C.J., Mnih, A., Teh, Y.W.: The concrete distribution: a continuous relaxation of discrete random variables. arXiv preprint arXiv:1611.00712 (2016)
30. Mahmood, R., Babier, A., Diamant, A., Chan, T.C.Y.: Interior point methods with adversarial networks. arXiv preprint arXiv:1805.09293 (2018)
31. Mahmood, R., Babier, A., McNiven, A., Diamant, A., Chan, T.C.Y.: Automated treatment planning in radiation therapy using generative adversarial networks. In: Doshi-Velez, F., et al. (eds.) Proceedings of the 3rd Machine Learning for Healthcare Conference, vol. 85, PMLR, Palo Alto, California, pp. 484–499, 17–18 August 2018
32. Pascanu, R., Mikolov, T., Bengio, Y.: Understanding the exploding gradient problem. CoRR, abs/1211.5063 (2012)
33. Paszke, A., et al.: Automatic differentiation in PyTorch. In: NeurIPS AutoDiff Workshop (2017)
34. QHull Library (2018). https://docs.scipy.org/doc/scipy-0.19.0/reference/generated/scipy.spatial.ConvexHull.html. Accessed 2 Nov 2018
35. Rumelhart, D.E., Hinton, G.E., Williams, R.J.: Learning representations by back-propagating errors. Nature **323**(6088), 533–536 (1986)
36. Saez-Gallego, J., Morales, J.M.: Short-term forecasting of price-responsiveloads using inverse optimization. IEEE Trans. Smart Grid (2017)
37. Sahoo, S.S., Lampert, C.H., Martius, G.: Learning equations for extrapolation and control. In: International Conference on Machine Learning (ICML) (2018)
38. Snoek, J., Larochelle, H., Adams, R.P.: Practical Bayesian optimization of machine learning algorithms. In: Advances in Neural Information Processing Systems, pp. 2951–2959 (2012)
39. Soudry, D., Hoffer, E.: Exponentially vanishing sub-optimal local minima in multilayer neural networks. arXiv preprint arXiv:1702.05777 (2017)
40. Sutskever, I.: Training recurrent neural networks. University of Toronto Toronto, Ontario, Canada (2013)
41. Sutskever, I., Martens, J., Dahl, G., Hinton, G.: On the importance of initialization and momentum in deep learning. In: International Conference on Machine Learning, pp. 1139–1147 (2013)
42. Tarantola, A.: Inverse Problem Theory: Methods for Data Fitting and ModelParameter Estimation. Elsevier, Amsterdam (1987)
43. Troutt, M.D., Brandyberry, A.A., Sohn, C., Tadisina, S.K.: Linear programming system identification: the general nonnegative parameters case. Eur. J. Oper. Res. **185**(1), 63–75 (2008)
44. Troutt, M.D., Tadisina, S.K., Sohn, C., Brandyberry, A.A.: Linear programming system identification. Eur. J. Oper. Res. **161**(3), 663–672 (2005)

A Study on the Traveling Salesman Problem with a Drone

Ziye Tang[1](✉), Willem-Jan van Hoeve[1], and Paul Shaw[2]

[1] Tepper School of Business, Carnegie Mellon University, Pittsburgh, USA
{ziyet,vanhoeve}@andrew.cmu.edu
[2] IBM, Valbonne, France
paul.shaw@fr.ibm.com

Abstract. A promising new model for future logistics networks involves the collaboration between traditional trucks and modern drones. The drone can pick up packages from the truck and deliver them by air while the truck is serving other customers. The operational challenge combines the allocation of delivery locations to either the truck or the drone, and the coordinated routing of the truck and the drone. In this work, we consider the scenario of a single truck and one drone, with the objective to minimize the completion time (or makespan). As our first contribution, we prove that this problem is strongly NP-hard, even in the restricted case when drone deliveries need to be optimally integrated in a given truck route. We then present a constraint programming formulation that compactly represents the operational constraints between the truck and the drone. Our computational experiments show that solving the CP model to optimality is significantly faster than the state-of-the-art exact algorithm. For larger instances, our CP-based heuristic algorithm is competitive with a state-of-the-art heuristic method.

1 Introduction

Vehicle routing problems have become increasingly important with the evolution of online shopping and fulfillment and a variety of delivery services. The use of unmanned aerial vehicles, or drones, for this purpose is actively explored by industry [12]. A common model is to equip a delivery truck with one or more drones to deliver packages in parallel to the truck [15]. Unlike the traditional setting where a fleet of vehicles have little operational constraints to each other, the drone operation is highly constrained to the truck operation because it needs to pick up packages for delivery from a truck. As a result the completion time also depends on the waiting time incurred due to the synchronization between the truck and the drone.

In this paper we study the design of optimal joint truck and drone routes under this scenario. We consider the elementary case where only one truck and one drone is available. Given a set of customers to be served either by a truck or

This work was supported by Office of Naval Research grant N00014-18-1-2129.

a drone, our objective is to minimize the completion time of the entire delivery task, *i.e.* the total time it takes to serve all customers. For operational simplicity, we assume the drone can only be dispatched at a customer location and the service time at each location is instant. In a feasible solution, the truck route forms a tour which starts from and returns to the depot with a subset of customers served along the tour. Each remaining customer is served by the drone which is dispatched from a customer location and returns to a (possibly different) location on the truck tour. We follow Agatz *et al.* [1] and call this problem the *traveling salesman problem with a drone* (TSP-D).

Contributions. Our first contribution is a proof that the TSP-D is strongly NP-hard even in the case when a truck route is given and we need to optimally integrate the remaining drone visits. Our second contribution is a new constraint programming (CP) formulation that relies on representing the TSP-D as a scheduling problem. We show experimentally that our CP approach outperforms the best exact method from the literature, and is competitive with a state-of-the art heuristic method in terms of solution quality.

2   Related Work

The hybrid truck and drone model was first studied by Murray *et al.* [11]. Agatz *et al.* [1] propose an exponential-sized integer programming model. Ha *et al.* [6] consider a variant where the objective is to minimize operational costs including total transportation cost and the cost incurred by vehicles' waiting time. Ham [7] considers a different integrated model where after one delivery task, the drone may return to the depot or fly to another customer to pick up a return order from a customer, with the truck traveling separately along a cycle. Bouman *et al.* [2] introduce a dynamic programming (DP) formulation of the TSP-D and solve it with A^* search. Yurek and Ozmutlu [17] propose a decomposition method; in the first stage, the truck nodes and the truck routes are generated and determined and the second stage solves a mixed-integer program to determine the optimal drone schedule. Lastly, Poikonen *et al.* [13] develop a specialized branch-and-bound procedure, which includes boosted lower bound heuristics to further speed up the solving process. Their method assumes insertion of a customer node into a sequence of nodes will not increase the optimal cost, which does not hold when the drone has a finite flight range. Other heuristic algorithms have also been proposed in the literature, see *e.g.* [1,3–5,11]. The best exact method for the TSP-D is the DP approach in [2], which can optimally solve instances with up to 15 locations in reasonable time.

The first theoretical study was performed by Wang *et al.* [16], who consider the more general vehicle routing problem with multiple trucks and drones. They study the maximum savings that can be obtained from using drones compared to truck-only deliveries (*i.e.* TSP cost) and derive several tight theoretical bounds under different truck and drone configurations. Poikonen *et al.* [14] extend [16] to different cases by incorporating cost, limited battery life and different metrics respectively.

3   Problem Definition

We are given an undirected graph $G = (V, E)$ with $V = C \cup \{r\}$ where r is the depot and C is the set of customers to be served by the truck or the drone. Let $n = |C|$. The travel time between a pair of nodes (i, j) is given by metric $w(i, j)$. $\rho \geq 0$ is the ratio between the truck's and the drone's travel time per unit distance. ρ is also called the *speed differential*. Every customer demands one parcel, which can be delivered by either a truck or a drone. A drone can only deliver one parcel at the time. We make the following assumptions about the behavior of the truck and the drone:

(a) The truck can dispatch and pick up a drone only at the depot or a customer location. The truck can continue serving customers after a drone is dispatched and reconnect with the drone at a possibly different node.
(b) The vehicle (truck or drone) that first arrives at the reconnection node has to wait for the other one, which we call *synchronization*.
(c) Upon returning to the truck, the time required to prepare the drone for another launch is negligible.

Our objective is to minimize the completion time, *i.e.* from the time the truck is dispatched from the depot with the drone to the time when the truck and the drone returns to the depot.

Notation. In a feasible solution to TSP-D, denote V_d as the set of nodes visited by the drone only. Denote $V_t := V \backslash V_d$ as the set of nodes including the depot visited by the truck either with or without the drone atop the truck. By a slight abuse of notation, we also call V_d the set of *drone nodes* and V_t the set of *truck nodes*. For each $i \in V_d$, let $p(i)$ be the dispatch node where the drone is dispatched right before visiting i, $q(i)$ be the pick up node where the drone returns immediately after visiting i, Let E_t be the set of edges in the truck tour. For $i, j \in V_t$, let T_{ij} denote the path induced by E_t and $w(T_{ij}) = \sum_{e \in T_{ij}} w_e$. Consider a partial drone schedule where the drone is dispatched from the truck at node j, visits node i and meets up with the truck at node k (we allow $j = k$). We call this partial drone schedule a *drone activity* and use a shorthand notation $j \rightarrow i \rightarrow k$ to represent this activity.

4   Computational Complexity

Solving the TSP-D to proven optimality is highly challenging, as witnessed by the performance of the best exact methods—they scale up to about 15 locations only. While the TSP-D is known to be NP-hard due to a reduction from the TSP, we aim to provide more insight in the computational difficulty by considering a restricted version, which we call the drone routing subproblem (DRS). We next prove strong NP-hardness of this restricted version.

We associate the drone activity $j \rightarrow i \rightarrow k$ with a cost c_{ijk} defined as

$$c_{ijk} = \max\{0, w(j, i) + w(i, k) - \rho w(T_{jk})\} \tag{1}$$

This is the marginal time a drone activity adds to the truck tour. Given V_t, V_d and E_t, we define a set of drone activities to be *feasible* if (1) each drone node in V_d appears in exactly one drone activity and (2) any pair of activities do not overlap in time (Fig. 1).

We now define the drone routing subproblem as follows:

Definition 1 (Drone routing subproblem). *Given V_t, V_d, E_t. The drone routing subproblem is to find a feasible set of drone activities with minimum total drone activity cost.*

We show that DRS is strongly NP-hard by a reduction from 3-partition.

Definition 2 (3-Partition). *Given positive integers m, B and $3m$ positive integers $x_1, \ldots, x_{3m}$ satisfying $\sum_{q=1}^{3m} x_q = mB$ and $\frac{B}{4} < x_q < \frac{B}{2}$ for $q = 1, \ldots, 3m$. Does there exist a partition of the set $Y = \{1, \ldots, 3m\}$ into m disjoint subsets $Y_1, \ldots, Y_m$ such that $\sum_{q \in Y_i} x_q = B$ for $i = 1, \ldots, m$?*

Theorem 1. *The drone routing subproblem is strongly NP-hard.*

Proof. We prove the theorem for $\rho = 1$. We give a pseudo-polynomial time reduction from 3-partition. Given an instance of 3-partition as in Definition 2, we construct a graph with $m(B + 1)$ truck nodes and $4m$ drone nodes. The truck route connects m paths P_i, each having $B + 1$ nodes and B unit edges. Edges that connect two paths are assigned $\epsilon = \frac{1}{2m}$. Direct all edges in the cycle counterclockwise. The tail of a directed path P_i is defined as the tail of the first arc in P_i. We similarly define the head of P_i. The drone nodes are partitioned into two disjoint sets A and B. A has $3m$ nodes $v_1, \ldots, v_{3m}$. For $i = 1, \ldots, 3m$, v_i is connected to each node on the cycle via an edge of weight $\frac{x_i}{2}$. B contains m dummy nodes $u_1, \ldots, u_m$. For $i = 1, \ldots, m$, each u_i is connected to the head of P_i and tail of P_{i+1} via two edges of weight $\frac{\epsilon}{2}$ (we assume $P_{m+1} = P_1$). Other edges connected to u_i are assigned a unit weight so metric inequality holds. Below we show Lemma 1, from which the theorem follows.   □

Lemma 1. *There exists a 3-partition if and only if there exists a feasible solution to the above DRS instance of zero total cost.*

Proof. 'if': connect each dummy node u_i to the head of P_i and tail of P_{i+1}. Without loss of generality assume the feasible partition is $\{x_{3k+1}, x_{3k+2}, x_{3k+3}\}$ for $k = 0, \ldots, m - 1$. Then $v_{3k+1}, v_{3k+2}, v_{3k+3}$ are connected to path P_k in the following way: v_{3k+1} is connected to the first node and $(x_{3k+1} + 1)$-th node on the path, v_{3k+2} is connected to $(x_{3k+1} + 1)$-th and $(x_{3k+1} + x_{3k+2} + 1)$-th nodes, v_{3k+3} is connected to $(x_{3k+1} + x_{3k+2} + 1)$-th and x_{B+1}-th nodes. It is easy to check that the total cost is zero.

'only if': we claim each dummy node u_i in any solution with zero total cost must be connected to the head of P_i and tail of P_{i+1}: suppose not, note for any $t \neq i$, u_i cannot be connected to the head of P_t and tail of P_{t+1} since otherwise such a drone activity has non-zero cost. As a result any drone activity which

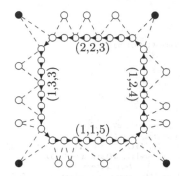

Fig. 1. An example of the reduction, with $m = 4$, $B = 7$ and feasible partitions $(1, 1, 5)$, $(1, 2, 4)$, $(1, 3, 3)$, $(2, 2, 3)$. Drone activities are shown as dashed lines and dummy nodes are marked as solid. While each drone node has the same distance to any node on the truck cycle, we put the drone nodes outside the cycle for visualization purposes.

visits u_i covers at least a unit-length edge on the cycle. Therefore after visiting u_i, remaining edges on the cycle have at most $mB - 1 + m\epsilon = mB - \frac{1}{2} < mB$ length to use for visiting the remaining drone nodes. Notice each visit of a node $v_l \in A$ must cover a path at least x_l to make the drone activity cost zero and each visit must cover non-overlapping path on the cycle, which is a contradiction to the fact that the remaining edge length on the cycle is less than mB. Therefore we've shown the claim. The result follows by reversing the steps in the 'if' part to partition drone nodes into m sets where each set contains 3 nodes that are visited by using edges in the same path. $\qquad\square$

5   Constraint Programming Formulation

An essential feature of a truck-drone schedule is the synchronization between truck and drone operations. This poses a significant challenge to construct a MIP model with a tight linear relaxation. Below we explain how to construct a compact CP with $O(n^2)$ variables and constraints, using the constraint-based scheduling formalism introduced in [8–10]. The CP solver IBM ILOG CP Optimizer [10] provides an expressive modeling language based on the notion of interval variables representing the execution of an activity. Its domain encodes the presence status (Boolean) (true if the activity is executed). When a is present, it is represented by variables $s(a)$ for its start time, $e(a)$ for its end time, and $d(a)$ for its duration, obeying the relationship $d(a) = e(a) - s(a)$. On the other hand, an absent interval variable is not considered by any constraint or expression on interval variables in which it is involved. An activity can be forced to present or declared 'optional', *i.e.* its presence status can be either true or false to be decided by the solver. Below we assume all interval variables are optional unless stated otherwise.

Recall the number of nodes including the depot is $n + 1$. Denote both node 0 and n as the depot (leaving and returning). For each node i define three

```
1    minimize
2        endOf(tVisit[n])
3    subject to {
4        forall (i in 0...n) setPresent(visit[i])
5        setPresent(tVisit[0])
6        setPresent(tVisit[n+1])
7        first(tVisit, tVisit[0])
8        last(tVisit, tVisit[n])
9        no_overlap(tVisit, w)
10       no_overlap(dVisit)
11       forall(i in 0...n) {
12           alternative(visit[i], [tVisit[i], dVisit[i]])
13           alternative(dVisit_before[i], [all (j in 0...n) tdVisit[i][j]])
14           alternative(dVisit_after[i], [all (j in 0...n) dtVisit[i][j]])
15           span(dVisit[i], [dVisit_before[i], dVisit_after[i]])
16           end_at_start(dVisit_before[i], dVisit_after[i])
17           if_then(presence_of(dVisit[i]), presence_of(dVisit_before[i]) & presence_of(dVisit_after[i]))
18       forall(i, j in 0...n) {
19           if_then(presence_of(tdVisit[i][j]), presence_of(tVisit[j]))
20           if_then(presence_of(dtVisit[i][j]), presence_of(tVisit[j]))
21           start_before_start(tVisit[j], tdVisit[i][j])
22           start_before_end(tdVisit[i][j], tVisit[j])
23           end_before_end(dtVisit[i][j], tVisit[j])
```

Fig. 2. A compact constraint programming formulation for TSP-D. Note we can enforce finite drone range by adding an upper bound on the duration of each dVisit[i].

interval variables: visit[i] (forced to be present), dVisit[i] and tVisit[i]. Each dVisit[i] represents the time period from the drone just leaving for node i to the drone arriving at the first truck node after serving i. Note we can enforce finite drone range by adding an upper bound on the duration of each dVisit[i]. Furthermore, for each node i, we create two interval variables dVisit_before[i] and dVisit_after[i] which represent splitting dVisit[i] by the time point of the drone visiting i. For each pair (i, j), we define two interval variables tdVisit[i][j] and dtVisit[i][j], where the former represents the drone leaving from truck node j to drone node i and the latter represents the drone leaving from drone node i to truck node j. Each tdVisit[i][j] is lower bounded by the drone travel time from j to i and similarly for dtVisit[i][j]. As an example, an activity $i \rightarrow j \rightarrow k$ is composed of tdVisit[i][j] and dtVisit[i][k], which are constrained to be equivalent to dVisit_before[i] and dVisit_after[i] respectively. The complete model is presented in pseudocode in Fig. 2.

Lines 7–8 require the truck tour to start and end at the depot. The next constraint requires that for each pair (i, j), their truck visits have to be at least w_{ij} apart if both of them are served by the truck. Similarly for line 10. The remaining constraints enforce logical constraints between different sets of interval variables. For example, alternative(*interval*, *array*) creates an alternative constraint between interval variable *interval* and the set of interval variables in *array*. If *interval* is present, then one and only one of the intervals in *array* will be selected by the alternative constraint to be present and the start and end values of *interval* will be the same as the one of the selected intervals. Line 21–23 implements the synchronization constraint between the truck and the drone. We refer the reader to the manual [8] for the definition and usage of each constraint.

Table 1. Comparison of our constraint programming (CP) approach with (a) the exact dynamic programming (DP) method of Bouman *et al.* [2] in terms of runtime (s), and (b) the heuristic branch-and-bound method (BAB) of Poikonen *et al.* [13] in terms of solution quality (average objective value).

Size	10	11	12	13	14	15	16	17	18
CP	6.79	5.71	16.66	15.66	50.83	120.59	216.46	375.49	564.22
DP	1.00	4.00	12.00	56.00	306.00	1568.00	9508.00	–	–

a. Runtime comparison (s) of CP and DP (exact).

Size	10	20	30	40	50	60	70	80	90	100	200
CP	116.60	136.64	160.12	198.88	237.4	276.96	316.20	407.36	515.80	679.64	–
BAB	149.53	171.64	200.95	226.15	241.36	267.54	283.30	299.09	322.37	337.91	465.63

b. Solution value comparison of CP and BAB (heuristic).

6   Computational Experiments

We implemented and solved our CP model with CP Optimizer version 12.8.0, using the Python interface DOcplex. Our experiments are run on a 2.2 GHz Intel Core i7 quad-core machine with 16 GB RAM. We compared our approach to the two best approaches from the literature: the exact dynamic programming (DP) algorithm in [2], and the branch-and-bound algorithm from [13]. The implementation of the latter relies on the assumption that the drone has a finite range, for which the method is not guaranteed to provide optimal solutions.

We first present the results on the exact comparison. Since benchmark instances used in [2] are not publicly available, we follow their approach to generate 10 uniform instances of each size. We use the same parameter $\rho = 2$ as speed differential. Table 1a reports the average runtime (in seconds) of our approach (CP) and the reported runtime from [2] (DP). While DP can solve the smaller problems faster than CP, our approach scales more gracefully.

Table 1b presents the comparison with the branch-and-bound method (BAB) of [13] in terms of solution quality. For this experiment, we apply a time limit of 10 min for each instance. As a benchmark, we use the same dataset as [13] (25 instances of each size). We use the same parameter values for the speed differential $\rho = 2$ and drone range $R = 20$. The table reports the mean objective value for each problem size. The results for BAB are the best solutions found among all branch-and-bound heuristic variants. We note that the runtime of the BAB approach is typically less than one minute. These results show that the time-limited CP approach can produce better solutions for smaller instances (up to 50 locations) but that the dedicated heuristic branch-and-bound outperforms CP on larger instances.

Acknowledgement. The first author would like to thank Thomas Bosman for helpful discussions on the complexity proof.

References

1. Agatz, N., Bouman, P., Schmidt, M.: Optimization approaches for the traveling salesman problem with drone. Transp. Sci. **52**(4), 965–981 (2018)
2. Bouman, P., Agatz, N., Schmidt, M.: Dynamic programming approaches for the traveling salesman problem with drone. Networks **72**(4), 528–542 (2018)
3. Daknama, R., Kraus, E.: Vehicle routing with drones. arXiv preprint arXiv:1705.06431 (2017)
4. Dorling, K., Heinrichs, J., Messier, G.G., Magierowski, S.: Vehicle routing problems for drone delivery. IEEE Trans. Syst. Man Cybern. Syst. **47**(1), 70–85 (2017)
5. Ha, Q.M., Deville, Y., Dung, P.Q., Hà, M.H.: Heuristic methods for the traveling salesman problem with drone. CoRR abs/1509.08764 (2015)
6. Ha, Q.M., Deville, Y., Pham, Q.D., Hà, M.H.: On the min-cost traveling salesman problem with drone. Transp. Res. Part C Emerg. Technol. **86**, 597–621 (2018)
7. Ham, A.M.: Integrated scheduling of m-truck, m-drone, and m-depot constrained by time-window, drop-pickup, and m-visit using constraint programming. Transp. Res. Part C Emerg. Technol. **91**, 1–14 (2018)
8. IBM ILOG CPLEX: CP Optimizer 12.7 Users Manual (2017)
9. Laborie, P.: IBM ILOG CP optimizer for detailed scheduling illustrated on three problems. In: van Hoeve, W.-J., Hooker, J.N. (eds.) CPAIOR 2009. LNCS, vol. 5547, pp. 148–162. Springer, Heidelberg (2009). https://doi.org/10.1007/978-3-642-01929-6_12
10. Laborie, P., Rogerie, J., Shaw, P., Vilím, P.: IBM ILOG CP optimizer for scheduling. Constraints **23**(2), 210–250 (2018)
11. Murray, C.C., Chu, A.G.: The flying sidekick traveling salesman problem: optimization of drone-assisted parcel delivery. Transp. Res. Part C Emerg. Technol. **54**, 86–109 (2015)
12. Otto, A., Agatz, N., Campbell, J., Golden, B., Pesch, E.: Optimization approaches for civil applications of unmanned aerial vehicles (UAVs) or aerial drones: a survey. Networks **72**(4), 411–458 (2018)
13. Poikonen, S., Golden, B., Wasil, E.: A branch and bound approach to the TSP with drone. INFORMS J. Comput. (to appear)
14. Poikonen, S., Wang, X., Golden, B.: The vehicle routing problem with drones: extended models and connections. Networks **70**(1), 34–43 (2017)
15. UPS: UPS Tests Residential Delivery Via Drone. Youtube (2017). https://www.youtube.com/watch?v=xx9_6OyjJrQ
16. Wang, X., Poikonen, S., Golden, B.: The vehicle routing problem with drones: several worst-case results. Optim. Lett. **11**(4), 679–697 (2017)
17. Yurek, E.E., Ozmutlu, H.C.: A decomposition-based iterative optimization algorithm for traveling salesman problem with drone. Transp. Res. Part C Emerg. Technol. **91**, 249–262 (2018)

Lower Bounds for Uniform Machine Scheduling Using Decision Diagrams

Pim van den Bogaerdt[✉] and Mathijs de Weerdt

Faculty of Electrical Engineering, Mathematics and Computer Science,
Delft University of Technology, Van Mourik Broekmanweg 6,
2628 XE Delft, The Netherlands
P.vandenBogaerdt@tudelft.nl

Abstract. We propose a relaxed decision diagram (DD) formulation for obtaining lower bounds on uniform machine scheduling instances, based on *separators* to separate jobs on different machines. Experiments on the total tardiness for instances with tight due times show that for obtaining nontrivial bounds, it is important to partition the DD nodes on a layer based on their machine finishing time. When the number of jobs is small, DDs provide stronger bounds in less time than a time-indexed LP relaxation.

Keywords: Multi-machine scheduling · Uniform machines · Lower bounds · Decision diagrams

1 Introduction

We consider machine scheduling on uniform machines with release times and sequence-dependent setup times. This problem models an environment where machines have different speeds, time is incurred between jobs depending on the pair of jobs and their machine assignment, and each job may only be scheduled after a given time depending on the job. This is an abstract model of, for example, scheduling a production factory (see [17, pp. 1–2]).

The aim is to find a schedule that minimizes a given objective function. This problem is hard in many cases, for example it is NP-complete if there are two machines with equal speed, setup times and release times are not present, and the objective is the maximum completion time [15]. We propose a decision diagram formulation for a uniform machine scheduling problem that provides lower bounds on an objective for a given instance.

Decision diagrams (DDs) have proven useful as a tool in optimization (see, e.g., [7]). A particular successful application of decision diagrams has been single-machine scheduling (see, e.g., [10, 12]) as well as multi-machine scheduling where the machines are considered identical [9]. To the best of our knowledge, there has so far been no extension to more general multi-machine scheduling problems where the machines are not exchangeable.

© Springer Nature Switzerland AG 2019
L.-M. Rousseau and K. Stergiou (Eds.): CPAIOR 2019, LNCS 11494, pp. 565–580, 2019.
https://doi.org/10.1007/978-3-030-19212-9_38

In this paper, we propose a DD formulation for uniform multi-machine scheduling; that is, each machine has a speed with which it processes a job. More precisely, we provide a formulation for a relaxed DD, which gives lower bounds on the optimal objective for instances of this problem. Lower bounds can be used to appreciate the quality of a feasible schedule, and may help in search algorithms such as branch-and-bound.

We also propose a merge heuristic based on partitioning nodes on their machine finishing time. An experiment for the total tardiness objective on instances with tight due times shows that this is important for obtaining non-trivial bounds. We also show that, at least when the number of jobs is small, bounds given by DDs are stronger and computed faster than those given by the linear programming relaxation of a time-indexed mixed integer program that [6] is based on, solved by IBM ILOG CPLEX.

The remainder of this paper is structured as follows. In Sect. 2, we introduce our problem formally and explain the basics of DDs. In Sect. 3, we briefly review previous work in the literature on DDs for single-machine scheduling by formulating a single-machine DD of our problem. In Sect. 4, we present our DD formulation for the multi-machine problem, and in Sect. 5, we elaborate on improving the bounds provided by our DD formulation. In Sect. 6, we present computational results. Finally, in Sect. 7, we conclude and present directions for future work.

2   Background

2.1   Problem Formulation

In a uniform machine scheduling problem, a number of *machines* is available that can each process one *job* at a time. The set of m machines is $\mathcal{M} = \{M_1, \ldots, M_m\}$ and the set of n jobs is $\mathcal{J} = \{j_1, \ldots, j_n\}$. Each job j has a *processing time* $p_{i,j}$ on machine i. We assume the ratio $p_{i,j}/p_{i',j}$ is constant over all jobs j for each pair of machines i, i'; that is, each machine has a *speed* with which it processes jobs.

Each job has a *release time* r_j and can only start after that time. We also allow a sequence-dependent *setup time* $\sigma_{M,j,j'}$ between each pair of jobs j, j' and for each machine M, which means job j' can only start $\sigma_{M,j,j'}$ time after job j has ended if both jobs are scheduled on M. The first job on a machine may have a setup time as well; let $\sigma_{M,\vdash,j}$ denote such a setup time for job j on machine M. Here, we let a "dummy job" $\vdash$ denote there is no job before j.

We assume the setup times satisfy the triangle inequality, that is, $\sigma_{M,j,j'} \leq \sigma_{M,j,j''} + \sigma_{M,j'',j'}$ for all jobs $j \in \mathcal{J} \cup \{\vdash\}, j', j'' \in \mathcal{J}$ and machines $M \in \mathcal{M}$. This is a reasonable assumption because a direct setup should not take more time than the time it takes performing the setup indirectly. The setup times need not be symmetric.

Release times look like setup times involving the dummy job, but are not superfluous. Namely, the release times are not involved in the above triangle inequality. A slight generalization of this problem has been formulated (but not in the context of lower bounds) in [16].

In a *schedule*, each job $j \in \mathcal{J}$ has a machine assignment $m(j)$, start time $s(j)$ and completion (end) time $e(j) = s(j) + p_{m(j),j}$. We seek a schedule that minimizes an *objective function* $z(e(j_1), \ldots, e(j_n))$. We assume that the objective function is non-decreasing in each completion time, and that it can be written as a sum $\sum_j z_j(e(j))$ of functions, each depending on the completion time of only one job. In this article, we let each job have a *due time* d_j, and consider the *total tardiness* $\sum_j \max\{0, e(j) - d_j\}$ as the objective function. This is a measure of the total delay of the jobs in a schedule.

2.2 Decision Diagrams

A *decision diagram* (DD) is a directed acyclic graph for which all edges go from one layer to the next. (See [7] for an elaborate introduction to decision diagrams.) The first and last layer consist of a single node, called the *root* and *terminal*, respectively. One can model the search space of a minimization problem in n variables as a DD of $n + 1$ layers: each layer (except the last) corresponds to a variable in the sense that each outgoing edge of that layer corresponds to a choice of that variable. By setting an appropriate cost to each edge, the smallest root-terminal path represents the optimal solution of the problem. Such a DD is called *exact*.

Exact DDs may be of exponential size. To ensure tractability, we consider *relaxed DDs* with a bounded number of nodes. Such DDs may also represent infeasible solutions, and each root-terminal path is only required to be a *lower bound* on the objective value of the corresponding solution. These two properties together imply that the shortest root-terminal path in a relaxed DD is a lower bound on the optimal solution. To bound the number of nodes, we let each layer contain at most $w \geq 1$ nodes, where w is a fixed parameter.

To build an exact DD, one can keep track of what choices have been made by means of a *state* S in each node. This state can be used to define the set $F(S)$ of outgoing edges (i.e., choices) of this node as well as the cost $c(S, j)$ of each such edge. The state of a node after following an edge for choice j is defined as $\varphi(S, j)$. This algorithm of building a DD is called *top-down compilation*.

To build a relaxed DD, one needs to additionally ensure that each layer contains at most w nodes. This is done by *merging* nodes. The state of a merged node is defined through an associative *merge operator* $\oplus$. A merged node represents multiple paths from the root but only has a single state. This state should "underapproximate" these paths in some sense. Intuitively, the shortest root-terminal path then has a length that is an underapproximation (i.e., lower bound) of the optimum.

The merge operator needs to be *valid*, i.e., a DD resulting from applying it must be relaxed. In [12], a theorem is presented to prove the validity of a merge operator. We present it in the following formulation, like [9].

Definition 1. *A binary operator $\preceq$ on states is a* state relaxation relation *if it adheres to the following properties:*

- *(R1) $\preceq$ is reflexive and transitive.*
- *(R2) If $S' \preceq S$, then $F(S') \supseteq F(S)$.*
- *(R3) If $S' \preceq S$, then for $j \in F(S), c(S', j) \leq c(S, j)$.*
- *(R4) If $S' \preceq S$, then for $j \in F(S), \varphi(S', j) \preceq \varphi(S, j)$.*

Theorem 1 (Hooker [12]). *Suppose a state relaxation relation $\preceq$ is given. If $S \oplus S' \preceq S, S'$ for all states S, S' for which $\oplus$ is defined, then $\oplus$ is a valid merge operator.*

This theorem as presented here is actually slightly stronger than [12] because we only consider pairs for which $S \oplus S'$ is defined. The same proof as in [12] is applicable, however. The purpose of this addition becomes clear when we prove our DD formulation in Sect. 4.

3   DDs for Single-Machine Scheduling

DDs have been used for single-machine scheduling before (see, e.g., [10,12]). In this section, we review existing work by modelling the single-machine variant of our problem as a DD. In the next section, we extend this model to uniform machines.

A single-machine schedule can be considered as a permutation of the jobs. A permutation can be transformed into a schedule by considering the jobs iteratively, scheduling each job as soon as possible. The set of schedules generated this way includes an optimal schedule because we assume the objective function is non-decreasing in each of the job completion times. It is thus reasonable to let the DD represent permutations (see, e.g., [10,12]). We now explain the details of this formulation, which we then extend to multiple machines in the next section.

In each node, we keep track of a set V of jobs we have certainly already scheduled; that is, the jobs that appear on all paths from the root to the node. These jobs are removed from the feasible set in the node, because all paths through the edge for this job would contain that job twice, and so would not represent a feasible solution. The root has $V = \varnothing$ and the transition for job j is $\overline{V} = V \cup \{j\}$. (We use a bar to denote the state variable after a transition.) The merge operator is the intersection: given two nodes v_1, v_2, the jobs that occur on all paths from the root to the merged node are precisely the jobs that appear on all paths from the root to v_1 and to v_2.

When we schedule a next job, we need (a relaxation of) the finishing time of the machine, so that we know when the job can start. To this end, we keep a number f in each node that represents this finishing time [12]. The root has $f = 0$; the transition will be discussed below. The merge operator is the minimum. Intuitively, by underestimating the finishing time of the machine, we also underestimate the end time of jobs and hence their costs, thereby obtaining a valid relaxation on the objective.

To incorporate setup times, we use a set L of jobs that may have been scheduled last (see [7, p. 143][1]). If we make a transition for job j, we need to take into account the setup time between j and the previous job (or the dummy job if j is the first job). A valid relaxation (underestimate) of this setup time is $\min_{j' \in L} \sigma_{j',j}$. Hence, we could define the transition for f as $f + \min_{j' \in L} \sigma_{j',j} + p_j$. Rather, to take into account the release time, we define

$$\overline{f} = \max\{r_j, f + \min_{j' \in L} \sigma_{j',j}\} + p_j.$$

Note that in an exact DD, L is a singleton set, and we simply take the setup time between that job and j. The root has $L = \{\vdash\}$ and the transition for job j is $\overline{L} = \{j\}$. The merge operator is the union.

4   DD Formulation for Uniform Machines

In this section, we propose a DD formulation for uniform machines, based on the single-machine DD formulation discussed in the previous section.

We represent a schedule by a list of n jobs and $m - 1$ *separators*, inspired by [11]. The schedule represented by such a list of $n + m - 1$ elements is as follows: the i'th machine has the jobs between the $(i - 1)$'th and i'th separator, in the order of the list. (We assume there is an implicit separator before and after the list.) Between two separators, we thus consider a single machine. We let $\ddagger$ denote a separator.

See Fig. 1 for an example with $n = 5$ jobs, $m = 2$ machines of equal speed, no release and setup times, and processing times $p_{j_i} = i$.

Fig. 1. Example of a list representation (left) of a schedule (right)

Our DD formulation is based on this list representation and has $n + m$ layers, where the outgoing edges of the i'th layer, $1 \leq i \leq n + m - 1$, correspond to making a choice for the i'th item in the list. Between two separators, the formulation is essentially the single-machine DD formulation described in the previous section. We keep track of the current machine i in each node, which is considered as a parameter for the single-machine formulation (for example, so that we can use $p_{i,j}$ as the processing time for job j on this machine). Also, we do not merge nodes with different values of i.

Figure 2 contains a sketch of a possible DD of our formulation for $n = 3$ jobs and $m = 2$ machines. Each of the two columns of nodes corresponds to a machine.

---

[1] We use a slightly simpler definition, where $L \subseteq \mathcal{J}$.

The four edges between the columns correspond to choosing a separator, whereas the other edges correspond to choosing a job. The DD has $n + m = 5$ layers, and the number of separators in any root-terminal path is $m - 1 = 1$.

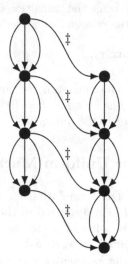

Fig. 2. Example decision diagram with three jobs and two machines. The separator edges are labeled as such.

We now consider our formulation formally. The state in a node is a tuple $S = (i, V, L, f)$ where V, L are sets and i, f are numbers. The root has state $(1, \varnothing, \{\vdash\}, 0)$. Recall that i is the current machine, V is the set of jobs that are certainly already scheduled, L is the set of jobs that may have been scheduled last, and f is a lower bound on the finishing time of the machine.

To define the feasible set $F(S)$, we first define $X = \mathcal{J} - V$. Then, we define $F(S)$ to be X, additionally with a separator if we are not on the last machine:

$$F(S) = \begin{cases} X \cup \{\ddagger\} & \text{if } i < m \\ X & \text{otherwise.} \end{cases}$$

When making a transition from a node, we define the new state $\bar{S} = (\bar{i}, \bar{V}, \bar{L}, \bar{f})$ as follows. If we choose a job j, we use definitions based on single-machine scheduling:

$$\bar{i} = i, \qquad \bar{V} = V \cup \{j\}, \qquad \bar{L} = \{j\}, \qquad \bar{f} = \max\{r_j, f + \min_{j' \in L} \sigma_{i,j',j}\} + p_{i,j}$$

If we instead choose a separator, we let:

$$\bar{i} = i + 1, \qquad \bar{V} = V, \qquad \bar{L} = \{\vdash\}, \qquad \bar{f} = 0$$

We pass V along as we still should not schedule these jobs on the new machine. Also, we reset $\bar{f}$ to 0 because the new machine does not contain any jobs yet.

The cost of an edge corresponding to job j is

$$c(S, j) = z_j(\max\{r_j, f + \min_{j' \in L} \sigma_{i,j',j}\} + p_{i,j}),$$

as the operand of z_j is the time job j finishes according to the state information. For a separator, we set $c(S, \ddagger) = 0$. In an exact DD (built top-down without merging), the cost of any root-terminal path is therefore equal to the objective value of the corresponding schedule.

We proceed with proving the validity of the merge operator using Theorem 1. Parts of the ideas below are similar to the single-machine case and can (to some extent) be found in [12].

We define a state relaxation relation $\preceq$ on states S, S' for which $i = i'$ as follows: $S' \preceq S$ means $V' \subseteq V \wedge L' \supseteq L \wedge f' \leq f$.

Theorem 2. *This relation satisfies the conditions of being a state relaxation relation (Definition 1).*

Proof. We show each of (R1)–(R4). Let S, S' be states such that $S' \preceq S$. In particular, assume $i = i'$.

For (R1), reflexivity and transitivity follow from that of $\subseteq, \supseteq, \leq$.

For (R2), we need to show $F(S') \supseteq F(S)$. Since $i = i'$, it suffices to show $X' \supseteq X$, where X' denotes the set used in the definition of $F(S')$. Since $V' \subseteq V$, indeed $\mathcal{J} - V' \supseteq \mathcal{J} - V$.

For (R3), let j be a job or separator in $F(S)$. We need to show that $c(S', j) \leq c(S, j)$. If j is a separator, both costs are zero, so the inequality holds. So assume j is a job.

Since the z_j are non-decreasing, we need to show that

$$\max\{r_j, f' + \min_{j' \in L'} \sigma_{j',j}\} + p_{i',j} \leq \max\{r_j, f + \min_{j' \in L} \sigma_{j',j}\} + p_{i,j}.$$

Since $i = i'$, also $p_{i,j} = p_{i',j}$ and hence we only need to show

$$\max\{r_j, f' + \min_{j' \in L'} \sigma_{j',j}\} \leq \max\{r_j, f + \min_{j' \in L} \sigma_{j',j}\}.$$

In turn, it is sufficient to show

$$f' + \min_{j' \in L'} \sigma_{j',j} \leq f + \min_{j' \in L} \sigma_{j',j}.$$

First, we have $f' \leq f$. Also, $L' \supseteq L$, so the minimum is taken over a superset. Hence the inequality holds.

For (R4), define $\bar{S} = \varphi(S, j)$ and $\bar{S'} = \varphi(S', j)$. We have to prove $\bar{S'} \preceq \bar{S}$, which we do by showing each of the relations that $\preceq$ entails.

First, consider the case where we select a job j in $F(S)$.

- $\bar{i}' = i' = i = \bar{i}$ so it is meaningful to prove the claim.
- $\overline{V}' = V' \cup \{j\} \subseteq V \cup \{j\} = \overline{V}$ because $V' \subseteq V$.
- $\overline{L}' = \{j\} = \overline{L}$, so also $L' \supseteq L$.
- $\bar{f}' \leq \bar{f}$ can be shown using the same calculation as for the costs (R3).

Next, consider the case where we select a separator in $F(S)$.

- $\bar{i}' = i' + 1 = i + 1 = \bar{i}$ so it is meaningful to prove the claim.
- $\overline{V}' = V'$ and $\overline{V} = V$ so by assumption, $\overline{V}' \subseteq \overline{V}$.
- $\overline{L}' = \{\vdash\} = \overline{L}$ so also $\overline{L}' \supseteq \overline{L}$.
- $\bar{f}' = 0 = \bar{f}$ so also $\bar{f}' \leq \bar{f}$.

$\square$

We define the merge operator $\oplus$ for states S, S' as for the single-machine case, but we only define it for $i = i'$. In that case,

$$S \oplus S' = (i, V \cap V', L \cup L', \min\{f, f'\}).$$

Using the state relaxation relation $\preceq$, we can utilize Theorem 1 to show that this merge operator is valid.

Theorem 3. *The merge operator $\oplus$ is valid.*

Proof. Consider states S, S' for which $S \oplus S'$ is defined, that is, $i = i'$. We need to show $S \oplus S' \preceq S, S'$. The merged state also has this value for i, so it is meaningful to prove the two $\preceq$ claims.

We indeed have $V \cap V' \subseteq V, V'$ and $L \cup L' \supseteq L, L'$, as well as $\min\{f, f'\} \leq f, f'$. $\square$

5   Merging Heuristics

In the previous section, we have proven the correctness of the proposed DD formulation for uniform machines. The *performance*, i.e., the quality of the bounds, depends on the choice of nodes to merge. Recall that, for correctness, we only merge nodes with the same value of i, i.e., corresponding to the same machine. Among such nodes, however, we have the freedom to choose what nodes to merge, impacting the performance. In this section, we first explain two existing merging heuristics (based on f or the partial objective) and their weakness in our model. We then explain the new merging heuristic that we use.

Ideally, one wants to merge nodes which are going to have identical subtrees, since this way no information is lost, and so would give the best bound (i.e., the optimal objective). However, identifying such nodes is NP-hard [20].

Instead, at least for single-machine scheduling, it seems reasonable to merge nodes which have a large f or a large partial objective [12]. Merging nodes with high partial objective likely does not affect the shortest path (i.e., the bound). Similarly, nodes with a relatively high f may correspond to the machine having

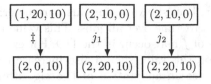

Fig. 3. Part of a DD with two consecutive layers and three nodes in each layer. The state variables depicted are (i, f, z) where z is the partial objective.

much idle time, and so the optimal solution likely does not go through such a node.

In our model, however, one may expect these heuristics to work poorly. Recall that a transition corresponding to a separator sets $\bar{f} = 0$ in the new node. Thus, merging based on f may give priority to such nodes, which does not necessarily make sense.

For example, a path consisting of separators on each of the first few layers would go through several nodes with $f = 0$, which would not get merged. However, selecting multiple separators after each other, thereby leaving machines empty, likely does not result in a good schedule.

Merging based on the partial objective alone may also not be the best choice. Namely, we might merge nodes for which $f = 0$ and nodes for which f is large. In the merged node, the f value is set to the minimum of these f values, which can result in a weak relaxation. We illustrate this with an example.

In Fig. 3, a layer with three nodes is depicted, and a transition from each of these nodes, resulting in three nodes in the next layer. Suppose we need to decide which nodes of this latter layer to merge. If we consider the partial objective values (z), then, because all these three values are equal (and all states have the same value for i), we might decide to merge the first two nodes. The merged node then has $(i, f, z) = (2, 0, 10)$. In particular, the f value is $0 = \min\{0, 20\}$, so that the (relaxed) cost of scheduling a job in this state is likely small. This may result in a weak bound.

Therefore, it seems reasonable to merge nodes with similar values of f. We do this by *partitioning* the nodes on a given layer such that we only merge nodes within the same partition. In the example, the first node (with $f = 0$) should be in a different partition than the other two (with $f = 20$). In that case, we would merge the last two nodes, resulting in state $(2, 20, 10)$, which has a stronger relaxation of f: it is $20 = \min\{20, 20\}$. Within each partition, we merge the nodes with the highest partial objective.

There are two factors influencing the partitioning: the number of partitions, and the range of f values of each partition. We let each partition have a range of equal size; hence, the latter factor reduces to finding a approximation of the type of values of f that may occur in the DD.

We use the following heuristic value for this parameter: construct a schedule and select the largest start time over all jobs. This value is a heuristic estimate of what order of values of f may occur in the DD. For example, if the time scale is multiplied by a factor, this value should also increase with that factor.

More precisely, we use an adaptation of the algorithm by [18], which is a heuristic to find a good schedule by transforming permutations of jobs into schedules. This heuristic is designed for a different scheduling problem; we use the idea by [19] by iteratively selecting the machine for which a job *completes* (rather than starts) earliest. We set $N = 1$ in the heuristic because we do not need the algorithm to spend much time on finding a good solution. Nodes with an f value above the approximation are put in the last partition; the last partition therefore has a larger range than the others.

6 Experiments

In this section, we investigate the effect of partitioning on the quality of the bound. We also compare the bound given by the DDs to that by a linear programming (LP) relaxation of a mixed integer program (MIP). We use the total tardiness $\sum_j \max\{0, e(j) - d_j\}$ as the objective function.

When we compute a gap (that is, the difference between an upper and a lower bound, divided by the upper bound), we use the upper bound of a constraint program (CP, see Fig. 4) found by IBM ILOG CP Optimizer 12.8 after 5 s with default settings. The CP is based on one of the samples provided with this solver. We supply the solution we use for partitioning (the adaptation of [18]) as a warm start to CP Optimizer. Our DD implementation was written in C# (x64, .NET 4.7) and the experiments were run on an Intel Xeon E5-1620 v2 (3.70 GHz, 8 threads), 8 GB RAM, Windows 7. We parallelized our DD implementation, where each thread builds and merges part of each layer, like [9]. We thus also partition the nodes based on the thread they are created in.

$$\text{Minimize} \quad \sum_j \max\{0, \text{EndOf}(I_j) - d_j\}$$

$$\text{s.t.} \ I_{j,M} \in \text{Intervals}([\max\{r_j, \sigma_{M,\vdash,j}\}, \infty), p_{j,M}) \qquad (j \in \mathcal{J}, \ M \in \mathcal{M})$$

$$I_j \in \{I_{j,M} \mid M \in \mathcal{M}\} \qquad (j \in \mathcal{J})$$

$$\text{NoOverlap}(\{I_{j,M} \mid j \in \mathcal{J}\}, \sigma_{M,\cdot,\cdot}) \qquad (M \in \mathcal{M})$$

Fig. 4. Constraint program

6.1 Instances

We first describe the instances we created. We construct processing times by first drawing an integer uniformly between 5 and 10 inclusive, and then multiplying this integer with $m - 1$. We set the speed of machine M, where $1 \leq M \leq m$, to

$1/(1 + q(M - 1)/(m - 1))$, so that the speeds are between $1/(1 + q)$ and 1. (A speed of $1/2$ means all jobs are processed twice as slowly compared to a speed of 1.) The parameter q controls the difference in processing speeds between the machines, and the factor $m - 1$ in the processing times ensures that all $p_{i,j}$ are integer.

Setup times are initially drawn uniformly between 0 and $10s$ where s is a parameter. The setup times are then modified by the Floyd-Warshall algorithm to make them satisfy the triangle inequality, as proposed in [14, p. 113]. We use the factor s to increase setup times if the Floyd-Warshall algorithm makes them small.

Release times are drawn uniformly between 0 and $10s$ inclusive for half of the jobs; for the other half, we set them to 0 (so as to make use of the dummy job setup times). For due times, we use the TF/RDD model [3,4], adapted to multiple machines much like [9]. Given TF, RDD $\in [0..1]$, due times are drawn uniformly integer between $\lfloor R_j + (1 - \text{TF} - \text{RDD}/2)S \rfloor$ and $\lfloor R_j + (1 - \text{TF} + \text{RDD}/2)S \rfloor$ inclusive (with negative values removed). Here, we let $R_j = \max\{r_j, \sum_M \sigma_{M,\vdash,j}/m\}$ be an approximation of when j can start, and $S = \sum_j p_j / \sum_M \text{speed}_M$ is the total processing time corrected for the speeds of the machines. For a machine M, speed_M is the speed of machine M as described above.

We set TF $= 0.8$, RDD $= 0.2$. The latter two parameters were chosen so that the partial objectives are likely often nonzero, thereby likely improving the merge heuristic (like [9]).

6.2 Partitioning

We first consider the performance of the partitioning based on f. Figure 5 shows the average gap of 25 instances with $n = 60$ jobs and $m = 5$ machines. We set $q = 3$, and $s = 7$ so that the average setup time is about 3.5. Further, we set the width to 8 threads $\cdot$ 5 machines $\cdot$ 500 $= 20,000$. If the width is not divisible by the number of partitions, we divided the width as evenly as possible, giving priority to partitions with a smaller value of f.

We see that partitioning is important: at least 8 partitions are necessary to obtain a nonzero bound. Increasing the number of partitions after that gives better bounds. In particular, not using partitions (i.e., using one partition) gives a zero bound. However, the gap is not monotonically decreasing in the number of partitions.

6.3 Comparison to LPs

We also compare our formulation to the LP relaxation of a time-indexed MIP, solved by IBM ILOG CPLEX 12.8 with default settings. The MIP is based on the one given in [6]. It has a binary variable for each tuple of job, machine and time step, indicating whether a job starts on a machine at a time step. See Fig. 6. We modified the MIP to incorporate setup times; the release times constraint is adapted from [2].

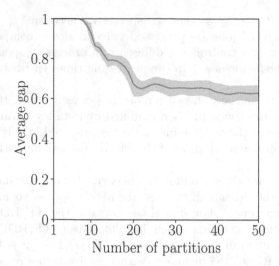

Fig. 5. Average gap with standard deviation of 25 instances as a function of the number of partitions, for $n = 60, m = 5, q = 3, s = 7$

The authors of [6] propose two improvements to the plain LP relaxation of the MIP: preprocessing and cutting planes. However, both require multiple LPs to be solved, whereas our DD model is a single relaxation. We therefore did not use these additions.

Instead, to improve the LP relaxation, we added constraint $(*)$, which is used in the identical machine MIP in [2]. Conceptually, this constraint states that at most m jobs can be processed at any time. While this constraint follows from the others in the MIP used here, we found that it improves the optimum of the LP.

The horizon H in the LP needs to be sufficiently large: if it is too small, the model may be infeasible because not all jobs can be scheduled within the time frame $\{0, \ldots, H-1\}$. However, choosing H too large may yield bad performance because of the large number of variables and constraints. We are not aware of a method to compute the best value of H for our problem.

However, there is a lower bound on H of 1, and an upper bound is induced by the schedule where all jobs are scheduled after each other on the slowest machine, starting at the latest release time (or dummy setup time). We increase horizons between these bounds (divided in 100 steps); when the LP for a horizon becomes feasible, its solution *might* be a lower bound for the MIP (i.e., the scheduling problem), but is not if the horizon is too small. Nevertheless, a larger horizon is guaranteed to result in a smaller LP optimum because the feasible set for such a horizon is a superset of that for a smaller horizon. Thus, any horizon for which the LP is feasible gives an *upper bound* on the lower bound that LP is able to give for the scheduling problem. Equivalently, it gives a lower bound on the gap.

Minimize $\displaystyle\sum_j \sum_M \sum_t x_{j,M,t} \max\{0, t + p_{j,M} - d_j\}$

s.t. $\displaystyle\sum_M \sum_t x_{j,M,t} = 1$ $\hspace{3cm} (\forall j)$

$x_{j,M,t} = 0$ $\hspace{2.5cm} (\forall j, \forall M, \forall t : t < \max\{r_j, \sigma_{M,\vdash,j}\})$

$x_{j,M,t} + \displaystyle\sum_{t \leq s < \min\{t+p_{j,M}+\sigma_{M,j,j'}, H\}} x_{j',M,s} \leq 1 \quad (\forall j, j' : j \neq j', \forall M, \forall t)$

$\displaystyle\sum_j \sum_M \sum_{\max\{0, t-p_{j,M}+1\} \leq s \leq t} x_{j,M,s} \leq m \hspace{1cm} (\forall t) \hspace{1cm} (*)$

$x_{j,M,t} \geq 0$ $\hspace{4cm} (\forall j, \forall M, \forall t)$

Fig. 6. LP relaxation of the time-indexed MIP. The notation t denotes a time step $0 \leq t < H$.

We report the gap and solve time of the LP with the first (i.e., smallest) horizon we tried that makes the LP feasible.

We tested instances with $n \in \{10, 15, 20\}, m \in \{2, 5\}, q \in \{3, 6\}$. The values of s for the various n are respectively $2, 2.5, 3$, so that the average setup time is about 4.4. For each parameter combination we generated 10 instances. We set a width of 8 threads $\cdot m$ machines $\cdot 500$ and fix the number of partitions to 100. We used relatively small values of n because the size of the model becomes too large otherwise. We measured time using the System.Diagnostics.Stopwatch class so that the LPs and DDs are measured in the same way; we did not count the time of building the LP models. We used the solver with and without presolving, reporting the best result. For the DDs, we also included the time of determining the value on which we base the partitioning, i.e., the time of running the adaptation of [18].

The average gap with standard deviation, with and without constraint $(*)$, is shown in Table 1. We see that the gap of the DDs is smaller than that of the LP. In particular, constraint $(*)$ improves the bound but requires quite some more time. We also provide the bound given by CP Optimizer that we used for obtaining the upper bound (i.e., after 5 s); these gaps are typically between LP without and with constraint $(*)$. It seems that the bound provided by the DDs becomes weaker as the instance grows, a phenomenon also mentioned by [12] (although they consider a harder problem).

We additionally compared to LP relaxations of non-time-indexed MIPs, based instead on binary variables denoting the relative order of pairs of jobs on a machine. Specifically, we used the LP relaxation of MIPs adapted from [13, 16] and [1]. While these were solved very quickly, the bounds were always worse than that of the time-indexed LP.

Table 1. Average gap and time (ms) with standard deviation

(n, m, q)	DD		Time-indexed LP without (∗)		Time-indexed LP with (∗)		CP
	Gap	Time	Gap	Time	Gap	Time	Gap
$(10, 2, 3)$	0.148 (0.079)	8 (1)	0.869 (0.027)	11 (4)	0.710 (0.041)	42 (9)	0.540 (0.376)
$(10, 2, 6)$	0.142 (0.050)	5 (0)	0.875 (0.037)	18 (3)	0.746 (0.036)	44 (13)	0.566 (0.311)
$(10, 5, 3)$	0.228 (0.053)	6 (0)	0.791 (0.035)	32 (7)	0.622 (0.036)	137 (46)	0.811 (0.031)
$(10, 5, 6)$	0.275 (0.054)	4 (0)	0.864 (0.021)	31 (3)	0.751 (0.033)	145 (20)	0.851 (0.017)
$(15, 2, 3)$	0.309 (0.070)	30 (3)	0.973 (0.014)	47 (14)	0.753 (0.038)	181 (70)	0.959 (0.008)
$(15, 2, 6)$	0.319 (0.063)	19 (1)	0.975 (0.011)	63 (16)	0.825 (0.030)	189 (67)	0.938 (0.005)
$(15, 5, 3)$	0.386 (0.072)	22 (10)	0.927 (0.014)	121 (47)	0.704 (0.022)	1125 (284)	0.935 (0.013)
$(15, 5, 6)$	0.366 (0.057)	35 (3)	0.957 (0.014)	140 (18)	0.848 (0.018)	790 (140)	0.946 (0.010)
$(20, 2, 3)$	0.451 (0.060)	67 (7)	0.990 (0.003)	115 (18)	0.742 (0.045)	701 (142)	0.978 (0.004)
$(20, 2, 6)$	0.458 (0.064)	39 (2)	0.994 (0.006)	146 (40)	0.818 (0.042)	982 (115)	0.966 (0.005)
$(20, 5, 3)$	0.369 (0.058)	138 (10)	0.971 (0.013)	362 (100)	0.739 (0.025)	7998 (1911)	0.969 (0.008)
$(20, 5, 6)$	0.393 (0.045)	109 (23)	0.989 (0.005)	423 (53)	0.882 (0.020)	3198 (935)	0.977 (0.004)

7   Conclusion, Discussion, and Future Work

We proposed a DD formulation for finding lower bounds on a uniform machine scheduling problem, based on separators to identify the machine assignment of jobs. Additionally, we proposed a merge heuristic based on partitioning nodes on a layer, and showed this is important for obtaining a nontrivial bound. Further, we compared to the LP relaxation of a time-indexed MIP, and found the DDs give stronger bounds in less time. We thus conclude that using DDs to obtain lower bounds can contribute to shorter solve times of uniform machine scheduling problems.

The DD formulation proposed in this paper schedules on an "assignment first" basis: we first decide which jobs to schedule on machine M_1, then which jobs to schedule on M_2, etc. In contrast, [9] schedules on a "time first" basis (and the machine assignment is implicit). The latter approach may have advantages, such as being able to model precedence constraints. In our formulation, a precedence constraint $i \to j$ can be enforced if i is scheduled on an earlier machine than j, but we are unsure how to model precedence constraints in general.

In order to use an efficient "time first" DD formulation, one ideally wants a machine dispatching rule, such as the SPTF rule as used in [9]. An interesting direction for future research is devising such a dispatching rule for uniform machines (i.e., one that is guaranteed to generate an optimal schedule for some permutation of jobs), if one exists.

Another possibility is applying the proposed DD formulation to *unrelated machines*, where the processing times $p_{i,j}$ can depend on i and j in an arbitrary way, and need not necessarily be factored by means of a machine speed. Our DD formulation is directly applicable to this setting; however, the performance is yet to be investigated.

In further performance analysis, it may be interesting to consider alternatives to the LP-relaxation benchmark, such as solving a MIP rather than its LP (e.g., using branch-and-bound). We expect that this will give better bounds, but will require significantly more computation time than the LP relaxations.

Given that DDs have proven useful in branch-and-bound scheme [8], incorporating our DD formulation in such a search procedure may also improve their performance.

Finally, our DD model may also be applicable to vehicle routing problems, which show similarities to machine scheduling problems [5].

Compliance with Ethical Standards
Declaration of interest: When this work was carried out, the first author had a paid PhD position, and the second author was a visiting scientist at the Dutch Railways (NS).

Role of funding source: The NS had no involvement in the conduct of research, this article, or the decision to submit for publication.

References

1. Balakrishnan, N., Kanet, J.J., Sridharan, V.: Early/tardy scheduling with sequence dependent setups on uniform parallel machines. Comput. Oper. Res. **26**(2), 127–141 (1999)
2. Baptiste, P., Jouglet, A., Savourey, D.: Lower bounds for parallel machine scheduling problems. Int. J. Oper. Res. **3**(6), 643–664 (2008)
3. Beasley, J.E.: Weighted tardiness (OR library) (2018). http://people.brunel.ac.uk/~mastjjb/jeb/orlib/wtinfo.html, originally described in [4]. Accessed 26 Feb 2018
4. Beasley, J.E.: OR-Library: distributing test problems by electronic mail. J. Oper. Res. Soc. **41**(11), 1069–1072 (1990)
5. Beck, J.C., Prosser, P., Selensky, E.: Vehicle routing and job shop scheduling: what's the difference? In: ICAPS, pp. 267–276 (2003)
6. Berghman, L., Spieksma, F., T'Kindt, V.: Solving a time-indexed formulation by preprocessing and cutting planes (2014). https://doi.org/10.2139/ssrn.2437371, Available at SSRN (First appeared as extended abstract in the 6th Multidisciplinary International Scheduling Conference: Theory & Applications)
7. Bergman, D., Cire, A.A., van Hoeve, W.J., Hooker, J.: Decision Diagrams for Optimization. Springer, Cham (2016). https://doi.org/10.1007/978-3-319-42849-9
8. Bergman, D., Cire, A.A., van Hoeve, W.J., Hooker, J.N.: Discrete optimization with decision diagrams. INFORMS J. Comput. **28**(1), 47–66 (2016)
9. van den Bogaerdt, P., de Weerdt, M.M.: Multi-machine scheduling lower bounds using decision diagrams. Oper. Res. Lett. **46**(6), 616–621 (2018)
10. Cire, A.A., Van Hoeve, W.J.: Multivalued decision diagrams for sequencing problems. Oper. Res. **61**(6), 1411–1428 (2013)
11. Driessel, R., Mönch, L.: Variable neighborhood search approaches for scheduling jobs on parallel machines with sequence-dependent setup times, precedence constraints, and ready times. Comput. Ind. Eng. **61**(2), 336–345 (2011)
12. Hooker, J.N.: Job sequencing bounds from decision diagrams. In: Beck, J.C. (ed.) CP 2017. LNCS, vol. 10416, pp. 565–578. Springer, Cham (2017). https://doi.org/10.1007/978-3-319-66158-2_36
13. Jain, V., Grossmann, I.E.: Algorithms for hybrid MILP/CP models for a class of optimization problems. INFORMS J. Comput. **13**(4), 258–276 (2001)

14. Jordan, C.: Batching and Scheduling: Models and Methods for Several Problem Classes. Lecture Notes in Economics and Mathematical Systems, vol. 437. Springer, Heidelberg (1996). https://doi.org/10.1007/978-3-642-48403-2
15. Lenstra, J.K., Kan, A.R., Brucker, P.: Complexity of machine scheduling problems. Ann. Discrete Math. **1**, 343–362 (1977)
16. Lin, Y.K., Hsieh, F.Y.: Unrelated parallel machine scheduling with setup times and ready times. Int. J. Prod. Res. **52**(4), 1200–1214 (2014)
17. Pinedo, M.L.: Scheduling: Theory, Algorithms, and Systems, 4th edn. Springer, New York (2016). https://doi.org/10.1007/978-3-319-26580-3
18. Rodrigues, R., Pessoa, A., Uchoa, E., Poggi de Aragão, M.: Heuristic algorithm for the parallel machine total weighted tardiness scheduling problem. Relat. Pesquisa Engenh. Produçao **8**(10), 1–11 (2008)
19. Schutten, J.M.J.: List scheduling revisited. Oper. Res. Lett. **18**(4), 167–170 (1996)
20. Van Hoeve, W.J.: Decision diagrams for sequencing and scheduling (2016). http://icaps16.icaps-conference.org/proceedings/tutorials/tutorial5.pdf, Tutorial ICAPS 2016. Accessed 8 Nov 2018

Extending Compact-Diagram to Basic Smart Multi-Valued Variable Diagrams

Hélène Verhaeghe[1]([✉]), Christophe Lecoutre[2], and Pierre Schaus[1]

[1] ICTEAM, UCLouvain, Place Sainte Barbe 2, 1348 Louvain-la-Neuve, Belgium
{helene.verhaeghe,pierre.schaus}@uclouvain.be
[2] CRIL-CNRS UMR 8188, Université d'Artois, 62307 Lens, France
lecoutre@cril.fr

Abstract. Multi-Valued Decision Diagrams (MDDs), and more generally Multi-Valued Variable Diagrams (MVDs), are instrumental in modeling constrained combinatorial problems. This has led to a number of algorithms for filtering constraints such as mddc, MDD4R and CD (Compact-Diagram). Many compressed forms of tables have also been proposed over the years, leading to a 'smart' hybridization between extensional an intentional representations, which was obtained by embedding simple arithmetic constraints in tuples (of tables). Interestingly, the state-of-the-art algorithm CT (Compact-Table) has been recently extended to deal efficiently with bs-tables, i.e., 'basic smart' tables containing expressions of the form '$*$', '$\neq v$', '$\leq v$', '$\geq v$' and '$\in S$'. In this paper, we introduce the concept of bs-MVDs by enabling arcs of diagrams to be labelled with similar expressions. We show how such diagrams can be naturally derived from ordinary tables and MDDs, and we extend the state-of-the-art algorithm CD in order to handle bs-MVDs (and bs-MDDs).

Keywords: Multi-Valued Decision Diagrams · Filtering · Compression · Compact-Table · Bitset

1 Introduction

Efficiently representing constraints under extensional forms such as tables and decision diagrams has been a hot research topic for the last decade; concerning MDDs (Multi-Valued Decision Diagrams), see e.g., [1,2,4,5,13–15,28]. Actually, two main lines of improvements have been followed when handling extensional forms of constraints. Firstly, high effective filtering techniques have been proposed over the years, such as those based on tabular reduction [18,19] and bitwise operations [11,16,32]. Secondly, compact representation techniques have been intensively studied, mainly by allowing simple constraints to be put in tables as in [17,20,30,31] or by directly using decision diagrams such as MDDs [10,22,23].

CD (Compact-Diagram), previously called Compact-MDD, has been recently introduced [29] for Multi-Valued Variable Diagrams (MVDs), which are diagrams

© Springer Nature Switzerland AG 2019
L.-M. Rousseau and K. Stergiou (Eds.): CPAIOR 2019, LNCS 11494, pp. 581–598, 2019.
https://doi.org/10.1007/978-3-030-19212-9_39

generalizing MDDs by authorizing non-determinism. By combining several ideas and techniques, in particular, those proposed in [29,30], we show how a bitwise filtering algorithm can be conceived for constraints represented by bs-MVDS, which are 'basic smart' MVDs accepting arcs to be labelled with unary constraints. These bs-MVDs can be seen as a very promising modeling tool.

As a direct application of bs-MVDs, we find the possible compact representation of regular constraints [3,6,25], imposing that a specified sequence of variables must be accepted by a given automaton (derived from a regular expression). As a user, it is quite natural to express an automaton using expressions on transitions, as illustrated in Fig. 1a for the regular language $[1-9].^*[\wedge 5].^*[05]$. As described in [24], a layered deterministic automaton graph basically defines an MDD constraint; a layered automaton graph obtained from a non-deterministic automaton then naturally leads to an MVD constraint. Figure 1b displays the bs-MVD that corresponds to the unfolding of the automaton over a sequence of 5 variables.

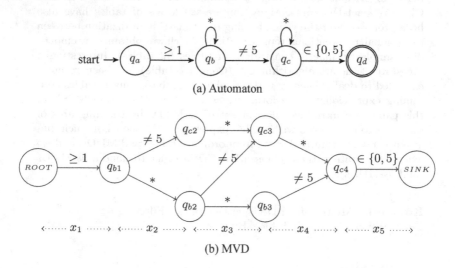

(a) Automaton

(b) MVD

Fig. 1. The regex $[1-9].^*[\wedge 5].^*[05]$ describes a number divisible by 5 with at least one of its inner digits being different from 5.

The main contributions of this paper are:

- Different strategies to create bs-MVD constraints from table constraints, with improved compression compared to classical MDDs.
- An efficient bitwise filtering algorithm that enforces the property known as Generalized Arc Consistency on bs-MVD constraints.
- An experimentation that demonstrates the practical interest of our approach.

2   Technical Background

A *constraint network* is composed of a set of variables and a set of constraints. Each *variable* x has an associated (ordered) domain $dom(x)$ containing the values that can be assigned to it; the *current* domain is included in the *initial* domain $dom^0(x)$. We respectively denote by $min(x)$ and $max(x)$ the smallest and greatest values in $dom(x)$. Each *constraint* c involves an ordered set of variables, called the *scope* of c and denoted by $scp(c)$, and is semantically defined by a *relation* $rel(c)$ containing the tuples allowed for the variables involved in c. The *arity* of a constraint c is $|scp(c)|$. When the domain of a variable x is (becomes) singleton, we say that x is *bound*.

Given a sequence $\langle x_1, \ldots, x_r \rangle$ of r variables, an r-tuple τ on this sequence of variables is a sequence of r values $\langle a_1, \ldots, a_r \rangle$, where the individual value a_i is also denoted by $\tau[x_i]$. An r-tuple τ is *valid* on an r-ary constraint c iff $\forall x \in scp(c), \tau[x] \in dom(x)$, and τ is *allowed* by c iff $\tau \in rel(c)$. A *support* of c is a tuple which is both valid on c and allowed by c. A *literal* is a pair (x, a) where x is a variable and a a value. A literal (x, a) is *Generalized Arc-Consistent* (GAC) on c iff there is a support τ on c such that $\tau[x] = a$. A constraint c is GAC iff any literal (x, a) such that $x \in scp(c)$ and $a \in dom(x)$ is GAC on c.

A directed graph is composed of a set of nodes and a set of arcs. Each arc has an orientation from one node, the *tail* of the arc, to another node, the *head* of the arc. For a given node ν, the set of arcs with ν as tail (resp., head) is called the set of *outgoing* (resp., *incoming*) arcs of ν. A labelled directed graph is a directed graph such that a label $l(\omega)$ is associated with each arc ω. A node is *in-d* (in-deterministic) iff it does not have two incoming arcs with the same label, *in-nd* otherwise. A node is *out-d* (out-deterministic) iff no two outgoing arcs have the same label, *out-nd* otherwise. A directed acyclic graph (DAG) is a directed graph with no directed cycles. An MVD (Multi-valued Variable Diagram) [1] *for* a constraint c (called an MVD constraint) is a layered DAG, with one special root node at level 0, denoted by ROOT, r layers of arcs, one layer $L(x_i)$ for each variable x_i of the scope $\langle x_1, \ldots, x_r \rangle$ of c, and one special sink node at level r, denoted by SINK. The arcs in $L(x_i)$ going from level $i - 1$ to level i are *on* the variable x_i: any such arc is labelled by a value in $dom^0(x_i)$. A *valid path* in such an MVD is a path p from the root to the sink such that for each variable x_i in $scp(c)$ the label of the arc going in p from level $i - 1$ to i is a value in $dom(x_i)$. The set of supports of an MVD constraint c corresponds to the valid paths in the MVD for c. One classical type of MVD is the Multi-valued Decision Diagram (MDD) [8], which guarantees that each node is out-d (each node at level i has at most $|dom^0(x_i)|$ outgoing arcs, labelled with different values), but possibly in-nd. Another type of MVD is the semi-MDD (sMDD) [29] which guarantees that each node at a level $< \lfloor \frac{r}{2} \rfloor$ is out-d and each node at a level $> \lfloor \frac{r}{2} \rfloor + 1$ is in-d.

Algorithm 1. Class ConstraintCD

---

1 **Method** enforceGAC()
2   updateGraph()
3   filterDomains()

4 **Method** updateGraph()
5   **foreach** *variable* $x \in$ scp **do**
6   mask$[x] \leftarrow 0^{64}$
7   updateMasks()
8   propagateDown(x_1, false)
9   propagateUp(x_r, false)

10 **Method** updateMasks()
11   **foreach** *variable* $x \in \{x \in scp : |\Delta_x| > 0\}$ **do**
12   **if** $|\Delta_x| < |dom(x)|$ **then**   // Incremental update
13   **foreach** *value* $a \in \Delta_x$ **do**
14   mask$[x] \leftarrow$ mask$[x]$ | supports$[x, a]$ // bitwise OR
15   **else**   // Reset-based update
16   **foreach** *value* $a \in dom(x)$ **do**
17   mask$[x] \leftarrow$ mask$[x]$ | supports$[x, a]$ // bitwise OR
18   mask$[x] \leftarrow \sim$ mask$[x]$ // bitwise NOT

19 **Method** propagateDown(x_i, localChange)
20   **if** $|\Delta_{x_i}| > 0$ *or* localChange **then**
21   currArcs$[x_i] \leftarrow$ currArcs$[x_i]$ & $\sim$ mask$[x_i]$
22   **if** currArcs$[x_i] = 0$ **then**
23   **throw** Backtrack
24   **if** $x_i \neq x_r$ **then**
25   localChange $\leftarrow$ false
26   **foreach** *node* $\nu \in \{\nu :$ currArcs$[x_{i+1}]$ & arcsT$[\nu, x_{i+1}] \neq 0^{64}\}$ **do**
27   **if** currArcs$[x_i]$ & arcsH$[x_i, \nu] = 0^{64}$ **then**
28   mask$[x_{i+1}] \leftarrow$ mask$[x_{i+1}]$ | arcsT$[\nu, x_{i+1}]$
29   localChange $\leftarrow$ true
30   propagateDown(x_{i+1}, localChange)
31   **else if** $x_i \neq x_r$ **then**
32   propagateDown(x_{i+1}, false)

33 **Method** propagateUp(x_i, localChange)
    /* Similar to propagateDown with x_1 instead of x_r, x_{i-1} instead of x_{i+1},
    inverted use of arcsT and arcsH.   */

34 **Method** filterDomains()
35   **foreach** *variable* $x \in \{x \in scp : |dom(x)| > 1\}$ **do**
36   **foreach** *value* $a \in dom(x)$ **do**
37   **if** currArcs$[x]$ & supports$[x, a] = 0^{64}$ **then**   // bitwise AND
38                 $dom(x) \leftarrow dom(x) \setminus \{a\}$

CD. Compact-Diagram, or CD (previously called Compact-MDD [29]), is a filtering algorithm (propagator) that uses bitwise operations for MVD constraints. It is based on some ideas behind both Compact-Table [11], a propagator for table constraints that uses bitwise operations as well, and MDD4R [23], a propagator for MDD constraints.

The idea is to keep track of the arcs that remain valid during the filtering process; namely by introducing (reversible sparse) bitsets, one per layer of the MVD (and so, per variable of the constraint). At layer i, one bit, in the bitset currArcs[x_i], is associated with each arc: when the bit is set to 1, it means that the arc is considered as valid. This way, the current MVD, which can be seen as a subgraph of the initial MVD, can be identified, and used to remove the values without any supports left.

To ease computations, at each level there are three types of precomputed bitsets: these bitsets are never modified. First, supports[x, a] indicates for each arc on the variable x whether or not the value a is initially supported by this arc (bit is set to 1 iff a is supported). Second, arcsT[ν, x] and arcsH[x, ν'] indicate for each arc on x whether ν and ν' are respectively the tail and the head of this arc. Finally, a temporary bitset mask[x_i] is associated with each variable x_i to store the results of intermediate computations.

The pseudo-code for enforcing GAC on an MVD constraint is given by Algorithm 1, which is, for simplicity, a simplified version of the one given in [29]. In method updateGraph(), after initializing all masks, all arcs that can be trivially removed are handled by calling updateMasks(). This method assumes access to the set of values Δ_x removed from $dom(x)$ since the last call to enforceGAC()[1]. There are two ways of updating the masks (before updating currArcs from these masks, later): either incrementally or from scratch after resetting. In case of an incremental update, we perform the union of the arcs to be removed, whereas in case of a reset-based update, we perform the union of the arcs to be kept, followed by a reverse operation. Next, we need to determine which arcs can be subsequently removed: this is achieved by calling the methods propagateDown() and propagateUp(), which, similarly to MDD4R, perform two passes on the diagram. During the downward (resp., upward) pass, each level is examined from the root (resp., sink) to the sink (resp., root). When there are no more valid arcs entering (resp., exiting) a node, it becomes unreachable and all arcs exiting (resp., entering) the node becomes invalid. Identifying unreachable nodes is done by testing if the intersection between currArcs and arcsT (for outgoing arcs) or arcsH (for incoming arcs) is empty.

The process of filtering domains is very similar to the one described in CT [11]. This is given by method filterDomains() in Algorithm 1. For each unbounded variable x and each value a in $dom(x)$, the intersection between the valid arcs on x, currArcs[x], and the arcs allowing value a, supports[x, a], determines if a is still supported. An empty intersection means that a can be deleted from $dom(x)$.

---

[1] In [27], a sparse-set domain implementation for obtaining Δ_x without overhead is described.

Let us mention an important fact about the bitwise operations performed at Lines 14, 17 and 28 of Algorithm 1. As described in [11], bitsets are implemented as arrays of words (long integers). Progressively, while arcs become invalid, some words of currArcs become equal to 0 (all bits set to 0). In practice, operations are only performed on the active (i.e., non-zero) words of currArcs, which are easily retrievable due to the use of a sparse set maintaining the indexes of active words (called validWords). Eventually, the mask is intended for being intersected with currArcs. Therefore, computing values of the words of mask corresponding to a non-valid word of currArcs is not needed and not done.

3   Transforming Tables into bs-MVDs

A bs-MVD, or basic smart MVD, is defined similarly to a bs-table, or basic smart table [30]. Namely, it is an MVD where each arc is labelled by one the following expressions: '$*$', '$\langle relop \rangle\, v$', '$\in S$' and '$\notin S$' where v is value, S is a set of values, and $\langle relop \rangle$ is an operator in $\{<, \leq, =, \neq, \geq, >\}$. The operator involved in the labeling expression of an arc ω is denoted by $op(\omega)$. There are two main strategies of generating bs-MVDs from (ordinary) tables:

1. Through MVDs: the table is first transformed into an MVD, using pReduce [24] (leading to an MDD) or sReduce [29] (leading to an sMDD). Then, the arcs with the same tail and head nodes are processed, targeting the general unary expressions given above. Note that the initial structure of the MVD is preserved. Namely, an MDD becomes a bs-MDD, while an sMDD becomes a bs-sMDD.
2. Through bs-tables: the table is first transformed into a bs-table, for example using the algorithms described in [17,30]. Then, the bs-table is transformed into a bs-MVD using a slightly modified version of an algorithm transforming tables into diagrams. However, as some (smart) tuples may overlap, the transformation may lead to a bs-MVD with nodes that are non deterministic (in-nd and out-nd) at any level. This is not an issue since CD can handle non-deterministic diagrams.

We now describe how to carry out the second step of both approaches.

3.1   From MVDs to bs-MVDs: Arc Merging

Generating a bs-MVD from an MVD is straightforward. At each level i, we simply process every group G of (at least two) arcs sharing the same tail and head nodes. Specifically, we can compare $V = \{l(\omega) : \omega \in G\}$ with $dom(x_i)$, and consequently apply some rules (given in order of priority) for merging some arcs of G:

1. if $V = dom(x_i)$, then G is replaced by a unique arc labelled with '$*$';
2. if $\exists a \in dom(x_i)$ s.t. $V \cup \{a\} = dom(x_i)$, then G is replaced by a unique arc labelled with '$\neq a$';

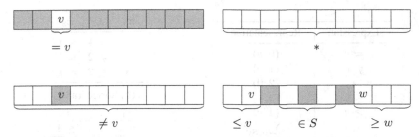

(a) Illustration of possible merging rules (on a domain of size 10).

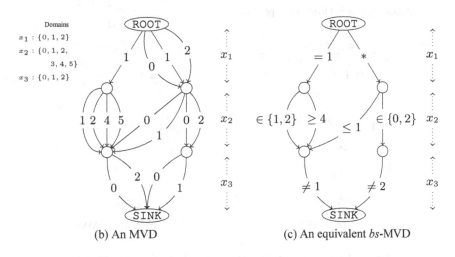

(b) An MVD   (c) An equivalent bs-MVD

Fig. 2. Transforming an MVD into an equivalent bs-MVD.

3. if m, defined as $\max\{v : \{v' \in dom(x_i) : v' \leq v\} \subseteq G\}$ is not equal to $min(x_i)$, then $G^m = \{\omega \in G : l(\omega) \leq m\}$ is replaced by a unique arc labelled with '$\leq m$' (otherwise, $G^m = \emptyset$); if M, defined as $\min\{v : \{v' \in dom(x_i) : v' \geq v\} \subseteq G \setminus G^m\}$, is not equal to $max(x_i)$, then $G^M = \{\omega \in G \setminus G^m : l(\omega) \geq M\}$ is replaced by a unique arc labelled with '$\geq M$' (otherwise, $G^M = \emptyset$); with $G' = G \setminus G^m \setminus G^M$, if $|G'| > 1$ then G' is replaced by a unique arc labelled with '$\in S$' where $S = \{l(\omega) : \omega \in G'\}$.

Figure 2a illustrates these merging rules. The variable of interest x_i has a domain (initially) composed of 10 values, and white cells represent the values that are present in G.

Note that our merging procedure keeps at most three arcs between any two nodes. An example is given in Fig. 2.

3.2 From bs-Tables to bs-MVDs: pReduce$_{bs}$

To create a bs-MVD from a bs-table, one can easily adapt the known procedure pReduce (initially introduced for creating MDDs from tables) so as to generate

Expression	Representation
$= v$	$(0, v)$
$\neq v$	$(1, v)$
$*$	$(2, 0)$
$\leq v$	$(3, v)$
$\geq v$	$(4, v)$
$\in S$	$(5, \sum_{i \in S} 2^i)$
$\notin S$	$(6, \sum_{i \notin S} 2^i)$

(a) Lexicographic Order on Expressions

x_1	x_2	x_3
$= 1$	$= 1$	≤ 1
$*$	$\neq 2$	≤ 1
$*$	≤ 2	$= 1$

(b) Sorted Table

x_1	x_2	x_3
$= 1$	$= 1$	≤ 1
	$\neq 2$	≤ 1
$*$	≤ 2	$= 1$

(c) Trie

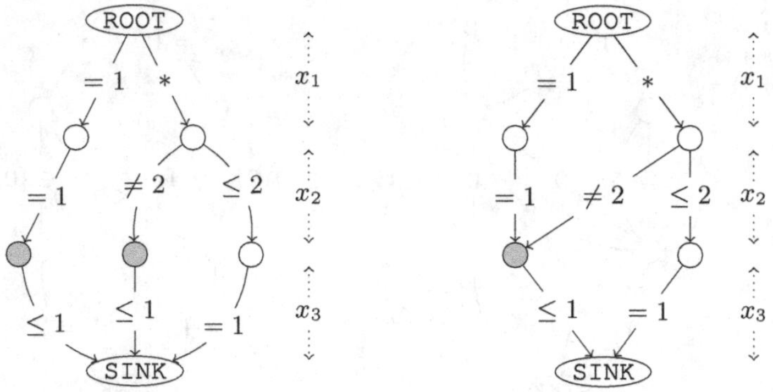

(d) MVD

(e) Reduced MVD

Fig. 3. Turning a bs-table into a bs-MVD using pReduce$_{bs}$.

MVDs; the adaptation is called pReduce$_{bs}$. The four steps of the procedure are the following. First, the tuples of the table are sorted using a lexicographic ordering. Second, the corresponding trie (i.e., prefix tree) is created by sharing common prefixes among the tuples. Third, a diagram is derived from the trie by merging all the leaves of the trie to form the sink node. Finally, the diagram is reduced by merging, in a bottom-up way, each pair of nodes having similar sets of outgoing edges. Two sets of outgoing arcs are similar if they have the same cardinality, and for each arc in one set, there is an arc in the other set with the same label (value) and the same head. Actually, for adapting it, we just need to impose a total order on expressions (unary constraints) involved in basic smart tuples. For example, we can simply associate a pair of integers with each expression (unary constraint) such that the first element of the pair represents the type (operator) of the expression and the second element the operand involved in the expression. Figure 3a illustrates the naturally derived lexicographic order.

Figure 3 illustrates through an example the four steps of pReduce$_{bs}$: going from a sorted bs-table (Fig. 3b) to a trie (Fig. 3c), then into an MVD (Fig. 3d) and finally into a reduced MVD (Fig. 3e, where the gray node is the result of merging two nodes with similar outgoing sets of edges). This example shows that pReduce$_{bs}$ does not necessarily generate a bs-MDD, because some nodes are not out-d, possibly leading to multiple paths for a same tuple as it is the case for $(1, 1, 1)$.

A similar adaptation exists for sReduce [29], the procedure that generates sMDDs, leading to sReduce$_{bs}$.

4 CD^{bs}: Compact-Diagram Handling bs-MVDs

CD and CT are quite similar in term of design. Basically, both of them use the bitsets called supports to respectively find the arcs and tuples that must be discarded. Recently, the CT^{bs} [30] algorithm, which can deal with bs-tables, was proposed as an extension of CT, by only modifying the update procedure. In the same spirit, we show how similar ideas can be reused to adapt the method updateMask() of CD in order to define CD^{bs}. We present first a simple version of CD^{bs}, before introducing an optimized version that strongly relies on a partition of the arcs at each level i, defined as follows:

- $C^{bas}(x_i) = \{\omega \in L(x_i) : op(\omega) \in \{=, \neq, *\}\}$,
- $C^{min}(x_i) = \{\omega \in L(x_i) : op(\omega) \in \{<, \leq\}\}$,
- $C^{max}(x_i) = \{\omega \in L(x_i) : op(\omega) \in \{>, \geq\}\}$,
- $C^{set}(x_i) = \{\omega \in L(x_i) : op(\omega) \in \{\in, \notin\}\}$

Simple Adaptation of CT^{bs}. As in CT^{bs}, in addition to bitsets supports, we introduce auxiliary bitsets:

- supports*$[x, a]$, the exclusive supports: for each arc for which the label of arc ω is exactly a ('$= a$'), the bit is set to 1,
- supportsMin$[x, a]$, the lower bound supports: for each arc which would be still valid if the minimum of the domain was a, the bit is set to 1,
- supportsMax$[x, a]$, the upper bound supports: for each arc which would be still valid if the maximum of the domain was a, the bit is set to 1.

Algorithm 2 displays the method updateMasks() for the simple version of CD^{bs}. This is for Compact-Diagram a simple adaptation of the modifications made to pass from CT to CT^{bs}. Resetting (and recomputing) is performed when the number of removed values (i.e., values in Δ_x) is too large by collecting the supports of every value in the current domain (lines 10–8). Otherwise an incremental update is performed. Notice that contrarily to the reset-based update, one needs to also collect invalid arcs for operators in $\{<, \leq, >, \geq\}$ using supportsMin and supportsMax at lines 7 and 9 of Algorithm 2. The time complexity of one call to updateMasks(), for a given variable x, is $\Theta(dt)$ where t is the number of valid words and d is $\min(|\Delta_x|, |dom(x)|)$ if $C^{set}(x) = \phi$ and $|dom(x)|$ if not.

Algorithm 2. Simple Version of CD^{bs}

---

1 **Method** updateMasks()
2   **foreach** *variable* $x \in \{x \in scp : |\Delta_x| > 0\}$ **do**
3   **if** $|\Delta_x| < |dom(x)|$ *and* $\mathsf{C}^{\text{set}}(x) = \phi$ **then**   // Incremental update
4   **foreach** *value* $a \in \Delta_x$ **do**
5                  $\text{mask}[x] \leftarrow \text{mask}[x] \mid \text{supports}^*[x, a]$ // bitwise OR
6   **if** $dom(x).\text{minChanged}()$ **then**
7                  $\text{mask}[x] \leftarrow \text{mask}[x] \mid {\sim} \text{supportsMin}[x, x.min]$
8   **if** $dom(x).\text{maxChanged}()$ **then**
9                  $\text{mask}[x] \leftarrow \text{mask}[x] \mid {\sim} \text{supportsMax}[x, x.max]$
10   **else**   // Reset-based update
11   **foreach** *value* $a \in dom(x)$ **do**
12                 $\text{mask}[x] \leftarrow \text{mask}[x] \mid \text{supports}[x, a]$ // bitwise OR
13             $\text{mask}[x] \leftarrow {\sim} \text{mask}[x]$ // bitwise NOT

Exploiting Partitions of Arcs. The time complexity of Algorithm 2 can be improved to reach $\Omega(t)$ and $\mathcal{O}(dt)$. For that, let us consider the hypothetical case of a variable with an operator in $\{<, \leq, >, \geq\}$ for each of its associated arc labels. In such a case, one can collect invalid arcs using lines 7 and 9 from Algorithm 2, and there is no need to iterate over the sets $dom(x)$ or Δ_x. This favorable situation can be partially forced by sorting arcs in bitsets supports so that the bits in a computer word only represent arcs from a given category ($\mathsf{C}^{\text{bas}}, \mathsf{C}^{\text{set}}, \mathsf{C}^{\text{min}}, \mathsf{C}^{\text{max}}$). If each computer word is filled with (bits for) arcs belonging to the same category (dummy invalid arcs are used to complete a word if necessary), then only the required specific operations can be systematically applied to this word. This leads to Algorithm 3 that iterates over the valid words and only applies the operations required by the category of the word (note that the category for the jth word is given by currArcs[x].category[j]). Arcs from C^{bas} are updated using supports* or supports (incremental or reset case). Arcs from C^{set} are updated using supports in all cases. Arcs from C^{min} and C^{max} are updated using supportsMin and supportsMax, respectively. It appears that the categories C^{min} and C^{max} are particularly cheap to treat as they only imply one value.

An Interesting Observation. In Algorithm 3, each valid word is associated with a (unique) category. From this fact, one can observe that supportsMin and supportsMax are useless.

Observation 1 *For any variable x, and any word index j of* currArcs[x]*, we have:*

$$\text{currArcs}[x].\text{category}[j] = \mathsf{C}^{\text{min}} \Rightarrow$$

$$\text{supportsMin}[x, a][j] = \text{supports}[x, a][j]$$

Algorithm 3. Optimized Version of CD^{bs}

---

1 **Method** updateMasks()
2   **foreach** *variable* $x \in \{x \in scp : |\Delta_x| > 0\}$ **do**
3   **foreach** *index* $j \in$ currArcs[x].validWords **do**
4   **switch** currArcs[x].category[j] **do**
5   **case** C^{bas} **do**
6   **if** $|\Delta_x| < |dom(x)|$ **then**   // Incremental update
7   **foreach** *value* $a \in \Delta_x$ **do**
8               $\lfloor$ mask[x][j] $\leftarrow$ mask[x][j] | supports*[x, a][j]
9   **else**   // Reset update
10   **foreach** *value* $a \in dom(x)$ **do**
11               $\lfloor$ mask[x][j] $\leftarrow$ mask[x][j] | supports[x, a][j]
12   mask[x][j] $\leftarrow \sim$ mask[x][j]
13   **case** C^{set} **do**
14   **foreach** *value* $a \in dom(x)$ **do**
15             $\lfloor$ mask[x][j] $\leftarrow$ mask[x][j] | supports[x, a][j]
16   mask[x][j] $\leftarrow \sim$ mask[x][j]
17   **case** C^{min} **do**
18   **if** $dom(x)$.minChanged() **then**
19             $\lfloor$ mask[x][j] $\leftarrow$ mask[x][j] | $\sim$ supportsMin[x, x.min][j]
20   **case** C^{max} **do**
21   **if** $dom(x)$.maxChanged() **then**
22             $\lfloor$ mask[x][j] $\leftarrow$ mask[x][j] | $\sim$ supportsMax[x, x.max][j]

Similarly,

$$\text{currArcs}[x].\text{category}[j] = C^{\mathrm{max}} \Rightarrow$$
$$\text{supportsMax}[x, a][j] = \text{supports}[x, a][j]$$

Proof (sketch for C^{min}) By restricting the scope of the definitions of the bitsets to the word (index) j whose bits are exclusively associated with arcs from C^{min}, supports[x, a][j] contains arcs represented by this word that accept the value a, i.e. arcs labelled by $\leq v$ with $v \geq a$, whereas supportsMin[x, a][j] contains arcs for which $\exists b \in dom(x)$ accepted by the arcs such as $a \leq b$, i.e., arcs labelled by $\leq v$ with $v \geq a$. The two words end up to be equal: the exact same bits are set for both supports[x, a][j] and supportsMin[x, a][j].

This observation is illustrated by Fig. 4. For any literal (x, a) and any word index j of category C^{min} (resp., C^{max}), the word supportsMin[x, v][j] (resp., supportsMax[x, v][j]) is equal to the word supports[x, v][j]. Therefore, we can simply use supports at lines 19 and 22. It means that the only required auxiliary bitset is supports* for words attached to C^{bas}.

(a) Labels of Arcs

	x
ω_0	$= 1$
ω_1	≤ 2
ω_2	≥ 1
ω_3	$\in \{1,3\}$
ω_4	$\neq 1$
ω_5	> 2
ω_6	$\notin \{0,3\}$
ω_7	< 2
ω_8	$\neq 2$
ω_9	$*$

(Category)	word 0 C^{bas}				word 1 C^{set}				word 2 C^{min}				word 3 C^{max}			
	ω_0	ω_4	ω_8	ω_9	ω_3	ω_6			ω_1	ω_7			ω_2	ω_5		
$[x,0]$	0	1	1	1	0	0	0	0	1	1	0	0	0	0	0	0
$[x,1]$	1	0	1	1	1	1	0	0	1	1	0	0	1	0	0	0
$[x,2]$	0	1	0	1	0	1	0	0	1	0	0	0	1	0	0	0
$[x,3]$	0	1	1	1	1	0	0	0	0	0	0	0	1	1	0	0
$[x,4]$	0	1	1	1	0	1	0	0	0	0	0	0	1	1	0	0

(b) Bitsets `supports` for literals on x

(Category) (From)	word 0 C^{bas} supports*				word 1 C^{set} no auxiliary				word 2 C^{min} supportsMin				word 3 C^{max} supportsMax			
	ω_0	ω_4	ω_8	ω_9	ω_3	ω_6			ω_1	ω_7			ω_2	ω_5		
$[x,0]$	0	0	0	0	-	-	-	-	1	1	0	0	0	0	0	0
$[x,1]$	1	0	0	0	-	-	-	-	1	1	0	0	1	0	0	0
$[x,2]$	0	0	0	0	-	-	-	-	1	0	0	0	1	0	0	0
$[x,3]$	0	0	0	0	-	-	-	-	0	0	0	0	1	1	0	0
$[x,4]$	0	0	0	0	-	-	-	-	0	0	0	0	1	1	0	0

(c) Auxiliary bitsets for literals on x

Fig. 4. Bitsets related to a variable x, assuming 10 associated arcs $\omega_0, \omega_1, \ldots$ in the bs-MVD. The size of computer words is assumed to be 4, for simplicity.

Overall Complexity of the Propagator. Regarding the time complexity of the propagator (and not only the updateMasks() method), CD is $\mathcal{O}(max(n,d)r\frac{a}{w})$ where r is the arity of the constraint, d the greatest domain size, n (resp. a) the maximum number of nodes (resp. arcs) per level and w the size of computer words ($w = 64$ for Java long integer type). CD^{bs} keeps the same complexity. Regarding the space complexity, the maximum number of words of one bitset is $\lceil\frac{a}{w}\rceil + 3$. Per level, there is one currArcs, d supports and supports* (its length is min 0 words, if $C^{bas} = \phi$ and $\lceil\frac{a}{w}\rceil$ max, if $|C^{set}| \leq w$, $|C^{min}| \leq w$ and $|C^{max}| \leq w$) and n arcsH and arcsT. The space complexity is thus $\mathcal{O}((d+n)r\frac{a}{w})$.

5   Experimental Results

All algorithms described in this paper are implemented in the Oscar solver [21], using 64-bit words (Long). Our implementation benefits from all optimization techniques described in [11,29]. Notably, we manage sparse sets in order to avoid handling zero computer words.

All the results of our experiments are displayed using performance profiles [12]. A performance profile is a cumulative distribution of the improved performance of an algorithm $s \in S$ compared to other algorithms of S over a set I

of instances: $\rho_s(\tau) = \frac{1}{|I|} \times |\{i \in I : r_{i,s} \leq \tau\}|$ where the performance ratio is defined as $r_{i,s} = \frac{t_{i,s}}{\min\{t_{i,s} | s \in S\}}$ with $t_{i,s}$ the value of the measured unit (here, either the number of nodes, the number of arcs or the CPU time) obtained with algorithm $s \in S$ on instance $i \in I$. A ratio $r_{i,s} = 1$ thus means that s is the best algorithm for instance i.

Depending on the main data structure (table or diagram) and possible transformation, we use different names to describe the benchmark suite:

- β_t: the initial benchmark. It is a set of roughly $4,000$ instances only containing (positive) table constraints, and available on the XCSP3 website [7].
- β_{bst}: instances of β_t have been transformed into instances where bs-tables replace (ordinary) tables. The compression algorithm detailed in [30] was used.
- β_{mdd}: instances of β_t have been transformed into instances where MDDs replace (ordinary) tables. The algorithm pReduce [24] was used.
- β_{bsmvd}: instances of β_{bst} have been transformed into instances where bs-MVDs replace bs-tables. The algorithm pReduce$_{bs}$ was used.
- β_{bsmdd}: instances of β_{mdd} have been transformed into instances where bs-MDDs replace MDDs.

Our experiments have two main objectives:

1. analyzing the compression quality of the different approaches, when transforming tables,
2. analyzing the speedup obtained by the new filtering algorithms over the existing ones.

5.1 Quality of Compression

To start, we consider the results depicted in Fig. 5. The three benchmarks involving MVDs are β_{mdd}, β_{bsmvd} and β_{bsmdd}. In term of compression, the clear winner is β_{bsmdd} with substantially less arcs than in the diagrams generated by the two other approaches. Let us recall that this approach consists of two main steps: (1) creating a graph, and (2) merging arcs greedily. The alternative approach β_{bsmvd} that creates first a bs-table, and then converts it into a bs-MVD is worse both in terms of the number of nodes and the number of arcs, even when compared to a standard generation of MDDs (β_{mdd}). One explanation is that, despite starting from smaller tables, there is less chance to merge nodes due to the proliferation of constraint labels in the compressed tables.

5.2 Speedup

Figure 6 shows the results of a comparison between CD and CD^{bs}. The new filtering algorithm CD^{bs}, as it could be expected, obtains a larger speedup when applied on graphs with fewer nodes and arcs, i.e., on instances from β_{bsmdd}.

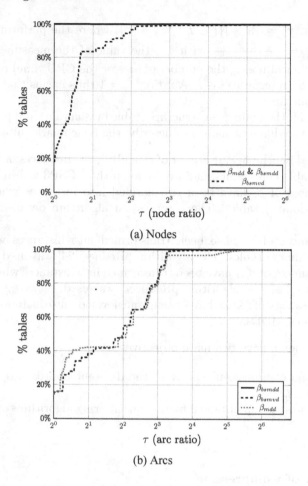

(a) Nodes

(b) Arcs

Fig. 5. Performance profile comparing the structure of the graphs from β_{bsmdd}, β_{bsmvd} and β_{mdd}.

In particular, we can see that on the benchmark β_{bsmvd} (based on a compression into bs-tables, followed by a generation of bs-MVDs) CD^{bs} performs worse than CD applied on β_{mdd} (standard MDDs). The reason is that graphs in β_{bsmvd} have generally a greater number of nodes than other equivalent graphs as shown before in Fig. 5a. This follows the same conclusions as in [29] regarding why CD was more efficient on sMDDs (having fewer nodes than MDDs).

An interesting remark is that, contrarily to CT^{bs} (Compact-Table for basic smart tables) [30], the presence of expressions '∈ S' does not induce any overhead for CD^{bs}. Since the arcs involving expressions of the form '∈ S' are gathered on the same bitwords, they don't prevent from doing an incremental update when considering the other categories of expressions, as it was the case for CT.

In [29], CT was shown to remain faster than CD despite the introduction of bitwise operations. We revisit the same experiment with the newly presented

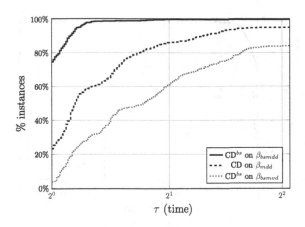

Fig. 6. Performance profile comparing CD^{bs} to CD on various basic smart MVD benchmarks CD on β_{mdd}, CT^{bs} on β_{bsmdd}, β_{bsmvd} and β_{bsmdd}.

algorithm. Figure 7 compares four scenarios, including the use of CT: CT on β_t, CT^{bs} (the extension of CT [30] handling directly bs-tables) on β_{bst}, CD (Compact-Diagram, the adaptation of CT to MVD [29]) on β_{mdd} and CD^{bs} on β_{bsmdd}.

One can see that CT is still the best approach, followed by CT^{bs}. Nevertheless, as it can be observed in the figure, the gap is shrinking when using the new algorithm CD^{bs}. Also, there is now around 10% of the instances where CD^{bs} is the fastest algorithm. A post analysis has shown that instances with larger domains are the more favorable for CD^{bs}. In such cases, we could observe for some tables a reduction by a factor of up to 8 on the number of arcs.

The main advantage of CD thus lies in the potential compactness of the diagrams, although this is really problem/constraint dependent. On the one side, some graphs, when expanded into tables, can't even fit in memory. On the other side, some constraints, like AllDifferent [22] are not well suited for an MDD representation because there is almost no compression. When CD can benefit from a large compression, it becomes faster.

For a fair comparison, the choice was made not to evaluate the new algorithm on a priori favorable problems, hence the benchmarks composed of problems that initially contain table constraints. Also the order of variables remained unchanged (order as described in the initial instances used). Optimizing this order may also have an impact on the size of the graphs [9].

In our opinion, having both CT and CD is useful: if, for a given constraint, a high compression (by an MDD or another diagram) is possible, CD should be used, otherwise CT is more suited. Also, the new algorithm should typically be used for solving combinatorial problems with complex constraints that can't even be represented in memory as simple tables. One good example of work in that direction is [26]. Another promising direction for applying this propagator is for solving combinatorial problems on Strings.

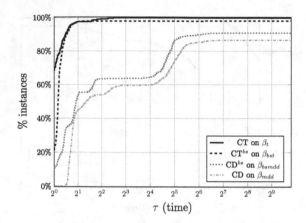

Fig. 7. Performance profile of the comparison of the best-case scenario of CD^{bs}, CD and the tables algorithms CT and CT^{bs}.

6   Conclusion

We have proposed to use a new general form of Multi-valued Variable Diagram (MVD) for representing constraints: the bs-MVD that accepts unary constraints as labels of arcs. We have also shown how to generate such diagrams from (ordinary) tables. Finally, we have adapted the propagator CD (Compact-Diagram) to bs-MVDs, by inspiring ourselves from the adaptation of CT to bs-tables. This new propagator is efficient and makes little closer graph-based approaches and table-based approaches.

Acknowledgements. The second author is supported by the project CPER Data from the "Hauts-de-France".

References

1. Amilhastre, J., Fargier, H., Niveau, A., Pralet, C.: Compiling CSPs: a complexity map of (non-deterministic) multivalued decision diagrams. Int. J. Artif. Intell. Tools **23**(04) (2014)
2. Andersen, H.R., Hadzic, T., Hooker, J.N., Tiedemann, P.: A constraint store based on multivalued decision diagrams. In: Bessière, C. (ed.) CP 2007. LNCS, vol. 4741, pp. 118–132. Springer, Heidelberg (2007). https://doi.org/10.1007/978-3-540-74970-7_11
3. Beldiceanu, N., Carlsson, M., Petit, T.: Deriving filtering algorithms from constraint checkers. In: Wallace, M. (ed.) CP 2004. LNCS, vol. 3258, pp. 107–122. Springer, Heidelberg (2004). https://doi.org/10.1007/978-3-540-30201-8_11
4. Bergman, D., Ciré, A., van Hoeve, W.: MDD propagation for sequence constraints. J. Artif. Intell. Res. **50**, 697–722 (2014)
5. Bergman, D., Ciré, A., van Hoeve, W., Hooker, J.: Decision Diagrams for Optimization. Springer, Cham (2016). https://doi.org/10.1007/978-3-319-42849-9

6. Bessiere, C., Hebrard, E., Hnich, B., Kiziltan, Z., Quimper, C.-G., Walsh, T.: Reformulating global constraints: the SLIDE and REGULAR constraints. In: Miguel, I., Ruml, W. (eds.) SARA 2007. LNCS (LNAI), vol. 4612, pp. 80–92. Springer, Heidelberg (2007). https://doi.org/10.1007/978-3-540-73580-9_9

7. Boussemart, F., Lecoutre, C., Piette, C.: XCSP3: an integrated format for benchmarking combinatorial constrained problems. Technical report. arXiv:1611.03398, CoRR (2016). http://www.xcsp.org

8. Bryant, R.: Graph-based algorithms for Boolean function manipulation. IEEE Trans. Comput. **35**(8), 677–691 (1986)

9. Cappart, Q., Goutierre, E., Bergman, D., Rousseau, L.M.: Improving optimization bounds using machine learning: decision diagrams meet deep reinforcement learning. In: Proceedings of AAAI 2019 (2019)

10. Cheng, K., Yap, R.: An MDD-based generalized arc consistency algorithm for positive and negative table constraints and some global constraints. Constraints **15**(2), 265–304 (2010)

11. Demeulenaere, J., et al.: Compact-table: efficiently filtering table constraints with reversible sparse bit-sets. In: Rueher, M. (ed.) CP 2016. LNCS, vol. 9892, pp. 207–223. Springer, Cham (2016). https://doi.org/10.1007/978-3-319-44953-1_14

12. Dolan, E.D., Moré, J.J.: Benchmarking optimization software with performance profiles. Math. Programm. **91**(2), 201–213 (2002)

13. Gange, G., Stuckey, P., Szymanek, R.: MDD propagators with explanation. Constraints **16**(4), 407–429 (2011)

14. Hadzic, T., Hooker, J.N., O'Sullivan, B., Tiedemann, P.: Approximate compilation of constraints into multivalued decision diagrams. In: Stuckey, P.J. (ed.) CP 2008. LNCS, vol. 5202, pp. 448–462. Springer, Heidelberg (2008). https://doi.org/10. 1007/978-3-540-85958-1_30

15. Hoda, S., van Hoeve, W.-J., Hooker, J.N.: A systematic approach to MDD-based constraint programming. In: Cohen, D. (ed.) CP 2010. LNCS, vol. 6308, pp. 266–280. Springer, Heidelberg (2010). https://doi.org/10.1007/978-3-642-15396-9_23

16. Ingmar, L., Schulte, C.: Making compact-table compact. In: Hooker, J. (ed.) CP 2018. LNCS, vol. 11008, pp. 210–218. Springer, Cham (2018). https://doi.org/10. 1007/978-3-319-98334-9_14

17. Le Charlier, B., Khong, M.T., Lecoutre, C., Deville, Y.: Automatic synthesis of smart table constraints by abstraction of table constraints

18. Lecoutre, C.: STR2: optimized simple tabular reduction for table constraints. Constraints **16**(4), 341–371 (2011)

19. Lecoutre, C., Likitvivatanavong, C., Yap, R.: STR3: a path-optimal filtering algorithm for table constraints. Artif. Intell. **220**, 1–27 (2015)

20. Mairy, J.-B., Deville, Y., Lecoutre, C.: The smart table constraint. In: Michel, L. (ed.) CPAIOR 2015. LNCS, vol. 9075, pp. 271–287. Springer, Cham (2015). https://doi.org/10.1007/978-3-319-18008-3_19

21. OscaR Team: OscaR: Scala in OR (2012). https://bitbucket.org/oscarlib/oscar

22. Perez, G.: Decision diagrams: constraints and algorithms. Ph.D. thesis, Université de Nice (2017)

23. Perez, G., Régin, J.-C.: Improving GAC-4 for table and MDD constraints. In: O'Sullivan, B. (ed.) CP 2014. LNCS, vol. 8656, pp. 606–621. Springer, Cham (2014). https://doi.org/10.1007/978-3-319-10428-7_44

24. Perez, G., Régin, J.C.: Efficient operations ON MDDs for building constraint programming models. In: Twenty-Fourth International Joint Conference on Artificial Intelligence (2015)

25. Pesant, G.: A regular language membership constraint for finite sequences of variables. In: Wallace, M. (ed.) CP 2004. LNCS, vol. 3258, pp. 482–495. Springer, Heidelberg (2004). https://doi.org/10.1007/978-3-540-30201-8_36
26. Roy, P., Perez, G., Régin, J.-C., Papadopoulos, A., Pachet, F., Marchini, M.: Enforcing structure on temporal sequences: the allen constraint. In: Rueher, M. (ed.) CP 2016. LNCS, vol. 9892, pp. 786–801. Springer, Cham (2016). https://doi.org/10.1007/978-3-319-44953-1_49
27. le Clément de Saint-Marcq, V., Schaus, P., Solnon, C., Lecoutre, C.: Sparse-sets for domain implementation. In: Proceeding of TRICS 2013, pp. 1–10 (2013)
28. de Uña, D., Gange, G., Schachte, P., Stuckey, P.J.: Compiling CP subproblems to mdds and d-DNNFs. Constraints **24**(1), 56–93 (2019)
29. Verhaeghe, H., Lecoutre, C., Schaus, P.: Compact-MDD: efficiently filtering (s)MDD constraints with reversible sparse bit-sets. In: IJCAI, pp. 1383–1389 (2018)
30. Verhaeghe, H., Lecoutre, C., Deville, Y., Schaus, P.: Extending compact-table to basic smart tables. In: Beck, J.C. (ed.) CP 2017. LNCS, vol. 10416, pp. 297–307. Springer, Cham (2017). https://doi.org/10.1007/978-3-319-66158-2_19
31. Verhaeghe, H., Lecoutre, C., Schaus, P.: Extending compact-table to negative and short tables. In: Proceedings of AAAI 2017 (2017)
32. Wang, R., Xia, W., Yap, R., Li, Z.: Optimizing Simple Tabular Reduction with a bitwise representation. In: Proceedings of IJCAI 2016, pp. 787–795 (2016)

Arc Consistency Revisited

Ruiwei Wang$^{(\boxtimes)}$ and Roland H. C. Yap

National University of Singapore, Singapore, Singapore
{ruiwei,ryap}@comp.nus.edu.sg

Abstract. Binary constraints are a general representation for constraints and is used in Constraint Satisfaction Problems (CSPs). However, many problems are more easily modelled with non-binary constraints (constraints with arity >2). Several well-known binary encoding methods can be used to transform non-binary CSPs to binary CSPs. Historically, work on constraint satisfaction began with binary CSPs with many algorithms proposed to maintain Arc Consistency (AC) on binary constraints. In more recent times, research has focused on non-binary constraints and efficient Generalized Arc Consistency (GAC) algorithms for non-binary constraints. Existing results and "folklore" suggest that AC algorithms on the binary encoding of a non-binary CSP do not compete with GAC algorithms on the original problem. We propose new algorithms to enforce AC on binary encoded instances. Preliminary experiments show that our AC algorithm on the binary encoded instances is competitive to state-of-the-art GAC algorithms on the original non-binary instances and faster in some instances. This result is surprising and is contrary to the "folklore" on AC versus GAC algorithms. We believe our results can lead to a revival of AC algorithms as binary constraints and resulting algorithms are simpler than the non-binary ones.

Keywords: Binary constraint · Binary encoding · Arc Consistency · Generalized Arc Consistency · CSP

1 Introduction

Binary constraint is a general representation for constraints and is used in Constraint Satisfaction Problems (CSPs) to model/solve any discrete combinatorial problem. Historically, work on constraint satisfaction began with binary CSPs, problems with at most two variables per constraint and many algorithms have been proposed to maintain Arc Consistency (AC) on binary constraints. The seminal work of Mackworth [16] proposed a basic local consistency, arc consistency, which has been the main reasoning technique used in constraint solvers for CSPs. However, many problems are more naturally modelled with non-binary constraints (constraints with arity >2). Several well-known binary encoding methods can be used to transform non-binary CSPs to binary CSPs. Non-binary CSPs can be also solved directly which would require non-binary constraint solvers, Generalized Arc Consistency (GAC) is the natural extension

© Springer Nature Switzerland AG 2019
L.-M. Rousseau and K. Stergiou (Eds.): CPAIOR 2019, LNCS 11494, pp. 599–615, 2019.
https://doi.org/10.1007/978-3-030-19212-9_40

of AC. In more recent times, research has focused on non-binary constraints and efficient Generalized Arc Consistency (GAC) algorithms using clever algorithms, representations and data structures [3,5,6,8–10,12,18,22–27]. Improvements in GAC algorithms have led to a "folklore belief" that AC algorithms on the binary encoding of a non-binary CSP do not compete with GAC.[1] It has also spurred major developments in GAC algorithms.

We first show with experimental comparisons of binary encoding with state-of-art GAC algorithms reasons why binary encoding with existing AC algorithms are outperformed by GAC on non-binary constraints. We propose new algorithms to enforce AC on binary encoded instances which address these factors: (i) a more efficient propagator for hidden variable binary constraints; and (ii) control the interaction between the search heuristic and the binary encoded model. Preliminary experiments show that our AC algorithm can be much faster than state-of-the-art AC algorithms for non-binary CSPs, CT [5] and STRbit [26], on their binary encoded instances. This result is surprising and is contrary to the "folklore" on AC vs GAC algorithms. We believe our results can lead to a revival of AC algorithms since binary constraints and resulting algorithms are simpler than the non-binary ones. For example, many stronger consistencies were proposed to handle binary constraints and these have been more extensively studied in the case of binary constraints. Many fundamental works studying properties of CSPs are often also studied in the binary case.

2   Background

A *CSP* (*Constraint Satisfaction Problem*) $\mathcal{P}$ is a pair $(\mathcal{X}, \mathcal{C})$ with n variables, $\mathcal{X} = \{x_1, x_2, \dots x_n\}$, and m constraints, $\mathcal{C}$ $\{c_1, c_2, \dots c_m\}$. The variable domains are finite, $D(x_i)$ is the domain of $x_i \in \mathcal{X}$. We distinguish the current domain of x_i, $dom(x_i) \subseteq D(x_i)$, the domain may shrink during search when solving the CSP. The variables in each constraint c_i is called the *constraint scope*, $scp(c_i) = \{x_{i_1}, x_{i_2}, \dots x_{i_r}\}$ and r is the (constraint) arity. The constraint is a relation defined over the constraint scope, $rel(c_i) \subseteq \prod_{j=1}^{r} D(x_{i_j})$. In this paper, we only consider non-trivial constraints, hence, $r > 1$. A constraint c is a *binary constraint* iff $r = 2$, i.e $scp(c) = \{x, y\}$, otherwise, c is a *non-binary constraint* iff $r > 2$. A *binary CSP* only has binary constraints; otherwise the CSP is a *non-binary CSP*. An assignment $\mathcal{A} = \{(x_1, a_1), (x_2, a_2), \dots (x_n, a_n)\}$ satisfies a constraint c iff $\mathcal{A}[scp(c)] \in rel(c)$ where the notation $[v]$ denotes projection on the set of variables v. Then $\mathcal{A}$ is a solution satisfying $(\mathcal{X}, \mathcal{C})$ iff $\mathcal{A}$ satisfies all constraints in $\mathcal{C}$ and $\mathcal{A} \in \prod_{i=1}^{n} dom(x_i)$. Following [19], we say a CSP $\mathcal{P}_1$ is equivalent to $\mathcal{P}$ if they are mutually reducible. A CSP $\mathcal{P}$ is *reducible* to another CSP $\mathcal{P}_1$ if the solution of $\mathcal{P}$ can be obtained from the solution of $\mathcal{P}_1$, by mapping the variable values in one CSP to variable values in the other.

---

[1] We focus on AC and GAC algorithms for the general finite domain CSPs with table constraints. Global constraints with special semantics and special GAC algorithms exploiting the semantics are outside our scope.

A tuple $\tau \in rel(c)$ is *valid* iff $\tau[x] \in dom(x)$ for all $x \in scp(c)$. We say (x, a) is a *support* of tuple $\tau \in rel(c)$ iff $\tau[x] = a$. A variable value (x, a) is *generalized arc consistent* (GAC) on c iff (x, a) has a valid support in $rel(c)$. A variable x is GAC on c iff for all value $a \in dom(x)$, (x, a) is GAC. A constraint c is GAC iff all variables in $scp(c)$ is GAC on c. A CSP $(\mathcal{X}, \mathcal{C})$ is GAC iff all constraints in $\mathcal{C}$ is GAC. A binary CSP $\mathcal{P}$ is *arc consistent* (AC) iff $\mathcal{P}$ is GAC, i.e. arc consistency is a special case of GAC. For a binary constraint c, arc consistency uses a simpler definition of support: a value $a \in dom(x)$ has a valid support in rel(c) iff (x, a) has a valid support b in $dom(y)$ such that $\{(x, a), (y, b)\} \in rel(c)$, where $scp(c) = \{x, y\}$. M(G)AC is used to denote maintaining (G)AC during search. In this paper, we focus on MGAC and MAC and simply say GAC or AC.

2.1 Binary Encodings

A non-binary CSP $\mathcal{P}_1 = (\mathcal{X}_1, \mathcal{C}_1)$ can be solved through transformation by encoding into an "equivalent" binary CSP $\mathcal{P}_2 = (\mathcal{X}_2, \mathcal{C}_2)$ such that $\mathcal{P}_2$ is reducible to $\mathcal{P}_1$. This means there are two options to solving a non-binary CSP P_1: (i) directly solving P_1; or (ii) indirectly by solving P_2. There are two well known binary encodings, namely, the *dual encoding* [4] and the *hidden variable encoding* (HVE) [19]:

- the dual encoding of P_1 is a binary CSP $(\mathcal{H}, \mathcal{DC})$
- the HVE of P_1 is a binary CSP $(\mathcal{X}_1 \cup \mathcal{H}, \mathcal{HC})$

with new variables $\mathcal{H} = \{hv_i | c_i \in \mathcal{C}_1\}$ where the domain of hv_i is the tuples of the c_i itself, $D(hv_i) = rel(c_i)$. Variables $\mathcal{H}$ are called *hidden variables* and also sometimes called *dual variables* [20,21]. In the dual encoding, the new constraints are $\mathcal{DC} = \{c_{ij} | s = scp(c_i) \cap scp(c_j) \neq \emptyset\}$, i.e. $scp(c_{ij}) = \{hv_i, hv_j\}$, and $rel(c_{ij}) = \{(\tau_1 \in rel(c_i), \tau_2 \in rel(c_j)) \mid \tau_1[s] = \tau_2[s]\}$. The hidden variable encoding has constraints $\mathcal{HC} = \{c_i^x | x \in scp(c_i)\}$, one new constraint per variable in c_i, i.e. $scp(c_i^x) = \{x, hv_i\}$, and $rel(c_i^x) = \{(a \in D(x), \tau \in rel(c_i)) | \tau[x] = a\}$.

Example 1. *Consider a CSP P $(\mathcal{X}, \mathcal{C})$, where $\mathcal{X} = \{x_1, \ldots, x_4\}$, $D(x_i) = \{0, 1\}$ and $\mathcal{C} = \{c_1 : x_1 + x_2 + x_3 = 1, c_2 : x_2 + x_3 + x_4 < 2, c_3 : x_1 + x_2 + x_4 < 2, c_4 : x_1 + x_3 + x_4 = 1\}$. Figure 1(a) gives the HVE CSP instance of P, and Fig. 1(b) is the dual encoding instance. Every node in the figure is a variable, each edge corresponds to a binary constraint, and the label of the edge denotes the relation of the binary constraint. E.g. $D(hv_1) = D(hv_4) = \{1', 2', 4'\}$ and $D(hv_2) = D(hv_3) = \{0', 1', 2', 4'\}$, where the values $0'$, $1'$, $2'$ and $4'$ represent $(0, 0, 0), (0, 0, 1), (0, 1, 0)$ and $(1, 0, 0)$ respectively (the figure uses the tuples notation). The constraint $r_1 = \{(0, 1'), (0, 2'), (1, 4')\}$ is the relation in the HVE constraints with scope $\{x_1, hv_1\}$ and $\{x_1, hv_3\}$ in the HVE while $r_{13} = \{(1', 2'), (2', 4'), (4', 0'), (4', 1')\}$ is the relation in the dual encoding with scope $\{hv_1, hv_2\}$.*

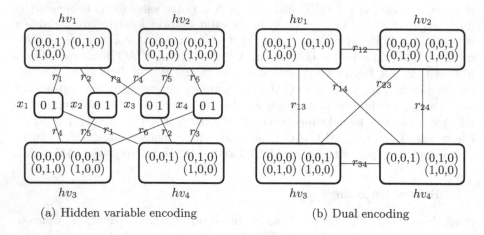

(a) Hidden variable encoding   (b) Dual encoding

Fig. 1. Binary encodings

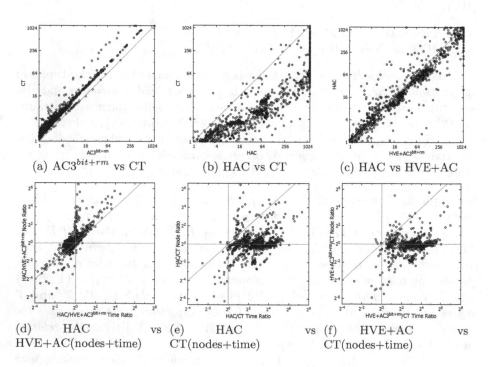

(a) $AC3^{bit+rm}$ vs CT   (b) HAC vs CT   (c) HAC vs HVE+AC

(d)   HAC   vs (e)   HAC   vs (f)   HVE+AC   vs
HVE+AC(nodes+time)   CT(nodes+time)   CT(nodes+time)

Fig. 2. AC vs GAC algorithm: (a)–(c) use time on the axis while (d)–(f) use time and node ratios

Table 1. Dual encoding memory size results: m/n means m instances were memory-out of n total, x% are fraction of original constraints

Series	100%	80%	60%	40%	20%
Rand-5-12	50/50	50/50	50/50	50/50	50/50
Rand-15-23	25/25	25/25	25/25	25/25	25/25
Rand-10-60	50/50	50/50	50/50	50/50	50/50
Nonogram	88/170	78/170	71/170	53/170	40/170
Tsp	15/45	15/45	15/45	15/45	15/45
Jnh	50/50	40/50	25/50	2/50	0/50
Mdd-7-25	25/25	25/25	25/25	25/25	25/25
Rand-3-20	30/50	0/50	0/50	0/50	0/50

3   History and the Problem

In this paper, we revisit the question whether non-binary CSPs are better solved directly using a non-binary solver or the non-binary CSP is encoded to a new binary CSP and solved by a binary constraint solver. We focus on the comparison between binary constraints and table constraints which are the most general representation for constraints in CSPs. We start with a chronology of binary encodings and corresponding consistency algorithms (if any). In 1998, [1] showed on some instances, forward checking (FC) with backtracking search on binary encoded instances can be faster than solving the non-binary instances directly. In 1999, experimental results in [21] showed that enforcing AC on the dual encoding instances is very expensive. In 2001, [17] proposed HAC showing that MHAC is competitive with M(G)AC, and M(G)AC algorithms can be mapped to the corresponding AC algorithms on binary encoded instances. In 2005, [20] showed that binary encodings are competitive with the non-binary representation. It also showed that a higher order consistency PW-AC can work well on binary encoding instances. MHAC-2001 and PW-AC can be faster than MGAC-2001 in some cases. However, they only tested some special cases for the dual encoding. In 2011, [10] showed that the dual and double encodings run out of memory on many instances, and STR2+ can outperform HAC and HVE+AC3^{bit+rm}.[2]

During the past decades, many AC and table GAC algorithms were proposed. AC algorithms check whether a variable value has a valid support on another variable domain, and many methods are proposed to reduce the cost of AC consistency algorithm during search. Over the period from 2007–2018, there has been considerable research efforts expanded on GAC algorithms, but little work on AC algorithms given the shift to GAC algorithms. Many of the GAC algorithms use special ideas to make GAC more efficient. An incomplete list is as follows: reducing the size of tables during search, e.g. algorithms using simple table reduction during search [10,12,22], algorithms using decision diagrams

---

[2] (HVE+AC3^{bit+rm} means using the AC3^{bit+rm} on the HVE binary instances).

[3,18,25], algorithms using compressed tables [6,8,9,27], and bit set representations [5,23,24,26]. Some of the state-of-the-art GAC algorithms are CT [5] and STRbit [26] which combine bit set with simple table reduction can be much faster than STR2+ [10].

In order to understand and revisit the results from existing papers such as the ones above, we test various kinds of instances to compare different existing algorithms. Experimental details are given in Sect. 5 to avoid repetition. The main drawback of dual encoding is the large constraints which lead to the solver running out of memory, which we call *memory-out*. This was also shown in [10]. We also tested with various instances, Table 1, where the 100% column is the original instance and a random set of constraints removed creating a smaller CSP. Many instances are simply memory-out, e.g. with jnh instances at 60% (40% constraint removal) still 25 instances were memory-out. In this paper, we focus on the hidden variable encoding as the dual encoding starts to become infeasible as the constraints become larger.

We revisit GAC vs binary encodings for hidden variables with results in Fig. 2. Each dot in the graphs is a problem instance. For AC, we employ $AC3^{bit}$ [15] with residues [11] denoted as $AC3^{bit+rm}$. It has been shown to be efficient in practice for AC because of the bit representation [13]. Figure 2(a) compares time (in seconds) of AC^{bit+rm} with CT on binary CSP instances. It shows that $AC3^{bit+rm}$ can be faster than a recent state-of-the-art GAC algorithm, CT, on binary CSP instances. While it might not be surprising that a binary AC algorithm is faster than a non-binary algorithm, it highlights the special nature of binary constraints. The special case of binary constraints is simpler than the non-binary case which typically has more complex algorithms and data structures. For example, the $AC3^{bit+rm}$ algorithm in Fig. 2(a) is much simpler than the CT algorithm and uses simpler data structures.

We now compare different ways of solving non-binary CSP instances. We first compare solving non-binary CSP instances with the HAC algorithm on the hidden variable encoding of the non-binary instances with the CT algorithm on the original non-binary instances. Figure 2(b) compares the time of HAC[3] with CT and shows that HAC is much slower than CT on non-binary CSP instances. This result is the opposite of the binary-only instances in Fig. 2(a) and suggests what is known in the folklore that encoding a non-binary constraint to binary form to be solved using a binary solver is much slower than using a (modern) GAC solver on the non-binary constraints directly. The HVE encoding with AC (using $AC3^{bit+rm}$) is also slow, shown in the comparison of time between HAC and HVE+AC in Fig. 2(c). Figure 2(d) compares time versus search nodes of HAC with HVE+AC with the y-axis giving the number of search nodes of HAC/(HVE+AC) and x-axis giving the time of HAC/(HVE+AC). It shows that the special propagator of HAC for HVE is more efficient than $AC3^{bit+rm}$ on HVE.

---

[3] In this paper, we use HAC to implicitly refer to HVE+HAC.

Figure 2(e) and (f) reveals important factors behind why the performance of HVE encoding + AC algorithm is worse than GAC (with CT) on the non-binary instances:

(i) The concentration of points around the x-axis to the right of the y-axis shows that for a similar number of search nodes, the search time is significantly slower than CT (yet Fig. 2(a) shows $AC3^{bit+rm}$ is faster than CT for binary constraints which do not come from the hidden variable encoding). This suggests that the CT propagator is more efficient than the AC propagators on HVE instances;

(ii) There are many differences in the search nodes for the binary encoding versus the original instance. Many instances have more search nodes in the binary encoding as shown by the density of points above the x-axis.

As many points are far to the right, we see that the propagator efficiency may be the factor for the superiority of CT though the difference in search nodes is also a factor.

Search heuristics and consistency propagators are the main components in a constraint solver. The results above show that with CT (and also other modern GAC propagators)[4] both the efficiency of the constraint propagator and effectiveness are reasons for the folklore superiority of GAC on the non-binary instance over AC on the encoded instance. Furthermore, we have seen that the binary constraints in HVE instances are very special. In this paper, we focus on improving two problems identified for binary encoding instances:

1. Designing a special AC propagator which is more efficient for the binary constraints in HVE instances.
2. Avoiding making the search heuristic on HVE instances worse than on the original instances.

4   The Hidable Model Transformation Propagator

We saw that in Sect. 3, results illustrating the folklore suggesting it is better to solve a non-binary CSP directly using GAC rather than with binary encoding. The goal in this paper is to dispel this folklore notion. We also saw that binary encoding can interact poorly with the search heuristic and that the binary constraints from the hidden variable encoding are special.

To deal with the search heuristic problem, we "virtualize" the binary encoding so that the interaction between the binary encoded constraints can be hidden from the search heuristic making it behave like the search heuristic for the original non-binary constraint. This allows us to investigate the search heuristic which behaves like in the non-binary instance but also have an alternative where the search heuristic works on the HVE instance. We incorporate ideas from modern GAC algorithms to get a more efficient AC propagator for the special kinds of binary constraints in HVE instances.

---

[4] Experimental results for STR2+ are also similar to CT and have not been shown.

4.1   The Hidable Model Transformation

The hidden variable encoding is a special transformation of a CSP P_1 to a P_2 where P_2 is reducible to P_1. We generalize this idea to allow different kinds of transformations and search strategies.

Definition 1. $(\mathcal{X}_1 \cup \mathcal{X}_2, \mathcal{C}_2)$ *is a GAC hidable transformation of* $(\mathcal{X}_1, \mathcal{C}_1)$ *iff* $\mathcal{C}_2$ *has a partition* $\{s_1, ..., s_m\}$ *such that for each* $c_i \in \mathcal{C}_1$, $scp(c_i) \subseteq scp(s_i)$ *and* c_i *is GAC iff all constraints in* s_i *are GAC, where* $scp(s_i) = \bigcup_{c \in s_i} scp(c)$.

Corollary 1. *If* $(\mathcal{X}_1 \cup \mathcal{X}_2, \mathcal{C}_2)$ *is a GAC hidable transformation of* $(\mathcal{X}_1, \mathcal{C}_1)$, *then* $(\mathcal{X}_1, \mathcal{C}_1)$ *is GAC iff* $(\mathcal{X}_1 \cup \mathcal{X}_2, \mathcal{C}_2)$ *is GAC.*

If $P_2 = (\mathcal{X}_1 \cup \mathcal{X}_2, \mathcal{C}_2)$ is an GAC hidable transformation of P_1, then the solver does not need to search the variables in $\mathcal{X}_2$, since GAC on P_2 can check whether an assignment on $\mathcal{X}_1$ is a solution of P_1. As such, the search algorithm only needs to consider $\mathcal{X}_1$ where the GAC propagators on P_2 are a "black box". In this paper, we only consider binary encodings, i.e. P_2 is a binary CSP.

Corollary 2. *HVE is a GAC hidable model transformation.*

Proof. For a non-binary constraint c_i, we can set $s_i = \{c_i^x \in \mathcal{HC} | x \in scp(c_i)\}$, since $(scp(s_i), s_i)$ is the HVE of $(scp(c_i), \{c_i\})$. The HVE encoding by construction of c_i^x already meets the requirements of Definition 1. The HVE transformation is only on non-binary constraints. For a binary constraint c_i, we set $s_i = \{c_i\}$, also meeting the definition.

4.2   A Propagator for the Hidable Model Transformation

Algorithm 1 gives the *HTAC* algorithm to enforce AC on hidable binary encoding instances. HTAC adds a variable $x \in \mathcal{X}_1 \cup \mathcal{X}_2$ to the propagation queue $\mathcal{Q}$ if x may be used to update the domains of other variables. Then HTAC iteratively picks a variable x from $\mathcal{Q}$, and then use AC algorithms to enforces AC on all constraints in every subset $s_i \in \mathcal{S}$ such that $x \in scp(s_i)$. For different s_i, we can use different AC algorithms. If c_i is a binary constraint in $\mathcal{C}_1$ and $s_i = \{c_i\}$, then HTAC can use any efficient AC algorithms, e.g. AC3^{bit+rm}, to enforce AC on c_i. For a GAC hidable model transformation, we give special AC-H algorithms exploiting the nature of constraints used in the transformation. In Sect. 4.3, we present a AC-H algorithm to handle the constraints used in HVE transformation. HTAC is different from HAC [20]: HTAC adds original variables to $\mathcal{Q}$ while HAC only adds hidden variables to $\mathcal{Q}$. When the domain of a variable x is changed, HAC directly updates the domains of all hidden variables constrained by x and does not add x to $\mathcal{Q}$. HTAC uses a reversible bit set to represent the domain of a hidden variable (see Sect. 4.3), but HAC does not; The revise functions used in AC-H are also different from HAC.

For a GAC hidable model transformation $(\mathcal{X}_1 \cup \mathcal{X}_2, \mathcal{C}_2)$, the solver only needs to search the variables in $\mathcal{X}_1$. Search heuristics which use information from the

Algorithm 1. HTAC $(\mathcal{X}_1, \mathcal{C}_1)$

let $(\mathcal{X}_1 \cup \mathcal{X}_2, \mathcal{C}_2)$ be the hidable model transformation of $(\mathcal{X}_1, \mathcal{C}_1)$;
let $\mathcal{S}$ be a partition of $\mathcal{C}_2$ making $(\mathcal{X}_1 \cup \mathcal{X}_2, \mathcal{C}_2)$ hidable;
$\mathcal{Q} \leftarrow \mathcal{X}_1$;
while $\mathcal{Q} \neq \emptyset$ **do**
    pick and delete x from $\mathcal{Q}$;
    for $s_i \in \mathcal{S}$ s.t. $x \in scp(s_i)$ **do**
        if $|s_i| = 1$ **then**
            // $s_i = \{c_i\}$
            if $\neg AC(s_i)$ **then**
                return false;
        else if $\neg AC\text{-}H(s_i)$ **then**
            return false;
return true;

constraint structure can choose to use the structure of the GAC hidable model transformation or the original model. For example, the wdeg/dom [2] heuristic records a weight w for each constraint, and increasing w by one if the constraint causes inconsistency. Variables are selected based on the weights of constraints. For the HVE transformation, we propose two alternatives for wdeg/dom:

A. using wdeg/dom with the original model, we record a weight w_i for each $s_i \in \mathcal{S}$. Thus, w_i is regarded as a weight for a virtual constraint representing the weight of $c_i \in \mathcal{C}_1$. Weight w_i is incremented if AC(-H)(s_i) is not consistent;
B. using wdeg/dom with the HVE transformation, we record a weight w_i^x for each $c_i^x \in \mathcal{C}_2$ and w_i^x is incremented if AC(-H)(s_i) is not consistent, where x is picked from $\mathcal{Q}$ and $x \in scp(s_i)$.

We call HTAC as *HTAC1* if the heuristics use the original non-binary model (A); and *HTAC2* if the heuristics use the transformation model (B).

4.3   The AC-H Algorithm for HVE

We first introduce the data structures used in the algorithm which incorporates data structures used by (modern) GAC and AC algorithms [5,10,14,15]:

1. For a original variable x:
   - $dom(x)$ uses an "ordered link" data structure proposed in [14] to represent the current domain of x.
   - $bitDom(x)$ uses a bit set to represent the domain of a variable x [15].
   - $bitSup(c, x, a)$ is used to represent all supports of variable value (x, a) in $bitDom(y)$, where $scp(c) = \{x, y\}$ [15].
   - $lastSize(c, x)$ is used to record the size of $dom(x)$ after the last update on the domain of x based on c [10].

2. For a hidden variable x:
   - $bitDom(x)$ uses a bit set to represent the domain of x, if $bitDom(x)$ is changed, the old states of $bitDom$ are recorded in a stack so that it can be undone on backtracking.
   - $wordDom(x)$ is a sparse set used to record the non-ZERO words in $bitDom(x)$. It uses the reversible sparse bit-set idea in [5].
   - $prevDom(x)$ is a copy of $bitDom(x)$, we use $prevDom(x)$ to check whether $bitDom(x)$ is changed.
   - $buf0$ is a bit set, where all words in $buf0$ are initialized as ZERO (ZERO is the bit set with all zeroes).

Algorithm 2 is to enforce AC on a set (s_i) of constraints for HVE instances, where $s_i = \{c_i^x \in \mathcal{HC} | hv_i \in scp(c_i^x)\}$. The HVE transformation has a star structure constraint graph (a special tree). This allows the AC-H algorithm to update the domains of variables in two passes: (i) from leaves ($x \in c_i$) to the root (hv_i), and (ii) from the root to the leaves. The first phase of revise is with function $revise2$ to (partially) update the domain of hv_i based on the current domains of variables in $scp(c_i)$. Then function $update$ updates $wordDom$ representation of hv_i. If $wordDom(hv_i) = \emptyset$, the instance is not AC. The second revise phase uses function $revise1$ to update the domains of all variables in $scp(c_i)$. If the domain of a variable x is changed, x is added to Q. We do not add hv_i to Q, since the domains of variables constrained with hv_i are updated in the AC-H algorithm.

Algorithm 2. AC-H (s_i)

---

let hv_i be the hidden variable constrained with binary constraints in s_i;
for each $c_i^x \in s_i$ **do**
   | revise2(c_i^x, hv_i);
if $\neg update(hv_i)$ **then**
   | **return** false;
for each $c_i^x \in s_i$ **do**
   | **if** $revise1(c_i^x, x)$ **then**
   | | $Q \leftarrow Q \cup \{x\}$;

return true;

Due to lack of space, we briefly sketch correctness of AC-H and associated functions. The overall structure is similar to any AC algorithm using revise except that we exploit the star constraint graph as explained above. The current domain of hv_i is only updated by the function $revise2$. Meanwhile, the function $revise2$ deletes the values which do not have valid supports in the current domains of the variables in $scp(c_i)$ from the current domain of hv_i. If the function $update$ return false, all words in $bitDom(hv_i)$ become ZERO, which means that it is not AC. Finally. the function $revise1$ is similar to AC3^{bit+rm}.

4.4   Revise Functions

In AC algorithms, the $revise(c, x)$ functions are used to update $dom(x)$ based on $dom(y)$, where $\{x, y\} = scp(c)$, i.e removing the values in $dom(x)$ which don't have valid supports in $dom(y)$. We give two revise functions used in AC-H to handle the reversible bit-set domains. The function $revise1$ updates original variable domains using function $seekSupport$ which is similar to that in AC3^{bit+rm} algorithm. The difference is that our $seekSupport$ only check the words in $wordDom$. For hidden variables, we do not use the $seekSupport$ function, since each value in hidden variable domains (by construction) only has one support in $rel(c)$, i.e. an hidden variable hv functionally determines the values in the domains of original variables constrained with hv (see [1]), which make $bitSup$ useless. Using the ideas from GAC algorithms CT [5] and STRbit [26], function $revise2$ applies the $delete$ and $reset$ functions to update $bitDom$. If the number of values deleted from $dom(x)$ is larger than $|dom(x)|$, then function $delete$ is used to remove all supports of the deleted values from $bitDom(hv)$. Otherwise, function $reset$ is used to build a new $bitDom(hv)$ based on current $dom(x)$. After updating $bitDom(hv)$ of a hidden variable hv, the function $update$ is used to check $bitDom(hv)$ for domain wipeout, for each word w in $bitDom(hv)$, if w is changed, the old value of $prcDom(w)$ in saved on a stack for backtracking, and if $w = ZERO$ it is removed from $wordDom$.

Function $revise1(c, x)$

$size \leftarrow |dom(x)|;$
for each $a \in dom(x)$ **do**
$\quad$ **if** $\neg seekSupport(c, x, a)$ **then**
$\quad\quad$ remove a from $dom(x);$

return $|dom(x)| \neq size;$

Function $seekSupport(c, x, a)$

Let y be the variable such that $scp(c) = \{x, y\};$
$w \leftarrow rm[c, x, a];$
if $(bitSup[c, x, a][w] \& bitDom[y, w]) \neq ZERO$ **then**
$\quad$ **return** true;

for each $w \in wordDom(y)$ **do**
$\quad$ **if** $(bitSup[c, x, a][w] \& bitDom[y, w]) \neq ZERO$ **then**
$\quad\quad$ $rm[c, x, a] \leftarrow w;$
$\quad\quad$ **return** true;

return false;

Function $revise2(c, hv)$

Let x be the variable such that $scp(c) = \{x, hv\}$;
$dn \leftarrow lastSize(c, x) - |dom(x)|$;
if $dn > |dom(x)|$ **then**
$\quad \rfloor\ reset(c, hv)$;
else if $dn > 0$ **then**
$\quad |\ delete(c, hv, dn)$;

Function $delete(c, hv, dn)$

Let x be the variable such that $scp(c) = \{x, hv\}$;
for $i = 0$ *to* dn **do**
$\quad$ Let a be the last i value deleted from $dom(x)$;
$\quad$ **for each** $w \in wordDom(hv)$ **do**
$\quad\quad \rfloor\ bitDom[hv, w] \leftarrow bitDom[hv, w]\&\neg bitSup[c, x, a]$;

Function $reset(c, hv)$

Let x be the variable such that $scp(c) = \{x, hv\}$;
for each $a \in dom(x)$ **do**
$\quad$ **for each** $w \in wordDom(hv)$ **do**
$\quad\quad \rfloor\ buf0[w] \leftarrow buf0[w]|bitSup[c, x, a]$;

for each $w \in wordDom(hv)$ **do**
$\quad bitDom[x, w] \leftarrow buf0[w]\&bitDom[hv, w]$;
$\quad buf0[w] \leftarrow ZERO$;

Function $update(x)$

for each $w \in wordDom(x)$ **do**
$\quad$ **if** $bitDom[x, w] \neq prevDom[x, w]$ **then**
$\quad\quad$ save $prevDom[x, w]$ in a stack;
$\quad\quad prevDom[x, w] \leftarrow bitDom[x, w]$;
$\quad\quad$ **if** $bitDom[x, w] = ZERO$ **then**
$\quad\quad\quad \rfloor\ $ remove w from $wordDom(x)$;

return $wordDom(x) \neq \emptyset$;

---

5   Experiments

We present experiments to evaluate HTAC on the hidden variable encoding (HVE+HTAC1 and HVE+HTAC2) compared with HVE+AC3^{bit+rm}, HAC, STR2+, CT and STRbit algorithms. HTAC, HVE+AC3^{bit+rm} and HAC maintain AC (MAC) on the hidden variable encoding instances. STR2+, CT and STRbit maintain GAC (MGAC) on the original non-binary instances. CT and

STRbit are the state-of-the-art GAC algorithms, and HAC is the best existing algorithm for binary encoding from Sect. 3. We used binary and non-binary instances from the XCSP3 website (http://xcsp.org). Instances from XCSP3 which timeout for all compared algorithms are ignored. In total, we evaluated 1328 binary and 2431 non-binary problem instances. The binary CSP series are:

QCP, QWH, Geometric, Rlfap, Driver, Lard, Queens, RoomMate, Prop-Stress, QueensKnights, KnightTour, Random

The binary instances are discussed in Sect. 3 in Fig. 2(a). The non-binary CSP series are:

Kakuro, Dubois, PigeonsPlus, MaxCSP,Renault, Aim, Jnh, Cril, Tsp, Various, Nonogram, Bdd-{15,18}-21, mdd-7-25-{5,5-p7,5-p7}, reg2ext,Rand-3-24-24, Rand-3-24-24f, Rand-5-12-12, Rand-5-{2,4,8}X, Rand-10-20-10, Rand-10-20-60, Rand-15-23-3, Rand-5-10-10, Rand-5-12-12t, Rand-7-40-8t, Rand-8-20-5, Rand-3-20-20, Rand-3-20-20f

This section focuses on the non-binary instances. Results on the non-binary series are also given in Sect. 3 comparing HAC, HVE+AC3^{bit+rm} and CT. The experiments were run on a 3.20 GHz Intel i7-8700 machine. We implemented HTAC in the Abscon solver[5] which has the other algorithms implemented. In addition, we optimized the Abscon CT and HAC implementation to be a little faster.[6] The variable search heuristic used is $wdeg/dom$ and the value heuristic used is lexical value order. The $wdeg/dom$ with restart is considered one of state-of-the-art heuristics in classic constraint solvers [7]. The restart policy was geometric restart (the initial $cutoff = 10$ and $\rho = 1.1$)[7]. CPU time is limited to 1200 s per instance and memory to 8 GB.

Figures 3 shows 9 scatter plots to compare HVE+HTAC with other algorithms. Each dot in the plots is an instance. Figures 3(a) to (f) compare the time[8] of different algorithms. Meanwhile, Figs. 3(g), (h) and (i) compare the time ratio and node ratio, where the time (node) ratio of A/B means the ratio "the time (number of search nodes) of algorithm A" to "the time (number of search nodes) of algorithm B". Figures 3(a), (b) and (c) show that HVE+HTAC2 and HVE+HTAC1 can outperform the other binary algorithms. From Fig. 3(g), giving the time ratio and node ratio of (HTAC2+HVE)/HAC (see the discussion of ratio graphs in Sect. 3), we see that HTAC is generally much faster than HAC, since most points around the x-axis are at the left of the y-axis. For most instances, the search nodes of HTAC2 is less than HAC.

---

[5] https://www.cril.univ-artois.fr/~lecoutre/#/softwares.

[6] While implementing HTAC, we found some optimizations for the existing CT and HAC code.

[7] $cutoff$ is the allowed number of failed assignments for each restart. After restart, $cutoff$ increases by ($cutoff \times \rho$).

[8] For binary encoding instances, the time includes solving time and model transformation time.

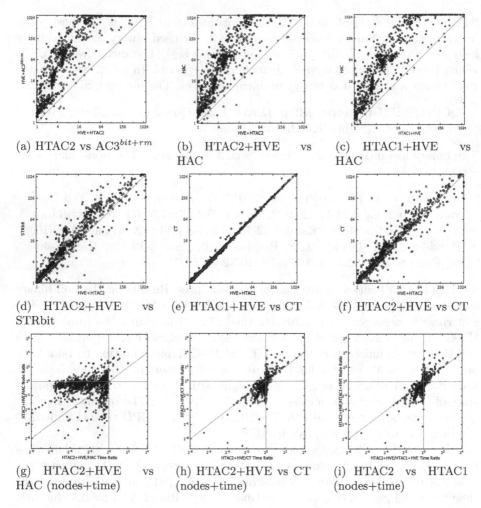

(a) HTAC2 vs AC3^{bit+rm}

(b) HTAC2+HVE vs HAC

(c) HTAC1+HVE vs HAC

(d) HTAC2+HVE vs STRbit

(e) HTAC1+HVE vs CT

(f) HTAC2+HVE vs CT

(g) HTAC2+HVE vs HAC (nodes+time)

(h) HTAC2+HVE vs CT (nodes+time)

(i) HTAC2 vs HTAC1 (nodes+time)

Fig. 3. HTAC vs other algorithms: (a)–(f) use time on the axis while (g)–(i) use node and time ratios in the axis

Figures 3(d), (e) and (f) show HTAC is competitive with the state-of-the-art GAC algorithms CT and STRbit. HTAC1+HVE using wdeg/dom on the original model is competitive with CT, being faster than CT on some instances. HTAC2+HVE using wdeg/dom on the HVE transformed model is faster than STRbit and CT on many instances. Figure 3(h) combines node ratio and time ratio to show the runtime and search nodes tradeoffs of HTAC2 with CT with more instances having less nodes and time. Figure 3(i) compares HTAC2 with HTAC1, the performance of HTAC1 is similar to CT.

Figure 4 shows the runtime distribution of the problem instances solved by the different algorithms. The y-axis is the solving time (in seconds) and the x-axis is the number of instances solved within the time limit. It shows firstly, the folklore that binary encodings are slower with HVE+AC3^{bit+rm} and HAC

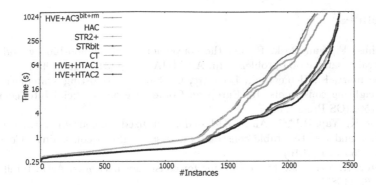

Fig. 4. Runtime Distribution: HTAC, HAC, HVE+AC3^{bit+rm}, CT, STRbit, STR2+

being behind STR2+ (also show in [10]). There is a separation between STR2+ and newer GAC algorithms (STRbit and CT). The performance of our HTAC algorithms is competitive with STRbit and CT, in particular, HTAC2 is faster on some instances.

6 Conclusion

We first show experimental results which can explain the folklore that it is better to solve a non-binary CSP instance directly with GAC than by a binary encoding of the instance and using AC. We show that this folklore is misleading, solving with the binary encoding can be improved by having a more efficient propagator for binary constraints from the HVE instances and preventing poor interaction of the HVE model with the search heuristic.

We propose a new propagator HTAC. By using the GAC hidable binary encoding with HTAC, we can address the differences in search nodes so that the search space on the binary instance behaves as in the non-binary instance but it also allows search on the binary encoded model. The HTAC propagator gains efficiency by using properties of binary constraints in the HVE. It is also efficient as we apply ideas from modern GAC algorithms. Experiments show that HTAC on the binary encoded instance is competitive with state-of-the-art GAC algorithms on the original instances, in some cases, HTAC is faster. Not only have we shown that solving with the binary encoding is viable and competitive, we believe that it opens new directions for modelling and solver algorithms while still retaining the original non-binary instance. Binary instances and constraints are special being simpler so the algorithms can also be simpler. Many transformations and higher consistencies can be applied directly to binary instances.

Acknowledgements. We acknowledge the support of grant MOE2015-T2-1-117.

References

1. Bacchus, F., Van Beek, P.: On the conversion between non-binary and binary constraint satisfaction problems. In: AAAI/IAAI, pp. 310–318 (1998)
2. Boussemart, F., Hemery, F., Lecoutre, C., Sais, L.: Boosting systematic search by weighting constraints. In: European Conference on Artificial Intelligence, pp. 146–150. IOS Press (2004)
3. Cheng, K., Yap, R.H.C.: An MDD-based generalized arc consistency algorithm for positive and negative table constraints and some global constraints. Constraints **15**(2), 265–304 (2010)
4. Dechter, R., Pearl, J.: Tree clustering for constraint networks. Artif. Intell. **38**(3), 353–366 (1989)
5. Demeulenaere, J., et al.: Compact-table: efficiently filtering table constraints with reversible sparse bit-sets. In: Rueher, M. (ed.) CP 2016. LNCS, vol. 9892, pp. 207–223. Springer, Cham (2016). https://doi.org/10.1007/978-3-319-44953-1_14
6. Gharbi, N., Hemery, F., Lecoutre, C., Roussel, O.: Sliced table constraints: combining compression and tabular reduction. In: Simonis, H. (ed.) CPAIOR 2014. LNCS, vol. 8451, pp. 120–135. Springer, Cham (2014). https://doi.org/10.1007/978-3-319-07046-9_9
7. Hebrard, E., Siala, M.: Explanation-based weighted degree. In: Salvagnin, D., Lombardi, M. (eds.) CPAIOR 2017. LNCS, vol. 10335, pp. 167–175. Springer, Cham (2017). https://doi.org/10.1007/978-3-319-59776-8_13
8. Jefferson, C., Nightingale, P.: Extending simple tabular reduction with short supports. In: Proceedings of the 23rd International Joint Conferences on Artificial Intelligence, pp. 573–579 (2013)
9. Katsirelos, G., Walsh, T.: A compression algorithm for large arity extensional constraints. In: Bessière, C. (ed.) CP 2007. LNCS, vol. 4741, pp. 379–393. Springer, Heidelberg (2007). https://doi.org/10.1007/978-3-540-74970-7_28
10. Lecoutre, C.: STR2: optimized simple tabular reduction for table constraints. Constraints **16**(4), 341–371 (2011)
11. Lecoutre, C., Hemery, F., et al.: A study of residual supports in arc consistency. In: IJCAI, vol. 7, pp. 125–130 (2007)
12. Lecoutre, C., Likitvivatanavong, C., Yap, R.H.C.: A path-optimal GAC algorithm for table constraints. In: Proceedings of the 20th European Conference on Artificial Intelligence, pp. 510–515 (2012)
13. Lecoutre, C., Likitvivatanavong, C., Yap, R.H.C.: STR3: a path-optimal filtering algorithm for table constraints. Artif. Intell. **220**, 1–27 (2015)
14. Lecoutre, C., Szymanek, R.: Generalized arc consistency for positive table constraints. In: Benhamou, F. (ed.) CP 2006. LNCS, vol. 4204, pp. 284–298. Springer, Heidelberg (2006). https://doi.org/10.1007/11889205_22
15. Lecoutre, C., Vion, J.: Enforcing arc consistency using bitwise operations. Constraint Programm. Lett. **2**, 21–35 (2008)
16. Mackworth, A.K.: Consistency in networks of relations. Artif. Intell. **8**(1), 99–118 (1977)
17. Mamoulis, N., Stergiou, K.: Solving non-binary CSPs using the hidden variable encoding. In: Walsh, T. (ed.) CP 2001. LNCS, vol. 2239, pp. 168–182. Springer, Heidelberg (2001). https://doi.org/10.1007/3-540-45578-7_12
18. Perez, G., Régin, J.-C.: Improving GAC-4 for table and MDD constraints. In: O'Sullivan, B. (ed.) CP 2014. LNCS, vol. 8656, pp. 606–621. Springer, Cham (2014). https://doi.org/10.1007/978-3-319-10428-7_44

19. Rossi, F., Petrie, C.J., Dhar, V.: On the equivalence of constraint satisfaction problems. ECAI **90**, 550–556 (1990)
20. Samaras, N., Stergiou, K.: Binary encodings of non-binary constraint satisfaction problems: algorithms and experimental results. J. Artif. Intell. Res. **24**, 641–684 (2005)
21. Stergiou, K., Walsh, T.: Encodings of non-binary constraint satisfaction problems. In: AAAI/IAAI, pp. 163–168 (1999)
22. Ullmann, J.R.: Partition search for non-binary constraint satisfaction. Inf. Sci. **177**, 3639–3678 (2007)
23. Verhaeghe, H., Lecoutre, C., Deville, Y., Schaus, P.: Extending compact-table to basic smart tables. In: Beck, J.C. (ed.) CP 2017. LNCS, vol. 10416, pp. 297–307. Springer, Cham (2017). https://doi.org/10.1007/978-3-319-66158-2_19
24. Verhaeghe, H., Lecoutre, C., Schaus, P.: Extending compact-table to negative and short tables. In: AAAI, pp. 3951–3957 (2017)
25. Verhaeghe, H., Lecoutre, C., Schaus, P.: Compact-MDD: efficiently filtering (s)MDD constraints with reversible sparse bit-sets. In: IJCAI, pp. 1383–1389 (2018)
26. Wang, R., Xia, W., Yap, R.H.C., Li, Z.: Optimizing simple tabular reduction with a bitwise representation. In: IJCAI, pp. 787–795 (2016)
27. Xia, W., Yap, R.H.C.: Optimizing STR algorithms with tuple compression. In: Schulte, C. (ed.) CP 2013. LNCS, vol. 8124, pp. 724–732. Springer, Heidelberg (2013). https://doi.org/10.1007/978-3-642-40627-0_53

Embedding Decision Diagrams
into Generative Adversarial Networks

Yexiang Xue[1(✉)] and Willem-Jan van Hoeve[2]

[1] Department of Computer Science, Purdue University, West Lafayette, USA
yexiang@purdue.edu
[2] Tepper School of Business, Carnegie Mellon University, Pittsburgh, USA
vanhoeve@andrew.cmu.edu

Abstract. Many real-world decision-making problems do not possess a clearly defined objective function, but instead aim to find solutions that capture implicit user preferences. This makes it challenging to directly apply classical optimization technology such as integer programming or constraint programming. Machine learning provides an alternative by learning the agents' decision-making implicitly via neural networks. However, solutions generated by neural networks often fail to satisfy physical or operational constraints. We propose a hybrid approach, DDGAN, that embeds a Decision Diagram (DD) into a Generative Adversarial Network (GAN). In DDGAN, the solutions generated from the neural network are filtered through a decision diagram module to ensure feasibility. DDGAN thus combines the benefits of machine learning and constraint reasoning. When applied to the problem of schedule generation, we demonstrate that DDGAN generates schedules that reflect the agents' implicit preferences, and better satisfy operational constraints.

1 Introduction

Traditional management science approaches to decision-making involve defining a mathematical model of the situation, including decision variables, constraints, and an objective function to be optimized. While common objectives such as cost minimization or profit maximization are widely applied, many operational decision-making processes depend on factors that cannot be captured easily by a single mathematical expression. For example, in production planning and scheduling problems one typically takes into account priority classes of jobs and due dates, but ideally also (soft) business rules or the preferences of the workers who execute the plan. Those rules and preferences may be observed from historic data, but creating a model that results in, say, a linear objective function is far from straightforward. Instead, one may represent the objective function, or even the constraints, using machine learning models that are then embedded into the optimization models; we refer to [24] for a recent survey.

In this work, we study the integration of machine learning and constraint reasoning in the context of sequential decision making. In particular, we aim

This work was supported by Office of Naval Research grant N00014-18-1-2129.

L.-M. Rousseau and K. Stergiou (Eds.): CPAIOR 2019, LNCS 11494, pp. 616–632, 2019.
https://doi.org/10.1007/978-3-030-19212-9_41

to extend (recursive) generative adversarial neural networks (GANs) with constraint reasoning. We assume that the context specifies certain physical or operational requirements, within which we need to find solutions that are similar to the decisions that were made historically, as in the following stylized example.

Example 1. Consider a routing problem in which we need to dispatch a service vehicle to perform maintenance at a set of locations. The set of locations differs per day and is almost never the same. Previous routes indicate that the driver does not follow a distance or time optimal sequence of visits, even though there are no side constraints such as time windows or precedence relations. Instead, the routes suggest that the driver has an underlying route preference, that is exposed by, e.g., visiting shopping and lunch areas at specific times of the day. Our decision making task is now: Given the historic routes and a set of locations to visit today, determine a route that (1) visits all locations, and (2) is most similar to the historic routes. In addition, we might add further restrictions such as a maximum time limit, for example 8 h.

In these contexts, traditional optimization methods such as integer programming or constraint programming cannot be applied directly since they are unable to represent an appropriate objective function. Instead, it is natural to represent the structure and preferences from the historic solutions using a machine learning model. For example, we could design and train a generative adversarial neural network (GAN) for this purpose, which will be able to produce sequences of decisions that aim to be similar to the historic data. However, because GANs are trained with respect to an objective function (loss function) to be minimized, hard operational constraints cannot be directly enforced. For example, in case of the routing problem above, the sequences produced by the GAN usually fail to visit all locations, visit some locations multiple times, or fail to recognize constraints such as the maximum time limit.

Contributions. We propose a hybrid approach, which we call DDGAN, that embeds a decision diagram (DD) into a generative adversarial neural network. The decision diagram represents the constraint set and serves as a filter for the solutions generated by the GAN, to ensure feasibility. As proof of concept, we develop a DDGAN to represent routing problems as in Example 1. We show that without the DD module, the GAN indeed produces sequences that are rarely feasible, while the DD filter substantially increases the feasibility. Moreover, we show that DDGAN converges much more smoothly than the GAN.

We note that, in principle, any constraint reasoning module could have been applied; e.g., we could embed an integer program or constraint program that contains all constraints of the problem. The variable/value assignments suggested by the GAN can then be checked for feasibility by running a complete search, but this is time consuming. By compiling a decision diagram offline, we can check for feasibility instantly during execution. Moreover, for larger problems we can apply relaxed decision diagrams of polynomial size that may not guarantee feasibility in all cases, but can produce much better solutions that those generated by the GAN alone.

2   Related Work

Within the broader context of learning constraint models, several works have studied automated constraint acquisition from historic data or (user-)generated queries, including [1,4,7,11,23]. These approaches use partial or complete examples to identify constraints that can be added to the model. The type of constraints that can be learned depends on the formalism that is used, for example linear constraints for integer programming or global constraints for constraint programming.

A different approach is to embed a machine learning model into the optimization formalism, e.g., by extending a constraint system with appropriate global constraints. For example, [22] integrate neural networks and decision trees in constraint programming, while [25,26] introduce a 'Neuron' global constraint that represents a pre-trained neural network. Another series of approaches based on Grammar Variational Autoencoders [12,19,21] use neural networks to encode and decode from the parse-tree of a context free grammar to generate discrete structures. Such approaches are used to generate chemical molecule expressions, which is also a structured domain. Compared to our work, their models are not conditional, and therefore cannot solve decision-making problems with varying contextual information.

Machine learning approaches have also been used to solve optimization problems. This includes the works [14,30], which use neural networks to extend partial solutions to complete ones. The authors of [5] handle the traveling salesman problem by framing it as reinforcement learning. Approaches based on Neural Turing Machines [16] employ neural networks with external memory for discrete structure generation. More recently, the authors of [20] solve graph combinatorial optimization problems, and employ neural networks to learn the heuristics in backtrack-free searching. The scopes of these works are different from ours, since they do not deal with optimization problems without clear objective functions. Recently, the work of [29] combine reasoning and learning using a max-margin approach in hybrid domains based on Satisfiability Modulo Theories (SMT). They also show applications in constructive preference elicitation [13]. Compared to our work, their approach formulates the entire learning and reasoning problem as a single SMT, while we combine reasoning and learning tools, namely the neural networks and decision diagrams, into a unified framework.

3   Preliminaries

3.1   Structure Generation via Generative Adversarial Networks

Finding structured sequences in presence of implicit preferences is a more complex problem compared to classical supervised learning, in which a classifier is trained to make single-bit predictions that match those in the dataset. The problem of finding such sequences is broadly classified as a *structure generation* problem in machine learning, which arises naturally in natural language processing [17,18], computer vision [8,28], and chemistry engineering [21].

Specifically, we are given common context R_c and a historical data set $\mathcal{D} = \{(R_1, S_1), (R_2, S_2), \ldots, (R_{t-1}, S_{t-1})\}$, in which R_i represents the contextual information specific to data point i and S_i the decision sequence for scenario $i = 1, \ldots, t - 1$. For our routing example, R_i represents the daily location requests while S_i is the ordered sequence of visits that covers the requested locations in R_i. R_c, in this case, represents the set of constraints that is common to all data points; for example, each location can be visited at most once. Notice that R_i can be also interpreted in the form of constraints, which we will leverage later in the paper. For example, in the routing problem, R_i can be interpreted as the constraint that sequence S_i can only contain locations from the requested set R_i, and all locations in R_i must be visited.

Given present contextual data R_t, our goal is to find a good sequence of decisions S_t. Using machine learning terminology, we would like to learn a conditional probability distribution of "good" sequences

$$Pr(S_t \mid R_t)$$

based on the historical dataset. More precisely, $Pr(S_t \mid R_t)$ should assign high probabilities to "good" sequences, i.e., sequences that satisfy operational constraints and look similar to the ones in the historical dataset. A "bad" sequence, for example, a sequence that violates key operational constraints, should ideally be assigned a probability zero. When we deploy our tool (i.e., for scenario t), we first observe the contextual data R_t, and then sample a sequence S_t according to $Pr(S_t \mid R_t)$, which should satisfy all operational constraints and the implicit preferences reflected by the historical dataset.

Generative Adversarial Networks (GAN). GAN is a powerful tool developed in the machine learning community for complex structure generation [15]. The original GAN is developed to learn the probability distribution $Pr(x)$ of complex structural objects x. We will briefly review GAN in this simple context.

Given a dataset X in which each entry $x \in X$ is independently and identically drawn from the underlying (unknown) data distribution $Pr_{data}(x)$, GAN fits a probability distribution $Pr(x)$ that best matches $Pr_{data}(x)$. Instead of directly fitting the density function, GAN starts with random variables z with a known prior probability distribution $Pr(z)$ (such as multi-variate Gaussian), and then learns a *deterministic mapping* $G : Z \to X$ in the form of a neural network, which maps every element $z \in Z$ to an element $x \in X$. The goal is to fit the function $G(.)$ so that the distribution of $G(z)$ matches the true data distribution $Pr_{data}(x)$ when z is sampled from the known prior distribution.

GAN also trains another *discriminator* neural network $D : x \to \mathcal{R}$ to determine the closeness of the generated and the true structures. D is trained to separate the real examples from the dataset with the fake examples generated by function G. Both the generator G and the discriminator D are trained in a competing manner. The overall objective function for GAN is:

$$\min_G \max_D \ \mathbb{E}_{x \sim \text{data}}[D(x)] + \mathbb{E}_z[1 - D(G(z))]. \tag{1}$$

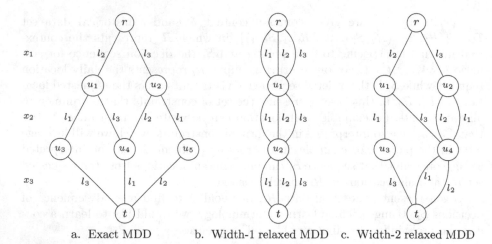

a. Exact MDD   b. Width-1 relaxed MDD   c. Width-2 relaxed MDD

Fig. 1. Multi-valued decision diagrams (MDDs) representing possible assignments of x_1, x_2, x_3. (a) Exact MDD representing the subset of permutations satisfying `alldifferent`(x_1, x_2, x_3) and $x_1 \neq l_1$. Each path from r to t represents a valid permutation satisfying the two constraints. (b) A width-1 relaxed MDD for the exact MDD in (a). (c) A width-2 relaxed MDD, which is formed by combining nodes u_4 and u_5 of the MDD in (a).

For our application, we extend the classical GAN into a conditional structure, as will be discussed in Sect. 4. We acknowledge that GAN is a recent popular probabilistic model for structural generation. Nevertheless, structural generation is challenging and many research questions still remain open. Our embedding framework is general and can be applied beyond the GAN structure.

3.2   Decision Diagrams

Decision diagrams were originally introduced to compactly represent Boolean functions as a graphical model [2,9], and have since been widely used, e.g., in the context of verification and configuration problems [31]. More recently decision diagrams have been used successfully as a tool for optimization, by representing the set of solutions to combinatorial optimization problems [6].

Decision diagrams are defined with respect to a sequence of decision variables $x_1, x_2, \ldots, x_n$. Variable x_i has a domain of possible values $D(x_i)$, for $i = 1, \ldots, n$. For our purposes, a decision diagram is a layered directed acyclic graph, with $n + 1$ layers of nodes; see Fig. 1 for an example. Layer 1 contains a single node, called the root r. Layer $n+1$ also contains a single node, called the terminal t. An arc from a node in layer i to a node in layer $i+1$ represents a possible assignment of variable x_i to a value in its domain, and is therefore associated with a label $l \in D(x_i)$. For an arc (v, w), we use $var(v, w)$ to represent the variable being assigned through this arc, and use $val(v, w)$ to represent its assigned value. For a node m in the MDD, we use $val(m)$ to represent the union of the values of each arc starting from node m, i.e., $val(m) = \{val(u, v) : u = m\}$. Notice that $val(m)$ represents the possible value assignments of the decision variable corresponding to node m. Each path from the root r to the terminal t represents a solution, i.e., a complete variable-value assignment. We can extend the arc definition to allow for "long arcs" that skip layers; a long arc out of a node in layer i still represents a value assignment to variable x_i and assigns the skipped layers to a default value (for example 0). In our case, we consider variables with arbitrary domains, which results in so-called multi-valued decision diagrams (MDDs).

Example 2. Let x_1, x_2, x_3 represent a sequence of decision variables, each with domain $\{l_1, l_2, l_3\}$. The constraint $\texttt{alldifferent}(x_1, x_2, x_3)$ restricts the values of x_1, x_2, x_3 to be all different; i.e., they form a permutation. Along with another constraint $x_1 \neq l_1$, it restricts the set of feasible permutations to be $\{(l_2, l_1, l_3), (l_2, l_3, l_1), (l_3, l_2, l_1), (l_3, l_1, l_2)\}$. Figure 1(a) depicts the exact MDD that encodes all permutations satisfying these two constraints.

Exact Decision Diagram. Given a set of constraints R, MDD $\mathcal{M}$ is said to be *exact* with respect to R if and only if every path that leads from the root node r to the terminal node t in $\mathcal{M}$ is a variable-value assignment satisfying all constraints in R. Conversely, every valid variable-value assignment can be found as a path from r to t in $\mathcal{M}$. For example, Fig. 1(a) represents an exact MDD that encodes the constraints $\texttt{alldifferent}(x_1, x_2, x_3)$ and $x_1 \neq l_1$.

Relaxed Decision Diagram. Because exact decision diagrams can grow exponentially large, we will also apply *relaxed* decision diagrams of polynomial size [3]. The set of paths in a relaxed decision diagram forms a *superset* of that of an exact decision diagram. For example, the set of paths in Fig. 1(a) is fully contained in the sets of paths in the Figs. 1(b) and (c). Therefore, the MDDs in Figs. 1(b) and (c) form two relaxed versions of the MDD in Fig. 1(a). Relaxed MDDs are often defined with respect to a maximum *width*, i.e., the number of nodes in its largest layer. For example, Fig. 1(b) is a width-1 relaxed MDD, which trivially forms the superset of any constrained set of x_1, x_2, x_3, while Fig. 1(c) is a width-2 relaxed MDD.

Decision Diagram Compilation. Decision diagrams can be constructed to encode constraints over the variables, by a process of node refinement and arc filtering [3,6]. In general, arc filtering removes arcs that lead to infeasible solutions, while node refinement (or splitting) improves the precision in characterizing the solution space. One can reach an exact MDD by repeatedly going through the filtering and the refinement process from a width-1 MDD. We refer to [10] for details on MDD compilation for sequencing and permutation problems.

Exact MDD Filtering. MDD filtering algorithms can also be applied without node refinement, to represent additional constraints in a given MDD. Generally, MDD filtering does not guarantee that each remaining r-t path is feasible. To establish that, we next introduce the notion of an exact MDD filter.

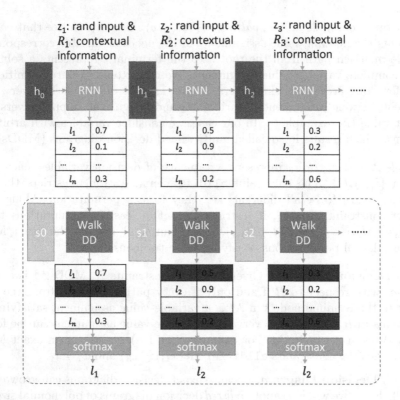

Fig. 2. DDGAN as the generator G within the GAN framework. On the top is a recursive neural network (RNN) structure. RNN takes as input the contextual information R_t and random variables z, and outputs scores for values for each variable. The values are filtered by the WALKDD module on the bottom that represents a constraint set. The values that lead to contradictions are filtered out (marked by the red color). Finally, a softmax layer decides the value of each variable by picking up the one with the largest score among all non-filtered values. (Color figure online)

Definition 1. *Let M be an exact MDD with respect to a constraint set R, and let C be an additional constraint. An MDD filter for C is exact if applying it to M results in an MDD M' that is exact with respect to R and C.*

Consider the following MDD filtering algorithm `FilterUnary` for unary constraints, i.e., constraints of the form $x \neq l$ for some variable x and value $l \in D(x)$. We first let `FilterUnary` remove any arc that violates the unary constraints. Then, working from the last layer up to the first layer, `FilterUnary` recursively removes any nodes and arcs that do not have path that lead to the terminal t. We have the following result.

Proposition 1. `FilterUnary` *is an exact MDD filter.*

4 Embedding Decision Diagrams into GANs

We next present our hybrid approach, DDGAN, which embeds a multi-valued decision diagram into a neural network, to generate structures that (i) satisfy a set of constraints, and (ii) capture the user preferences embedded implicitly in the historical dataset. The structure of DDGAN is shown in Fig. 2.

To achieve this, DDGAN has a recursive neural network (RNN) as its first layer. The RNN module generates scores of possible assignments to variables $x_1, \ldots, x_n$ in a sequence of n steps. We refer the entire recursive neural network (the upper part of Fig. 2) as a RNN and refer to the one-step unrolling of the network as a RNN cell. In the i-th step, one RNN cell takes the hidden state of the previous step as input, outputs the hidden state of this step, and a table of dimension $|D(x_i)|$. The table corresponds to the score for each value of variable x_i. In general, the higher the score is, the more likely the RNN believes that x_i should be assigned to the particular value. RNN is trained to capture implicit preferences which give higher scores to the structures in the historical dataset. Because of the link through hidden states among RNN cells of different steps, RNN is able to capture the correlations among variables.

However, the structures generated by the RNN module may not satisfy key (operational) constraints. Therefore, we embed a WALKDD neural network as a second layer (Fig. 2 bottom, within the dashed rectangular) to filter out actions that violate the constraint set. WALKDD simulates the process of descending along a particular path of the MDD. During this process, WALKDD marks certain assignments generated by the RNN module as infeasible ones (as shown in the entries with red background). Finally, a softmax layer takes the action with highest score among all feasible actions. In this way, WALKDD filters out structures that violate operational constraints.

Full WALKDD Filtering. We assume access to an MDD $\mathcal{M}$, which is compiled with respect to the common constraint set R_c. We first focus on the case where $\mathcal{M}$ is exact, i.e., each r-t path in $\mathcal{M}$ represents an assignment to $x_1, \ldots, x_n$ that satisfies all constraints in R_c. We also assume we are given an filtering scheme Filter for the additional contextual constraints R_t for data point t. Let $\mathcal{M}_t$ be the MDD resulted from $\mathcal{M}$ filtered through constraints R_t using Filter.

WALKDD is executed as follows. It keeps the current MDD node of $\mathcal{M}_t$ as its internal state. Initially, the internal state is at the root node r. In each step, WALKDD moves to a new MDD node in the next layer once one variable is set to a particular value. Suppose WALKDD is at step i, the MDD node m_i, and the corresponding decision variable x_i. Recall that $val(m_i)$ represents the set of values that can be assigned to variable x_i according to the labels of the arcs out of m_i. WALKDD blocks all assignments to x_i outside of $val(m_i)$ (shown as the entries with the red background in Fig. 2). After this step, a softmax layer picks up the variable assignment with the largest score among all non-blocked entries and set x_i to be the corresponding value. WALKDD then moves its state to the corresponding new node in the MDD following the variable assignment.

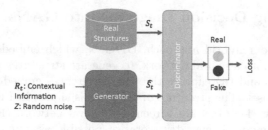

Fig. 3. GAN structure for solving the structure generation problem with implicit preferences. The contexts R_t and random variables are fed into the generator G, which will produce a structure S_t. The Discriminator D are trained to separate the generated structures $\hat{S}_t$ with the real structures S_t. D and G are trained in a competing manner.

Proposition 2. *Let $\mathcal{M}$ be an exact MDD with respect to constraint set R_c and let* Filter *be an exact filter for constraint set R_t. Then* WALKDD *is guaranteed to produce sequences that satisfy the constraints in both R_c and R_t.*

Overall Conditional GAN Architecture. During training, the aforementioned DDGAN structure is fed as the generator function G in the conditional GAN architecture [27], which is the classical GAN network, modified to take into account contextual information R_t. We use conditional GAN as a example to host DDGAN. Nevertheless, DDGAN can be accommodated by other structures as well. For example, infeasible actions can be filtered out by WALKDD in a similar fashion in the Viterbi algorithm for Hidden Markov Models. The overall conditional GAN architecture is shown in Fig. 3. In this structure, R_t and random variables z are arguments of the generator G, which in turn produces a structure (or sequence) $\hat{S}_t$. The discriminator D is trained to separate the generated structure $\hat{S}_t$ with the real one S_t. D and G are trained in a competing manner using stochastic gradient descent. The overall objective function we optimize is:

$$\min_G \max_D \quad \mathrm{E}_t \left[D(S_t, R_t) + \mathrm{E}_z[1 - D(G(R_t, z), R_t)] \right]. \tag{2}$$

Note that, compared to the objective function of the classical GAN (Eq. 1), the discriminator and the generator in Equation (2) both take the contextual information R_t as an additional input.

Implementation. WALKDD heavily uses matrix operations, most of which can be efficiently carried out on GPUs. We have a state-transition matrix of the MDD, which are hard-coded prior to training and in which infeasible transitions are labeled with a unique symbol. During execution, we maintain the *state* tensor, which contains the current MDD node of each data point in the mini-batch. We also maintain the *mask* tensor, which indicates values that variables cannot take.

Backpropagate the Gradients. We heavily rely on non-differentiable *gather* and *scatter* operations offered by Pytorch to maintain the state and the mask tensors. As a filter, WALKDD can have non-differentiable components because it is not updated during training. We pass gradients through the non-blocked entries of WALKDD into the fully-differentiable RNN. We leave it as a future research direction to make WALKDD fully differentiable following the work of Neural Turing Machine [16].

5 Case Study: Routing with Implicit Preferences

As proof of concept, we apply DDGAN to a routing problem similar to Example 1. We consider a set of n locations $\mathcal{L} = \{l_1, l_2, \ldots, l_n\}$. At each day t, a service person (an agent) receives the request to visit a subset of locations $R_t \subset \mathcal{L}$. For this work, we assume that the agent can visit at most M locations in a day; i.e., $|R_t| \leq M$ for all t. The agent has a starting location $s \in \mathcal{L}$ and an ending location $e \in \mathcal{L}$. When $s = e$, the agents's route is a Hamiltonian circle.

The agent's actual visit schedule for day t, $S_t = (s_{1,t}, s_{2,t}, \ldots, s_{|R_t|,t})$, is a permutation (or, an ordered list) of locations in R_t. Notice that it is sufficient to specify the schedule fully using a permutation. Other information, such as the earliest time and latest time to visit locations in the schedule, can be inferred from the permutation. For this work, we assume the agent's schedules are subject to the following constraints:

1. **All-different Constraint.** S_t must be a subset of $\mathcal{L}$, in which no location is visited more than once.
2. **Full-coverage Constraint.** S_t must visit all and only the locations in R_t.
3. **Total Travel Distance Constraint.** Let $d_{i,j}$ be the length of the shortest path between l_i and l_j. The total travel distance for the agent is: $dist_t = d_{s,s_1} + \sum_{i=1}^{|R_t|-1} d_{s_i,s_{i+1}} + d_{s_{|R_t|},e}$. Suppose the agent has a total travel budget B. We must have $dist_t \leq B$.

Observe that only the Full-coverage constraint requires the contextual data R_t. Therefore, the common constraints R_c contains the All-different and the Total Travel Distance constraint. Lastly, the agent has **implicit preferences**. As a result, the schedule S_t may deviate from the shortest path connecting the locations in R_t. Moreover, the agent cannot represent these preferences as a clear objective function. Instead, we are given the travel history $S_1, S_2, \ldots, S_{t-1}$ and the request locations R_t for day t. The goal is to find a valid schedule S_t, which satisfies all constraints, but also serves his preferences reflected implicitly in the travel history.

Constructing MDD for the Routing Problem. For this application, the set of nodes of the MDD $\mathcal{M}$ is partitioned into $M + 2$ layers, representing variables $x_1, \ldots, x_{M+2}$. The first layer contains only one root node, representing the agent at the starting location, s. The last layer also contains only one terminal node, representing the agent at the ending location, e. The nodes in the layer of variable x_i correspond to the agent making stops at the i-th position of the schedule. As initial domain we use $D(x_i) = \mathcal{L}$, i.e., the set if all possible locations. There are two types of arcs. An arc $a = (u, v)$ of the first type is always directed from a source node u in one layer x_i to a target node v in the subsequent layer x_{i+1}. Each arc a of the first type in the i-th layer is associated with a label $val(a) \in \{l_1, \ldots, l_n\}$, meaning that the agent visits location l_i as the i-th stop. The second type of arcs b, whose values $val(b)$ are always e, connect every node to the terminal node. These arcs are used to allow the agent to travel back to the end location at any time. The terminal node is connected to itself with an arc of the second type. This allows the agent to stay at the end location for the rest of his day, once arrived.

We follow the procedure in [10] for constructing the (relaxed) MDD for the routing problem, with respect to both the alldifferent and the maximum distance constraint in R_c. That is, we start with a width-1 MDD, and then repeatedly apply the filter and the refine operations until the MDD is exact or a fixed width limit has reached. These operations make use of specific state information that is maintained at each node of the MDD. Since we will re-use some of these, we revisit them here. For an MDD node v, (i) $All^{\downarrow}(v)$ is the set of locations that *every* path from the root node r to the current node v passes, while (ii) $Some^{\downarrow}(v)$ is the set of locations that *at least* one path from the root node r to the current node v passes. (iii) $All^{\uparrow}(v)$ and (iv) $Some^{\uparrow}(v)$ are defined similarly, except that they consider locations from the current node v to the terminal node t. An arc in the MDD of a routing problem corresponds to visiting one location. For an arc a, define (v) $st^{\downarrow}(a)$ as the *shortest travel distance from the root*, which is the shortest distance that the location $val(a)$ can be reached along any path from the root r. Similarly, define (vi) $st^{\uparrow}(a)$ as the *shortest travel distance from the target*, which is the shortest distance to travel to the target node t from the location $val(a)$ along any path in the MDD. When the MDD is exact, the sum $st^{\downarrow}(a) + st^{\uparrow}(a)$ represents the shortest distance to travel from the root r to target t along any valid path passing arc a. When the MDD is relaxed, the computation of $st^{\downarrow}(a)$ and $st^{\uparrow}(a)$ can be along any path, regardless its validity. The sum $st^{\downarrow}(a) + st^{\uparrow}(a)$ therefore becomes the lower bound of the shortest distance. More details on how this information is used for filtering and refinement can be found in [10].

Full WALKDD Filtering for the Routing Problem. The daily requests R_t for the routing problem can be translated into the following two constraints in addition to the all-different and maximum distance requirements in R_c: (i) only locations in the requested set R_t are allowed to be visited other than the starting and ending locations. (ii) The trip length is exactly $|R_t| + 2$. Notice that these two constraints can be realized by imposing unary constraints to the MDD. To enforce the first constraint, we can remove all first-type arcs in the MDD whose corresponding location is outside of R_t. To enforce the second constraint, we remove all second-type arcs which imply the wrong trip length. Because we have access to an exact filter for unary constraints (Proposition 1), the schedules produced by the full WALKDD scheme presented in Sect. 4 satisfy all constraints, if the MDD is exact with respect to R_c (Proposition 2).

Local WALKDD Filtering for the Routing Problem. The full WALKDD filtering scheme requires a pass over the entire MDD for each data point (filtering unary constraints). While this guarantees to produce structures that satisfy all constraints, when the MDD is exact, it is also relatively computationally expensive. We next discuss a more efficient *local* WALKDD *filtering scheme* that removes infeasible transitions only from the information that is *local* to the current MDD node. This local scheme is not comprehensive, i.e., local WALKDD filter may generate structures that do not fully satisfy all constraints, even though the MDD is exact. In practice, however, exact MDDs often become too large, in which case we apply relaxed MDDs for which the guarantees in Proposition 2 no longer hold. In other words, the computational benefit of local filtering often outweighs theoretical guarantees: it only requires to visit *a single path* from the root to the terminal for local filtering, which is substantially cheaper than visiting the entire MDD as in the full case.

The local WALKDD filter rules out actions of the RNN in each step by only examining information that is local to the current MDD node, as follows. As before, we assume that MDD $\mathcal{M}$ represents the common constraints R_c. For day t, local WALKDD keeps its own internal state: $W_{t,i} = (u_{t,i}, l_{t,i}, V_{t,i}, R_{t,i}, Time_{t,i})$ after deciding the first i locations, where $u_{t,i}$ is the MDD node in $\mathcal{M}$ representing the current state, $l_{t,i}$ is the current location of the agent, $V_{t,i}$ is the set of locations that the agent already visited, $R_{t,i}$ is the set of locations that remains to be visited, and $Time_{t,i}$ is the time elapsed after visiting the first i locations. Local WALKDD applies the following local filters based on its internal state:

- **Next Location Filter:** The location to be visited should follow one of the arcs that starts from the current node $u_{t,i}$. Otherwise, the location is filtered.
- **Locations Visited Filter:** If the location to be visited is in the set of visited locations $V_{t,i}$, then the location is filtered out, except for the end location e.
- **Locations to be Visited Filter:** If the location to be visited is not in the set $R_{t,i}$, then the location is filtered out. We guarantee that the end location e is always in $R_{t,i}$.
- **Future Location Set Filter:** Suppose the next location to be visited is $l_{t,i+1}$ and the MDD node following the arc of visiting $l_{t,i+1}$ is $u_{t,i+1}$. If $R_{t,i} \setminus \{l_{t,i+1}\}$ is not a subset of $Some^{\uparrow}(u_{t,i+1})$, then we cannot cover all the locations remaining to be visited following any paths starting from $u_{t,i+1}$. Therefore, $l_{t,i+1}$ should be filtered out.
- **Total Travel Time Filter:** Let $l_{t,i+1}$ be the next location to visit and $st^{\uparrow}(l_{t,i+1})$ be the shortest time to reach the end location e from $l_{t,i+1}$. If $Time_{t,i} + d_{l_{t,i}, l_{t,i+1}} + st^{\uparrow}(l_{t,i+1}) > B$, this suggests that no path from $l_{t,i+1}$ to the end location satisfies the total distance constraint. Therefore, $l_{t,i+1}$ should be filtered out.

6 Experiments

The purpose of our experiments is to evaluate the performance of the GAN with and without the DD module. We first describe the implementation details of the GAN. We use a LSTM as the RNN module. The dimension of the hidden state of the LSTM in DDGAN is set to 100. The dimension of the random input z is

20. During training, DDGAN is used as the generator of the conditional GAN infrastructure. The discriminator of the conditional GAN is also a RNN-based classifier, whose hidden dimension is 100. The batch size is set to be 100. We compare our DDGAN with the same neural network structure except without the WALKDD module as the baseline. The entire conditional GAN is trained using stochastic gradient descent. The learning rate of both the generator and the discriminator are both set to be 0.01. Every 10 epochs, the performance of DDGAN and the baselines are tested by feeding in 1,000 scheduling requests into the generator part of the neural network (the structure shown in Fig. 2) and examining the schedules it generates.

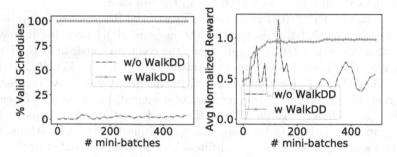

Fig. 4. DDGAN on a small scheduling problem with 6 locations. DDGAN has access to an exact MDD and the full filtering scheme. (Left) The percentage of valid schedules generated along training progress. DDGAN always generates valid schedules, while the same neural network without the WALKDD component cannot. (Right) the normalized reward for the schedules generated by DDGAN and the competing approach. The normalized rewards converge to 1 for DDGAN, suggesting that DDGAN is able to fully recover the implicit preference of the agent. (Color figure online)

We assume the agent's implicit preferences are reflected by a hidden reward function $r_{i,j}$, which is the reward that the agent visits location j in the i-th position of his schedule. For our experiments, this reward function is generated uniformly randomly between 0 and 1. The agent's optimal schedule is the one that maximizes the total reward while observing all the operational constraints. The reward function is hidden from the neural network. In our application, the goal of the neural networks is to generate schedules subject to operational constraints, which also score high in terms of this hidden reward function.

We first test DDGAN on a synthetic small instance of 6 locations. In this case, we embed an exact MDD into the DDGAN structure and we apply the full filtering scheme as discussed in Sect. 4. The result is shown in Fig. 4. As we can see, the schedules generated by DDGAN always satisfy operational constraints (red curve, left panel), while the schedules generated by the same neural network without the WALKDD module are often not valid (blue curve, left panel). On the right we plot the total reward of the schedules generated by the two approaches. Because the number of locations are small, we compute offline the

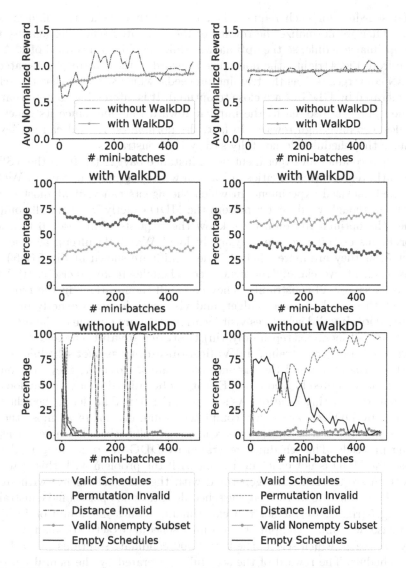

Fig. 5. DDGAN on a medium-sized scheduling problem in TSPLib consisting of 29 locations, where DDGAN can only access a relaxed MDD of width at most 1,000. The agent is allowed to visit 6 locations (left) or 12 locations (right) maximally. (Top) The normalized reward for the schedules generated with DDGAN and the same neural network without the WALKDD module. DDGAN learns to generate schedules that are close to the optimal ones (with normalized reward close to 1). (Middle, lower) The percentage of different types of schedules generated by DDGAN (middle) and the baseline (lower). Legends on the bottom. The schedules generated by DDGAN (middle) always satisfy the alldifferent and distance constraints. The percentage of fully valid schedules increase in (middle). However, the same neural network without WALKDD cannot satisfy major constraints (lower).

optimal schedule for each request, i.e., the one that yields the highest total reward. Then we normalize the reward of the generated schedules against that of the optimal schedule, so the optimal schedule should get a reward of 1. As we can see from Fig. 4 (right), the normalized reward of the schedules generated by DDGAN converges to 1 as the training proceeds, which suggests that the schedules generated by DDGAN are close to optimal. It is interesting to note that the baseline approach also learns the implicit reward function, since its generated schedules also have high reward. In fact, the normalized reward can go beyond 1 because the schedules do not fully satisfy the constraints.

We then conduct an experiment using instance bayg29.tsp from the TSPLib benchmark, representing 29 cities in Bavaria with geographic distances. We first run a medium-sized experiment, in which the agents can visit at most 6 locations. Even though we do not represent the MDD exactly, we can still compute (offline) the optimal route once we know the requests of at most 6 locations. For problems of this size, we only apply local WALKDD filters, as discussed in Sect. 5, as they are more efficient. The results are shown in Fig. 5 (left). For this experiment, we classify the generated schedules more precisely: (i) **Valid schedules** satisfy all constraints. They cover all the locations in the requested set, meet the travel distance budget, and visit each location exactly once. (ii) **Permutation invalid** schedules visit locations that are outside of the requested set and/or visit locations repeatedly. (iii) **Distance invalid** schedules break the total travel distance constraint. (iv) **Valid non-empty subset** schedules satisfy both the permutation and the distance constraints. However, they visit only a subset of the requested locations. (v) **Empty schedules** do not visit any location other than the starting location. As shown in Fig. 5 (middle, left), the schedules generated by DDGAN are either completely valid, or violate at most the full coverage constraint. Moreover, the percentage of valid schedules increases as the training proceeds. The schedules generated by DDGAN are no longer all valid because the MDD is not exact in dealing with the problem with this scale. On the bottom are the schedules generated with the same neural net without the WALKDD module. As we can see, the schedules break all operational constraints.

In Fig. 5 (right), we further run experiments with a maximum of 12 visits. In this case, we cannot compute the optimal schedule exactly. Instead, we use a greedy approach, which selects the best 1,000 candidate solutions for each stop in the schedule. The reward of the schedules generated by the neural networks are normalized with respect to that of the greedy approach. We can see that for larger problems we can apply relaxed decision diagrams of polynomial size that may not guarantee feasibility in all cases, but can produce much better solutions that those generated by the GAN.

7   Conclusion

In this work, we study the integration of machine learning and constraint reasoning in the context of sequential decision-making without clear objectives. We propose a hybrid approach, DDGAN, which embeds a decision diagram (DD) into a generative adversarial network (GAN). The decision diagram represents the constraint set and serves as a filter for the solutions generated by the GAN, to ensure feasibility. We demonstrate the effectiveness of DDGAN to solve routing problems with implicit preferences. We show that without the decision diagram module, the GAN indeed produces sequences that are rarely feasible, while the decision diagram filter substantially increases the feasibility. Moreover, we show that DDGAN converges much more smoothly.

References

1. Addi, H.A., Bessiere, C., Ezzahir, R., Lazaar, N.: Time-bounded query generator for constraint acquisition. In: van Hoeve, W.-J. (ed.) CPAIOR 2018. LNCS, vol. 10848, pp. 1–17. Springer, Cham (2018). https://doi.org/10.1007/978-3-319-93031-2_1
2. Akers, S.B.: Binary decision diagrams. IEEE Trans. Comput. C–27, 509–516 (1978)
3. Andersen, H.R., Hadzic, T., Hooker, J.N., Tiedemann, P.: A constraint store based on multivalued decision diagrams. In: Bessière, C. (ed.) CP 2007. LNCS, vol. 4741, pp. 118–132. Springer, Heidelberg (2007). https://doi.org/10.1007/978-3-540-74970-7_11
4. Beldiceanu, N., Simonis, H.: A model seeker: extracting global constraint models from positive examples. In: Milano, M. (ed.) CP 2012. LNCS, pp. 141–157. Springer, Heidelberg (2012). https://doi.org/10.1007/978-3-642-33558-7_13
5. Bello, I., Pham, H., Le, Q.V., Norouzi, M., Bengio, S.: Neural combinatorial optimization with reinforcement learning. CoRR abs/1611.09940 (2016). http://arxiv.org/abs/1611.09940
6. Bergman, D., Cire, A.A., van Hoeve, W.J., Hooker, J.N.: Decision Diagrams for Optimization. Artificial Intelligence: Foundations, Theory, and Algorithms, 1st edn, p. 254. Springer, Cham (2016). https://doi.org/10.1007/978-3-319-42849-9
7. Bessiere, C., Koriche, F., Lazaar, N., O'Sullivan, B.: Constraint acquisition. Artif. Intell. 244, 315–342 (2017)
8. Brock, A., Donahue, J., Simonyan, K.: Large scale GAN training for high fidelity natural image synthesis. CoRR abs/1809.11096 (2018)
9. Bryant, R.E.: Graph-based algorithms for Boolean function manipulation. IEEE Trans. Comput. C–35, 677–691 (1986)
10. Cire, A.A., van Hoeve, W.J.: Multivalued decision diagrams for sequencing problems. Oper. Res. 61(6), 1411–1428 (2013)
11. Coletta, R., Bessiere, C., O'Sullivan, B., Freuder, E.C., O'Connell, S., Quinqueton, J.: Constraint acquisition as semi-automatic modeling. In: Coenen, F., Preece, A., Macintosh, A. (eds.) SGAI 2003, pp. 111–124. Springer, London (2004). https://doi.org/10.1007/978-0-85729-412-8_9
12. Dai, H., Tian, Y., Dai, B., Skiena, S., Song, L.: Syntax-directed variational autoencoder for structured data. CoRR abs/1802.08786 (2018)
13. Dragone, P., Teso, S., Passerini, A.: Constructive preference elicitation. Front. Robot. AI 4, 71(2018)
14. Galassi, A., Lombardi, M., Mello, P., Milano, M.: Model agnostic solution of CSPs via deep learning: a preliminary study. In: van Hoeve, W.-J. (ed.) CPAIOR 2018. LNCS, vol. 10848, pp. 254–262. Springer, Cham (2018). https://doi.org/10.1007/978-3-319-93031-2_18

15. Goodfellow, I.J., et al.: Generative adversarial nets. In: Proceedings of the 27th International Conference on Neural Information Processing System, NIPS 2014, vol. 2, pp. 2672–2680 (2014)
16. Graves, A., et al.: Hybrid computing using a neural network with dynamic external memory. Nature **538**(7626), 471–476 (2016)
17. Hu, Z., Yang, Z., Liang, X., Salakhutdinov, R., Xing, E.P.: Toward controlled generation of text. In: Proceedings of the 34th International Conference on Machine Learning, vol. 70, pp. 1587–1596 (2017)
18. Hu, Z., et al.: Deep generative models with learnable knowledge constraints. CoRR abs/1806.09764 (2018)
19. Jin, W., Barzilay, R., Jaakkola, T.S.: Junction tree variational autoencoder for molecular graph generation. In: ICML (2018)
20. Khalil, E., Dai, H., Zhang, Y., Dilkina, B., Song, L.: Learning combinatorial optimization algorithms over graphs. In: Advances in Neural Information Processing Systems, pp. 6348–6358 (2017)
21. Kusner, M.J., Paige, B., Hernández-Lobato, J.M.: Grammar variational autoencoder. In: Proceedings of the 34th International Conference on Machine Learning, vol. 70, pp. 1945–1954 (2017)
22. Lallouet, A., Legtchenko, A.: Building consistencies for partially defined constraints with decision trees and neural networks. Int. J. Artif. Intell. Tools **16**(4), 683–706 (2007)
23. Lallouet, A., Lopez, M., Marti, L., Vrain, C.: On learning constraint problems. In: Proceedings of IJCAI, pp. 45–52 (2010)
24. Lombardi, M., Milano, M.: Boosting combinatorial problem modeling with machine learning. In: Proceedings of IJCAI, pp. 5472–5478 (2018)
25. Lombardi, M., Milano, M., Bartolini, A.: Empirical decision model learning. Artif. Intell. **244**, 343–367 (2017)
26. Lombardi, M., Gualandi, S.: A Lagrangian propagator for artificial neural networks in constraint programming. Constraints **21**(4), 435–462 (2016)
27. Mirza, M., Osindero, S.: Conditional generative adversarial nets. CoRR abs/1411.1784 (2014)
28. Radford, A., Metz, L., Chintala, S.: Unsupervised representation learning with deep convolutional generative adversarial networks. CoRR abs/1511.06434 (2015)
29. Teso, S., Sebastiani, R., Passerini, A.: Structured learning modulo theories. Artif. Intell. **244**, 166–187 (2017)
30. Vinyals, O., Fortunato, M., Jaitly, N.: Pointer networks. In: Cortes, C., Lawrence, N.D., Lee, D.D., Sugiyama, M., Garnett, R. (eds.) Advances in Neural Information Processing Systems, pp. 2692–2700 (2015)
31. Wegener, I.: Branching Programs and Binary Decision Diagrams: Theory and Applications. SIAM Monographs on Discrete Mathematics and Applications. Society for Industrial and Applied Mathematics (2000)

Time Table Edge Finding with Energy Variables

Moli Yang[1], Andreas Schutt[1,2], and Peter J. Stuckey[2,3(✉)]

[1] University of Melbourne, Melbourne, Australia
[2] Data61, CSIRO, Melbourne, Australia
[3] Monash University, Melbourne, Australia
peter.stuckey@monash.edu

Abstract. Cumulative resource constraints can model scarce resources in scheduling problems or a dimension in packing and cutting problems. In order to efficiently solve such problems with a constraint programming solver, it is important to have strong and fast propagators for cumulative resource constraints. In this paper, we develop a time-table edge-finding energy propagator for cumulative constraint which can reason more strongly based on energy. We give results using this propagator in a lazy clause generation system on rectangle packing and evacuation scheduling problems. We are able to prune the search space and reduce solve time compared with a time-table or time-table edge-finding propagator.

1 Introduction

The `cumulative` constraint models the use of a limited resource over time in executing a series of tasks requiring the resource. The resource may be a set of machines, a group of workers, entities like power supply or even a dimension in a packing or cutting problem. Because of its broad modelling capability the `cumulative` constraint has been widely used in many industrial scheduling problems. Hence it is important to have strong and fast propagation techniques for the `cumulative` constraint so that constraint programming (CP) solvers can detect inconsistency and remove invalid values for the domains of the variables involved more efficiently. Moreover, for CP solvers that incorporate nogood learning [11], it is also important to generate strong reusable explanations for the reasoning of the `cumulative` constraint.

Vilím [18] developed TTEF propagation combining time-table propagation [1], which is usually superior for *highly disjunctive* problems, and edge-finding propagation [2], which is more appropriate for *highly cumulative* problems, in order to perform stronger propagation while having a low runtime overhead. Vilím shows that on a range of highly disjunctive project scheduling problems, TTEF propagation can generate lower bounds on the project deadline that are superior to previous methods. He used a CP solver without nogood learning.

Schutt *et al.* [14] extended TTEF propagation for use in a lazy clause generation (LCG) CP solver [11] by showing how to explain its propagation.

© Springer Nature Switzerland AG 2019
L.-M. Rousseau and K. Stergiou (Eds.): CPAIOR 2019, LNCS 11494, pp. 633–642, 2019.
https://doi.org/10.1007/978-3-030-19212-9_42

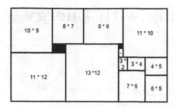
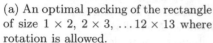

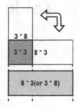

(a) An optimal packing of the rectangle of size 1×2, 2×3, ...12×13 where rotation is allowed.

(b) The two rotations of a rectangle and its energy usage

Fig. 1. (a) Rectangle packing and (b) the loss of information when only using duration and resource usage variables

LCG solvers are state of the art for solving many problems involving cumulative constraints. Schutt *et al.* [14] show that TTEF performs well in both lowering runtime and reducing search space for highly cumulative scheduling problems. However, the stronger propagation does not generally pay off for highly disjunctive problems.

An example of the usage of the `cumulative` constraint is in optimal rectangle packing [10], which is, given a set of rectangles, find the minimum area of a rectangle containing all rectangles without overlap. The cumulative constraint is used as a redundant constraint to constrain the maximal usage of height, when considering each rectangle as a task of duration length, and resource usage height; and similarly to constraint the maximum usage of width, when considering each rectangle as a task of duration height, and resource usage length. Note that the `cumulative` constraint provides very strong propagation in the case that the orientation of the rectangle is fixed, so the length and the height are known. But if we allow rectangles to be rotated, then we do not know the length and height of the rectangle, since each has two possibilities (unless it is a square).

Example 1. Consider a set of rectangles of sizes 1×2, 2×3, ... , up to 12×13, where rectangles may be rotated by 90-degree. Figure 1a shows the optimal solution in a 21×35 bounding box. □

If we consider the rectangle packing problem we can immediately see a weakness of a cumulative constraint that reasons using only start times, durations and resources usages. When we consider packing a rectangle of dimensions $w \times h$ whose orientation is not fixed then the minimum duration is $\min(w, h)$ and the minimum resource usage is $\min(w, h)$ and hence the overall minimum energy utilization is $\min(w, h)^2$, whereas we know the energy utilization is always exactly $w \times h$. With explicit energy usage variables, we can make use of the much larger lower bound on energy usage, and hence hope to propagate more.

Example 2. Figure 1b illustrates this phenomena explicitly by packing a 3×8 rectangle into an interval which is at least 8 long. Without knowing the orientation of the rectangle the lower bound on duration and resource usage are both 3, for a minimum resource usage of 9. But since the entire rectangle fits in the interval whatever rotation we know the energy usage is exactly 24. □

In this paper we define a `cumulative` propagator that uses energy variables in a TTEF propagation algorithm; we show how to explain its propagation; and we compare it against time-table and TTEF propagators.

2   Cumulative Resource Constraint with Energy Variables

In cumulative resource scheduling, a set of (non-preemptive) tasks $\mathcal{V}$ and one cumulative resource with a (constant) resource limit L is given where a *task i* is specified by its *start time* S_i, its *duration* D_i, its *resource usage* R_i, its *energy* $E_i = D_i \cdot R_i$. In this paper we assume each of S_i, D_i, R_i and E_i may be an integer variable and L is assumed to be an integer constant.

We assume a set of integer times τ and use notation $[t_1, t_2)$ to indicate the period starting at time t_1 and finishing (non-inclusive) at time t_2. We define est_i (d_i^{min}, r_i^{min}, e_i^{min}) and lst_i (d_i^{max}, r_i^{max}, e_i^{max}) as the current *lower* and *upper* bounds of start time (duration, resource usage, energy respectively) of i. Further, we define the *earliest completion time* $ect_i \leftarrow est_i + d_i^{min}$, and the *latest completion time* $lct_i \leftarrow lst_i + d_i^{max}$.

The cumulative resource constraint with energy `cumulative`(S, D, R, E, L) is characterized by the set of tasks $\mathcal{V}$ and a cumulative resource with resource capacity L. The constraint is satisfied by finding a solution that assigns values to each of the start time variables S_i, duration variables D_i, resource usage variables R_i and energy usage variables E_i ($i \in \mathcal{V}$), so that the following conditions hold.

$$est_i \leq S_i \leq lst_i, \qquad\qquad \forall i \in \mathcal{V}$$
$$d_i^{min} \leq D_i \leq d_i^{max}, \qquad\qquad \forall i \in \mathcal{V}$$
$$r_i^{min} \leq R_i \leq r_i^{max}, \qquad\qquad \forall i \in \mathcal{V}$$
$$e_i^{min} \leq E_i \leq e_i^{max}, \qquad\qquad \forall i \in \mathcal{V}$$
$$R_i \times D_i = E_i, \qquad\qquad \forall i \in \mathcal{V}$$
$$\sum_{i \in \mathcal{V}:\tau \in [S_i, S_i + D_i)} R_i \leq L \qquad\qquad \forall \tau$$

where τ ranges over the time periods considered. Note that this problem is NP-hard [6].

3   Time-Table Edge-Finding Propagation with Energy Variables

The basic idea of TTEF propagation is to treat a task as a fixed part (used in time-table propagation) and a free part and to determine the range of start times based on the energy available from the resource and the energy required for the tasks in specific time windows.

Time-table edge-finding [14, 18] calculates the amount of energy $e_i(a, b)$ that must be used by a task i in the time window between two time points a and b. The TTEF calculation for $e_i(a, b)$ without energy variables is given by

$$e_i(a, b) := \begin{cases} d_i^{min} \times r_i^{min}, & a \leq est_i \wedge lct_i \leq b \\ \max(0, b - lst_i) \cdot r_i^{min} & a \leq est_i \wedge lct_i > b \\ \max(0, \min(b, ect_i) - \max(a, lst_i)) \cdot r_i^{min} & otherwise \end{cases}$$

The first case is when the entire task must occur in the time window, here we can use the lower bound on the total energy of the task given by $d_i^{min} \times r_i^{min}$. The second case is when the task partially overlaps and some parts might run after the time window, here we use the minimum duration of the overlap times the minimum resource usage. The third case is for all others for which we only consider the minimum energy from the overlapping compulsory part of the task.

The weakness of the usual TTEF formulation without energy variables is that the lower bound of energy of a task $d_i^{min} \times r_i^{min}$ can be very weak, as shown in Example 2. When we have energy variables we can calculate minimum energy usage within a time window more accurately.

$$e_i(a, b) := \begin{cases} e_i^{min} & a \leq est_i \wedge lct_i \leq b \\ \max(0, b - lst_i) \cdot r_i^{min} & a \leq est_i \wedge lct_i > b \\ \max(0, \min(b, ect_i) - \max(a, lst_i)) \cdot r_i^{min} & otherwise \end{cases}$$

Note that only the first case for the calculation changes. We assume that the product constraint $E_i = D_i \times R_i$ is separately propagated so $e_i^{min} \geq d_i^{min} \times r_i^{min}$.

3.1   Consistency Check with Energy Variables

The consistency check is the part of TTEF energy propagation that checks if there is a resource overload in any task interval. Time-table edge finding splits the total energy of a task into a *fixed part* $e_i^{fix} \leftarrow \max(0, r_i^{min} \cdot (lst_i - ect_i))$ and a *free part* $e_i^{free} \leftarrow e_i^{min} - e_i^{fix}$. Let $\mathcal{V}^{Free}$ be the set of tasks with a non-empty free part $\{i \in \mathcal{V} \mid e_i^{free} > 0\}$. The use of energy variables simply allows us to have a better estimation of the least energy used by a task within a time window $[a, b)$.

Proposition 1 (Consistency Check). *The cumulative resource scheduling problem is inconsistent if $R \cdot (b - a) - energy(a, b) < 0$ where $energy(a, b) := \sum_{i \in \mathcal{V}} e_i(a, b)$*

This check can be done in $\mathcal{O}(l^2 + n)$ runtime [18], where $l = |\mathcal{V}^{Free}|$, if the resource profile is given.

The algorithm for the consistency check is shown in Algorithm 1. The difference from TTEF is that for a task i, if lct_i is later than the end time, we can take all its free energy into account; if not, we use the part of free energy lying between the time interval. The differences from the algorithm of Schutt *et al.* [14] are shown in blue.

In order to use the cumulative propagator within a CP solver with nogood learning [11] the propagator needs to be able to explain the *reason* for failures

---

Algorithm 1. TTEF_En consistency check algorithm

Input: X activity array sorted in non-decreasing order of the earliest start time.
Input: Y activity array sorted by (non-decreasing) latest completion time.

1: **procedure** TTEF_EN CONSISTENCY CHECK
2:    $end \leftarrow \infty; minAvail \leftarrow \infty$
3:    **for** $y \leftarrow n$ **down to** 1 **do**
4:        $b \leftarrow Y[y]$
5:        **if** $lct_b = end$ **then** *continue*
6:        **if** $end \neq \infty$ **and** $minAvail \neq \infty$ **and** $minAvail \geq R \cdot (end - lct_b) - ttEn(lct_b, end)$ **then** *continue*
7:        $end \leftarrow lct_b; En^{free} \leftarrow 0; minAvail \leftarrow \infty$
8:        **for** $x \leftarrow n$ **down to** 1 **do**
9:            $a \leftarrow X[x]$
10:            **if** $end \leq est_a$ **then** *continue*
11:            $begin \leftarrow est_a$
12:            $eMin = max(e_a^{min}, d_a^{min} \times r_a^{min})$
13:            **if** $lct_a \leq end$ **then**
14:                $En^{free} \leftarrow En^{free} + eMin - e_a^{fix}$
15:            **else**
16:                $enIn^{fix} \leftarrow max(0, min(end, ect_a) - lst_a) \times r_a^{min}$
17:                $enIn^{free} \leftarrow min(e_a^{free}, max(0, r_a^{min} \times (end - lst_a) - enIn^{fix}))$
18:                $En^{free} \leftarrow En^{free} + enIn^{free}$
19:            $En^{avail} \leftarrow R \cdot (end - begin) - En^{free} - ttEn(a, b)$
20:            **if** $En^{avail} < 0$ **then**
21:                $explainOverload(begin, end)$
22:                **return** *false*
23:            **if** $En^{avail} < minAvail$ **then** $minAvail \leftarrow En^{avail}$

---

of the consistency check. That is it needs to determine a set of facts true about the current domain D which ensure that the consistency check leads to failure.

We need to explain for each task i where $e_i(a, b) > 0$ why its energy usage was at least $e_i(a, b)$. Hence, for task i, the start time should be larger than $a - \left\lfloor \frac{e_i^{min} - e_i(a,b)}{r_i^{max}} \right\rfloor$ and less than $b - \left\lceil \frac{e_i(a,b)}{r_i^{min}} \right\rceil$. And also, the resource usage should be less than r_i^{max} and larger than r_i^{min} because if not, the energy of task i which lie in the time window could be less $e_i(a, b)$. In summary, the explanation should be of the form.

$$\bigwedge_{i \in \mathcal{V}: e_i(a,b) > 0} \llbracket a - \left\lfloor \frac{e_i^{min} - e_i(a,b)}{r_i^{max}} \right\rfloor \leq S_i \rrbracket \wedge \llbracket S_i \leq b - \left\lceil \frac{e_i(a,b)}{r_i^{min}} \right\rceil \rrbracket$$

$$\wedge \llbracket r_i^{min} \leq R_i \rrbracket \wedge \llbracket R_i \leq r_i^{max} \rrbracket \wedge \llbracket e_i^{min} \leq E_i \rrbracket \rightarrow \bot$$

To further generalize the explanation, we can make use of the overload energy $\Delta := energy(a, b) - R \cdot (b - a) - 1$, if $\Delta > 0$. Since the overload occurs even if some tasks use less energy we can give some allowable reduction δ_i in the energy used in the time window for each task i and still use up too much energy in the time window. We choose δ_i such that $\sum_{i\in\mathcal{V}:e_i(a,b)>0}\delta_i = \Delta$. For task i, if $\delta_i \geq e_i(a,b)$ then we can remove the task i completely from the explanation. Otherwise the start time lower bound and upper bound for the explanation can be relaxed to $a - \frac{e_i^{min}-e_i(a,b)+\delta_i}{r_i^{max}}$ and $b - \frac{e_i(a,b)-\delta_i}{r_i^{min}}$, respectively. By default we generalise the tasks in order. We experimented with different policies, but found any reasonable generalization policy was equally effective.

3.2 Start Time Lower Bound Propagation with Energy Variables

Propagation on the lower and upper bounds of the start time variables S_i are basically symmetric; consequently we only discuss the case for the lower bounds' propagation.

To prune the start time lower bound of a task u, TTEF_EN checks if there is an overload when task u starts at its earliest start time in a time window $[a, b)$. Vilím [18] considers four positions of u relative to the time window. In our case, the four positions should be defined as *right* ($a \leq est_u < b < ect_u$), *inside* ($a \leq est_u < ect_u \leq b$), *through* ($est_u < a \wedge b < ect_u$), and *left* ($est_u < a < ect_u \leq b$).

For *right* and *inside* task u, we define $e_u^{est}(a,b) := min(e_u^{min}, r_u^{min} \times (min(est_u + d_u^{max}, b) - est_u))$ as the minimum energy used when u starts at its earliest start time and $e_u^{max}(a,b)$ as the maximum available energy remaining in the time window when u is left out. The update rule for the *right* and *inside* task u, illustrated in Fig. 1, is

$$R \cdot (b - a) - (energy(a,b,u) + e_u^{est}(a,b)) < 0 \rightarrow b - \left\lfloor \frac{e_u^{max}(a,b)}{r_u^{min}} \right\rfloor \leq S_u \quad (1)$$

where $energy(a,b,u) := energy(a,b) - e_u(a,b)$ and $e_u^{max}(a,b) := R \times (b - a) - energy(a,b,u)$. We omit the propagation algorithm for space reasons, it is similar to that shown in [14].

To explain the propagation of the new start time lower bound est_u' for task u, the principle is that we decrease the lower bound on the left hand side as much as possible so that the same propagation holds. For the task u, we can push the explanation of lower bound to the left until the minimum energy lying in the time window just equals $e_u^{max}(a,b)$. And for all other tasks we can perform similar generalization as discussed in the case of resource overload.

$$\bigwedge_{i\in\mathcal{V}\setminus\{u\}:e_i(a,b)>0} (\llbracket a - \left\lfloor \frac{e_i^{min} - e_i(a,b)}{r_i^{max}} \right\rfloor \leq S_i \rrbracket \wedge \llbracket S_i \leq b - \left\lceil \frac{e_i(a,b)}{r_i^{min}} \right\rceil \rrbracket)$$

$$\wedge \; \llbracket r_i^{min} \leq R_i \rrbracket \wedge \llbracket R_i \leq r_i^{max} \rrbracket \wedge \llbracket e_i^{min} \leq E_i \rrbracket \wedge \llbracket e_u^{min} \leq E_u \rrbracket$$

$$\wedge \; \llbracket a - \left\lfloor \frac{e_u^{min} - e_u^{max}(a,b)}{r_u^{max}} + 1 \right\rfloor \leq S_u \rrbracket \wedge \llbracket r_u^{min} \leq R_u \rrbracket \wedge \llbracket R_u \leq r_u^{max} \rrbracket$$

$$\rightarrow \llbracket est_u' \leq S_u \rrbracket$$

4 Experimental Evaluation

We now compare our solution approach TTEF_EN to both time-table (tt), time-table edge-finding TTEF propagation. We compare average conflicts (conf) and average time (in seconds) for 10 runs, ∞ indicates all runs fail to prove optimality in time. The experiments were run on a X86-64 running MacOS 10.13 and a Intel Core m3 CPU processor at 1.2 GHz. We set the timeout for each run to 1800 s. All models and data are available at people.eng.unimelb.edu.au/pstuckey/ttefen.

Rectangle Packing problems [10] are highly cumulative and hence good examples for TTEF propagation. We compare three different versions: (a) consecutive rectangle packing [10], where instance N is the set of rectangles of size $1 \times 2, 2 \times 3, ...,$ up to $N \times (N+1)$ that may be rotated. (b) double-perimeter rectangle packing [10], where instance N is the set of rectangles of size $1 \times (2N-1), 2 \times (2N-2), ...,$ up to $N \times (N+1)$ that may be rotated. (c) free rectangle packing, where instance N is a set of rectangles constrained to take areas to be $1 \times (2N-1), 2 \times (2N-2), ...,$ up to $N \times (N+1)$ with any height and width giving the correct area. The results using default activity based search are shown in Table 1. Clearly TTEF_EN propagation is superior to the alternatives, and its advantage grows with problem size. We also compared using fixed search (not shown) where TTEF_EN was also superior, but not by as much.

Evacuation Planning problems [5] try to schedule evacuation tasks so everyone is evacuated as quickly as possible. Cumulative constraints constrain the flow rates r_i of evacuation tasks on road segments. The total energy of a task i is

Table 1. Results for rectangle packing

Consecutive	12		13		14		15		16	
	conf	time	conf	time	conf	time	conf	time	conf	time
tt	11250	9.21	33556	23.91	111749	71.85	306233	191.01	∞	∞
TTEF	13912	9.76	30705	20.91	72670	45.64	216442	144.92	885743	581.41
TTEF_EN	**8367**	**7.23**	**23836**	**15.36**	**59998**	**37.57**	**157539**	**113.56**	**806538**	**503.17**
Double perimeter	7		8		9		10		11	
	conf	time	conf	time	conf	time	conf	time	conf	time
tt	2251	0.90	12195	3.55	41203	16.88	106326	62.34	∞	∞
TTEF	2200	1.18	10550	5.08	39329	17.06	93773	61.69	817887	815.26
TTEF_EN	**1979**	**0.91**	**8035**	**3.40**	**25127**	**9.57**	**55212**	**34.56**	**454654**	**361.18**
Free	6		7		8		9		10	
	conf	time	conf	time	conf	time	conf	time	conf	time
tt	1896	1.71	6965	6.88	155992	119.72	594878	488.63	∞	∞
TTEF	**1700**	**1.65**	8001	7.74	153885	118.31	546442	457.52	∞	∞
TTEF_EN	1712	1.68	**6237**	**5.76**	**96074**	**80.68**	**446776**	**311.91**	∞	∞

Table 2. Evacuation problem.

	9		10		11		20	30	40	50
	conf	time	conf	time	conf	time	evac	evac	evac	evac
tt	2069.8	**218.81**	6962.7	684.59	1830.5	490.84	9326.4	17348.3	21941.5	27926.0
TTEF	2306.2	228.42	7108.7	675.45	2437.4	551.64	8995.8	16968.1	21909.9	27925.3
TTEF_EN	**1993.9**	219.21	**5817.8**	**578.43**	**1646.1**	**448.05**	**8959.7**	**16851.6**	**21884.9**	**27879.7**

the number n_i of evacuees constrained so that the $d_i \times r_i \geq n_i$. The results using default search are shown in Table 2, where N is the number of evacuation zones, we use 10 randomly generated instances for each N and show average results. For small examples where all methods can prove optimality, we compare conflicts and time. For larger examples we simple compare minimal evacuation time (evac) at time out. The smaller results show that energy variables improve the number of conflicts and time (except the smallest example). Interestingly here TTEF does not beat tt in terms of conflicts or time. For larger results we see TTEF is superior to tt and bettered by TTEF_EN.

5   Conclusion and Related Work

The addition of energy variables to the cumulative constraint allows us to improve any energy based reasoning approach for cumulative. The experiments show that in problem classes where TTEF propagation is effective, the version using energy variables TTEF_EN is even more effective. We expect the same would occur if we added energy variables to other cumulative propagators that reason about energy, e.g., [7–9,12,13,15,16].

Note that a number of versions of the cumulative constraint appearing in the CHIP system [4] included energy variables (there called "surface" variables). How these variables are used in propagation is not described in any detail; we do not believe they are combined with TTEF propagation. Interestingly, no version of cumulative with energy variables appears in the Global Constraint Catalog, even though the CHIP developers are key contributors.

However, Beldiceanu [3], one of the key contributors, describes a function called *ask_what_if* that can be passed to global constraint propagators and the propagator can query about bounds on, e.g., a product of two variables. This could be used to imitate energy variables, but is not implemented in any system we are aware of.

The most common used CP solvers (Gecode, Choco, JaCoP, SICStus Prolog, CP Optimizer, OR Tools) do not have an implementation, which considers the product/ energy variable. To best of our knowledge, there is no publication of filtering algorithms on the energy variable. Note that Vilim [17] determines the maximal available energy in certain time windows according to the edge-finder rule in order to propagate max duration and max resource usage, but does not consider energy variables which would improve the propagation.

References

1. Aggoun, A., Beldiceanu, N.: Extending CHIP in order to solve complex scheduling and placement problems. Math. Comput. Modell. **17**(7), 57–73 (1993)
2. Baptiste, P., Le Pape, C.: Constraint propagation and decomposition techniques for highly disjunctive and highly cumulative project scheduling problems. Constraints **5**(1–2), 119–139 (2000)
3. Beldiceanu, N.: Global constraints as graph properties on structured network of elementary constraints of the same type. SICS Technical report T2000/01, SICS, Uppsala, Sweden (2000)
4. COSYTEC SA: CHIP v5.12.2.0 finite domain constraints reference manual. Technical report, COSYTEC SA (2017)
5. Even, C., Schutt, A., Van Hentenryck, P.: A constraint programming approach for non-preemptive evacuation scheduling. In: Pesant, G. (ed.) CP 2015. LNCS, vol. 9255, pp. 574–591. Springer, Cham (2015). https://doi.org/10.1007/978-3-319-23219-5_40
6. Garey, M.R., Johnson, D.R.: Computers and Intractability. W.H Freeman and Co, San Francisco (1979)
7. Gingras, V., Quimper, C.G.: Generalizing the edge-finder rule for the cumulative constraint. In: IJCAI, pp. 3103–3109 (2016)
8. Kameugne, R., Fotso, L.P.: A cumulative not-first/not-last filtering algorithm in $O(n^2 \log(n))$. Indian J. Pure Appl. Math. **44**(1), 95–115 (2013)
9. Kameugne, R., Fotso, L.P., Scott, J.: A quadratic extended edge-finding filtering algorithm for cumulative resource constraints. Int. J. Plan. Sched. **1**(4), 264–284 (2013)
10. Korf, R.: Optimal rectangle packing: initial results. In: Giunchiglia, E., Muscettola, N., Nau, D. (eds.) Proceedings of the Thirteenth International Conference on Automated Planning and Scheduling (ICAPS 2003), pp. 287–295. AAAI PRess (2003)
11. Ohrimenko, O., Stuckey, P.J., Codish, M.: Propagation via lazy clause generation. Constraints **14**(3), 357–391 (2009)
12. Ouellet, P., Quimper, C.-G.: Time-table extended-edge-finding for the cumulative constraint. In: Schulte, C. (ed.) CP 2013. LNCS, vol. 8124, pp. 562–577. Springer, Heidelberg (2013). https://doi.org/10.1007/978-3-642-40627-0_42
13. Ouellet, Y., Quimper, C.-G.: A $O(n \log^2 n)$ checker and $O(n^2 \log n)$ filtering algorithm for the energetic reasoning. In: van Hoeve, W.-J. (ed.) CPAIOR 2018. LNCS, vol. 10848, pp. 477–494. Springer, Cham (2018). https://doi.org/10.1007/978-3-319-93031-2_34
14. Schutt, A., Feydy, T., Stuckey, P.J.: Explaining time-table-edge-finding propagation for the cumulative resource constraint. In: Gomes, C., Sellmann, M. (eds.) CPAIOR 2013. LNCS, vol. 7874, pp. 234–250. Springer, Heidelberg (2013). https://doi.org/10.1007/978-3-642-38171-3_16
15. Schutt, A., Wolf, A.: A new $\mathcal{O}(n^2 \log n)$ not-first/not-last pruning algorithm for cumulative resource constraints. In: Cohen, D. (ed.) CP 2010. LNCS, vol. 6308, pp. 445–459. Springer, Heidelberg (2010). https://doi.org/10.1007/978-3-642-15396-9_36
16. Tesch, A.: Improving energetic propagations for cumulative scheduling. In: Hooker, J. (ed.) CP 2018. LNCS, vol. 11008, pp. 629–645. Springer, Cham (2018). https://doi.org/10.1007/978-3-319-98334-9_41

17. Vilím, P.: Max energy filtering algorithm for discrete cumulative resources. In: van Hoeve, W.-J., Hooker, J.N. (eds.) CPAIOR 2009. LNCS, vol. 5547, pp. 294–308. Springer, Heidelberg (2009). https://doi.org/10.1007/978-3-642-01929-6_22
18. Vilím, P.: Timetable edge finding filtering algorithm for discrete cumulative resources. In: Achterberg, T., Beck, J.C. (eds.) CPAIOR 2011. LNCS, vol. 6697, pp. 230–245. Springer, Heidelberg (2011). https://doi.org/10.1007/978-3-642-21311-3_22

Quadratic Reformulation of Nonlinear Pseudo-Boolean Functions via the Constraint Composite Graph

Ka Wa Yip$^{(\boxtimes)}$, Hong Xu$^{(\boxtimes)}$, Sven Koenig, and T. K. Satish Kumar

University of Southern California, Los Angeles, CA 90089, USA
{kawayip,hongx,skoenig}@usc.edu, tkskwork@gmail.com

Abstract. *Nonlinear pseudo-Boolean optimization* (nonlinear PBO) is the minimization problem on *nonlinear pseudo-Boolean functions* (nonlinear PBFs). One promising approach to nonlinear PBO is to first use a *quadratization algorithm* to reduce the PBF to a quadratic PBF by introducing intelligently chosen auxiliary variables and then solve it using a quadratic PBO solver. In this paper, we develop a new quadratization algorithm based on the idea of the *constraint composite graph* (CCG). We demonstrate its theoretical advantages over state-of-the-art quadratization algorithms. We experimentally demonstrate that our CCG-based quadratization algorithm outperforms the state-of-the-art algorithms in terms of both effectiveness and efficiency on randomly generated instances and a novel reformulation of the uncapacitated facility location problem.

1 Introduction

Nonlinear pseudo-Boolean optimization (nonlinear PBO) refers to the minimization problem on *nonlinear pseudo-Boolean functions* (nonlinear PBFs). Formally, a PBF is a mapping $f : \mathbb{B}^n \to \mathbb{R}$ that maps each assignment of values to a set of Boolean variables to a real number. The Boolean variables are restricted to take a value in $\mathbb{B} = \{0, 1\}$. A PBF is nonlinear iff it cannot be reformulated as a linear combination of the Boolean variables. (Nonlinear) PBO asks for an assignment of values to the Boolean variables that minimizes the value of a (nonlinear) PBF, i.e., it is the task of computing arg $\min_{x \in \mathbb{B}^n} f(x)$. Nonlinear PBO is known to be NP-hard and subsumes many classic optimization problems, such as MAX-SAT and MAX-CUT [10]. It has been used in many real-world applications, such as computer vision [26], operations research and traffic planning [15,18,37], chip design [11], evolutionary computation [38], and spin-glass models [25].

While a lot of research has concentrated on linear PBO (with linear constraints) [1,8,13,23,31,33,35], nonlinear PBO has not been very well studied. There only exist a handful of techniques dedicated to nonlinear PBO, such as

The research at the University of Southern California was supported by NSF under grant numbers 1724392, 1409987, 1817189, and 1837779.

© Springer Nature Switzerland AG 2019
L.-M. Rousseau and K. Stergiou (Eds.): CPAIOR 2019, LNCS 11494, pp. 643–660, 2019.
https://doi.org/10.1007/978-3-030-19212-9_43

reformulation to 0/1 *integer linear programming* (ILP) [16], constraint integer programming [6], constraint logic programming [7], and graph cuts [26]. Among these techniques, the most viable approach involves *quadratization algorithms*, i.e., reformulating the PBF as a *quadratic* PBF [3,21]. A PBF is quadratic iff it is a sum of monomials in which each monomial is a product of at most two Boolean variables. Several authors have pointed out the benefits of quadratization algorithms over algorithms based on linearization and ILP [4,10,17].

Quadratization algorithms are not only the most viable approach to PBO, but also the only viable approach that serves some fundamental purposes. For example, quantum annealers—such as the D-Wave chips—can solve only quadratic unconstrained Boolean optimization problems [22]. Therefore, a quadratization algorithm is indispensable for solving nonlinear PBO instances on quantum annealers. In addition, no existing weighted MAX-SAT or ILP solver can solve nonlinear PBO instances with arbitrary lengths of monomials. They, including BiqMac [36], are only applicable to unconstrained binary quadratic programs. Quadratization algorithms are therefore required for reformulating nonlinear PBO instances to make them amenable to such solvers. In general, quadratization algorithms are useful due to the existence of more efficient algorithms that are dedicated to minimizing quadratic PBFs, i.e., to *quadratic PBO* (QPBO). For example, a QPBO solver can make use of the peculiar properties of quadratic PBFs, such as their *roof duality* [24] and the existence of polynomial-time algorithms for finding partial solutions even if the PBFs are not submodular [21].

Formally, a *quadratization* of a PBF $f(\boldsymbol{x})$ is a quadratic PBF $g(\boldsymbol{x}, \boldsymbol{y})$ such that

$$f(\boldsymbol{x}) = \min_{\boldsymbol{y} \in \mathbb{B}^m} g(\boldsymbol{x}, \boldsymbol{y}) \quad \forall \boldsymbol{x} \in \mathbb{B}^n, \tag{1}$$

where $\boldsymbol{y}$ is a set of m auxiliary Boolean variables. Since minimizing a quadratic PBF is also NP-hard, quadratization algorithms should preferably be achieved in polynomial time using a small number of auxiliary variables. Since the number of variables largely determines the size of the search space, existing algorithms focus on minimizing the number of auxiliary variables [3,21].

In this paper, we develop a new polynomial-time quadratization algorithm based on the *constraint composite graph* (CCG) [28–30]. We show that our CCG-based quadratization algorithm has a theoretical advantage over the state-of-the-art algorithms proposed in [21]. We also experimentally demonstrate that our CCG-based quadratization algorithm outperforms the state-of-the-art quadratization algorithms in terms of the required number of auxiliary variables, the number of terms in the quadratization, and the runtime. We conduct experiments on both randomly generated instances and a novel reformulation of the uncapacitated facility location problem.

2   Preliminaries

In this section, we give a brief background on quadratizations and the CCG.

2.1   Quadratizations

A PBF is a function that maps n Boolean variables to a real number. As proved in [18], any PBF f of n Boolean variables $x = \{x_1, \ldots, x_n\}$ can be uniquely represented as a polynomial of the form

$$f(x) = \sum_{S \subseteq x} c_S \prod_{x \in S} x, \tag{2}$$

where $c_S \in \mathbb{R}$. Throughout this paper, we specify all PBFs in this form. We let d denote the degree of a PBF, i.e., the maximum degree of all its monomials.

A quadratization of a PBF $f(\cdot)$ is a quadratic PBF $g(\cdot)$ that satisfies Eq. (1). For any given PBF, its quadratizations exist but are not necessarily unique. Since a quadratization algorithm can be seen as a preprocessing algorithm, its effectiveness can be evaluated using two metrics: the number of auxiliary variables and the number of terms in $g(\cdot)$, which usually are good indicators of the time required to solve the resulting quadratic PBF using a QPBO solver.

In terms of the number of auxiliary variables in $g(\cdot)$, some current state-of-the-art algorithms are given in [21]. The first algorithm is called *polynomial expansion* and is polynomial-time. It first *quadratizes*, i.e., finds a quadratization of, each monomial in $f(\cdot)$ individually and then combines all like quadratic terms. Polynomial expansion quadratizes a monomial $ax_1 \ldots x_d$ of degree $d > 2$ to

$$\begin{cases} \min_{w \in \mathbb{B}} aw\left(S_1 - (d-1)\right) & \text{if } a < 0 \\ \min_{\{w_1, \ldots, w_{n_d}\} \in \mathbb{B}^{n_d}} \sum_{i=1}^{n_d} w_i \left(c_{i,d}(-S_1 + 2i) - 1\right) + aS_2 & \text{if } a > 0, \end{cases} \tag{3}$$

where

$$S_1 = \sum_{i=1}^{d} x_i \qquad S_2 = \sum_{i=1}^{d-1} \sum_{j=i+1}^{d} x_i x_j = \frac{S_1(S_1 - 1)}{2}$$

$$n_d = \left\lfloor \frac{d-1}{2} \right\rfloor \qquad c_{i,d} = \begin{cases} 1 & \text{if } i = n_d \text{ and } d \text{ is odd} \\ 2 & \text{otherwise.} \end{cases}$$

Therefore, if $a < 0$, quadratizing this monomial requires 1 auxiliary variable; if $a > 0$, it requires n_d auxiliary variables[1].

The second algorithm is called γ *flipping*. Let $\gamma = \{\gamma_1, \ldots, \gamma_n\} \in \mathbb{B}^n$ and

$$x_i^{(\gamma)} = \gamma_i x_i + \bar{\gamma}_i \bar{x}_i = \begin{cases} x_i & \text{if } \gamma_i = 1 \\ \bar{x}_i & \text{if } \gamma_i = 0. \end{cases} \text{ Then, we have}$$

$$f(x) = \sum_{\gamma \in \mathbb{B}^n} f(\gamma) x_1^{(\gamma)} \ldots x_n^{(\gamma)} = \lambda + \sum_{\gamma \in \mathbb{B}^n} (f(\gamma) - \lambda) x_1^{(\gamma)} \ldots x_n^{(\gamma)}. \tag{4}$$

---

[1] For a single positive monomial, the smallest possible number of auxiliary variables achievable is $\lceil \log d \rceil - 1$, as proven in [9].

When using

$$\lambda = \max_{\gamma \in \mathbb{B}^n} f(\gamma), \tag{5}$$

every monomial in Eq. (4) is non-positive. Then, by following the first case in Eq. (3), γ flipping requires exactly $2^n - 1$ auxiliary variables. This is superpolynomial with respect to the size of input if the number of terms in Eq. (2) is, for example, polynomial with respect to n.

The computation of Eq. (5) is the bottleneck. Its time complexity is superpolynomial with respect to n. Hence, γ flipping is superpolynomial-time if the number of terms in Eq. (2) is, for example, polynomial with respect to n.

2.2   Constraint Composite Graph

The CCG [28–30] is a combinatorial structure associated with an optimization problem posed as the *weighted constraint satisfaction problem* (WCSP). It simultaneously represents the graphical structure of the variable interactions in the WCSP and the numerical structure of the constraints in it. The task of solving the WCSP can be reformulated as the task of finding a *minimum weighted vertex cover* (MWVC) (called the MWVC problem) on its associated CCG. CCGs can be constructed in polynomial time and are always tripartite. A subclass of the WCSP has instances with bipartite CCGs. This subclass is tractable since an MWVC can be found in polynomial time on bipartite graphs using a maxflow algorithm [27]. The CCG also facilitates kernelization, message passing [14,40], and an efficient encoding of the WCSP as an integer linear program [39].

Given an undirected graph $G = \langle V, E \rangle$, a vertex cover of G is a set of vertices $S \subseteq V$ such that every edge in E has at least one of its endpoint vertices in S. A *minimum vertex cover* (MVC) of G is a vertex cover of minimum cardinality. When G is vertex-weighted—i.e., each vertex $v_i \in V$ has a non-negative weight w_i associated with it—its MWVC is defined as a vertex cover of minimum total weight of its vertices. The MWVC problem is the task of computing an MWVC on a given vertex-weighted undirected graph.

For a given graph G, the concept of the MWVC problem can be extended to the notion of projecting MWVCs onto a given *independent set* (IS) $U \subseteq V$. (An IS of G is a set of vertices in which no two vertices are adjacent to each other.) The input to such a projection is the graph G as well as an IS $U = \{u_1, u_2, \ldots, u_k\}$. The output is a table of 2^k numbers. Each entry in this table corresponds to a k-bit vector. We say that a k-bit vector t imposes the following restrictions: (i) if the ith bit t_i is 0, the vertex u_i has to be excluded from the MWVC; and (ii) if the ith bit t_i is 1, the vertex u_i has to be included in the MWVC. The projection of the MWVC problem onto the IS U is then defined to be a table with entries corresponding to each of the 2^k possible k-bit vectors $t^{(1)}, t^{(2)}, \ldots, t^{(2^k)}$. The value of the entry corresponding to $t^{(j)}$ is equal to the weight of the MWVC conditioned on the restrictions imposed by $t^{(j)}$. [28, Fig. 2] illustrates this projection.

The table of numbers produced above can be viewed as a weighted constraint over $|U|$ Boolean variables. Conversely, given a (Boolean) weighted constraint, we design a lifted representation for it so as to be able to view it as the projection of an MWVC onto an IS of some intelligently constructed vertex-weighted undirected graph [28,29]. The benefit of constructing these representations for individual constraints lies in the fact that the lifted representation for the entire WCSP, i.e., the CCG of the WCSP, can be obtained simply by "merging" them.

[28, Fig. 5] shows an example WCSP instance over 3 Boolean variables to illustrate the construction of the CCG. Here, there are 3 unary and 3 binary weighted constraints. Their lifted representations are shown next to them. The figure also illustrates how the CCG is obtained from the lifted representations of the weighted constraints: In the CCG, vertices that represent the same variable are simply "merged"—along with their edges—and every "composite" vertex is given a weight equal to the sum of the individual weights of the merged vertices. Computing the MWVC for the CCG yields a solution for the WCSP instance; namely, if X_i is in the MWVC, then it is assigned value 1 in the WCSP instance, otherwise it is assigned value 0 in the WCSP instance.

3   The CCG-Based Quadratization Algorithm

PBO is a special case of the WCSP and is therefore equivalent to solving the MWVC problem on its associated CCG. In turn, we show that the MWVC problem itself can be reformulated as QPBO. This leads to the CCG-based quadratization algorithm presented in this section.

Given a vertex-weighted graph $G = \langle V, E, w \rangle$ and one of its independent sets T, the projection of the MWVC problem onto T is a table of weights of MWVCs with all combinations of vertices in T imposed to be included in or excluded from the MWVC [28]. More formally:

Definition 1. *Let $T_+ \cup T_- = T$ and $T_+ \cap T_- = \emptyset$. S is a para-vertex cover on $\langle G, T_+, T_- \rangle$ iff S is a vertex cover on G, $T_+ \subseteq S$, and $T_- \cap S = \emptyset$. S is a para-MWVC on $\langle G, T_+, T_- \rangle$ iff S is a para-vertex cover on $\langle G, T_+, T_- \rangle$ and the sum of weights of all vertices in S is no greater than that of any other para-vertex cover on $\langle G, T_+, T_- \rangle$. The projection of the MWVC problem onto T (on G) is a function that maps $\langle T_+, T_- \rangle$ to the weight of a para-MWVC on $\langle G, T_+, T_- \rangle$.*

The following theorem is inspired by [12, Theorem 3].

Theorem 1. *Let us consider the finite graph $G = \langle V, E, w \rangle$ and an independent set $T = T_+ \cup T_-$ on it. Let $\boldsymbol{x} = (x_r : r \in V)$ and*

$$C(\boldsymbol{x}) = \sum_{p \in V} w_p x_p + \sum_{(p,q) \in E} J_{pq}(1 - x_p)(1 - x_q). \tag{6}$$

(i) If $\forall (p, q) \in E : J_{pq} \geq \max\{w_p, w_q\}$, then the projection of the MWVC problem onto an independent set T equals the function

$$h(\langle T_+, T_- \rangle) = \min_{x_j \in \mathbb{B} : j \in V \setminus T} C(\boldsymbol{x})\Big|_{\substack{x_i = 1 \quad \text{if } i \in T_+ \\ x_i = 0 \quad \text{if } i \in T_-}}. \tag{7}$$

(ii) If further $\forall (p,q) \in E : J_{pq} > \max\{w_p, w_q\}$, *then any* $S^* \subset V$ *that satisfies*

$$T_+ \subseteq S^* \tag{8}$$

$$T_- \cap S^* = \emptyset \tag{9}$$

$$\mathscr{C}(S^*) = h(\langle T_+, T_- \rangle) \tag{10}$$

is a para-MWVC on $\langle G, T_+, T_- \rangle$, *with* $\mathscr{C}(\cdot)$ *defined as* $\mathscr{C}(S) = C(\boldsymbol{x})\big|_{\substack{x_i=1 \quad if\ i \in S. \\ x_i=0 \quad if\ i \notin S}}$

Proof. Let us consider a given $\langle T_+, T_- \rangle$. We first prove (ii), then (i).

For (ii): We first prove by contradiction that, if $J_{pq} > \max\{w_p, w_q\}$, then S^* is a vertex cover. Let $x_i^* = 1$ if $i \in S^*$ and $x_i^* = 0$ if $i \in V \backslash S^*$, and $\boldsymbol{x}^* = (x_i^* : i \in V)$. We assume that there exists an edge (a, b) such that $x_a^* = x_b^* = 0$. Neither a nor b can be in T_+ because $T_+ \subseteq S^*$. Since T is an independent set, a and b cannot be both in T_-. If we hold either of the rest cases, i.e.,

– if only one of a and b is in T_- (without loss of generality, we let $a \in T_-$), or
– if neither a nor b is in T_-,

then $a \notin S^*$ and $\mathscr{C}(S^*) - \mathscr{C}(S^* \cup \{b\}) = \sum_{k \notin S^* : (b,k) \in E} J_{bk} - w_b \geq J_{ab} - w_b > 0$, which contradicts Eq. (10).

In addition, S^* is also a para-MWVC on $\langle G, T_+, T_- \rangle$ because S^* being a vertex cover implies $\sum_{(p,q) \in E} J_{pq}(1 - x_p^*)(1 - x_q^*) = 0$. Therefore, (ii) holds.

For (i): Let $S^{*\prime}$ be a para-MWVC on $\langle G, T_+, T_- \rangle$. Because $S^{*\prime}$ is a vertex cover, the second summation of Eq. (6) in $\mathscr{C}(S^{*\prime})$ equals zero and thus the weight of $S^{*\prime}$ equals $\mathscr{C}(S^{*\prime})$. Therefore, it is sufficient to prove that there exists such an $S^{*\prime}$ that satisfies $\mathscr{C}(S^{*\prime}) = h(\langle T_+, T_- \rangle)$, or, equivalently, $\mathscr{C}(S^{*\prime}) = \mathscr{C}(S^*)$.

If $C(\cdot)$ is a constant function, then it is obvious that $\mathscr{C}(S^{*\prime}) = h(\langle T_+, T_- \rangle)$. We now consider the case where $C(\cdot)$ is not a constant function. Let $E' = \{(p,q) \in E : J_{pq} = \max\{w_p, w_q\}\}$. Let

$$C'(\boldsymbol{x}) = \sum_{p \in V} w_p x_p + \sum_{(p,q) \in E} J'_{pq}(1 - x_p)(1 - x_q). \tag{11}$$

Here, $J'_{pq} = \begin{cases} J_{pq} & \text{if } (p,q) \notin E' \\ J_{pq} + \epsilon_{pq} & \text{if } (p,q) \in E' \end{cases}$, where $\forall (p,q) \in E' : \epsilon_{pq} > 0$ and they satisfy

$$\sum_{(p,q) \in E'} \epsilon_{pq} < \epsilon_0, \tag{12}$$

where ϵ_0 is the smallest positive value that $C(\boldsymbol{x}) - C(\boldsymbol{y})$ can be for all $\boldsymbol{x}, \boldsymbol{y} \in \mathbb{B}^{|V|}$, i.e., $\epsilon_0 = \min\{C(\boldsymbol{x}) - C(\boldsymbol{y}) \in \mathbb{R}_{>0} : \boldsymbol{x}, \boldsymbol{y} \in \mathbb{B}^{|V|}\}$. Here, the operand of min cannot be $\emptyset$ because $C(\cdot)$ is not a constant function.

($\star$) Let $S^{*\prime} \subseteq V$ satisfy Eqs. (8) to (10) except that $\mathscr{C}(\cdot)$ in Eq. (10) is replaced by $\mathscr{C}'(\cdot)$, defined as $\mathscr{C}'(S) = C'(\boldsymbol{x})\big|_{\substack{x_i=1 \quad if\ i \in S, \\ x_i=0 \quad if\ i \notin S}}$ and all occurrences of $C(\cdot)$ are replaced by $C'(\cdot)$. According to (ii), $S^{*\prime}$ is a para-MWVC on $\langle G, T_+, T_- \rangle$.

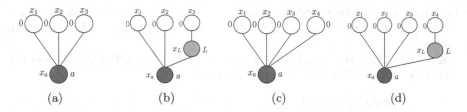

Fig. 1. The graph gadgets for the construction of the CCG. Each vertex is associated with a weight and a label. "x_a" and "x_L" are the labels of the auxiliary variables.

Let $x_i^{*\prime} = 1$ if $i \in S^{*\prime}$ and $x_i^{*\prime} = 0$ if $i \in V \setminus S^{*\prime}$, and $\boldsymbol{x}^{*\prime} = (x_i^{*\prime} : i \in V)$. Now we only need to prove $\mathscr{C}(S^{*\prime}) = \mathscr{C}(S^*)$, or, equivalently, $\mathcal{C}(\boldsymbol{x}^{*\prime}) = \mathcal{C}(\boldsymbol{x}^*)$.

According to Eq. (10), $\mathcal{C}(\boldsymbol{x}^*) = h(\langle T_+, T_- \rangle)$. Therefore, $\mathcal{C}(\boldsymbol{x}^{*\prime}) \geq \mathcal{C}(\boldsymbol{x}^*)$. We now only need to prove that $\mathcal{C}(\boldsymbol{x}^{*\prime}) > \mathcal{C}(\boldsymbol{x}^*)$ cannot hold. We prove by contradiction. Assume $\mathcal{C}(\boldsymbol{x}^{*\prime}) > \mathcal{C}(\boldsymbol{x}^*)$. Then, according to the definition of ϵ_0, $\mathcal{C}(\boldsymbol{x}^{*\prime}) - \mathcal{C}(\boldsymbol{x}^*) \geq \epsilon_0$. According to Eqs. (11) and (12), $\mathcal{C}(\boldsymbol{x}^*) - \mathcal{C}'(\boldsymbol{x}^*) \geq -\sum_{(p,q)\in E'} \epsilon_{pq} > -\epsilon_0$. According to Eq. (11), $\mathcal{C}'(\boldsymbol{x}^{*\prime}) - \mathcal{C}(\boldsymbol{x}^{*\prime}) \geq 0$. Adding these three inequalities, we have $\mathcal{C}'(\boldsymbol{x}^{*\prime}) - \mathcal{C}'(\boldsymbol{x}^*) > 0$, and thus $\mathscr{C}'(S^{*\prime}) > \mathscr{C}'(S^*)$. This contradicts Eq. (10) after the replacements in (✫). □

Based on Theorem 1, we outline a quadratization algorithm for a (nonlinear) PBO—or more generally for a WCSP—as follows: (i) Using the polynomial-time algorithm proposed in [28], reformulate the input PBF $f(\boldsymbol{x})$ to the projection of the MWVC problem on its CCG; and (ii) using Theorem 1, convert this projection to a quadratic PBF and output it as its quadratization.

3.1   A Full Example

Consider the PBF

$$P(x_1, x_2, x_3, x_4) = 3x_2x_3 + 5x_1x_2x_3 + 6x_1x_2x_3x_4 - 3x_1x_3x_4. \tag{13}$$

The CCG is a composition of graph gadgets, each of which represents a monomial [28]. Each monomial is related to an MWVC of a particular graph gadget (Fig. 1). Assume $a > 0$ (throughout this subsection), for a monomial $-ax_1x_2x_3$, MWVC $\{$Fig. 1a$\} = a - ax_1x_2x_3$, where MWVC $\{$Fig. 1a$\}$ is the weight of the MWVCs of Fig. 1a, i.e.,

$$-ax_1x_2x_3 = \text{MWVC}\,\{\text{Fig. 1a}\} - a. \tag{14}$$

For $ax_1x_2x_3$, MWVC $\{$Fig. 1b$\} = L(1-x_3) + a - a(x_1x_2(1-x_3))$, for a sufficiently large constant L, i.e.,

$$ax_1x_2x_3 = \text{MWVC}\,\{\text{Fig. 1b}\} - L(1 - x_3) - a + ax_1x_2. \tag{15}$$

For $ax_1x_2x_3x_4$, MWVC $\{$Fig. 1d$\} = L(1 - x_4) + a - a(x_1x_2x_3(1 - x_4))$, i.e.,

$$ax_1x_2x_3x_4 = \text{MWVC}\,\{\text{Fig. 1d}\} - L(1 - x_4) - a + ax_1x_2x_3, \tag{16}$$

where $ax_1x_2x_3$ can be rewritten as in Eq. (15). Here, all monomials of degree > 2 have been rewritten as quadratic PBFs and weights of MWVCs of graph gadgets.

Applying Theorem 1 and setting $J \geq L > a > 0$, we further express the weights of MWVCs in algebraic quadratic forms as

$$
\text{MWVC} \{\text{Fig. } 1a\} = \min_{x_a}[ax_a + J(1 - x_1)(1 - x_a) \\
+ J(1 - x_2)(1 - x_a) + J(1 - x_3)(1 - x_a)]
\tag{17}
$$

$$
\text{MWVC} \{\text{Fig. } 1b\} = \min_{x_a, x_L}[ax_a + Lx_L + J(1 - x_1)(1 - x_a) \\
+ J(1 - x_2)(1 - x_a) + J(1 - x_3)(1 - x_L) + J(1 - x_L)(1 - x_a)]
\tag{18}
$$

$$
\text{MWVC} \{\text{Fig. } 1c\} = \min_{x_a}[ax_a + J(1 - x_1)(1 - x_a) \\
+ J(1 - x_2)(1 - x_a) + J(1 - x_3)(1 - x_a) + J(1 - x_4)(1 - x_a)]
\tag{19}
$$

$$
\text{MWVC} \{\text{Fig. } 1d\} = \min_{x_a, x_L}[ax_a + Lx_L + J(1 - x_1)(1 - x_a) \\
+ J(1 - x_2)(1 - x_a) + J(1 - x_3)(1 - x_a) \\
+ J(1 - x_4)(1 - x_L) + J(1 - x_L)(1 - x_a)].
\tag{20}
$$

Here, we have quadratized all monomials of degree > 2 using the algebraic expression of the weight of MWVCs on its graph gadget. The auxiliary variables are named uniquely for each graph gadget. For the PBF in Eq. (13), we therefore need 5 auxiliary variables: x_a and x_L for the degree-4 term, $x_{a'}$ and $x_{L'}$ for the degree-3 term with positive coefficient that combines the existing degree-3 term with the degree-3 term that comes from the reduction of the degree-4 term, and $x_{a''}$ for the degree-3 term with negative coefficient.

3.2   Details of the CCG-Based Quadratization Algorithm

The CCG-based quadratization algorithm is an iterative algorithm. Let $f(x)$ be the input PBF. It initializes a polynomial $f'(x)$ to $f(x)$. In each iteration, let d be the degree of $f'(x)$. It substitutes each degree-d negative monomial and positive monomial in $f'(x)$, respectively, using

$$
-ax_1 \ldots x_d = \min_{x_a} \left[ax_a + J\sum_{i=1}^{d}(1 - x_i)(1 - x_a) \right] - a
\tag{21}
$$

$$
ax_1 \ldots x_d = \min_{x_a, x_L} \left[ax_a + Lx_L + J\sum_{i=1}^{d-1}(1 - x_i)(1 - x_a) \right.
$$

$$
\left. + J(1 - x_d)(1 - x_L) + J(1 - x_L)(1 - x_a) \right]
\tag{22}
$$

$$
- L(1 - x_d) - a + ax_1 \ldots x_{d-1},
$$

where $J \geq L > a > 0$. It then combines all like terms in $f'(x)$. Because the right-hand sides of Eqs. (21) and (22) are of degrees that are lower than the left-hand sides, the degree of $f'(x)$ decreases by at least 1 after each iteration. The iterating procedure terminates until the degree of $f'(x)$ is no larger than 2. Finally, the algorithm outputs $f'(x)$ as the quadratization.

4   Evaluation

In this section, we evaluate our CCG-based quadratization algorithm both theoretically and experimentally, and illustrate its uses and advantages on a real-world problem.

4.1   Theoretical Results

In this subsection, we theoretically compare our CCG-based quadratization algorithm with the state-of-the-art algorithms in [21] in terms of the number of auxiliary variables. For any PBF of degree d on n variables, the maximum number of non-zero monomials of each degree $i \leq d$ is $\binom{n}{i}$. For polynomial expansion, the worst case occurs when all coefficients of non-zero monomials are positive[2]. From Eq. (3), each positive monomial of degree $i \geq 3$ generates $\lfloor \frac{i-1}{2} \rfloor$ auxiliary variables. Therefore, the number of auxiliary variables in the worst case is $N^{\text{all}+}(n, d) = \sum_{i=3}^{d} \binom{n}{i} \lfloor \frac{i-1}{2} \rfloor = O\left(\lfloor \frac{\hat{d}-1}{2} \rfloor \frac{n!}{\hat{d}!(n-\hat{d})!} \right) = O\left(\lfloor \frac{\hat{d}-1}{2} \rfloor \frac{n^{\hat{d}}}{\hat{d}!} \right)$, where $\hat{d} = \min\{\lceil n/2 \rceil, d\}$ and the expression is with respect to asymptotically large n.

For the CCG-based quadratization algorithm, the worst case also occurs when all monomials are positive. Each positive monomial of degree $i \geq 3$ generates 2 auxiliary variables (i.e., X_a and X_L in Eq. (22)) when it is reduced to the sum of a quadratic polynomial and a monomial of degree $i - 1$, which can then be combined with existing monomials of degree $i-1$ if they are composed of the same variables. This combination of monomials can take place in each iteration, until the whole PBF becomes quadratic. In the worst case, only positive monomials remain after the combining step of each iteration, and therefore the number of auxiliary variables is $\sum_{i=3}^{d} 2\binom{n}{i} = O\left(\frac{n!}{\hat{d}!(n-\hat{d})!} \right) = O\left(\frac{n^{\hat{d}}}{\hat{d}!} \right)$.

In the best case, where all monomials are negative (assuming that all monomials up to degree d are present), both polynomial expansion and the CCG-based quadratization algorithm need just one auxiliary variable for each monomial, i.e., $N^{\text{all}-}(n, d) = \sum_{i=3}^{d} \binom{n}{i} = O\left(\frac{n!}{\hat{d}!(n-\hat{d})!} \right) = O\left(\frac{n^{\hat{d}}}{\hat{d}!} \right)$.

Table 1 summarizes our theoretical results. It shows that the CCG-based quadratization algorithm is advantageous over both polynomial expansion and γ flipping in terms of the required number of auxiliary variables. γ flipping has the same complexity regardless of the number of monomials in the input PBF, which is undesirable for PBFs that do not have an exponential number of monomials. In the best case, the numbers of auxiliary variables required by

---

[2] We follow the worst case definition in [21].

polynomial expansion and the CCG-based quadratization algorithm are only polynomial in n, while that of γ flipping is exponential in n.

4.2 Experimental Results

In this subsection, we focus on an experimental comparison of polynomial expansion and the CCG-based quadratization algorithm, since both of them have time complexities that are polynomial in d. We implement both algorithms in Python 2.7. Although *pseudo-Boolean* (PB) competitions have been regularly held [32], none of their instances have objective functions that are nonlinear PBFs. Therefore, we generate our own instances. We experiment with random instances and instances that model real-world facility location problems. These instances have

Table 1. Number of auxiliary variables for different quadratization algorithms as a function of the number of variables n and $\hat{d} = \min\{\lceil n/2 \rceil, d\}$, where d is the degree of the PBF.

	if $n \neq d$	if $n = d$
Polynomial expansion (worst case)	$O\left(\left\lfloor \frac{\hat{d}-1}{2} \right\rfloor \frac{n^{\hat{d}}}{\hat{d}!}\right)$	$2^{d-2}(d-3)+1$
CCG-based (worst case)	$O\left(\frac{n^{\hat{d}}}{\hat{d}!}\right)$	$2^{d+1} - 2 - 2d - d(d-1)$
Polynomial expansion and CCG-based (best case)	$O\left(\frac{n^{\hat{d}}}{\hat{d}!}\right)$	$2^d - 1 - \frac{d(d+1)}{2}$
γ flipping	$2^n - 1$	$2^d - 1$

Table 2. Number of auxiliary variables, number of terms in the quadratization, runtime of the quadratization algorithm, and runtime of the QPBO solver for the minimization of the quadratization. All reported numbers are averaged over 10 instances with $n = d$ and $m = 2^d$ monomials. Numbers after $\pm$ are standard deviations. The monomial coefficients are integers chosen randomly from $[1, 300]$. The smaller numbers of auxiliary variables and terms of quadratizations of each column are highlighted.

d		3	10	11	12	13	14	15
Original number of terms		8	1024	2048	4096	8192	16384	32758
Number of auxiliary variables	Poly	1	**1793**	4097	9217	20481	45057	>24 h
	CCG	2	1936	**3962**	**8034**	**16200**	**32556**	65294
Number of terms	Poly	11	12089	29508	70736	167005	389227	>24 h
	CCG	14	**7988**	**17162**	**36572**	**77482**	**163444**	343610
Quadratization time (s)[a]	Poly	0.0006	10.87 ±0.348	58.84 ±2.43	341.19 ±27.87	3435.50 ±39.34	17241.10 ±98.35	>24 h
	CCG	0.0006	1.828 ±0.007	5.133 ±0.006	23.16 ±0.10	104.58 ±1.94	584.33 ±21.90	3396.50 ±9.63
QPBO (s)[b]	Poly	0.0197 ±0.003	0.0835 ±0.0023	0.5086 ±0.0090	2.7316 ±0.081	14.5621 ±1.01	88.362 ±5.12	>24 h
	CCG	0.0080 ±0.00049	0.0043 ±0.00024	0.0132 ±0.00082	0.0616 ±0.0018	0.2968 ±0.0065	1.3018 ±0.0333	5.5763 ±0.5098

[a] Intel Xeon 4-core 2.3 GHz/6-core 2.6 GHz
[b] Intel Core i7-4960HQ Processor 6M Cache 2.60 GHz 8 GB SDRAM

a range of d wider than that of the problem of image denoising used in [21], which always has $d = 4$.

To generate random instances with a PBF on n variables, we generate each monomial with degree i randomly chosen from $\{0, \ldots, d\}$. Such a monomial has i unique variables randomly chosen from $\{1, \ldots, n\}$ along with a non-zero random integer coefficient. If a newly generated monomial is on the same variables as those of already generated monomials, it is rejected and a new one is generated. We also check that at least one of the m terms generated in this way is of degree d. For polynomial expansion, we also add up all quadratic like terms in the resulting quadratization (which can be expensive since polynomial expansion can potentially generate a lot of like terms in Eq. (3)). We use the QPBO solver from [26] for the minimization of the quadratizations. We run it via the open-source MATLAB wrapper qpboMex [34] on MATLAB R2016a and measure the actual wall-clock time. For the CCG-based quadratization algorithm, all wall-clock times include the runtime of the CCG construction. For each monomial coefficient a, the CCG-based quadratization algorithm for all experiments uses $J = L + 1$ and $L = a + 1$. The exact values of these parameters do not matter insofar as the condition $J \geq L > a > 0$ holds.

Table 3. Similar to Table 2, except that the monomial coefficients are non-zero integers chosen randomly from $[-300, 300]$.

d		3	10	11	12	13	14	15
Original number of terms		8	1024	2048	4096	8192	16384	32758
Number of auxiliary variables	Poly	1	1371.7 ±13.50	3022.1 ±28.75	6630.3 ±39.16	14317 ±43.35	30580 ±111	65623 ±193
	CCG	1.5 ±0.52	1545 ±14.5	3182 ±18.3	6503 ±25.9	13091 ±49.9	26397 ±41.1	52908 ±70
Number of terms	Poly	11	9002.7 ±120.2	21205.8 ±227.1	49758.7 ±379.4	114482 ±493.3	259370 ±1055	588832 ±2111
	CCG	12.5 ±1.58	7205.4 ±28.2	15601 ±36.3	33504 ±51.9	71251 ±81.3	151089 ±75.8	318781 ±150
Quadratization time (s)[a]	Poly	0.0006	4.01 ±0.18	23.22 ±0.91	133.5 ±0.71	1048.8 ±3.20	5724.5 ±747.7	84538 ±788.1
	CCG	0.0006	0.9516 ±0.011	4.21 ±0.022	19.14 ±0.057	133.9 ±0.58	594.8 ±73	3046 ±475
QPBO (s)[b]	Poly	0.0196 ±0.0012	0.0420 ±0.0094	0.27 ±0.0536	1.67 ±0.3289	8.43 ±0.0380	43.2 ±7.67	376.1 ±15.2
	CCG	0.0071 ±0.00021	0.0049 ±0.00052	0.0155 ±0.0025	0.058 ±0.011	0.37 ±0.010	1.10 ±0.26	4.38 ±0.90

[a] Intel Xeon 4-core 2.3 GHz/6-core 2.6 GHz
[b] Intel Core i7-4960HQ Processor 6M Cache 2.60 GHz 8 GB SDRAM

For polynomial expansion and the CCG-based quadratization algorithm, we first compare the numbers of auxiliary variables, the numbers of terms in the quadratizations, the runtime of the quadratization algorithms, and the runtime of the QPBO solver for minimization of the quadratizations. The runtime of the quadratization algorithm, referred to as its *quadratization time*, also includes the time for combining like terms. While depending mostly on the number of auxiliary variables, the quadratization time also depends on the number of like terms combined in the quadratization algorithm and the number of terms in the quadratization. Table 2 reports these comparisons for PBFs with all positive monomials. In each instance, the number of monomials is maximized, with integer coefficients chosen randomly from the interval $[1, 300]$. Table 2 shows the average and standard deviation for the results of the worst-case scenario over 10 instances where $n = d$ and the number of monomials is $m = 2^d$. The CCG-based quadratization algorithm significantly outperforms polynomial expansion in all four metrics as d increases.

Table 3 reports a comparison similar to Table 2 for the "average" cases, i.e., in each instance, $n = d$ and the number of monomials is 2^d but the coefficient of each monomial is a non-zero integer chosen randomly from the interval $[-300, 300]$. Here, too, the CCG-based quadratization algorithm significantly outperforms polynomial expansion in all four metrics as d increases.

Table 4. Similar to Table 2, except that all reported numbers are averaged over 100 instances, $n = 15$, $m = 500$, and the monomial coefficients are non-zero integers chosen randomly from $[-300, 300]$.

d		3	4	5	6	7	8	9	10	11
Original number of terms		500	500	500	500	500	500	500	500	500
Number of auxiliary variables	Poly	394.05 ±0.2179	405.71 ±2.7	491.56 ±9.23	538.27 ±10.43	614.47 ±17.73	666.57 ±19.45	733.76 ±22.15	789.70 ±23.12	859.77 ±26.37
	CCG	584.11 ±9.91	720.21 ±20.62	905.18 ±29.41	1088.23 ±43.52	1271.9 ±55.54	1438.61 ±61.74	1591.95 ±63.92	1742.11 ±70.80	1873.79 ±75.93
Number of terms	Poly	1637.18 ±0.887	1896.20 ±15.02	2596.18 ±61.81	3121.74 ±75.32	3954.80 ±154.1	4662.54 ±181.1	5607.75 ±214.2	6498.62 ±249.0	7630.98 ±314.8
	CCG	2029.31 ±19.92	2642.21 ±58.01	3473.31 ±98.33	4374.41 ±155.8	5346.31 ±220.1	6331.16 ±262.8	7343.51 ±275.9	8397.41 ±323.4	9430.44 ±374.6
Quadratization time (s)[a]	Poly	0.1147 ±0.0096	0.1783 ±0.0122	0.3170 ±0.0261	0.5284 ±0.0624	0.8172 ±0.0928	1.1943 ±0.1459	1.6922 ±0.1574	2.3604 ±0.2247	3.2838 ±0.3174
	CCG	0.1126 ±0.0066	0.1730 ±0.0121	0.2754 ±0.0211	0.4289 ±0.0428	0.5937 ±0.0534	0.8082 ±0.0741	1.0397 ±0.0799	1.3350 ±0.1003	1.6727 ±0.1312
QPBO (s)[b]	Poly	0.0013 ±0.0003	0.0019 ±0.0002	0.0032 ±0.0004	0.0048 ±0.0006	0.0076 ±0.0010	0.0105 ±0.0018	0.0148 ±0.0032	0.0205 ±0.0045	0.0291 ±0.0063
	CCG	0.0011 ±0.0001	0.0015 ±0.0002	0.0020 ±0.0001	0.0027 ±0.0002	0.0035 ±0.0003	0.0043 ±0.0003	0.0052 ±0.0007	0.0062 ±0.0006	0.0071 ±0.0006

[a] Intel Xeon 4-core 2.3 GHz/6-core 2.6 GHz
[b] Intel Core i7-4960HQ Processor 6M Cache 2.60 GHz 8 GB SDRAM

Table 4 reports a comparison similar to Table 2 for the case where $n = 15 > d$. Here, the degree i of each monomial is randomly chosen from $\{0, \ldots, d\}$. Then, i unique variables are chosen randomly from $\{1, \ldots, n\}$ to construct this monomial along with a non-zero integer coefficient for it chosen randomly from the interval $[-300, 300]$. $m = 500$ such monomials are generated and we report averages over 100 instances. Table 4 shows that the CCG-based quadratization algorithm continues to outperform polynomial expansion in the quadratization time and the runtime of the QPBO solver as d increases, although it uses more auxiliary variables and results in a quadratization with more terms. The reason is that the CCG-based quadratization algorithm derives its advantage from the recursive combinations of monomials, and the probability that these combinations take place decreases as the gap between $\sum_{i=0}^{d} \binom{n}{i}$, the maximum number of terms a degree-d PBF can have, and m, the actual number, increases. (As this gap increases, it is more difficult to encounter monomials that are of the same variables during each reduction process in the CCG-based quadratization algorithm). Nonetheless, the CCG-based quadratization algorithm is more efficient than polynomial expansion in its quadratization time since polynomial expansion not only generates more quadratic like terms for each monomial but also considers each monomial individually and altogether accumulates many quadratic like terms to be added up.

Table 5. Similar to Table 2, except that $n = d = 12$, m varies in density $100m/2^d$, and the monomial coefficients are non-zero integers drawn randomly from $[-300, 300]$.

d		12	12	12	12	12
Original number of terms (density)		10%	20%	80%	90%	100%
Number of auxiliary variables	Poly	671 ±20.1	1295 ±23.8	5287 ±35.6	6062 ±40.3	6641 ±43.3
	CCG	1219 ±28.3	2031 ±35.6	5612 ±49.8	6005 ±40.3	6510 ±46.2
Number of terms	Poly	5645 ±20.9	10769 ±30.0	39996 ±320.9	45019 ±380.1	49812 ±410.3
	CCG	6025 ±24.1	10142 ±28.6	28733 ±95.5	31264 ±102.2	33519 ±106.8
Quadratization time (s)[a]	Poly	1.74 .±0.03	6.31 ±0.05	108.07 ±10.1	136.05 ±12.2	186.50 ±20.16
	CCG	0.68 ±0.06	1.92 ±0.05	16.80 ±2.80	20.51 ±3.23	24.24 ±5.75
QPBO (s)[b]	Poly	0.0157 ±0.017	0.0825 ±0.018	1.1389 ±0.210	1.50 ±0.232	1.66 ±0.243
	CCG	0.0029 ±0.0005	0.0068 ±0.0015	0.0397 ±0.0079	0.042 ±0.0079	0.054 ±0.0082

[a] Intel Xeon 4-core 2.3 GHz/6-core 2.6 GHz
[b] Intel Core I7-4960HQ Processor 6M Cache 2.60 GHz 8 GB SDRAM

We finally investigate the role of m in comparison to the worst-case number of monomials. Table 5 reports a comparison similar to Table 2 for the case of varying density, i.e., $100m/2^d$ in percentage. We set $n = d = 12$ and report averages over 10 instances. We observe that the advantages of the CCG-based quadratization algorithm become more pronounced as the density increases. While the CCG-based quadratization algorithm is advantageous in quadratization time and runtime of the QPBO solver for all densities, it becomes more useful in the number of auxiliary variables and the number of terms in the quadratizations as the density increases.

4.3 Case Study: The Uncapacitated Facility Location Problem

We consider a real-world problem called the *uncapacitated facility location problem* (UFLP), also known as the *simple plant location problem*. The UFLP can also be used to model other real-world problems such as vehicle dispatching. This problem is NP-hard and can be reformulated as a nonlinear PBO [2,5,19,20].

Formally, the UFLP is characterized by a set of locations $I = \{1, \ldots, M\}$ and a set of users $J = \{1, \ldots, N\}$. Let f_i be the fixed cost of opening and operating a facility at location $i \in I$. Each user $j \in J$ is required to be served by exactly one facility. An $M \times N$ matrix $C = [c_{ij}]$ specifies the transportation cost of delivering products from a facility at location i to user j. The goal is to open facilities at a subset $S \subseteq I$ of locations that minimizes the sum of fixed costs and transportation costs, i.e.,

$$\sum_{i \in S} f_i + \sum_{j \in J} \min_{i \in S} c_{ij} . \tag{23}$$

In [5,19], the following method is used for reformulating the UFLP as a nonlinear PBO. For each column j in C, we assume a non-decreasing ordering of its elements as

$$c_{i_1^j} \leq c_{i_2^j} \leq \cdots \leq c_{i_M^j} . \tag{24}$$

We denote the difference between consecutive elements as

$$\Delta c_{0j} = c_{i_1^j} , \tag{25}$$

$$\Delta c_{lj} = c_{i_{l+1}^j} - c_{i_l^j} , \quad 1 \leq l < M . \tag{26}$$

Let $z_i = \begin{cases} 0 & \text{if } i \in S \\ 1 & \text{otherwise} \end{cases}$ for each $i \in \{1, \ldots, M\}$ and $\mathbf{z} = (z_1, \ldots, z_M)$. For any valid solution S, we have $\mathbf{z} \neq (1, \ldots, 1)$, and therefore $\forall j \in J : \min_{i|z_i=0} c_{ij} = \Delta c_{0j} + \sum_{l=1}^{M-1} \Delta c_{lj} z_{i_1^j} \ldots z_{i_l^j}$. Therefore, according to Eq. (23), the transportation cost is $\sum_{j \in J} \min_{i \in S} c_{ij} = \sum_{j=1}^{N} \left\{ \Delta c_{0j} + \sum_{l=1}^{M-1} \Delta c_{lj} z_{i_1^j} \ldots z_{i_l^j} \right\}$ and the fixed cost is $\sum_{i \in S} f_i = \sum_{i=1}^{M} f_i (1 - z_i)$. Hence, the total cost is

$$\sum_{i=1}^{M} f_i (1 - z_i) + \sum_{j=1}^{N} \left\{ \Delta c_{0j} + \sum_{l=1}^{M-1} \Delta c_{lj} z_{i_1^j} \ldots z_{i_l^j} \right\} . \tag{27}$$

The UFLP is equivalent to computing

$$\arg \min_{\mathbf{z}} \left\{ \sum_{i=1}^{M} f_i (1 - z_i) + \sum_{j=1}^{N} \left\{ \Delta c_{0j} + \sum_{l=1}^{M-1} \Delta c_{lj} z_{i_1^j} \ldots z_{i_l^j} \right\} \right\}$$

subject to: $\mathbf{z} \neq (1, \ldots, 1)$,

which, in turn, is equivalent to computing

$$\arg \min_{\mathbf{z}} \left\{ \sum_{i=1}^{M} f_i (1 - z_i) + \sum_{j=1}^{N} \left\{ \Delta c_{0j} + \sum_{l=1}^{M-1} \Delta c_{lj} z_{i_1^j} \ldots z_{i_l^j} \right\} + \lambda \prod_{i=1}^{M} z_i \right\} ,$$

where $\lambda > \max_{i \in I} f_i$. All nonlinear terms have positive coefficients because of Eq. (24). The degree of the resulting PBF is determined by the number of facilities. While different columns of C potentially use different orderings, the same ordering can be applicable to different columns. The number of unlike terms in the PBF is determined by the number of different orderings, which is generally affected by the number of users.

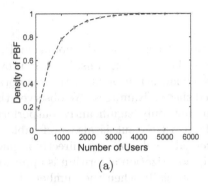
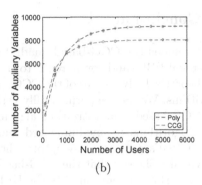

(a)   (b)

Fig. 2. UFLP experimental results. (a) shows $M = 12$ facility locations and N users. 10 instances are generated for each N. For each instance, the fixed cost for location i is an integer chosen randomly from $[1, 10000]$. The transportation cost from location i to user j is an integer chosen randomly from the interval $[1, 100]$. The density of the PBF is defined as the number of terms in the PBF divided by 2^M. (b) compares the number of auxiliary variables in the quadratizations of polynomial expansion and the CCG-based quadratization algorithm on all PBFs in (a). The y-axis indicates the number of auxiliary variables for each quadratization algorithm. (Color figure online)

We perform experiments to compare the performance (in terms of the number of auxiliary variables) of polynomial expansion and the CCG-based quadratization algorithm on PBFs resulting from UFLP instances. As reported in [21] and Tables 2, 3, 4 and 5, the number of auxiliary variables is the most important comparison parameter and is also indicative of the quadratization time and the runtime of the QPBO solver. We set the number of locations $M = 12$, which results in PBFs of degree 12. We vary the number of users N, which results in PBFs of varying densities. Except that the linear term coefficients are negative, this resembles the case of $d = 12$ in Table 2. For each instance, we randomly select the fixed costs of the facilities and the transportation costs from location i to user j. As the number of users increases, we plot the number of terms in the PBFs in Fig. 2a and the number of auxiliary variables for each quadratization algorithm in Fig. 2b.

Figure 2a shows that, as more users are added, the density increases rapidly at first and quickly approaches 1. Figure 2b shows that polynomial expansion is preferable when the number of users is small but is quickly outperformed by the CCG-based quadratization algorithm as the number of users increases. This observation is consistent with the results in Table 5, which show that the

CCG-based quadratization algorithm is more beneficial for higher densities. For the UFLP, the superior performance of the CCG-based quadratization algorithm with respect to the number of auxiliary variables is due to three possible reasons: (i) The degrees of the resulting PBFs are usually high, (ii) the coefficients of the nonlinear terms in these PBFs are all positive, and (iii) the PBFs become denser with a higher number of users.

5   Conclusion

We developed the CCG-based quadratization algorithm for the nonlinear PBO on general PBFs and compared it to state-of-the-art algorithms. We first proved the theoretical advantages of the CCG-based quadratization algorithm over other algorithms. We then experimentally verified these advantages. We observed that our CCG-based quadratization algorithm not only significantly outperforms other algorithms on medium-sized and large PBFs but is also preferable for smaller PBFs, to which asymptotic theoretical results are not directly applicable. We also showed that the CCG-based quadratization algorithm is applicable to real-world problems such as the UFLP, especially when the number of users to deliver products to is large.

References

1. Abío, I., Nieuwenhuis, R., Oliveras, A., Rodríguez-Carbonell, E., Mayer-Eichberger, V.: A new look at BDDs for Pseudo-Boolean constraints. J. Artif. Intell. Res. **45**(1), 443–480 (2012)
2. AlBdaiwi, B.F., Goldengorin, B., Sierksma, G.: Equivalent instances of the simple plant location problem. Comput. Math. Appl. **57**(5), 812–820 (2009)
3. Anthony, M., Boros, E., Crama, Y., Gruber, A.: Quadratization of symmetric Pseudo-Boolean functions. Discrete Appl. Math. **203**, 1–12 (2016). https://doi.org/10.1016/j.dam.2016.01.001
4. Anthony, M., Boros, E., Crama, Y., Gruber, A.: Quadratic reformulations of nonlinear binary optimization problems. Math. Program. **162**(1–2), 115–144 (2017). https://doi.org/10.1007/s10107-016-1032-4
5. Beresnev, V.: On a problem of mathematical standardization theory. Upr. Sistemy **11**, 43–54 (1973). (in Russian)
6. Berthold, T., Heinz, S., Pfetsch, M.E.: Nonlinear Pseudo-Boolean optimization: relaxation or propagation? In: Kullmann, O. (ed.) SAT 2009. LNCS, vol. 5584, pp. 441–446. Springer, Heidelberg (2009). https://doi.org/10.1007/978-3-642-02777-2_40
7. Bockmayr, A.: Logic programming with Pseudo-Boolean constraints. In: Constraint Logic Programming, pp. 327–350 (1993)
8. Bofill, M., Coll, J., Suy, J., Villaret, M.: Compact MDDs for Pseudo-Boolean constraints with at-most-one relations in resource-constrained scheduling problems. In: International Joint Conference on Artificial Intelligence, pp. 555–562 (2017). https://doi.org/10.24963/ijcai.2017/78

9. Boros, E., Crama, Y., Rodríguez-Heck, E.: Quadratizations of symmetric Pseudo-Boolean functions: Sub-linear bounds on the number of auxiliary variables. In: International Symposium on Artificial Intelligence and Mathematics (2018). http://isaim2018.cs.virginia.edu/papers/ISAIM2018_Boolean_Boros_etal.pdf
10. Boros, E., Gruber, A.: On quadratization of Pseudo-Boolean functions (2014). arXiv preprint: arXiv:1404.6538
11. Boros, E., Hammer, P.L., Minoux, M., Rader Jr., D.J.: Optimal cell flipping to minimize channel density in VLSI design and Pseudo-Boolean optimization. Discrete Appl. Math. **90**(1–3), 69–88 (1999)
12. Choi, V.: Minor-embedding in adiabatic quantum computation: I. The parameter setting problem. Quantum Inf. Process. **7**(5), 193–209 (2008). https://doi.org/10.1007/s11128-008-0082-9
13. Eén, N., Sörensson, N.: Translating Pseudo-Boolean constraints into SAT. J. Satisf. Boolean Model. Comput. **2**, 1–26 (2006)
14. Fioretto, F., Xu, H., Koenig, S., Kumar, T.K.S.: Solving multiagent constraint optimization problems on the constraint composite graph. In: Miller, T., et al. (eds.) PRIMA 2018. LNCS (LNAI), vol. 11224, pp. 106–122. Springer, Cham (2018). https://doi.org/10.1007/978-3-030-03098-8_7
15. Freeman, R.J., Gogerty, D.C., Graves, G.W., Brooks, R.B.: A mathematical model of supply support for space operations. Oper. Res. **14**(1), 1–15 (1966)
16. Glover, F., Woolsey, E.: Converting the 0-1 polynomial programming problem to a 0-1 linear program. Oper. Res. **22**(1), 180–182 (1974)
17. Gruber, A.G.: Algorithmic and complexity results for Boolean and Pseudo-Boolean functions. Ph.D. thesis, Rutgers University-Graduate School-New Brunswick (2015)
18. Hammer, P.L., Rudeanu, S.: Boolean Methods in Operations Research and Related Areas, vol. 7. Springer, Heidelberg (2012)
19. Hammer, P.: Plant location – a Pseudo-Boolean approach. Isr. J. Technol. **6**, 330–332 (1968)
20. Hansen, P., Kochetov, Y., Mladenovi, N.: Lower bounds for the uncapacitated facility location problem with user preferences. Groupe d'études et de recherche en analyse des décisions, HEC Montréal (2004)
21. Ishikawa, H.: Transformation of general binary MRF minimization to the first-order case. IEEE Trans. Pattern Anal. Mach. Intell. **33**(6), 1234–1249 (2011). https://doi.org/10.1109/TPAMI.2010.91
22. Johnson, M.W., et al.: Quantum annealing with manufactured spins. Nature **473**, 194–198 (2011). https://doi.org/10.1038/nature10012
23. Joshi, S., Martins, R., Manquinho, V.: Generalized totalizer encoding for Pseudo-Boolean constraints. In: Pesant, G. (ed.) CP 2015. LNCS, vol. 9255, pp. 200–209. Springer, Cham (2015). https://doi.org/10.1007/978-3-319-23219-5_15
24. Kahl, F., Strandmark, P.: Generalized roof duality for Pseudo-Boolean optimization. In: International Conference on Computer Vision, pp. 255–262 (2011). https://doi.org/10.1109/ICCV.2011.6126250
25. Kirkpatrick, S., Gelatt, C.D., Vecchi, M.P.: Optimization by simulated annealing. Science **220**(4598), 671–680 (1983)
26. Kolmogorov, V., Rother, C.: Minimizing nonsubmodular functions with graph cuts – a review. IEEE Trans. Pattern Anal. Mach. Intell. **29**(7), 1274–1279 (2007). https://doi.org/10.1109/TPAMI.2007.1031
27. Kumar, T.K.S.: Incremental computation of resource-envelopes in producer-consumer models. In: Rossi, F. (ed.) CP 2003. LNCS, vol. 2833, pp. 664–678. Springer, Heidelberg (2003). https://doi.org/10.1007/978-3-540-45193-8_45

28. Kumar, T.K.S.: A framework for hybrid tractability results in boolean weighted constraint satisfaction problems. In: Stuckey, P.J. (ed.) CP 2008. LNCS, vol. 5202, pp. 282–297. Springer, Heidelberg (2008). https://doi.org/10.1007/978-3-540-85958-1_19

29. Kumar, T.K.S.: Lifting techniques for weighted constraint satisfaction problems. In: International Symposium on Artificial Intelligence and Mathematics (2008). http://isaim2008.unl.edu/PAPERS/TechnicalProgram/ISAIM2008_0004_d1de5114b3cb94de7e670ab2905c3b3d.pdf

30. Kumar, T.K.S.: Kernelization, generation of bounds, and the scope of incremental computation for weighted constraint satisfaction problems. In: International Symposium on Artificial Intelligence and Mathematics (2016)

31. Manquinho, V., Marques-Silva, J., Planes, J.: Algorithms for weighted Boolean optimization. In: Kullmann, O. (ed.) SAT 2009. LNCS, vol. 5584, pp. 495–508. Springer, Heidelberg (2009). https://doi.org/10.1007/978-3-642-02777-2_45

32. Manquinho, V., Roussel, O.: Pseudo-Boolean competition (2016). http://www.cril.univ-artois.fr/PB16

33. Manquinho, V.M., Marques-Silva, J.: On using cutting planes in Pseudo-Boolean optimization. J. Satisf. Boolean Model. Comput. **2**, 209–219 (2006)

34. Osokin, A.: MATLAB wrapper to the QPBO algorithm by V. Kolmogorov (2014). https://github.com/aosokin/qpboMex

35. Philipp, T., Steinke, P.: PBLib – a library for encoding Pseudo-Boolean constraints into CNF. In: Heule, M., Weaver, S. (eds.) SAT 2015. LNCS, vol. 9340, pp. 9–16. Springer, Cham (2015). https://doi.org/10.1007/978-3-319-24318-4_2

36. Rendl, F., Rinaldi, G., Wiegele, A.: Solving max-cut to optimality by intersecting semidefinite and polyhedral relaxations. Math. Program. **121**(2), 307 (2010)

37. Rhys, J.: A selection problem of shared fixed costs and network flows. Manag. Sci. **17**(3), 200–207 (1970)

38. Wegener, I., Witt, C.: On the analysis of a simple evolutionary algorithm on quadratic Pseudo-Boolean functions. J. Discrete Algorithms **3**(1), 61–78 (2005)

39. Xu, H., Koenig, S., Kumar, T.K.S.: A constraint composite graph-based ILP encoding of the Boolean weighted CSP. In: Beck, J.C. (ed.) CP 2017. LNCS, vol. 10416, pp. 630–638. Springer, Cham (2017). https://doi.org/10.1007/978-3-319-66158-2_40

40. Xu, H., Satish Kumar, T.K., Koenig, S.: The Nemhauser-Trotter reduction and lifted message passing for the weighted CSP. In: Salvagnin, D., Lombardi, M. (eds.) CPAIOR 2017. LNCS, vol. 10335, pp. 387–402. Springer, Cham (2017). https://doi.org/10.1007/978-3-319-59776-8_31

Author Index

Printed in the United States
By Bookmasters